MW00636655

FLORA OF THE
CAYMAN
ISLANDS

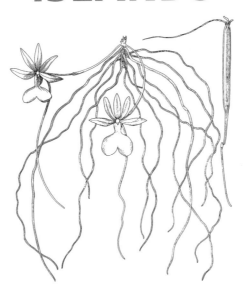

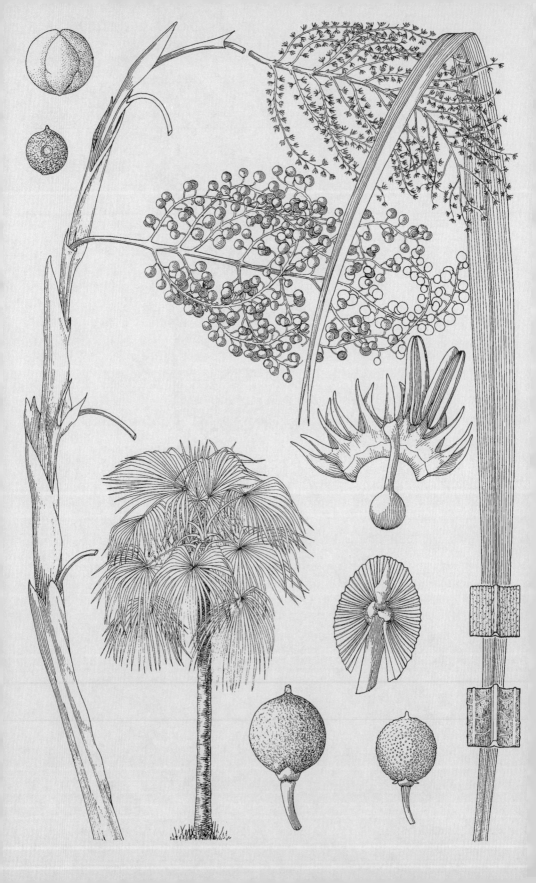

FLORA OF THE CAYMAN ISLANDS

Second Edition

George R. Proctor

Kew Publishing
Royal Botanic Gardens, Kew

ROYAL BOTANIC GARDENS

Second Edition published in 2012
by Royal Botanic Gardens, Kew
Richmond, Surrey, TW9 3AB, UK
in association with the Cayman Islands Department of the Environment

www.kew.org

Distributed on behalf of the Royal Botanic Gardens, Kew in North America by the University of Chicago Press, 1427 East 60th Street, Chicago, IL 60637, USA.

ISBN 978-1-84246-403-8

British Library Cataloguing in Publication Data
A catalogue record for this book is available from the British Library.

PROJECT EDITOR: Sharon Whitehead

DESIGN: Jill Bryan

TYPESETTING AND PAGE LAYOUT: Margaret Newman

COVER ILLUSTRATIONS: Front: main picture, 'Private Island' (Patrick Broderick Photography); left to right, *Tribulus cistoides* (George R. Proctor); *Hymenocallis latifolia* (Mat Cottam) and *Passiflora cupraea* (Kristan D. Godbeer). Spine: *Plumeria obtusa* (P. Ann van Battenburg-Stafford). Back cover drawing: *Agave caymanensis* (D. Erasmus).

Printed and bound in Malta by Melita

Funding for the second edition of this book has been provided by the Government of the Cayman Islands.

For information or to purchase all Kew titles please visit
www.kewbooks.com or email publishing@kew.org

Kew's mission is to inspire and deliver science-based plant conservation worldwide, enhancing the quality of life.

Kew receives half of its running costs from Government through the Department of Environment, Food and Rural Affairs (Defra). All other funding needed to support Kew's vital work comes from members, foundations, donors and commercial activities including book sales.

CONTENTS

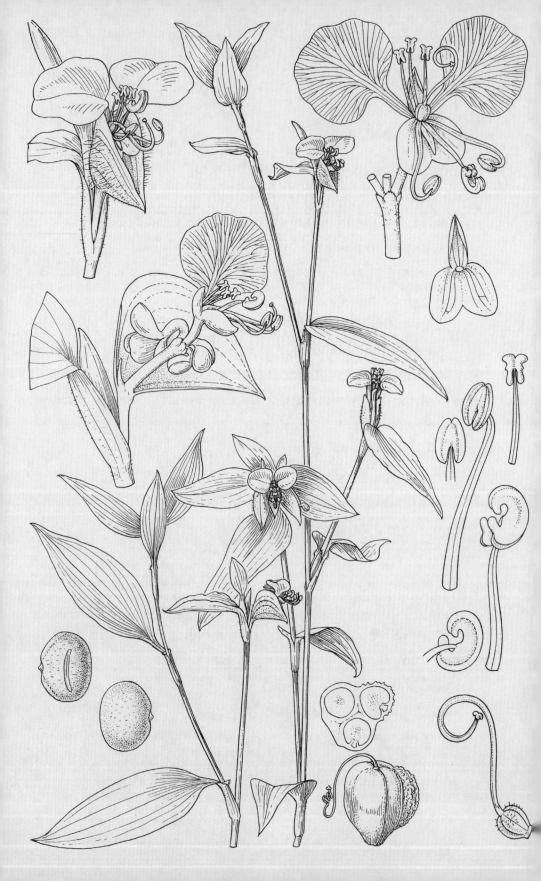

EDITORS' NOTE ON SECOND EDITION

Most modern taxonomic treatments follow the Angiosperm Phylogeny Group III (APG III) System, a molecular-based system of plant classification at family level and above (APG III, 2009) that is the standard for all Kew publications. The APG III System does not, of course, update classification of the Pteridophyta or Gymnospermae. Here, Dr Proctor has largely retained the arrangement established in the First Edition of his Flora (1986), which is based on Cronquist's *Evolution and Classification of Flowering Plants* (1968), with some updating based on APG III.

Under different circumstances, we would have liked to have worked with Dr Proctor to update the Second Edition of *The Flora of the Cayman Islands* to be fully compliant with APG III, but this was not possible and the editors have respected Dr Proctor's taxonomic viewpoint. To aid users who are familiar with APG III, we have included a table (pp. 720–724) that maps all of the orders, families and genera used by Dr Proctor in this Flora against the accepted order and family names currently used in APG III.

During editing, it became evident that a small part of the nomenclature retained from the First Edition was incorrect on the basis of recent taxonomic treatments. These areas of taxonomy have been updated, either by a family specialist based at Kew or by cross-reference to *The Plant List* (http://www.theplantlist.org/), and the resulting changes incorporated into the Flora, in some cases without a full treatment of the new taxa (e.g. see treatments for *Senna*, *Cassia* and *Chamaecrista*). The main aim of these updates was to enable users to apply the keys to identify plants in the Cayman Islands, and to get to the correct name. Some new records, verified after completion of Dr Proctor's manuscript, have been inserted at the most appropriate place in the Flora without a major renumbering of families, genera and so on (e.g. see insertion of first record for family Aristolochiaceae p. 668).

The following people at the Royal Botanic Gardens, Kew are thanked for their valuable contributions to and comments on this Second Edition: Catherine Challis (Latin descriptions for the new species described herein), Colin Clubbe (overall co-ordination), Phillip Cribb (Orchidaceae), Martin Hamilton (nomenclature), Nicholas Hind (Compositae), Gwilym Lewis (Leguminosae), Steve Renvoize (Gramineae), David Simpson (Cyperaceae) and Nigel Taylor (Cactaceae).

ACKNOWLEDGEMENTS

I gratefully acknowledge the many organisations and individuals who have assisted in the compilation of this Flora. I acknowledge especially the contributions made by the Cayman Islands authorities, the United Kingdom Overseas Development Agency (ODA), now the Department for International Development (DFID), and the Institute of Jamaica. Thanks are due also to the custodians of plant materials and illustrations: among the latter, I particularly thank the Trustees of the British Museum (Natural History) and the Rachel McMasters Miller Hunt Botanical Library for their kind permission to use important drawings. Grateful acknowledgement is also made to the former ODA for commissioning original drawings and for arranging for permission to use the others; Mr Martin Brunt of the former ODA's Land Resources Development Centre was largely responsible for the success of these arrangements.

NOTES ON THE ILLUSTRATIONS AND PHOTOGRAPHS

The illustrations have been obtained from a number of sources as indicated by bracketed initials inserted at the end of each caption. The explanation of these initials is as follows:

(D.E.), drawn by Mr D. Erasmus; (F. & R.), reproduced with the permission of the Trustees of the British Museum (Natural History) from Fawcett & Rendle, *Flora of Jamaica*; (G.), drawn by Miss V. Goaman; (H.), reproduced with the permission of the Office of the Secretary, United States Department of Agriculture, from Hitchcock, *Manual of the Grasses of the West Indies* and collateral publications, the original drawings kindly made available by the Rachel McMasters Miller Hunt Botanical Library, Carnegie-Mellon University, Pittsburgh; (J.C.W.), drawn by Mrs D. Erasmus; (St.), made available by Dr W. T. Stearn from drawings prepared for volume 6 of *Flora of Jamaica* (unpublished) — these drawings are held in the *Flora of Jamaica* collection in the Department of Botany at the British Museum; (R.R.I.), the drawing of *Pennisetum purpureum* by W. E. Trevithick was originally in a book by R. Rose-Innes; (H.U.), a drawing of *Epidendrum cochleatum* is reproduced with the permission of L. G. Garay, Curator of the Orchid Herbarium of Oake Ames Botanical Museum, Harvard University.

A significant number (116) of the illustrations in this Flora are original illustrations of species either not depicted elsewhere, or else seldom or never presented in as much detail in other publications. The high quality of the drawings by Mr & Mrs D. Erasmus and by Miss V. Goaman provide this Flora with a value beyond that which the text alone could support. The important role of Dr W. T. Stearn and Mr A. C. Jermy (ferns) in supervising or monitoring the preparation of these drawings is also acknowledged.

Mrs N. Cruikshank, of the Mosquito Research and Control Unit, Cayman Islands, drew the diagrams in the section on 'Environment and plant communities'. All photographs were taken by the author except where otherwise indicated by the following initials: JFB, John F. Binns; DB, Denise Bodden; FJB, Frederic J. Burton; MC, Mat Cottam; JDM, João de Deus Medeiros; KDG, Kristan D. Godbeer; WP, Wallace Platts; CR, Carla Reid; JR, Joanne Ross; FR, Frank Roulstone; and AS, P. Ann van Batenburg-Stafford.

ACKNOWLEDGEMENTS FOR THE SECOND EDITION

The gathering of information and materials for the Second Edition of this book has received a wide spectrum of encouragement and support from several organizations and many individuals within the Cayman Islands. First and foremost has been the decision of the Cayman Islands Government, through its Department of Environment, to support both the preparation and publication costs of the book. Thanks are also due to this agency for the purchase of special equipment such as a computer and an 'Optelec' reading machine; the latter, essential because of the author's failing eyesight, was also partially contributed by the Lions Club of the Cayman Islands. A substantial cash donation was generously made by the Garden Club of the Cayman Islands, and this is gratefully acknowledged. A much larger contribution was received from the Dart Foundation, expressing its keen interest in the Flora project and emphasizing its dedication to improving scientific and horticultural knowledge of Cayman plant life. Sincere thanks are due to Mrs. Lois Blumenthal, who has worked tirelessly to promote and coordinate the Flora project.

Many people have helped in the pursuit of new Cayman plant records, and I fear that no list of such individuals can ever be complete. However, the following alphabetical summary includes at least a majority of the persons whose help has been vital to the project: Margaret Barwick, Patricia Bradley, Fred Burton, Penny Clifford, Gladys Howard (Little Cayman), Wallace Platts (Cayman Brac), Carla Reid, Joanne Ross, Frank Roulstone III and P. Ann van B. Stafford.

Very special thanks are also due to Wayne Ross, who arranged to have the entire First Edition photocopied in an enlarged format on one side of a page to make easier the editing and interpolation of new information. This proved to be of immense help to the project.

Funding for the Second Edition has been provided by the Cayman Islands Department of the Environment.

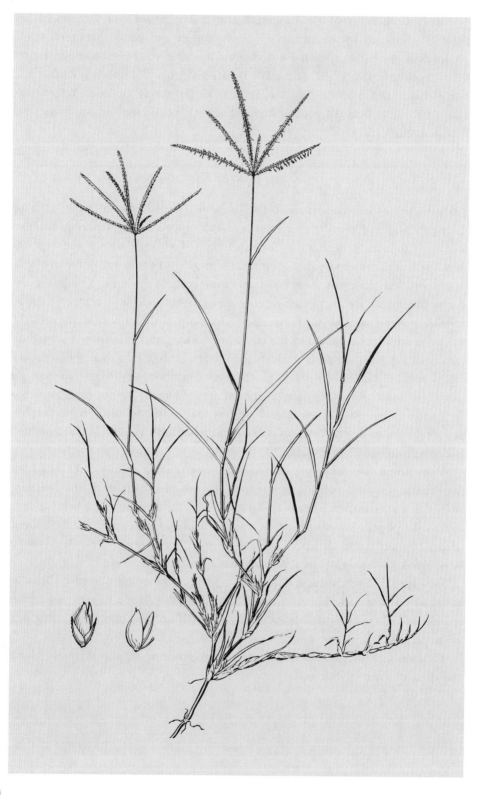

PLAN OF THE FLORA

The development of plant classification since the time of Linnaeus has been marked by a series of more or less competing systems. Each system was intended to express contemporary ideas about species and their relationships, and to arrange groups of species in hierarchies according to their supposed phylogeny. It has been customary for writers of Floras to use whatever sequence of taxa was considered orthodox for their particular region or nationality, and this has often tended to delay and restrict the spread of new taxonomic ideas. For example, it has long been customary for Floras in the N. American and West Indian region to be arranged more or less according to the system of Engler and Prantl, first proposed in Germany in the late nineteenth century (Engler & Prantl, 1895). In recent years, however, the accumulation of much new information has led to profound changes in taxonomic concepts. In particular (though not exclusively), writers such as Cronquist and Thorne in the U.S.A. and Takhtajan in the U.S.S.R., reinforced by the work of numerous monographers, have convincingly demonstrated that many details of the older systems have become obsolete. Unfortunately, although these developments are usually taught to modern university students, the writing of floristic books has lagged behind. The families of flowering plants in this book are arranged according to Arthur Cronquist's *The Evolution and Classification of Flowering Plants* (1968). Even though modifications of this system will undoubtedly continue to occur as knowledge improves, at the present time it reflects much current thinking about plant classification, and therefore deserves to be exercised at the practical level (see Editors' Note on Second Edition: p. 7).

The chief exceptions to the original Cronquist system in this book are: (1) the monocotyledon series is placed before the dicotyledons, more out of habit than conviction, as there seems no doubt that the monocotyledons as a group were derived from primitive dicotyledonous ancestors; (2) the Viscaceae are separated as a family from the Loranthaceae; (3) the name Surianaceae is used in place of Stylobasiaceae (on the basis of being a *nomen conservandum*); and (4) *Picrodendron* is included in the Euphorbiaceae.

Unlike Engler & Prantl (1895), Cronquist does not deal with classification below the family level, so in this regard the writer has simply followed his own preferences. For the most part, a broad concept of genus has been followed in the belief that the binomial system of nomenclature works better as a means of recognition and information-retrieval when the number of generic names is kept to a minimum consistent with what is known or believed about natural relationships. It is the writer's opinion that the trend toward more and more splitting at the generic and family levels results in an unbalanced taxonomy that unduly exaggerates the significance of differences while obscuring resemblances, and at the same time impairs the usefulness of names as a means of recognition.

Descriptions of families, genera and species are provided in the text, together with keys for the identification of taxa at each level. The identification keys are dichotomous; users should note that characters used in the keys are frequently not repeated in the relevant descriptions. Thus the keys should be referred to as an integral part of the descriptions.

All taxa believed to be not indigenous to the Cayman Islands (especially those supposedly or certainly introduced by human agencies) are enclosed in square brackets.

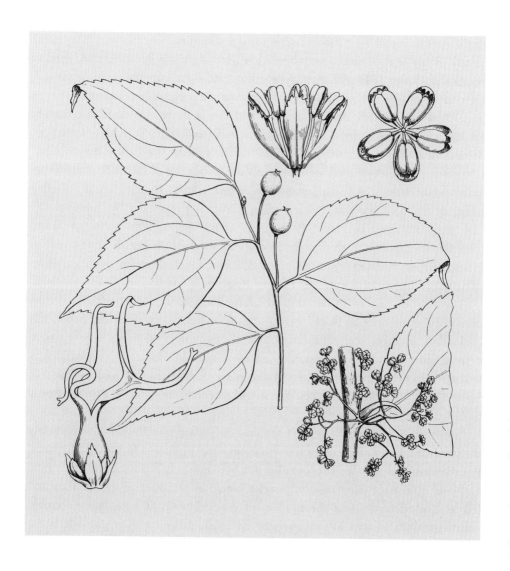

INTRODUCTION

GEOGRAPHY AND PHYTOGEOGRAPHY

BY FREDERIC J. BURTON AND COLIN CLUBBE

Geographically and floristically, the Cayman Islands are closely associated with the Greater Antilles, although in terms of land area, they are very small in relation to neighbouring Cuba and Jamaica. At 197 km², Grand Cayman is the largest of the three Cayman Islands. Little Cayman and Cayman Brac are 28.5 and 38 km², respectively. The three islands represent the emergent peaks of a largely submerged mountain range that extends from the Sierra Maestra in Cuba to the Gulf of Honduras, which was created by ancient uplifting of the southern rim of the N. American tectonic plate.

The islands have never been connected by land bridge to any mainland, and last emerged from a period of total submergence some 2–3 million years ago. Since then, the modern flora and fauna of the islands has become established and has evolved to suit local conditions. Some 415 taxa (species and varieties) of vascular plant are now believed to be truly native to the Cayman Islands. Of these, 28 taxa are regarded as endemic (Burton, 2008a).

At latitude 19–20° N, the Cayman Islands have a distinctly seasonal climate, with a sub-humid to humid wet season between May and November, and a semi-arid to arid dry season between December and April. Average daily minimum–maximum temperatures are 21–28°C during January and February, rising to 25–32°C in July and August. Extremes recorded between 1973 and 1987 spanned from 11.2°C to 36.5°C. Annual rainfall averages between 1,049 mm and 1,595 mm at various locations on the islands, with a strong trend to increased rainfall from east to west on Grand Cayman. Occasional hurricanes are a major feature of the climate, and they exert a powerful influence on both landforms and biological communities.

The surface rock strata of all three islands are carbonate rocks, primarily dolostone karst at higher elevations, and limestone karst or pavement rock in areas that were submerged during the Sangamon highstand. There are no rivers, and fissures in the bedrock allow tidal fluctuations to influence groundwater throughout the interior. Rainfall percolation locally forms fresh water lenses, with the fresh water floating over the saline groundwater.

Soil is scarce, and chemically unrelated to the bedrock. It is rich in iron and aluminium oxides that are derived, at least partially, from fallout from Saharan dust plumes that are entrained in stratospheric currents across the Atlantic.

The primary natural vegetation of the Cayman Islands, before human impacts, consisted of semi-deciduous dry forests on the highest land, where trees can root into several metres of rock above the permanent water table. Dry land at lower elevations, and any dry land under extreme stress from salt spray, was blanketed in semi-deciduous xerophytic shrubland. Seasonally rain-flooded forests and shrublands blended into extensive seasonally and tidally flooded mangrove forests and shrublands, growing on the autochthonous peat that is still accumulating with rising sea levels.

Since the first colonization by humans some 300 years ago, there has been large-scale deforestation, especially on Grand Cayman and Cayman Brac. A large array of non-native plants has been introduced, including some extremely invasive species. New man-made landscapes, including grasslands, second growth woodlands, and coastal *Casuarina* forests, are now widespread.

The natural vegetation communities of the Cayman Islands have been described in detail by Burton (2008b). They include forests, woodlands, shrublands, dwarf-shrublands, and herbaceous and sparsely vegetated communities. Of these, the dry forests and dry shrubland communities show the greatest biodiversity. The dry forests of each of the three islands are floristically distinct.

The dry forests of Grand Cayman and Little Cayman are lowland semi-deciduous forests, typically dominated by *Bursera simaruba* and *Guapira discolor*. They range in canopy height from 4.5 to 16 m. Cayman Brac's dry forests are mainly xeromorphic semi-deciduous forests, usually dominated by *B. simaruba* and the cactus *Pilosocereus royenii*.

Seasonally flooded semi-deciduous forests are characterized by *Roystonea regia* on Grand Cayman, *Metopium toxiferum* on Little Cayman, and *Thrinax radiata* and *Swietenia mahogani* on both islands. These forests are restricted in distribution and are absent from Cayman Brac.

Mangrove forests are widespread on Grand Cayman and Little Cayman. They are dominated by one or any combination of the mangroves *Rhizophora mangle*, *Avicennia germinans* and *Laguncularia racemosa*, with *Conocarpus erectus* in less saline settings. These forests occur in both seasonally and tidally flooded settings.

Natural woodlands are rare on the Cayman Islands, usually forming an intermediate between forest and shrubland, and sharing some species characteristic of each. The transition to shrubland occurs when the main canopy falls below 4.5 m.

Xerophytic shrublands on all three islands tend to be dominated by *Savia erythroxyloides* and *Agave caymanensis*. On sandy and coral rubble coasts, *Coccoloba uvifera* shrublands form dense salt-sheared hedges on the crests of the beach ridges. Seasonally flooded *Conocarpus erectus* shrublands and extensive mangrove shrublands are the reduced-canopy counterparts of the corresponding forests, stunted by nutrient limitations or at an early stage of succession after a hurricane.

Dwarf shrublands are characteristic of the 'ironshore' coasts on the three islands, consisting mainly of *Rhachicallis americana* with stunted, prostrated *Conocarpus erectus*.

Herbaceous communities are found naturally not only on the sandy beaches seaward of the *Coccoloba* hedge but also as sedge wetlands, as tidally flooded succulents bordering *Avicennia* forest or shrubland, and as aquatic vegetation, including sea grass beds.

As a result of ongoing deforestation and the impacts of invasive species, some 46% of the Cayman Islands' native plant species are now under some threat of extinction (Vulnerable or worse under International Union for Conservation of Nature (IUCN) Red List Criteria). Of the islands' 28 endemics, 15 are 'Critically Endangered'. Meanwhile, the introduction and establishment of alien species in the wild continues apace, now reaching 38% of the flora listed in this volume.

These statistics give great urgency to the National Trust for the Cayman Islands' and the Cayman Islands Department of Environment's efforts to expand the terrestrial protected area system on all three islands. By 1998, some 37% of Grand Cayman, 26% of Cayman Brac and 19% of Little Cayman had already been deforested. By the end of the 21st century, deforestation seems likely to be complete, save only for protected areas, which currently account for only 7.1% of the Cayman Islands' combined land area.

FLORISTIC BACKGROUND
SPECIES AND VARIETIES

The following 28 species and varieties are considered endemic to the Cayman Islands; the symbols 'GC', 'LC', and 'CB' refer to Grand Cayman, Little Cayman, and Cayman Brac, respectively; an asterisk (*) means that a close congener is also present.

Aegiphila caymanensis Moldenke (VERBENACEAE) GC*

Agalinis kingsii Proctor (SCROPHULARIACEAE) GC

Agave caymanensis Proctor (AGAVACEAE) GC, LC, CB

Allophylus cominia var. *caymanensis* Proctor (SAPINDACEAE) GC, LC, CB

Argythamnia proctorii Ingram (EUPHORBIACEAE) GC, LC, CB

Banara caymanensis Proctor (SALICACEAE) LC, CB

Casearia staffordiae Proctor (SALICACEAE) GC

Chamaesyce bruntii Proctor (EUPHORBIACEAE) LC

Chionanthus caymanensis Stearn (OLEACEAE) GC, LC, CB

Coccothrinax proctorii Read (PALMAE) GC, LC, CB

Cordia sebestena var. *caymanensis* (Urb.) Proctor (BORAGINACEAE) GC, LC, CB

Crossopetalum caymanense Proctor (CELASTRACEAE) GC, LC, CB*

Dendropemon caymanensis Proctor (LORANTHACEAE) LC

Dendrophylax fawcettii Rolfe (ORCHIDACEAE) GC

Encyclia kingsii (C. D. Adams) Nir (ORCHIDACEAE) LC, CB

Epiphyllum phyllanthus var. *plattsii* Proctor (CACTACEAE) CB

Hohenbergia caymanensis Britton ex L. B. Smith (BROMELIACEAE) GC

Myrmecophila thomsoniana var. *minor* (Starchan ex Fawc) Dressler (ORCHIDACEAE) LC, CB

Myrmecophila thomsoniana var. *thomsoniana* (ORCHIDACEAE) GC

Pectis caymanensis var. *robusta* Proctor (COMPOSITAE) GC

Phyllanthus caymanensis Webster & Proctor (EUPHORBIACEAE) GC, LC, CB

Pilostyles globosa var. *caymanensis* Proctor (APODANTHACEAE) GC, LC, CB

Pisonia margaretiae Proctor (NYCTAGINACEAE) GC

Salvia caymanensis Millsp. & Uline (LABIATAE) GC

Scolosanthus roulstonii Proctor (RUBIACEAE) GC

Terminalia eriostachya var. *margaretiae* Proctor (COMBRETACEAE) GC

Turnera triglandulosa Millsp. (TURNERACEAE) LC, CB

Verbesina caymanensis Proctor (COMPOSITAE) CB

The Jamaican element in the Cayman flora consists of 17 species and varieties. All of these are otherwise confined to Jamaica except for five that also occur on the Swan Islands, and one on Swan and Cozumel.

Astrocasia tremula (Griseb.) Webster (EUPHORBIACEAE) GC

Bourreria venosa (Miers) Stearn (BORAGINACEAE) GC, LC, CB (Swan)

Casearia odorata Macf. (SALICACEAE) GC*

Cestrum diurnum var. *venenatum* (Mill.) O. E. Sch. (SOLANACEAE) GC, CB (Swan)

Cionosicyos pomiformis Griseb. (CUCURBITACEAE) GC

Cordia brownei (Friesen) I. M. Johnst. (BORAGINACEAE) GC, LC, CB

Daphnopsis occidentalis (Sw.) Krug & Urban (THYMELAEACEAE) GC, CB

Jacquinia proctorii Stearn (THEOPHRASTACEAE) GC, LC, CB

Jatropha divaricata Sw. (EUPHORBIACEAE) GC

Vernonia divaricata Sw. (COMPOSITAE) GC, LC, CB

Peperomia simplex Ham. (PIPERACEAE) GC

Phyllanthus angustifolius (Sw.) Sw. (EUPHORBIACEAE) GC, LC, CB (Swan)

Phyllanthus nutans ssp. *nutans* (EUPHORBIACEAE) GC, LC, CB (Swan)

Tabernaemontana laurifolia L. (APOCYNACEAE) GC

Tournefortia astrotricha var. *astrotricha* (BORAGINACEAE) GC, CB (Swan)

Tournefortia astrotricha var. *subglabra* Stearn (BORAGINACEAE) GC

Trichilia glabra L. (MELIACEAE) GC, LC, CB (Swan)

The strictly Cuban element in the Cayman flora consists of only 14 species and varieties, but this figure can be reinforced by numerous others of wider distribution that occur in Cuba but not in Jamaica. Cayman plants found otherwise only in Cuba are as follows:

Cestrum diurnum var. *marcianum* Proctor (SOLANACEAE) GC

Chamaesyce torralbasii (Urban) Millsp. (EUPHORBIACEAE) CB

Clerodendrum aculeatum var. *gracile* Griseb. ex Moldenke (VERBENACEAE) GC

Ipomoea passifloroides House (CONVOLVULACEAE) GC

Malpighia cubensis Kunth (MALPIGHIACEAE) GC

Pectis caymanensis var. *caymanensis* (COMPOSITAE) GC, LC, CB

Peperomia pseudoperskiifolia C. DC. (PIPERACEAE) GC, CB

Phyllanthus nutans ssp. *grisebachianus* (Muell. Arg.) Webster (EUPHORBIACEAE) LC

Pleurothallis caymanensis Adams (ORCHIDACEAE) GC

Polygala propinqua (Britton) Blake (POLYGALACEAE) GC, LC, CB

Portulaca tuberculata Leon (PORTULACACEAE) LC, CB

Roystonea regia (Kunth) O.F. Cook (PALMAE) GC

Tillandsia fasciculata var. *clavispica* Mez in DC. (BROMELIACEAE) GC, LC, CB

Encyclia phoenicia (Lindl.) Neum. (ENCYCLIA)

The vegetation of the Cayman Islands shows virtually no direct affinity with that of C. America. Hence, the discovery in 1975 of *Phyllanthus caymanensis* Webster & Proctor on Little Cayman and Cayman Brac was quite surprising, as it is closely related to two C. American species and represents the only known endemic Antillean element in its section of the genus. Another surprise has been the discovery of an endemic variety of *Epiphyllum phyllanthus* (Cactaceae) on Cayman Brac. This species is otherwise confined to S. America and Panama; no other representative of this genus has been found in the West Indies.

The finding of a variety of the parasitic *Pilostyles globosa* (Apodanthaceae) on all three Cayman Islands provides an unusual phytogeographic linkage between Mexico and Jamaica that is not easy to explain.

There are, on the other hand, 40 Cayman species whose distribution includes Cuba, the Bahamas and, in some cases, Hispaniola (Hispaniola alone in three cases: *Chascotheca domingensis*, *Metastelma picardae* and *Neoregnellia cubensis*) but that do not occur in Jamaica.

A further group of 8 species represents plants endemic to the Greater Antilles, which occur in both Cuba and Jamaica:

Bunchosia media Griseb. (MALPIGHIACEAE) GC, LC

Capparis ferruginea L. (CAPPARACEAE) GC, CB

Chamaecrista lineata (Sw.) Greene (LEGUMINOSAE-CAESALPINIOIDEAE) GC (Note: this entity is not very distinct from a wider-ranging species)

Celtis trinervia Lam. (ULMACEAE) GC

Cleome procumbens Jacq. (CAPPARACEAE) GC

Evolvulus squamosus Britton (CONVOLVULACEAE) LC (Note: closely related forms in the Bahamas may not be distinct)

Paspalum distortum Chase (GRAMINEAE) GC

Solanum havanense Jacq. (SOLANACEAE) GC

This list can be augmented by at least 17 other species of primarily Greater Antillean distribution that extend slightly beyond this area (chiefly into the Bahamas). These occur in both Cuba and Jamaica, and also usually in Hispaniola but not in Puerto Rico. The great majority of these have been found within the Cayman Islands only on Grand Cayman.

The remainder of the species considered indigenous to the Caymans (i.e. those not introduced by human agencies) consist chiefly of wide-ranging tropical American species, pantropical strand plants, and virtually pantropical weeds. An unknown number of these occurred in the Cayman Islands before the earliest human settlement, and we can assume that a certain percentage have introduced themselves since that time. A further 90 species, now more or less established or naturalized, owe their presence to intentional (or sometimes unintentional!) introduction by humans, and this number will no doubt continue to grow.

The whole list of vascular plants now known to grow without cultivation in the Cayman Islands totals 716 species (27 pteridophytes, 1 gymnosperm, 173 monocotyledons and 515 dicotyledons). It is quite probable that others will eventually be found as a few relatively inaccessible portions of all three islands have still not been investigated adequately in all seasons. On the other hand, there is real danger that the building of new roads, hotels, and housing settlements will bring about the total extermination of many interesting or unique plant species unless enough representative habitats can be saved from the bulldozers.

HISTORY OF BOTANICAL COLLECTIONS

Botanical investigation began relatively late in the Cayman Islands and has involved comparatively few people. The earliest known plant collections[1] were made in May 1888, by William Fawcett, then Director of Public Gardens and Plantations in Jamaica, and later co-author of *Flora of Jamaica*. His visit, of just a few days' duration, resulted in the publication of a short report dealing with the natural and agricultural resources of the islands. He also reported on such subjects as the disease of coconut palms (presumably what is now called Lethal Yellowing) then ravaging Grand Cayman. Added to this report was a list of 112 plants that he had collected, including both indigenous and introduced species. The specimens on which this list was based were deposited in the herbarium of the Royal Botanic Gardens, Kew. Among Fawcett's discoveries was the endemic orchid *Dendrophylax fawcettii*, described by Rolfe in the *Gardener's Chronicle* of Nov. 10, 1888. Alluding to this plant, Fawcett later commented, "As I saw but a small portion of the Islands, and that chiefly on the sea shore, I feel little doubt that a complete collection of the plants would be of very great interest, and that perhaps other endemic species would be found."

Following Fawcett, several American botanists paid short visits to the Cayman Islands during the 1890s. Of these, the most important were Albert S. Hitchcock (in 1891), who later became a leading authority on grasses, and Charles F. Millspaugh (in 1899). Hitchcock's specimens (mostly miserable scraps, unfortunately) were collected while he was a member of a party of naturalists led by J. T. Rothrock; they are preserved in the herbarium of the Missouri Botanical Garden in St. Louis. Millspaugh made his collections while a guest of Allison V. Armour (the Chicago meat-packing millionaire) on a West Indian cruise aboard the yacht 'Utowana'; they visited the Cayman Islands during February 1899. The chief set of Millspaugh's specimens is in the herbarium of the Field Museum of Natural History in Chicago. Both Hitchcock and Millspaugh published lists of their collections.

After Millspaugh, there was an interval of 39 years during which there was virtually no botanical activity in the Cayman Islands. A few plants only were gathered on Cayman Brac by the geologist C. A. Matley in 1924.

The Oxford University Biological Expedition to the Cayman Islands, a party of five under the leadership of W. G. Alexander, carried out fieldwork from April 17 to August 27 1938. The primary objects of attention were plants, insects, reptiles, and fishes, but nearly all animal taxa received some attention. The official botanist of this group was Wilfred W. Kings, who joined the expedition about a month later than the others; he had been especially recruited from Lawrence Sheriff School, Rugby, because Oxford had no available botanist at that time. Before his arrival, some plant-collecting was done by C. Bernard Lewis, whose interests were otherwise chiefly zoological. Kings gathered a large collection of material from all three islands; until recently, these excellent specimens constituted the major basis of our knowledge of the Cayman flora. The main set of the Kings collection is deposited at the British Museum (Natural History) in London, while duplicate material can be found in several other herbaria.

Lewis, then an Oxford student (a Rhodes Scholar from the United States), later became Director of the Institute of Jamaica in Kingston; he collected further Cayman plant specimens during the 1940s. His continued interest in the Cayman Islands has been a constant source of encouragement during the writing of this book.

Coincident with the Oxford Expedition was a brief visit, during June 1938, of the Cap Pilar Expedition. The 'Cap Pilar' was a French sailing vessel (a 'barquentine') that made a two-year trip around the world, beginning in September 1936, under the captaincy of Adrian Seligman, then of Wimbledon, England. The purpose of the trip was partly adventure and partly to collect plants for the Royal Botanic Gardens, Kew. Originally, the botanist on this expedition was A. F. Roper, but when the ship reached South Africa on its outward journey, his place was taken (for unexplained reasons) by C. M. Maggs, then a horticulturist at the Kirstenbosch Botanical Garden. The 'Cap Pilar' visited Australia, and after making collections on various Pacific islands (including the Galapagos) and at Panama, made a final stop at Grand Cayman, where a few plants were collected.

I first visited Grand Cayman on April 19, 1948 while a member of the Catherwood–Chaplin West Indies Expedition of the Academy of Natural Sciences of Philadelphia. This one-day stop was followed eight years later by a longer collecting trip on behalf of the Institute of Jamaica and the British Museum (Natural History), traveling to the islands as a guest of the British survey ship H.M.S. Vidal. At that time, we spent nearly three weeks on Grand Cayman. I made a one-day visit to Cayman Brac on May 2, 1956. My next Cayman trip was made during June 1967 at the invitation of

Dr. M. E. C. Giglioli, Director of the Cayman Government Mosquito Research and Control Unit (M.R.C.U.), following a suggestion by Martin Brunt. The latter, then an ecologist on the staff of the Land Resources Development Centre of the British Overseas Development Administration (O.D.A.), was at the time making ecological studies of the Cayman swamps for M.R.C.U. He had already collected numerous Cayman plant specimens, which he and his colleagues were having difficulties indentifying, partly because no descriptive flora of the islands had ever been written. This stimulating collaboration with Brunt led me to agree to the request of Mr J. A. Cumber, then Administrator of the Cayman Islands, that I should write the First Edition of this Flora. Brunt's continued interest was largely responsible for making its publication possible.

Later, I made short collecting trips to Little Cayman (July 7–12, 1967 and August 2–12, 1975), Cayman Brac (August 6–10 and November 9–11, 1968), and again to Grand Cayman (September 6–7, 1969), in order to augment the information available on the botany of these islands. Particular mention should be made of the hospitality of the late Dr Logan Robertson, which made possible the earlier work on Little Cayman. My second trip to that island in 1975 was sponsored by the Cayman Islands Government as part of their joint study of the island with the Royal Society of London. This complemented the ODA-funded Natural Resources Study of all the Cayman Islands. I am indebted to Dr David Stoddart of Cambridge University for his invitation to participate in this program.

The First Edition of this Flora was published in 1984, but any complacency about its completeness was shattered in 1991 by the discovery by Margaret Barwick of *Terminalia eriostachya* and by my renewed field work on all three islands. Between 1991 and 2004, I carried out 15 mostly rather short collecting trips, gathering 610 specimens including numerous new records and identifying several species new to Science. Some of this new information was presented in *Kew Bulletin* (vol. 51, no. 3, pp. 483–507) in 1996. A further list of more new records by G. F. Guala *et al.* appeared in the same journal in 2002 (vol. 57, no. 1, pp. 235–237). Still further species have been revealed since 2002; Guala was right in characterizing the Cayman flora as "dynamic".

The prime set of the writer's collections is preserved in the herbarium of the Institute of Jamaica; most of this material is duplicated at the British Museum (Natural History), and smaller sets have been distributed to other herbaria. Collections made between 1991 and 1996 are located primarily at SJ (Department of Natural and Environmental Resources, San Juan, Puerto Rico).

In recent years, an unknown number of American collectors have visited the Cayman Islands (usually only Grand Cayman), but for the most part, it has not been possible to take their material into account. The Flora does, however, take into account the collection of Marie-Helène Sachet (1958), some of the specimens collected by Jonathan Sauer (1962 and 1967), and the collection of Donovan S. and Helen B. Correll (1979). Sauer's gatherings totalled 243 specimens, according to lists he kindly supplied, but only a small fraction of these were actually seen in connection with the present study. A single species, *Atriplex pentandra*, is recorded from the Cayman Islands solely on the basis of a Sauer specimen. Sauer's material is deposited in the herbarium of the University of Wisconsin at Madison; duplicate material is filed at the Field Museum of Natural History in Chicago.

The Corrells' specimens, 65 numbers in all, are deposited in the herbarium of the Fairchild Tropical Botanic Garden, Miami. Included in the collection are the first examples of *Corchorus hirsutus* and *Gomphrena globosa* from the Cayman Islands, and the first specimens of *Citharexylum spinosum*, *Passiflora cupraea* and *Phoradendron quadrangulare* from Grand Cayman.

I visited all the cited institutions during the preparation of this Flora, and I thank the relevant officials, curators, and staff for much help and much kindness received while studying the Cayman plant specimens under their care.

The Cayman Islands specimens that I have seen are summarized in Table 1 below. They are listed chronologically by dates of collection. The herbarium symbols signify the following: A, Arnold Arboretum; BM, British Museum (Natural History); F, Field Museum of Natural History; FTG, Fairchild Tropical Garden; GH, Gray Herbarium of Harvard University; IJ, Institute of Jamaica; K, Royal Botanic Gardens, Kew; MO, Missouri Botanical Garden; SJ, Department of Natural and Environmental Resources, San Juan, Puerto Rico; US, U.S. National Herbarium, Washington, D.C.

Table 1. Collectors of Cayman Islands plants, in chronological order.			
Name	**Collection dates**	**Herbaria where specimens are deposited**	**Number of specimens seen**
William Fawcett	May 1888	K	16
John T. Rothrock	"Winter of 1890–1891"	F	10
A. S. Hitchcock	Jan. 1891	MO, F	166
Charles F. Millspaugh	Feb. 1899	F	216
C. A. Mately	Jan. 1924	K	3
Wilfrid W. Kings	May–Aug. 1938	BM, GH, MO	645
C. Bernard Lewis	Apr. 1938, Dec. 1944, Mar. 1945, Dec. 1945	BM, IJ	43
C. M. Maggs	June 1938	K	18
George R. Proctor	Feb. 1948, Apr.–May 1956, Jun.–Jul. 1967, Aug. 1968, Nov. 1968, Sep. 1969, Aug. 1975, Nov. 1991–Apr. 2004 (15 short collecting trips)		1617
Marie-Helène Sachet	Sep. 1958	US	95
Robert A. Dressler	May 1964	IJ	2
Richard A. Howard & B. Wagenknecht	Jan. 1969	A	11
Martin Brunt	May–Jun. 1967	BM, IJ	503
Jonathan Sauer	Jun. 1967	F	15
John Popenoe	Apr. 1969	FTG IJ	1
Donovan S. & Helen B. Correll	Nov. 1979	FTG	65
G. F. Guala	Jun. 1998	FTG	Not seen
		Total	**2816**

DESCRIPTION OF THE FLORA

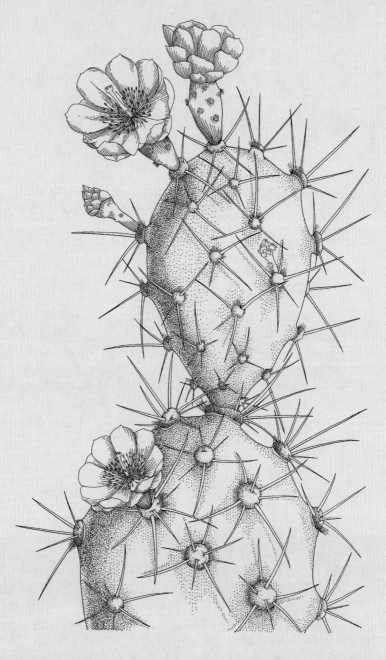

Pteridophyta

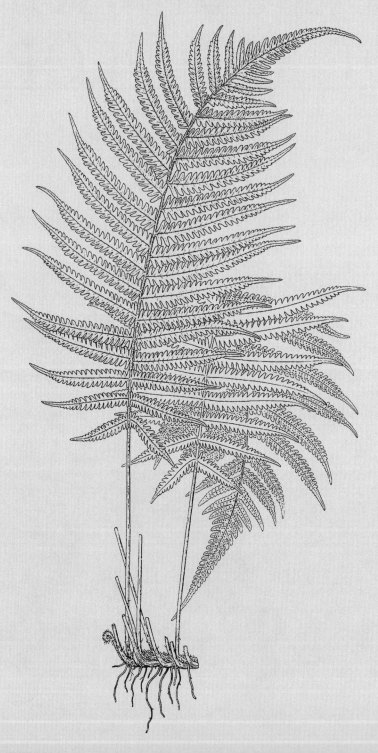

Vascular plants without flowers, fruits or seeds, having a life-cycle of two distinct, independent phases, called the sporophyte and gametophyte, which are normally produced in an alternating sequence. The sporophyte, nearly always the more conspicuous phase, has parts differentiated into roots, stem (rhizome or caudex) and leaf (frond in the ferns). These structures always contain special hardened (vascular) tissues for the conduction of water and various water-soluble substances. In leafy tissues, the vascular strands are normally evident as veins. The sporophyte produces dust-like asexual spores (reproductive granules), which may be all similar (in homosporous plants) or differing in size and form (in heterosporous plants). (All Cayman pteridophytes are homosporous.) Spores are borne in very small or minute capsules (sporangia), which are isolated or variously clumped on the sporophyte leaves or in their axils. A germinating spore grows by normal cell division into a gametophyte (prothallus) characteristic in form according to taxonomic group; those of true ferns usually scale-like or semi-filamentous, green and lacking vascular tissue except in rare cases. The prothalli of homosporous species are all alike in general appearance, but may be either monoecious (organs of both sexes on the same individual) or dioecious (male and female organs on separate individuals). The male reproductive cells (gametes or spermatozoids) of all pteridophytes are motile by means of coiled cilia. Fertilization of the egg cell (female gamete or archegone) by a spermatozoid requires the presence of a droplet of water so that the spermatozoid can reach its destination; the resulting zygote immediately initiates the growth of a new sporophyte plant, which soon assumes the characteristic appearance of a fern or other pteridophyte. In many cases, both the sporophyte and the gametophyte are also capable of reproduction by vegetative means.

A world-wide group of more than 12,000 species, classified in a number of families whose relationship is remote in many cases. Only three of these families occur in the Cayman Islands.

KEY TO FAMILIES

1. Plant without true roots or leaves, the plant body consisting chiefly of a dichotomously branched system of erect green stems set with scattered minute pointed scales, the fertile branches bearing isolated 3-locular sporangial capsules: **Psilotaceae**

1. Plant with true roots, stems and leaves, these clearly differentiated; sporangia minute, variously clustered on the backs or margins of fertile leaves:

   2. Spores produced on specialized non-leafy pinnae arising from the base of a sterile green blade; sporangia with cap-like annulis, opening by a vertical slit down one side: **Schizaeaceae**

   2. Spores produced on the undersides of green leafy fronds, the sporangia variously arranged; sporangial annulus vertical and elastic, rupturing transversely: **Polypodiaceae**

PSILOTACEAE

Plants mostly epiphytic, the stems either simple and provided with small 2-ranked leaves (*Tmesipteris*) or many times dichotomous and appearing leafless, the leaves being remote, minute, and scale-like. Sporangia 2-locular (*Tmesipteris*) or 3-locular and dehiscing vertically, attached on adaxial base of minute bifid sporophylls. Spores reniform, all similar. Gametophytes subterranean or embedded in humus, tuberous.

A very primitive group with two genera, *Tmesipteris* of the South Pacific islands and *Psilotum*, which is pantropical.

Psilotum Sw.

Plants epiphytic or on old masonry, sometimes terrestrial, the rhizomatous part of the stem short-creeping, beset with small brownish hairs. Aerial stems loosely clustered or solitary, the lower unbranched part more or less elongate, dichotomous above into several to numerous 2- or 3-angled divisions. Scale-like leaves few, scattered, minutely awl-shaped, alternate and distichous or else 3-ranked. Sporangia depressed-globose, sessile, 3-lobed, 3-celled; spores hyaline.

A pantropical genus of two species.

Psilotum nudum (L.) Beauv., *Prodr. Fam. Aethéog.* 112 (1805).

Mostly erect, 20–60 cm high, the main stalk 2–4 mm thick. Branches more or less 3-angled. Scale-like leaves remote, mostly 1–2 mm long, rarely to 3 mm. Sporangia ca. 2 mm thick, yellow or brownish.

GRAND CAYMAN: *Brunt* 1920; *Kings* GC 174, GC 408.
— Pantropical.

SCHIZAEACEAE

Terrestrial (rarely epiphytic) ferns of widely diverse habit and form. Rhizome erect or decumbent, usually small, sometimes branched or long creeping, clothed with hairs (or scales in one genus). Fronds small and erect or else (in *Lygodium*) elongate (to many meters) and vine-like, rarely simple and filiform, more often dichotomously or pinnately divided, varying from glabrous to pubescent or (in one genus) scaly, of circinnate growth. Sporangia mostly obovoid or pyriform, with subapical annulus contracted below, opening by a longitudinal fissure, borne singly or in rows, usually on more or less specialized lobes or segments (sporangiophores), or on slender ultimate divisions of special non-leafy pinnae; indusium present or absent. Homosporous, the spores trilete or monolete, without chlorophyll. Gametophytes epigeal, flat and green, or filamentous and partly epigeal, the exposed parts with chlorophyll, or else tuberous and subterranean, with associated mycorrhizae and lacking chlorophyll.

A chiefly pantropical group of four distinctive genera, sometimes treated as separate families, with a total of about 150 species. One genus with one species has been found in the Cayman Islands.

Anemia Sw.

Small terrestrial or epiphytic ferns; rhizome creeping or ascending, rarely erect, usually short, clothed with articulate (pluricellular) hairs. Fronds usually long-stipitate, partially or wholly dimorphic; pinnatifid to pinnately compound, either entirely fertile or entirely sterile, or else with a basal pair of contracted fertile pinnae on otherwise sterile fronds; veins free, or sometimes reticulate without free included veinlets. Sporangia borne in a row on either side of more or less contracted ultimate divisions (the fertile divisions sometimes reduced to the axis only). Spores tetrahedral-globose, trilete, the surface with prominent parallel ridges, or ridged and coarsely echinate, or else broadly reticulate without spines. Chromosomes: n = 38 or 76.

A pantropical genus of about 90 species, the majority occurring in the neotropics. The name *Anemia* is of greek derivation (*aneimon*, naked) and alludes to the naked sporangia.

Anemia adiantifolia (L.) Sw., *Syn. Fil.* 157 (1806). PLATE 1.

Rhizome creeping, branched, densely clothed toward apex with brown articulate hairs. Fronds closely distichous, the sterile ones arching or ascending, 11–70 cm long, with stipes as long as or exceeding the blades; fertile fronds slightly longer and more erect; all with strongly pubescent stipe and rhachis, the leaf-tissue sparingly hispidulous on both sides. Sterile blades ovate-deltate or subpentagonal, 5–35 cm long, 4–28 cm broad at base; pinnae stalked and inequilateral, close or distant; ultimate segments mostly oblong-obovate, obtuse, of subcoriaceous texture, the margins erose-denticulate and thickened, the venation flabellate, free. Fertile pinnae rigidly erect, usually shorter than to slightly exceeding the sterile portion of the blade. Spores cristate, the ridges undulate.

GRAND CAYMAN: George Town Quarry, *Walls* s.n., 19 Apr. 1992 (BM), a single sterile leaf det. by C. D. Adams. George Town Quarry, 10 Jul. 2006, fertile plants found by *P. Ann van B. Stafford* s.n.

— Florida, Bahamas, Greater and Lesser Antilles, Trinidad, and from Mexico to Costa Rica. Very rare in the Cayman Islands.

POLYPODIACEAE

Small to rather large leafy plants of varying habit, terrestrial or epiphytic, the usually hairy or scaly rhizome creeping to erect, producing few or many roots and sometimes also stoloniferous. Fronds usually stalked (stipitate), uniform or dimorphic, the lamina (blade) simple to several-times pinnately compound, or rarely digitately or palmately divided. Sporangia long-stalked, dehiscing by means of an annulus of elastic cells interrupted by thin-walled cells at one side, and arranged in lines or clusters (sori), or completely covering the underside of fronds or frond-divisions (pinnae or segments). Sori naked or protected by scale-like indusia that develop either from the veins or from the margins of the lamina (rarely both). Spores bilateral to globose or tetrahedral, often surrounded by a husk-like outer covering called the perispore, which is smooth or variously sculptured or even spinulose, the form usually characteristic for genera or often for species. Gametophytes green, flat, glabrous or hairy, sometimes glandular.

More that 170 genera (up to 240 or more recognized by some authors) and nearly 10,000 species, as here defined. Some specialists subdivide the group into numerous smaller families, but there is wide divergence of opinion as to whether and how this should be done. I do not favour fragmentation of the ferns at the family level, preferring a concept of subfamilies. In any case, the relatively very few Cayman fern genera are more conveniently arranged under a single family heading. A summary of the habitats of the Cayman ferns is given in Figure 2A.

KEY TO GENERA

1. Sporangia completely covering the backs of fertile pinnae; large, coarse ferns of brackish swamps:	**Acrostichum**
1. Sporangia clustered in linear or roundish sori, never completely covering the lamina tissue; small to medium-sized ferns of various habitats, but not occurring in brackish or saline situations:	
2. Sori linear or marginal (or both):	
3. Sori linear along veins beneath:	**Pityrogramma**
3. Sori not along veins:	
4. Sori marginal:	
5. Stipes wiry, black or nearly so, glossy:	
6. Ultimate frond-divisions mostly sessile and less than 3 mm broad; sporangia arising from the lamina, protected by a recurved marginal indusium:	**Cheilanthes**
6. Ultimate frond-divisions distinctly black-stalked and up to 10 mm broad; sporangia arising from the underside of the indusial flap:	**Adiantum**
5. Stipes neither black, wiry, nor particularly glossy:	
7. Rhizome elongate, without scales; fronds scattered, long-stalked, the lamina deltate in outline and 3–4-pinnate:	**Pteridium**
7. Rhizome short-creeping, densely scaly; fronds clustered, 1-pinnate:	**Pteris**
4. Sori not marginal, elongate and parallel to midveins of pinnae:	**Blechnum**
2. Sori round or nearly so, not marginal:	
8. Lamina 3–4 pinnate:	**Macrothelypteris**
8. Lamina less than 2-pinnate:	
9. Lamina fully pinnate, at least at the base (i.e. at least the lowest pinnae attached to the rhachis by a junction much narrower than the width of the pinna); indusium present or apparently absent:	
10. Veins all free, none joined or connivent in the tissue; pinnae entire to serrate, jointed to the rhachis:	**Nephrolepis**
10. At least some veins joined or connivent in the tissue; pinnae more or less lobed, and not jointed to the rhachis:	
11. Pinnae few (3–6 pairs), 1–6 cm broad with entire margins, the lowest with basal lobes; veins reticulate:	**Tectaria**
11. Pinnae numerous (7–25 or more pairs), 0.5–1.5 cm broad with margins lobed throughout; only basal veins joined or connivent:	**Thelypteris**
9. Lamina simple or pinnatisect (i.e. joined to the rhachis by a broad base, as broad as the segment itself); sori without any trace of indusium:	**Polypodium**

Pityrogramma Link

Small to medium-sized terrestrial or epipetric ferns. Rhizomes short-creeping to erect, clothed with narrow, more or less attenuate, non-clathrate scales. Fronds fasciculate, all similar or sub-dimorphic, not articulate to the rhizome; stipes usually dark, hard, and lustrous. Blades 1–3- pinnate, narrowly oblong or lanceolate to ovate or deltate, usually covered on abaxial side with white or yellow to nearly orange wax-like powder, otherwise naked or rarely with scales; veins free, pinnately forked. Sori indefinite and indusium absent; sporangia produced along the course of the veins and often at maturity forming confluent lines or masses; annulus of 20–24 cells; spores tetrahedral-globose, trilete, the surface coarsely tuberculate, reticulate, or ridged, rarely smooth and minutely granulate.

A small genus of fewer than 20 species, most of them occurring in the neotropical region. The generic name is derived from the Greek *pityron*, meaning scurf or farina, and *gramme*, meaning line; the undersurface of the fronds is covered with farina-like powder, and the sporangia are in lines along the veins.

Pityrogramma calomelanos (L.) Link, *Handbuch* 3: 20 (1833).

Rhizome ascending to erect, clothed at apex with dark golden-brown, narrowly lanceolate scales mostly 3–5.5 mm long, these long filamentous at apex (a long portion one cell wide below terminal cell). Fronds closely tufted, up to 1 m long; stipes dark lustrous reddish-brown, about as long as the blades or shorter. Blades lanceolate to oblong-deltate, 20–95 cm long, 10–30 cm broad below the middle, 2-pinnate to 2-pinnate-pinnatifid (rarely 3-pinnate); segments lanceolate to elliptic, usually at least sharp-pointed and often more or less sharply serrate or deeply incised; tissue firmly herbaceous. Sporangia numerous, often nearly concealing the undersurface of the pinnules; spores prominently ridged.

GRAND CAYMAN: *Proctor 50746*.

Tropical and subtropical America and Africa; naturalized elsewhere in warm regions. Often weedy in disturbed ground.

Cheilanthes Sw.

Small terrestrial or epipetric ferns. Rhizomes short-creeping, scaly. Fronds clustered, erect or spreading, often glandular, hairy or scaly (but not in the single Cayman species), the lamina 1–4-pinnate, or 1-pinnate and variously pinnatifid, the ultimate divisions in some species minute. Rhachis and costae terete, black; veins free, simple or forked. Sori marginal, arising from the enlarged tips of the veins, usually numerous and narrowly confluent, more or less protected by the reflexed, membranous, indusioid margin. Spores globose or tetrahedral and smooth, granulose, or sometimes corrugated.

A world-wide genus of more than 125 species, many of them characteristically occurring in the crevices of cliffs or among rocks. The name is of Greek origin and alludes to the lip-like or marginal indusia.

Cheilanthes microphylla (Sw.) Sw., *Syn. Fil.* 127 (1806). **FIG. 1**.

Rhizome clothed with soft, narrowly linear scales, these pale yellow or reddish-brown. Fronds mostly 8–20 cm long (becoming larger in other regions), the wiry black stipes much shorter than or equal in length to the lamina, and bearing a few hair-like scales near the base. Lamina oblong-lanceolate, seldom exceeding 2 cm in width in Cayman plants, 2–3-pinnate;

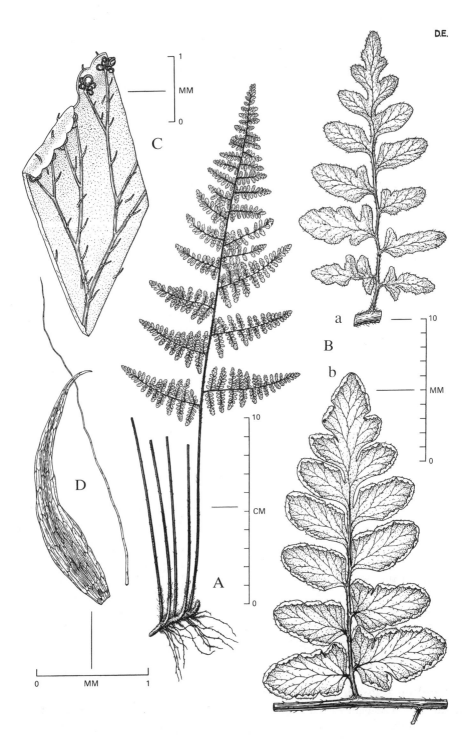

FIG. 1 **Cheilanthes microphylla**. A, habit. B, single pinna; a, upper side; b, lower side. C, portion of fertile pinnule. D, scale and hair of rhizome. (D.E.)

rhachis and costae sparsely clothed with lax, jointed, whitish hairs. Pinnules simple, rarely lobed or pinnate, the ultimate divisions always more or less oval or rounded-oblong.

GRAND CAYMAN: *Proctor* 48307; CAYMAN BRAC: *Proctor* 29007.

— Southern United States and Mexico, the West Indies (common), and northern S. America, usually on shaded limestone ledges or similar habitats. The Cayman Brac plants are frequent in certain places along the northern side of The Bluff.

Adiantum L.

Terrestrial ferns of forest floors, shady ravines and rocky banks or cliffs. Rhizomes slender and wide-creeping to short and suberect, bearing numerous narrow scales chiefly near the apex. Fronds usually distichous, apart or clustered with firm, dark to black, usually highly glossy stipes. Lamina 1–5-pinnate (rarely simple), of various patterns of dissection, most often glabrous or apparently so, sometimes glaucous beneath. Ultimate divisions sessile or stalked, often articulate and deciduous; veins free, often flabellately branched. Sori borne on the underside of the reflexed margin (or marginal lobe), the sporangia arising along the ends of (and sometimes between) the ultimate veinlets. Spores most often tetrahedral, dark, smooth.

A large pantropical genus of about 200 species, most numerous in S. America; a few also occur in temperate regions. Often called maidenhair ferns.

KEY TO SPECIES

1. Lamina 1-pinnate, or (if 2–3-pinnate) with an elongate terminal pinna essentially like the lateral ones; ultimate divisions mostly curved-oblong with one entire side, the stalks not articulate:	A. melanoleucum
1. Lamina compoundly dissected without a distinct terminal pinna; ultimate divisions more or less rhombic or flabellate-cuneate with 2 entire sides, the stalks articulate at the apex:	A. tenerum

Adiantum melanoleucum Willd. in L., *Sp. Pl.* 5: 443 (1810). **FIG. 2.**

Rhizome short-creeping, densely clothed with dull brownish, concolorous scales. Fronds closely distichous, in Cayman plants seldom longer than 30 cm, the stipes lustrous purple-black, scabrous. Lamina variable, in juvenile or depauperate forms linear, 1-pinnate, and mostly less than 20 cm long; those of well-developed mature plants 2-pinnate (or 3-pinnate at the base); pinnules or ultimate divisions oblong, close, dimidiate. Sori oblong-lunate or linear and deeply curved, occurring only along the distal margins.

GRAND CAYMAN: *Kings* F 24.

— Florida, Bahamas, and the Greater Antilles, on shaded limestone ledges or cliffs. The Grand Cayman plants were found on the sides of a 'well'.

Adiantum tenerum Sw., *Nov. Gen. & Sp. Pl.* 135 (1788). **FIG. 2.**

Rhizome short-creeping, densely clothed with lustrous dark brown scales having pale, lacerate margins. Fronds few, distichous, erect and spreading or sometimes pendent, reaching 70 cm long or more, the stipes lustrous black or purple-black, smooth. Lamina deltate-ovate, subpentagonal, nearly as broad as long, 3–5-pinnate at the base; pinnae alternate, stalked; ultimate divisions more or less rhombic-oblong to flabellate-cuneate, on

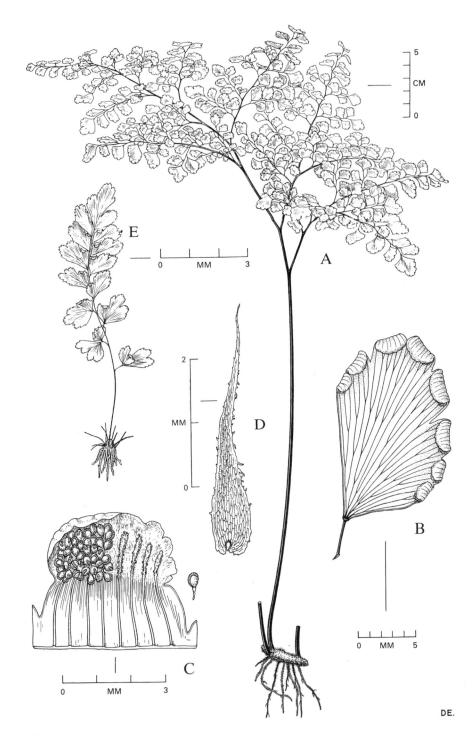

FIG. 2 **Adiantum**. A, **A. tenerum**, habit. B, pinnule, C, fertile lobe (dissected). D, scale of rhizome. E, **A. melanoleucum**, depauperate plant. (D.E.)

2–4 mm stalks, these distinctly jointed at the apex. Sori retuse-oblong, borne in pairs on each of the lightly bifid lobes of the distal margin.

GRAND CAYMAN: *Kings* F 40; *Proctor* 15274.

— Southern United States, the West Indies, and Mexico to northern S. America. Cayman plants occur on shaded limestone banks and ledges, and in moist, shaded 'wells', not common.

Pteridium Gled. ex Scop., nom. cons.

Terrestrial ferns; rhizomes deeply subterranean, extensively creeping, repeatedly branched, lightly clothed toward the apex with reddish-brown, pluricellular, articulate hairs; scales lacking. Fronds of coarse, hard texture, tripartite and mostly 3–4-pinnate, the lower pinnae with glandular so-called 'nectaries' at base when young; ultimate segments variable in size and shape, but always with revolute margins; veins free. Sori marginal, usually continuous, the sporangia borne between the outer reflexed margin and an inner indusial flange, on a vascular strand connecting the vein-ends. Spores globose-tetrahedral, very finely spinulose.

Usually considered to consist of a single, variable, world-wide species, a concept that unfortunately results in a cumbersome hierarchy of trinomial and quadrinomial names. In many countries, this fern is (or these ferns are) known as bracken. It is considered a pernicious weed throughout most of its distribution.

Pteridium aquilinum (L.) Kuhn in v.d. Decken, *Reisen in Ost-Afrika* 33: 11 (1879) var. *caudatum* (L.) Sadebeck in *Jahrb. Hamb. Wiss. Anst.* 14, Beiheft 3: 5 (1897). PLATE 1.

Fronds often 1 m tall or more, the greenish to straw-colored stipes nearly as long as the deltate lamina, glabrous throughout except for the lamina tissue beneath. Ultimate segments mostly linear, 1–5 cm long, 1–4 mm broad, distant and distinct.

GRAND CAYMAN: *Kings* F 39.

— Florida, the West Indies, Mexico and C. America (this variety). Apparently rare in the Cayman Islands; it occurs chiefly in old fields and secondary thickets.

Pteris L.

Terrestrial ferns; rhizomes often stout and woody, creeping to erect, clothed with scales at least near the apex. Fronds various in outline, the veins free or areolate. Sori marginal, usually linear, the sporangia borne in a continuous line on a slender inframarginal vein connecting the ultimate vein-ends; paraphyses usually present; indusium linear, formed by the modified reflexed margin, opening inwardly. Spores tetrahedral (rarely bilateral), and smooth, tuberculate or otherwise sculptured.

A large, world-wide genus of at least 280 species, most abundant in tropical regions.

KEY TO SPECIES

1. Pinnae articulate to the rhachis	P. longifolia var. bahamensis
1. Pinnae not articulate to the rhachis, the costae decurrent:	[2. P. vittata]

33

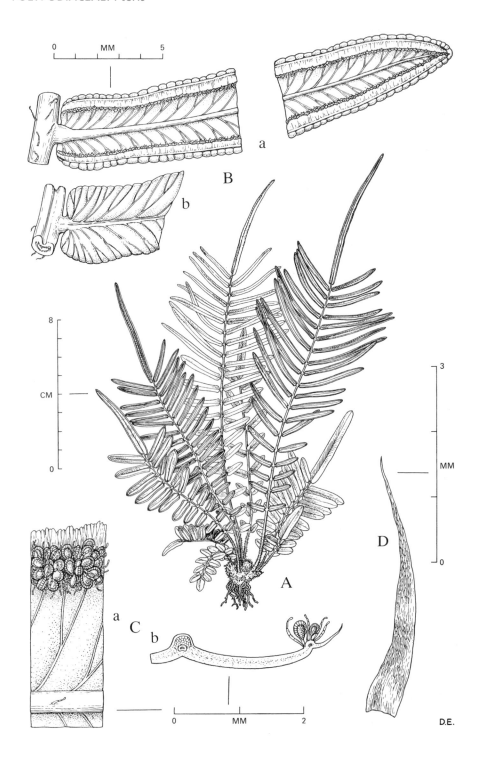

FIG. 3 **Pteris longifolia** var. **bahamensis**. A, habit. B, pinna; a, lower side; b, upper side. C, details of fterile pinna; a, with indusium removed; b, section. D, scale of rhizome. (D.E.)

Pteris longifolia L., *Sp. Pl.* 2: 1074 (1753) var. **bahamensis** (Ag.) Hieron. in *Hedwigia* 54: 289 (1914). **FIG. 3.** PLATE 1.

Pteris bahamensis (Ag.) Fée (1852).

Rhizome decumbent, densely clothed with yellowish to rusty-brown hair-pointed scales. Fronds ascending or spreading (forming an open rosette), usually less than 50 cm long, including the short stipe; stipes 5–15 cm long, bearing soft scales that leave a smooth surface on falling; lamina glabrous except for a few scattered minute hairs and glands along the rhachis, oblanceolate or rarely elliptic-oblong, 3–12 cm broad, 1-pinnate, terminating abruptly in a linear segment; pinnae narrowly linear, ascending, 2–7 mm broad, the sterile ones with finely crenulate margins; veins simple or once-forked.

GRAND CAYMAN: *Proctor* 15304. CAYMAN BRAC: *Kings* CB 97; *Proctor* 29363.

— Florida, Bahamas, and Cuba, the Cayman population occurring on moist limestone ledges, the sides of old excavations, and on talus slopes below cliffs, not common. Differs from 'typical' *P. longifolia* in its smaller fronds with smooth stipes and ascending pinnae, which are truncate or nearly so at the base. Grand Cayman *P. longifolia* plants have a diploid chromosome count, in contrast to the tetraploid Jamaican population (T. Walker, personal communication).

[*Pteris vittata* L., *Sp. Pl.* 2: 1074 (1753).

Rhizome decumbent to erect, densely clothed with pale greenish to pale brownish, linear, hair-pointed scales to 5 mm long. Fronds prostrate to ascending or arching , mostly 20–100 cm long; stipes pale brown, grooved adaxially, much shorter than the blades and densely clothed with pale hair-like scales. Blades usually 6–25 cm broad or more, oblanceolate to narrowly obovate, attenuate at base, at apex terminating abruptly in a long, linear pinna usually larger than the lateral ones; lateral pinnae narrowly lance-linear, the bases often slightly dilated, the margins (where not soriferous) finely and sharply serrulate; tissue thin-herbaceous, bright green, glabrous, the veins raised, mostly 1-forked near base. Indusium firm, greenish-brown, entire or slightly undulate.

GRAND CAYMAN: *Proctor* 50745.

— Old World tropics, first described from China; introduced and naturalized in Florida and other parts of the southern United States, Bahamas, Puerto Rico, Virgin Islands, Lesser Antilles, Trinidad, and probably other areas. Occurs on old walls, shaded waste ground, rocky banks, and sometimes on cliffs.]

Blechnum L.

Mostly rather coarse terrestrial ferns; rhizomes ascending to erect and often stout or even trunk-like, or else elongate and scandent, densely scaly, sometimes stoloniferous. Fronds 1-pinnate (rarely simple), usually glabrous, all similar (as in the Cayman species) or dimorphic; veins of sterile pinnae all free. Sori elongate-linear, usually continuous, borne near or against the pinna midvein (costa) on an elongate transverse veinlet connecting the main veins and parallel to the costa; indusium narrowly linear, continuous or nearly so, firm, opening toward the costa. Spores bilateral, usually smooth.

A world-wide genus of at least 180 species, the majority in the southern parts of both hemispheres.

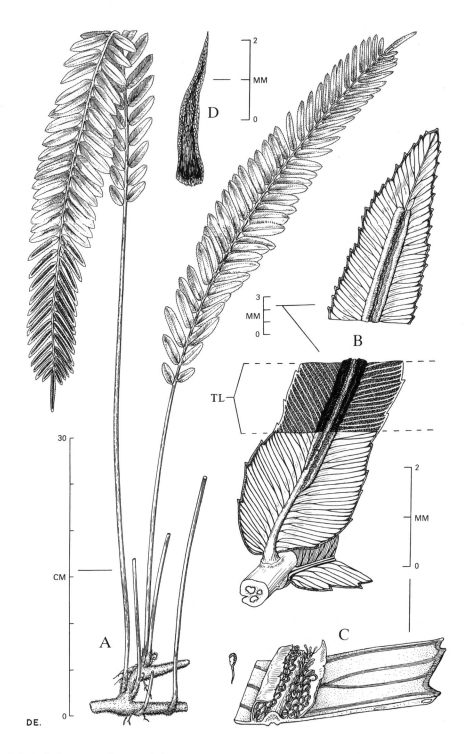

FIG. 4 **Blechnum serrulatum**. A, habit. B, pinna (TL, by transmitted light). C, enlarged portion of fertile pinna. D, scale of rhizome. (D.E.)

Blechnum serrulatum Rich., *Act. Soc. Hist. Nat. Paris* 1: 114 (1792). **FIG. 4.**

Rhizome subterranean, wide-creeping and stolon-like, with erect, subwoody branches bearing the fronds, and clothed with small, rigid, dark brown scales. Fronds rigidly erect, up to 1 m long or more, the stipes shorter than the lamina; lamina narrowly to broadly oblong, mostly 10–20 cm broad, terminating in an apical pinna similar to the lateral ones; pinnae linear-ligulate, 5–15 mm broad, sessile and articulate at the base, eventually deciduous; midvein (costa) bearing a few small yellowish scales beneath; veins close, 1–3-forked; tissue of hard texture. Sori as described for the genus; indusium becoming minutely and irregularly lacerate.

GRAND CAYMAN: *Brunt* 1951; *Kings* F 37, F 43; *Lewis* P 39; *Proctor* 3241.

— Florida, the West Indies, and Mexico to S. America; Cayman plants form tufts in grassy swamplands and in roadside ditches.

Acrostichum L.

Large, coarse ferns of brackish or saline swamps; rhizomes stout, woody, erect, scaly at the apex. Fronds erect, 1-pinnate with large, simple pinnae; venation closely areolate without included free veinlets. Sporangia densely covering the under-surface of fertile pinnae, copiously mingled with sterile paraphyses; indusium lacking. Spores tetrahedral, minutely tuberculate.

A small pantropical genus with several species of wide distribution.

KEY TO SPECIES

1. Fertile fronds with only the upper pinnae bearing sporangia; paraphyses capitate-stellate, dark glistening brown:	A. aureum
1. Fertile fronds with (usually) all the pinnae bearing sporangia, the sterile and fertile fronds thus completely dimorphic; paraphyses sausage-shaped, pale brown:	A. danaeifolium

Acrostichum aureum L., *Sp. Pl.* 2: 1069 (1753). **FIG. 5.** PLATE 1.

Rhizome bearing at apex a dense tuft of rigid linear, dark brown scales ca. 1.5 cm long or less. Fronds up to 3 m long (usually much shorter); stipes much shorter than the lamina, subterete, light brown, bearing at base a cluster of ovate, thinly papery scales, these pale brown with a blackish midrib, the margins minutely lacerate, up to 1.5 cm long; upward the stipes bearing several alternate hard spurs. Lamina oblong, 20–40 cm broad, with 10–14 ascending pinnae on each side and a terminal one; tissue of hard texture, glabrous; fertile pinnae (when present) 1–4 on a side.

GRAND CAYMAN: *Correll & Correll* 51049; *Maggs* II 66; *Proctor* 15270. LITTLE CAYMAN: *Kings* LC54; *Proctor* 28090.

— Pantropical; common in more or less brackish swamps, also occasionally in wet pockets of limestone rock.

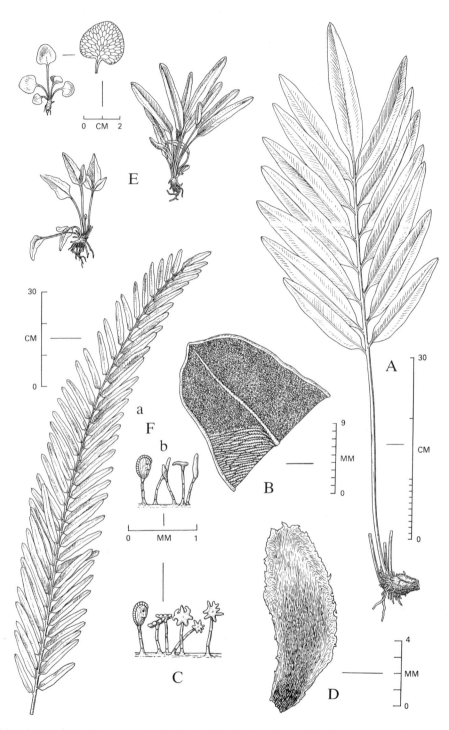

FIG. 5 **Acrostichum**. A, **A. aureum**, habit. B, tip of fertile pinna, lower side. C, sporangium and paraphyses. D, scale of rhizome. E, juvenile sporophytes. F, **A. danaeifolium**; a, frond; b, sporangium and paraphyses. (D.E.)

Acrostichum danaeifolium Langsd. & Fisch., *Ic. Fil.* 1: 5, t. 1 (1810). **FIG. 5**.

Rhizome massive, clothed at apex with rigid, linear, dark brown scales up to 2 cm long. Fronds up to 4 m tall; stipes very stout, grooved, dark brown, bearing at the base a cluster of broadly linear scales up to 2.5 cm long, these with a dark brown longitudinal band in the middle and paler, retrorsely fibrillose margins. Sterile fronds erect to somewhat spreading, the fertile ones taller and rigidly erect; pinnae very numerous, 20–40 or more on each side plus a terminal pinnae; tissue of sterile pinnae sometimes finely pubescent beneath; fertile pinnae of somewhat fleshy texture.

GRAND CAYMAN: *Proctor* 15269.

— Western Hemisphere tropics, in the same habitats as *Acrostichum aureum*, but usually much less common.

Nephrolepis Schott

Plants terrestrial or epiphytic, the rhizomes mostly erect and short, scaly and usually extensively stoloniferous. Fronds clustered, crowded, persistent, with relatively short stipes; lamina usually oblong-linear, more or less elongate, the apex of slow growth (thus often appearing indeterminate), normally 1-pinnate (some horticultural forms are more finely dissected). Pinnae articulate to the rhachis and eventually deciduous; margins entire to more or less serrate; veins 1–4-forked, all but the fertile branches reaching nearly to the margins, with enlarged tips often secreting a small white calcareous scale on the upper (adaxial) side of the pinna. Sori terminal on the first distal vein-branches, medial to submarginal in relative position, in a single row on either side of the midvein, each protected by an orbicular to lunate indusium, which is attached at the sinus. Spores bilateral.

A pantropical genus of about 30 species, some of them variable and difficult to define. Two indigenous and one naturalized species occur outside of cultivation in the Cayman Islands. Several horticultural forms of other species are sometimes grown for ornament.

KEY TO SPECIES

1. Costae glabrous (or nearly so) on adaxial side; indusium reniform or broadly U-shaped; plants often growing in axils of palm leaves: **N. exaltata**

1. Costae clothed on adaxial side with few to many short or longer hairs; indusium nearly orbicular with closed sinus; plants mostly terrestrial:

    2. Midvein of pinnae glabrous on upper side, or else bearing a few scattered flexuous hairs 0.5–1 mm long; scales of stipe spreading and rather loose; pinnae broadly cuneate at base, some of them short-stalked: **N. biserrata**

    2. Midvein of pinnae on upper side densely clothed with very short, straight, pale brown pluricellular hairs 0.2–0.3 mm long; scales of stipe (or many of them) closely appressed and blackish-brown with pale minutely fibrillose margins; pinnae subcordate and auricled at the sessile base: **[N. multiflora]**

Nephrolepis exaltata (L.) Schott, *Gen. fil.* under t. 3 (1834).

Plants terrestrial or occasionally epiphytic. Rhizome short, suberect, concealed by the stout stipe bases, the apex clothed with a dense tuft of narrowly lance-attenuate, glabrous, light orange-brown scales up to 8 mm long and ca. 0.8 mm broad above the base, terminating in a long hair-like apex. Fronds closely clustered, suberect or spreading, up to 2.5 m long (but usually much shorter), the stipes much shorter than the blades (mostly 6–20 cm long), deciduously fibrillose-scaly, the scales spreading, linear-filiform, concolorous, pale orange-brown, basally attached. Blades linear, mostly 50–100(–200) cm long, 6–14 cm broad near the middle, slightly narrowed toward the base, the apex apparently of indeterminate growth; rhachis light brown, deciduously fibrillose-scaly like the stipe, the scales glabrous; pinnae numerous, close, narrowly oblong or narrowly deltate-oblong and subfalcate, 3–7 cm long, 0.8–1.3 cm broad at the middle, acutish to sub-acuminate at the apex, subcordate and obtusely auriculate at the base, the auricle (on acroscopic side) often overlapping the rhachis, the margins bluntly serrulate to lightly crenate; tissue deciduously fibrillose or else apparently glabrate; veins 1- or 2-forked. Sori supramedial, rather close; indusium variable, orbicular-cordate to subreniform, the sinus usually open and U-shaped.

GRAND CAYMAN: *Proctor 48248.*

— Often stated to be pantropical, but many specimens with this name in herbaria have been misidentified, or else originated from cultivated or escaped plants. Its probable true natural distribution includes Florida, the Bahamas, Greater Antilles, and Mexico.

Nephrolepis biserrata (Sw.) Schott, *Gen. Fil.* under t. 3 (1834). FIG. 6. PLATE 2.

Rhizome woody, densely clothed at apex with light brown, sparingly ciliate and denticulate, hair-pointed scales; stolons numerous, slender, elongate, scaly. Fronds 1–5 m long, often arching or pendent, with fibrillose to nearly naked rhachis; pinnae 1.2–2.5 cm broad, the margins varying from finely dentate-serrulate in sterile pinnae to bicrenate in fertile ones, the crenations often minutely toothed.

GRAND CAYMAN: *Kings F–33; Millspaugh 1376; Proctor 47268.*

— Pantropical. Rare in the Cayman Islands, where its precise habitat is not recorded. Elsewhere, it occurs both on the ground, on rocky ledges, and as an epiphyte on trees especially in swamps.

[*Nephrolepis multiflora* (Roxb.) Jarrett ex Morton in *Contr. U.S. Nat. Herb.* 38: 309 (1974). PLATE 2.

Rhizome short, woody, erect, clothed especially at the apex with lance-attenuate, mahogany to dark-brown adpressed scales, these with pale, finely fimbriate-ciliate margins; similar scales on stipe and rhachis, becoming progressively smaller above; stolons numerous, wiry. Fronds 15–120 cm long, the stipes relatively very short. Lamina narrowed at base; pinnae 2–7.5 cm long, 0.5–1.2 cm broad above the auricled base, the base sessile and subcordate with the acroscopic auricle often overlapping the rhachis, the margins bluntly serrulate to finely crenate. Sori submarginal, with nearly orbicular indusium.

GRAND CAYMAN: *Brunt 1854, 1864, 2034; Proctor 15285.* CAYMAN BRAC: Proctor 29364.

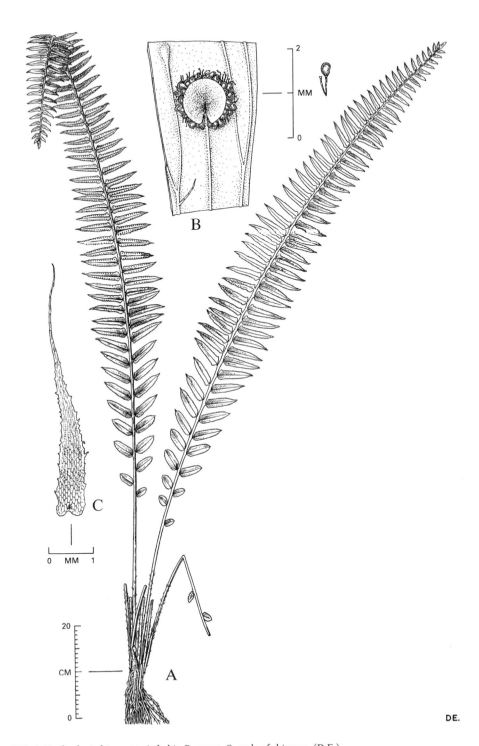

FIG. 6 **Nephrolepis biserrata**. A, habit. B, sorus. C, scale of rhizome. (D.E.)

41

— Native to India, this species has become widely naturalized in tropical areas, including southern Florida and the West Indies. Although elsewhere often a roadside weed. Cayman plants occur on shaded limestone ledges or on the moist sheltered sides of 'wells', and are not common.]

The following may be found under cultivation:

(a) *Nephrolepis cordifolia* (L.) Presl, with narrow fronds, the stolons usually bearing small, scaly tubers. Tuberless clones are also known.

(b) *N. cordifolia* cv. 'Duffii' (*N. cordifolia* f. *duffii* (T. Moore) Proctor), with small orbicular or doubly orbicular pinnae and repeatedly forked rhachises, the stolons without tubers. This form is nearly always sterile. Its alleged relationship with *N. cordifolia* may be incorrect.

(c) *N. exaltata* (L.) Schott cv. 'Boston Fern' (*N. exaltata* var. *bostoniensis* Davenp.), a variable, more or less finely dissected form that occasionally reverts to the typical 1-pinnate condition. A more extreme form is cv. 'Verona', the most finely dissected of all.

(d) *N. falcata* (Cav.) C. Chr. cv. 'Furcans' (*N. falcata* f. *furcans* (T. Moore in Nicholson) Proctor), the fishtail fern often cultivated in the West Indies, with pinnae and sometimes the rhachis forked 1–several times. It is often incorrectly described as a form of *N. biserrata*. The ancestral stock of cv. 'Furcans' apparently originated in Australia or New Guinea.

(e) *N. hirsutula* (Forst.) Presl cv. 'Superba', a hairy, always sterile form with deeply and irregularly lobed pinnae. Typical *N. hirsutula* is native to tropical Asia and islands of the western Pacific Ocean. It has been reported to be occasionally adventive in the American tropics.

Tectaria Cav.

Medium to rather large terrestrial ferns; rhizomes woody, short-creeping to erect, scaly at the apex. Fronds clustered, simple to 1-pinnate or pinnately dissected, the divisions usually broad with entire or sparingly incised margins; venation reticulate, the areoles with or without free included veinlets. Sori mostly round (sometimes somewhat elongate or irregular), scattered or borne in open rows; indusium peltate, reniform, or sometimes lacking. Spores bilateral with thick perispore, becoming tuberculate or spinulose.

A pantropical genus of perhaps nearly 100 species, many of them poorly understood.

Tectaria incisa Cav., *Descr. Pl.* 249 (1802). FIG. 7.

Rhizome woody, erect, at maturity 1.5–3 cm thick, the apex clothed with dark brown, lance-deltate, denticulate-fimbriate scales up to 7 mm long. Fronds scarcely more than 50 cm long in Cayman plants, elsewhere up to 1.5 m long, the grooved stipe about as long as the lamina; lamina oblong or ovate-oblong, 1-pinnate with basal pinna deeply 2-lobed, the lower division shorter than the upper; undivided pinnae narrowly or broadly oblong with long-acuminate apex, usually several cm broad, the terminal pinna confluent with the uppermost lateral ones; tissue normally glabrous; vein-areoles often with a free included veinlet. Sori round; indusium round-reniform, persistent, often appearing peltate because of the overlapping of the basal lobes.

GRAND CAYMAN: *Kings* F 46.

— West Indies and Mexico to S. America. Rare in the Cayman Islands, where it occurs on the moist, shaded sides of 'wells'.

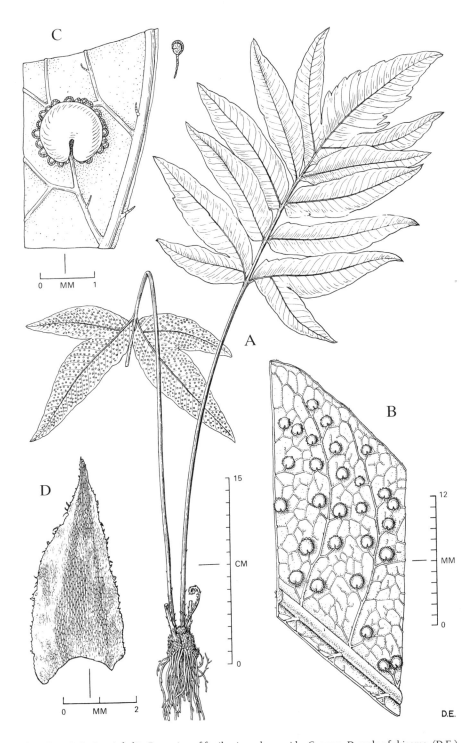

FIG. 7 **Tectaria incisa**. A, habit. B, portion of fertile pinna, lower side. C, sorus. D, scale of rhizome. (D.E.)

Macrothelypteris (H. Itô) Ching

Medium-sized terrestrial ferns. Rhizomes creeping, clothed with more or less ciliate scales. Fronds subdistichous to loosely clustered, long-stipitate; blades of ample size and mostly 3–4-pinnate, clothed throughout with long unbranched pluricellular hairs (or rarely glabrescent), these white or colorless; veins all free, with enlarged tips not reaching the margins, the ultimate ones simple or with a short spur bearing a sorus. Sori small; indusium lacking, or else rudimentary and quickly deciduous; paraphyses lacking; sporangia often with short-stipitate glands; spores ellipsoidal and monolete, the surface reticulate with coarse, more or less connected, perforate ridges.

A genus of about nine species as defined by Holttum (1969), all indigenous to tropical and subtropical Asia and adjacent islands. Of these, a single weedy, aggressive species has established itself in the Western Hemisphere, apparently within the past century. Its advent into the Cayman Islands must have been very recent, because I have found it at but a single site on Grand Cayman in the Beach Bay area (in February 1994).

Macrothelypteris torresiana (Gaudichaud) Ching, *Acta Phytotax. Sin.* 8: 310 (1963).

Rhizome stout, short-creeping or ascending, densely clothed at green apex with brown, linear-attenuate, ciliate scales, these up to 8 mm long. Fronds close, arching, variable in size, reaching 2 m in length but commonly fertile when very much smaller; stipes mostly 20–35 cm long (rarely to 75 cm), pale green and glaucous at first, becoming straw-colored, densely scaly toward base, the scales like those of rhizome apex, otherwise glabrate and minutely stipate-glandular. Blades deltate-ovate to deltate-lanceolate, longer than the stipes, to 1.5 m long or longer, mostly 20–60 cm broad, long-acuminate, the apex pinnatifid; tissue thin, membranous, with veins 3–5 pairs in the ultimate divisions, simple or (when fertile) 1-forked; smallest fertile blades 2-pinnate-pinnatifid, the largest 3-pinnate-pinnatifid to 4-pinnate; rhachis 2-grooved on adaxial side, the grooves separated by a rounded, pubescent ridge, otherwise sparsely and minutely stipitate-glandular; pinnae deltate-lanceolate or lanceolate, attenuate, short-stalked; ultimate divisions oblong-falcate or linear-oblong, mostly 1.5–2 mm broad, subentire to crenate-denticulate; costae narrowly green-winged; costules whitish short pubescent adaxially, all vascular parts beneath bearing numerous long, colorless, pluricellular hairs. Sori medial; indusium minute, soon deciduous, the sorus usually appearing to lack an indusium; sporangia glabrous.

GRAND CAYMAN: *Proctor* 49212.

— Tropical southeast Asia and islands of the southwest Pacific; now naturalized in southeastern United States, Bahamas (Crooked Island), Greater and Lesser Antilles, Trinidad, parts of S. America, and doubtless elsewhere.

In Grand Cayman, *Macrothelypteris torresiana* has been found in moist sheltered hollows and grottoes among jagged limestone rocks.

Thelypteris Schmid.

Small to moderately large terrestrial ferns. Rhizomes slender and wide-creeping to thick and erect, clothed with scales. Fronds scattered or clustered, the lamina usually 1-pinnate (rarely simple), the pinnae with margins variously lobed or pinnatifid, glabrous or commonly pubescent, the hairs simple, forked, or stellate. Veins free, or the lowermost connivent at the sinuses of lobes, or adjacent pairs joined in the tissue, at their junction producing a short excurrent veinlet; in some species (not occurring in the Cayman Islands) all the adjacent veins are joined, the venation thus being completely reticulate. Sori roundish or elliptic, dorsal on veins, with or without indusium; indusium (if present) usually roundish-reniform, attached at the sinus. Spores bilateral, monolete, with variously wrinkled or sculptured perispore.

One of the largest genera of ferns, world-wide in general distribution, with probably at least 900 species. Some authors have preferred to subdivide the group into several or many smaller genera, a procedure I do not favor. In areas where many species occur, however, it is convenient to arrange them in subgenera that emphasize lines of closer affinity within the genus. It is hardly necessary to do this when describing the few Cayman Islands species.

KEY TO SPECIES

1. Lowest veins of adjacent lobes always free or connivent to a membrane at the sinus, not truly joined in the tissue:

  2. Median pinnae with mostly fewer than 25 lobes on a side (on mature fronds), these mostly longer than broad, the basal ones of lower pinnae usually enlarged and overlapping the rhachis: **T. kunthii**

  2. Median pinnae with up to 40 lobes on a side, these mostly as broad as long and distinctly triangular, the basal ones of lower pinnae, if slightly enlarged, not overlapping the rhachis: **T. augescens**

1. Lowest veins of adjacent lobes always or at least sometimes clearly joined in the tissue:

  3. Hairs always simple; fronds never arching and rooting at the tip; plant of moist, swampy, or boggy habitats:

    4. Rhizome wide-creeping, black, appearing naked, the appressed scales widely scattered; fronds not tuffed: **T. interrupta**

    4. Rhizome short-creeping, densely scaly at apex; fronds tufted: **T. hispidula var. versicolor**

  3. Hairs both branched and simple (stellate hairs abundant on vascular parts); sterile fronds often elongate, arching, and rooting at the tip; plant of limestone ledges, with short decumbent rhizome: **T. reptans**

Thelypteris kunthii (Desv.) Morton in *Contr. U. S. Nat. Herb.* 38: 53 (1967). **FIG. 8.**

  Thelypteris normalis (C. Chr.) Moxley (1920). PLATE 2.

Rhizomes creeping, clothed at the apex with narrow, pale brown, ciliate scales. Fronds erect, with glabrate or sparsely pubescent stipes shorter than the lamina; lamina lance-oblong to broadly ovate-oblong, up to 50 cm long or more and 25 cm broad (often less), the lowest pinnae often deflexed at an angle to the plane of the lamina; vascular parts obliquely fine-hairy above, the underside finely whitish-hairy and also minutely capitate-glandular. Pinnae linear with attenuate apex, usually not more than 2 cm broad; segments more or less oblong-falcate with 8–10 pairs of veins. Sori medial, with persistent hairy indusium.

  GRAND CAYMAN: *Brunt* 1821, 1925; *Kings* F 10, F 17, F 18, F 20, F 22, F 28; *Proctor* 15271. CAYMAN BRAC: *Proctor* 29365.

  — Southeastern United States, the West Indies, and Mexico to C. America; rare in the Lesser Antilles and northern S. America. Cayman plants occur in moist marly swales and depressions, on ledges and the sides of excavations, and in 'wells'; frequent.

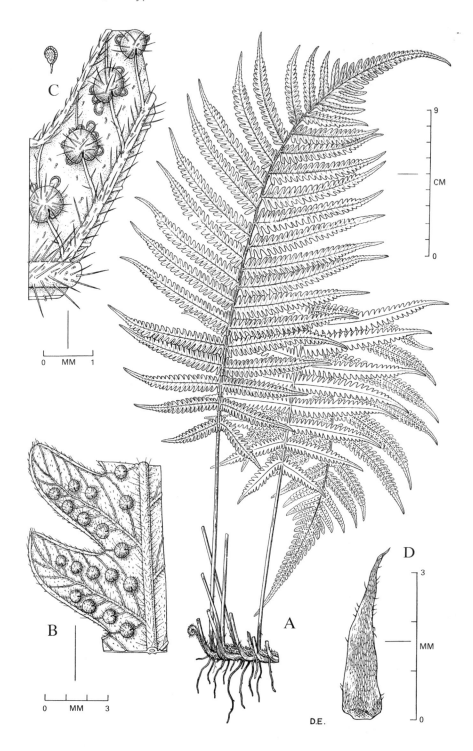

FIG. 8 **Thelypteris kunthii**. A, habit. B, portion of fertile pinna, lower side. C, enlargement of same. D, scale of rhizome. (D.E.)

Thelypteris augescens (Link) Munz & Johnston, *Amer. Fern J.* 12: 75 (1922). PLATE 2.

Rhizome creeping, the apex clothed with lance-attenuate, yellow-brown, ciliate scales. Fronds few, spaced 1–3.5 cm apart, erect to arching, with quadrangular glabrous stipes equal in length to or a little shorter than the lamina; lamina broadly oblong to roundish-ovate, up to 60 cm long or more (usually less in Cayman plants), mostly 20–40 cm broad, rather abruptly ending in a hastate-pinnatifid apex or else a somewhat distinct terminal pinna wider than the lateral ones and 7–17 cm long; underside finely and rather densely pubescent (hairs mostly 0.2–0.4 mm long). Pinnae linear with attenuate apex, usually not over 1 cm broad and of firm texture, incised halfway or a little more to the costa; segments oblique, deltate or subfalcate, with about 5–8 pairs of veins. Costae bearing small, linear, ciliate scales beneath. Sori slightly supramedial, the persistent indusium pubescent.

GRAND CAYMAN: *Proctor* 15272, 48244a.

— Florida, Bahamas, Cuba, and Guatemala. Cayman plants were found only at one locality on moist limestone ledges overhanging an open depression.

Thelypteris interrupta (Willd.) Iwatsuki, *J. Jap. Bot.* 38(10): 314 (1963).

Thelypteris gongylodes (sometimes spelled '*goggilodus*') (Schkuhr) Small (1938).
Thelypteris totta (Thunb.) Scheipe (1963).

Rhizome wide-creeping, branched, black, 2–5 mm thick, bearing scattered, lanceolate, dark purple-brown, sparsely ciliate scales. Fronds erect, distant, with glabrous stipes of equal length to or longer (sometimes much longer) than the lamina; lamina narrowly oblong, up to 50 cm long or more and 20 cm broad (commonly narrower), with abruptly acuminate apex, glabrous on the upper (adaxial) side, sparsely pubescent beneath. Pinnae linear with bluntly acuminate apex, 5–15 mm broad, lobed less than halfway to the costa; costae bearing a few small brown scales beneath, the costules minutely resinous-glandular; lobes debate or rounded, with 7–15 pairs of veins, the basal adjacent ones united in the tissue. Sori medial, with glabrous persistent indusium.

GRAND CAYMAN: *Brunt* 1975; *Proctor* 48247.

— Pantropical, usually in mats of floating vegetation on small freshwater ponds or in boggy marshes; rare in the Cayman Islands.

Thelypteris hispidula (Decaisne) Reed var. **versicolor** (R. St. John) Lellinger in *Amer. Fern Jour.* 71: 94 (1981).

Rhizome short-creeping, tiped with brownish, linear-attenuate, ciliate scales. Fronds crowded, numerous, 40–80 cm long, with slender stipes up to about 20 cm long; blades lanceolate or broadly lanceolate, mostly 8–15 cm broad below the middle, short-tapering at base and acuminate at apex. Pinnae about 12–19 pairs, linear-lanceolate, sessile, not narrowed or auriculate at base, narrowed and slightly falcate at apex, bluntly lobed with mostly 5–6 pairs of single veins in each lobe, the basal veins of adjacent lobes variably joined in the tissue or ending at or above the sinus. Sori medial; indusium minutely hairy, persistent. Tissue light green, rather thin, minutely hairy throughout.

GRAND CAYMAN: *Proctor* 48244 (SJ).

— The variety widely distributed in southeastern United States from South Carolina and Florida to eastern Texas; also recorded in Cuba. The Cayman plants were growing in a moist roadside ditch S.E. of Halfway Pond, S. of George Town.

Thelypteris reptans (J. F. Gmel.) Morton in *Fieldiana* (*Bot.*) 28: 12 (1951); *Amer. Fern Jour.* 41: 87 (1951). **FIG. 9.**

Rhizome short, decumbent, at the apex bearing lance-attenuate, brown, stellate-pubescent scales. Fronds clustered, variable in form, procumbent and often attenuate to a proliferous tip to shorter and laxly ascending or erect, the sterile and fertile ones often of different form; stipes much shorter than the lamina. Lamina oblong to linear-attenuate, mostly 10–30 cm long, 2–5 cm broad, pinnate throughout or pinnatifid toward the apex; rhachis stellate-pubescent and usually often bearing longer simple hairs. Pinnae few or numerous, close or distant, linear to oval, the apex rounded to acute, the margins more or less crenately lobed, minutely stellate-pubescent on both sides; veins 2–7 pairs per lobe, the basal ones usually united and sending an excurrent veinlet to the sinus between the lobes. Sori inframedial in relation to the midveins of the lobes; indusium minute, bearing numerous long, simple or forked hairs.

GRAND CAYMAN: *Brunt* 1820; *Kings* F 47; *Proctor* 15273.

— Florida, the West Indies, and Mexico to Venezuela. Cayman plants occur along the shaded bases of limestone ledges and are frequent.

Polypodium L.

Small to moderately large, chiefly epiphytic or epipetric ferns of varied habit. Rhizomes creeping, clothed with often clathrate scales, rarely nearly naked. Fronds articulate to the rhizome, all similar or somewhat dimorphic, glabrous or variously hairy or scaly; lamina simple to pinnatisect or 1-pinnate, sometimes further subdivided, the margins entire or rarely crenate-toothed; veins branched, free or variously reticulate, the areoles often with free included veinlets. Sori round or oval (rarely oblong to linear), terminal on vein-branchlets, not marginal, usually in 1 or more regular rows, always lacking an indusium. Spores bilateral, with smooth to tuberculate perispore.

A world-wide genus of about 225 species, often subdivided into smaller genera on the basis of frond-outline and variations in the pattern of venation.

KEY TO SPECIES

1. Lamina pinnatisect:
  2. Rhizome-scales orange, mostly 10–15 mm long; frond-segments more than 15 mm wide: P. aureum
  2. Rhizome-scales dark brown or bicolorous, mostly 2–4 mm long; frond-segments less than 5 mm wide:
    3. Fronds essentially naked except for scattered minute scales along the rhachis beneath; veins all free: P. dispersum
    3. Fronds densely scaly, especially beneath; veins more or less reticulate: P. polypodioides
1. Lamina simple:
  4. Rhizome elongate, less than 1.5 mm thick, bearing scattered fronds of less than 12 cm long, of thin texture; sori uniseriate: P. heterophyllum
  4. Rhizome short-creeping, 5–8 mm thick, embedded in a mass of roots, bearing clustered fronds often more than 20 cm long (to 50 cm or more), of stiff, somewhat cartilaginous texture; sori multiseriate: P. phyllitidis

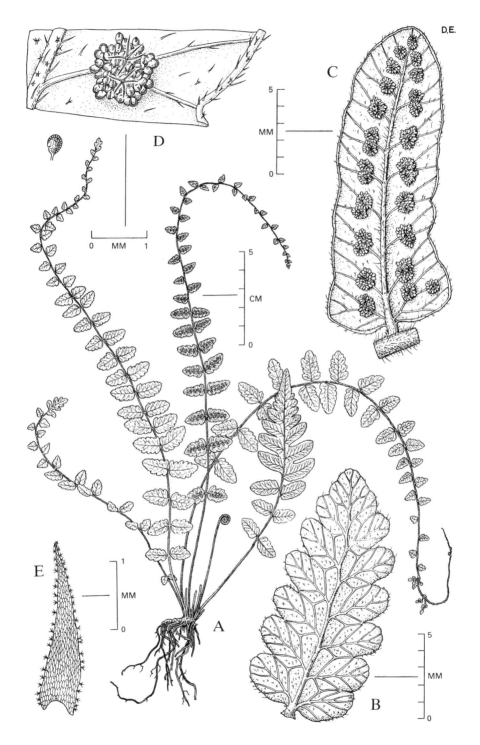

FIG. 9 **Thelypteris reptans**. A, habit. B, pinna, upper side. C, fertile pinna, lower side. D, sorus. E, scale of rhizome. (D.E.)

Polypodium aureum (L.) *Sp. Pl.* 2: 1087 (1753).

Rhizome creeping, 9–15 mm thick (excluding scales), densely clothed with numerous pale orange or tawny, denticulate-ciliate scales mostly 10–15 mm long, these linear attenuate or filiform from slightly enlarged and darker base attached peltately by a minute brown stalk. Fronds arching or pendent, 55–170 cm long, seasonally deciduous; stipes shorter than the blades, 20–70 cm long, lustrous dark brown to light reddish-brown, glabrous. Blades ovate-oblong or broadly oblong, 35–100 cm long, 22–45 cm broad, deeply and coarsely pinnatifid, with a long, often much larger terminal segment; segments 6–22 pairs, ligulate or lance-ligulate, 2–3.8 cm broad, obtuse to acuminate at apex, joined at base by a rhachis-wing 2–9 mm wide; margins subentire to minutely and distantly crenulate; veins mostly reticulate, forming a row of oblong to narrowly obdeltate costal areoles without included veinlets, then two uneven and irregular series of larger areoles with included veinlets mixed with smaller areoles lacking included veinlets; veinlets 1–3 (if 2 or 3 usually joined at the tips, or 2 joined and one free), then (toward margin) many smaller areoles of irregular shape and unequal size, some with a single included free veinlet, terminating along the margin in an irregular row of small adaxial hydathodes, these on very short free veinlet tips or at the junction of two veinlets, and each secreting a minute white scale of calcium carbonate; tissue usually green, rarely somewhat glaucous. Sori round or oval, located at the junction of two intra-areolate veinlets or sometimes at tip of a single free veinlet; very rarely a partial third row of sori present.

GRAND CAYMAN: *Proctor* 48249.

Florida, Bahamas, Greater Antilles (but rare or nearly absent in Jamaica), Lesser Antilles, Trinidad, and continental tropical America. The sole Cayman record came from the rocky forest behind University College of the Cayman Islands. It occurs on trees and mossy ledges.

This species has been shown to have originated as a fertile allotetraploid hybrid of *Polypodium pseudoaureum* and *P. decumanum* (both diploid). It is more common and widespread than either of its parent species, and often occurs where both are absent.

Polypodium dispersum A. M. Evans in *Amer. Fern Jour.* 58: 173, t. 27. (1969). FIG. 10.

Rhizome short-creeping, 4–8 mm thick, densely clothed with narrow, attenuate, dark brown, denticulate scales to 3 mm long. Fronds elastic-herbaceous, with blackish puberulous stipes much shorter than the lamina; lamina narrowly elliptic-oblong, ovate-oblong, or linear-oblong, 12–40 cm long, mostly 4–6 cm broad; rhachis black, bearing scattered small brown scales beneath, these flat with denticulate margins; segments numerous, close, horizontal, ligulate, with ciliolate margins, the underside minutely septate-puberulous; veins 2-forked, obscure. Sori terminal on distal vein-forks, superficial.

GRAND CAYMAN: *Kings* F 3, F 34; *Proctor* 47263.

— Florida, the West Indies, and continental tropical America, usually on more or less shaded limestone rocks. This species is apogamous, i.e. it produces diploid spores and gametophytes that give rise to new sporophyte plants vegetatively, without the intervention of a sexual process.

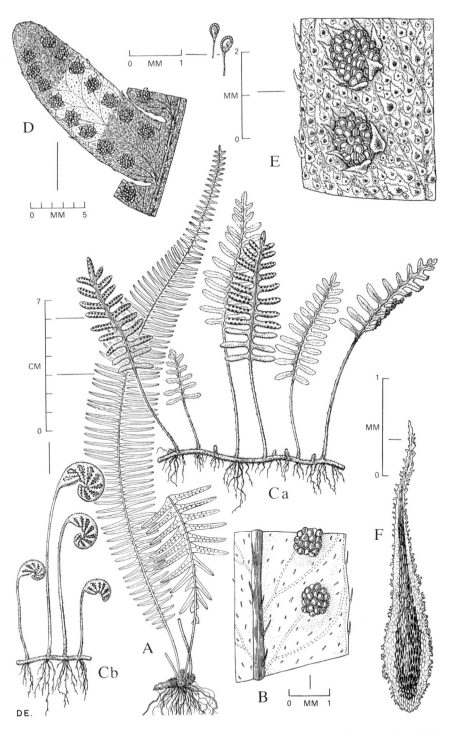

FIG. 10 **Polypodium**. A, **P. dispersum**, habit. B, portion of fertile pinna, lower side. C, **P. polypodioides**; a, habit (expanded after rain); b, habit during dry weather. D, fertile segment, lower side. E, scale of rhizome. (D.E.)

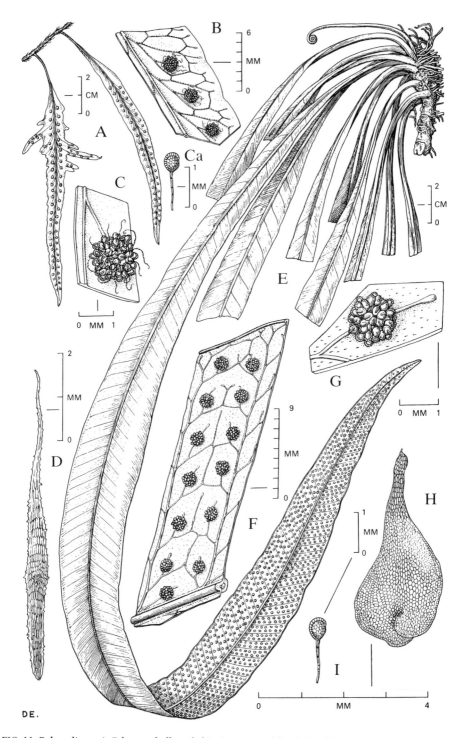

FIG. 11 **Polypodium**. A, **P. heterophyllum**, habit. B, portion of fertile frond, lower side. C, sorus; a, sporangium. D, scale of rhizome. E, **P. phyllitidis**, habit. F, portion of fertile frond, lower side. G, sorus. H, scale of rhizome. I, sporangium. (D.E.)

Polypodium polypodioides (L.) Watt in *Canadian Nat. II*, 13: 158 (1867). **FIG. 10.** PLATES 2 & 3.

Rhizome wide-creeping, 1–2 mm thick, clothed with lance-subulate, dark brown scales with pale denticulate-ciliolate margins. Fronds scattered, hygroscopic (curling up in dry weather, opening out after rain), the densely appressed-scaly stipes shorter than the lamina; lamina chiefly oblong or linear-oblong, 4–15 cm long, 1.5–6 cm broad, densely clothed beneath with roundish or deltate-ovate, peltate scales that entirely conceal the tissue, on the upper side bearing a few scattered, slender, pale, fimbriate scales; segments 2.5–5 mm broad, rounded at the apex, dilated at the base; veins obscure. Sori supramedial, protruding from pockets in the lamina tissue.

GRAND CAYMAN: *Brunt* 2146; *Kings* F 32, F 32-a, F 36, F 45; *Proctor* 15034, 47313, 47846, 48211. CAYMAN BRAC: *Proctor* 29084.

— Widespread in the warmer parts of the Western Hemisphere from Ohio and Virginia to Argentina; common in the West Indies. Cayman plants occur on mossy logs and tree-trunks in sheltered woodlands, and are rather rare.

Polypodium heterophyllum L., *Sp. Pl.* 2: 1083 (1753). **FIG. 11.** PLATE 3.

Rhizome elongate, densely clothed with linear-attenuate, tawny to reddish-brown, denticulate scales, these peltately attached far above the narrowed base. Fronds distant, variable in form, glabrous, the stipes very short or nearly lacking; sterile lamina oval or elliptic and 1–3 cm long, or lanceolate to linear and 3–12 cm long, narrowed at both ends, with margins undulate, sinuate, or (rarely) irregularly crenate to incised; fertile lamina mostly narrower than the sterile and somewhat longer; veins reticulate, forming a row of areoles on either side of the costa, producing numerous short free excurrent veinlets near the margins, and a single free or closed veinlet within each areole. Sori borne singly on the infra-areolar veinlets, forming a single row on either side of the costa; sporangia accompanied by brown hair-like paraphyses.

GRAND CAYMAN: *Brunt* 2174.

— Florida and the West Indies, often climbing on the stems of shrubs or small trees in dense woodlands; sterile plants sometimes grow on limestone rocks. Rare in the Cayman Islands.

Polypodium phyllitidis L., *Sp. Pl.* 2: 1083 (1753). **FIG. 11.** PLATE 3.

Rhizome short-creeping, subwoody, usually enveloped in a mass of brown-tomentose rootlets, bearing at the apex a few appressed-imbricate, more or less ovate scales, these 2–6 mm long, acute to acuminate, gray-brown, clathrate, glabrous. Fronds with stipes very short or nearly lacking, passing gradually upwardly into the long-decurrent lamina; lamina naked, glabrous, narrowly lance-oblong, 4–8 cm broad near the middle; principal veins ascending, joined by arching cross-veins, the areoles mostly with 2 or 3 included free excurrent veinlets, one of these sometimes prolonged and joined to the next cross-vein above. Sori in 2 rows between the main lateral veins.

GRAND CAYMAN: *Brunt* 1865; *Kings* F 2, F 2-a, F 26, F 41, F 42; *Proctor* 27973, 48313.

— Florida, the West Indies, and continental tropical America, on trunks or mossy bases of trees in sheltered woodlands, uncommon in the Cayman Islands.

Spermatophyta

Vascular plants that produce seeds containing an embryo, i.e. a rudimentary plant that remains in a dormant condition until germination. An alternation of sporophyte and gametophyte generations (cf. discussion above under Pteridophyta) occurs, but in this case, the gametophyte is much reduced and wholly dependent on the sporophyte for its nutrition. The sporophylls (equivalent to the sporangium-bearing structures of the Pteridophyta) are arranged in groups (cones or strobili in Gymnosperms and flowers in Angiosperms) of definite or indefinite number and are heterosporous, those bearing microsporangia (anther sacs) termed stamens, and those producing megasporangia termed pistils. The megasporangium (nucellus) itself is enclosed along with the embryo sac inside an integument (testa), the whole structure termed an ovule. Each ovule, when fertilized, becomes a seed. The megasporangiate sporophyll, containing one to many ovules (open and often scale-like in Gymnosperms, closed in Angiosperms) is called a carpel. The female gametophyte is confined within the megasporangium, where its egg cell is fertilized by a non-motile (motile only in Cycadaceae) male gamete that is introduced by the growth of a pollen tube. The cellular contents of a pollen grain together with the pollen-tube constitute the male gametophyte. Pollen is normally transmitted from an anther sac to the receptive point or surface (stigma) of a carpel or fused group of carpels by the agency of wind, insects, birds, bats, or other means. The evolution of flowering plants is closely associated with that of various active agents of pollination. After the fertilization of an ovule, the young sporophyte begins its development, forming a seed, while still attached to the sporophyte of the preceding generation. The production of seeds, allowing a period of dormancy during times unfavourable for growth, and also providing a wide variety of devices for dispersal, gives seed-bearing plants a competitive advantage over the pteridophytes, whose spores and gametophytes must fend for themselves without any special protective or dispersal mechanism. This is the chief reason why the seed plants now dominate the world's land vegetation.

Seed-bearing plants, numbering an estimated total of more than 250,000 species, are subdivided into two classes, the Gymnospermae and the Angiospermae.

GYMNOSPERMAE

More or less woody plants, monoecious or dioecious, the ovules not enclosed in an ovary, typically borne on scale-like carpels that are arranged in a cone or strobilus, or else sometimes terminal on naked or bracteate stalks. Stamens solitary or clustered on the scales of a strobilus, likewise the ovules. Male and female strobili separate and dissimilar. Pollen transferral always by means of wind. Seeds nut-like, winged or unwinged, or else berry-like or resembling a drupe; endosperm present. The endosperm of gymnosperm seeds is residual female-gametophyte (embryo-sac) tissue (cf. Angiospermae, for different origin of endosperm).

The vascular tissues of gymnosperms differ from those of most angiosperms in lacking xylem vessels.

A world-wide taxon of about 700 species, mostly trees or shrubs (a few vines, a few others with subterranean stems or almost acaulescent), which represent the remnant of a group that was more abundant in past geologic periods. The gymnospermae include many trees of great economic value, such as various species of pines (*Pinus*). Only one of the several gymnosperm families occurs naturally in the Cayman Islands.

ZAMIACEAE

Woody or subwoody plants with a usually subterranean caudex that is sometimes short and partly or wholly buried in the ground, growing only from the apex and thus often unbranched; leaves rather large and of leathery texture, pinnately compound without a terminal leaflet, forming a fern-like or palm-like apical crown. Dioecious, the stamens and carpels in terminal strobili. Scales of staminate strobili bearing several anther sacs on the lower (abaxial) side; pollen distributed by wind but fertilization effected by means of motile antherozoids. Scales of carpellate strobili bearing 2 or more naked ovules. Seeds drupe-like or nut-like.

A tropical and subtropical family of 9 genera and at least 75 species, occurring chiefly in America, Australia and Africa.

Zamia L.

Caudex subwoody or toughly succulent, more or less starchy, wholly or partly buried in the ground; leaves stiff and coriaceous, the segments parallel-veined and entire or toothed, often serrulate at the apex; petiole unarmed in Cayman species. Strobili dense, many-scaled, oblong-cylindric to subglobose, the female larger and thicker than the male; scales stalked, peltate, more or less hexagonal, eventually deciduous. Anther sacs several per scale, sessile. Ovules 2 per scale, sessile. Seeds drupe-like, angled.

A tropical American genus of several variable species. West Indian forms have been used as a source of starchy flour, from which a poisonous substance must be extracted before it is edible.

Zamia integrifolia L. f. ex Aiton. **FIG. 12.** PLATE 3.

> *Zamia media* var. *commeliniana* Schuster in Engler & Diels (1932).
> *Zamia pumila* of the 1ˢᵗ ed. of this Flora, not L. (1753).

BULRUSH, BULL RUSH

Caudex stout, up to 20 cm long or more and 6 cm thick, usually completely buried in the ground, clothed at the apex with silky villous, narrowly acuminate scales, 2–3.5 cm long. Leaves usually 4–6 in a crown, erect-arching, up to 75 cm long; leaflets 12–20 on a side, alternate or subopposite, linear to narrowly oblanceolate, 9–14 cm long, 0.7–2 cm broad, serrulate at the obtuse apex. Peduncles villous-pubescent, 1–3 cm long; strobili with dark red tomentose scales. Male strobili oblong, 3.5–7.5 cm long at anthesis. Carpellate strobili broadly oblong-ellipsoid, 6–9 cm long. Seeds red, angled, ca. 2 cm long.

GRAND CAYMAN: *Brunt* 1868, 1876; *Kings* GC 256; *Proctor* 15008, 50750, 52117, 52118. LITTLE CAYMAN: *Proctor* 35122. CAYMAN BRAC: *Fawcett*; *Kings* PCB 1; *Proctor* 15325.

— Florida, Jamaica, and Puerto Rico, in rocky woodlands at low elevations.

[*Cycas revoluta* Thunb., SAGO, occurs under cultivation in Grand Cayman (*Kings* PGC 258).]

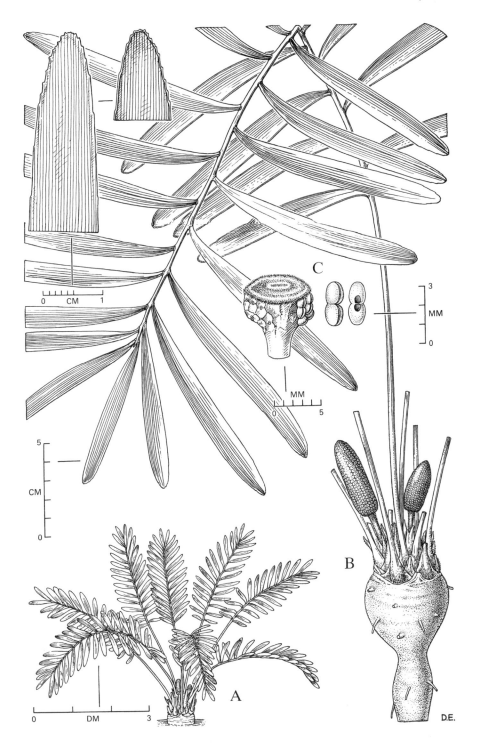

FIG. 12 **Zamia integrifolia**. A, habit. B, male plant with details of leaf-tips. C, scale of male cone and pollen-sacs. (D.E.)

ANGIOSPERMAE

Plants of very diverse habit, structure, form, size, and habitat; ovules (and seeds) borne enclosed in carpels located in the center of flowers, these carpels interpreted as modified fertile leaves (megasporophylls) infolded along a median line or zone so that the margins form a more or less firmly sealed ventral (adaxial) suture. Carpels either free or often several united into a compound pistil, the ovule-bearing portion (the ovary) ripening into a fruit. Fertilization is usually a double process in which (1) the egg cell is fertilized by a male gamete reaching it by means of the pollen tube, and (2) other female gametophyte cells within the ovule are united with additional pollen-tube nuclei, and subsequently often develop into a triploid or polyploid nutritive tissue called endosperm, which is closely associated with the embryo in the seed.

The flowering plants dominate the land vegetation of most regions of the world, and in all form a vast array of probably nearly 250,000 species, including the majority of plants of economic value and nearly all those used as a source of food or fiber. They are customarily divided into two subclasses, as follows:

KEY TO SUBCLASSES

1. Stem (if present) with vascular bundles scattered through a large solid or spongy pith, or sometimes hollow with the bundles scattered in the surrounding cylinder; leaves usually parallel-veined (reticulate in Araceae and Smilacaceae); parts of the flower nearly always in 3s; embryo with 1 cotyledon: **MONOCOTYLEDONES**

1. Stem with a single continuous or interrupted vascular cylinder, in woody members this cylinder increasing in diameter by growth of a cambium, the pith (if any) usually small and with no vascular bundles scattered through it; leaves usually reticulate-veined; parts of the flower most often in 5s or 4s, sometimes more numerous, seldom in 3s; embryo with 2 cotyledons (rarely more): **DICOTYLEDONES**

MONOCOTYLEDONES

The monocotyledonous families are arranged in the following sequence of orders and families, in accordance with the taxonomic sequence of Cronquist:

ALISMATALES	Alismataceae
HYDROCHARITALES	Hydrocharitaceae
NAJADALES	Ruppiaceae, Cymodoceaceae
COMMELINALES	Commelinaceae
CYPERALES	Cyperaceae, Gramineae
TYPHALES	Typhaceae
BROMELIALES	Bromeliaceae
ARECALES	Palmae
ARALES	Araceae, Lemnaceae
LILIALES	Pontederiaceae, Dracaenaceae, Asphodelaceae, Amaryllidaceae, Agavaceae, Smilacaceae
ORCHIDALES	Orchidaceae

KEY TO THE MONOCOTYLEDONOUS FAMILIES

1. **Aquatic plants, at least the base normally growing in water:**

  2. Plants entirely submerged:

    3. Plants of brackish water; stems trailing loosely in the water, not subterranean: **Ruppiaceae**

    3. Plants of sea water; stems (rootstocks) wide-creeping, subterranean:

      4. Leaves strap-like, ca. 10mm wide: **Hydrocharitaceae**

      4. Leaves narrowly linear, less than 1.5 mm wide: **Cymodoceaceae**

  2. Plants floating or else rooted in mud, often with emergent leaves:

    5. Plants floating on fresh water pools or tanks, minute, lacking leaves, the plant-body a flat disk usually less than 4 mm long, multiplying by budding: **Lemnaceae**

    5. Plants usually rooted in mud and always with conspicuous leaves:

      6. Each flower with numerous separate carpels; petals 3, white, conspicuous: **Alismataceae**

      6. Each flower with 1 carpel, or if more than one, these joined in a common ovary; flowers not white:

        7. Flowers minute, unisexual, aggregated into dense brown cylindric spikes: **Typhaceae**

        7. Flowers not as above:

          8. Leaves with inflated petioles; flowers with showy perianth with 6 parts: **Pontederiaceae**

          8. Leaves linear and lacking petioles; flowers without a perianth, or this structure represented by minute bristles or scales:

            9. Leaves (when present) 3-ranked, the basal sheaths tubular and not split; aerial stems usually 3-angled, not hollow; anthers attached at the base: **Cyperaceae**

            9. Leaves 2-ranked, the basal sheaths usually split to the point of attachment; aerial stems usually cylindric and hollow; anthers attached by the middle: **Gramineae**

1. **Terrestrial or epiphytic plants, not normally growing in water:**

  10. Ovary superior or naked:

    11. Perianth lacking or represented by minute bristles or scales; flowers aggregated in spikes or spikelets:

      12. Leaves linear, without differentiated petiole and blade; veins parallel:

        13. Leaves (when present) 3-ranked, the basal sheaths tubular and not split; aerial stems usually 3-angled, not hollow; anthers attached at the base: **Cyperaceae**

        13. Leaves 2-ranked, the basal sheaths usually split to the point of attachment; aerial stems usually cylindric and hollow; anthers attached by the middle: **Gramineae**

      12. Leaves differentiated into petiole and flat expanded blade; veins reticulate: **Araceae**

    11. Perianth obviously present:

      14. Plants tree-like with solid woody trunks; leaves large, pinnately compound or palmately divided into numerous segments: **Palmae**

      14. Plants herbaceous or vine-like, the leaves always simple:

15. Perianth segments all petal-like, equal:
    16. Woody vine with unisexual flowers, the plants dioecious;
        fruit a berry: **Smilacaceae**
    16. Terrestrial herbs with bisexual flowers; fruit a dehiscent
        or indehiscent capsule: **Amaryilldaceae**
15. Perianth segments in two series, the inner petal-like, the
    outer sepal-like and usually green:
    17. Plants epiphytic or terrestrial; leaves often clothed with
        very minute scales, and usually forming a stiff rosette
        (except in *Tillandsia recurvata*) **Bromeliaceae**
    17. Plants always terrestrial; leaves naked, alternate on
        elongate trailing or ascending stems (forming a rosette in
        Tradoscantia spathacea, with leaves red-purple beneath): **Commelinaceae**
10. Ovary inferior:
  18. Flowers regular (actinomorphic) or nearly so; stamens 6, free from
    the style; seeds with endosperm; always terrestrial:
    19. Leaves with entire margins and lacking an apical spine; bulbous
        herbs with solitary or umbellate flowers: **Amaryllidaceae**
    19. Leaves with prickly margins or at least with an apical spine;
        plants coarse and rigid, not bulbous at base:
        20. Perianth segments all similar, petaloid, united below to
            form a tube; flowers in compound panicles much exceeding
            the leaves: **Agavaceae**
        20. Perianth segments of two kinds, consisting of dissimilar
            sepals and petals; flowers in simple or compound head-
            like inflorescences shorter than or barely equal in length
            to the leaves: **Bromeliaceae**
  18. Flowers irregular (zygomorphic); stamens 1 or 2, adnate to the style;
    seeds powdery, without endosperm; epiphytic or sometimes terrestrial: **Orchidaceae**
    Dracaenaceae
    Asphodelaceae

ALISMATACEAE

Annual or perennial acaulescent herbs, chiefly of marshes or wet places. Leaves basal, the elongate petioles sheathing the base, the blades flat and several-ribbed. Scapes erect, simple or branched. Flowers perfect or unisexual, regular, whorled, borne in terminal racemes or panicles. Calyx of 3 persistent green sepals. Petals 3, white, soon falling. Stamens 6 or more, the filaments distinct, the anthers 2-celled. Carpels distinct and free, few or many, capitate in our species, each with usually 1 ovule; style usually persistent, appearing as a beak on the fruit. Fruit (in the Caymanian species) a head of achenes; seeds curved, the embryo horseshoe-shaped.

A widely distributed family of about 12 genera and 75 species.

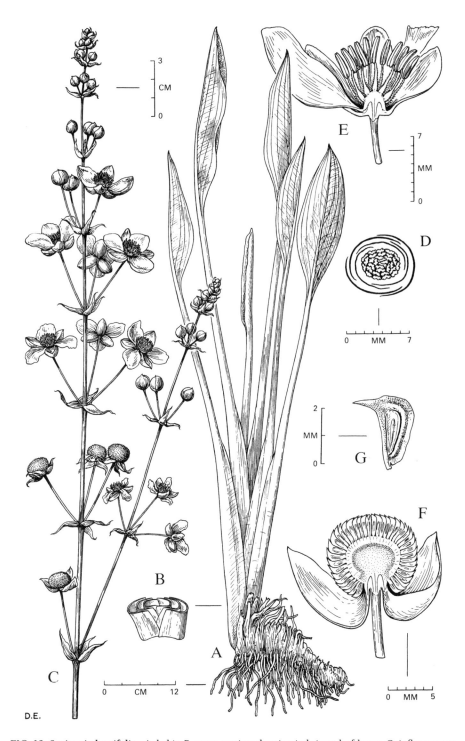

FIG. 13 **Sagittaria lancifolia**. A, habit. B, cross-section showing imbricate leaf-bases. C, inflorescence. D, floral diagram. E, staminate flower, longitudinal section. F, pistillate flower, longitudinal section. G, achene, longitudinal section. (D.E.)

61

Sagittaria L.

Perennial herbs growing in shallow fresh water or wet soil, arising from thick, somewhat tuberous horizontal rhizomes. Leaves erect, with spongy petioles. Scapes simple or branched, the flowers monoecious or dioecious, in whorls of 3, the flowers of the upper whorls usually staminate, of the lower pistillate. Stamens few or many; carpels very numerous, crowded on a head-like receptacle.

A cosmopolitan genus of about 20 species; one species is recorded from the Cayman Islands.

Sagittaria lancifolia L., *Pl. Jam. Pug.* 27 (1759). **FIG. 13**.

Plants coarse, glabrous, up to 1 m tall or sometimes more. Leaves with lance-linear to elliptic blades usually 12–35 cm long, 3–11 cm broad, acute or acuminate at apex, acuminate at base, conspicuously nerved. Scapes simple or often branched, with flowers on long, slender, spreading peduncles. Flowers 2–4 cm broad with 3 delicate pure white petals. Fruiting heads 1–1.5 cm in diameter; achenes more or less obovate, dorsally winged, with a short horizontal beak at the apex.

GRAND CAYMAN: *Brunt* 1658, 2158; *Kings* GC 189; *Lewis* 4289; *Proctor* 27963, 31024. Chiefly occurring in the George Town area.

— Southern United States, Mexico and C. America, West Indies, S. America.

HYDROCHARITACEAE

Aquatic herbs, floating or submerged, with short to elongate leafy stem. Leaves linear or strap-shaped, sessile or stalked. Flowers regular, unisexual or rarely perfect, usually solitary within a stalked, tubular, 2-lipped spathe or else subtended by 2 overlapping bracts. Calyx of 3 sepals; corolla of 3 thin petals or lacking. Stamens 3–12, with linear 2-celled anthers. Carpels 3–15, united, the ovary 1–15-locular with numerous ovules; styles or stigmas as many as the ovary-locules. Fruit a berry or capsule.

A widely distributed marine and fresh-water family of about 17 genera and 130 species.

KEY TO GENERA

1. Leaves less than 4 cm long, petiolate, with oblong to elliptic blades:	Halophila
1. Leaves 6–60 cm long, strap-shaped and without petiole:	Thalassia

Halophila Thouars.

Submerged (often deeply), rooted, monoecious or dioecious, marine herbs with slender, wide-creeping rhizomes bearing erect short shoots at nodes, these shoots bracteate at the base or above. Leaves two or more in pairs or pseudowhorls, sessile or petiolate, linear to elliptic or ovate, entire or serrulate, glabrous or minutely pubescent. Flowers unisexual, solitary or sometimes 1-staminate and 1-pistillate flower together in a single spathe; spathes sessile, comprised of 2 overlapping membranous bracts, usually axillary to erect shoots. Staminate flowers pedicellate, with 3 tepals and 3 stamens; stamens alternating with the tepals, the anthers 2–4-locular; pollen grains globose, attached together in moniliform chains. Pistillate flowers sessile, the ovary ellipsoid, unilocular, the hypanthium elongate with reduced tepals; styles 3–5, linear. Fruit ovoid, rostrate, with membranous pericarp; seeds globose, few to several.

A cosmopolitan genus of 8 species, at least some occurring in all tropical and subtropical seas, a few extending into warm-temperate areas. Two species have been found in the marine waters of the Cayman Islands.

KEY TO SPECIES

1. Individual leaves distinctly petiolate, obtuse at apex, with 3–5 pairs of cross veins:	H. baillonis
1. Leaves subsessile, acute at apex, with 6–8 pairs of cross veins:	H. engelmannii

Halophila baillonis Ascherson ex Dickie in Hook. f., *J. Linn. Soc. Bot.* 14: 317 (1874).

Dioecious marine herb; rhizomes slender, branching, 0.8–1 mm in diameter, with a single unbranched root at each node. Lateral erect shoots 0.6–2(– 4) cm long, with 2 or 3 pairs of petiolate leaves in a pseudowhorl at the apex; petioles 2–5 mm long; blades elliptic, 5–22 mm long, 2–5(–8) mm wide, the margins finely spinulose; cross-veins ascending at an angle of 60–80°. Spathe bracts lanceolate, 5–8 mm long. Staminate flower on pedicel ca. 3 mm long, with anther 4 mm long. Pistillate flower 6–7 mm long with minute perianth; ovary sessile, extended into a hypanthium with 2–5 styles, each 10–30 mm long. Fruit globular, 2–3 mm in diameter with a beak 4–5 mm long; pericarp membranous; seeds 10–20, subglobose, apiculate at both ends.

GRAND CAYMAN: East End Lagoon, *Swain* s.n. (BM).

— Panama, Cayman Islands, Jamaica, Puerto Rico, Virgin Islands, Guadeloupe, Curacao and Brazil.

Halophila engelmannii Ascherson in Neumayer, *Anl. wiss. Beoob. Reisen* ed. 1, 368 (1875) (nomen nudum); op. cit. ed. 3, 2: 395 (1906).

Dioecious marine herb; rhizome slender, branching, 1–1.3 mm in diameter, with a single unbranched root at each node. Lateral erect shoots mostly 2–4(–10) cm long bearing 2 lanceolate to obovate scales about half way to the top, and at the apex 2–4 pairs of subsessile leaves arranged in a pseudowhorl, the blades oblong or elliptic-oblong, 10–25(–30) mm long, 3–6 mm wide, of rather thick texture, with obtuse to acute apex and cuneate base, the margins minutely serrulate, the cross-veins ascending at a 30–45-degree angle. Spathe bracts lanceolate, acuminate, sessile in the axil of a leaf, enclosing one flower only. Staminate flowers unknown. Pistillate flowers consisting of a minute 3-parted perianth, a sessile or subsessile ovoid ovary 3–4 mm long and a hypanthium 3–5 mm long; styles 3, ca. 3 cm long. Fruit and seeds not described.

GRAND CAYMAN: Palm Heights Canal, North Sound, *Swain* s.n. (BM).

— Coasts of Florida and the Gulf of Mexico to Texas; Bahamas, northern Cuba, Cayman Islands and Puerto Rico.

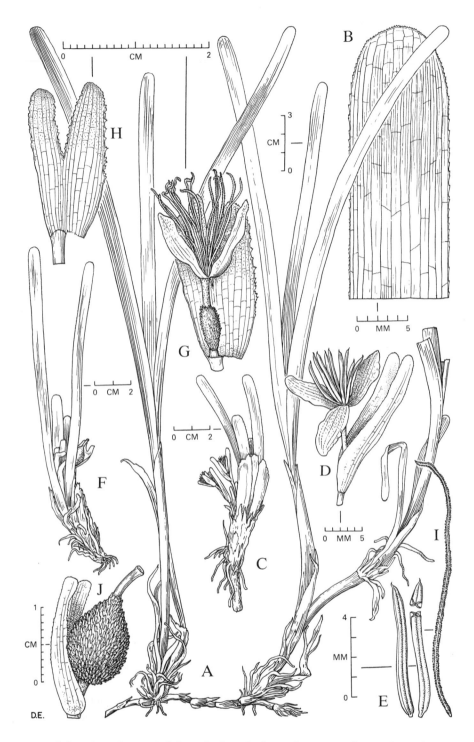

FIG. 14 **Thalassia testudinum**. A, habit. B, leaf-tip. C, plant with staminate flowers. D, staminate flower. E, anthers. F, plant with pistillate flower. G, pistillate flower. H, spathe of pistillate flower. I, stigma. J, fruit. (D.E.)

Thalassia Solander

Marine herbs with thick, creeping rootstocks, the strap-like leaves tufted on short erect branches at the nodes. Scapes elongate, arising singly among the leaves, terminating in a 1-flowered inflorescence enclosed in a tubular, 2-cleft spathe. Staminate flowers stalked, consisting of 3 petaloid sepals and 8–13 distinct stamens with very short filaments. Pistillate flowers nearly sessile within the spathe, consisting of an ellipsoid inferior ovary terminated by 12–18 linear-spindle-shaped, pilose stigmas, these always arranged in pairs. Ovary incompletely 6–12-celled by the projection of irregular placental lobes. Pollination entirely submarine. Fruit a stalked capsule with warty surface, opening by valves.

A genus of 2 species, one occurring in the Pacific and Indian Oceans, the other Caribbean.

Thalassia testudinum Konig ex Konig & Sims in *Ann. Bot.* 2: 96 (1805). **FIG. 14.**

TURTLE GRASS

Plants forming extensive colonies; leaves 2–5 in a cluster, sheathing at base, up to 30 cm long or more, 6–11 mm wide, rounded and minutely denticulate at the apex. Flowers seldom observed; sepals 10–12 mm long; anthers linear, 8 mm long; stigmas 10 mm long. Fruit ellipsoid or spindle-shaped, short-stalked and short-beaked.

GRAND CAYMAN: *Kings* GC 20, GC 158, GC 160, GC 184; *Proctor* 15133; *Sachet* 435. LITTLE CAYMAN: *Kings* LC 62a, LC 98, LC 121; *Proctor* 28109. CAYMAN BRAC: *Proctor* 29373.

— Florida to northern S. America, commonly occurring in shallow sandy bays and lagoons, and especially in areas protected by reefs.

RUPPIACEAE

Submerged aquatic herbs with long, thread-like, forking stems; leaves alternate, narrowly linear, 1-veined, and sheathing at the base. Flowers perfect, without bracts, in 2-flowered spikes on short peduncles which elongate after flowering; perianth lacking. Stamens 4, the anthers sessile, 2-celled, soon deciduous. Carpels 4, each with 1 pendulous ovule, sessile at first, each carpel eventually long-stalked; stigma minutely peltate. Fruit of long-stalked drupelets each crowned with the persistent stigma.

One genus of 2 somewhat variable species, widely distributed.

Ruppia L.

Characters as given for the family; both species occur in the Cayman Islands. They are not easy to differentiate.

KEY TO SPECIES

1. Peduncle becoming elongate in fruit and often spiral; leaves of juvenile plants obtuse or rounded:	R. cirrhosa
1. Peduncle not much longer in fruit than in flower; leaves always sharply acute:	R. maritima

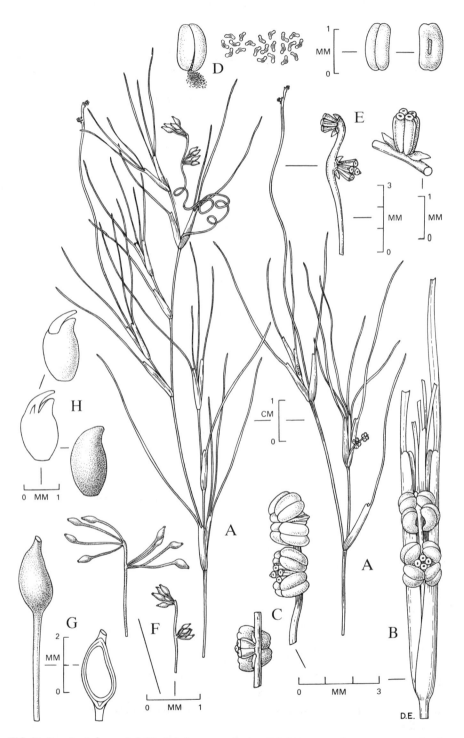

FIG. 15 **Ruppia cirrhosa**. A, habit. B, inflorescence *in situ*. C, inflorescence showing position of anthers. D, anther and pollen. E, inflorescence with early development of fruits. F, stages in development of fruits. G, single fruit. H, germination of seed. (D.E.)

Ruppia cirrhosa (Petagna) Grande in *Bull. Orto Bot. Univ. Nap.* 5: 58 (1918). **FIG. 15.**
Ruppia spiralis L. ex Dumort. (1827).

Stems flaccid, slender, repeatedly forked, up to 1 m long or more, forming loose, intricate, submerged masses. Leaves linear-capillary, up to 12 cm long, 0.3–0.8 mm wide, broadly sheathing at the base, the sheaths 6–12 mm long. Fruiting peduncles capillary, elongate, and usually curved, flexuous or spiral and up to 10 cm long or more. Drupelets ovoid, 2–3 mm long, on slender stalks (carpophores) 1–3 cm long.

GRAND CAYMAN: *Kings* GC 405.
— Cosmopolitan in suitable habitats. The Cayman plants were found "submerged in pond at Battle Ground, Batabano".

Ruppia maritima L., *Sp. Pl.* 1: 127 (1753).
Similar in habit to the preceding species, differentiated by the key characters.

LITTLE CAYMAN: *Proctor* 47304. CAYMAN BRAC: *Proctor* 29129.
— Cosmopolitan. The Cayman plants were found "in warm shallow brackish pools", and in a brackish lagoon.

CYMODOCEACEAE

Submerged marine perennial herbs with slender creeping stems (rootstocks). Leaves linear with sheathing bases, tufted at nodes on the rootstock. Flowers unisexual or (rarely) perfect, solitary or clustered, usually naked, rarely with minute bracts; perianth lacking. Staminate flower consisting of 1 sessile or 2 long-stalked, 2-celled anthers bearing thread-like pollen. Pistillate flower of 1 or 2 fused carpels, sessile or stalked, with hair-like style. Fruit a 1-seeded drupelet.

Four genera with about 22 species, widely distributed in tropical, subtropical and temperate regions.

KEY TO GENERA

1. Leaves flat; stigma 1:	Halodule
1. Leaves terete; stigmas 2:	Syringodium

Halodule Endl.

Dioecious herbs with creeping jointed rootstocks, rooting at the nodes. Leaves linear, grass-like, flat, 2- or 3-toothed at the apex, 2 or several tufted on short spurs at nodes of the rootstock, each tuft protected at the base by a membranous, ligulate sheath. Flowers solitary, unisexual. Staminate flower consisting of 2 anthers attached at different levels at the end of a long pedicel, the pedicel sheathed at the base. Pistillate flower consisting of a single naked carpel; style short, ending in a slender stigma. Fruit a small globose drupelet.

Four species were attributed to the Caribbean area (and Bermuda) in a monographic study (Den Hartog in *Blumea* 12(2): 289–313 (1964)). This publication did not ascribe any species to the Cayman Islands, but the same author's later book 'The Sea-grasses of the World' (1970) cites *Proctor* 15134 as *Halodule wrightii*.

Halodule wrightii Aschers. in *Sitz. ber. Ges. Nat. Freunde Berlin*. 1868: 19 (1868).
Halodule beaudettei (Hartog) Hartog, *Blumea* 12: 303 (1964).
Diplanthera wrightii (Aschers.) Aschers. (1897).

EEL GRASS

Leaves 0.3–0.8 mm wide, 2-toothed at the apex without secondary projections on the teeth, the inner side of these teeth concave. Flowers rarely observed; anthers 2-celled, about 6 mm long; carpel ca. 3 mm long. Mature fruit black.

GRAND CAYMAN: *Kings* GC 156, GC 159; *Proctor* 15134. LITTLE CAYMAN: *Kings* LC 120a; *Proctor* 28108.

— Coasts of Florida and West Indian islands to Puerto Rico; apparently absent from Jamaica, where it is allegedly replaced by *Halodule beaudettei* (Den Hartog) Den Hartog, which is said to differ in having broader leaves with 3-toothed tips. None of the Cayman collections cited above is fertile. They were found in sandy salt-water lagoons and bays.

Syringodium Kütz.

Dioecious herbs with creeping jointed rootstocks, rooting at nodes. Leaves linear, terete, pointed, 2 or several tufted on small erect branchlets at the nodes, each tuft protected at the base by an auricled membranous sheath. Flowers solitary or in cymose clusters arising from the leaf-axils. Staminate flowers consisting of 2 anthers attached at the same level on a slender stalk. Pistillate flowers of 2 fused carpels; pistil with a slender style terminating in 2 stigmas.

A genus of 2 species, one occurring in the Indian and western Pacific Oceans, the other in the Caribbean region.

Syringodium filiforme Kütz. in Hohenack, *Alg. Marin. Sico. IX*, no. 426 (1860).

Cymodocea manatorum Aschers. (1868). For discussion of this species, see Dandy & Tandy in *Jour. Bot. Lond.* 77: 116 (1939).

MANATEE GRASS

Rootstock red-brown, wide-creeping; leaves mostly 10–25 cm long, 0.5–1 mm thick, in tufts of 2 or 3 separated by internodes 1–5 cm long; basal sheaths 2.5–6 cm long. Flowers rarely observed. Fruit ellipsoid, flattened, 5–6.5 mm long, 2.5 mm broad, beaked by the persistent style; seed 2.5–3 mm long.

GRAND CAYMAN: *Kings* GC 157; *Sachet* 437. LITTLE CAYMAN: *Kings* LC 120, MA 93a; *Proctor* 28107. Undoubtedly occurs at Cayman Brac.

— Southern coasts of the United States; Bermuda and West Indian islands, in sandy salt-water lagoons and around reefs.

COMMELINACEAE

Annual or perennial herbs, often rather succulent, with alternate entire leaves, these usually with sheathing or clasping bases. Flowers perfect, regular or irregular, in cymes or umbels usually subtended by spathe-like or leafy bracts. Sepals 3, free and distinct, usually green and herbaceous. Petals usually 3, delicate and soon withering, free or united into a tube. Stamens typically 6, in 2 whorls of 3 each, but in some genera dimorphic or reduced in number; anthers 2-celled. Ovary superior, 2- or 3-celled; ovules 1 to several in each cell; sometimes only 1 or 2 cells fertile. Fruit a 2- or 3-valved capsule, or indehiscent.

A tropical and warm-temperate family of about 30 genera and 650 species.

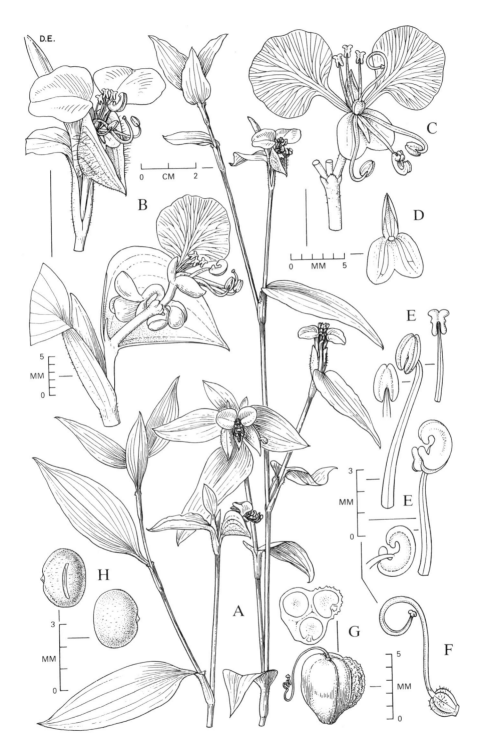

FIG. 16 **Commelina elegans**. A, habit. B, inflorescence. C, flower. D, sepals. E, stamens. F, pistil. G, capsule. H, seeds. (D.E.)

69

KEY TO GENERA

1. Plants trailing or suberect, with elongate slender stems; leaves less than 10 cm long:

    2. Inflorescences pedunculate; corolla irregular, one petal much smaller than the other two: **Commelina**

    2. Inflorescences sessile in leaf axils; corolla regular, of 3 equal petals: **Callisia**

1. Plants with short erect stem ca. 1.5 cm thick; leaves usually 20–35 cm long, red-purple beneath: **[Tradescantia]**

Commelina L.

Annual or perennial branching herbs, the stems trailing or ascending, somewhat succulent. Leaves sessile or short-petioled. Inflorescence a small pedunculate cyme or group of cymules, more or less enclosed by a folded, heart-shaped spathe-like bract; flowers blue or white. Sepals unequal, 2 usually being united toward the base. Petals markedly unequal, 2 of them being much longer than the third, clawed. Fertile stamens 3, two with oblong anthers, the other incurved with a longer anther; sterile stamens 3, smaller than the fertile ones, bearing small, cross-shaped, non-functional anthers; filaments glabrous. Ovary 3-celled, one of the cells often reduced or abortive; fertile cells with 1 or 2 ovules. Seeds smooth, rough, or reticulate.

A large genus of about 230 tropical or warm-temperate species, several of them pantropical weeds; one species occurs in the Cayman Islands.

Commelina erecta L., *Sp. Pl.* 2: 41 (1753). **FIG. 16.**

Commelina elegans Kunth in H.B.K. (1816).

WATER GRASS

Stems more or less decumbent, glabrous. Leaves lanceolate to narrowly elliptic or oblong-lanceolate, 3–8 cm long, acute or acuminate at apex; sheaths minutely ciliate on margins. Spathes 1–2 cm long, glabrous or minutely pubescent, acute, the margins joined near the base (in the closely related pantropical weed *Commelina longicaulis* Jacq., which may yet be found in the Cayman Islands, the spathe-margins are not joined at all). Petals blue, pale bluish, or sometimes white. Capsule 4–5 mm long, 3-seeded; seeds smooth, 3–3.5 mm long (in *C. longicaulis* the seeds are reticulated and 2–3 mm long).

GRAND CAYMAN: *Brunt* 2089; *Kings* GC 312, GC 313, GC 366; *Proctor* 27994. CAYMAN BRAC: *Proctor* 290557.

— Widespread throughout tropical America, common along damp roadside ditches and in other low moist situations.

Callisia Loefl.

Creeping, trailing or ascending herbs. Flowers in sessile axillary glomerules, or umbellate on filiform exserted peduncles; sepals 2 or 3; petals 2 or 3, equal; stamens 1–3, with glabrous filaments, the anther cells separated by a broad, hastate connective. Ovary 2- or 3-celled, compressed or angled, with 2 ovules in each cell; stigma more or less deeply 3-lobed. Capsule membranous, 2- or 3-valved.

A tropical American genus of 12 species, one occurring in the Cayman Islands.

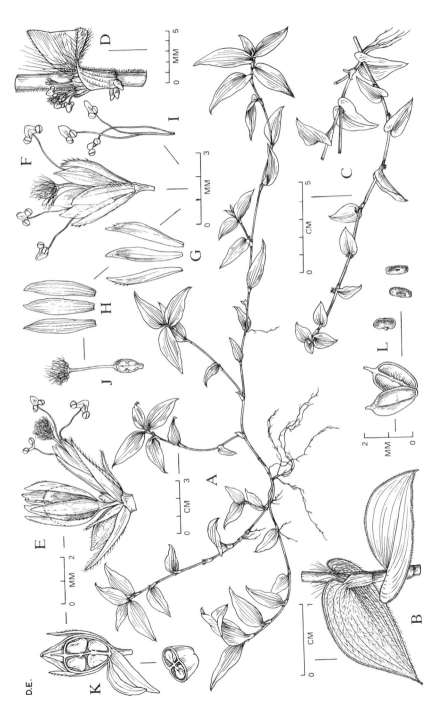

FIG. 17 **Callisia repens.** A, habit. B, leaves. C, flowering branch. D, inflorescence in situ. E, inflorescence with leaf-like bract removed. F, single flower. G, sepals. H, petals. I, stamens. J, pistil. K, capsule, dissected. L, dehisced capsule and seeds. (D.E.)

71

Callisia repens (Jacq.) L., *Sp. Pl.* 2, 1: 62 (1762). **FIG. 17.**

Stems prostrate, creeping, glabrous, with short erect flowering branches. Leaves ovate, 1–4 cm long, 0.6–1.3 cm wide, rather fleshy or succulent, often minutely speckled, apex sharply acuminate, base clasping and margins minutely ciliate. Flowers in a dense, sessile, axillary glomerule more or less enclosed by the clasping, spathe-like leaf base; sepals 3–4 mm long, pilose on back; petals 3, translucent white, shorter than the sepals. Ovary pilose at the apex; stigma trifid, long-pilose. Capsule oblong, 1.5 mm long; seeds dark brown, wrinkled, 1 mm long.

GRAND CAYMAN: *Brunt* 1775, 1777; *Kings* GC 201, GC 341; *Proctor* 27974.

— West Indies, C. and S. America, in partly shaded, often rocky situations. Various populations of this species are somewhat different in appearance, but the variations seem to be merely clonal.

[*Tradescantia* L.

Perennial herbs of diverse habit, the single Cayman species with erect succulent stems, leaves somewhat fleshy, attached in a dense close spiral to form a rosette-like cluster much longer than the inflorescences. Peduncles axillary, short; flowers in contracted scorpioid clusters enclosed within an erect, firm, boat-shaped spathe formed by 2 overlapping bracts. Sepals free, somewhat petaloid; petals 3, equal, free; stamens 6, subequal, with hairy filaments. Ovary 3-locular, with a single ovule in each locule. Fruit a 3-valved capsule; seeds rugose.

An American genus of more than 60 species occurring both as widely cultivated and as wild species, some naturalized throughout tropical and temperate areas.

Tradescantia spathacea Sw., *Prodr.* 57 (June–July) (1788).

Rhoeo discolor (L'Herit.) Hance (1853).

BOAT LILY

Plants often forming colonies; stems up to 20 cm tall, or old plants more elongate, becoming procumbent-ascending, ca. 1.5 cm thick. Leaves nearly erect, oblong-lanceolate, easily broken, mostly 20–35 cm long, 3–5 cm broad, long-acuminate at apex, dark green above and red-purple beneath in the common form. Peduncles 2–4 cm long; inflorescences several- to many-flowered; pedicels ca. 1 cm long. Petals white, 5–8 mm long. Capsule 4–4.5 mm long; seeds 3–4 mm long.

GRAND CAYMAN: *Kings* GC 188. CAYMAN BRAC: *Proctor* (sight), cult. and escaping.

— Type from Nicaragua; widely cultivated and naturalized.]

CYPERACEAE

Grass-like or rush-like herbs (rarely somewhat woody), annual or perennial, often with more or less elongate rhizomes. Culms usually erect, solid (rarely hollow), more or less 3-angled (rarely terete). Leaves linear or nearly so, 3-ranked, with closed sheaths at base, this lacking a ligule or the ligule very small; sometimes lacking a blade, the entire leaf then consisting of a tubular sheath. Inflorescences simple in heads, or variously compound; flowers bi-sexual or unisexual, arranged in dense or loose spikelets, the individual flowers always solitary in the axils of deciduous or persistent papery scale-like glumes. Spikelets 1–many-flowered, clustered or numerous, rarely solitary. Perianth composed of bristles or small scales, rarely calyx-like, sometimes lacking. Stamens usually 1–3, the anthers 2-celled. Ovary superior, 1-celled, with a single erect ovule; style usually 2- or 3-cleft (rarely more divided). Fruit an achene.

A large, world-wide family of about 106 genera and over 5,000 species, some of which are of economic importance. Members of this family are often mistaken for grasses. Most are wind-pollinated; exceptions include the white-bracted species of *Rhynchospora*; one of these (*R. stellata*) commonly occurs in Grand Cayman.

KEY TO GENERA

1. Culms terminating in a single spikelet:

2. Leaves all reduced to bladeless sheaths; spikelets conic-cylindric, the glumes spirally arranged:	**Eleocharis**
2. Leaves with free blades present; spikelets flattened, the glumes 2-ranked:	**Abildgaardia**

1. Culms terminating in more than 1 spikelet, the inflorescence compound or condensed:

3. Glumes 2-ranked, the spikelets thus evidently flattened:	**Cyperus**
3. Glumes spirally imbricate:	
4. Bracts below the inflorescence white at base; spikelets condensed into a dense, head-like cluster:	**Rhynchospora**
4. Bracts below the inflorescence green throughout, or else minute or lacking:	
5. Flowers all unisexual (plants monoecious); achenes white, shining, enamel-like:	**Scleria**
5. At least the fertile flowers bisexual; achenes not white and enamel-like:	
6. Empty glumes at base of spikelet 1 or 2:	
7. Base of style swollen; perianth bristles absent:	**Fimbristylis**
7. Base of style not swollen; perianth bristles present:	**Schoenoplectus**
6. Empty glumes at base of spikelet 3 or more:	
8. Fertile flower lateral in the spikelet; spikelets in loose panicles; culms erect, usually 1–3 m tall, distantly leafy:	**Cladium**
8. Fertile flower terminal in the spikelet; spikelets condensed into a dense, head-like cluster; culms decumbent or ascending, less than 30 cm long, densely leafy:	**Remirea**

Cyperus L.

Annuals or perennials, the leaves and culms solitary or tufted, sometimes with corm-like thickenings at the base, often proliferating by slender creeping subterranean rhizomes that in some cases may bear tuber-like nodules, all the underground parts frequently aromatic. Culms simple, erect, leafy near the base (the leaves shorter than or exceeding the culms), at apex bearing a capitate or branched umbelliform inflorescence, subtended by 1–many more or less leafy bracts. Spikelets flattened or angular, few–many in dense or loose spikes, or, by a shortening of the rhachis, in a capitate cluster. Glumes 2-ranked, deciduous or persistent. Flowers bisexual, without perianth, sometimes only 1 or 2 of a spikelet fertile. Stamens 1–3; style 2- or 3-cleft. Achenes flattened or 3-angled, neither beaked not tuberculate.

A cosmopolitan genus of about 820 species, most numerous in tropical and warm-temperate regions. The five major subdivisions (*Kyllinga*, *Pycreus*, *Cyperus*, *Mariscus* and *Torulinium*) are treated by some authors as separate genera, but I prefer to view them as subgenera. All five are represented in the Cayman Islands. In the following treatment, the correct assignment with respect to these taxa is indicated in the synonymy of each species.

Few species of *Cyperus* have any economic value. Among the Cayman representatives, *C. compressus* and *C. rotundus* have found minor usefulness elsewhere, the former as a source of fiber, the latter for perfume and medicine.

KEY TO SPECIES

1. Styles 2; achenes flattened:
   2. Spikelets aggregated into a single dense ovoid or globose head; rhachillas jointed at base, the whole spikelet falling when ripe:
      3. Leaves reduced to bladeless sheaths; bracts not exceeding the inflorescence: *C. peruvianus*
      3. Leaves with foliaceous blades; bracts leaf-like, much exceeding the inflorescence: *C. brevifolius*
   2. Spikelets fasciculate, forming penicillate clusters either on a branched or contracted lobed or head-like inflorescence; rhachillas not jointed at base, persistent after the achenes and glumes have fallen: *C. polystachyos*
1. Styles 3; achenes 3-angled:
   4. Rhachillas not jointed at all, persistent after the achenes and glumes have fallen:
      5. Culms mostly solitary; spikelets red-brown; rhizomes elongate, slender, bearing aromatic nut-like tubers: *C. rotundus*
      5. Culms tufted; spikelets green; rhizomes absent or very short:
         6. Plants annual with fibrous roots only; culms and leaves neither viscid nor septate-nodulose; glumes appressed, 3–3.5 mm long: *C. compressus*
         6. Plants perennial with short rhizomes:
            7. Culms and leaves viscid, the latter septate-nodose: *C. elegans*
            7. Culms and leaves not viscid, the latter flat, not septate: *C. surinamensis*
   4. Rhachillas jointed at least at the base:
      8. Rhachillas jointed at base only, the entire spikelets falling as a unit at maturity:
         9. Spikelets compressed, more than 4 mm long:
            10. Spikes cylindric, the spikelets not over 6 mm long: *C. ligularis*
            10. Spikes flattened, the spikelets more than 6 mm long (often much more):
               11. Spikelets 1.5–1.7 mm wide; glumes acutely keeled, ca. 1 mm wide, usually light brown; achenes narrowly obovate in outline: *C. planifolius*
               11. Spikelets 2.5–2.8 mm wide; glumes obtusely keeled, 1.5–1.8 mm wide, deep red-brown; achenes broadly elliptic in outline: *C. brunneus*
         9. Spikelets inflated, less than 3 mm long: *C. swartzii*
      8. Rhachillas jointed both at base and between flowers, the spikelets breaking into internode-segments when falling:

12. Leaves 4–13 mm wide; culms 2–5 mm thick; annual with fibrous roots only:		C. odoratus
12. Leaves not over 2 mm wide; culms 0.3–0.5 mm thick; perennials with knotted rhizomes:		
	13. Leaves 1–2 mm wide; culms up to 35 cm tall; spikelets mostly 9–20 mm long, the glumes 2.5–3 mm long:	C. filiformis
	13. Leaves 0.6–0.9 mm wide; culms less than 9 cm tall; spikelets not over 8 mm long, the glumes 1.8–2 mm long:	C. floridanus

Cyperus peruvianus (Lam.) F. N. Williams in *Bull. Herb. Boiss.* ser. 2, 7: 90 (1907).

Kyllinga peruviana Lam. (1789).

Perennial by short creeping rhizomes covered with conspicuous brown ovate scales. Leaves reduced to tubular green sheaths, membranous at the apex. Culms mostly 20–50 cm tall or sometimes taller, appearing naked. Bracts triangular, inconspicuous. Spikelets ca. 4 mm long, arranged in a dense globular head ca. 1 cm in diameter, each spikelet with about 3 glumes, only 1 of these fertile. Achenes flattened, yellowish-brown, ca. 1 mm long.

GRAND CAYMAN: *Brunt* 2085; *Proctor* 15286; *Sachet* 453.

— Widespread in tropical America and Africa, in wet soils, not common.

Cyperus brevifolius (Rottb.) Radlk. ex Hassk., *Cat. Pl. Hort. Bogor.* 24 (1844).

Kyllinga brevifolia Rottb. (1773).

Perennial by creeping rhizomes; leaves shorter than the culms, mostly 2–3 mm wide. Culms slender, mostly 10–30 cm tall. Inflorescence a single small (5–10 mm long), ovoid or globose head of densely aggregated spikelets, subtended by 3 elongate but unequal leafy bracts. Spikelets each with 3–4 glumes, only 1 of these fertile. Achenes flattened, brown, ca. 1 mm long.

GRAND CAYMAN: *Kings GC* 182; *Proctor* 47365.

— Pantropical, also extending somewhat into warm-temperate regions, in damp soils, often a weed of lawns and gardens.

Cyperus polystachyos Rottb., *Descr. Pl.* 21 (1772).

Pycreus polystachyos (Rottb.) Beauv. (1807).

Cyperus odoratus sensu Britton & Wilson (1923), not L. (1753).

Annual, or short-lived perennial with short rhizomes. Culms tufted, slender, mostly 20–60 cm tall, usually nearly equalled by the basal leaves. Inflorescence compound-umbellate (rarely capitate), subtended by several spreading, elongate, leafy bracts. Spikelets linear-lanceolate, chiefly 8–12 mm long, 1–1.5 mm wide, gray- or yellow-brown, densely fasciculate. Achenes flattened, brown to black, 1 mm long.

GRAND CAYMAN: *Brunt* 2092; *Kings GC* 182b, GC 194, GC 425; *Lewis GC* 9, GC 33; *Proctor* 15241, 27950, 31036.

— Pantropical, often a weed of roadside ditches.

Cyperus rotundus L., *Sp. Pl.* 1: 45 (1753).

NUT GRASS

Perennial with long, slender, fragile, deeply subterranean rhizomes bearing aromatic, nut-like tubers at wide intervals. Culms mostly 15–30 cm tall (rarely taller). Inflorescence compound with 2 or 3 erect, simple rays subtended by shorter leafy bracts. Spikelets few, 1–2 cm long (rarely much longer), loosely spreading. Achenes 3-angled, shining black, 1.5–2 mm long.

GRAND CAYMAN: *Proctor 31798.*

— Pantropical, also extending into many warm-temperate regions. A persistent weed of lawns, gardens and open waste ground, this species easily regenerates from any tubers left in the ground. These tubers contain an aromatic oil that has been used in some countries in perfume and in medicines as a remedy for digestive disorders.

Cyperus compressus L., *Sp. Pl.* 1: 46 (1753).

Annual, with fibrous roots only; leaves 1–3 mm wide. Culms tufted, 10–40 cm tall, wiry. Inflorescence a simple capitate umbel of spikelets or more often compound, having 2 or 3 slender rays up to 1.5 cm long, each bearing an umbel of spikelets at the apex, the whole inflorescence subtended by 2 or 3 bracts, one of these longer than the inflorescence. Spikelets 3–10 in a cluster, 8–25 mm long, 3–5 mm wide, much flattened; glumes light green with a yellowish band along each side, deciduous beginning from base of the spikelet. Achenes 3-angled, brown to black, shining, 1–1.3 mm long.

CAYMAN BRAC: *Proctor 29130.*

— Pantropical. The culms contain fibers that have been used in India for weaving small mats.

Cyperus elegans L., *Sp. Pl.* 1: 45 (1753).

Cyperus viscosus Sw. (1788).

Perennial by short tangled rhizomes. Culms tufted, mostly 30–60 cm long, sometimes overtopped by the more or less convolute wiry basal leaves. Inflorescence decompound-umbellate, subtended and exceeded by several narrow, elongate bracts; rays up to 15 cm long. Spikelets mostly 4–10 mm long, up to 3 mm wide, borne in numerous small heads; glumes greenish-brown, sharply mucronate. Achenes 3-angled, nearly black, 1.5 mm long.

GRAND CAYMAN: *Brunt 1848, 2040; Proctor 15065, 15300, 48278.* CAYMAN BRAC: *Proctor 29002.*

— Florida, West Indies, and continental tropical America, chiefly growing in moist subsaline or brackish soils near seacoasts.

Cyperus surinamensis Rottb., *Descr. Icon. Rar. Nov. Pl.* 35, t.6, f.5 (1773).

Rhizome short with rather short roots. Culms tufted, 30–60 cm tall, triquetrous, retrorsely scabrid on angles except at base, light green. Leaves shorter than or nearly equal in length to the culm, narrowly linear, 2–4 mm wide, flattish, herbaceous, gradually narrowed to a long acute apex. Inflorescence compound, 5–11 cm long, 4–10 cm wide; leafy bracts 4–7, mostly exceeding the inflorescence; corymb-rays 6–15, very unequal, up to 7 cm long; secondary corymbs bearing 3–9 heads. Spikelets lance-oblong to ovate, acute, 3–6 mm long, flattened, 10–24-flowered, light yellow-green; rhachilla not winged; glumes 1.2–1.4 mm long, incurved, thinly membranous, minutely reticulate. Achenes oblong, trigonous, 0.5–0.6 mm long, densely punctulate, maturing black.

GRAND CAYMAN: *Proctor* 52098; known only from wet swales near West Bay.

— Widely distributed in the Western Hemisphere from southern United States to Argentina.

Cyperus ligularis L., *Syst. Nat.* ed. 10, 2: 867 (1759).

Mariscus ligularis (L.) Urban (1900).

CUTTING GRASS

Perennial, forming large, dense clumps. Leaves conspicuous, pale green, often 1 cm wide or more, with harsh, rough margins. Culms stout, up to 1 m tall or more. Inflorescence compound, subtended by long leafy scabrous bracts. Spikelets mostly 4–6 mm long, ca. 1 mm wide, borne in dense cylindric spikes up to 4 cm long and 1 cm thick, these often aggregated closely; glumes light brown. Achenes 3-angled, brown, 1.2–1.4 mm long.

GRAND CAYMAN: *Brunt* 1796; *Kings GC* 168, GC 278; *Maggs* II 62; *Proctor* 15104. LITTLE CAYMAN: *Proctor* 28072. CAYMAN BRAC: *Proctor* 28983.

— Tropical America and Africa, usually at low elevations near the sea.

Cyperus planifolius L. C. Rich. in *Act. Soc. Hist. Nat. Paris* 1: 106 (1792). **FIG. 18.**

Mariscus planifolius (L. C. Rich.) Urban (1900).

CUTTING GRASS

Perennial, densely tufted from short rhizomes. Leaves many, glaucous green and of stiff texture, much longer than the culms and 8–10 mm wide. Culms 60–90 cm tall, 3–4 mm thick toward base. Inflorescence compound, subtended and much exceeded by 5–8 leafy bracts; rays very unequal, bearing corymbose clusters of spikes, these ovoid, 1.5–3 cm long, densely composed of many spikelets. Spikelets linear, 6–16 mm long, 1.5–1.7 mm wide, light brown sometimes tinged with red, 8–14-flowered; glumes rather loosely overlapping. Achenes 3-angled with shallowly concave sides, 1.2–1.5 mm long.

GRAND CAYMAN: *Sauer* 4231.

— West Indies and French Guiana, usually in coastal sands.

Cyperus brunneus Sw., *Fl. Ind. Occ.* 1: 116 (1797).

Cyperus brizaeus Vahl (1806).
Mariscus brunneus (Sw.) C. B. Clarke in Urban (1900).

CUTTING GRASS

Perennial, densely tufted from short rhizomes. Leaves many, glaucous green and of rather stiff texture, shorter than to equal in length to the culms and 3–8 mm wide. Culms 20–80 cm tall, 1.5–3 mm thick toward base. Inflorescence compound, either open with elongate rays or congested in a single head-like lobed cluster, much overtopped by 3–5 leafy bracts; rays (when developed) 1–7 cm long, bearing 1–4 spikes, these broadly ovoid, up to 3 cm long and 2.5 cm thick. Spikelets divergent, linear-oblong or lance-oblong, 10–20 mm long, mostly 2.5–2.8 mm wide (rarely less), compressed, usually deep red-brown, 6–17-flowered; glumes rather densely overlapping. Achenes 3-angled with convex sides, 1.2–1.5 mm long.

GRAND CAYMAN: *Proctor* 15130, 48267.

— Florida, the West Indies, and southern Mexico, frequent in coastal sands.

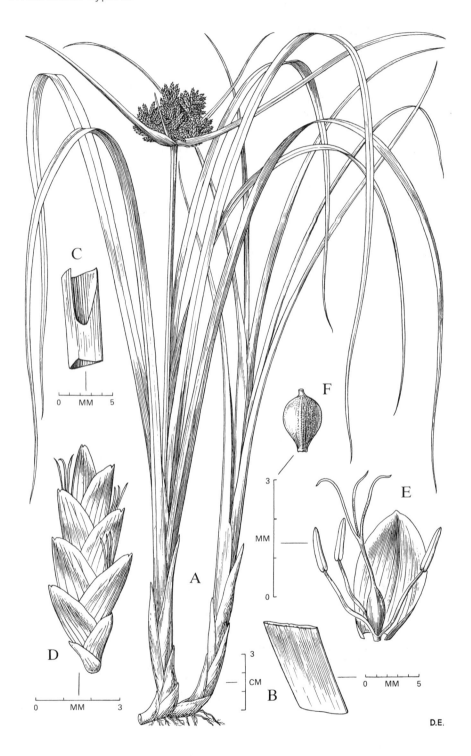

FIG. 18 **Cyperus planifolius**. A, habit. B, section of leaf. C, section of stem. D, spikelet. E, floret. F, achene. (D.E.)

The preceding two species have customarily been combined under the name *Cyperus planifolius* because of their close similarity. Since their distinctness was first recognized (T. Koyama in Howard, *Flora of the Lesser Antilles* 3: 268–269 (1979)), it has not been possible to re-study the numerous specimens of this complex in order to cite them under their correct names. They can be listed as follows: GRAND CAYMAN: *Brunt* 1705; *Kings* GC 28, GC 267; *Lewis* GC 40; *Millspaugh* 1248; *Sachet* 387a, 387b. LITTLE CAYMAN: *Kings* LC 3, LC 92, LC 101; *Proctor* 28071. CAYMAN BRAC: *Brunt* 1675; *Kings* CB 10, CB 62; *Proctor* 28924.

Cyperus swartzii (Dietr.) Boeck ex Kük. in *Fedde Repert.* 23: 186 (1926).

Mariscus gracilis Vahl (1806).
Cyperus caymanensis Millsp. (1900).

Perennial with short, tangled, woody rhizomes, forming dense tufts. Leaves few, or sometimes apparently lacking, shorter than the culms, 1–4 mm wide, slightly rough on the margins; basal sheaths red-brown. Culms slender and wiry, usually 15–40 cm tall or more. Inflorescence capitate, consisting of up to 3 densely aggregated spikes each bearing 30–40 crowded spikelets, the whole subtended by 3–5 unequally elongate bracts, these 1–2 mm wide. Spikelets 1–2.2 mm long, asymmetrically inflated, 1-flowered; fertile glumes pale greenish, longitudinally 11–15-nerved. Achenes 3-angled with concave sides, purplish-brown, 1–1.2 mm long.

GRAND CAYMAN: *Brunt* 1822; *Kings* GC 177, GC 286; *Millspaugh* 1334 (type of *Cyperus caymanensis*); *Proctor* 27964, 48304.

— Greater and Lesser Antilles, also reported from southern Mexico, often growing in damp shaded glades.

Cyperus odoratus L., *Sp. Pl.* 1: 46 (1753).

Cyperus ferax L. C. Rich. (1792).
Torulinium odoratum (L.) Hooper (1972).

Annual or perhaps sometimes a short-lived perennial, solitary or loosely tufted, with fibrous roots. Leaves shorter than the culms, up to 13 mm wide. Culms stout, smooth, up to 1 m tall. Inflorescence loosely compound, ample, subtended and overtopped by several long leafy bracts; rays 5–12, up to 10 cm long or more (rarely to 20 cm), bearing loose oblong-cylindric spikes 2–3 cm long, with 20–40 spikelets. Spikelets linear, 10–20 mm long, less than 1 mm wide, the glumes light yellowish-brown, 2 mm long or more, only slightly overlapping; at maturity each rhachilla separating into 1-fruited joints. Achenes 3-angled, brown to black, 1–1.5 mm long.

GRAND CAYMAN: *Brunt* 1845.
— Widespread in tropical and warm-temperate regions, commonly a weed of roadside ditches, damp fields and marshes.

Cyperus filiformis Sw., *Nov. Gen. & Sp. Pl.* 20 (1788).

Torulinium filiforme (Sw.) C. B. Clarke in Urban (1900).

Perennial, densely tufted, the rhizomes short and knotted. Leaves few (1–3 per culm), (less than) half as long as the culms, filiform, 0.5–1.5 mm wide, slightly rough-margined, the sheaths red-brown. Culms usually 20–35 cm tall, 0.3–0.5 mm thick. Inflorescence a solitary, lateral-appearing, loose cluster of 2–8 spikelets of unequal length, subtended by 2

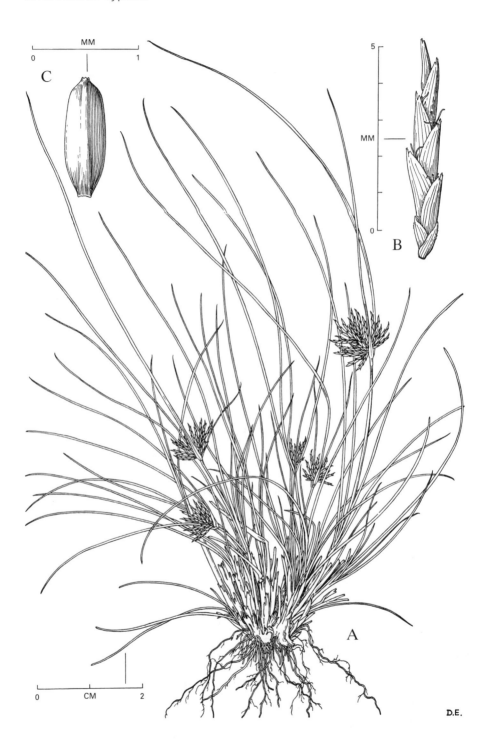

FIG. 19 **Cyperus floridanus**. A, habit. B, spikelet. C, achene. (D.E.)

or 3 bracts one of which is much longer than the others and appears like an extension of the culm. Spikelets 9–20 mm long, 0.7–1 mm wide; glumes yellow-brown, 2.5–3 mm long, appressed. Achenes 3-angled, brown to black, 2 mm long.

GRAND CAYMAN: *Kings* GC 384. CAYMAN BRAC: *Proctor* 28936.

— Bahamas, Greater Antilles and the Virgin Islands, in shaded sandy or rocky soils. Reports from Florida pertain to *Cyperus floridanus*.

Cyperus floridanus Britton in Small, *Fl. Southeastern U. S.* 170, 1327 (1903). **FIG. 19.**

Mariscus floridensis C. B. Clarke (1908) illegit. (*C. floridanus* Brit. has not so far been validly transferred to *Torulinium* with its relatives).

Cyperus filiformis var. *densiceps* Kük. (1926).

Cyperus kingsii C. D. Adams (1977).

Perennial with short knotted aromatic rhizomes; leaves and culms densely tufted. Leaves numerous, filiform, often exceeding the culms, 0.6–0.9 mm wide. Culms 4–8 cm tall, 3-angled and more slender than the leaves. Inflorescence a solitary, lateral-appearing loose cluster of usually 10–16 spikelets, the aggregate 1 cm in diameter or less. Spikelets 6–8 mm long, 0.6–0.8 mm wide, 3–7-flowered; glumes yellow-brown, striate, broadly green-ribbed, 1.8–2 mm long, mucronate. Achenes oblong-ellipsoid, 3-angled, tawny to black, 1–1.1 mm long.

GRAND CAYMAN: *Kings* GC 410 (type of *Cyperus kingsii*). LITTLE CAYMAN: *Proctor* 35098, 35205, 47285. CAYMAN BRAC: *Proctor* 28910.

— Florida Keys, Bahamas, Cuba, Jamaica, Hispaniola (Isla Saona) and Mona Island, growing in pockets of flat limestone, in loose dry calcareous sand, or in thin bauxitic soil of pastured slopes.

Eleocharis R. Br.

Annual or perennial herbs of watery habitats or wet soil, solitary or tufted, often with long rhizomes. Leaves reduced to bladeless sheaths enclosing base of culm. Culms angled, flattened, grooved or terete, sometimes hollow and septate. Inflorescence a single spikelet; subtending bracts minute or absent. Spikelet terminal, erect, few- or many-flowered; glumes spirally imbricate, deciduous. Flowers bisexual, with a perianth of normally 6 bristles, these usually retrorsely barbed; stamens 1–2; style 2–3-cleft, the expanded base persisting on the apex of the ripe achene as a cap-like tubercle. Achene flat, 3-angled or turgid.

A world-wide genus of about 200 species.

KEY TO SPECIES

1. Spikelet about same diameter as the culm or only a little thicker, 1–5 cm long; culms 1–5 mm thick; plants perennial:

    2. Culms nodose-septate: **E. interstincta**

    2. Culms not nodose-septate:

        3. Culms terete or nearly so: **E. cellulosa**

        3. Culms rather sharply 3-angled: **E. mutata**

1. Spikelet obviously much thicker than the culm:

4. Spikelets 6–20 mm long; culms more than 0.7 mm thick; plants perennial, forming tussocks: **E. rostellata**
4. Spikelets less than 6 mm long; culms less than 0.5 mm thick; plants annual:
   5. Achenes jet black, smooth and shining, more or less flattened; glumes ovate, faintly bicolorous:
      6. Perianth bristles brown; achenes 1 mm long; culms mostly 5–20 cm tall: **E. geniculata**
      6. Perianth bristles nearly white; achenes 0.5 mm long; culms mostly 2–7 cm tall: **E. atropurpurea**
   5. Achenes whitish to pale brown, reticulate-striate, acutely 3-angled; glumes ovate-lanceolate, strongly bicolorous: **E. minima**

Eleocharis interstincta (Vahl) Roem. & Schult. in L., *Syst. Veg.* ed. nov. 2: 149 (1817). FIG. 20.

Perennial, spreading by stout rhizomes. Culms mostly 40–100 cm tall, 4–5 mm thick, terete, hollow and nodose-septate; basal sheaths membranous, oblique at apex. Spikelets cylindric, up to 4 cm long; glumes rigid, oblong or ovate, often acute, pale yellowish or greenish, striate. Bristles 6, retrorsely barbed, about as long as the achene. Style 2- or 3-cleft. Achenes 1.5–2 mm long, brown or yellowish-brown, with minute transverse ridges; tubercle conic, acute.

GRAND CAYMAN: *Brunt* 1794, 1842; *Kings* GC 191, GC 192.

— Southern United States, West Indies, C. and S. America.

Eleocharis cellulosa Torr. in *Ann. Lyc. N. Y.* 3: 298 (1836).

Perennial, with deep-seated creeping rhizomes, often forming dense colonies. Culms mostly 30–70 cm tall, 1–3.5 mm thick, terete in upper part, obscurely 3-angled toward base; sheaths membranous, oblique at apex, usually purple. Spikelets cylindric, 1.5–4 cm long, thicker than the culm; glumes rigid, orbicular or obovate, obtuse, yellowish with brown border and whitish membranous margins, striate. Bristles about 6, not barbed, about as long as the achenes. Style 3-cleft. Achenes 2–2.5 mm long, olive-brown, nearly smooth; tubercle broadly conic, whitish, tipped by the blackish style-base.

CAYMAN BRAC: *Proctor* 29130. The single colony seen was growing densely in brackish sandy mud.

— Southern United States, Bermuda, West Indies, Mexico and Belize.

Eleocharis mutata (L.) Roem. & Schult. in L., *Syst. Veg.* ed. nov. 2: 155 (1817). FIG. 21.

Perennial, spreading by stout rhizomes. Culms up to 1 m tall or more, sharply 3- or 4-angled; sheaths membranous, oblique at apex. Spikelets cylindric, 2–5 cm long; glumes firm, ovate or obovate, pale yellowish. Bristles 6, brown, retrorsely barbed, about as long as the achene. Style 3-cleft. Achenes 1.5–2.3 mm long, yellowish-brown, lustrous, obscurely longitudinally striate; tubercle a low, annular depressed cap surmounted from the middle by the acuminate style-base.

GRAND CAYMAN: *Brunt* 1799, 1967, 2093; *Kings* GC 102; *Proctor* 27947, 50078.

— Widespread in tropical America; also in Africa.

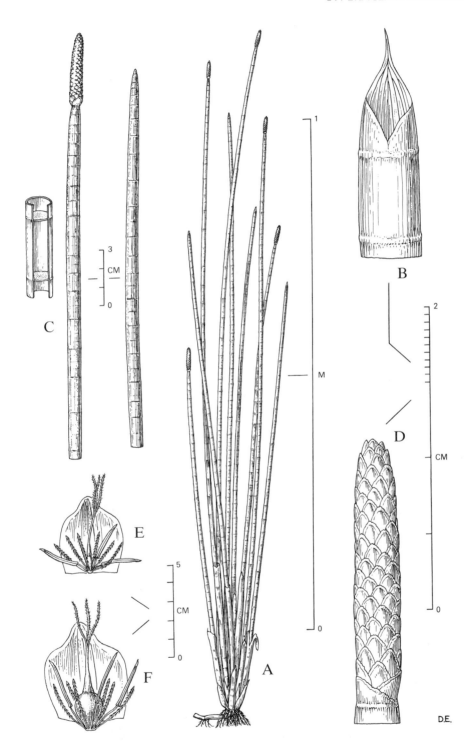

FIG. 20 **Eleocharis interstincta**. A, habit. B, basal sheath. C, sterile and fertile culms (upper end). D, spikelet. E, young floret. F, fruiting floret. (D.E.)

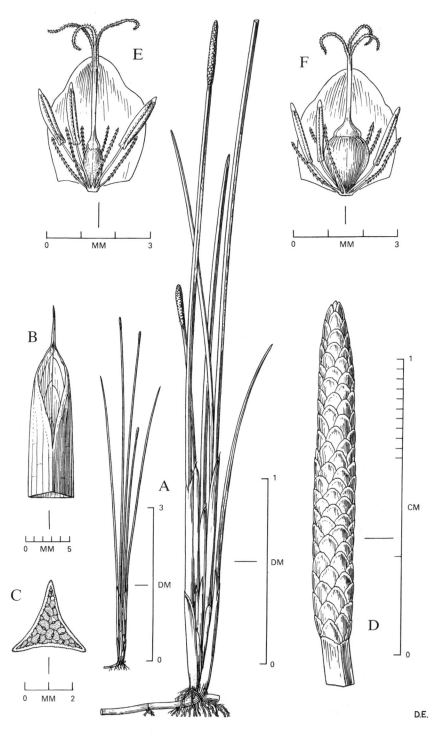

FIG. 21 **Eleocharis mutata**. A, habit. B, basal sheath. C, cross-section of culm. D, spikelet. E, young floret. F, fruiting floret. (D.E.)

Eleocharis rostellata (Torr.) Torr., *Fl. New York* 2: 347 (1843).

Rhizomatous perennial forming dense tussocks, the rhizomes thickened and erect with coarse roots. Culms mostly erect or ascending, sometimes arching and rooting at tips, 10–60 cm tall, mostly 0.7–1.8 mm thick, somewhat flattened and sulcate; sheath light to chocolate brown, obtuse to subacute, the orifice suboblique to nearly truncate. Spikelets ovoid to fusiform, acute at apex, 6–20 mm long, 2–35 mm thick, many-flowered; fertile scales ovate to ovate-lanceolate, 3–5 mm long, 1.8–3 mm wide, rather stiff, drab or brown, the margins thickly scarious, obtuse at apex. Stamens 3, style 3-branched. Achenes acutely trigonous, obovoid to ellipsoid-obovoid, 1.4–2 mm long, narrowed at summit to the confluent pyramidal style base, obtuse at apex, yellowish-brown to olive brown, smoothish, finely and indistinctly cellular-reticulate, lustrous; bristles 6, subulate, closely retrorsely spinulose, equalling or exceeding the style base in length.

GRAND CAYMAN: *Proctor* 50721, found on wet peaty margin of fresh water pond N. of the Fire Station, Frank Sound Road.

Eastern U.S.A, Puerto Rico.

Eleocharis geniculata (L.) Roem. & Schult. in L., *Syst. Veg.* ed. nov. 2: 150 (1817). FIG. 22.

Eleocharis caribaea (Rottb.) Blake (1918).

Eleocharis capitata R. Br. (1810).

Annual, with fibrous roots. Culms densely tufted, mostly 5–20 cm tall; basal sheaths firm, oblique at apex. Spikelets ovoid, obtuse, 2–6 mm long; glumes ovate, obtuse, yellowish or pale brown. Bristles 5–8, brown, longer than the achene. Style 2-cleft. Achenes ca. 1 mm long, shining black, with a short whitish tubercle.

GRAND CAYMAN: *Brunt* 1965, 2162; *Kings GC* 206; *Proctor* 48287.

— Pantropical, extending also into temperate regions. Usually a weed of wet ditches and damp open ground.

Eleocharis atropurpurea (Retz.) Kunth, *Enum. Pl.* 2: 151 (1837).

A miniature annual. Culms densely tufted, very slender, 2–7 cm tall. Spikelets ovoid, obtuse or subacute, 2.5–4 mm long; glumes ovate-oblong, light purple-brown with green midvein and narrow scarious margins. Bristles 2–4, whitish, retrorsely hairy, nearly as long as the achene. Style 2- or 3-cleft. Achenes 0.5 mm long, shining black, with minute whitish tubercle constricted at base.

GRAND CAYMAN: *Kings GC* 248.

— Of scattered but wide distribution in the warmer parts of both hemispheres.

Eleocharis minima Kunth, *Enum. Pl.* 2: 139 (1837).

A miniature annual. Culms tufted, capillary, 2–7 cm tall, longitudinally grooved; sheaths short, conspicuous, with inflated apex. Spikelets lanceolate or lance-ovate, 3–7 mm long; glumes ovate-lanceolate, usually acute, dark brown with greenish midvein and broad white hyaline margins. Bristles several, white, obscurely toothed, much shorter than the achene. Style 3-cleft. Achenes 0.7–1 mm long, narrowed at both ends; tubercle short-pyramidal, brownish or gray.

GRAND CAYMAN: *Brunt* 1954.

— Widespread in tropical Americam, but in the West Indies known otherwise only from Cuba.

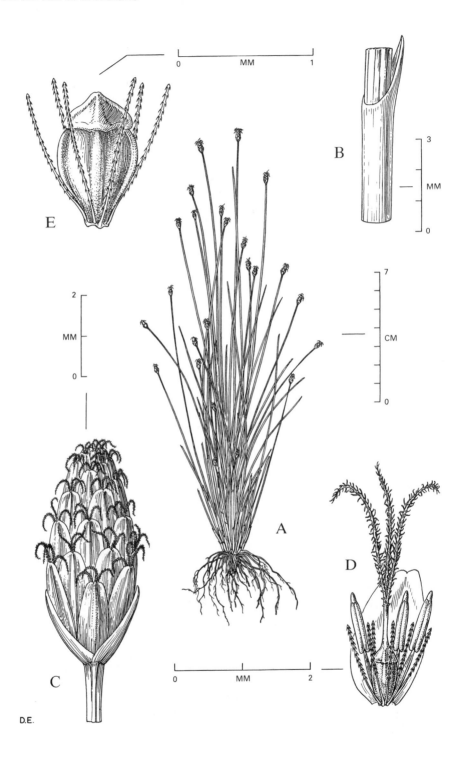

FIG. 22 **Eleocharis geniculata**. A, habit. B, basal sheath. C, spikelet. D, floret. E, achene with bristles. (D.E.)

Abildgaardia Vahl

Glabrous tufted perennials. Leaves basal, flat, sharp-pointed. Culms slender, numerous in a tuft. Inflorescence a solitary spikelet (rarely more than 1), subtended by 1 or 2 very small inconspicuous bracts. Spikelet few–many-flowered, the deciduous scales imbricated in 2 rows. Bristles lacking. Stamens 1–3. Style pubescent, deciduous, with enlarged base; stigmas 3. Achenes 3-angled.

A small genus of about 15 species, chiefly occurring in the Old World tropics.

Abildgaardia ovata (Burm. f.) Kral in *Sida* 4: 72 (1971).

Abildgaardia monostachya (L.) Vahl (1805).
Fimbristylis monostachya (L.) Hassk. (1848).
Fimbristylis ovata (Burm. f.) J. Kern (1967).

A low, tufted plant with a hard, knotted base. Leaves up to 20 cm long (usually less), 0.5–1.5 mm wide, with roughened, minutely bristly margins. Culms wiry, about twice as long as the leaves. Bracts much shorter than the pale, solitary spikelets. Spikelets ovate or ovate-lanceolate, 5–15 mm long, up to 5 cm broad, flattened; glumes ovate, keeled, mucronate, pale greenish-brown with whitish margins. Achenes 2–2.5 mm long, constricted near the base, yellowish, tuberculate.

LITTLE CAYMAN: *Proctor* 28190, 35133. CAYMAN BRAC: *Proctor* 35153.

— Pantropical, chiefly at low to medium elevations. The Cayman specimens were collected in pockets of soil on flat, shaded limestone pavements.

Fimbristylis Vahl

Annual or perennial plants, usually tufted, with culms leafy toward base. Inflorescence a simple or compound cyme (often appearing umbellate by reduction), subtended by 1–several narrow bracts. Spikelets loosely or densely aggregated, cylindric, the glumes spirally imbricate and deciduous. Flowers perfect (but apparently sometimes functionally unisexual and monoecious), without perianth. Stamens 1–3. Style 2- or 3-cleft, often pubescent near the enlarged base, completely deciduous. Achenes flattened or 3-angled, often finely reticulate or striate.

A large genus of about 150 tropical or subtropical to warm-temperate species, most abundant in the Indonesian–Australian region.

KEY TO SPECIES

1. Spikelets more than 5 mm long (up to 15 mm), in loose inflorescences (or contracted in *Fimbristylis ferruginea*); achenes various but not granular:
  2. Glumes glabrous; achenes not smooth:
    3. Achenes minutely reticulate in lines but not ribbed; sheaths at base of culm dark brown to nearly black:
      4. Spikelets narrowly oblong- to lance-cylindric; longest bract of inflorescence equal in length to or longer than the entire inflorescence: F. spadicea
      4. Spikelets ovoid or broadly ellipsoid; longest bract of inflorescence much shorter than the entire inflorescence: F. castanea
    3. Achenes longitudinally ribbed as well as minutely reticulate; sheaths at base of culm green or pale brown, inconspicuous: F. dichotoma

2. Glumes minutely pubescent; achenes appearing smooth:	**F. ferruginea**
1. Spikelets usually less than 4 mm long, aggregated in dense heads; achenes with granular surface:	**F. cymosa** subsp. **spathacea**

Fimbristylis spadicea (L.) Vahl, *Enum. Pl.* 2: 294 (1805).

Perennial, densely tufted, spreading by elongate stolons. Leaves shorter than or nearly equal in length to the culms, erect, more or less linear-involute with roughened margins. Culms mostly 30–70 cm tall, wiry, the broad basal sheaths dark brown to black. Inflorescences with 3–8 unequal rays, these less than 1–3 cm long, often branched. Spikelets usually 1–1.5 cm long; glumes ovate, obtuse to apiculate, shiny dark brown to nearly black with pale margins. Stamens 2 or 3, with dark brown subulate anthers 1.5–2 mm long. Achenes flattened-obovoid, brown, 1.5–1.8 mm long, the surface finely reticulate due to regular rows of deeply pitted quadrangular cells.

GRAND CAYMAN: *Brunt* 1642; *Proctor* 15242.

— Widespread in tropical and temperate America, occurring as far north as southern Canada. Many variants occur, and *Fimbristylis castanea* is often combined with it. In the West Indies, *F. spadicea* occurs chiefly in damp brackish soils at low elevations.

Fimbristylis castanea Vahl, *Enum. Pl.* 2: 292 (1805).

Perennial, rather loosely tufted, the bases deeply set in the substratum, the outer and older leaf-bases persistent as imbricated scales. Leaves and culms similar to those of *Fimbristylis spadicea*. Inflorescences with 3–6 slender rays, these 2–9 cm long and often unbranched. Spikelets relatively fewer than in *F. spadicea*, and rarely more than 1 cm long; glumes broadly ovate with rounded apex, dull brown. Stamens as in *F. spadicea*, likewise the achenes.

GRAND CAYMAN: *Kings GC* 193.

— Seacoasts of eastern N. America from New York to the Yucatan peninsula; also the Bahamas and Cuba, always in coastal marshes or swales.

Fimbristylis dichotoma (L.) Vahl, *Enum. Pl.* 2: 287 (1805).

Fimbristylis diphylla (Retz.) Vahl (1805).
Fimbristylis annua (All.) Roem. & Schult. (1817).

Annual or short-lived perennial, tufted. Leaves shorter than the culms, flat, to 3 mm wide. Culms slender, 10–60 cm tall, glabrous or pubescent. Inflorescence loose, compound-umbellate with slender unequal rays, subtended by several very short rough-margined bracts, these ciliate at the sheathing base. Spikelets usually numerous, oblong to ovoid, mostly 5–10 mm long, ca. 2 mm thick or less, solitary or in clusters of 2 or 3; glumes ovate, 2–3 mm long, brown with green midrib, acute or apiculate. Stamens 1 or 2; style 2-forked. Achenes flattened-obovoid, whitish to pale yellow-brown, 1 mm long, longitudinally ribbed, the ribs minutely tuberculate.

GRAND CAYMAN: *Brunt* 2175; *Kings GC–193*; *Proctor* 27949, 47870.

— Cosmopolitan; one of the most widely distributed plants in the world, but often known by other names. One authority estimates that about 400 synonyms apply to this species.

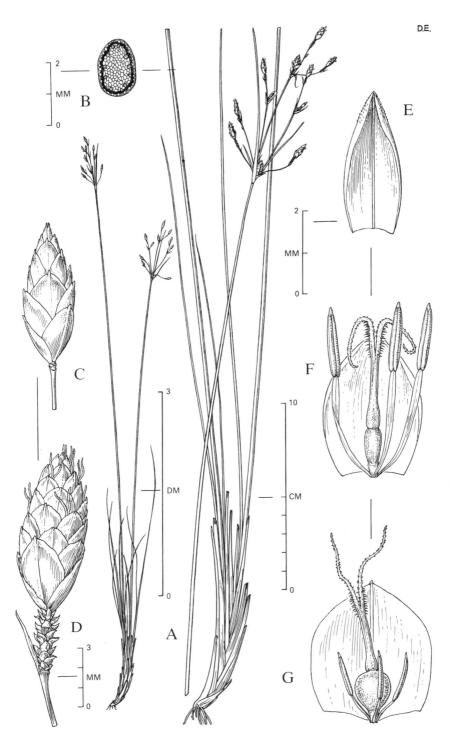

D.E.

FIG. 23 **Fimbristylis ferruginea**. A, habit. B, cross-section of culm. C, young spikelet. D, mature spikelet. E, bract. F, functionally staminate floret. G, fruiting floret. (D.E.)

Fimbristylis ferruginea (L.) Vahl, *Enum. Pl.* 2: 291 (1805). **FIG. 23**.

Perennial, in slender tufts. Leaves inconspicuous, much shorter than the culms or often lacking. Culms 20–70 cm tall, narrowly brown-sheathed at base. Inflorescence rays 0.5–2 cm long, forming a small rather compact cluster. Spikelets 5–10 (rarely 1) per inflorescence, ovoid-oblong, 8–20 mm long; glumes ovate, minutely apiculate, appressed puberulous toward the apex. Stamens 3, anthers 1–1.5 mm long. Achenes flattened-obovoid, 1.5 mm long, the surface very minutely reticulate (visible under high magnification), ordinarily appearing smooth.

GRAND CAYMAN: *Brunt* 1774, 1922, 1945; *Kings* GC 169, GC 356; *Proctor* 15059. LITTLE CAYMAN: *Kings* LC 74. CAYMAN BRAC: *Proctor* 29131.

— Pantropical.

Fimbristylis cymosa R. Br. subsp. **spathacea** (Roth) T. Koyama in *Micronesica* 1: 83 (1964).

Fimbristylis spathacea Roth (1821).
Fimbristylis cymosa sensu Adams (1972), not R. Br. (1810).

Perennial, forming dense tufts often aggregated in a turf. Leaves much shorter than the culms, stiff, flat, spreading, 1.5–3 mm wide, mucronate at tip. Culms 10–40 cm tall, stiffly erect. Inflorescence a dense head of small spikelets, subtended by short glabrous entire bracts. Spikelets ellipsoid to short-cylindric, 2–3(–6) mm long, 1.5–2 mm thick; glumes ovate, 1–1.3 mm long, light to dark brown, broadly keeled and obtuse to emarginate. Stamens 1 or 2, the anthers 0.6 mm long; style 2-forked (very rarely 3-forked). Achenes dark brown, ca. 0.75 mm long.

GRAND CAYMAN: *Brunt* 1851, 2024; *Proctor* 15287. LITTLE CAYMAN: *Proctor* 35177. CAYMAN BRAC: *Proctor* (sight).

— Pantropical, chiefly near sea-coasts in brackish or subsaline soils. The similar *Fimbristylis cymosa* subsp. *cymosa* differs chiefly in having 3-branched styles and slightly different achenes; it does not occur in the West Indies. The two are maintained as separate species by some authors, but the differentiating characters are reported not to be constant.

Schoenoplectus (Rchb.) Palla.

Usually perennials, spreading by stolons or sometimes caespitose. Culms terete or angulate, leafy or the leaves reduced to tubular sheaths. Inflorescence umbellate, paniculate, or reduced to 1 spikelet or a small cluster of sessile spikelets, subtended by 1 to many bracts. Spikelets terminal or lateral, sessile or pedunculate, 3–many-flowered; glumes spirally imbricate, all fertile or the lowest sometimes empty. Flowers bisexual, with perianth of 1–6 bristles, or bristles sometimes lacking. Stamens 2–3; style 2- or 3-cleft, not enlarged at base, usually deciduous. Achenes flattened or 3-angled.

A cosmopolitan genus of about 300 species, usually of wet habitats. A single species is reported from the Cayman Islands.

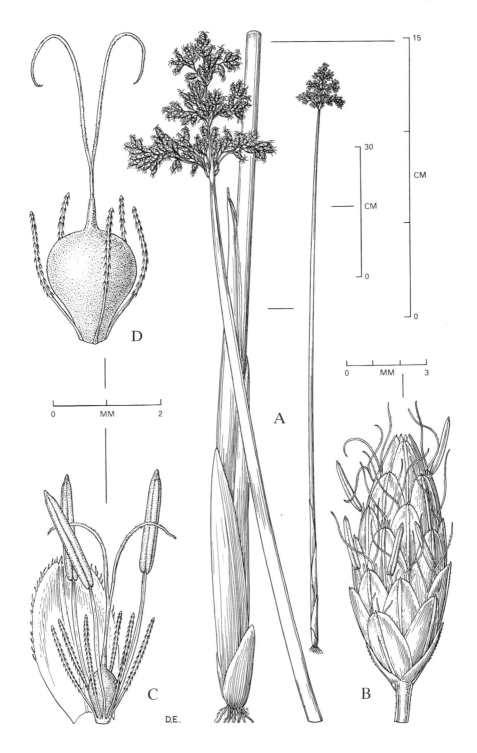

FIG. 24 **Schoenoplectus validus**. A, habit. B, spikelet. C, floret. D, achene with bristle and persistent style. (D.E.)

Schoenoplectus validus (Vahl), A. & D. Love in *Bull. Torr. Bot. Club* 81: 33 (1954). FIG. 24.

Rhizomes elongate, creeping, ca. 1 cm in diameter, covered with brown papery scales. Culms erect, up to 2 m tall (often less), terete, reddish- or brownish-sheathed at base, the sheaths terminating in a narrow free blade 0.5–10 cm long. Inflorescence compound-umbellate, subtended by a narrow rigid green bract 1–7 cm long. Spikelets ovoid, 5–10 mm long, 4–5 mm thick, numerous, in clusters of 2–4 on scabrous-margined peduncles; glumes mucronate; bristles 6, reddish, tortuous, retrorsely barbed. Style 2-cleft. Achenes flattened, 2.5 mm long, dark gray when ripe.

GRAND CAYMAN: *Kings GC 411.*

— Widely distributed in chiefly temperate regions; also common in the Greater Antilles, usually in open marshes or swales at low elevations.

Rhynchospora Vahl

Perennial sedges, spreading by means of creeping rhizomes. Culms usually solitary, leafy below. Inflorescence (in the only Caymanian species) a terminal head of sessile spikelets, subtended by long leafy bracts that are white toward base. Spikelets somewhat compressed; scales spirally arranged in 2 or 3 rows, several usually empty or with abortive flowers. Flowers perfect, with or without perianth. Stamens 3. Style 2-cleft. Achenes flattened, transversely rugose, capped with the beak-like persistent base of the style.

A nearly cosmopolitan genus of more than 250 species, best represented in the tropics. The Cayman species belongs to a group that is often maintained as a separate genus *Dichromena*, and the generic description given above applies more especially to this taxon.

Rhynchospora colorata (L.) H. Pfeiff (see Thomas, 1984)

Rhynchospora stellata (Lam.) Griseb. in *Gött. Abh.* 7: 271 (1857). **FIG. 25.**
Dichromena colorata (L.) Hitchc. (1893).

Rhizomes elongate (to 30 cm) and very slender, stolon-like. Leaves several, shorter than the culm, 1–3 mm wide, usually flat. Culms 20–50 cm tall. Bracts 4–6, leafy, 1–12 cm long, spreading or reflexed, white at base, glabrous. Spikelets numerous, 6–8 mm long; scales broadly ovate-lanceolate, 3–4 mm long. Achenes ca. 0.8 mm long, with a flat tubercle.

GRAND CAYMAN: *Brunt* 1646, 1921; *Kings GC* 100; *Proctor* 15288, 48280.
— Southeastern United States; Mexico to Belize; Bermuda and throughout the West Indies to Martinique. One of the few sedges pollinated by insects.

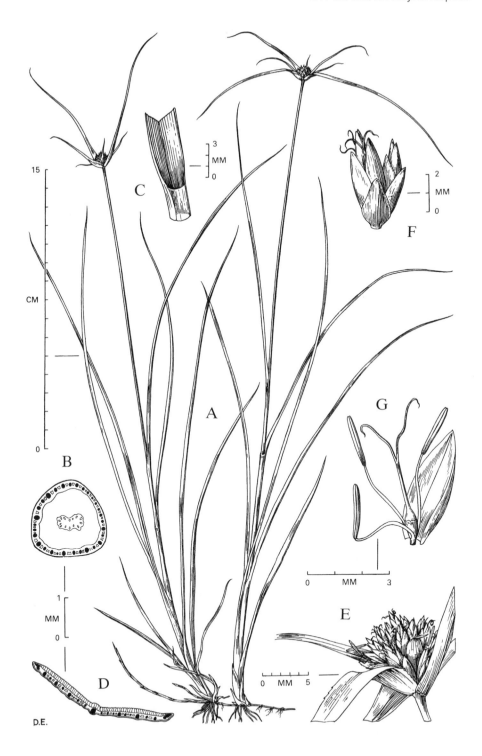

FIG. 25 **Rhynchospora stellata**. A, habit. B, cross-section of culm. C, junction of leaf-blade with sheath. D, cross-section of leaf-blade. E, inflorescence. F, two spikelets. G, floret with scale. (D.E.)

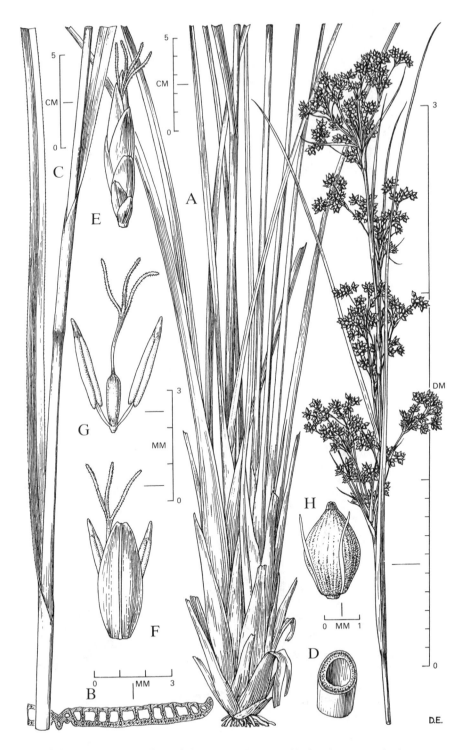

FIG. 26 **Cladium jamaicense**. A, base of plant. B, cross-section of leaf. C, lower part of culm. D, cross-section of culm. E, spikelet. F, floret. G, floret with glume removed. H, achene. (D.E.)

Cladium R. Br.

Tall coarse perennials with thick rhizomes. Leaves long and narrow, with rough margins. Culms tall, the inflorescence a compound terminal panicle with numerous small clusters of sessile spikelets, or a series of axillary panicles. Spikelets mostly 1–3-flowered with usually only the lowest flower fertile; scales imbricate, the lower ones empty. Flowers perfect, with perianth of bristles or lacking. Stamens 2–3. Style 3-cleft, the thickened base often persistent on the apex of the achene as a tubercle. Achenes somewhat flattened or 3-angled, smooth or rugose.

A genus of 4 species of pantropical and warm-temperate distribution. A single species occurs in the Cayman Islands.

Cladium jamaicense Crantz, *Inst. R. Herb.* 1: 362 (1766). **FIG. 26.**

Mariscus jamaicensis (Crantz) Britton (1913).

CUTTING GRASS, SAWGRASS

Rhizomes elongate, woody, 4–6 mm thick, clothed with brown papery overlapping scales. Culms 1–3 m tall, up to 2.5 cm thick near base, obscurely 3-angled, leafy; cauline leaves numerous, up to 1 m long or more, 6–20 mm wide, of hard texture, the margins very scabrous. Panicles several in leaf-axils along upper part of the culms, densely or laxly compound-umbelliform, mostly 2–12 cm long. Spikelets narrowly ovoid, 4–5 mm long, brown, acute. Perianth lacking. Stamens 2, anthers 3 mm long. Achenes ovoid, 2 mm long, acute, brown and somewhat rugose.

GRAND CAYMAN: *Brunt* 1690, 1863, 1948; *Kings* GC 256, GC 363.

— Pantropical as a species, but subsp. *jamaicense* confined to the American tropics. The tough leaves are used in Jamaica for weaving baskets.

Remirea Aubl.

A perennial with elongate rhizomes and solitary or clustered densely leafy culms. Leaves rigid, linear-lanceolate, spinulose-tipped, the basal sheaths imbricate. Inflorescence a dense solitary head of 1-flowered spikelets; rhachilla jointed above the base. Flowers bisexual, without perianth. Stamens 3; style 3-cleft. Achenes smooth, sessile.

A pantropical genus of 1 species. Some authors have included it in *Cyperus* or its segregate *Mariscus* on narrow technical grounds, despite its distinctive habit and unique features.

Remirea maritima Aubl., *Pl. Guian.* 1: 45 (1775). **FIG. 27.**

Remirea pedunculata R. Br. (1810).
Cyperus pedunculatus (R. Br.) Kern (1958); Adams (1972).
Mariscus pedunculatus (R. Br.) T. Koyama (1977).

Culms 2–30 cm long, decumbent to erect, leafy throughout. Leaves flat, 2–8 cm long, 2–7 mm wide near base. Heads 1–2 cm long, subtended by leaf-like bracts. Spikelets many, ovoid, brownish, 3–5 mm long, each with 4 glumes. Glumes ovate, several-veined, 3–4.5 mm long. Achene trigonous, grayish-brown, 2.5 mm long.

GRAND CAYMAN: *Proctor* 15216. LITTLE CAYMAN: *Kings* LC 81; *Proctor* 35084.

— Of wide but sporadic distribution, usually found in loose sand at the top of sea-beaches.

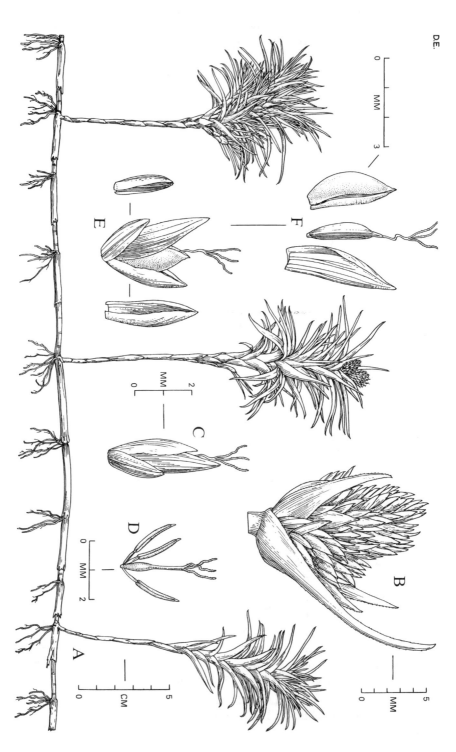

FIG. 27 **Remirea maritima**. A, habit. B, inflorescence. C, flowering spikelet. D, floret. E, fruiting spikelet. F, achene and scales. (D,E)

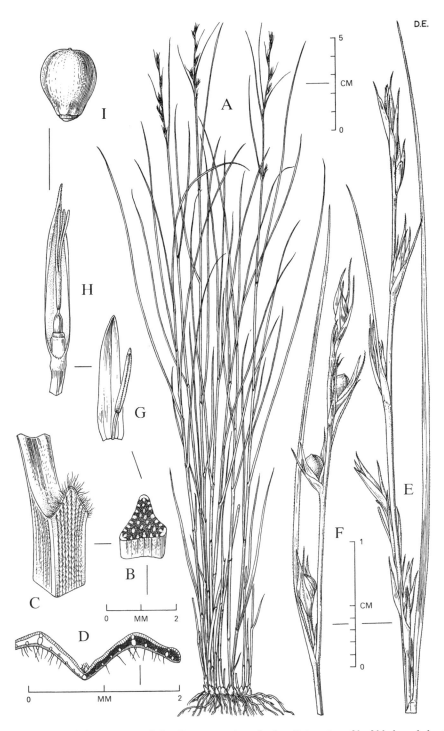

D.E.

FIG. 28 **Scleria lithosperma**. A, habit. B, cross-section of culm. C, junction of leaf-blade and sheath. D, cross-section of leaf. E, staminate inflorescence. F, pistillate inflorescence. G, fstaminate floret and scale. H, pistillate floret and scale. I, achene. (D.E.)

Scleria Bergius

Chiefly perennial, often with rhizomes, and with sharply 3-angled more or less leafy culms, these short and erect to tall and erect, or sometimes greatly elongate and vine-like. Spikelets small, few-flowered, clustered in terminal or axillary fascicles, or in interrupted spikes; glumes spirally attached. Flowers monoecious, without perianth; staminate spikelets several-flowered; pistillate ones 1-flowered. Stamens 1–3. Ovary sometimes supported on a disc (hypogynium); stigmas 3. Achenes globose to ovoid, usually white, with gleaming, enamel-like surface.

A pantropical genus of 200 species, a few extending into temperate regions.

Scleria lithosperma (L.) Sw., *Nov. Gen. & Sp. Pl.* 18 (1788). **FIG. 28.**

Perennial, the culms 20–60 cm tall, glabrous, more or less clustered from short, hard, knotted rhizomes. Leaves involute, 10–20 cm long, 0.5–3 mm broad, the lowest ones reduced to finely pubescent, papery sheaths. Inflorescence of several small, distant clusters of spikelets, each subtended by a more or less filiform bract. Hypogynium lacking. Achenes obovoid-ellipsoid, ca. 2 mm long.

GRAND CAYMAN: *Brunt* 2175; *Kings GC* 235, *GC* 245. LITTLE CAYMAN: *Kings LC* 58; *Proctor* 28120. CAYMAN BRAC: *Proctor* 28998.

— Pantropical, mostly found in rather dry woodlands.

GRAMINEAE (POACEAE)

Annual or perennial herbs, rarely woody or tree-like. Stems (culms) usually terete and hollow (rarely solid), with closed nodes. Leaves 2-ranked, usually elongate, with a distinct basal sheath split on one side to the point of attachment, and a small appendage (the ligule) at the junction of sheath and blade. Inflorescence a spike, raceme or panicle of densely or loosely aggregated spikelets, each spikelet consisting of a rhachilla bearing 1–several flowers subtended by small bractlets; the lowest 2 bractlets (glumes) sterile; each succeeding bractlet (lemma) enclosing a flower, with another bractlet (palea) subtending the flower on the upper side. Flowers perfect or unisexual. Perianth none, or reduced to 2 or 3 minute scales (lodicules). Stamens 1–6 (usually 3), separate; anthers 2-locular, attached at the middle. Ovary superior, 1-locular with 1 ovule; styles 1–3 (usually 2), more or less plumose. Fruit a seed-like grain (caryopsis); endosperm copious, starchy.

A large, world-wide family of about 750 genera and more than 9000 species (some authorities estimate as many as 10,000). This is economically the most important family of plants. Because of the complexity of classification, it is customary to arrange grasses in groups called tribes, and this practice was followed in the First Edition of this Flora, following the system presented by Hitchcock (1936). However, grass taxonomy has undergone so much change in recent years that in the present book, the genera are arranged alphabetically without reference to tribal position.

KEY TO THE GRASS GENERA OF THE CAYMAN ISLANDS

1. Stems woody, plants of tree-like proportions: **[Bambusa]**

1. Stems herbaceous, if woody then not of tree-like proportions (Lasiacis):

  2. Plant monoecious with female spikelets condensed in a cob and male spikelets in racemes on an axis forming a pyramidal inflorescence: **[Zea (mays)]**

  2. Plant not as above:

    3. Spikelets 1–many-flowered, if 2-flowered then both bisexual or the upper barren:

      4. Spikelets unisexual, leaf-blades with oblique veins: **Pharus**

      4. Spikelets bisexual, leaf-blades with parallel veins:

        5. Reed-like plants with large plumose panicles: **[Arundo]**

        5. Plants not reed-like:

          6. Spikelets 1-flowered:

            7. Spikelets falling entire at maturity:

              8. Inflorescence a solitary, cylindrical raceme: **[Zoysia]**

              8. Inflorescence of several racemes dispersed along an axis: **Spartina**

            7. Spikelets breaking up at maturity, inflorescence a contracted or open panicle or composed of digitate racemes:

              9. Spikelets fusiform, inflorescence a contracted or open panicle: **Sporobolus**

              9. Spikelets laterally compressed, inflorescence composed of digitate racemes: **Cynodon**

          6. Spikelets 2–many-flowered:

            10. Lemmas 7–11-nerved, inflorescence spicate to ovoid, plant dioecious: **Distichlis**

            10. Lemmas 3-nerved, plant bisexual:

              11. Inflorescence an effuse or contracted panicle: **Eragrostis**

              11. Inflorescence composed of racemes:

                12. Racemes digitate or subdigitate:

                  13. Spikelets many-flowered, all florets fertile or the uppermost reduced:

                    14. Raceme axis terminating in a spikelet: **Eleusine**

                    14. Raceme axis terminating in a bristle: **Dactyloctenium**

                  13. Spikelets with only one floret fertile, with or without additional male or barren florets: **Chloris**

                12. Racemes arranged along an axis: **Leptochloa**

    3. Spikelets 2-flowered, lower floret male or barren, upper floret pistilate or perfect:

      15. Spikelets solitary or in clusters:

        16. Inflorescence an open panicle:

          17. Upper lemma coriaceous to bony at maturity, with inrolled margins:

            18. Upper lemma with an excavation and small tuft of wool at the tip: **Lasiacis**

            18. Upper lemma entire at the tip, without an excavation or tuft of wool: **Panicum**

          17. Upper lemma cartilaginous, laterally compressed, with flat margins: **[Melinis]**

        16. Inflorescence a contracted spike or composed of racemes:

          19. Inflorescence an interrupted or dense spike:

20. Spikelets in clusters of 1–11 surrounded by an involucre of bristles, spines or bracts:
   21. Involucre of stiffly coriaceous, narrowly elliptic several-nerved bracts: **Anthephora**
   21. Involucre of bristles or spines:
      22. Bristles slender and free to the base: **[Pennisetum]**
      22. Bristles flattened and spinous or slender, often in separate whorls and fused at the base to form a cupule: **Cenchrus**
20. Spikelets solitary, each subtended by one or two bristles: **Setaria**
19. Inflorescence of racemes, these solitary, digitate or scattered on an axis:
   23. Racemes very short and appressed to or embedded in the axis: **Stenotaphrum**
   23. Racemes long or short, not embedded in the axis:
      24. Glumes or lemmas awned:
         25. Spikelets laterally compressed: **Oplismenus**
         25. Spikelets dorsally compressed: **Echinochloa**
      24. Glumes or lemmas awnless or at most cuspidate:
         26. Upper lemma coriaceous to crustaceous, the margins inrolled:
            27. Lower glume absent:
               28. Spikelets plano-convex: **Paspalum**
               28. Spikelets biconvex: **Axonopus**
            27. Lower glume present:
               29. Spikelets paired:
                  30. Tip of upper palea reflexed and protruding: **Echinochloa**
                  30. Tip of upper palea enclosed by the lemma: **[Urochloa]**
               29. Spikelets solitary:
                  31. Spikelets plano-convex: **Paspalum**
                  31. Spikelets biconvex: **[Urochloa]**
         26. Upper lemma chartaceous to cartilaginous, the margins flat: **Digitaria**
15. Spikelets paired, one sessile and the other pedicelled or both pedicelled, similar or dissimilar, the pedicelled sometimes much reduced:
   32. Inflorescence a panicle of plumose racemes on an elongated axis, pedicelled spikelet similar to the sessile:
      33. Panicle large and spreading, raceme rachis fragile, one spikelet of the pair sessile: **[Saccharum]**
      33. Panicle contracted and spiciform, raceme rachis tough, both spikelets of a pair pedicelled: **Imperata**
   32. Inflorescence of paniculate, solitary, paired or digitate racemes, the sessile spikelet bisexual, the pedicelled male or barren:
      34. Rachis internodes and pedicels slender, upper lemma usually awned:
         35. Pedicels and internodes with a translucent median line: **[Bothriochloa]**
         35. Pedicels and internodes without a translucent median line:
            36. Inflorescence of paired racemes aggregated into a large, dense, feathery compound panicle: **Andropogon**
            36. Inflorescence a panicle of racemes: **[Sorghum]**
      34. Rachis internodes and pedicels stout, upper lemma awnless: **[Rottboellia]**

Andropogon L.

Annual or perennial herbs, the culms erect or decumbent. Leaves chiefly long and narrow, the ligules fringed or not, the blades flat or involute. Inflorescence often of one to numerous palmately arranged racemes arising from the middle and upper nodes of the culm; raceme bases unequal and terete; rachis internodes slender to club-shaped, often ciliate, disarticulating along the rachis with spikelets falling as pairs (sessile and pedicellate together) or just the pedicellate spikelets falling first. Sessile spikelets bisexual or pistillate, 2-flowered. Glumes subequal, about as long as the spikelet, membranous, the lower glume keeled, 1–3-nerved, usually awnless; florets hyaline, shorter than the spikelet; lower floret sterile, lacking a palea; upper floret with an awned or unawned lemma, the palea often well developed. Pedicellate spikelets absent to well developed, usually similar to sessile spikelets when present; lower glume convex with many nerves. Seed with a large embryo.

A world-wide genus of about 100 species, occurring in both tropical and temperate areas.

Andropogon glomeratus (Walter) Britton, Sterns & Poggenb., *Prel. Cat. N.Y.* 67 (1888). **FIG. 29**.

Densely tufted perennial; culms compressed, 1–1.5 m tall. Leaves with flat or folded blades 3–5 mm broad; lower sheaths crowded, keeled. Inflorescence densely corymbose, villous, with paired racemes 1.5–3 cm long; rhachis slender, flexuous; sessile spikelet 3–4 mm long, with awn 1.5 cm long.

GRAND CAYMAN: *Brunt* 1656, 1974; *Hitchcock*; *Kings* GC 164; *Lewis* GC 21.

— Southeastern United States, Mexico and the West Indies to northern S. America. Frequent in moist open ground. The very similar *Andropogon bicornis* (not yet recorded from the Cayman Islands) can be distinguished by having all its spikelets awnless.

FIG. 29 **Andropogon glomeratus**, × 1. (H.)

FIG. 30 **Anthephora hermaphrodita**, × $\frac{1}{2}$. (H.)

Anthephora Schreb.

Loosely tufted annuals, the inflorescence narrow and spike-like. Spikelets in small clusters of 4, the indurate first glumes united at base, forming a pitcher-shaped pseudo-involucre, the clusters subsessile and erect on a slender flexuous continuous axis; glumes rigid, acute, or produced into short awns.

A small, pantropical genus of 20 species.

Anthephora hermaphrodita (L.) Kuntze, *Révis. Gen. Pl.* 2: 759 (1891). **FIG. 30.**

Culms leafy, branched, ascending or decumbent; leaf-blades flat, up to 8 mm broad, velvety pubescent on the upper surface, the margins more or less wavy; sheaths glabrous with ciliate margins; ligule prominent, 1.5 mm long. Inflorescence with numerous approximate, deciduous groups of spikelets, each enclosed by leathery lower glumes.

GRAND CAYMAN: *Brunt* 2066; *Proctor* 15219. LITTLE CAYMAN: *Proctor* 28156. CAYMAN BRAC: *Proctor* 28920.

— Throughout the West Indies and other parts of tropical America; common at lower elevations.

[*Arundo* L.

Tall perennial reeds, with broad linear leaf-blades and large plume-like terminal panicles. Spikelets several-flowered, the florets successively smaller, the summits of all about equal; rhachillas glabrous, disarticulating above the glumes and between the florets; glumes somewhat unequal, membranaceous, 3-nerved, narrow, tapering into a slender point, about as long as the spikelet; lemmas thin, 3-nerved, densely long-pilose, gradually narrowed at the apex, the nerves ending in slender teeth, the middle tooth extending into a straight awn.

A genus of 12 temperate and tropical species of wide distribution.

Arundo donax L., *Sp. Pl.* 1: 81 (1753). **FIG. 31.**

Culms 2–6 m tall, growing in large clumps, sparingly branched from thick, knotty rhizomes. Leaf-blades numerous, elongate, 3–5 cm broad or more, with a characteristic light convoluted area at the junction of the sheath, and slightly pubescent there; sheath glabrous; ligule 1 mm long, with toothed margin. Inflorescence a large, tawny panicle to 70 cm long; spikelets 1 cm long, 3-flowered; glumes with 3 prominent nerves.

GRAND CAYMAN: *Brunt* 1712; *Proctor* 48273.

— Indigenous to the Mediterranean region and widely distributed in warmer parts of the Old World; cultivated chiefly for ornament in America, and occurring as an escape in the southern United States through the West Indies to S. America. Grows well in sandy soils. The hard, hollow culms are used for making fishing-rods, flutes and bagpipes.]

FIG. 31 **Arundo donax**. A, inflorescence, × ¹/₃. B, culm with leaves, × ¹/₃. C, rhizome, × ¹/₃. D, spikelet, × 3. E, floret × 3. (H.)

Axonopus P. Beauv.

Annual to robust perennial herbs, erect or decumbent, branched or unbranched, the internodes solid, the leaf-sheaths smooth or scabrous; ligule a ciliate membrane or a fringe of hairs. Leaf-blades filiform to lanceolate, flat, conduplicate or involute, the margins smooth or scabrous. Inflorescence with spreading primary branches having 1-sided or distichous spikelets; secondary branches (if present) appressesed; rachis terminating in a spikelet; callus not differentiated. Spikelets disarticulating at base, dorsiventrally compressed or plano-convex; lower glume absent, upper glume as long as the spikelet. Lower floret lemma membranous; upper floret lemma 0.5–1-times the length of the lower floret lemma, cartilaginous and smooth or muricate, shiny or dull, mostly with involute margins. Seed with white or grayish embryo.

A tropical and subtropical genus with about 110 species.

Axonopus compressus (Sw.) P. Beauv., *Ess. Agrostogr.* 154 (1812).

Perennial, stoloniferous herb. Flowering culms 10–60 cm tall, erect from the base, unbranched. Leaf-sheaths glabrous; ligule 0.4–0.6 mm long; leaf-blades linear to lanceolate, flat or conduplicate, mostly 3–8 (rarely to 25) cm long and 0.4–1 cm broad, glabrous, smooth on the upper surface. Inflorescence ovate, main axis present or absent, to 2 cm long; primary branches 3–8 cm long. Spikelets solitary, ovate to elliptic, 2.1–2.8 mm long, 0.6–0.9 mm broad; upper glume 2.1–2.8 mm long, 4–5-nerved; in the lower floret the lemma lanceolate to ovate, 4-nerved; palea absent. In the upper floret the lemma oblong to ovate, membranous, white, 1.7–2.4 mm long.

GRAND CAYMAN: *Guala* 1870, 1896, "dispersing along forest trails and roadsides".

— A pantropical weed, often used as a lawn grass.

[*Bambusa* Schrad.

Large woody perennials with conspicuously jointed, hollow culms. Spikelets arranged in branched leafy or leafless panicles or in panicled spikes, often oblong or ovate, 1–many-flowered; paleas 2-keeled. Stamens 6.

A primarily Old World genus of about 70 species, a few of them now widely naturalized.

BAMBOO

Bambusa vulgaris Schrad. ex Wendl., *Collect. Pl.* 2: 26, t. 47 (1810).

Bambusa sieberi Griseb. (1864).

A giant bamboo up to 16 m tall or more (usually much shorter in the Cayman Islands), forming large open clumps from branched, creeping, woody rhizomes. Culms 10–12 cm in diameter, bright green or yellow. Leaf-blades linear-lanceolate, acuminate, 10–25 cm long, mostly 1–3 cm broad, usually glabrous at least on the surfaces. Flowers seldom seen. Spikelets 15–20 mm long, closely 6–10-flowered.

GRAND CAYMAN: *Brunt* 1686; *Kings* GC 163.

— Pantropical in cultivation; probably of Asiatic origin but now naturalized in many countries of both hemispheres. The culms form durable poles and have many uses. The very young shoots, 15–30 cm long, are edible both cooked or pickled and are considered a delicacy by many people.]

[*Bothriochloa* Kuntze

Annual or perennial herbs, caespitose or sometimes rhizomatous or stoloniferous, with erect culms; leaf-blades flat, the ligules membranous, fringed or not. Inflorescence of several to many long-pedunculate racemes, borne along a short to elongate rachis. Spikelets paired, falling attached to the rachis internodes; rachis internodes filiform, hairy, brittle, with a thin central longitudinal groove appearing as a translucent stripe. Sessile spikelets awned, 2-flowered; lower glume dorsally compressed, about as long as the spikelet, many-nerved, the nerves evident near the apex; upper glume delicately membranous, about as long as the spikelet, 1–3 nerved, keeled; lower floret sterile, without a palea; upper floret fertile, bisexual, the lemma reduced to a geniculate awn, the palea absent. Pedicellate spikelets sterile or staminate, similar to the sessile spikelets or reduced.

A pantropical genus of 30–40 species of which perhaps a dozen occur in the neotropics.

Bothriochloa pertusa (L.) Camus in *Ann. Sci. Linn. Lyon* ser. 2, 76: 164 (1931).

Andropogon pertusus of Adams (1972) and this Flora (1984).

SEYMOUR GRASS

Stoloniferous perennial with laxly erect culms 20–100 cm tall when flowering. Leaf-blades 10–20 cm long, 1–4 mm broad. Racemes villous, 2–6 cm long; spikelets 4.5–5 mm long, the lower glumes pitted; awns ca. 1.5 cm long, twice geniculate, the distal half black.

GRAND CAYMAN: *Brunt* 1839, 1852, 1894a; *Proctor* 15069. LITTLE CAYMAN: *Proctor* 35200. CAYMAN BRAC: *Proctor* 29338, 47331.

— Now pantropical but native to tropical Asia. Common weed of open fields and roadsides; eaten by livestock but not very palatable.]

Cenchrus L.

Annuals or perennials, with terminal racemes of spiny burs. Spikelets sessile, 1 to several together, permanently enclosed in a bristly or spiny involucre or bur, composed of more or less coalesced sterile branchlets; burs sessile or nearly so on a slender, compressed or angled axis, its apex produced into a short point beyond the uppermost bur, the burs falling entire, the grains germinating within them; involucre somewhat oblique, its body irregularly cleft, the lobes rigid, in most species resembling the spines, the cleft on the side of the bur next to the axis reaching to the tapering, abruptly narrowed or truncate base, the bristles or spines barbed, at least toward the summit; spikelets mostly glabrous or nearly so; first glume 1-nerved, usually narrow, sometimes lacking; second glume and sterile lemma 3–5-nerved, the lemma enclosing a well developed palea and usually a staminate flower. Fruit usually turgid, indurate, the lemma acuminate, the nerves visible toward the summit, the margins thin, flat.

About 25 tropical and warm-temperate species; most are troublesome weeds because of their prickly burs.

KEY TO SPECIES

1. Plants perennial; burs glabrous except at base of spines:	**C. gracillimus**
1. Plants annual; burs more or less pubescent (the hairs sometimes visible only under a strong lens):	
2. Involucre with a ring of slender bristles at base:	
3. Burs, excluding the bristles, not more than 4 mm wide, numerous, crowded in a long spike; lobes of involucre interlocking, not spine-like:	**C. brownii**
3. Burs, excluding the bristles, ca. 5.5 mm wide, not densely crowded; lobes of the involucre erect or nearly so (or rarely one or two lobes loosely interlocking), the tips spine-like:	**C. echinatus**
2. Involucre beset with flattened spreading spines, no ring of slender bristles at base:	
4. Burs, including spines, 7–8 mm wide, finely and minutely pubescent:	**C. incertus**
4. Burs, including spines, 10–12 mm wide, usually densely woolly pubescent, some hairs as long as the width of the spines near the base:	**C. tribuloides**

Cenchrus brownii Roem. & Schult. in L., *Syst. Veg.* 2: 258 (1817). **FIG. 32.**

Culms erect from a sparingly branched, more or less geniculate base, 30–100 cm tall; leaf-blades flat, thin, mostly 10–30 cm long, 6–12 mm broad. Spike 4–10 cm long, dense; burs with numerous slender outer bristles, the inner usually exceeding the bur, erect or spreading; lobes of the bur usually 6–8, interlocking at maturity; spikelets usually 3.

GRAND CAYMAN: *Hitchcock*; *Millspaugh* 1268; *Proctor* 15174. CAYMAN BRAC: *Kings* CB 36.

— Florida and Mexico to Brazil; a common weed of fields and waste places throughout the West Indies.

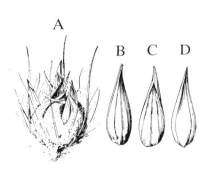

FIG. 32 **Cenchrus brownii.** A, bur, × 5. B, C, two views of spikelet, × 5. D, floret, × 5. (H.)

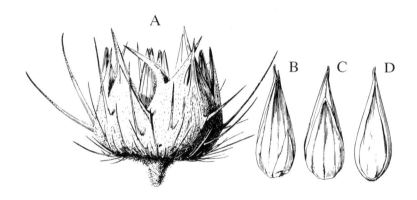

FIG. 33 **Cenchrus echinatus**. A, bur, × 5. B, C, two views of spikelet, × 5. D, floret, × 5. (H.)

Cenchrus echinatus L., *Sp. Pl.* 2: 1050 (1753). **FIG. 33**.

SOFT BUR

Culms ascending from a branched decumbent base, often rooting at the lower nodes, 25–60 cm long or longer. Leaf-blades 6–20 cm long, 3–9 mm broad; sheath with marginal silky hairs. Spikes 3–7(–10) cm long; burs truncate at base, the body 4–7 mm high and as wide or wider; longest bristles about equal in length to the lobes of the bur in length; lobes of the bur about 10, erect or bent inward, the tips hard and spine-like; spikelets about 4.

GRAND CAYMAN: *Brunt 1707*; *Hitchcock*; *Kings GC 170*; *Proctor 15256*. LITTLE CAYMAN: *Proctor 28153*. CAYMAN BRAC: *Proctor 29098*.

— Common throughout tropical and subtropical America, a weed of fields and open waste ground.

Cenchrus gracillimus Nash in *Bull. Torr. Bot. Club* 22: 299 (1895). **FIG. 34**.

Plants tufted, at length forming dense clumps, glabrous throughout. Culms erect or ascending, 20–80 cm tall, slender and wiry; leaf-blades 5–20 cm long, 2–3 mm broad, usually folded. Spikes 2–6 cm long, the burs not crowded; burs 3.5–5 mm wide (excluding the spines), somewhat tapering at the base; spines spreading or reflexed, all flat and broadened at base, retrorsely scabrous toward the apex; lobes of the bur 5–6 mm long; spikelets 2 or 3, 5.5–7 mm long.

GRAND CAYMAN: *Lewis 3853*; *Proctor 15146, 31046*.

— Florida, Cuba, Jamaica and Hispaniola, on sandy open ground.

Cenchrus incertus M. A. Curtis in *Boston J. Nat. Hist.* 1: 135 (1837). **FIG. 35**.

Cenchrus pauciflorus Benth. (1840).

Plants sometimes forming large mats; culms 20–90 cm long or more, spreading to ascending from a branched decumbent base. Leaf-blades 3–15 cm long, 2–7 mm broad, flat or folded. Spikes 3–10 cm long; burs mostly 4–6 mm wide, often densely (but very minutely) pubescent, rarely nearly glabrous; spines numerous, spreading or reflexed, flat and broadened at base, some of the upper ones 5 mm long; lobes of the bur about 8, rigid and spine-like; spikelets usually 2.

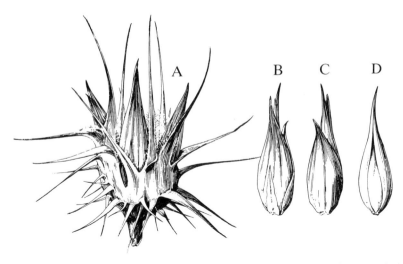

FIG. 34 **Cenchrus gracillimus**. A, bur, × 5. B, C, two views of spikelet, × 5. D, floret, × 5. (H.)

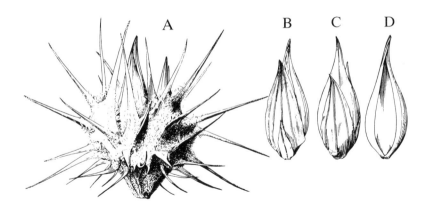

FIG. 35 **Cenchrus incertus**. A, bur, × 5. B, C, two views of spikelet, × 5. D, floret, × 5. (H.)

GRAND CAYMAN: *Sachet* 384. LITTLE CAYMAN: *Proctor* 28157. CAYMAN BRAC: *Proctor* 28991.

— United States to Argentina, in the West Indies common in sandy open ground.

Cenchrus tribuloides L., *Sp. Pl.* 2: 1050 (1753). **FIG. 36.**

Plants with stout, branching, radiate-decumbent culms 15–60 cm long or more, rooting at the nodes and with numerous ascending branches 10–30 cm tall; leaf-sheaths usually overlapping, sharply keeled, those below the spikes inflated. Spikes 3–9 cm long; burs 5–6 mm wide and 8–9 mm high (excluding the spines); spines more or less spreading, flat, the upper sometimes as much as 3 mm broad at base; spikelets usually 2, 7–8 mm long.

GRAND CAYMAN: *Brunt* 1741; *Kings GC 26*; *Millspaugh* 1249; *Proctor* 15069. LITTLE CAYMAN: *Kings* LC 107. CAYMAN BRAC: *Kings* CB 39; *Millspaugh* 1152.

— Eastern United States, the West Indies, and Brazil, in loose coastal sands.

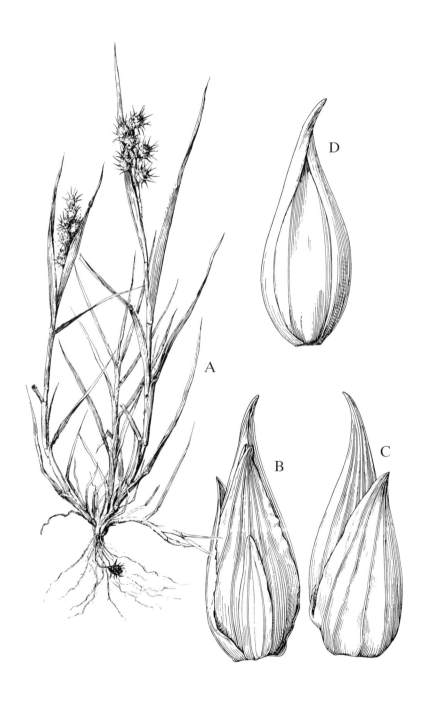

FIG. 36 **Cenchrus tribuloides**. A, plant, × ¹/₂. B, C, two views of spikelet, × 10. D, floret, × 10. (H.)

Chloris Sw.

Perennials or annuals, with flat or folded leaf-blades and 2–many often showy digitate spikes. Spikelets with 1 perfect floret, sessile in 2 rows along one side of a continuous rhachis, the rhachilla disarticulating above the glumes, produced beyond the perfect floret and bearing 1–several reduced florets consisting of empty lemmas, these often truncate, and, if more than one, the smaller ones enclosed in the lower, forming a usually club-shaped rudiment; glumes somewhat unequal, the first shorter, narrow, acute; lemma keeled, usually broad, 1–5-nerved, often villous on the callus and villous or long-ciliate on the keel or marginal nerves, awned from between the short teeth of a bifid apex, the awn slender or sometimes reduced to a mucro, the sterile lemmas awned or awnless.

About 40 species of tropical and warm temperate regions. Several are useful pasture grasses.

KEY TO SPECIES

1. Lemmas awnless; spikes dark brown: — **C. petraea**

1. Lemmas awned; spikes pale or purplish:

  2. Plants perennial, the culms commonly more than 1 m tall; spikes flexuous, 8–10 cm long: — **C. barbata**

  2. Plants annual, the culms usually less than 75 cm tall; spikes less than 8 cm long: — **C. inflata**

Chloris petraea Sw., *Nov. Gen. & Sp. Pl.* 25 (1788). **FIG. 37.**

Eustachys petraea (Sw.) Desv. (1810).

Stoloniferous perennial, glabrous and glaucous. Culms flat, averaging about 30 cm tall; leaf-blades 6–20 cm long, 3–6 mm broad, folded lengthwise and markedly keeled; ligule very reduced, difficult to distinguish. Inflorescence of 4–7 spikes, these 5–10 cm long, erect or ascending; spikelets 2 mm long, 2-flowered, only one of the florets maturing; glumes green, lemmas dark brown.

GRAND CAYMAN: *Brunt* 1802, 2039; *Hitchcock*; *Kings GC* 266; *Lewis* 3849, GC 26a; *Millspaugh* 1255; *Proctor* 15160; *Sachet* 388b. LITTLE CAYMAN: *Kings LC* 103; *Proctor* 28083. CAYMAN BRAC: *Millspaugh* 1181.

— Southern United States, West Indies, and eastern Mexico south to Panama and Trinidad; common on sandy beaches and similar inland habitats.

Chloris barbata Sw., *Fl. Ind. Occ.* 1: 200 (1797); Fosberg in *Taxon* 25(1): 176–178 (1976). **FIG. 38.**

Andropogon barbatus L. (1759).

Chloris polydactyla Sw. (1788), illegit.

C. dandyana C. D. Adams (1971).

BITTER GRASS

Culms rather stout; leaf-blades up to 15 mm broad, scabrous toward the long-acuminate tip. Spikes 5–10, pale; spikelets closely imbricate, silky; first lemma ca. 2.5 mm long, pubescent on the keel, long-ciliate on the margins, the hairs 2 mm long, the awn about as long as the lemmas; second lemma shorter and narrower than the first, the awn ca. 2 mm long.

GRAND CAYMAN: *Brunt* 2036; *Hitchcock*; *Kings GC* 262; *Millspaugh* 1271; *Proctor* 15129.
— Florida and the West Indies south to Brazil, chiefly in grassy fields.

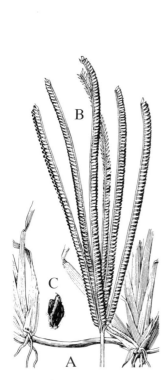

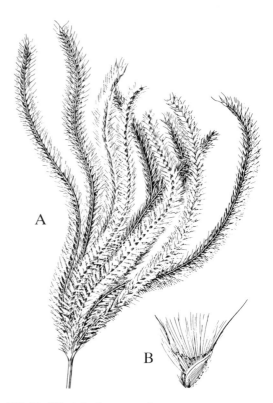

FIG. 37 **Chloris petraea**. A, portion of plant with stolons, plant, × 5. B, inflorescence × 1. C, florets × 5. (H.)

FIG. 38 **Chloris barbata**. A, inflorescence, × 1. B, florets, × 5. (H.)

Chloris inflata Link, *Enum. Pl.* 1: 105 (1821).

> *Chloris barbata* Sw. (1797), based on *Andropogon barbatus* L. (1771) not *A. barbatus* L. (1759).

Tufted annual, with compressed culms mostly 30–75 cm tall; leaf-blades up to 20 cm long or more, 2–5 mm broad, glabrous except at the junction of the sheath where a few long scattered hairs occur; sheath glabrous and markedly keeled; ligule small, ciliate. Inflorescence of mostly 5–8 reddish-purple spikes, 4–6 cm long. Spikelets 3-flowered with one fertile and two sterile florets; lemmas of the three florets each with an awn ca. 7mm long.

GRAND CAYMAN: *Brunt* 1785, 2035; *Correll & Correll* 51055; *Maggs* II 61; *Proctor* 15057. CAYMAN BRAC: *Proctor* 29090.

— Mexico and the West Indies south to Argentina. A common weed of roadsides, pastures and sandy waste ground.

Cynodon L. C. Rich.

Perennials, with creeping stolons or rhizomes, short leaf-blades, and several slender digitate spikes. Spikelets 1-flowered, awnless, sessile in 2 rows along one side of a slender continuous rhachis, the rhachilla disarticulating above the glumes and prolonged behind the palea as a slender naked bristle, this sometimes bearing a rudimentary lemma; glumes narrow, acuminate, 1-nerved, about equal in size, shorter than the floret; lemma strongly compressed, pubescent on the keel, firm, 3-nerved, the lateral nerves close to the margins.

About 10 tropical and subtropical species.

KEY TO SPECIES

1. Plants with underground rhizomes; leaf-blades mostly 2–5 cm long; ligules of 1 or 2 rows of small hairs; flowering culms mostly 10–25 cm tall: **C. dactylon**

1. Plants with surface stolons but without underground rhizomes; leaf-blades mostly 5–16 cm long; ligule a scarious rim 0.3 mm long; flowering culms 30–60 cm tall: **[C. nlemfuensis]**

Cynodon dactylon (L.) Pers., *Synops. Pl.* 1: 85 (1805). **FIG. 39.**

BAHAMA GRASS, BERMUDA GRASS

Stoloniferous perennial with wiry underground creeping rhizomes. Flowering culms erect, usually 10–30 cm tall; leaf-blades 2–5 cm long or more, 1.5–2.5 mm broad, glabrous or finely pubescent; ligule in some strains with a double row of hairs, one row being shorter than the other, other strains with only a single line of hairs. Inflorescence of 3–5 spikes digitate at the apex of the culm; spikes 2.5–5 cm long. Spikelets 2–2.5 mm long.

GRAND CAYMAN: *Hitchcock*; *Proctor* 15132. CAYMAN BRAC: *Proctor* 29099.

— Widespread in the warmer parts of both hemispheres. Commonly used for lawns and pastures throughout the West Indies; it especially thrives in sandy soils.

[*Cynodon nlemfuensis* Vanderyst in *Bull. Agric. Congo Belg.* 11: 121 (1920).

AFRICAN STAR GRASS

Stoloniferous perennial herb without rhizomes; stolons stout and woody, creeping on ground surface. Leaf-blades flat, linear-lanceolate, 5–16 cm long, 2–6 mm broad. Inflorescence with 1 (sometimes 2) whorls of 4–9 slender racemes 4–7 cm long; spikelets 2–3 mm long, green or sometimes reddish or purple; glumes narrowly lanceolate, the upper 0.5–0.75 times the length of the spikelet; lemma silky pubescent on keel, the palea glabrous.

GRAND CAYMAN: *Guala* 1886, in pastures.

— Native to Kenya and other parts of East Africa, now introduced in many tropical countries.]

FIG. 39 **Cynodon dactylon**. A, plant, × ¹/₂. B, spikelet, × 5. C, floret, × 5. (H.)

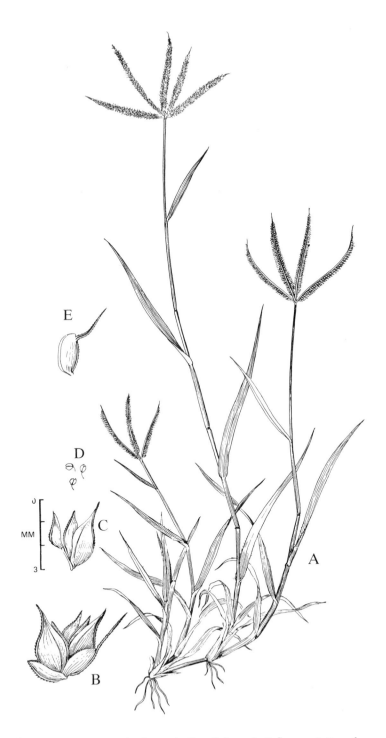

FIG. 40 **Dactyloctenium aegyptium**. A, plant, × ½. B, spikelet, × 5. C, floret, × 5. D, anthers, × 5. E, second glume, × 5. (H.)

Dactyloctenium Willd.

Annuals or perennials, with flat leaf-blades and 2–several short thick spikes, which are digitate and widely spreading at the summit of the culms. Spikelets 3–5-flowered, compressed, sessile and closely imbricate, in 2 rows along one side of the rather narrow flat rhachis, the end projecting in a point beyond the spikelets; rhachilla disarticulating above the first glume and between the florets; glumes somewhat unequal, broad, 1-nerved, the first persistent upon the rhachis, the second mucronate or short-awned below the tip, deciduous; lemmas firm, broad, keeled, acuminate or short-awned, 3-nerved, the lateral nerves indistinct, the upper floret reduced; palea about as long as the lemma. Seed subglobose, ridged or wrinkled, enclosed in a thin, early disappearing pericarp.

A small warm-climate genus of 10 species, chiefly African.

Dactyloctenium aegyptium (L.) P. Beauv., *Ess. Nouv. Agrost. Expl. Planch.* 10 (1812). FIG. 40. PLATE 4.

Creeping stoloniferous annual, up to 15 cm tall. Leaf-blades 6–14 cm long, to 5 mm broad, hairy on both sides; sheath sometimes glabrous, sometimes hairy; ligule 1 mm long, ciliate. Inflorescence of 3–4 spikes arranged digitately on the end of the culm; spikes to 4 cm long with rhachis extending beyond the spikelets. Spikelets 4–5 flowered, ca. 3 mm long and broad.

GRAND CAYMAN: *Millspaugh* 1267; *Proctor* 15067.
— Pantropical, in various habitats.

Digitaria Heist. ex Fabr.

Erect or prostrate annuals or perennials, the slender racemes glabrous to silky-pubescent, digitate, or if paniculate then aggregated along the upper part of the culms. Spikelets solitary or in 2s or 3s, subsessile or short-stalked, alternate in 2 rows on one side of a 3-winged or wingless rhachis, lanceolate or elliptic, plano-convex; first glume minute or lacking; second glume equal in length to or shorter than the sterile lemma, glabrous or silky-hairy; fertile lemma cartilaginous with pale hyaline margins.

A large pantropical genus of nearly 400 species, many extending into warm-temperate regions.

KEY TO SPECIES

1. Racemes paniculately arranged, silky-pubescent; fruit lance-acuminate: **D. insularis**

1. Racemes digitately or subdigitately arranged, glabrous or with inconspicuous short pubescence; fruit elliptic:

  2. Rachis of inflorescence branches sparsely pubescent; spikelets 2.5 mm long; first glume obsolete or lacking: **D. horizontalis**

  2. Rachis of inflorescence branches glabrous; spikelets 3 mm long; first glume present: **D. ciliaris**

FIG. 41 **Digitaria insularis**. A, B, C, parts of plant, × $\frac{1}{2}$. D, spikelet, × 10. E, floret, × 10. (H.)

Digitaria insularis (L.) Mez ex Ekman in *Ark. f. Bot.* 11(4): 17 (1912). **FIG. 41.**

Panicum insulare (L.) Meyer (1818).
Trichachne insularis (L.) Nees (1829).
Valota insularis (L.) Chase (1906).

Culms tufted, coarse, erect, 50–150 cm tall; leaf-blades flat, usually scabrous, up to 25 cm long and 15 mm broad; sheaths sparsely silky-hairy chiefly along the margins; ligule distinct, 1 mm long. Inflorescence a greenish to pale tawny panicle mostly 15–20 cm long; spikelets 4–5 mm long, silky-hairy.

GRAND CAYMAN: *Brunt* 2167; *Hitchcock*; *Kings* GC 252; *Proctor* 15284. LITTLE CAYMAN: *Kings* LC 97; *Proctor* 28158. CAYMAN BRAC: *Millspaugh* 1153; *Proctor* 29081.

— Widely distributed in the warmer parts of America, chiefly occurring in open ground and waste places throughout the West Indies at low elevations.

Digitaria horizontalis Willd., *Enum. Hort. Berol.* 92 (1809). **FIG. 42.**

Stoloniferous annual with decumbent branching culms; leaves flat, pubescent, the sheaths pilose. Inflorescence of 5–15 very slender, lax, sparsely hairy racemes up to 8 cm long, subdigitate or in fascicles along a slender axis. Spikelets narrow, ca. 2.5 mm long; first glume obsolete or lacking.

GRAND CAYMAN: *Hitchcock*; *Kings* GC 344. LITTLE CAYMAN: *Proctor* 28152.
— Pantropical, common in fields, open ground, and waste places.

Digitaria ciliaris (Retz.) Koeler, *Descr. Gram.* 27 (1802).

Annual herb lacking stolons; flowering culms decumbent at the base, sparingly branched, 10–100 cm tall; leaf-sheaths glabrous or pubescent; ligules 1–4 mm long; leaf-blades

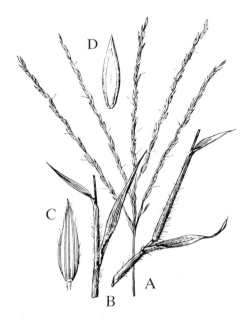

FIG. 42 **Digitaria horizontalia**. A, portions of plant, × 1. B, spikelet, × 10. C, spikelet, × 10. D, spikelet, × 10. (H.)

linear, flat, 2–20 cm long, 3–10 mm broad, scabrous, mostly glabrous on the upper surface. Primary branches of the inflorescence 4–24 cm long, distinctly winged. Spikelets paired, lanceolate; lower glume nerveless, less than 0.8 mm long; upper glume 3–5-nerved, up to 2.7 mm long.

GRAND CAYMAN: *Guala* 1873, 1887. Found in "pasture, lawns, and roadsides, and many other disturbed areas".

— A pantropical weed.

Distichlis Raf.

Dioecious perennials with extensively creeping scaly rhizomes. Culms erect, stiff, leafy, terminating in a small dense panicle. Spikelets several- to many-flowered, the rhachilla of the pistillate ones disarticulating above the glumes and between the florets. Glumes unequal, broad, acute, keeled, mostly 3-nerved; lemmas closely imbricate, firm, 3-nerved with intermediate striations; pistillate palea enclosing the grain.

A genus of 4 N. American, 8 S. American and 1 Australian species.

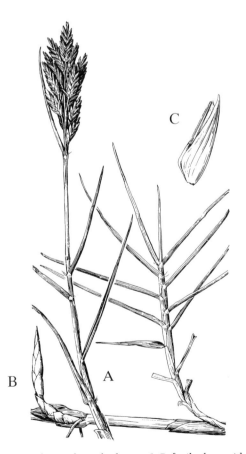

FIG. 43 **Distichlis spicata**. A, plant with sterile shoot, × 1. B, fertile shoot with inflorescence, × 1. C, floret, × 5. (H.)

Distichlis spicata (L.) Greene in *Bull. Calif. Acad. Sci.* 2: 415 (1887). **FIG. 43.**

Sometimes forming extensive colonies of sterile plants. Leaves strongly distichous, with stiff, glabrous, involute blades. Spikelets ca. 1 cm long, aggregated in a small, compact panicle.

GRAND CAYMAN: *Brunt* 1824, 1866, 1886; *Kings GC* 424; *Proctor* 52186.

— United States to Mexico and the northern West Indies, in saline soils.

Echinochloa P. Beauv.

Coarse, often succulent, annuals or perennials, with compressed sheaths, linear flat leaf-blades, and rather compact panicles composed of short, densely flowered racemes along the main axis. Spikelets plano-convex, often stiffly hispid, subsessile, solitary or in irregular clusters on one side of the panicle branches; first glume about half the length of the spikelet, pointed; second glume and sterile lemma of equal length, pointed, mucronate, or the glume short-awned and the lemma long-awned, sometimes conspicuously so, enclosing a membranous palea and sometimes a staminate flower; fertile lemma plano-convex, smooth and shining, acuminate-pointed, the margins inrolled below, flat above, the apex of the palea not enclosed.

A genus of about 30 species occurring widely in tropical and subtropical regions.

KEY TO SPECIES

1. Racemes simple, 1–2 cm long; awn of the sterile lemma reduced to a short point; leaves 3–6 mm broad:	E. colonum
1. Racemes often branched, more than 2 cm long; awn of the sterile lemma 1–2 cm long, conspicuous; leaves more than 10 mm broad:	E. walteri

Echinochloa colona (L.) Link, *Hort. Bot. Berol.* 2: 209 (1833). **FIG. 44.**

Weedy annual, usually much branched at base; culms decumbent-ascending or erect, 20–40 cm long, glabrous; leaf-blades rather lax, 5–10 cm long, 3–6(–10) mm broad, somewhat scabrous on the margins; sheaths glabrous; ligule lacking. Inflorescence a compressed panicle of 4–6 appressed or ascending racemes; spikelets crowded, 3 mm long, in 4 rows.

GRAND CAYMAN: *Brunt* 2098; *Kings GC* 399.

— Pantropical, often in roadside ditches and moist fields.

Echinochloa walteri (Pursh) Heller, *Cat. N. Amer. Pl.* ed. 2, 21 (1900). **FIG. 45.**

Annual; culms erect, often succulent and rooting at the lower nodes, 1–2 m tall, up to 2.5 cm thick at base; leaf-blades 10–15(–30) mm broad, scabrous on both sides; sheaths papillose-hairy or papillose only, rarely glabrous; ligule lacking or represented by a line of hairs. Panicle dense, up to 30 cm long, erect or nodding; racemes oppressed or ascending, up to 10 cm long; spikelets close, long-awned, often purple, ca. 3 mm long.

GRAND CAYMAN: *Brunt* 1659, 1772; *Kings GC* 279; *Proctor* 15021.

— Eastern United States, Cuba, Jamaica, and Hispaniola, in marshes, lagoons and wet places.

FIG. 44 **Echinochloa colona**, × 1. (H.)   FIG. 45 **Echinochloa walteri**, × 1. (H.)

Eleusine Gaertn.

Annuals, with 2–several rather stout spikes digitately arranged at the summit of the culms, rarely with 1–2 (or a whorl) a short distance below, or a single terminal spike. Spikelets few- to many-flowered, compressed, sessile and closely imbricate, in 2 rows along one side of a rather broad rhachis, the latter not prolonged beyond the spikelets; rhachilla disarticulating above the glumes and between the florets; glumes unequal, rather broad, acute, 1-nerved, shorter than the first lemma; lemmas acute, with 3 stong green nerves close together forming a keel, the upper floret somewhat reduced. Seed dark brown, loosely enclosed in the thin pericarp.

A small genus of 9 tropical or subtropical species, occurring in both hemispheres. Several are useful fodder plants, and one (*Eleusine coracana*) is cultivated as a cereal and for an alcoholic beverage in Ceylon, India and Africa.

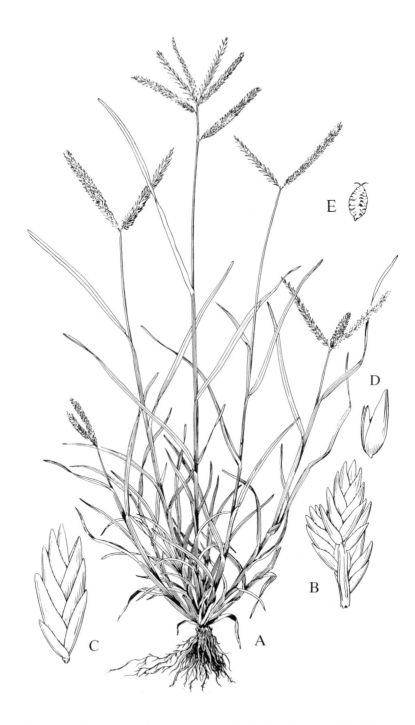

FIG. 46 **Eleusine indica**. A, plant, × ¹/₂. B, C, spikelets, × 5. D, floret × 5. E, seed, × 5. (H.)

Eleusine indica (L.) Gaertn., *Fruct. & Sem. Pl.* 1: 8 (1788). **FIG. 46.**

Loosely tufted, branching at base, the compressed culms ascending or prostrate, less than 50 cm long. Leaf-blades usually flat (sometimes folded), 3–8 mm broad. Spikes mostly 2–6 in number, 4–15 cm long.

GRAND CAYMAN: *Brunt* 2037; *Millspaugh* 1270; *Proctor* 15097; *Sachet* 391. LITTLE CAYMAN: *Proctor* 28154. CAYMAN BRAC: *Proctor* 29106.

— A common pantropical weed.

Eragrostis P. Beauv.

Mostly tufted annuals or perennials with open or contracted panicles. Spikelets few- to many-flowered, the florets usually imbricate, the rhachilla disarticulating above the glumes and between the florets, or continuous, the lemmas deciduous, the paleas persistent; glumes somewhat unequal, acute or acuminate, 1-nerved, or the second rarely 3-nerved; lemmas acute or acuminate, keeled or rounded on the back, 3-nerved, the nerves usually prominent; paleas 2-nerved, the keels sometimes ciliate.

A cosmopolitan genus of about 300 species, the majority occurring in subtropical regions.

KEY TO SPECIES

1. Annuals; paleas ciliate on the keels, the cilia conspicuous, usually as long as the width of the lemma; culms mostly 10–30 cm tall:	
2. Pedicels as long as the spikelets or longer; panicle open, oblong:	[E. tenella]
2. Pedicels very short; panicle spike-like, more or less interrupted:	E. ciliaris
1. Perennials; paleas scaberulous on the keels but not prominently long-ciliate; culms 40–150 cm tall:	E. domingensis

[*Eragrostis tenella* (L.) P. Beauv. ex Roem. & Schult. in L., *Syst. Veg.* ed. nov. 2: 76 (1817).

Eragrostis amabilis (L.) Wight & Arn. (1841).

Small tufted annual usually 15 cm tall or less. Leaf-blades 3–8 cm long, ca. 2 mm broad, glabrous, with inconspicuous ligule. Panicle open and rather delicate, ca. 6 cm long or sometimes longer. Spikelets 2 mm long.

GRAND CAYMAN: *Correll & Correll* 51056. LITTLE CAYMAN: *Proctor* 28192. CAYMAN BRAC: *Proctor* 15310.

— Native of the Old World tropics, originally described from India. This small weed now occurs throughout the West Indies.]

Eragrostis ciliaris (L.) R. Br. in Tuckey, *Narr. Exped. Riv. Zaire* 478 (1818). **FIG. 47.**

Small tufted annual seldom taller than 20 cm. Leaf-biades averaging 6 cm long, 5 mm broad, glabrous except for a tuft of long hairs (3 mm long) just above the junction with the sheath; sheath glabrous except for a ring of hairs on the back just below where the 3 mm hairs occur; basal sheath purplish, ligule reduced to a trace. Inflorescence a ragged spike-like panicle averaging 7 cm long. Spikelets 8-flowered; cilia of the paleas with a characteristic comb-like appearance when viewed through a hand-lens.

FIG. 47 **Eragrostis ciliaris**. A, plant, × $^1/_2$. B, spikelet, × 5. C, floret × 10. (H.)

GRAND CAYMAN: *Brunt* 1708, 2179; *Proctor* 11965. LITTLE CAYMAN: *Proctor* 28155. CAYMAN BRAC: *Millspaugh* 1190; *Proctor* 29060.

— A pantropical weed.

Eragrostis domingensis (Pers.) Steud., *Syn. Pl. Glum.* 1: 278 (1854).

Tufted perennial with strong erect culms up to 1.5 m tall. Leaf-blades elongate, 2–5 mm broad; sheaths glabrous or sparsely pilose at summit. Panicles 25–50 cm long, loosely flowered. Spikelets short-pedicellate, 5–10 mm long, 10–25-flowered; lemmas 1.5–1.8 mm long.

GRAND CAYMAN: *Brunt* 1660, 1704, 2067, 2178; *Hitchcock*; *Millspaugh* 1240; *Proctor* 11966. CAYMAN BRAC: *Proctor* 29003.

— West Indies.

Imperata Cyrillo

Slender erect perennials with terminal narrow silky panicles. Spikelets all alike, awnless, in pairs, unequally pedicellate on a slender continous rhachis, surrounded by long silky hairs; glumes about equal, membranous; sterile lemma, fertile lemma, and palea thin and hyaline.

A small genus of 10 tropical and subtropical species.

Imperata contracta (Kunth) Hitchc. in *Rep. Mo. Bot. Gard.* 4: 146 (1893).

Tufted, spreading by scaly rhizomes; culms erect, often 1 m tall or more, leafy; leaf-blades flat, elongate, 6–10 mm broad, glabrous but with sharply scabrous margins. Inflorescence a narrow whitish panicle up to 40 cm long; spikelets 3 mm long, densely silky-hairy from the base, the hairs much longer than the spikelet.

GRAND CAYMAN: "Hitchcock in 1891", as recorded in *Manual of the Grasses of the West Indies*, p. 379. No Cayman material seen in preparing this Flora.

— Southern Mexico and the West Indies to Brazil, in swamps and moist open ground.

Lasiacis (Griseb.) Hitchc.

Large branching perennials with woody culms, often clambering, the mostly firm leaf-blades narrowed into a minute petiole, and with open or somewhat compact panicles. Spikelets subglobose, placed obliquely on their pedicels; first glume broad, somewhat inflated-ventricose, usually not more than one-third the length of the spikelet; second glume and sterile lemma about equal in length, broad, abruptly apiculate, papery-chartaceous, shining, many-nerved, glabrous or lanose at the apex only, the lemma enclosing a membranous palea and sometimes a staminate flower; fertile lemma white, bony-indurate, obovoid, obtuse, this and the palea having the same texture, bearing at the apex in a slight crateriform depression a tuft of woolly hairs, the palea concave below, gibbous above, the apex often free at maturity.

About 30 species of tropical and subtropical America.

Lasiacis divaricata (L.) Hitchc. in *Contr. U.S. Nat. Herb.* 15: 16 (1910). **FIG. 48.**

Woody perennial climber, usually glabrous throughout (except margins of sheaths); culms much-branched, reaching 3 m or more in length; leaf-blades narrowly lanceolate, 5–12 cm long, 5–15 mm broad, with scabrous margins; ligule inconspicuous, ciliate. Panicles terminating the main culm and fertile branches, 5–20 cm long, loosely flowered; spikelets ovoid, ca. 4 mm long.

GRAND CAYMAN: *Brunt* 1694, 2031, 2191; *Kings GC* 166, GC 166a; *Lewis* 3859, *Sachet* 372. LITTLE CAYMAN: *Kings LC* 94. CAYMAN BRAC: *Kings CB* 44; *Millspaugh* 1172, 1226; *Proctor* 28962.

— Florida, West Indies, C. and S. America, common in thickets and woodlands at low to middle elevations.

FIG. 48 **Lasiacis divaricata**. A, portion of plant, × $^1/_2$. B, spikelet, × 10. C, floret × 10. (H.)

Leptochloa P. Beauv.

Annuals or perennials with flat leaf-blades and numerous narrow racemes scattered along a common axis, forming a rather ample panicle. Spikelets 2–several-flowered, sessile or short-stalked, approximate or somewhat distant along one side of a slender rhachis; rhachilla disarticulating above the glumes and between the florets; glumes unequal or nearly equal, awnless or mucronate, 1-nerved, usually shorter than the first lemma; lemmas obtuse or acute, sometimes 2-toothed and mucronate or short-awned from between the teeth, 3-nerved, the nerves sometimes pubescent.

A genus of 27 tropical or warm-temperate species, the majority in the Western Hemisphere.

KEY TO SPECIES

1. Plants annual:

    2. Spikelets 1–2 mm long, the lemmas awnless, 1.5 mm long; leaf-sheaths papillose-hispid: **L. filiformis**

    2. Spikelets 7–12 mm long, the lemmas distinctly awned, 4–5 mm long; leaf-sheaths glabrous or scabrous: **L. fascicularis**

1. Plants perennial; lemmas 1.5–2mm long, awnless or short-awned; leaf-sheaths glabrous (usually somewhat glaucous):   **L. virgata**

Leptochloa filiformis (Lam.) P. Beauv., *Ess. Nouv. Agrost.* 166 (1812). **FIG. 49.**

Tufted annual with flat leaf-blades 4–10 mm broad; ligule long and membranaceous. Inflorescence up to 25 cm long, usually lax and open with numerous very slender racemes, slightly viscid. Spikelets 3–4-flowered; glumes acuminate, longer than the first floret; lemmas pubescent on the nerves.

GRAND CAYMAN: *Brunt* 2176.

— Virginia to California, south to S. America. A common weed of fields and cultivated ground throughout the West Indies.

Leptochloa fascicularis (Lam.) A. Gray, *Man.* 588 (1848). **FIG. 50.**

Somewhat succulent annual with freely branching culms 30–100 cm long, erect to spreading or prostrate. Leaf-blades flat to loosely involute, 1–3 mm broad; sheaths often purple. Inflorescence often partly enclosed by the upper leaves, the several to numerous racemes ascending or appressed, or spreading at maturity. Spikelets overlapping, 6–12-flowered; awn of lemma up to the same length as the lemma.

GRAND CAYMAN: *Brunt* 1631, 1966.

— United States, Mexico, and the West Indies, mostly in ditches and wet places.

Leptochloa virgata (L.) P. Beauv., *Ess. Nouv. Agrost.* 166 (1812).

Tufted perennial with wiry culms 50–100 cm tall. Inflorescence ample, the slender racemes several to many, often laxly spreading, the lower distant. Spikelets 3–5-flowered.

GRAND CAYMAN: *Hitchcock*.

— Mexico and the West Indies to S. America, common.

FIG. 49 **Leptochloa filiformis**. A, plant, × $\frac{1}{2}$. B, inflorescence, × $\frac{1}{2}$. C, spikelet × 10. D, floret, × 10. (H.)

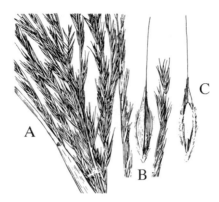

FIG. 50 **Leptochloa fascicularis**. A, portion of inflorescence, × 1. B, C, floret (two views), × 10. (H.)

[*Melinis* P. Beauv.

Perennials or annuals, with rather open panicles of silky spikelets. Spikelets on short capillary pedicels; first glume minute; second glume and sterile lemma of equal length, raised on a stipe above the first glume, emarginate or slightly lobed, short-awned, covered (except toward the apex) with long silky hairs, the palea of the sterile lemma well-developed; fertile lemma shorter than the spikelet, cartilaginous, smooth, boat-shaped, obtuse, the margins thin, not inrolled, the palea not enclosed.

A genus of 20 or more species occurring chiefly in tropical Africa, Madagascar, and from Arabia to Vietnam; *Melinis repens* is widely naturalized elsewhere.

Melinis repens (Willd.) Zizka in *Biblioth Bot.* 138: 55 (1988).

Rhynchelytrum roseum (Nees) Stapf & Hubb. ex Bews (1929).
R. repens sensu C. E. Hubb. (1934), in part, not *Saccharum repens* Willd. (1797).
Tricholaena repens sensu Hitchc. (1936), not *S. repens* Willd. (1797).
Tricholaena rosea of this Flora, 1st edition.

Tufted annual; culms often decumbent at base, rooting freely at the lower nodes, 30–100 cm tall; leaf-blades flat, 5–15 cm long, 2–7 mm broad, nearly glabrous; sheath hairy especially just below the junction with the blade and on the nodes; ligule represented by a line of hairs 1 mm long. Inflorescence a showy red-purple panicle 10–15 cm long; spikelets ca. 5 mm long, covered with long silky hairs.

GRAND CAYMAN: *Brunt* 2131; *Proctor* 15306. LITTLE CAYMAN: *Proctor* 28143. CAYMAN BRAC: *Proctor* 28995.

— Native of Africa; introduced into the West Indies, now widespread and common along roadsides and in fields. A somewhat useful pasture grass on poor soils, it is also cultivated in some countries to make dry bouquets.]

Oplismenus P. Beauv.

Annual or weak-stemmed perennial herbs; culms decumbent, sparingly or moderately branched. Leaf-sheaths smooth, the ligule a small ciliate membrane; leaf-blades linear, lanceolate or ovate, flat. Main axis of inflorescence with alternating 1-sided appressed or spreading short primary branches; secondary branches usually reduced to a fascicle of spikelets; rachis terminating in a spikelet that disarticulates at its base. Spikelets adaxial, more or less compressed; lower glume awned; upper glume 0.5–0.8 times the length of spikelets; sterile lemma of lower floret membranous; lemma of the upper floret nearly the length of the lower floret, cartilaginous, smooth and shiny, with involute margins, mucronate.

A small pantropical genus of about 5–7 species.

Oplismenus hirtellus (L.) P. Beauv., *Ess. Agrostogr.* 54 (1812).

Annual or perennial stoloniferous herb. Leaf-sheaths glabrous or hairy; ligules 0.4–1.1 mm long; leaf-blades linear to ovate, flat, 1–10 cm long, 0.4–1.7 cm broad, scabrous, glabrous or hairy on the upper surface. Flowering culms 5–50 cm long, decumbent, sparingly branched; primary branches of the inflorescence 5–15 mm long; spikelets paired, lanceolate to elliptic, 2.1–3 mm long; lower glumes 3–5-nerved, with awns 5–10 mm long.

GRAND CAYMAN: *Burton* 202, "dispersing along forest trails".
— Pantropical.

Panicum L.

Annuals or perennials of various habit. Inflorescence paniculate, rarely racemose; spikelets more or less compressed dorsiventrally; glumes herbaceous, nerved, usually very unequal, the first often minute, the second typically equal in length to the sterile lemma, the latter of the same texture and simulating a third glume, bearing in its axil a membranous or hyaline palea and sometimes a staminate flower, or rarely the palea lacking; fertile lemma chartaceous-indurate, typically obtuse, the nerves obsolete, the margins inrolled over an enclosed palea of the same texture.

A large genus of about 500 species, widely distributed in tropical and warm-temperate regions. Several species, known collectively as millets, are important cereal crops, especially in India and southern Europe. Others are important pasture grasses. The genus has been subject to considerable taxonomic fragmentation in recent years.

KEY TO SPECIES

1. Spikelets arranged in spike-like racemes along one side of the panicle branches:

    2. Nodes glabrous; racemes erect, 0.5–3 cm long: — **P. geminatum**

    2. Nodes bearded; racemes ascending, 3–9 cm long: — **P. purpurascens**

1. Spikelets in open or condensed panicles, the branches not raceme-like:

    3. Panicles open, diffuse:

        4. Plants with extensively creeping rhizomes; leaf-blades 2–5 mm wide; spikelets ca. 2.5 mm long: — **P. repens**

        4. Plants densely tufted, without extensively creeping rhizomes; leaf-blades usually more than 10 mm wide; spikelets 3 mm long: — **[P. maximum]**

    3. Panicles (or panicle-branches) narrow and condensed, 10–25 cm long, mostly less than 5 cm broad; spikelets 2.5 mm long, the glumes with a scabrous keel: — **P. rigidulum**

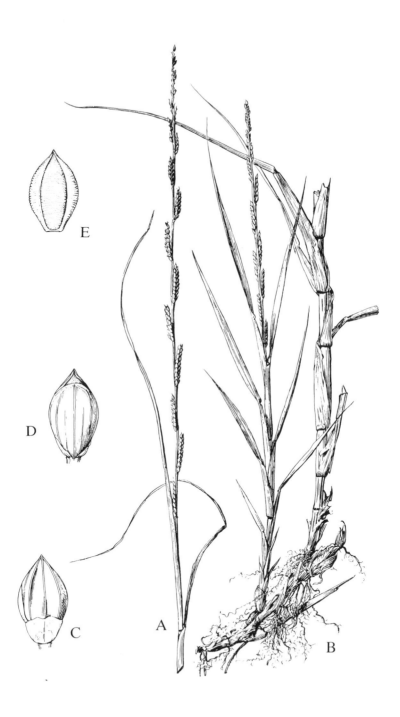

FIG. 51 **Panicum geminatum**. A, B, inflorescence and portion of plant plant, × ¹/₂. B, inflorescence, × ¹/₂. C, D, two views of spikelet × 10. E, floret, × 10. (H.)

FIG. 52 **Panicum maximum**. A, B, C, vegetative portions of plant, × ½. D, inflorescence, × ½. E, F, two views of spikelet × 10. G, floret, × 10. (H.)

Panicum geminatum Forsk., *Fl. Aegypt.-Arab. LX*, 18 (1775). **FIG. 51.**

Paspalidium geminatum (Forsk.) *Stap fin Prain* (1920).

Stoloniferous perennial; culms 25–80 cm tall, spreading from a decumbent base; leaf-blades 10–20 cm long, 3–8 mm wide, glabrous; ligule represented by a line of hairs just under 1 mm long. Inflorescence a narrow panicle, the main axis bearing 12–18 erect, subdistant racemes less than 3 cm long; spikelets 2–3 mm long, glabrous.

GRAND CAYMAN: *Brunt* 2090, 2099; *Proctor* 15283.

— Pantropical, occurring in swamps and damp places at low elevations.

Panicum purpurascens (Raddi) Henrard, *Agrost. Bras.* 47 (1823).

Brachiaria purpurascens (Raddi) Heur. (1940).
Panicum muticum sensu Adams (1972), not Forsk. (1775).

PARA GRASS

Widely creeping stoloniferous perennial; culms decumbent at base, rooting at the lower nodes, 2–6 m long, sometimes clambering on bushes or trees; leaf-blades 15–25 cm long, 10–15 mm wide, glabrous with scabrous margins; sheaths pubescent; nodes densely villous; ligule ca. 1 mm long, membranous or represented by a line of hairs. Inflorescence an open panicle; spikelets 3 mm long, glabrous.

GRAND CAYMAN: *Brunt* 7795, 1841, 1846, 1855, 1903.

— Widespread in tropical America, especially in shallow ponds and moist places; also in Africa. It is a valuable forage grass.

Panicum repens L., *Sp. Pl.* ed. 2, 1: 87 (1762).

Perennial from extensively creeping rhizomes. Culms erect, 10–80 cm tall, unbranched, with glabrous nodes; sheaths glabrous or sometimes pubescent, the ligule a ciliate membrane 0.4 –1.0 mm long; leaf-blades linear, 4–15 cm long, 2–8 mm broad, stiff, often distichous, sometimes glaucus, flat or involute, glabrous or pubescent. Panicles up to 19 cm long with ascending branches; spikelets oblong- to lance-elliptic, glabrous, sharply acute; upper glume 7–9-nerved; lower floret staminate, upper floret bisexual, 1.7–2.2 mm long, elliptic, smooth and shiny, the anthers orange.

GRAND CAYMAN: *Guala* 1890, "roadside and waste areas". This species is salt tolerant and may occur on sea beaches.

— Sporadically pantropical, extending into warm-temperate areas. Rare in the West Indies, recorded only from the Bahamas and Cuba.

[*Panicum maximum* Jacq., *Collect.* 1: 76 (1787). **FIG. 52.**

GUINEA GRASS

Robust tufted perennial; culms erect, 1–2.5 m tall, glabrous except for hairy nodes; leaf-blades flat, to 1 m long or more, 10–35 mm wide, usually glabrous or nearly so and with scabrous margins; sheath with scattered long hairs toward base. Inflorescence an open, diffuse panicle, the lower branches whorled; spikelets 3 mm long, oblong, acute.

GRAND CAYMAN: *Brunt* 1747; *Kings* GC 251; *Proctor* 15222; *Sachet* 375. LITTLE CAYMAN: *Proctor* 28151. CAYMAN BRAC: *Proctor* 29337.

— Native of Africa, now cultivated and naturalized in most tropical countries. It is an important pasture grass that is often cut and fed green.]

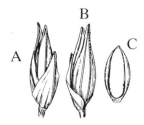

FIG. 53 **Panicum rigidulum**. A, B, two views of spikelet × 10. C, floret, × 10. (H.)

Panicum rigidulum Bosc ex Nees, *Agrost. Bras.* 163 (1829). **FIG. 53**.

Panicum condensum Nash in Small (1903); Adams (1972).

Tufted perennial, essentially glabrous; culms erect, 0.5–2 m tall; leaf-blades flat from a folded base, 15–30 cm long, 3–10 mm wide, margins slightly scabrous; sheaths (especially the lower ones) compressed-keeled. Panicles terminal and axillary, bearing appressed branchlets with crowded spikelets; spikelets ca. 2.5 mm long, narrowly oblong, acuminate, on scabrous pedicels.

GRAND CAYMAN: *Proctor* 15282. CAYMAN BRAC: *Proctor* 29131.

— Eastern United States, West Indies and C. America south to Brazil, in marshes and wet places.

Paspalum L.

Annuals or perennials with 1–many spike-like racemes, these single, paired at the summit of the culms, or racemosely arranged along the main axis. Spikelets plano-convex, usually obtuse, subsessile, solitary or in pairs, in 2 rows on one side of a narrow or dilated rhachis, the back of the fertile lemma toward it; first glume usually lacking; second glume and sterile lemma commonly about equal in length, the former rarely lacking; fertile lemma usually obtuse, chartaceous-indurate, the margins inrolled.

A nearly cosmopolitan genus of about 400 species, especially common in tropical regions.

KEY TO SPECIES

1. Racemes 2, conjugate or nearly so at the apex of the culm; perennials with creeping rhizomes or also stoloniferous:

    2. Spikelets narrowly elliptic, pointed at apex, 3.5–4 mm long; plants stoloniferous: **P. vaginatum**

    2. Spikelets broadly ovate to obovate, blunt at apex, 2.5–3.8 mm long; plants with creeping rhizomes but not stoloniferous: **P. notatum**

1. Racemes 1–many (never 2), if many not paired; plants tufted, not stoloniferous:

    3. Plants annual; spikelets with broad, firm, notched margin: **P. fimbriatum**

    3. Plants perennial; spikelets with entire margins:

        4. Racemes solitary; spikelets solitary (not paired) on the rhachis: **P. distortum**

        4. Racemes usually more than 2:

            5. Plants slender, the culms usually less than 1 m long:

                6. Racemes many, usually 10 or more; spikelets subhemispheric, 1.3–1.5 mm long, pubescent, the hairs not glandular: **P. paniculatum**

6. Racemes few, usually 3–8; spikelets elliptic to oval:
　　7. Spikelets oval, 1.3 mm long, glandular-pubescent:　　　**P. blodgettii**
　　7. Spikelets more than 1.3 mm long, not glandular-pubescent:
　　　　8. Ligule membranous with long white hairs behind it; leaf-
　　　　　　blades often villous-pubescent; spikelets suborbicular:　　**P. setaceum**
　　　　8. Ligule obsolete or apparently absent; leaf-blades usually
　　　　　　glabrous; spikelets elliptic:　　　　　　　　　　　**P. caespitosum**
5. Plants robust, the culms usually 1–2 m tall; spikelets more than
　　2 mm long:
　　9. Spikelets suborbicular, very crowded; rhachis of racemes ciliate:.　**P. millegranum**
　　9. Spikelets obovate-elliptic; rhachis of racemes not ciliate:　　**P. arundinaceum**

Paspalum vaginatum Sw., *Nov. Gen. & Sp. Pl.* 21 (1788).

Extensively creeping perennial with horizontal rhizomes and long leafy stolons, often forming large colonies. Flowering culms 8–60 cm tall; leaf-blades 1.5–15 cm long, 3–8 mm wide, glabrous, flat or folded with involute margins; ligule very short, truncate. Inflorescence usually a pair of racemes at apex of culm; racemes 1–7.5 cm long; spikelets 3–4.5 mm long, pale, glabrous.

GRAND CAYMAN: *Brunt* 1825, 1858, 1862, 1867, 1961, 2041; *Proctor* 15265; *Sachet* 396. LITTLE CAYMAN: *Kings* LC 84a; *Proctor* 28067. CAYMAN BRAC: *Kings* CB 99a.

— Pantropical, extending along seacoasts into warm-temperate areas; common in the West Indies. The closely related *Paspalum distichum* L., which may also occur in the Cayman Islands, can be distinguished by the appressed-pubescent second glume and lemma of the lower floret, and by hispid pubescence on the lower culm-nodes. This species is less salt-tolerant than *P. vaginatum*.

Paspalum notatum Flügge, *Monogr. Pasp.* 106 (1810).

Perennial with short or somewhat elongate stout subsurface rhizomes, from which the erect culms arise singly or in small clumps; rhizomes clothed with closely adhering, shiny, persistent, imbricated, scale-like sheaths. Leaf-blades linear, 6–25 cm long, 0.4–1 cm broad; ligule 0.3–1.5 mm long; flowering culms mostly 20–75 cm tall with usually a pair of racemes at the apex, these usually 4–12 cm long (rarely longer); occasionally a third raceme is produced on the culm below the apex. Spikelets solitary, glabrous, ovate or obovate, 2.7–3.8 mm long, broadly acute or obtuse at the apex.

GRAND CAYMAN: *Guala* 1900, 1930, found in "roadsides, pastures, and lawns".

— Greater and Lesser Antilles, eastern Mexico, C. America, and eastern S. America; introduced into the southern United States as a pasture grass known as bahia grass.

Paspalum fimbriatum Kunth in H.B.K., *Nov. Gen.* 1: 93, t. 28 (1816). **FIG. 54.**

Loosely tufted annual, the erect culms 25–100 cm tall; leaf-blades flat, mostly 10–20 cm long, 5–12 mm broad. Racemes 3–8, ascending or spreading, 2.5–8 cm long; spikelets ca. 2.3 mm long, with a broad, firm, notched wing, the wing and spikelets together ca. 3 mm long and wide.

GRAND CAYMAN: *Kings* GC 25; Proctor 15071.

— West Indies, Panama, and northern S. America, a common weed of open waste ground.

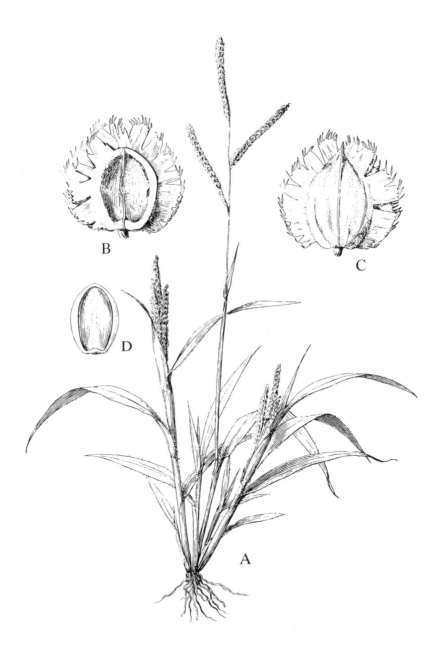

FIG. 54 **Paspalum fimbriatum**. A, plant, × ¹/₂. B, C, two views of spikelet × 10. D, floret, × 10. (H.)

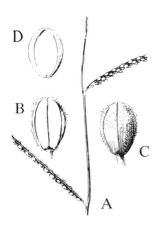

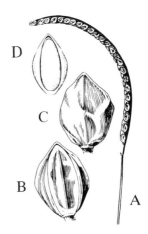

FIG. 55 **Paspalum blodgettii**. A, portion of inflorescence × 1. B, C, two views of spikelet × 10. D, floret, × 10. (H.)

FIG. 56 **Paspalum distortum**. A, inflorescence, × 1. B, C, two views of spikelet × 10. (H.)

Paspalum distortum Chase in *Contr. U.S. Nat. Herb.* 28: 142, f. 35 (1929). **FIG. 56**.

Culms slender, 15–50 cm tall; leaf-blades 15–40 cm long, 10–15 mm broad, involute, often somewhat tortuous. Inflorescence a solitary raceme 2.5–6 cm long, arcuate; spikelets 2 mm long, ovate to somewhat rhomboid, the glume and sterile lemma irregularly crumpled, glabrous.

GRAND CAYMAN: *Kings* GC 234; *Proctor* 15255.

— Greater Antilles, in grassy clearings and savannas.

Paspalum paniculatum L., *Syst. Nat.* ed. 10, 2: 855 (1759).

Culms erect or decumbent at base, mostly 50–100 cm tall; leaf-blades flat, mostly 10–25 cm long, 10–20 mm broad, more or less hairy; sheaths papillose-hispid. Panicle usually 8–20 cm long, of several to many arched-spreading, somewhat fascicled racemes, the lower ones 4–12 cm long; spikelets ca. 1.3 mm long, subhemispheric, pubescent.

GRAND CAYMAN: *Millspaugh* 1406 (syntype of var. *minus* Scribn.).

— Mexico and the West Indies to Argentina; a weed of cultivated and waste places, also occurring in ditches and moist open ground.

Paspalum blodgettii Chapm., *Fl. South. U.S.* 571 (1860). **FIG. 55**.

Culms erect, 40–100 cm tall; leaf-blades flat, 5–25 cm long, 5–10 mm wide; sheaths pubescent. Racemes 3–8, remote (or the upper approximate), 2–8 cm long; spikelets ca. 1.3 mm long, bearing numerous small gland-tipped hairs.

GRAND CAYMAN: *Brunt* 1918, 1953, 2038; *Kings* GC 136; *Proctor* 15068, 15257, 15265. LITTLE CAYMAN: *Proctor* 28082. CAYMAN BRAC: *Kings* CB 72; *Proctor* 28992.

— Florida, C. America, and the West Indies, in pastures, sandy clearings, and in soil pockets or rocky thickets and cultivated ground.

Paspalum setaceum Michx., *Fl. Bor. Am.* 1: 44 (1803).

Tufted perennial, the culms erect or spreading from a knotty base, up to 90 cm tall. Leaf-sheaths glabrous or pubescent; ligule a membrane with long white hairs behind it. Leaf-blades usually flat, 5–30 cm long, 1–20 mm broad, glabrous to pubescent, sometimes papillose-ciliate. Racemes up to 5 along the culm, often with a single axillary raceme arising within a leaf base; racemes 3–17 cm long, slender; spikelets plano-convex, usually in pairs, 1.4–1.9 mm long, suborbicular to elliptic, glabrous or appressed-pubescent; first glume usually absent.

GRAND CAYMAN: *Guala* 1891, 1915, 1932, found in "roadsides, pastures, and lawns".

— Central and eastern United States, Mexico, C. America and the West Indies, variable and polymorphic through much of its range.

Paspalum caespitosum Flügge, *Monogr. Pasp.* 161 (1810).

Densely tufted, the culms erect, slender, 30–60 cm tall, the base hard and slightly enlarged; leaf-blades flat or commonly folded or involute, 5–20 cm long, 4–10 mm wide. Racemes 3–5, remote, rather thick, ascending or somewhat spreading, usually 1.5–4 cm long; spikelets crowded, ca. 1.5 mm long or a little more.

GRAND CAYMAN: *Hitchcock.*

— Southern Florida, C. America and the West Indies, usually in partly shaded humus in pockets of limestone rock.

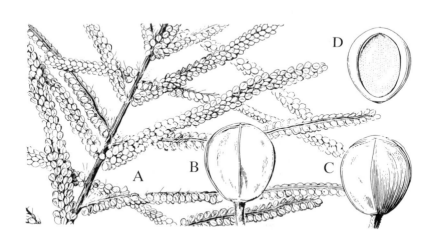

FIG. 57 **Paspalum millegranum.** A, portion of inflorescence × 1. B, C, two views of spikelet × 10. D, floret, × 10. (H.)

Paspalum millegranum Schrad. in Schult., *Mant.* 2: 175 (1824). **FIG. 57**.

Robust tufted perennial; culms 1–2 m tall; leaf-blades elongate, 7–15 mm broad, glabrous but with sharply scabrous margins; ligule 3 mm long, with a tuft of hairs just behind. Racemes numerous (usually 12–25), rather thick, 6–16 cm long; spikelets 2–2.4 mm long, suborbicular, flattened, pale to leaden-purplish.

GRAND CAYMAN: *Brunt* 1897.

— C. America and the West Indies to Brazil, in moist savannas and along ditches.

Paspalum arundinaceum Poir. in Lam., *Encyl. Méth. Bot. Suppl.* 4: 310 (1816).

Robust perennial with culms to 2 m tall or more, usually tufted in dense clumps. Leaf-blades flat or involute, of firm texture up to 1 m long or more, 5–10 mm broad. Panicles large, with usually 12–18 fruiting racemes, these stiffly ascending and mostly 10–20 cm long. Racemes rachises glabrous or scabrous, ca. 1 mm thick. Spikelets glabrous, obovate-elliptic, mostly 2.3–2.8 mm long; first glume occasionally developed.

GRAND CAYMAN: *Guala* 1908; found in a pasture.

— Greater and Lesser Antilles, C. America, and French Guyana, a rather uncommon species of low marshy habitats. This species belongs to a complex of closely related taxa.

[*Pennisetum* L. C. Rich.

Annuals or perennials, with usually flat leaf-blades and dense spike-like panicles. Spikelets solitary or in groups of 2 or 3, surrounded by an involucre of bristles, these not united except at the very base, often plumose, falling attached to the spikelets; first glume shorter than the spikelet, sometimes minute or lacking; second glume shorter than or equal in length to the sterile lemma; fertile lemma chartaceous, smooth, the margin thin, enclosing the palea.

A genus of 130 species widely distributed in warm regions (chiefly Old World). *Pennisetum typhoïdeum* L. C. Rich., pearl millet, is extensively cultivated in India.

Pennisetum purpureum Schum. in K. Dansk., *Vid. Selsk. Naturv. & Math. Afh.* 3: 64 (1828). **FIG. 58A**.

ELEPHANT GRASS, NAPIER GRASS

Robust tufted leafy perennial 2–4 m tall; leaf-blades elongate, 20–30 mm broad, the lower ones glabrous, the upper ones with long hairs on dorsal surface; sheath with scattered hairs; ligule represented by a line of hairs 3–4 mm long; nodes conspicuously hairy. Inflorescence a tawny spike-like panicle up to 30 cm long; spikelets 7 mm long, unequally pedicelled in fascicles, these subtended by long hair-like bristles.

GRAND CAYMAN: *Brunt* 1735; *Kings GC* 354; *Proctor* 15289.

— Native of tropical Africa; introduced into the West Indies and other areas as a forage grass and becoming naturalized.]

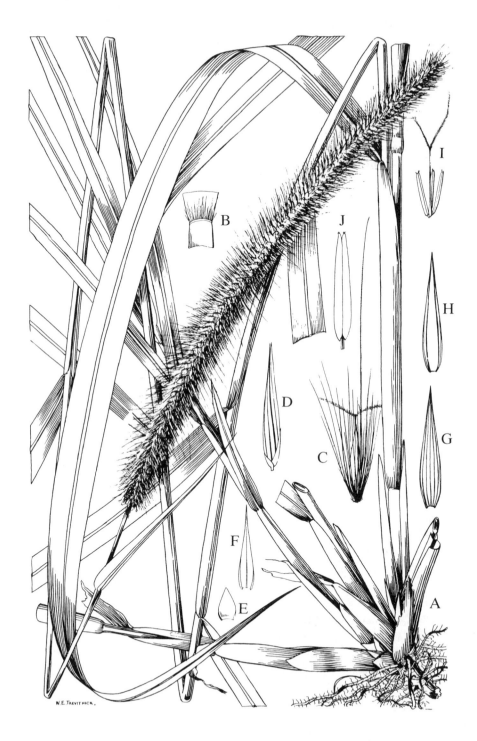

FIG. 58A **Pennisetum purpureum**. A, habit × ½. B, ligule, × 1. C, spikelet surrounded by bristles, × 3. D, spikelet, × 5. E, lower glume, × 5. F, upper glume, × 5. G, upper lemma, × 5. H, palea, × 5. I, flower, × 5. J, stamen, × 12. (R.R.I.)

Pharus L.

Monoecious perennials with erect or spreading herbaceous culms arising from an erect or decumbent base. Leaves with a petiole-like constriction between sheath and blade; blades flat, broad, tapering to both ends, with numerous fine transverse veinlets connecting the longitudinal nerves. Inflorescence a panicle of unisexual, 1-flowered spikelets in pairs of one large pistillate and sessile, and one small staminate and pedicellate, the spikelet pairs rather widely spaced and appressed along the unbranched or sparingly branched primary branches of the panicle. Pistillate spikelets with thin, nearly equal several-nerved glumes, these shorter than the lemma. Lemma densely pubescent with uncinate hairs at tip or all over, with a minute bent beak, becoming hardened and inrolled over the seed in fruits; stigmas 3. Staminate spikelets with 6 stamens.

A neotropical genus of about 8 species.

Pharus glaber Kunth in H.B.K., *Nov. Gen. Sp.* 1: 196 (1816). PLATE 4.

Culms spreading from an erect base, up to 75 cm long. Leaf-blades oblanceolate, usually 15–25 cm long, 3–5 cm broad above the middle, acuminate at the apex. Panicle branches few, stiffly ascending or spreading, strigose pubescent. Pistillate spikelets slender, ca. 1 cm long, the staminate spikelets mostly 3–5 mm long, the glumes dark brown. Fruit 2 to 3 times as long as the glumes, pubescent all over, readily disarticulating at maturity and clinging by the hooked hairs to passing objects or persons.

GRAND CAYMAN: *van B. Stafford* (photo and detached spikelet), found in forest area near the Mastic Trail.

— Greater and Lesser Antilles, and from Mexico to Brazil.

[*Rottboellia* L.f.

Tall robust annuals with flat leaf-blades; ligule a very short membrane. Inflorescence of pedunculate racemes, either single and terminal or several aggregated in a spathate false panicle; racemes cylindric, fragile, bearing paired spikelets; internodes thickened, short-clavate, fused to the adjacent pedicel; sessile spikelet 2-flowered, falling entire, sunk in the internode; callus truncate, with prominent central peg. Lower glume coriaceous, convex, 2-keeled, smooth, narrowly winged at the apex. Lower floret staminate, upper floret bisexual; lemma awnless. Pedicellate floret a little smaller than the sessile one.

A small genus of 4 species of the Old World tropics.

Rottboellia cochinchinensis (Lour.) W. D. Clayton in *Kew Bull.* 35: 817 (1981).

Rottboellia exalta of Adams (1972: 197), not (L.) L.f. (1779).

Tall tufted annual supported below by stilt roots, the basal leaf-sheaths painfully hispid. Culms up to 2 m tall or more, the leaf-blades broadly linear, up to 45 cm long and 2 cm broad. Racemes 3–15 cm long, glabrous, terminating in a tail of reduced spikelets gathered in a spathate false panicle. Sessile spikelets oblong-elliptic, pallid, the lower glume 3.5–5 mm long; upper glume with its keel narrowly winged toward the tip. Pedicellate spikelet narrowly ovate, 3–5 mm long, herbaceous, green.

GRAND CAYMAN: *Guala* 1887, found on "roadsides, abandoned lots and waste areas".
— Throughout the Old World tropics, introduced into Florida and the West Indies.]

[*Saccharum* L.

Tall, coarse perennial grasses with large plume-like inflorescences of panicled racemes. Culms solid, juicy (not hollow as in most grasses). Spikelets all alike, in pairs, one sessile, the other stalked, surrounded at base by long silky hairs, the rhachis readily disarticulating below the spikelets. Glumes firm, 1–3-nerved, acute or acuminate; sterile lemma similar to the glumes but hyaline; fertile lemma shorter than the glumes, hyaline, awnless, sometimes lacking.

Five or more species of the Old World tropics.

Saccharum officinarum L., *Sp. Pl.* 1: 54 (1753).

SUGAR CANE

Loosely tufted, with solid nodose leafy culms 2–3 m tall or more; leaf-blades 1.5–2 m long, 30–50 mm broad, with scabrous margins; sheaths deciduous from lower part of the culm; ligule 2 mm long, membranous, with tuft of hairs just behind. Panicles conspicuous, plume-like, silvery-white, with abundant woolly hairs, the whole inflorescence 50 cm long or more.

Cayman specimens not collected; sometimes cultivated and persisting.

— Cultivated in nearly all tropical and subtropical countries; origin somewhere in southeastern Asia.]

Setaria P. Beauv.

Annuals or perennials, with narrow (often spike-like) terminal panicles. Spikelets subtended by 1–several bristles (sterile branchlets), when ripe falling free from the persistent bristles, awnless; first glume broad, usually less than half the length of the spikelet, 3–5 nerved; second glume and sterile lemma equal, or the glume shorter, several-nerved; fertile lemma coriaceous-indurate, smooth or rugose.

A genus of 140 species widely occurring in tropical and warm-temperate regions. *Setaria italica* (L.) P. Beauv., Italian millet, is cultivated as a cereal from southern Europe to Japan.

Setaria geniculata (Lam.) P. Beauv., *Ess. Nouv. Agrost.* 178 (1812). **FIG. 58B.**

Chaetochloa geniculata (Lam.) Millsp. & Chase (1903).

Perennial with knotty branching rhizomes up to 4 cm long; culms usually erect, 30–100 cm tall, hard and wiry at base; leaf-blades flat, 10–25 cm long, 4–9 mm broad, nearly glabrous or somewhat hairy toward base on upper surface; sheath glabrous; ligule represented by a conspicuous line of hairs. Inflorescence a cylindric spike-like panicle 1–10 cm long (usually about 5 cm), 4–8 mm thick (excluding the bristles); spikelets mostly 2–2.5 mm long, subtended by bristles 1–3 times longer than the spikelets.

GRAND CAYMAN: *Brunt* 1640, 1896, 1927; *Hitchcock*; *Kings* GC 167, GC 325; *Lewis* GC 31; *Proctor* 15254.

— United States to Argentina, common throughout the West Indies in pastures, cultivated areas, and moist ground. Hitchcock notes that "this species is exceedingly variable in general appearance due to the variation in the color and length of the bristles. . . .".

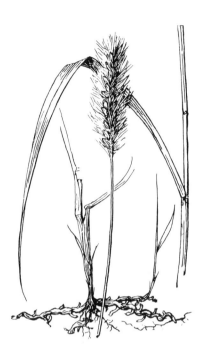

FIG. 58B **Setaria geniculata**, × 1. (H.)

[*Sorghum* Moench

Annuals or perennials, with elongate leaves and open, often large panicles of short racemes. Spikelets in pairs, one sessile and fertile, the other stalked and usually staminate, these pairs attached at the nodes of a tardily disarticulating rhachis of a short, few-jointed raceme, the terminal sessile spikelet with 2 pedicellate spikelets. Glumes of the fertile spikelet indurate, the first rounded but somewhat keeled at the summit; fertile lemma awnless or with a short, usually geniculate, twisted awn. Pedicellate spikelets herbaceous, lanceolate, the first glume several-nerved, 2-keeled in the upper half.

A genus of about 35 species, chiefly found in tropical Africa.

KEY TO SPECIES

1. Plants perennial, with creeping rhizomes; panicles loose, open; leaves less than 2 cm broad:	S. halepense
1. Plants annual, tufted (creeping rhizomes lacking); panicles usually dense, heavy; leaves up to 5 cm broad:	S. saccharatum

[*Sorghum halepense* (L.) Pers., *Syn. Pl.* 1: 101 (1805).

JOHNSON GRASS

Perennial with numerous strong rhizomes. Culms erect, usually 1–1.5 m tall, with appressed-pubescent nodes. Leaves elongate, mostly 1–1.5 cm broad, whitish-scabrous on the margins; sheaths shorter than the internodes, glabrous; ligule membranous, ciliate, ca. 2 mm long. Panicles 15–30 cm long; spikelets 5 mm long, acute, entirely pale or often with a large red spot at base, the awn (when present) 1–1.5 cm long, deciduous.

GRAND CAYMAN: *Hitchcock*.

— Widely naturalized in all warm countries, introduced from the Mediterranean region. A weed of fields and waste places.]

[*Sorghum saccharatum* (L.) Moench, *Meth. Pl.* 207 (1794).

Sorghum vulgare Pers. (1805).

SORGHUM, GUINEA CORN

Tufted annual, often 2 m tall or more (but dwarf races exist). Culms coarse, erect, the ample leaves with broad flat blades to 5 cm broad, the midrib white; ligule membranous, ciliate, 4 mm long. Panicles commonly dense and compact, with numerous turgid persistent spikelets.

CAYMAN BRAC: *Millspaugh* 1224. Also occasionally cultivated on Grand Cayman, but no preserved specimens have been seen.

— Pantropical in cultivation, with numerous varieties, some very different in appearance; occasionally escaping or persisting after cultivation. The species probably originated in Africa.]

Spartina Schreb.

Stout perennials with tough leaves and 2–many spikes racemose on a main axis. Spikelets 1-flowered, much flattened laterally, sessile and usually closely imbricate on one side of a continuous rhachis, disarticulating below the glumes, the rhachilla not produced beyond the floret. Glumes keeled, 1-nerved, acute or short-awned, the first shorter, the second often exceeding the lemma; lemma firm, keeled, the lateral nerves obscure, narrowed to a rather obscure point; palea 2-nerved, keeled and flattened, the keel between or at one side of the nerves.

A chiefly temperate-climate genus of 16 species, all halophytes.

Spartina patens (Ait.) Muhl., *Descr. Gram.* 55 (1817).

Plants forming loose (often sterile) colonies by means of hard, scaly, long-creeping, deep-seated rhizomes. Culms erect, tough and wiry, up to 1 m tall or more. Leaf-blades hard, involute, extending into a long fine point. Spikes mostly 2–8, ascending or spreading, remote, 4–6 cm long; spikelets closely imbricate, 6–8 mm long, the second glume scabrous on the keel.

GRAND CAYMAN: *Brunt* 2065; *Proctor* 31045. LITTLE CAYMAN: *Kings* LC 20; *Proctor* 28194.

— Southeastern United States and the West Indies, on sandy sea-beaches and salt marshes.

Sporobolus R. Br.

Perennials (or some species annual) with narrow to open panicles having numerous very small spikelets. Spikelets 1-flowered, the rhachilla disarticulating above the glumes; glumes usually unequal, the second often as long as the spikelet; lemma membranous, 1-nerved, awnless; palea usually prominent and as long as the lemma or longer. Seed often free from the pericarp.

A genus of about 150 species, widespread in tropical and warm-temperate regions.

KEY TO SPECIES

1. Plants tufted, without creeping rhizomes; leaves loosely or rather densely arranged from base of culm:

    2. Glumes nearly equal, both much shorter than the spikelet: **S. jacquemontii**

    2. Glumes very unequal, the second about as long as the spikelet: **S. domingense**

1. Plants with creeping rhizomes; leaves stiffly distichous: **S. virginicus**

Sporobolus jacquemontii Kunth, *Révis. Gram.* 2: 427, t. 127 (1831).

Sporobolus indicus of Hitchc. (1936, etc.), not (L.) R. Br. (1810).

Tufted perennial; culms 40–100 cm tall. Leaves elongate, to 50 cm long, mostly 1–4 mm wide, flat or somewhat involute, glabrous, the midrib dorsally marked as a white line; ligule very small, ciliate. Inflorescence a dense narrow panicle 15–30 cm long. Spikelets a little over 1 mm long.

    GRAND CAYMAN: *Proctor* 15070, 15258.

    — West Indies; Mexico to Brazil, common in open waste places.

Sporobolus domingensis (Trin.) Kunth, *Révis. Gram.* 2: 427 (1831). **FIG. 60.**

Tufted glabrous perennial; culms 30–80 cm tall. Leaves much shorter than the culms, to 20 cm long, 3–8 mm wide, more or less involute. Panicles contracted, densely flowered, 7–15 cm long. Spikelets ca. 2 mm long or slightly less.

    GRAND CAYMAN: *Brunt* 1683, 1726.

    — Florida, Bahamas, and the Greater Antilles, along sandy borders of salinas and open waste places near the sea.

Sporobolus virginicus (L.) Kunth, *Révis. Gram* 1: 67 (1829). **FIG. 59.**

Culms stiffly erect, 15–50 cm tall or more, from extensively creeping tough scaly perennial rhizomes. Leaves stiff, much shorter than the culms, 1–5 mm wide, involute on drying; ligule minute, ciliate. Panicles spike-like, usually 2–10 cm long; spikelets glabrous, 1.8–3.2 mm long, the glumes unequal or nearly equal.

    GRAND CAYMAN: *Brunt* 1624, 1641, 1861, 1910, 2105, 2120; *Hitchcock*; *Proctor* 27944. LITTLE CAYMAN: *Kings* LC 84; *Proctor* 28081. CAYMAN BRAC: *Brunt* 1682; *Kings* CB 42; *Proctor* 28928.

FIG. 59 **Sporobolus virginicus**, × ¹/₂. (H.)

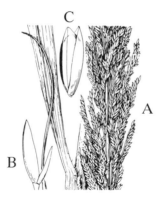

FIG. 60 **Sporobolus domingensis**. A, portion of leaf and inflorescence, × 1. B, glumes, × 10. C, floret, × 10. (H.)

— Eastern United States south to Brazil; common in saline soil or coastal sands throughout the West Indies. Often forms extensive colonies with mostly or entirely sterile culms, the flowering narrowly seasonal.

Stenotaphrum Trin.

Stoloniferous perennials with short flowering culms, rather broad obtuse leaf-blades, and flat terminal and axillary racemes. Spikelets enbedded in one side of an enlarged flattened corky rhachis disarticulating at maturity, the spikelets remaining attached. First glume small; second glumes and sterile lemma about equal in size, the latter with a palea or staminate flower; fertile lemma chartaceous.

A small tropical and subtropical genus of seven species.

Stenotaphrum secundatum (Walt.) Ktze., *Révis. Gen. Pl.* 2: 794 (1891). **FIG. 61**.

Plants extensively creeping, the flat stolons with long internodes and short erect leafy branches. Flowering culms 10–30 cm tall; leaf-blades mostly 10–15 cm long, up to 8 mm wide, glabrous except for a few short hairs on the margins just above junction with the sheath; ligule represented by a line of hairs. Racemes 5–10 cm long; spikelets remote, 4–6 mm long.

GRAND CAYMAN: *Brunt* 1768; *Hitchcock*; *Kings* GC 214; *Proctor* 15038. LITTLE CAYMAN: *Kings* LC 44. CAYMAN BRAC: *Brunt* 1680; *Kings* CB 20.

— Southern United States south to Argentina, common at low to medium elevations. An excellent pasture grass; sometimes used for lawns, especially in poor sandy soils.

FIG. 61 **Stenotaphrum secundatum**. A, plant, × 1. B, portion of spike. C, D, two views of spikelet. (H.)

[*Urochloa* P. Beauv.

Low annual or perennial herbs, mostly with spreading, decumbent or stoloniferous culms. Ligule a short, ciliate membrane. Inflorescence usually a few-flowered panicle; spikelets on short, more or less spreading spicate branches, these scabrous on the rachis and often with a few long stiff hairs. Spikelets solitary or sometimes paired, in 2 rows on either side of an angled rachis, 2-flowered, awnless, the lower floret staminate or sterile, the upper floret bisexual. Glumes unequal, the first short and broad, the second almost as long as the lemma of the lower floret. Lemma of upper floret indurated, usually rugose as in transverse lines or sometimes smooth with the margins inrolled over the palea.

A pantropical genus of about 50 species.

KEY TO SPECIES

1. Culms creeping or spreading; leaf margins entire; spikelets 3–4 mm long, glabrous: **U. subquadripara**

1. Culms tall-erect from a geniculate base; leaf margins serrate; spikelets 4.6–5.2 mm long, sparingly pilose: **U. decumbens**

Urochloa subquadripara (Trin.) R. D. Webster in *Austral. Paniceae* 252 (1987).

Rather slender perennial with weak, decumbent or creeping culms to 45 cm long or more, with upturned tips. Culm nodes glabrous or somewhat hairy, the sheaths ciliate on the margins; ligule a ciliate membrane or collar usually less than 1 mm long. Leaf-blades thin, glabrous on surfaces but sometimes ciliate toward the base, 5–20 cm long, 4–12 mm broad, tapering to apex. Panicle branches usually 3–6, widely spaced, mostly 2.5–9 cm long. Spikelets glabrous, solitary and loosely spaced on the flattened rachis, narrowly elliptic, 3.2–3.8 mm long, sharply acute at apex.

GRAND CAYMAN: *Guala* 1873, 1881, 1922, "spreading on roadsides, pastures and lawns".

— Native of China, India and Malaysia, also occurring on many Pacific islands and on Ceylon (Sri Lanka); introduced into Florida as a forage grass and now occurring sporadically on many West Indian islands, including the Bahamas, Cuba and Puerto Rico.

Urochloa decumbens (Stapf) R. D. Webster, *The Australian Paniceae* (Poaceae) 224 (1987).

Brachiaria decumbens Stapf (1919).

Robust perennial up to 1.5 m tall or more, tufted but sometimes with short rhizomes. Culms geniculate and rooting at the lower nodes and branching from the lower and middle nodes, the internodes flattened and solid or the lower becoming hollow, glabrous or often pilose just below the nodes. Sheaths compressed, shorter than the internodes or flowering culms, papillose-hirsute and ciliate; ligule 0.8–2 mm long, ciliate. Leaf-blades linear to linear-lanceolate, 5–19 cm long, 1–1.9 cm broad, acuminate at apex asymmetric at base, papillose-hirsute, the margins wrinkled and strongly serrate with even teeth. Panicle 6–14 cm long (rachis 2–9 cm long, hirsute), ending in a rudimentary branch; racemes 2–4, each 3–8 cm long, somewhat curved; spikelets 4.6–5.2 mm long, imbricate, strongly secund, solitary, 2-rowed, sparsely pilose, acute and short-apiculate at the apex. Lower glume 1.8–2.4 mm long, ovate and broadly acute, glabrous, 9–11-nerved, the nerves anastomosing. Anthers 2.5–3 mm long; stigmas purple.

GRAND CAYMAN: *Guala* 1936, found at edge of pasture off the Spotts–Newlands road.
— Native of tropical Africa; apparently not otherwise reported from the West Indies.]

[*Zea* L.

A coarse annual with broad drooping leaf-blades. Spikelets unisexual; staminate spikelets 2-flowered, in pairs on one side of a continuous rhachis, one flower nearly sessile, the other stalked; glumes membranous, acute; pistillate spikelets sessile, in pairs, consisting of 1 fertile and 1 sterile floret, the latter sometimes developed as a second fertile floret; glumes broad, rounded or emarginate at apex; sterile lemma similar to the fertile, with palea present; style very long and slender, stigmatic along both sides well toward the base.

A single variable species of probable hybrid origin, existing almost wholly as a cultivated plant.

Zea mays L., *Sp. Pl.* 2: 971 (1753).

CORN, MAIZE

A tall annual, extremely variable in size of plant and character of the pistillate inflorescences. Leaves conspicuously distichous, the blades up to 7 cm broad or more. Inflorescences monoecious, the staminate flowers in spike-like racemes, these numerous, forming large spreading panicles terminating the culms; pistillate inflorescences in the axils of the leaves, the spikelets in 8–16 (or even to 30) rows on a thickened, almost woody axis (cob), the whole enclosed in numerous large foliaceous bracts (husks), the long styles (silk) protruding from the top as a silky mass of threads.

No specimens have been collected in the Cayman Islands, but it is commonly cultivated.
— Grown in all tropical and temperate regions; apparently of tropical American origin.]

[*Zoysia* Willd.

Low perennials, often with rhizomes; culms decumbent or stoloniferous with erect leafy flowering branches. Leaf-blades narrowly linear, flat or involute; ligule very short, ciliolate. Inflorescence a small, pedunculate, spiciform raceme. Spikelets solitary, laterally compressed, falling entire, pedicellate, the pedicels flattened and widened upward. Lower glume absent; upper glume as long as the spikelet, sometimes mucronate. Floret 1, bisexual; lemma contained within the upper glume, hyaline; palea present or absent; stamens 3; styles connate at base. Seeds oblong.

A small genus of 10 species occurring in tropical and subtropical Asia and Australasia.

Zoysia tenuifolia Willd. ex Trin. in *Acad St. Petersb. Mem. VI, Sci. Nat.* **2**: 96 (1836).

Low sod-forming perennial with flowering culms up to 18 cm tall from creeping rhizomes. Leaf-sheaths with hyaline margins and sometimes a few cilia; ligule minute. Leaf-blades capillary-involute, up to 8 cm long. Inflorescence a few-flowered terminal spike; spikelets 2–2.3 mm long; first glume absent, the second glume firm, acute to short awned, enclosing the membranous lemma, palea absent.

GRAND CAYMAN: *Guala* 1924, 1930, 2220, on "roadsides and lawns".

— Introduced from the Asiatic tropics to Florida and the West Indies as a lawn grass, it has persisted and often become naturalized.]

TYPHACEAE

Erect, perennial, gregarious herbs with creeping rhizomes, erect terete flowering stems, and long linear leaves sheathing at the base. Flowers unisexual, densely crowded in elongate, cylindric, terminal spikes, the staminate above, the pistillate below. Perianth lacking. Staminate flowers composed of 1–7 stamens (usually 3), subtended by hairs; filaments very short, connate; anthers linear, 4-celled. Pistillate flowers consisting of a 1-celled ovary elevated on a short hairy stalk, containing 1 anatropous ovule, and terminated by a short style and a linear to spatulate stigma. Fruit a linear or fusiform achene, elevated on a long hairy stalk, and terminating in the persistent but fragile, greatly elongated style.

A small but cosmopolitan family with a single genus.

Typha L.

Characters as given for the family. It should be noted that the pistillate flower-spikes contain numerous sterile flowers mingled with the fertile ones; these enlarge after anthesis and become dilated at the apex.

A genus of about 10 species, all marsh plants, widely distributed in tempearate and tropical regions. In some countries, the fibrous stems and leaves are used for many purposes, such as for thatch, soft matting, ropes and baskets. The downy wool of the inflorescence can be applied like cotton to wounds and ulcers. The young shoots are edible, and are said to taste like asparagus. The pollen can be used as flour to make bread, and is also highly inflammable.

Typha domingensis Pers., *Syn. Pl.* **2**: 532 (1807). **FIG. 62.**

CAT-TAIL, RUSH, BULRUSH, BULL RUSH

Stems 1–3 m tall; leaves to 1.5 m long or more, 3–20 mm broad, nearly flat, glabrous. Spikes pale brown, the staminate and pistillate portions separated, each 10–40 cm long; hairs of pistillate pedicels usually minutely club-shaped at apex; pollen-grains 1-celled. GRAND CAYMAN: *Brunt* 1689; *Hitchcock*; *Kings* CG 216.

— Southern United States and throughout the West Indies, south to southern S. America, chiefly in coastal marshes.

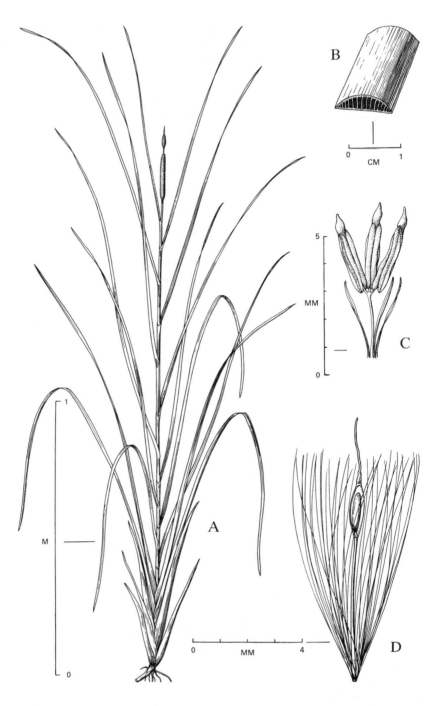

FIG. 62 **Typha domingensis**. A, habit. B, partial cross-section of stem. C, staminate flower. D, pistillate flower. (D.E.)

BROMELIACEAE

Mostly perennial herbs (rarely woody), epiphytic or terrestrial. Leaves spirally arranged, usually forming a basal tuft, dilated-sheathing toward the base, simple, the margins entire or spiny, the surface nearly always bearing minute peltate scales at least when young. Flowers perfect or functionally dioecious; inflorescence simple or paniculate, often with brightly colored bracts. Sepals and petals dissimilar, free or connate; stamens 6, in 2 series; filaments free, or joined to the petals or to each other; ovary superior to inferior, 3-celled; stigmas 3 or style 3-parted. Fruit a capsule or berry; seeds often plumose or winged, and containing plentiful mealy endosperm.

A chiefly tropical American family of about 50 genera and more than 1,500 species. The pineapple (*Ananas comosus*) is the most important economic plant of this family. Many of the species of this family are commonly called wild pine.

KEY TO GENERA

1. Leaves with entire margins; ovary superior or nearly so; fruit a dry capsule; seeds with plumose appendage:	**Tillandsia**
1. Leaves with spiny margins; ovary wholly inferior; fruit a berry; seeds naked:	
2. Petals naked, 3 cm long; ovaries free from each other; leaves bearing long curved spines on the margins:	**Bromelia**
2. Petals appendaged, less than 1.5 cm long:	
3. Ovaries remaining distinct from each other in fruit; leaves 8–12 cm broad with minutely spiny margins; inflorescence not foliaceous at apex:	**Hohenbergia**
3. Ovaries fused with each other and with the fleshy bracts to form a head-like syncarp; leaves less than 4 cm broad with densely spiny margins; inflorescence crowned by a tuft of leaves:	**[Ananas]**

Tillandsia L.

Caulescent or acaulescent herbs of varied habit, usually epiphytic. Leaves tufted or distributed along a stem, linear, ligulate or narrowly triangular. Inflorescence on a distinct scape, simple or compound, usually of one or more distichous-flowered spikes or sometimes reduced to a single polystichous-flowered spike by the reduction of lateral spikes to single flowers, or rarely the whole inflorescence reduced to a single flower. Flowers perfect; sepals usually symmetric, free, or equally or posteriorly connate; petals free or joined at the base, unappendaged. Stamens of various lengths relative to the petals and pistil. Ovary superior, glabrous; ovules usually many, caudate. Capsule septicidal. Seeds erect, narrowly cylindric or fusiform, the plumose appendage basal, straight, white.

A large genus of about 350 species widely distributed in the warmer parts of the Western Hemisphere.

KEY TO SPECIES

1. Stems elongate, loosely clothed with scattered, gray, linear leaves; inflorescence 1–2-flowered; stamens shorter than the petals: _____ **T. recurvata**

1. Stems very short, the leaves tufted, often with more or less broadly sheathing bases; inflorescence with several to numerous flowers; stamens longer than the petals:

  2. Floral bracts much less than twice the length of the internodes:

    3. Flowers appressed to the rhachis; inflorescence up to 1 m long, 2–3-pinnate with long slender branches; leaves numerous, not twisted or marked with transverse stripes: _____ **T. utriculata**

    3. Flowers more or less spreading; inflorescence less than 40 cm long, simple or with 2 or 3 short simple branches; leaves few, twisted and marked with transverse stripes: _____ **T. flexuosa**

  2. Floral bracts twice the length of the internodes or more:

    4. Leaf-blades linear-setiform, straight, less than 1 mm in diameter for most of their length, the sheaths not inflated: _____ **T. setacea**

    4. Leaf-blades flat or narrowly triangular, more than 1 mm broad (at least in the lower third, sometimes much more), the basal sheaths broadly expanded, erect, and more or less clasping:

      5. Inflorescence many-branched; leaves with basal sheaths essentially flat, not recurved-inflated:

        6. Inflorescence branches (spikes) 5–8 cm long; apical half of leaves filiform: _____ **T. festucoides**

        6. Inflorescence branches (spikes) 15–30 cm long; apical half of leaves long-acuminate, not filiform: _____ **T. fasciculata var. clarispica**

      5. Inflorescence simple or few-branched, the branches less than 10 cm long; leaves with basal sheaths more or less recurved-inflated:

        7. Basal leaves equalling or exceeding the inflorescence in size; plants not silvery-canescent:

          8. Swollen base of plant ellipsoid or spindle-shaped; leaf-sheaths loosely imbricate, gradually narrowing into the blades; floral bracts coriaceous: _____ **T. balbisiana**

          8. Swollen base of plant ovoid or bulbous; leaf-sheaths tightly imbricate, abruptly narrowing into the inrolled blades; floral bracts thin and papery: _____ **T. bulbosa**

        7. Basal leaves shorter than the inflorescence; plants densely silvery-canescent throughout: _____ **T. paucifolia**

**Tillandsia recurvata** (L.) L., _Sp. Pl._ ed. 2, 1: 410 (1762). **FIG. 63.**

OLD MAN'S BEARD

Stems several or many in a clump, simple or few-branched, 1–10 cm long, mostly shorter than the leaves; leaves distichous, 3–17 cm long, densely and minutely pruinose-scaly, the thin, many-nerved sheaths overlapping each other and concealing the stem, the blades recurved, linear, terete, 0.5–2 mm in diameter and of soft or lax texture. Scape terminal, up to 13 cm long, ca. 0.5 mm thick, bearing 1 or rarely 2 narrow scape-bracts immediately below

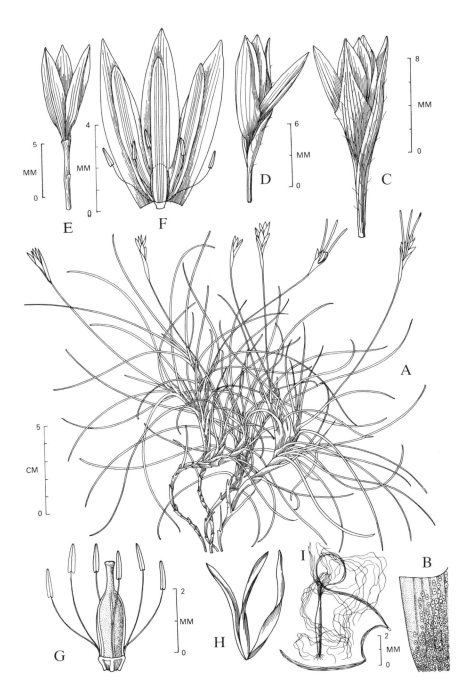

FIG. 63 **Tillandsia recurvata**. A, habit. B, portion of leaf-base greatly enlarged, showing scales. C, flower. D, flower with one sepal removed. E, flower with all sepals removed. F, flower spread out to show the stamens. G, stamens and pistil alone. H, fruit after dehiscence. I, seed with plumose fibrils. (G.)

the inflorescence (or sometimes one of them remote); inflorescence usually 1- or 2-flowered; floral bracts resembling the scape-bracts; sepals lanceolate, 4–9 mm long; petals narrow, pale violet or white; stamens deeply included but longer than the pistil. Capsule up to 3 cm long.

GRAND CAYMAN: *Kings* GC 108; *Proctor* 15172. LITTLE CAYMAN: *Kings* LC 76; *Proctor* 28134.

— Southern United States south to Argentina and Chile, always epiphytic, or in some countries (notably Jamaica) often growing on electric and telephone wires.

Tillandsia utriculata L., *Sp. Pl.* 1: 286 (1753). PLATE 4.

Plants stemless; leaves many in a dense rosette, up to 1 m long, densely and finely pale-appressed-scaly throughout; sheaths rather large, subovate; blades linear-triangular, long-acuminate, 2–7 cm broad at the base, the outer ones usually recurving. Scape erect, equalling or often exceeding the leaves in length, 8–14 mm thick, glabrous; scape-bracts erect, tubular-involute, barely overlapping or the uppermost sometimes remote. Inflorescence 2–3-pinnate (rarely simple) with lax branches, glabrous; primary bracts like the upper scape-bracts, less than 4 cm long; racemes up to 35 cm long with an elongate sterile base bearing several bracts, the rhachis slender, undulate, sulcate, and strongly flattened next to the flowers; floral bracts erect, enfolding the base of the flower but very little of the rhachis, much exceeded by the sepals, and equal in length to or shorter than the internodes, closely and prominently nerved, the margin often purple. Flowers appressed to the rhachis, the pedicels up to 5 mm long; sepals obtuse, 14–18 mm long; petals linear, 3–4 cm long, white; stamens and pistil exserted. Capsule 4 cm long, narrowly cylindric, acute.

GRAND CAYMAN: *Brunt* 1832, 1833, 1834; *Kings* GC 407; *Proctor* 15004. LITTLE CAYMAN: *Kings* LC 52; *Proctor* 35079, 35192. CAYMAN BRAC: *Kings* CB 34; *Proctor* 29345.

— Florida, West Indies, Mexico, Belize and Venezuela, common on trees at low elevations.

Tillandsia flexuosa Sw, *Nov. Gen. & Sp. Pl.* 56 (1788). PLATE 4.

Tillandsia aloifolia Hook. (1827).

Plant stemless; leaves 10 or more in a dense, twisted, often semi-bulbous rosette, 15–30 cm long or more, densely pale-appressed-scaly and usually marked with broad pale transverse stripes, the sheaths ovate, very large but passing into the blade without clear distinction, long-acuminate, stiff and curved at the apex. Scape erect, slender, glabrous; scape-bracts erect, tubular-involute, at least the upper ones shorter than the internodes. Inflorescence simple or with few branches; primary bracts like the upper scape-bracts; racemes very laxly flowered; rhachis slender, flexuous, sharply angled, glabrous; floral bracts 2–3 cm long, equal in length to or shorter than the sepals and about equal in length to the internodes. Flowers with pedicels to 7 mm long; sepals 2–3 cm long; petals tubular-erect, linear, to 4 cm long, usually pale rose; stamens exserted; capsule narrowly cylindric, up to 7 cm long.

GRAND CAYMAN: *Brunt* 1835; *Proctor* 31026. LITTLE CAYMAN: *Kings* LC 116; *Proctor* 28132. CAYMAN BRAC: *Kings* CB 32; *Millspaugh* 1219; *Proctor* 15317.

— Florida, Bahamas and Greater Antilles, and Panama to Guayana.

Tillandsia setacea Sw., *Fl. Ind. Occ.* 1: 593 (1797). PLATE 4.

Tillandsia tenuifolia L. (1753), in part, not as to type.

Plant stemless; leaves numerous in a dense fasciculate rosette, 15–30 cm long or more, often longer than the inflorescence, densely and finely lepidote throughout; sheaths ca. 2 cm long, often keeled, brownish; blades setiform, usually less than 1 mm thick. Scape erect, very slender, sparsely lepidote; scape-bracts erect, closely involute, densely overlapping. Inflorescence simple and distichous-flowered or shortly branched, mostly 2.5–5 cm long; primary bracts like the upper scape-bracts; spikes subsessile; floral bracts densely overlapping, mostly 8–14 mm long, longer than the sepals. Flowers subsessile; sepals 7–12 mm long, glabrous; petals tubular-erect, linear, 2 cm long, violet; stamens and pistil exserted. Capsule narrowly cylindric, 2.5 cm long.

GRAND CAYMAN: *Brunt* 1836; *Hitchcock*; *Kings* GC 180, GC 197; *Proctor* 15003.
— Florida, Greater Antilles, northern C. America, Venezuela, Brazil.

Tillandsia festucoides Brongn. ex Mez in C. DC., *Mongr. Phan.* 9: 678 (1896).

Plants epiphytic, forming clumps but non-stoloniferous; leaves with flattened expanded base and long-filiform apex. Infloresence compound, up to 17 cm long, with up to 10 or more arching-divergent spikes usually 5–8 cm long; floral bracts scarcely more than twice as long as the internodes; scape and floral bracts often pink or red; sepals up to 17 mm long; corolla pink to mauve.

CAYMAN BRAC: *Burton* 242, found in primary dry forest.
— Florida, Mexico to Costa Rica, Greater Antilles.

Tillandsia fasciculata Sw., *Nov. Gen. & Sp. Pl.* 56 (1788) var. **clavispica** Mez in DC., *Monog. Phan.* 9: 682 (1896).

Plant stemless; leaves numerous, up to 70 cm long, forming a stiff, rather dense rosette; sheaths ovate, nearly flat, dark brown near the base; blades narrowly triangular, long-acuminate, rigid, not more than 3 cm broad, finely and densely brown-appressed-lepidote throughout (but the surface appearing smooth and rather lustrous if examined without a lens). Scape erect, stout, shorter than the leaves, clothed with densely overlapping bracts, the lower of these leaf-like. Inflorescence compound with simple branches, overtopping the leaves; primary bracts like the upper scape-bracts; spikes ascending, clavate with an elongate sterile base, the apical fertile portion densely flowered. Floral bracts 2–2.8 cm long, bright pink or red at anthesis. Flowers erect, subsessile, with violet petals, the stamens and pistil exserted. Capsule acuminate, 4 cm long.

LITTLE CAYMAN: *Proctor* 35171.
— Cuba. In Little Cayman this plant grows on the ground or on the lower parts of trees in very dense woodland not near the sea. Other variants of *Tillandsia fasciculata* are widespread in the West Indies.

Tillandsia balbisiana Schult. f. in R. & S., *Syst. Veg.* 7, pt. 2: 1212 (1830). **FIG. 64.** PLATE 5.

Plant nearly stemless, sometimes several together in a loose clump; leaves many in a dense bulbous rosette, often longer than the inflorescence if extended (but typically recurved), densely and minutely pale-appressed-lepidote throughout; sheaths ovate, large, inflated, forming a more or less ellipsoid 'pseudobulb' up to 12 cm long, the blades abruptly

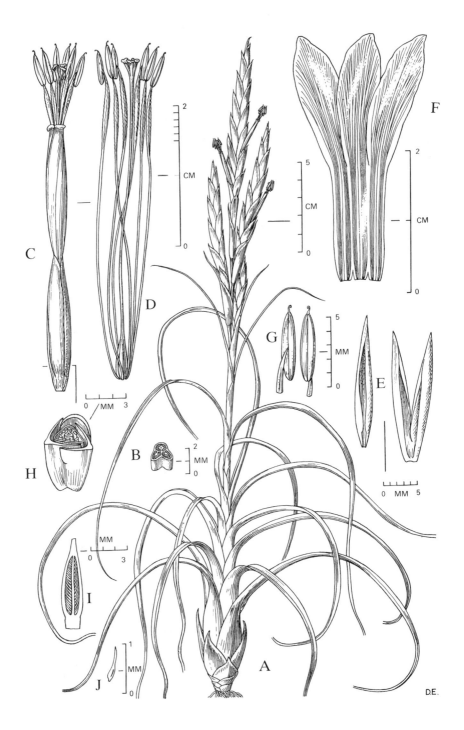

FIG. 64 **Tillandsia balbisiana**. A, habit. B, cross-section of stem. C, flower. D, flower with sepals and petals removed. E, sepals. F, corolla spread to show petals joined in the basal half. G, anthers. H, cross-section of ovary. I, longitudinal section of ovary. J, seed. (D.E.)

spreading or recurved from the tops of the sheaths, of hard texture and with involute margins, long-attenuate at the apex. Scape erect, slender, subglabrous; scape-bracts erect, overlapping, with long, linear, spreading or reflexed-contorted apices. Inflorescense simple and distichous-flowered or more often densely pinnate, 6–20 cm long; primary bracts like the upper scape-bracts, at least their sheaths shorter than the axillary spikes. Spikes sessile, strict, linear, 3–12 cm long, ca. 1 cm wide at anthesis; floral bracts overlapping, 1.5–2.2 cm long, longer than the sepals, often bright red at anthesis. Flowers erect, subsessile; sepals coriaceous; petals 3–4.5 cm long, violet or purple; stamens and pistil exserted. Capsule narrowly cylindric, 4 cm long.

GRAND CAYMAN: *Brunt* 1873; *Kings* GC 110, GC 195, GC 381. LITTLE CAYMAN: *Kings* LC 2, LC 113, LC 115; *Proctor* 28131. CAYMAN BRAC: *Kings* CB 31; *Proctor* 15316.

— Florida, Bahamas, Greater Antilles except Puerto Rico, and from Mexico south to Colombia and Venezuela, always epiphytic (occasionally on electric or telephone wires).

Tillandsia bulbosa Hook., *Exot. Fl.* 3: t. 173 (1826). PLATE 5.

Plants stemless, solitary or few in a clump; leaves 8–15, often exceeding the inflorescence, clothed with minute, closely adhering, cinereous scales; sheaths orbicular, inflated, forming a dense ovoid pseudobulb 2–5 cm long, green or greenish-white, abruptly contracted into narrow, contorted, involute blades, these up to 30 cm long and 2–7 mm thick. Scape erect; scape-bracts foliaceous with elongate blades exceeding the inflorescence. Inflorescence simple or subdigitate with few spikes, red or green; primary bracts ovate with foliaceous blades; spikes spreading, 2–5 cm long, 2–8-flowered; floral bracts overlapping, 15 mm long, longer than the sepals, 2 or 3 times as long as the internodes, keeled. Flowers sessile; sepals 13 mm long, glabrous; petals linear, 3–4 cm long, blue or violet; stamens and pistil exserted. Capsule cylindric, up to 4 cm long.

LITTLE CAYMAN: *Kings* LC 114.
— West Indies and Mexico to Colombia and Brazil.

Tillandsia paucifolia Baker in *Gard. Chron.* II, 10: 748 (1878).

Plants stemless, solitary or a few in a clump; leaves mostly 8–10, much shorter than the inflorescence, covered throughout with coarse, closely appressed, cinereous scales; sheaths broadly ovate, inflated, forming a narrowly ovoid or ellipsoid pseudobulb 5–15 cm long, gradually contracted into the blades, the outer ones reduced and bladeless; blades involute-subulate, shorter than or not much exceeding the sheaths, 3–7 mm thick, curved, contorted or coiled. Scape erect or decurved; scape-bracts overlapping, with curved, rigid, foliaceous blades. Inflorescence simple or digitately or pinnately compound, the spikes few; primary bracts like the scape-bracts. Spikes 5–12 cm long, 2–10 flowered; floral bracts overlapping, pink, 2–3 cm long, longer than the sepals, 2 or 3 times as long as the internodes, not keeled. Flowers sessile; sepals ca. 2 cm long, glabrous; petals linear, 4 cm long, purple; stamens and pistil exserted. Capsule narrowly cylindric, 4 cm long.

LITTLE CAYMAN: *Kings* LC 114a; *Proctor* 28133.
— Florida, Bahamas, Cuba, Hispaniola and Mexico. The name *T. circinnata* has been incorrectly applied to this species.

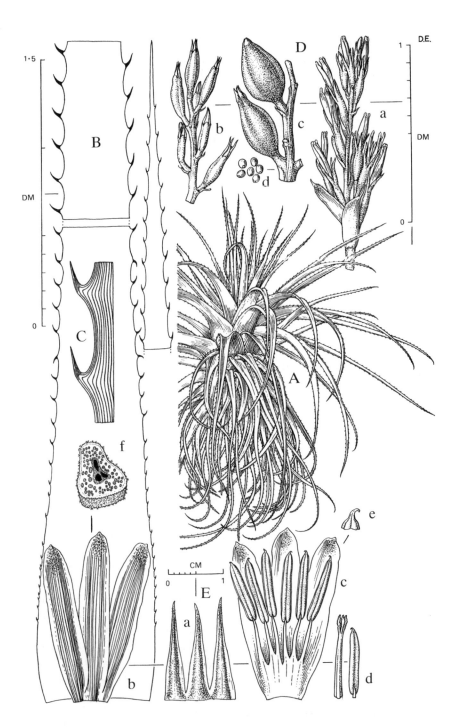

FIG. 65 **Bromelia pinguin**. A, habit, much reduced. B, outline of leaf segments showing marginal spines. C, marginal spines, enlarged. D, portions of inflorescence; a, with flowers; b, after fall of perianths; c, fruits; d, seeds. E, floral details; a, floral bracts; b, sepals; c, petals with attached stamens; d, style and anther; e, apex of anther; f, cross-section of ovary.

Bromelia L.

Coarse terrestrial herbs, spreading by underground stolons. Leaves arranged in a stiff, ascending rosette; margins armed with large curved spines. Inflorescence scapose or sessile, paniculate. Flowers pedicellate; sepals free or rarely somewhat united, obtuse or acute, rarely mucronulate; petals rather fleshy, with narrow free blades, the basal claws centrally united to the filament-tube but with free margins, unappendaged; stamens included, the anthers narrow and acute; ovary passing gradually into the pedicel, the epigynous tube conspicuous to nearly lacking. Berry succulent, relatively large; seeds few to many, flattened, naked.

A genus of nearly 50 species, widely distributed from Mexico and the West Indies to Argentina.

Bromelia pinguin L., *Sp. Pl.* 1: 285 (1753). **FIG. 65.**

PINGWING

Leaves numerous, usually 1–2 m tall, with broad, coarsely tomentose-lepidote sheaths at the base; blades linear, acuminate, up to 5 cm broad, armed with stout curved teeth up to 1 cm long. Scape stout, white-farinose; scape-bracts foliaceous with red, subinflated sheaths. Inflorescence many-flowered, narrowly pyramidal, white-farinose; primary bracts like the scape-bracts but the upper ones entire; branches up to 12-flowered; floral bracts linear-subulate from a short broad base, 3 cm long. Flowers up to 6 cm long, distinctly stalked; sepals 15–30 mm long; petals 3 cm long, rose with white base and margins, densely white-tomentose at the apex; ovary 2 cm long. Berry ovoid, ca. 3.5 cm long, yellowish, edible but very acidic.

GRAND CAYMAN: *Brunt* 1957; *Kings* GC 185; *Proctor* 27992.
— Mexico and the West Indies to Guyana and Ecuador.

Hohenbergia Schult. f.

Stemless, coarse, usually epiphytic herbs, sometimes growing on rocks. Leaves rosulate, the sheaths tightly overlapping and holding water, the blades broadly ligulate with more or less spiny margins. Scape well developed. Inflorescence bipinnate or tripinnate, composed of dense cone-like spikes. Flowers perfect, sessile, compressed; sepals nearly or quite free, distinctly asymmetric, mucronate; petals nearly or quite free, the claw bearing 2 scales on the inner surface; stamens shorter than the petals, the first series free, the second adnate to the petals; pollen with 2 or 4 pores. Ovary wholly inferior, compressed and more or less winged, usually changing little in fruit; ovules several in each locule.

A genus of about 35 species, concentrated chiefly in Jamaica and eastern Brazil, with a few species scattered elsewhere in the West Indies, C. America and northern S. America.

Hohenbergia caymanensis Britton ex L. B. Smith in *Proc. Amer. Acad.* 70: 150 (1935). PLATE 5.

OLD GEORGE

Leaves large, 1 m or more long and up to 13 cm broad, minutely brown-punctulate throughout, broadly rounded and apiculate at the apex, and with finely serrulate margins (the teeth less than 1 mm long). Scape 8–12 mm in diameter, densely and minutely brown-lepidote; scape-bracts overlapping, linear-lanceolate, minutely serrulate. Inflorescence rather laxly bipinnate, ca. 40 cm long; primary bracts like the scape-bracts, the lower ones much longer than the spikes; spikes densely ellipsoid, 2.5–4 cm long, stalked; floral bracts 12 mm long, with a mucro as long as the triangular-ovate base, strongly nerved. Flowers 14 mm long; sepals 6 mm long; petals greenish-white, 10 mm long; stamens included.

GRAND CAYMAN: *Rothrock* 495 (type) F!, *Kings GC* 196; *Proctor* 31031.

— Endemic, mostly confined to a small area of rocky woodland about a half mile southeast of George Town. The plants mostly grow on rocks, but also occasionally on the lower parts of tree-trunks. This species is not very different from *Hohenbergia penduliflora* of Cuba and Jamaica.

[*Ananas* Mill.

A terrestrial herb with numerous narrow, stiff leaves in a rosette, their margins plentifully armed with sharp, curved spines. Scape erect, stout, spiny-bracted. Inflorescence dense, head-like, crowned with a tuft of sterile leafy bracts. Flowers sessile, usually violet; sepals free, 5–7 mm long, slightly asymmetric; petals free, ca. 15 mm long, each bearing 2 slender scales; stamens included; pollen-grains ellipsoid with 2 pores. Ovaries coalescing with each other and with the bracts and axis to form a fleshy compound fruit. Berry sterile in the cultivated varieties, no seeds being formed.

A monotypic genus, probably of S. American origin.

Ananas comosus (L.) Merrill, *Interpr. Rumph. Amb.* 133 (1917).

PINEAPPLE

Characters of the genus. Leaves up to 1 m long, 1–3 cm broad, the marginal prickles ca. 2 mm long. Flowering spikes 4–10 cm long, enlarging in fruit, the fruits variable in shape according to the variety.

Often cultivated and sometimes persisting after cultivation; no Cayman specimens examined.]

[The Zingiberales are represented by cultivated plants of several families, notably Musaceae. *Musa sapientum* L., the BANANA, and *Musa paradisiaca* L., the PLANTAIN, are often grown.]

PALMAE (ARECACEAE)

Mostly trees with unbranched, erect trunks and a terminal bud. Leaves usually large, pinnately or palmately divided, forming a crown. Flowers perfect or unisexual; if the latter, then either on the same or on different plants. Inflorescence (spadix) paniculate, subtended or at first enclosed by a spathe usually of 2 valves, one of these usually much longer than the other. Sepals and petals 3 each, free or connate. Stamens 6–12; filaments distinct or joined toward the base; in pistillate flowers the stamens may be reduced to staminodes or lacking. Ovary usually 1–3-celled, each cell with a single ovule; style short or lacking. Fruit a 1-seeded berry (drupe); seed containing a horny or cartilaginous endosperm that is frequently rich in oil.

A large, economically important family of more than 180 genera and 2,500 species, represented in nearly all tropical and many warm-temperate regions. Many exotic palm species are now cultivated in the Cayman Islands.

KEY TO GENERA

1. Leaves pinnate:

  2. Petioles forming a smooth green crownshaft; inflorescences produced at base of crownshaft: **Roystonea**

  2. Petioles not forming a crownshaft; inflorescences produced among the leaves:

    3. Lower pinnae slender, stiff, and spine-like; trunk with persistent leaf-bases or conspicuous geometric scars; suckers produced at base of trunk; fruit less than 7 cm long: **[Phoenix]**

    3. Lower pinnae not spine-like; leaf-bases not persistent; trunk with rather smooth ring-like scars; no suckers produced; fruit up to 30 cm long: **[Cocos]**

1. Leaves palmate:

  4. Leaves green on both sides; fresh ripe fruits white; seed smooth: **Thrinax**

  4. Leaves silvery beneath; fresh ripe fruits black; seed grooved and fissured: **Coccothrinax**

Roystonea O. F. Cook

Tall, erect, simple, unarmed, monoecious trees with columnar, light-colored trunks that are often bulged or swollen. Leaves with tubular green petioles forming a crownshaft at the top of the trunk, this 1–2 m long, surmounted by a spreading crown of large pinnate leaf-blades; pinnae (leaflets) 1- or 2-seriate. Inflorescences springing from base of the crownshaft, each protected in bud by a pointed, glabrous, long-fusiform spathe; spathe-parts unequal, deciduous at anthesis. Spadix a whitish compound panicle, erect or ascending in flower, deflexed in fruit; flowers superficial on the rhachillae, unisexual, the staminate and pistillate ones randomly scattered on the same rhachillae. Sepals of staminate flowers imbricate and much shorter than the petals; stamens 6, bearing conspicuous anthers; pistillate flowers much smaller than the staminate, with staminodes represented by scales or a cup or ring; pistil of 3 carpels but usually only 1 developing, forming a 1-seeded fruit with a stigmatic scar on frontal side near the base. Fruit drupe-like, less than 2 cm long, oblong to globular, sessile; seed with raphe making a branched or lacerated pattern on its face; micropyle and embryo basal; albumen hard.

A chiefly Antillean genus of fewer than 10 species; several of these are prized as ornamentals and are planted in all tropical countries because of their tall, stately habit.

Roystonea regia (Kunth in H.B.K.) O. F. Cook in *Science* ser. 2, 12: 479 (1900); *Bull. Torr. Bot. Club* 28: 554 (1901). **FIG. 66.** PLATE 6.

  Oreodoxa regia Kunth in H.B.K. (1815).

Tree to 18 m tall or more, with trunk usually tapered at both ends, slightly bulged toward the middle. Leaf-blades rather lax and down-curved; leaflets biseriate and standing irregularly in 2 planes, giving the whole leaf a somewhat shaggy, plumose appearance, usually less than 1 m long and ca. 2–3 cm broad. Spadix 1 m long or more at anthesis, glabrous; staminate flowers with petals 5 mm long, the anthers concealed in bud; pistillate flowers ca. 3 mm long. Fruit oblong-globose, 8–13 mm long and 8–10 mm thick, dark purplish at maturity.

  GRAND CAYMAN: *Brunt* 1753, 1947; *Kings* PGC 259; *Proctor* (photo).

  — Florida and Cuba; also widely cultivated. The Grand Cayman population has the appearance of being indigenous; the trees are scattered or in small groups in swampy woodlands and clearings in the central part of the island.

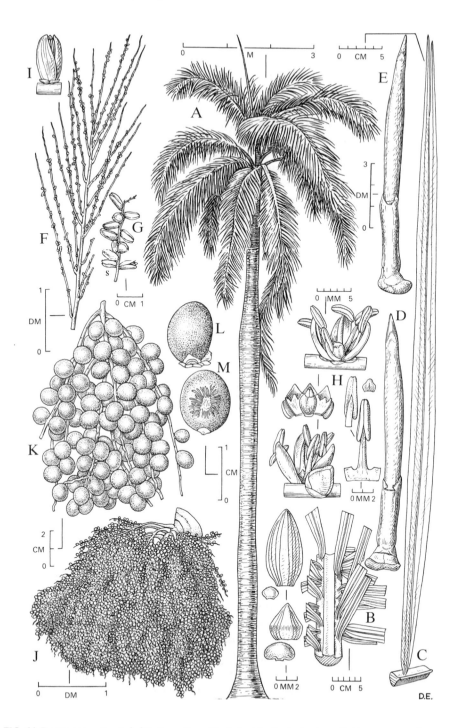

FIG. 66 **Roystonea regia**. A, habit. B, portion of leaf-rhachis. C, single pinna. D, E, two views of inflorescence-bud. F, portion of expanded inflorescence. G, portion of single rhachilla; s = staminate flowers (buds only). H, staminate flowers and anthers in various aspects. I, pistillate flower. J, fruiting inflorescence. K, enlarged portion of same. L, single fruit. M, seed showing branched raphe. (D.E.)

[*Phoenix* L.

The DATE-PALM, *Phoenix dactylifera* L., has long been cultivated in the Cayman Islands, presumably just as an ornamental, as it seldom if ever produces fruit here. It is not common, but being long-lived tends to persist near the sites of former dwellings. This species, which is native to N. Africa and Arabia, has been cultivated for many thousands of years, especially in desert oases and in the Tigris–Euphrates valley. It is an important source of food and revenue in these and some other regions.

GRAND CAYMAN: *Proctor* (sight). LITTLE CAYMAN: *Kings* LC 104a.]

[*Cocos* L.

The COCONUT-PALM, *Cocos nucifera* L., is economically the most important of all cultivated palms, but unfortunately was largely wiped out in the Cayman Islands by a mycoplasmic disease ('lethal yellowing'). It has been re-introduced and at this writing appears to be disease-free and thriving. In other tropical parts of the world, coconuts are often grown as a commercial crop. The oil extracted from the 'copra' or dried endosperm of the large seeds has many culinary and industrial uses.

GRAND CAYMAN: *Proctor* (sight). LITTLE CAYMAN: *Proctor* (sight). CAYMAN BRAC: *Kings* CB 52, CB 53.]

Thrinax Sw.

Mostly rather slender, unarmed trees. Leaves long-petiolate, the chiefly orbicular blades palmately cleft with narrow lobes, these 2-parted at the apex and obliquely folded; hastula thick, concave, pointed, often tomentose within when young. Spadix interfoliar, paniculate, partly enclosed toward the base by numerous tubular tomentose spathes. Flowers perfect, subequally 6-parted; stamens mostly 6, the filaments connate at the base; ovary superior, the stigma flat or concave; ovule solitary with lateral micropyle. Fruit 1-seeded, globose, usually white at maturity; seed with smooth endosperm.

A chiefly West Indian genus of 8 or 10 species or fewer.

Thrinax radiata Lodd. ex J. A. & J. H. Schultes in L., *Syst. Veg.* 7(2): 1301 (1830). FIG. 67. PLATE 6.

Thrinax multiflora of Adams (1972), not Mart. (1838).

Thrinax excelsa sensu L. H. Bailey (1938), not Griseb. (1864).

A small tree rarely more than 8 m tall (usually much shorter), the trunk 8–13 cm in diameter; leaf-blades 120–160 cm broad, the segments broadest at the point of fusion, beneath bearing scattered small fimbriate scales each having a conspicuous translucent central portion. Inflorescence and flowers creamy-white at anthesis, the axes glabrous. Fruits white and smooth at maturity, 7–8 mm in diameter.

GRAND CAYMAN: *Brunt* 1905, 1943; *Kings* PGC 255, PGC 257, PGC 257a; *Proctor* 27966. LITTLE CAYMAN: *Kings* LC 47, LC 104.

— Florida, Bahamas, Greater Antilles except Puerto Rico, Yucatan and Belize; frequent in rocky coastal thickets.

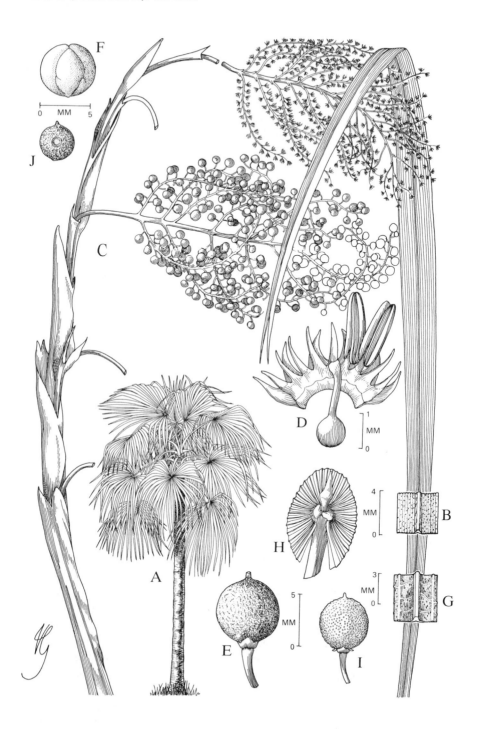

FIG. 67 **Coccothrinax proctorii**. A, habit (from photograph). B, section underside of leaf. C, portion of inflorescence with flowers and young fruits. D, dissected flower. E, fruit. F, seed showing grooved endosperm. **Thrinax radiata**. G, section underside of leaf. H, attachment of leaf-blade to petiole, showing hastula. I, fruit. J, seed with smooth endosperm. (G.)

Coccothrinax Sarg.

Slender unarmed trees. Leaves long-petiolate and palmately cleft, with segments bifid at the apex and obliquely folded; hastula free, thin, erect, concave, and usually pointed. Spadix interfoliar, paniculate, bearing numerous papery 2-cleft spathes. Flowers perfect, 6-parted; stamens 9–12, with subulate filaments barely united at the base. Ovary superior, the stigma funnel-shaped; ovule solitary with subbasal micropyle. Fruit 1-seeded, globose, black or dark brown at maturity; seed with deeply grooved endosperm.

A chiefly West Indian genus of about 30 species.

Coccothrinax proctorii Read, *Phytologia* 46 (5): 285 (1980). **FIG. 67.** PLATE 6.

THATCH PALM, SILVER THATCH

Slender tree, the trunk 2–5(–10) m tall. Leaf-blades orbicular in outline, with 39–48 narrowly trullate segments, these mostly 61–80 cm long and 3.2–4.2 cm at their widest, tapering to a very slightly bifid apex; in unexpanded blades the segment-tips are connected by a thread-like strand; adaxial surface dark green; abaxial (lower) surface appearing silvery to golden, covered with a dense indumentum of persistent irregularly shaped and interlocked fimbriate scales, the stalk or central portion of each scale conspicuous as a dark-colored dot; hastula variable in outline and width (2–3.3 cm wide), its free adaxial extension 0.5–1 cm long, conspicuously ciliate; petiole (32–)75–80 cm long, the abaxial surface usually densely covered with white scales that are soon lost. Inflorescence with 5–7 or more primary branches; flowers stalked, fragrant, white at anthesis, soon turning creamy; stamens about 10 in number, about equal in length to the pistil. Fruit purple-black at maturity, the mature fruiting pedicels mostly 1–4 mm long; seeds mostly 5–5.4 mm in diameter.

GRAND CAYMAN: *Brunt* 1906; *Proctor* 27991 (type); *Sachet* 380; *Sauer* 4091 (WIS, cited by Read). LITTLE CAYMAN: *Proctor* 28033, 35082. CAYMAN BRAC: *Brunt* 1674; *Proctor* 29045.

— Endemic, in rocky thickets and woodlands. The manufacture of rope from the leaves of the thatch palm was formerly an important cottage industry on Grand Cayman and Cayman Brac. This rope was especially prized by fishermen, both locally and in Jamaica, because of its durability in contact with sea-water. Synthetic substitutes have now replaced this commodity, and increased prosperity among the Cayman people has reduced the incentive to do this arduous and relatively low-paying work.

ARACEAE

Terrestrial, epiphytic, or aquatic herbs with thick, fleshy, tuberous rhizomes or corms, or sometimes rather slender creeping or climbing stems. Leaves alternate, petiolate, simple or compound, usually with reticulate venation. Inflorescence a fleshy cylindric or clavate spike (spadix) usually subtended by a conspicuous flat or hooded bract (spathe) which may be green, white or otherwise colored. Spadix densely beset with minute usually unisexual flowers, the staminate on the apical portion and the pistillate toward the base, or the plants dioecious; a few genera with bisexual flowers. Perianth, when present, minute, of 4–6 or more free or connate segments. Stamens 4–6, mostly without filaments or sometimes with broad flat filaments. Ovary 1–several-locular; ovules 1–several in each locule. Fruit a berry with 1–many seeds; seeds usually with endosperm.

A chiefly tropical and warm-temperate family of more than 100 genera and 2,000 species. Many species find a useful place in horticulture, and a few are commonly used for food. *Caladium bicolor* (Aiton) Vent., with spotted or variegated leaves, and *Dieffenbachia maculata* (Lodd.) G. Don, the dumb cane, are commonly cultivated ornamentals. *Colocasia esculenta* (L.) Schott (dasheen) and *Xanthosoma sagittifolium* (L.) Schott (coco) are widely cultivated in the West Indies for their edible starchy tubers. The latter is recorded as being grown on Grand Cayman (*Kings GC–84*).

KEY TO GENERA

1. Leaf-blades entire; sap clear:	Philodendron
1. Leaf-blades pedately divided; sap milky:	Syngonium

Philodendron Schott

Mostly scandent or vine-like, often high-climbing and emitting long aerial roots; internodes mostly elongate. Leaves with terete or vaginate petioles; blades entire or variously lobed or parted, the lateral nerves all parallel. Peduncles usually short; spathe persistent, convoluted into a tube at the base and closely surrounding the pistillate portion of the spadix, the apical part more or less hooded, whitish or green. Spadix erect, nearly equal in length to the spathe, sessile or short-stipitate, the basal pistillate portion cylindric and densely many-flowered, becoming fleshy in fruit, the apical staminate portion sterile toward base, fertile above, withering and recurved in fruit. Flowers unisexual, naked, the staminate with 2–6 stamens, these obpyramidal-prismatic, truncate at apex, the anthers oblong or linear; ovary of the pistillate flowers 2–several-celled, ovules 1–several; stigma sessile. Fruits crowded, 1–several-seeded.

A large genus of about 275 species, widely distributed in warm countries.

Philodendron hederaceum (Jacq.) Schott, *Wiener Z. Kunst* 3: 780 (1829).

A vine with slender, terete stems ca. 5 mm in diameter. Leaves with slender terete petioles up to 16 cm long; blades triangular-hastate, up to 11 cm long from attachment of the petiole to the shortly acuminate or merely acute apex, and mostly 5–9 cm broad, narrowly to broadly cordate at the base with rather open sinus, the rounded lobes somewhat divergent. Inflorescence unknown (in the Cayman Islands).

GRAND CAYMAN: *Brunt* 2170; *Kings GC* 259; *Proctor* 27998, 47349, 48295, 52119.

Related to *Philodendron scandens*, but its precise identity is still not fully certain as long as only sterile material is known. Kings' specimen was gathered at Forest Glen, the Brunt and Proctor material in dense woodlands about a mile due north of Breakers; plants from the latter locality were grown in Jamaica for several years without showing any sign of flowering.

Syngonium Schott

Epiphytes or hemiepiphytes, usually with long, root-climbing stems, producing milky sap. Leaves simple or variously divided, with 3–11 leaflets; petioles sheathed toward the base. Inflorescences 1–11 per axil; peduncles erect in flower, pendant fruit; spathe fleshy, convolute, conspicuously constricted medially, the tube ellipsoid, the blade whitish, greenish or variously colored, broadly spreading at anthesis; spadix much shorter than spathe, erect, with pistillate flowers on the basal portion. Flowers unisexual, the perianth wanting; stamens 3–4, united into a synandrium; ovary (1–)2(–3)-locular, with 1(–2) ovules per locule, the stigma discoid or bilabiate. Fruit a 1-seeded berry, connate into an ovoid syncarp; seeds obovoid or ovoid.

A neotropical genus of 33 described species most common in the region of Costa Rica and Panama. A single species has been found in the Cayman Islands.

Syngonium podophyllum Schott in *Bot. Zeitung* (Berlin) 9: 85 (1851).

Vine up to 10 m long, rooting at nodes; stem cylindric, glaucous, 1–2 cm diam., producing abundant milky sap. Leaves pedately divided; leaflets 3–11, united or free to base, coriaceous, the apex acuminate, the base variously auriculate, the margins sinuate, outermost leaflets smaller, the medial leaflets 16–38 × 6–17 cm, obovate, elliptic or lanceolate; petioles 15–60 cm long, nearly cylindric, sheathed $^2/_3$ of their length. Inflorescences 4–11 per axil, ascending; peduncles 8–9 cm long, slender; spathe ca. 10 cm long, convolute into an ellipsoid tube at the base, the blade cream-colored, concave, ephemeral; spadix whitish, sessile, cylindric with a constriction between the pistillate and the staminate areas. Syncarp ovoid, red, reddish orange or yellow, 3–5.5 cm long.

GRAND CAYMAN: *Proctor* 48296. Known only from tall old-growth forest in the Forest Glen area, where this plant appears to be indigenous. This species has not been observed under cultivation in the Cayman Islands.

Mexico and northern C. America; widely cultivated in the West Indies and Florida and often becoming naturalized. Its presence in a remote forest habitat in the Cayman Islands is somewhat enigmatic.

LEMNACEAE

Monoecious, very rarely dioecious, small to minute aquatic leafless annual herbs, floating on or submerged in fresh water. Plant body (frond) either solitary or connected in small groups, symmetric or asymmetric, flat or inflated, varying in shape from reniform, round, elliptic, lanceolate and linear to globose, usually green but red or brown pigments sometimes also present; roots several, 1 or none. Vegetative propagation from budding pouches; when there is one budding pouch there is also a flowering cavity bearing a spatheless inflorescence (except in *Wolffiopsis* where there are 2 flowering cavities); when there are two budding pouches one of these may give rise to an inflorescence surrounded by a spathe. Inflorescence consisting of 1 pistillate and 1 or 2 staminate flowers; perianth none. Staminate flower consisting of a single stamen, the anther 1- or 2-locular; pistillate flower consisting of 1 sessile, globular, 1-celled ovary with a short apical style; ovules 1–4. Fruit a 1–4-seeded utricle; seeds ribbed or smooth.

A small world-wide family of 6 genera and about 29 species, representing the world's smallest flowering plants. The present treatment is based on the synopsis presented by den Hartog & van der Plas in *Blumea* 18(2): 355–368 (1970).

Lemna L.

Small water plants, floating on the surface or submerged (in which case they rise to the surface during flowering periods). Fronds solitary or in small connected groups of 2–10, symmetric or slightly asymmetric, round, elliptic, oblong, obovate or lanceolate, flat or slightly swollen, often with a median row of papillae on the dorsal side; tissues containing raphides; stomata present only on the dorsal side of floating plants; margin entire or rarely denticulate; nerves 1–3 (rarely 5); root 1 per frond (rarely absent), terminated by a root cap of distinctive shape. Budding pouch opening at margin of frond, rarely slightly dorsal or ventral in position. Seeds longitudinally ribbed, rarely smooth.

A cosmopolitan genus of about 9 species.

Lemna aequinoctialis Welw. in *Anaes Conselho Ultramar*. 55: 543 (1858).

Lemna perpusilla sensu Adams (1972), not Torrey (1843).

DUCK WEED

Fronds solitary or 2–5 connected, obovate to elliptic, slightly asymmetric, 1.5–3.3 mm long, 0.8–2.5 mm wide, the upper surface minutely keeled with a median line of small papillae, with a more prominent papilla opposite the attachment of the root and often another at the apex; lower surface flat to slightly convex; root sheath with lateral wings; nerves usually 3 but often indistinct. Ovary with 1 ovule. Seed oblong-ovoid, prominently ribbed.

GRAND CAYMAN: *Kings GC 215, Proctor 49358.*

— Of cosmopolitan distribution, occurring widely in both tropical and temperate regions. The *Kings* collection was gathered at a "small cow well 1 mile S.E. of George Town" on 6 July 1938.

[PONTEDERIACEAE

Perennial aquatic herbs, rooting in mud or sometimes floating. Leaves chiefly basal, often with more or less succulent petioles, the blades with finely parallel venation. Flowers bisexual, irregular (zygomorphic), solitary or in spikes, racemes or panicles subtended by leaf-like bracts (spathes). Perianth corolla-like, 6-parted, the lobes usually joined toward the base. Stamens 3 or 6, the filaments adnate to the perianth tube; anthers 2-celled, sometimes dimorphic. Ovary superior, 1–3-celled; ovules 1–several; style 1; stigma 3–6-lobed. Fruit a capsule with several seeds, or achene-like with 1 seed; seeds with mealy endosperm.

A small family of 8 genera and about 25 species, chiefly of tropical and warm-temperate regions.

Eichhornia Kunth

Plants floating or rooted in mud. Leaves with slender, stout or inflated petioles and flat blades, or sometimes completely narrow and grass-like when submerged. Flowers solitary to numerous; perianth conspicuous, somewhat 2-lipped, the 6 parts in 2 series, all united into a tube below. Stamens 6, 3 included and 3 exserted. Ovary 3-celled; ovules numerous. Fruit a many-seeded capsule.

About 5 species, natives of tropical America.

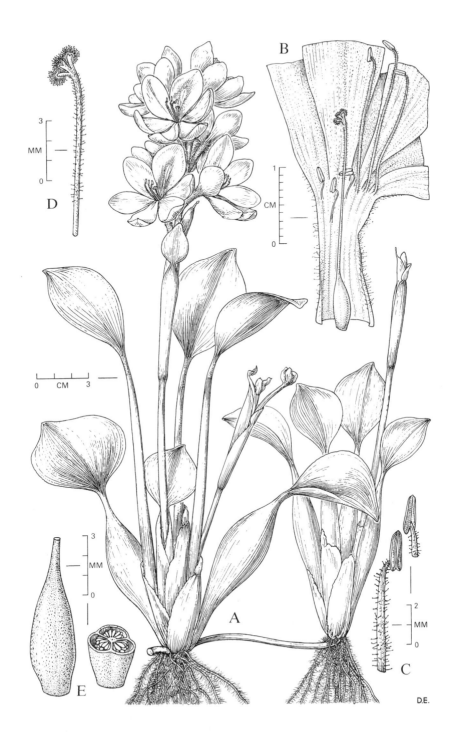

FIG. 68 **Eichhornia crassipes**. A, habit. B, dissected flower. C, stamen with wo views of an anther. D, stigma and upper portion of style. E, ovary, whole and in cross-section. (D.E.)

171

Eichhornia crassipes (Mart.) Solms in A. & C. DC., *Monog. Pha.* 4: 527 (1883). **FIG. 68.**

Piaropus crassipes (Mart.) Britton (1923).

WATER HYACINTH

Plants floating or the rhizomes rooting in mud. Leaves 5–40 cm long, in a basal tuft, glabrous, the petioles usually inflated, the blades ovate to orbicular or reniform, faintly many-nerved, 4–12 cm broad. Inflorescence a contracted panicle 4–15 cm long, subtended by a pair of unequal spathe-like bracts. Perianth 5–7 cm broad when fully expanded, lilac or rarely white, the upper lobe bearing a central violet blotch with yellow center. Stigma 3-lobed, glandular-hairy.

GRAND CAYMAN: *Brunt* 1840; *Kings* GC 230.

— Originally from Brazil, but introduced in most warm countries and widely naturalized, often becoming a pest of waterways and fresh-water lakes.]

THE 'LILIOID' FAMILIES

The family Liliaceae as presented in the first edition of this Flora is now considered to be an oversimplified classification of these plants. Modern analysis has revealed the polyphyletic nature of the old, broadly construed 'Liliaceae', resulting in the establishment of many smaller more homogenous taxa. In the Cayman Islands, we no longer recognize true Liliaceae. Two of the Liliaceous subfamilies (Amaryllidoideae and Agavoideae) are now treated as separate families (Amaryllidaceae and Agavaceae, respectively) whereas two exotic genera of African origin (*Sansevieria* and *Aloë*) are placed in two other families, Dracaenaceae and Asphodelaceae, respectively.

KEY TO 'LILIOID' SPECIES

1. Ovary superior; flowers in racemes or very narrow panicles:
    2. Leaves erect with entire margins; perianth whitish: **[Dracaenaceae]**
    2. Leaves coarsely toothed; perianth yellow: **[Asphodelaceae]**
1. Ovary inferior; flowers solitary, umbellate or in very large spreading panicles:
    3. Bulbous herbs; flowers solitary or umbellate, subtended by a spathe; leaves entire, not spine-tipped: **Amaryllidaceae**
    3. Coarse herbs of hard texture, not bulbous, sometimes developing a subwoody trunk; leaves with prickly margins or at least spine-tipped: **Agavaceae**

[DRACAENACEAE

Trees or shrubs with woody stems or else short-stemmed or stemless succulent herbs with fleshy creeping rhizomes; woody stems usually with secondary thickening. Leaves entire, lanceolate or linear-lanceolate to ovate and flat, channeled, cylindric or laterally compressed, sessile, glabrous, fleshy or leathery, rigid or flexible, often clustered at growing tips or in rosettes at the ends of branches. Inflorescence an axillary raceme or panicle; flowers bisexual, hypogynous, often fragrant; pedicels articulate; perianth tubular or funnel-shaped, the lobes 6, narrow, subequal, spreading or reflexed when fully expanded. Stamens 6, inserted at base of the lobes, the anthers dorsifixed, versatile, introrse. Ovary superior, 3-locular, with a single or many ovules in each locule; style slender, terminated by a 3-lobed or capitate stigma. Fruit a fleshy berry or sometimes hard and woody, usually red or orange; seeds 1–3, bony, globose or elongate.

Genus type: *Dracaena* Vand. ex L.

A family here construed as having 3 genera and about 225 species occurring primarily in the Old World tropics and subtropics, and on islands of the Pacific Ocean.

A single genus represented by one species occurs in the Cayman Islands.

Sansevieria Thunb.

Perennial herbs forming large colonies by means of creeping stolons. Leaves entire, leathery, erect, clustered, flat or terete. Flowers greenish-white, the inflorescence a raceme or panicle on an unbranched nearly naked scape. Perianth with slender tube and narrow spreading lobes; stamens 6, the filaments inserted at the base of perianth lobes; anthers oblong or linear. Ovary 3-celled, each cell with a single, erect ovule; style 1, long and filiform; stigma capitate. Fruit a thin-walled capsule; seeds 1–3, fleshy.

A genus of more than 60 species, indigenous to Africa and Asia. The leaves of many species yield valuable fibres; some are also widely cultivated for the ornamental value of their curious, sword-like leaves.

Sansevieria hyacinthoides (L.) Druce in *Rep. Bot. Exch. Cl. Brit. Isles* 1913 (3): 423 (1914).

BOWSTRING HEMP, LION'S TONGUE

Leaves lanceolate, nearly flat, mostly 30–100 cm long, 5–9 cm broad at the middle, narrowed at both ends, dark green mottled with white, the margin bordered with a fine red line. Inflorescence about as long as the leaves or slightly longer; flowers in clusters of 2 or 3 along the main axis; pedicels mostly less than 5 mm long; perianth greenish-white, the lobes about 1.5 cm long.

GRAND CAYMAN: *Brunt* 1942, 1956; *Kings* GC 82; *Proctor* 15206. LITTLE CAYMAN: *Proctor* (sight). CAYMAN BRAC: *Kings* CB 90; *Proctor* 29086.

The original habitat of this species was somewhere in Africa. It is rather widely naturalized in the West Indies, and can readily be distinguished from the similar *Sansevieria trifasciata* Prain (not yet recorded from the Cayman Islands, but may be present) by the red-lined margins of the leaves. The two species have often been confused under the name *S. guineensis*. The somewhat fleshy leaves of these and other related species contain numerous fine, white, silky fibers that can be used for many purposes; their strength is said to be about the same as those in sisal (*Agave sisalana*).]

[ASPHODELACEAE

Perennial often succulent or leathery herbs with rhizomatous roots, or sometimes trees with woody trunks. Leaves primarily basal, spirally arranged or sometimes 2-ranked, sheathing at base. Leaves more or less fleshy, often internally gelatinous, flat or cylindric, subulate to linear-lanceolate or elliptic, the apex often spine-tipped, the margins smooth, toothed or serrate. Inflorescence terminal or axillary, of simple or branched spikes or racemes; scapes leafless or beset with small bacteal leaves. Flowers bisexual with 6 perianth segments. Stamens 6, with free filaments, the anthers dorsifixed, introrse, longitudinally dehiscent. Ovary 3-locular, with 2–many ovules in each locule, the placentation axile; style simple, with small stigma. Fruit a loculicidal capsule; seeds ovoid or elongate, usually with aril.

A family with about 17 genera and 814 species, widely distributed in tropical and subtropical regions of the Old World, especially common in Africa. In older classifications, these plants were considered part of the Liliaceae. A single genus, *Aloë*, has been introduced into the Cayman Islands.

Aloë L.

Succulent herbs with basal leaves, or else with a woody caudex more or less well developed (a few species become large branching trees). Leaves rosulate or rarely distichous, fleshy, with toothed margins, containing bitter sap that turns black on exposure or drying. Inflorescence axillary, racemose or paniculate from a naked scapose base; flowers usually nodding; perianth a cylindric tube, the lobes coherent except at the spreading or recurved tips, often brightly colored. Ovary sessile, ovoid to oblong, usually 3-angled, the style filiform with a small capitate stigma. Fruit a coriaceous, dehiscent capsule; seeds numerous, flattened, black.

A genus of more than 300 species, chiefly occurring in Africa and Madagascar.

Aloë vera (L.) Burm.f., *Fl. Ind.* 83 (1768).

Aloë barbadensis Mill. (1768).
Aloë vulgaris Lam. (1783).

BITTER ALOES, SEMPERVIRENS

Stemless or with a very short upright stem, spreading by creeping stolons. Leaves narrowly deltate-lanceolate, 30–60 cm long, acuminate, turgid and watery within, pale glaucuous-green; marginal teeth usually less than 2 cm apart. Scape 60–120 cm tall, bearing distant, broad, acute scales; raceme dense, 10–30 cm long; bracts longer than the short pedicels. Flowers yellow, about 2.5 cm long; stamens about as long as the perianth, the style longer.

GRAND CAYMAN: *Kings* GC 122.

— Native of the Mediterranean region; widely naturalized in Florida, the West Indies and C. America. In addition to various medicinal uses, the plant can produce a valuable natural dye and also fibre.]

AMARYLLIDACEAE

Perennial herbs with usually a bulbous or rarely a rhizomatous base. Leaves few, basal, entire or nearly so, usually of rather soft or thinly subsucculent texture. Flowers bisexual, actinomorphic or subregular, often showy, solitary or umbellate and subtended by spathaceous bracts at the top of a simple naked peduncle. Perianth petaloid, of 6 segments in 2 series often united below into a tube. Stamens 6, opposite the perianth segments, hypogynous or epipetalous; filaments free or joined together, often curved upward in laterally directed flowers; anthers 2-locular, dorsi- or medifixed, opening lengthwise or rarely with terminal pores. Ovary usually inferior but sometimes superior, usually 3-locular with axile placentas; style slender; ovules often numerous, anatropous. Fruit a capsule or fleshy berry. Seeds with fleshy endosperm and a small embryo.

A widespread family of about 90 genera and 1,100 species.

KEY TO GENERA

1. Flowers solitary:	**Zephyranthes**
1. Flowers usually 2 or more in an umbel:	
2. Filaments free to the base:	
3. Flowers red; perianth tube with scales or a corona at the throat; seeds black:	**Hippeastrum**
3. Flowers white (or white with rose stripes); perianth tube lacking scales or a corona at the throat; seeds green, fleshy:	**[Crinum]**
2. Filaments joined below by a cup-like membrane; flowers pure white:	**Hymenocallis**

Zephyranthes Herbert

Bulbous herbs with narrow, rather grass-like, glabrous leaves. Flowers solitary at the apex of a leafless scape, the pedicel subtended by a papery bract (spathe). Perianth funnel-shaped, erect or inclined, of various colors, with 6 subequal lobes. Stamens 6, the filaments adnate to the throat of the perianth tube. Ovary 3-celled, with numerous ovules in 2 rows in each cell; style filiform, with a 3-lobed or nearly capitate stigma. Capsule 3-lobed, 3-celled, 3-valved; seeds blackish, usually flattened.

A genus of about 35 species occurring widely in the warmer parts of America. Many of these species are popular in cultivation. Because they tend to escape and become naturalized in fields and along roadsides, it is often hard to determine if a given species is truly indigenous to a particular area.

KEY TO SPECIES

1. Perianth yellow:	**Z. citrina**
1. Perianth pink:	**Z. rosea**
1. Perianth white:	**Z. tubispatha**

Zephyranthes citrina Baker in *Curt.*, *Bot. Mag.* 108: t. 6605 (1882).

Zephyranthes eggersiana Urban (1907).

YELLOW CROCUS

Bulb ovoid-globose, about 2–3 cm in diameter. Leaves mostly 15–30 cm long, 1.5–2.5 mm broad, narrowly linear. Scape 15–25 cm tall; spathe 2–2.5 cm long, about half as long as the pedicel or more. Perianth yellow, 3–4 cm long, the lobes elliptic; stamens shorter than the perianth; style 2 cm long, with a slightly trifid stigma.

GRAND CAYMAN: *Brunt* 2123; *Kings GC* 414b; *Proctor* 27933.
— West Indies.

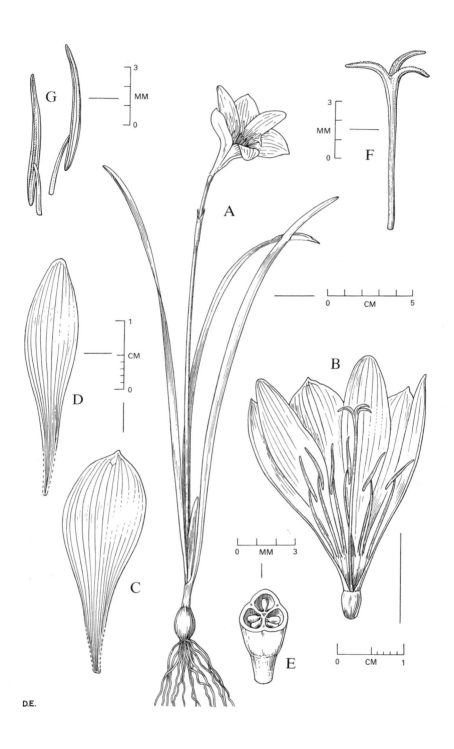

FIG. 69 **Zephyranthes rosea**. A, habit. B, dissected flower. C, D, outer and inner perianth-segments. E, cross-section of ovary. F, stigmas and upper portion of style. G, anthers. (D.E. & G.)

Zephyranthes rosea Lindl., *Bot. Reg.* t. 821 (1824). **FIG. 69**.

Bulb subglobose, 2–2.5 cm in diameter. Leaves linear, 10–25 cm long, 2–5.5 mm broad. Scape 10–25 cm tall, usually longer than the leaves; pedicel longer than the 2 cm spathe. Perianth bright pink, 3–4 cm long; stamens much shorter than the perianth; style nearly equal in length to the perianth, with a deeply trifid stigma.

GRAND CAYMAN: *Brunt* 2124; *Kings GC* 414; *Proctor* 27962.

— Cuba; naturalized in Bermuda, the Bahamas and Jamaica.

Zephyranthes tubispatha (L'Herit.) Herbert in Edw., *Bot. Reg.* 7, App. 36 (1821).

Bulb subglobose, 1.5–2.5 cm in diameter. Leaves narrowly linear, 16–30 cm long, 2–7 mm broad. Scape 15–25 cm tall (rarely less), shorter than or equal in length to the leaves; spathe 2–3.5 cm long, shorter than the pedicel. Perianth white, tinged greenish toward base, 3.5–5 cm long; stamens shorter than the perianth; style with deeply trifid stigma, overtopping the anthers.

GRAND CAYMAN: *Brunt* 2125; *Proctor* 27961.

— West Indies and northern S. America; often cultivated.

Hippeastrum Herbert

Bulbous plants with many, less strap-shaped, flat, entire leaves. Flowers large, showy, usually 2 or more in an umbellate cluster at the apex of a hollow, leafless scape. Perianth funnel-shaped, more or less nodding, the lobes nearly equal, the throat with scales or a corona; stamens inserted in the throat, the filaments separate; anthers linear or linear-oblong; ovary inferior, 3-celled; ovules many; style long, with a capitate or trifid stigma. Capsule globose, 3-valved; seeds flattened, with a thick black testa.

A genus of about 75 species, widely distributed in tropical and subtropical America.

Hippeastrum puniceum (Lam.) Kuntze, *Rev. Gen. Pl.* 2: 703 (1891) (err. *purpureum*).

Hippeastrum equestre Herbert (1821).

RED LILY

Bulb globose or ovoid-globose, stoloniferous, 4–5 cm long, the outer papery coats brown. Leaves mostly 25–50 cm long, 2.5–5 cm broad, gradually narrowed to the blunt apex, absent at flowering time. Scape terete, glaucous, 25–60 cm tall; umbel 2–4-flowered, subtended by papery, lanceolate bracts; pedicels 3.5–7 cm long. Perianth about 9 cm long or sometimes shorter, about the same in width when expanded, the segments bright red or rose-red with green at the base; stamens shorter than the perianth segments.

GRAND CAYMAN: *Brunt* 2061.

— West Indies, C. and S. America, often cultivated and persisting after cultivation.

[*Crinum* L.

Plants arising from bulbs that are narrowed at the apex into a short or long neck; leaves narrowly or broadly strap-shaped with entire, toothed or wavy margins. Flowers rather large, white or striped pink, in an umbellate cluster at the apex of a solid leafless scape; pedicels short or none. Perianth with subequal lobes broadly spreading or connivent-funnel-shaped, from a long, slender tube; stamens inserted in the throat of the perianth tube, the filaments long and filiform, the anthers linear; ovary 3-celled with few (sometimes only 2) ovules in each cell; style long and filiform with a minute capitate stigma. Capsule irregular in shape, tardily dehiscent; seeds large, green and bulb-like, with very thick endosperm.

An imperfectly known genus of about 100 species, at least a few occurring in nearly all warm countries. Three species are known to occur in the Cayman Islands; all have been introduced for their fragrant, ornamental flowers, and all tend to persist or become naturalized. It is not certain that any of them are correctly named here, as the preserved material is scanty, and a few discrepancies with published descriptions have not been resolved. However, until the genus as a whole is better understood, greater precision in identification probably is not possible. It should be noted that many crinums hybridize readily, and it is likely that many plants labelled with a particular name may, in fact, represent a mixture of two or more natural species. The descriptions given here are based primarily on Cayman specimens, reinforced by observations of what appear to be the same species in Jamaica.

KEY TO SPECIES

1. Bulb with a long, stalk-like massive neck rising 30 cm or more above ground; leaves 9–15 cm broad; flowers numerous, 18 or more per umbel: [C. amabile]

1. Bulb with neck, if any, subterranean; leaves less than 8 cm broad; flowers few, 8 or fewer per umbel:

2. Perianth lobes white or pale pink with pink median stripe, connivent and ca. 2 cm broad; leaf-margins entire but undulate: [C. zeylanicum]

2. Perianth lobes pure white, divergent and ca. 1 cm broad; leaf-margins roughened or minutely toothed: [C. americanum]

Crinum × amabile Donn, *Hort. Cantab*, ed. 6, 83 (1811).

GIANT LILY

Bulb with a rather massive, stem-like neck rising 30 cm or more above ground. Leaves numerous, spreading, narrowly lanceolate, entire, 70–100 cm long, 9–15 cm broad, clasping at the slightly narrowed base and gradually tapering to the apex. Scape 40–80 cm tall, 2-keeled. Spathe-valves deltate, reddish, 12–18 cm long. Umbel of 18–30 pedicellate flowers; perianth with slender cylindric deep rose or red tube 7–12 cm long; lobes ca. 2 cm broad, equalling or exceeding the tube in length, pink outside, and whitish with deep pink median stripe within. Fruit not seen.

GRAND CAYMAN: *Proctor* 15208, collected in sandy soil along the seacoast near Gun Bay. — Native of Sumatra; widely cultivated and perhaps of hybrid origin.

Crinum zeylanicum (L.) L., *Syst. Nat.* ed. 12, 1: 236 (1767).

Bulb said to be subglobose, 10–15 cm in diameter, the neck short (but similar plants naturalized in Jamaica have a slender neck up to 10 cm long). Leaves 6–10, strap-shaped, 30–60 cm long, 3–5 cm broad, the margins entire but more or less undulate. Scape 30–50 cm tall, not keeled, often tinged red. Spathe-valves deltate, thinly papery, 6–8 cm long. Umbel of 4–8 subsessile or short-pedicellate flowers; perianth with slender curved greenish or reddish tube 7–14 cm long; lobes ca. 2 cm broad or more, oblong-lanceolate with partly connivent margins, white with a pink to deep rose median stripe. Fruit not seen.

GRAND CAYMAN: *Brunt* 2062, growing by roadside in shallow soil over limestone pavement near Red Bay; *Sachet* 445. LITTLE CAYMAN: *Proctor* (sight).

— Native of tropical Africa and Asia.

Crinum americanum L., *Sp. Pl.* 1: 292 (1753).

SEVEN SISTERS

Bulb soft-succulent, cylindric, 7–11 cm thick, subterranean and stoloniferous. Leaves several, strap-shaped, 50–90 cm long, 4–7 cm broad (rarely more), the margins minutely roughened or irregularly denticulate. Scape 30–40 cm tall, not keeled. Spathe-valves narrowly to broadly deltate, papery, 5–7 cm long. Umbel of 3–8 sessile fragrant flowers; perianth pure white with slender cylindric tube 11–15 cm long, and wide-spreading narrow lobes ca. 1 cm broad; filaments rose. Fruit not seen, said to be irregularly subglobose and 4–6 cm thick.

GRAND CAYMAN: *Proctor* 15307, from sandy yards around George Town.

— Southeastern United States and Jamaica; cultivated in Cuba. Wild plants of this species customarily grow in swamps.]

Hymenocallis Salisb.

SPIDER LILIES

Bulbous plants with usually linear or strap-shaped leaves, sometimes contracted below into a petiole. Flowers pure white (or greenish in some continental species), few or many in an umbel at the apex of a solid, leafless scape, commonly fragrant chiefly at night; pedicels lacking or short. Perianth with an elongate narrow tube and equal, spreading or recurved narrow lobes; stamens inserted in the throat of the perianth tube, united in their lower part to form a rather conspicuous membranous cup, usually with an entire or bifid process between each pair of filaments; anthers linear. Ovary 3-celled, with 1 or 2 ovules in each cell; style long and filiform with a small capitate stigma. Capsule fleshy and somewhat irregular, splitting open as the seeds ripen; seeds large, fleshy, green.

A tropical American genus of about 30 species, doubtfully separable from the Old World genus *Pancratium*.

Hymenocallis latifolia (Mill.) M. J. Roem., *Syn. Monogr.* 4: 168 (1847). **FIG. 70.** PLATE 7.

Pancratium latifolium Mill. (1768).
Hymenocallis caymanensis Herbert (1837).

LILY, WILD WHITE LILY, EASTER LILY

Bulb ovoid-conic, up to 15 cm long or more and 6–8 cm thick, clothed with brownish papery-membranous epidermis. Leaves strap-shaped, 20–80 cm long, 3.5–6 cm broad, acute at the apex. Peduncle up to 60 cm long; flowers 6–15 in a cluster, greenish in bud, white at anthesis. Perianth tube 10–15 cm long, the linear-lobes 9–12 cm long, somewhat recurved; staminal cup mostly 2–3 cm long; anthers 1–1.3 cm long. Ovary sessile. Capsule 2–4 cm in diameter.

GRAND CAYMAN: *Brunt* 1713, 2181; *Kings* GC 111. LITTLE CAYMAN: *Kings* LC 105; *Proctor* 28079. CAYMAN BRAC: *Kings* CB 89; *Proctor* 28023.

— Florida, Cuba, Jamaica and Hispaniola, chiefly in sandy clearings near the sea.

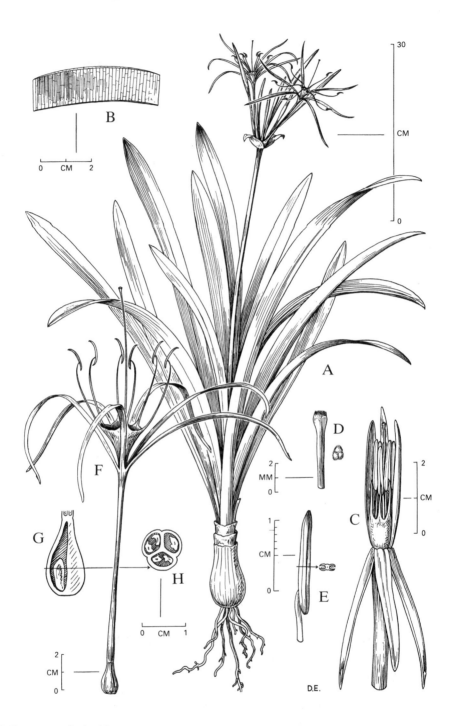

FIG. 70 **Hymenocallis latifolia.** A, habit. B, section across leaf. C, mature bud with three perianth-segments pulled down. D, apex of style and stigma. E, anther. F, flower. G, H, sections through ovary. (D.E.)

AGAVACEAE

Chiefly large, more or less succulent, leathery or fibrous rosette to very large rosette shrubs or trees with stout woody trunks often with secondary tissue growth. Roots fibrous and sometimes rhizomatous; plants sometimes proliferating by means of stolons. Leaves spirally arranged and with margins entire, spiny or toothed, usually broadest near the base and gradually tapering to a sharp apex; the vascular bundles in the leaves are accompanied by thick, strong fibers, thus various species are used as textile plants. Inflorescence stalks (scapes) stout, terminal, clothed with few or many more or less appressed bracts; inflorescences large, paniculate with more or less cymose branches; many species are monocarpic, producing a giant inflorescence and then dying. Flowers usually bisexual, trimerous, hypogynous or epigenous, actinomorphic or slightly zygomorphic, and usually arising from the axils of well-developed bracts; tepals (segments) most often white or yellow, free or more or less fused into a tubular or campanulate perianth, sometimes abruptly widened and urceolate in the outer part. Stamens 6, inserted at or near the base of the perianth, the filaments sometimes short and relatively stout (*Yucca*) or else filiform and long-exserted (*Agave*); anthers introrse, often peltate, opening longitudinally; pollen with reticulate exine, the grains dispersed singly or in tetrads. Ovary 3-locular, each cavity with several to many ovules; style short to rather long, with punctiform to trilobate stigma. Fruit a dry or berry-like capsule containing several to many seeds; seeds often flattened or compressed.

An entirely New World family of about 8 genera and more than 300 species. Many species have been introduced into all warmer parts of the world, either for ornament or for commercial production of textiles, and some of these have become naturalized in various countries.

KEY TO GENERA

1. Ovary superior; leaf-margins roughened but not spiny; plant normally producing a simple or branched trunk:	**Yucca**
1. Ovary inferior; leaf-margins usually spiny (smooth in some forms of *Agave sisalana*); plants not producing a trunk (except mature individuals of *Agave caymanensis*):	
2. Flowers yellow or greenish, directed upward on panicle branches:	**Agave**
2. Flowers white, pendant and directed downward on panicle branches:	**Furcraea**

Yucca L.

Large coarse plants, the majority with a woody trunk, often tall and tree-like, simple or branched. Leaves crowded toward the apex of the trunk or its branches, linear-lanceolate, thick and rigid, usually spine-tipped, the margins entire or fibrous. Flowers white or creamy, in large terminal panicles; perianth long-persistent after withering, the segments separate or nearly so but more or less connivent, rather fleshy. Stamens 6, hypogynous, free, much shorter than the perianth, with thick filaments and small sessile anthers. Ovary sessile or rarely stipitate, 3-celled, the cells incompletely partitioned in two; ovules numerous; style short, stout, divided at the apex into 3 or 6 stigmatic lobes. Fruit either indehiscent and pulpy or spongy within, or dry and splitting open by 6 valves; seeds black, flat.

A genus of more than 25 species, most of these occurring in the southwestern United States and Mexico, a single one being found in the West Indies.

Yucca aloifolia L., *Sp. Pl.* 1: 319 (1753).

A shrub or small tree up to 4 m tall, unbranched or with a few short branches, often growing in clumps or colonies. Leaves numerous, dark green, rigid, mostly 30–60 cm long 2–3.5 cm broad, minutely roughened on the margins and terminating in a sharp brown spine. Panicle compact, erect, up to 60 cm long; flowers numerous, nodding, creamy-white (rarely tinged purple), 3–5 cm long; ovary short-stipitate. Fruit an indehiscent pulpy capsule 7–9 cm long, the pulp dark purple; seeds roundish-flattened, up to 7 mm in greatest diameter.

GRAND CAYMAN: *Brunt* 1998. LITTLE CAYMAN: *Kings* LC 62.

— Southeastern United States and eastern Mexico, Bermuda, Bahamas and the Greater Antilles. Often cultivated, so that its true natural distribution is hard to determine; the Cayman records may represent escapes from cultivation. The flowers are edible, either raw (in salads) or cooked as a vegetable; those of *Yucca guatemalensis* in Guatemala are often dipped in egg batter and fried.

Another species of *Yucca*, not *aloifolia* but not yet identified, has been planted in George Town, Grand Cayman.

Agave L.

Large, slow-growing, fleshy herbs with massive leaves forming a basal rosette; rarely developing a short trunk. Leaves stiff, persistent, armed with a sharp spine at the apex and usually prickly along the margins. Flowers paniculate on tall scapes (or spicate in a few continental species). Perianth 6-parted, more or less funnel-shaped, rather fleshy. Stamens 6, exserted. Ovary 3-celled, with numerous ovules forming 2 rows in each cell; style 1; stigma capitately 3-lobed. Capsule oblong to globose, many-seeded; seeds flat, thin, black. Reproduction often by means of small vegetative 'bulbils', which are produced in large numbers on the inflorescence after flowering. The parent plant always dies after flowering, whether or not seeds or bulbils are formed.

A genus of perhaps 300 species distributed from the southern United States to tropical S. America, especially numerous in Mexico. About 50 indigenous species are recorded from the West Indies, but this number may be excessive, as many are distinguished on the basis of very small differences. These plants are characteristic of rocky or arid habitats, and are often abundant. Several Mexican species, notably *Agave sisalana* (sisal) and *A. fourcroydes* (henequen) are economically important for their useful fibres; these and others yield a copious flow of sap from their cut young inflorescences, which can be fermented to form 'pulque', a Mexican drink. Distillation of pulque yields 'mescal' and 'tequila', which are highly intoxicating. In some countries, *Agave* species are often planted to form living fences, and several (especially variegated forms of *A. americana*) are widely used in horticulture.

KEY TO SPECIES

1. Leaves curved at base; flowers yellow; indigenous species:	**A. caymanensis**
1. Leaves straight not curved at base; flowers green or yellowish-green; introduced species, somewhat naturalized:	
2. Leaves gray-green, the margins usually without prickles:	**A. sisalana**
2. Leaves dark green, the margins with numerous prickles:	**[A. vivipara]**

Agave caymanensis Proctor, sp.nov. **FIG. 71.** PLATE 7.

CORATO

Large, fleshy rosette plants, at first acaulescent but at maturity developing a woody caudex up to 1 m high. Leaves massive, medium green, elliptic-oblanceolate, up to 1.5 m long, 20–25 cm broad above the middle, acuminate at the spine-tipped apex; marginal prickles numerous, 5–14 mm apart, glossy dark brown, curved or reflexed-triangular (rarely straight), 1–4 mm long, often growing from green prominences of the undulate margin. Inflorescence up to 6 m tall or more, densely paniculate toward the apex; flowers bright yellow on pedicels 5–10 mm long; perianth 15–20 mm long, the tepals expanded at base, narrowing to an elongate apex; array narrowly fusiform 30–40 mm long, much longer than the perianth; style 30–35 mm long, long exserted, bifurcate at the expanded apex. Capsules not seen; bulbils often produced.

GRAND CAYMAN: *Brunt* 1765; *Proctor* 15160. LITTLE CAYMAN: *Proctor* 49338, 52171 (type, IJ). CAYMAN BRAC: *Kings* CB79, CB88; *Proctor* 47801.

Endemic; grows in dry, rocky, exposed situations. Differs from *Agave sobolifera* (with which it was formerly confused) by the development of a woody caudex at maturity; by having leaves broader in proportion to their length than those of *A. sobolifera*; and by having a much longer ovary (30–40 mm vs 15–20(–25) mm), shorter perianth, and longer exserted style.

Agave caymanensis Proctor, sp.nov. ab *A. sobolifera* caudice in maturitate lignoso, foliis in proportione latioribus, ovario multo longiore (30– 40 mm non 15–20 neque 25 mm), perianthio breviore et stylo longiore exserto differt.

[*Agave sisalana* Perrine in *House Rep. Doc.* 564: 8 (1838).

SISAL

Leaves at first somewhat glaucous, eventually green and rather glossy, linear-lanceolate, nearly flat, mostly up to ca. 1.5 m long and 10 cm broad near the base; apical spine dark brown, straight or slightly recurved, 2–2.5 cm long; marginal prickles often absent, when present slender and widely spaced. Inflorescence up to 6 m tall, the upper half loosely oblong-paniculate; flowers greenish or greenish-yellow with maroon speckles on the filaments and style; pedicels 5–10 mm long; ovary 20–25 mm long, shorter than the perianth. Perianth segments erect, 15–20 mm long. Capsules rarely produced, oblong, ca. 60 mm long; seeds 7–10 mm.

GRAND CAYMAN: *Proctor* 15207. CAYMAN BRAC: *Proctor* (sight).

— Originally from Yucatan; often cultivated in the hot drier parts of tropical countries, and tending to persist after cultivation.]

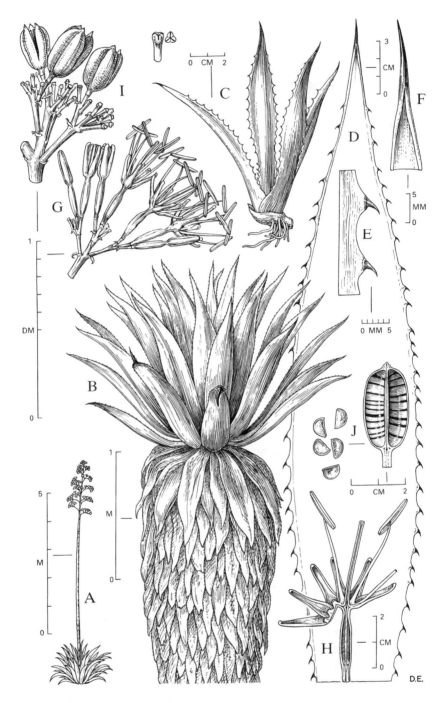

FIG. 71 **Agave caymanensis**. A, general habit. B, single very old plant with woody trunk. C, young plant growing from bulbil. D, outline of leaf. E, marginal spines. F, apical spine. G, portion of inflorescence, H, dissected flower. I, portion of fruiting inflorescence. J, long.-section of capsule and seeds. (D.E.)

[*Agave vivipara* L., *Sp. Pl.* 1: 323 (1753).

Agave angustifolia Haworth (1812); Hummelinck (1984).

Acaulescent rosette plant rarely developing a short caudex, usually proliferating freely by stoloniferous offshoots. Leaves narrowly linear to lanceolate, 40–75 cm long, 3.5–8 cm broad below the middle; apical spine 2.5–4 cm long; upper surface smooth but concave; marginal prickles numerous, black, 2–5 mm long, upcurved, sinuous or recurved from a black deltoid base. Inflorescence up to 6 m tall, the scape-bracts spreading; flowers pale green, the tepals shading to lighter yellowish toward the base, the perianth ca. 2 cm long. Anthers ca. 2 cm long, light brown with numerous minute maroon and silvery dots. Capsules subglabose to broadly ovoid, 3–5 cm long. Proliferous bulbils, instead of fruits, often produced on the inflorescence after flowering.

GRAND CAYMAN: *Hummelinck* 113, 114, 115, 116, 117, 118. LITTLE CAYMAN: *Hummelinck* 119; *Proctor* (sight). CAYMAN BRAC: *Hummelinck* 120, 121. The Hummelinck vouchers are deposited in the herbarium of the State University of Utrecht, The Netherlands.

Mexico to Nicaragua; planted in many warm countries and sometimes becoming naturalized.]

Furcraea Vent.

Large, coarse, rosette-forming monocarpic herbs, stem-less or with a very short caudex. Leaves large, thick, more or less oblong-lanceolate or oblanceolate, spine-tipped, commonly ridged longitudinally, the margins smooth or prickly. Inflorescence stout, paniculate, the elongate axis bearing broad triangular bracts. Flowers clustered, the perianth funnel form with 6 spreading segments, white or greenish-white. Stamens shorter than the perianth, borne on the base of tepals, the filaments swollen at their base. Ovary 3-locular with numerous ovules; style angled and thickened at base; stigma capitate. Commonly reproducing by bulbils formed in the inflorescence. Capsules (not often produced) oblong, 3-angled; seeds numerous, flat.

Furcraea hexapetala (Jacq.) Urban, *Symd. Ant.* 4: 152 (1903).

Leaves up to 150 cm long, straight, with upwardly hooked marginal spines. Inflorescence up to 7 m tall or more, with numerous drooping lateral branches to 2 m long. Flowers pendulous, solitary or clustered on the branches; perianth segments ca. 3 cm long, elliptic, free; ovary ca. 2 cm long. Fruit (rarely formed) oblong-ovoid, 3-lobed-sulcate, ca. 4.5 cm long and nearly 4 cm in diameter, beaked at apex with persistent withered perianth attached; seeds numerous, flat. Usually glossy, dark green, ovoid bulbils are produced instead of fruits; these fall to the ground and develop into new plants.

GRAND CAYMAN: *Proctor* 47267, found in thickets near Newlands.

— Bermuda, Bahamas, Cuba, Hispaniola and Jamaica. ("El género *Furcraea* (Agavaceae) en Cuba", Alberto Álvarez de Zayas, *Ann. Inst. Biol. Univ. Nac. Auton. Mexico ser. Bot.* 67 (2): 320–346 (1996).)

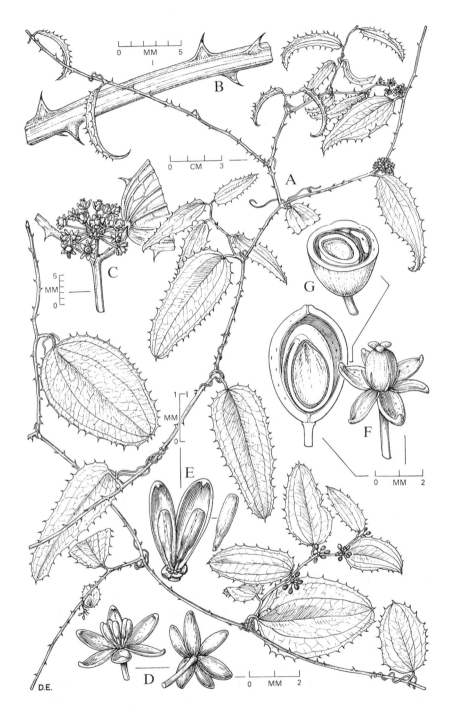

FIG. 72 **Smilax havanensis.** A, habit. B, enlarged portion of stem. C, inflorescence (pistillate). D, two views of a staminate flower. E, dissected portion of a staminate flower. F, pistillate flower. G, fruit, cross- and long.-sections. (D.E.)

SMILACACEAE

Shrubs or commonly vines, the latter climbing with the aid of petiolar tendrils. Leaves usually alternate and of leathery or hard texture, with 3 or more longitudinal nerves and reticulate venation between the nerves. Flowers usually dioecious, small, in axillary umbels, racemes or spikes. Perianth regular, 6-parted, the lobes all similar. Stamens usually 6, with confluent anther-cells. Ovary superior, 3-celled; ovules 1 or 2 in each cell. Fruit a berry; seeds with hard endosperm and small embryo.

A widely distributed family in tropical and temperate regions, with 4 genera and about 375 species.

Smilax L.

More or less woody vines, often armed with prickles on stems and leaves. growing from woody or fleshy tubers or long creeping rhizomes. Leaves petiolate, with entire, lobed, or prickly margins; petioles sheathing at base, and bearing a pair of coiling tendrils. Flowers umbellate on a globose or convex receptacle, terminating in an axillary peduncle; dioecious, usually small and greenish; perianth lobes separate to the base. Pistillate flowers usually smaller than the staminate. Fruit a red, blue or black berry.

A widespread genus of about 350 species. The drug known as sarsaparilla, formerly believed to have medicinal value, is obtained from the roots of various tropical species of *Smilax*. The roots of several species found in southeastern United States yield a refreshing jelly like condiment.

Smilax havanensis Jacq., *Enum. Syst. Pl. Carib.* 33 (1760). **FIG. 72.**

WIRE WISS

Trailing or climbing slender woody vine, armed with short hooked prickles, or sometimes nearly unarmed, up to 8 m long, the branches striate or angled, often zigzag. Leaves coriaceous, narrowly lanceolate or lance-elliptic to broadly ovate or suborbicular, 2–10 cm long, 0.6–4.5 cm broad or more, 3–7-nerved, the margins usually spinulose or rarely entire, the base rounded or subcordate. Peduncles about equal in length to the petioles; inflorescence 4–30-flowered; pedicels ca. 4 mm long; flowers green, 2–3 mm broad. Fruit black, 4–6 mm in diameter.

GRAND CAYMAN: *Brunt* 1788, 2084; *Correll & Correll* 51040; *Kings* GC 187; *Proctor* 15022, 15281, 47269, 48271, 50725, 52073; *Sachet* 379. Frequent in thickets and woodlands.

— Florida, Bahamas, Cuba and Hispaniola.

ORCHIDACEAE

by James D. Ackerman

Perennial herbs of diverse habit, terrestrial, lithophilic or epiphytic, with tuberous, fleshy or otherwise specialized, velamentous roots. Shoots monopodial or sympodial. Leaves convolute or conduplicate, simple, entire, thin and plicate to fleshy and often leathery, sometimes reduced to scales. Inflorescences of solitary flowers, corymbs, spikes, racemes or panicles. Flowers usually perfect, epigynous, zygomorphic, bracteate; perianth of 3 sepals, similar or nearly so, 3 petals, the lateral similar and the median third petal dissimilar, usually larger, sometimes spurred or otherwise modified, in the superior position but usually inferior by the resupination (twisting) of the pedicellate ovary. Stamens much reduced, usually 1, more or less united with the style and stigma to form a column (gynostemium); pollen aggregated in pollinia and often attached to caudicles, stipes and a viscidium to form a pollinarium. Stigmas 3, one of which is sterile and forms a structure known as the rostellum, a barrier between the stigma and anther, a portion of which often serves as part of the pollinarium apparatus; the lateral lobes of the stigma are viscid and receptive, united or separated. Ovary usually 1-celled, ovules numerous on 3 parietal placenta. Fruit a 3-valved capsule with numerous, dust-like seeds, or rarely a many-seeded indehiscent berry; seeds lack endosperm. After seed germination, growth of the plant is depends on symbiotic mycorrhizal fungi.

One of the largest families of flowering plants with about 800 genera and 20,000 species. A cosmopolitan family with greatest abundance and diversity in the moist and wet tropics.

KEY TO GENERA

1. Plants climbing vines with elongate, green to greenish-brown, fleshy stems:	Vanilla
1. Plants not vine-like:	
2. Plants terrestrial, rooting in soil:	
3. Base of plant a hard, swollen corm or pseudobulb; inflorescences lateral:	
4. Leaves thick, hard, conduplicate, mottled dark green; sepals and lateral petals tan:	Oeceoclades
4. Leaves thin, plicate, concolorous; sepals and lateral petals purple or rose:	Bletia
3. Base of plant not forming a swollen corm or pseudobulb; inflorescences terminal; flowers variously colored:	
5. Leaves cauline, sometimes reduced to scales or sheaths:	
6. Leaves thin, cartaceous, plicate; flowers cleistogamous; inflorescences paniculate:	Tropidia
6. Leaves reduced to scales or sheaths, fleshy, soft; flowers chasmogamous; inflorescences racemose:	Triphora
5. Leaves basal or absent:	
7. Lip in the dorsal position, less than 5 mm long:	Prescottia
7. Lip in the ventral (lowermost) position:	
8. Leaves absent or disintegrating at anthesis:	Sacoila
8. Leaves basal and not disintegrating at anthesis??:	
9. Lip fimbrate:	Eltroplectris
9. Lip entire:	
10. Rostellum longer than broad:	Cyclopogon
10. Rostellum broader than long:	Beloglottis
2. Plants epiphytic or rarely on rocks, never rooted in soil:	
11. Plants with evident leaves:	
12. Plants with distinct pseudobulbs:	
13. Pseudobulbs conspicuous, all leaves being attached to their apices:	
14. Pseudobulbs composed mostly of 1 internode; pollinia 4:	
15. Ovaries and fruits triangular in cross-section:	Prosthechea
15. Ovaries and fruits rounded in cross-section:	Encyclia
14. Pseudobulbs composed of 2–several somewhat equal internodes; pollinia 2 or 8:	

16. Inflorescences racemose; floral bracts inconspicuous, not colored as the perianth; pseudobulbs hollow; pollinia 8: **Myrmecophila**
16. Inflorescences paniculate; floral bracts large, colored as the flowers, yellow with reddish brown markings; pseudobulbs solid; pollinia 2: **Cyrtopodium**
13. Pseudobulbs inconspicuous, very small; leaves lateral: **Ionopsis**
12. Plants without pseudobulbs:
  17. Leaves scattered along stem:
    18. Leaves fleshy; floral bracts conspicuous, largely concealing the ovary; inflorescence racemose; pollinia 4: **Epidendrum**
    18. Leaves thin; floral bracts inconspicuous; inflorescence racemose or paniculate; pollinia 2: **Polystachya**
  17. Leaves solitary or else all basal:
    19. Leaves solitary on short erect stems:
      20. Leaf-blades elliptic-oblanceolate, not over 1.5 cm long; flowers very small, lip ca. 2 mm long: **Pleurothallis**
      20. Leaf-blades linear to linear-oblong, more than 6 cm long; flowers large lip over 3.5 cm long: **Brassavola**
    19. Leaves several, basal, the blades more than 2 cm long: **Tolumnia**
11. Plant leafless, consisting of a consipicuous cluster of elongate roots and a short naked scape bearing 1 or 2 showy cream-white flowers, each with an elongate spur at the base: **Dendrophylax**

Vanilla Plumier ex Miller

Fleshy green to yellow-brown vines, often high-climbing, with or without leaves, often producing roots at the nodes. Leaves of various textures from membranous to leathery, large to small and scale-like, persistent or early deciduous. Inflorescences short, axillary racemes or spikes. Flowers rather large, subtended by ovate bracts; sepals about equal, free, spreading; petals similar to the petals; lip and claw adnate to the column, the limb broad, concave, embracing the column at the base, disc variously ornamented with scales, bristles, hairs or callus ridges. Column long, without a foot; clinandrium short or obliquely raised. Anther attached to the margin of the clinandrium, incumbent; pollinia granular, composed of sticky monads, free or at length sessile on the rostellum. Stigma transverse under the short rostellum. Berries usually capsules, elongate, fleshy and leathery.

A pantropical genus of about 90 species. The spice called 'vanilla' is obtained from the fermented berries of *Vanilla planifolia* and (to a lesser extent) *V. pompona*.

Vanilla claviculata (W. Wr.) Sw. in *Nov. Act. Reg. Soc. Sci. Upsal.* 6: 66 (1799). PLATE 8.

Main stem leafless, long-trailing or climbing, 1–2 cm thick when fresh (shrinking by about one half when dried); internodes 10–13 cm long; adventitious roots often twistir.g like tendrils. Leaves early deciduous, 2–8 cm long, linear-lanceolate, rigid, involute, recurved, acuminate at the apex, half clasping at the base. Raceme with 8 or more sessile flowers, appearing stalked by the long narrow ovary; bracts reflexed, 2–11 mm long. Flowers fragrant; perianth 3.5–4.5 cm long, glaucous green; lip 4–4.5 cm long, its apex rounded, white or white with two reddish, lateral blotches near the base, more or less curled or ruffled on the upper margin; disc with a broad thickened mid-rib running the length of the lip and with erect, fleshy, often forked hairs, and a hinged beard nearly 1 cm long. Column 2.5–3 cm long. Berry ellipsoid-cylindric, 7–11 cm long.

GRAND CAYMAN: *Brunt* 2186; *Kings* GC 412.

— Bahamas and Greater Antilles, rather common.

Prescottia Lindl.

Terrestrial herbs with clustered fibrous or fleshy roots. Leaves basal, non-articulate, sessile or petiolate, membranous. Inflorescences a slender, erect spike with numerous flowers. Flowers non-resupinate, small; sepals membranous, connate at base to form a short cup or tube, spreading or revolute at the apex; petals membranous, narrow, adnate to the sepal-cup; lip with claw adnate to the sepal-cup, auriculate at the base, the apical part entire, deeply concave or galeate and often enclosing the column. Column very short, adnate to the sepal-cup. Pollinia 4, granular or powdery. Capsule small, suberect, ovoid or ellipsoid.

A genus of about 24 species found in tropical and subtropical America.

Prescottia oligantha (Sw.) Lindl., *Gen. & Sp. Orch. Pl.* 454 (1840).

Small, terrestrial herbs to 40 cm tall with a fascicle of thick, fleshy roots. Stems slender, erect, mostly 15–35 cm tall, glabrous, bearing several narrow, tubular sheaths. Leaves 2–3, basal, petiolate, cuneate, elliptical to suborbicular, acute to rounded, 1.5–13 cm long. Inflorescence a very slender, dense, many-flowered spike. Flowers pinkish or whitish, glabrous, minute, less than 2.5 mm long; floral bracts ovate to lanceolate, clasping the base of the ovary, 1.5–4 mm long; sepals arising from a broad connate base, 1–2.2 mm long; dorsal sepal recurved, ovate, usually obtuse, lateral sepals triangular to deltoid, connate with the lip to form a mentum; petals recurved, more or less linear, 1–1.5 mm long; lip white, erect, basally auriculate, suborbicular, concave-saccate, 1–2 mm long. Column adnate to sepaline tube, laterally winged near the apex, minute. Capsule ellipsoid, ca. 4 mm long, narrowly 6-keeled.

GRAND CAYMAN: *Kings* GC 427 (BM).

— Florida, West Indies, Mexico, south through C. America and tropical and subtropical S. America, including the Galapagos Islands. Rather common and widespread, growing in the humus of rocky woodlands. Although the Cayman specimen is sterile, there is no doubt of its identity. The description is based on material from the Greater Antilles.

Beloglottis Schltr.

Small terrestrial, erect herbs with fasciculate, fleshy roots. Leaves several in a basal rosette, petiolate. Flowers many in a loosely to densely flowered raceme. Sepals subparallel, free to connate at base, apices spreading, lateral sepals with an oblique, subdecurrent base; petals parallel and adnate to dorsal sepal; lip clawed, lamina canaliculate, sagittate, lateral margins adnate to sides of column. Column short, clinandrium inflated. Stigmas 2 and adjacent to each other, rostellum short, erect, bilobed, bifid or bidentate, broader than long. Anther ovate, concave, acute, slightly cordate at base, pollinia clavate, mealy, viscidium narrowly elliptic. Capsules erect, ellipsoial.

A subtropical and tropical genus of 10 species.

Beloglottis costaricensis (Rchb. f.) Schltr., *Beih. Bot. Centralbl.* 37(2): 365 (1920).

Terrestrial, erect herbs with few thick, fleshy, roots 1–6 cm long. Leaves 2–3, fugacious, present or absent at the time of flowering, delicate, glabrous; petioles erect, slender, 2–4 cm long; blades spreading, elliptic, acute, 2.5–4.5(–9) cm long, 1.3–2.0(–3.3) cm wide. Inflorescence a many-flowered, slender, pubescent raceme; floral bracts erect, clasping base of ovary, ovate, acuminate, 5–9 mm long. Flowers white with dark green midveins, resupinate, subsalverform; sepals dorsally and basally pubescent, united at their base, dorsal sepal 1-nerved, lanceolate, acute, 4–6 mm long, 1.5 mm wide; lateral sepals 1-nerved, oblique, lanceolate, acute to acuminate, 4.2–6 mm long, 1.2 mm wide; petals glabrous and parallel to the dorsal sepal, 1-nerved, linear, slightly falcate, acute, 4 mm long, 0.6 mm wide; lip glabrous, attached to base of column foot, 3-nerved, clawed, auriculate, pandurate, acute-acuminate, 4.5 mm long, 1.5 mm wide, the claw 1.5 mm long, the auricles subulate, 0.7–0.8 mm long, projecting basally and flanking the column and column foot. Column slender, 2.5 mm long with a foot that extends obliquely along the ovary 1.5 mm long without forming a mentum. Capsules erect, ellipsoidal, 4–6 mm long.

GRAND CAYMAN: *Proctor* 49218. Rocky woodland S.S.W. of Old Man Bay near the centre of the island.

— Florida, Dominican Republic, Mexico, Guatemala, Belize, Honduras, El Salvador, Nicaragua, Costa Rica, Panama and Trinidad.

Cyclopogon Presl

Small, terrestrial herbs with fleshy, fasciculate roots. Stems erect, simple, more or less concealed by leaf-sheaths. Leaves few to many, basal, non-articulate, petiolate. Inflorescences terminal, erect; scapes bracteate; spikes or racemes many-flowered. Flowers resupinate, erect; sepals subparallel, free or fused at base and forming an obscure mentum with base of column or a conspicuous sepaline nectar tube; petals connivent with dorsal sepal; lip unguiculate, sagittate to cordate, lateral margins adpressed to sides of column. Column erect. Anther dorsal, pollinia 2, clavate-oblong with an apical constriction, mealy with a relatively large disc-shaped viscidium. Stigma lobes 2, free to approximate, rostellum soft, longer than wide. Capsule ellipsoid to obovoid.

A neotropical genus of about 70 species and a segregate of *Spiranthes*. Although the dismemberment of *Spiranthes* has been suggested for decades on the basis of comparative morphology, circumscriptions of the different genera have been rigorously debated. Molecular systematic evidence strongly supports the break up of the genus into a number of smaller genera, including *Cyclopogon*, *Beloglottis* and *Sacoila* in the Cayman Islands (G. A. Salazar, M. W. Chase, M. A. Soto Arenas & M. Ingrouille (2003). Phylogenetics of Cranichideae with emphasis on Spiranthinae (Orchidaceae, Orchidoideae): evidence from plastid and nuclear DNA sequences. *Amer. J. Bot.* 90: 777–795).

KEY TO SPECIES

1. Lip broadest below apical constriction; leaf-blades dark, velvety purplish green, spreading; petioles 1–3.5 cm long:	C. cranichoides
1. Lip equal or broadest above apical constriction; leaf-blades green, erect-spreading; petioles 1.5–10 cm long:	C. elatus

Cyclopogon cranichoides (Griseb.) Schltr., *Beih. Bot. Centralbl.* 37(2): 387 (1920).

Small, terrestrial herbs with fleshy, fusiform roots. Leaves 1–5(–8), spreading and forming a loose, basal rosette; petioles 11–35(–55) mm long; blades broadly elliptic to ovate, 2.5–8 cm long, purplish below, dark green above, sometimes with whitish markings. Inflorescence a many-flowered raceme; floral bracts lanceolate, longer than the ovary, mottled. Sepals pubescent, greenish-brown, dorsal sepal midvein and margins dark near the apex, linear-elliptical to narrowly pandurate, acute, 3.5–5 mm long, lateral sepals often reflexed, linear-elliptical to linear-pandurate, acute, 4–6 mm long; petals greenish-brown proximally, greenish white distally with midvein and apical margins dark brown, linear spatulate-oblanceolate, acute, oblique, loosely adnate to the dorsal sepal, to 4.5 mm long; lip white, to 5 mm long, 2.5 mm wide, basally gibbose, oblong, constricted above the middle, slightly flared at the apex, provided with a pair of basal tubercles. Column slender, clavate, ca. 3.5 mm long. Capsules erect, ellipsoidal, 6–8 mm long.

GRAND CAYMAN: *Ebaules & Roulstone* 1 (K), *Proctor* 48675, collected for the first time in 1992 from near the center of the island.

— Florida, Bahamas, Cuba, Cayman Islands, Jamaica, Dominican Republic, Puerto Rico, United States Virgin Islands, Dominica, Guadeloupe, Grenada, Mexico, Guatemala, Belize, Honduras and Venezuela.

Cyclopogon elatus (Sw.) Schltr., *Repert. Spec. Nov. Regni Veg. Beih.* 6: 53 (1919).

Terrestrial herbs to 75 cm tall with fleshy, fasciculate roots. Leaves green, 2–6, basal; petioles erect, slender, 3–10 cm long; blades narrowly to broadly elliptic, to elliptic-lanceolate, obtuse to acute-acuminate, 3.7–13 cm long. Inflorescences pubescent racemes, many-flowered; floral bracts immaculate or minutely speckled, glabrous or sparsely pubescent at base, erect, lanceolate, often exceeding the length of the flowers, 8–19 mm long. Flowers green to coppery brown, sometimes purplish; sepals dorsally pubescent, basally connate for ca. 1 mm, dorsal sepal elliptic-lanceolate, acuminate, 4.5–7 mm long, 1.5–2 mm wide, lateral sepals slightly sinuate-falcate, linear-oblong, 5–7.5 mm long; petals glabrous, entire, linear to linear-spatulate, adnate to the dorsal sepal, 4–6.5 mm long; lip canaliculate, laterally thickened at base which may or may not have two erect callus horns, pandurate to oblong-ovate then constricted above the middle, usually broadest at flabellate to reniform apex, entire to crenulate, 5–7 mm long. Column slender, 3.5–4.5 mm long. Anther ca. 1.5 mm long. Capsules ellipsoidal, 6–12 mm long.

GRAND CAYMAN: *Roulstone s.n.* (K).
— Florida, Greater and Lesser Antilles, Mexico through C. America and into tropical and subtropical S. America.

Eltroplectris Raf.

Terrestrial herbs with fleshy, fasciculate roots. Stems rhizomatous from which flowering shoots emerge. Leaves non-articulate, basal, 1–many, long-petiolate. Inflorescence a slender, scapose raceme with few to many flowers. Flowers resupinate; sepals free, spreading, lateral sepals longer than the dorsal, attached along the column foot forming a spur; petals adpressed to the dorsal sepal, basally decurrent on the column; lip membranous, clawed, arcuately recurved, shorter than the sepals. Column slender, short, erect, column foot elongate, partially adnate to the ovary. Anther persistent, pollinia 2, clavate, mealy, attached to an oblong viscidium. Stigmas distinct, rostellum rigid, subulate to linear, pointed. Fruit a capsule.

A neotropical genus of about 12 species. Before detailed studies of the column morphologies and molecular phylogenetics of the Spiranthinae, this group was placed by various authors into a number of other genera, including *Spiranthes*, *Pelexia* and *Stenorrhynchos*. The data are now unequivocal and show that *Eltroplectris* is worthy of generic standing.

Eltroplectris calcarata (Sw.) Garay & Sweet, *J. Arnold Arb.* 53: 390 (1972).

Spiranthes calcarata (Sw.) Jiménez (1962).

Pelexia setacea Lindl. (1840).

Stems glabrous or minutely glandular-puberulous upwardly, 30–55 cm tall, sparingly clothed with a few adpressed membranous sheaths having attenuate tips, and accompanied from the base by a single leaf (rarely 2) with a reddish petiole, 8–20 cm long, and a lanceolate to broadly elliptic, acute to acuminate blade, 10–15 cm long. Inflorescences a loose spike of 2–11 flowers; bracts lanceolate, attenuate, exceeding the ovary. Sepals pale green, linear-lanceolate, 2.1–3 cm long, the lateral ones connate at the base into a spur-like appendage; petals similar to the sepals, attached their whole length to the median sepal, 1.9–2.1 cm long; lip white, oblong, many-nerved, the margins fimbriate in the median part, attached to the tip of the column foot within the sepal-spur, 1.2–1.5 cm long. Column ca. 10 mm long. Capsule broadly ellipsoid, 1.8–2.5 cm long, with 6 narrow keels.

GRAND CAYMAN: *Kings* GC 274 (BM).

— Florida, Bahamas, Greater Antilles and northern S. America, usually growing in shaded humus.

Sacoila Raf.

Terrestrial herbs with fleshy, tuberous, fasciculate roots. Stems erect. Leaves green to pale green, glossy, several, basal, senescent or absent at anthesis, subsessile, cuneate, lanceolate-elliptical to oblanceolate. Inflorescences scapose, pubescent racemes with numerous flowers; floral bracts green to red, glabrous to pubescent, lanceolate, attenuate. Flowers variously colored, odorless, pubescent, tubular; sepals free, erect, subsimilar, connivent, externally pubescent, dorsal sepal concave, lanceolate, lateral sepals connate at base, decurrent on column foot forming a mentum or spur-like extension, slightly spreading above the middle, obliquely lanceolate to oblong-lanceolate; petals adnate to the dorsal sepal, falcate, oblong; lip sessile or indistinctly clawed, united at base to the lateral sepals in the spur, basally canaliculate with linear, marginal or submarginal nectar glands, margins at middle adnate to column, slightly arcuate, upper half triangular-lanceolate to ovate, decurved or deflexed. Column short, stout with a long, column foot decurrent on sides of ovary and protruding below to a free tip. Anther ovate, acute to acuminate, filament adnate to column, pollinia white, narrowly clavate with a sheathing, linear viscidium. Stigmas 2, terminal, confluent, oriented at an oblique plane toward apex of column with its lower border prominent in side view, rostellum remnant rigid, linear-acicular. Capsules ellipsoid.

A neotropical genus of 7–10 species of Florida, West Indies, Mexico, Central and S. America. *Sacoila* is one of those genera of the Spiranthinae that has appeared and disappeared from synonymies a number of times. However, recent karyological and molecular evidence support its generic status (G. A. Salazar, M. W. Chase, M. A. Soto Arenas & M. Ingrouille (2003). Phylogenetics of Cranichideae with emphasis on Spiranthinae (Orchidaceae, Orchidoideae): evidence from plastid and nuclear DNA sequences. *Amer. J. Bot.* 90: 777–795).

Sacoila lanceolata (Aubl.) Garay, *Bot. Mus. Leafl. Harvard Univ.* 28: 352 (1982).

Stenorrhynchos lanceolatum (Aubl.) Rich. ex Spreng. (1826).
Spiranthes lanceolata (Aubl.) León (1946).
Stenorrhynchos orchioides (Sw.) Lindl. (1840).
Stenorrhynchos squamulosum (Kunth) Spreng (1826).
Spiranthes squamulosa (Kunth) León (1946).
Sacoila squamulosa (Kunth) Garay (1982).

Stems erect, minutely whitish-scurfy, 17–70 cm tall, clothed with about 8 acuminate sheaths 2–4 cm long. Basal leaves several, broadly lanceolate to elliptic, 11–41 cm long, usually absent at flowering time. Inflorescence more or less scurfy, spike loosely arranged or crowded with 6–14 flowers; bracts narrowly lanceolate, exceeding the ovary. Flowers pubescent like the stem, somewhat tubular; sepals brick red, salmon, coppery, greenish brown, or light buff, lanceolate, acuminate, 5-nerved, the lateral ones very oblique and 1.5–2.5 cm long, prolonged at base into a chin-like spur; median sepal 1–2 cm long; petals colored like the sepals, about as long as the adnate median sepal, oblong-lanceolate to elliptic-lanceolate, subacute, 1–1.2 cm long; lip lighter colored than sepals, often whitish, attached to the column foot, 1.4–2.35 cm long, dilated and saccate about the middle, lanceolate, acuminate above and linear-convolute below, more or less pubescent. Column straight, ca. 6 mm long with an elongate column foot. Anther ca. 5 mm long. Capsule scurfy, 1–1.5 cm long, 3-keeled.

GRAND CAYMAN: *Brunt* 2190; *Kings* GC 244.

— Florida, Greater and Lesser Antilles, and Mexico south to tropical S. America. *Sacoila lanceolata* is an extraordinarily variable species in pubescence, color, leaf form and size of spur; a number of segregates have been proposed, including *Sacoila squamulosa*. Nir (2000) (Symbolae Antillanae. DAG Media Publishing) recognizes *S. squamulosa* and distinguishes it from *S. lanceolata* by its scurfy pubescence and a spur about 2/3 the length of the ovary. Although local populations may be relatively uniform in such characteristics, these traits occur in all combinations so that variation is continuous across its entire distribution.

Triphora Nutt.

Small terrestrial, sparsely leafy (or mycotrophic and leafless) herbs, growing from a subterranean corm-like tuber, and spreading by delicate stolons. Leaves 2–4, alternate on the stem, sessile, relatively small or even reduced to mere sheaths. Inflorescence a 1–several-flowered raceme or corymb. Flowers small, axillary, pedicellate, erect or nodding; sepals and petals similar, free or somewhat connivent; lip sessile or clawed, parallel to the column, almost entire to 3-lobed, 1–3-crested, decurved at apex. Column slender, straight, apically entire or lobed. Anther rigidly attached to top of column; pollinia 2, mealy. Capsule ellipsoid to obovoid.

A genus of about 25 temperate and tropical American species.

Triphora gentianoides (Sw.) Ames & Schltr., *Orchidaceae* 7: 5 (1922).

Plant with erect glabrous dark red stem 8–26 cm tall; leaves 3–10, ovate, rounded to obtuse, sheathing, 8–16 mm long. Inflorescences in a short terminal corymb, 3–10-flowered, erect or somewhat spreading; bracts similar to leaves, ovate, acuminate, concave, 5–9 mm long. Flowers non-resupinate on pedicels up to 4 cm long or more; sepals narrowly oblong, 9–11 mm long, pale yellow-green and suffused with dark red-purple, 3-nerved; petals about as long as the sepals, white to pale yellow-green; lip 7.5–10 mm long, 3-lobed, white to pale green. Column 5–7 mm long, yellowish. Capsules 8–20 mm long, ellipsoid with 6 narrow keels.

GRAND CAYMAN: *Kings* GC 242 (BM). Recorded only from the Forest Glen area.

— Florida, Cuba, Hispaniola, Jamaica and Mexico southward at scattered localities to Venezuela and Ecuador, rare or uncommon.

Tropidia Lindl.

Erect terrestrial, caespitose herbs with short, rigid rhizomes and slender, simple or branched, hard erect stems. Leaves few, sheaths nonarticulate, blades thin, convolute and plicate. Inflorescences terminal or occasionally from axils of upper leaves, sessile or pedunculate, racemose or paniculate, rarely glomerate. Flowers few to many, small, resupinate; sepals subequal, dorsal sepal oblong-lanceolate, lateral sepals oblong, connate at base (except ours), forming a small mentum; petals slightly falcate, similar to sepals but shorter; lip sessile, saccate at base, oblong, slightly shorter than the petals. Column short, fleshy, with a short foot. Anther dorsal, erect, subequal to the rostellum, more than half the length of the column; pollinia 2, sectile, attached to a terminal viscidium. Stigma entire. Capsules oblong.

A genus of about 7 species in India, Sri Lanka, Malaysia, New Caledonia, Fiji, China and Japan, and one species in the neotropics.

Tropidia polystachya (Sw.) Ames, *Orchidaceae* 2: 262 (1908).

Plants to 70 cm tall with terete, branched stems, 31–55 cm high, 2–4 mm diam. Leaves several with clasping sheaths, 4–4.5 cm long; blades oblong-lanceolate, acuminate, 6–27 cm long, 2.4–6.7 cm wide. Inflorescences on peduncles 3.5–14 cm long; panicles densely flowered, subglobular, 2.5–9 cm long, 3–6.5 cm diam.; floral bracts linear to lanceolate, acuminate to subulate, 4–10 mm long. Flowers white to pale red, 3–48, numerous, 8–12 mm long; dorsal sepal concave, oblong to elliptic, acute to obtuse, 6.5–7 mm long, 1.7–2.2 mm wide, lateral sepals somewhat gibbous, lanceolate-oblong, acute, slightly falcate, 5.5–6.5 mm long, 2–2.5 mm wide; petals slightly falcate, lanceolate to elliptic-oblanceolate, acute, 5–6 mm long, 1.5–2 mm wide; lip very concave at base and embracing the column, disc 2-lamellate and flabellate above, oblong, constricted at the middle, apex apiculate, sometimes obscurely 4-lobed, 4.5–6 mm long, 2 mm wide. Column 4–4.5 mm long. Anther ovate, apiculate, 1.5 mm long, 1 mm wide. Ovary cylindrical, ca. 8 mm long on short pedicels. Capsules oblong, ribbed, 1–1.4 cm long, 3–6 mm diam.

GRAND CAYMAN: *Proctor* 48322.

— Florida, Cuba, Jamaica, Hispaniola, Mexico, Guatemala, Costa Rica, Venezuela, and Galapagos. Usually encountered in moist and wet montane forests, the Grand Cayman record of *Tropidia polystachya* is peculiar. First found on the island by Frank Roulstone III.

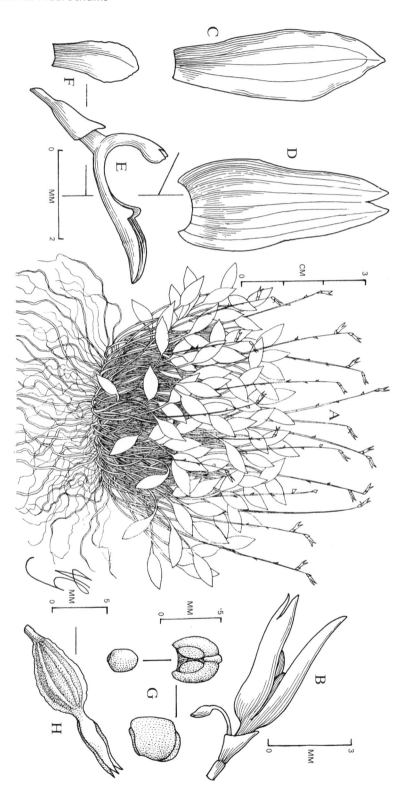

FIG. 73A **Pleurothallis caymanensis.** A, habit. B, single flower subtended by a small bud. C, dorsal sepal. D, synsepal. E, flower with sepals and lateral petals removed showing floral bract, ovary, column and lip. F, lateral petal. G, views of anther with pollinia. H, fruit. (G.)

Pleurothallis R. Br.

Small to large epiphytic herbs, with stems clustered or else branched from a creeping primary stem or rhizome; pseudobulbs absent. Leaves solitary on a stem, blades coriaceous, sessile or short-petiolate. Inflorescences subtended by a conspicuous or inconspicuous spathe, terminal (rarely lateral), racemose or sometimes bearing a solitary flower; floral bracts tubular. Flowers mostly small, yellowish, greenish or maroon; sepals equal or nearly so, erect or spreading, the median free or very shortly connate with the lateral; lateral sepals slightly or often completely connate; petals shorter or narrower than the sepals, sometimes minute; lip shorter or rarely a little longer than the petals, simple or 3-lobed, usually contracted at the base and jointed with the base of the column. Column semiterete, winged or wingless, often with a small column foot, this nearly obsolete to nearly as long as the column. Anther terminal, operculate, incumbent, 1- or 2-celled; pollinia 2, 6 or 8, hard, waxy. Ovary articulate with persistent pedicels. Capsule subglobose or ellipsoid.

A very large genus of hundreds of species (well over 2,000 epithets have been published), all of which are confined to the Western Hemisphere and most commonly found in the moist montane regions of the tropics and subtropics. The circumscription of the genus is unsettled as molecular data indicate that the genus includes a number of disparate elements. Very few species occur in the hot, dry lowland tropics, so it is rather surprising to find a species occurring near sea-level in the Cayman Islands.

Pleurothallis caymanensis C. D. Adams, *Orquideología* 6: 146 (1971). **FIG. 73A**. PLATE 8.

A small, caespitose epiphyte, the slender branched rhizomes concealed by the deflexed leaves; secondary stems with a single internode 2–3 cm long. Leaves sessile, thick, fleshy-coriaceous, mottled gray-green and minutely white-dotted, elliptic-oblanceolate, sharply acuminate, 6–15 mm long. Inflorescence solitary, scape filiform, articulate, 20–35 mm long, bearing 1–3 flowers successively. Flowers greenish-white or yellowish with purple veins; median sepal oblong-oblanceolate, 6.5–7 mm long; lateral sepals almost wholly connate, 6 mm long, the free apices acute; petals broadly obovate, 1.8–2.1 mm long; lip hinged to the tip of the column foot, ovate-oblong with broadly acute apex, entire, 2–2.25 mm long; column deeply curved. Column semiterete, broadly winged and minutely denticulate above the middle, 1.8 mm long. Ovary rugulose; capsule obliquely obovoid, ca. 7 mm long.

GRAND CAYMAN: *Kings* GC 250; *Proctor* 27983 (type).

— Recently discovered on Peninsula Guanahacabibes at the western tip of Cuba. Found in dense woodlands in the central part of Grand Cayman.

Bletia Ruiz & Pav.

Terrestrial erect herbs growing from more or less globose corms with few fleshy, villous roots. Leaves few, growing from top of corm, plicate, lanceolate, rather long and grass-like, usually withering at flowering time. Inflorescence scape slender, arising from side of corm, bearing small, distant scale-like bracts, the racemes or panicles with few to many, somewhat distant flowers. Flowers showy, resupinate, rather distant; sepals separate, the dorsal usually with a recurved apex and longer than the lateral sepals, elliptic, oblong or lanceolate; petals forming a hood over the column, sometimes partially covering the lateral lobes of the lip, similar to the sepals but slightly broader; lip trilobed, lateral lobes upturned, at least at the margins, midlobe declined, disc 3–7 lamellate, lamellae usually extending through the length of the lip, variously colored. Column elongate, arcuate, winged at apex, auriculate at base, footless. Anther terminal, operculate, pollinia 8, hard or soft. Capsule ellipsoid, erect or nodding.

A tropical American genus of about 35 species, some of them very common and widespread, often attracting the attention of travelers because they frequently grow on roadside banks.

Bletia florida (Salisb.) R. Br. in Aiton, *Hort. Kew.* ed. 2, 5: 206 (1813).

Terrestrial, erect herbs, up to 93 cm tall. Corms erect, globose, sometimes flattened, 1.2–3.7 cm diam, covered by 2–3 basal, scarious sheathing bracts. Leaves 3–5, present or absent during flowering, lanceolate, acute or acuminate, 40–82 cm long. Inflorescences scapose racemes or panicles 80–93 cm long; floral bracts ochraceous, acuminate, 3–12 mm long. Flowers mostly rose or purple colored; dorsal sepal reflexed, obovate-elliptic, 2.2–2.9 cm long, lateral sepals extended, ovate-elliptic or falcate, 2–2.5 cm long; petals ovate-elliptic, margin undulate, curved and partially enclosing the column, 2.1–2.7 cm long; lip trilobed, 1.7–2.4 cm long, obovate in outline, lateral lobes elliptic, partially covering the column, 0.8–1.4 cm long, midlobe horizontal to curved downwards, elliptic, 0.8–1.2 cm long, with an isthmus of 3.5–6 mm, margin crispate, disc with 5–7 yellow lamellae, extending from the base to the middle of the midlobe, elevated. Column arched, winged, 1.2–1.8 cm long. Capsules oblong, ascending, 2.8–3.5 cm long, 6.4–8.5 mm wide. GRAND CAYMAN: Presumed to be the identity of a cultivated species in George Town, Grand Cayman. It has deep purple flowers on branched inflorescences up to nearly 1 m tall, and was probably introduced from Jamaica. The terrestrial habit, ovoid pseudobulbs, and broad, somewhat grass-like leaves render identification easy. No herbarium material has been seen.

— Cuba, Jamaica, Haiti and Dominican Republic.

Polystachya Hook.

Epiphytic, lithophytic or terrestrial herbs with slender or pseudobulbous stems clustered or distant along the rhizome, 1–several-leaved. Leaves conduplicate, coriaceous, thin or fleshy. Inflorescences terminal, peduncle enclosed by scarious sheaths, racemose or paniculate with more or less secund branches, few- to many-flowered. Flowers generally small, mostly non-resupinate; sepals free; lateral sepals attached to the column foot forming a mentum; petals smaller than the sepals, free; lip simple or trilobed, attached to and articulate with the column foot, forming part of the mentum, disc usually callose and pubescent. Column short, semiterete, foot distinct. Anther terminal, operculate, incumbent, pollinia 4, globose or ellipsoidal, hard, waxy, attached to a single short stipe and an ovate to elliptic viscidium. Stigmas confluent, transverse under the rostellum. Capsules ellipsoid to cylindrical.

A pantropical genus of about 150 species, particularly species rich in Africa.

Polystachya concreta (Jacq.) Garay & Sweet in Howard, *Fl. Lesser Antil. Orch.* 178 (1974).

Caespitose epiphytes, lithophytes or terrestrial herbs. Stems pseudobulbous to slightly swollen at base, hidden by leaf-sheaths, composed of several subequal intenodes, 1–4 cm long. Leaves 2–4, sheathing at base; blades conduplicate and petiole-like at base, thin, narrowly oblong to elliptic, mostly acute, 3.5–21 cm long. Inflorescences erect, generally paniculate but racemose in small plants; peduncle compressed, covered by scarious sheaths, 12–17 cm long; branches of panicles secund, rachises many-flowered, 12–24 cm long. Flowers yellow or pale green with a white lip, non-resupinate; dorsal sepal ovate-elliptic, acute, 2.5–4 mm long, lateral sepals triangular, 4.5–5.0 mm in its longest dimension, attached to the column foot to form a mentum; petals narrowly oblanceolate, acute-obtuse, 2–3 mm long; lip trilobed, attached to the column foot, cuneate, rhombic, 3–5 mm long, 3 mm wide, lateral lobes falcate, apices almost reaching the apex of the midlobe, midlobe quadrate to oblong, retuse, subapiculate, disc with a linear, densely farinose callus (pseudopollen), 2 mm long, ending at the base of the midlobe. Column short, cylindrical, 1.5 mm long, column foot 2–3 mm long. Capsules brown at maturity, ellipsoid, 6–12 mm long.

LITTLE CAYMAN: *Proctor* 47314 (det. P. Cribb, Kew).

— Pantropical(?). African and Asian synonyms may represent different species even though some of them have been used for neotropical plants (e.g. *Polystachya flavescens*). Like *Polystachya foliosa*, morphological variation in vegetative parts of *P. concreta* is substantial with quite small plants producing inflorescences, and sometimes populations consist of only such small plants. However, floral morphology is relatively invariant. Over the centuries, there have been a number of attempts to recognize segregate taxa from both of these species, which may make sense locally but do not appear to hold up across the range of the species.

Epidendrum L.

Plants epiphytic, on rocks, or rarely terrestrial, extremely diverse in size, erect, pendent, or creeping and with or without a conspicuous rhizome. Stems either thickened into a 'pseudobulb' bearing leaves only at the apex, or cane-like, slender and more or less leafy, simple or branched. Leaves 1–numerous, conduplicate, terete or flattened, varing in outline from linear to oval. Inflorescence usually terminal (rarely lateral from leafy stem), 1-flowered, racemose, subumbellate or diffusely paniculate. Flowers variously colored, frequently green, usually fragrant, minute to rather large; perianth segments spreading; petals usually much narrower than the sepals; lip more or less adnate to the column, simple or lobed, smooth to warty or callose, often conspicuous. Column straight to arching, swollen toward apex, wingless, winged or auricled, short to elongate. Anther terminal, operculate, incumbent, 2-celled; pollinia 4 or rarely 2, waxy, equal, more or less flattened. Stigma entire; rostellum longitudinally slit or perpendicular to axis of column. Capsule usually ellipsoid.

As here defined this is one of the largest genera of neotropical orchids, with about 800 species.

KEY TO SPECIES

1. Flowers entirely green or yellowish green; lip entire, less than 1 cm long:	E. rigidum
1. Flowers with a white lip and column; lip deeply 3-lobed, more than 2 cm long:	E. nocturnum

Epidendrum rigidum Jacq., *Enum. Syst. Pl. Carib.* 29 (1760).

Epiphytic, with creeping, branched, compressed rhizome that give rise to scattered erect stems, these covered by the leaf-sheaths and mostly 10–20 cm long. Leaves distichous, the blades articulate to the tubular sheaths, usually 5–9 per stem, the lower ones lacking blades, the upper with coriaceous, oblong or elliptic-oblong blades mostly 2–7 cm long, obliquely notched at the apex. Inflorescence terminal, erect, distichous raceme of 2–7 flowers; floral bracts greenish, folded and enclosing the ovary, 8–15 mm long. Flowers green to yellowish-green, sessile, resupinate; sepals ovate, 4.5–7 mm long; petals linear to linear-oblanceolate, obtuse, denticulate on the margins, 4–6 mm long; lip adnate to the column, the free apex broadly rounded, entire to crenulate-denticulate, 2.5–4.5 mm long, 2.1–3.4 mm wide. Column straight, 2–3 mm long, erose-dentate at the apex. Capsule ellipsoid, beaked, mostly 1.1–2 cm long.

GRAND CAYMAN: *Dressler* 2906 (IJ), collected in May 1964.

— Widely distributed in tropical America from Florida south to Brazil, Peru and Bolivia. The sole Cayman record came from rocky woodland near George Town.

Epidendrum nocturnum Jacq., *Enum. Syst. Pl. Carib.* 29 (1760).

Epiphytic or epilithic herbaceous plants, 30–90 cm tall. Stems erect, slender, cane-like, terete, 10–80 cm long. Leaves 5–9, generally distributed along the upper half of the stem; blades coriaceous, linear-elliptical, elliptic-lanceolate to lanceolate, sometimes mucronate, dorsally keeled, 6–18 cm long. Inflorescence apical, erect; peduncle short, covered by a tubular, ancipitose, subfoliaceous, acute bract; racemes producing several flowers in succession. Flowers resupinate; sepals and petals pale green, olive green, dull yellow-orange to pinkish brown; sepals revolute, narrowly lanceolate, acute to shortly acuminate, 31–70 mm long, median sepal 7-nerved, lateral sepals 9-nerved; petals 3-nerved, linear-elliptic, acute, 33–60 mm long; lip white, united to the column, trilobed, 21–42 mm long, bicallose; lateral lobes spreading, lanceolate to lanceolate-triangular, acute, 19–33 mm long, midlobe narrowly ensiform, acute, 24–36 mm long. Column white, arcuate, conical in the basal half, abruptly dilated above and laterally flattened, 16–19 mm long, clinandrium short, erose-dentate. Anther subspherical, 4-locular, pollinia 4, subequal, laterally compressed. Capsules on pedicels 10–36 mm long, ellipsoid, 28–50 mm long, held approximately midway between the base of the pedicel and the tip of the beak, beak 6–17 mm long, overall length 45–80 mm long.

GRAND CAYMAN: Unlocalised material confirmed by Phillip Cribb at Kew.

— Widespread from Florida, Bahamas, throughout the West Indies, Mexico south to tropical and subtropical parts of S. America.

Encyclia Hook.

Epiphytes or epiliths most of which have pyriform pseudobulbs. Leaves articulate, 1–several from the apex of the pseudobulb, conduplicate. Inflorescence a terminal raceme or panicle. Flowers mostly resupinate, often showy; sepals and petals free, subsimilar, spreading; lip simple or trilobed, attached to the column near its base or rarely free. Column semiterete. Anther terminal, operculate, clinandrium lobed; pollinia 4, hard, waxy. Stigma entire, rostellum transverse, viscidium absent. Capsules ellipsoidal or fusiform.

A neotropical genus of about 140 species. The generic name refers to the labellum, which encloses the column.

Most students of neotropical Orchidaceae have accepted the distinctions between *Encyclia* and *Epidendrum* simply on morphological grounds, but there is a trend to recognize additional segregates of *Encyclia*. Recent molecular data suggest that the concept of the genus was too broad and at least paraphyletic if not polyphyletic. This is why *Prosthechea* has been segregated from *Encyclia* in this treatment.

KEY TO SPECIES

1. Sepals and petals 25–35 mm long:	**E. phoenicia**
1. Sepals and petals c. 10 mm long:	
2. Mid lobe of lip elliptic to obovate; column wingless;	**E. fucata (in addendum)**
2. Mid lobe of lip broadly ovate; column winged:	**E. kingsii**

Encyclia kingsii (C. D. Adams) Nir, *Lindleyana* 9: 147 (1994). PLATE 8.

Epidendrum kingsii C. D. Adams (1971).

Epiphytes with short rhizomes and ovoid-conicle pseudobulbs 5–6 cm long, 1.2–1.5 cm thick. Leaves solitary from the apex of the pseudobulb, conduplicate at the base, rigid, coriaceous, linear to oblong-oblanceolate, acute-obtuse to rounded, 27–34 cm long. Inflorescence from the apex of pseudobulbs, up to 80 cm long including the scape, diffuse, suberect panicle with branches up to 12 cm long, 24–60-flowered. Flowers brownish-yellow, resupinate; sepals elliptic-oblanceolate, obtuse, 10 mm long; petals oblanceolate, rounded 10 mm long; lip trilobed, 9–10 mm long, lateral lobes falcate, oblong, truncate, 6 mm long, midlobe broadly ovate, emarginate, apiculate, 7.5 mm long, 6 mm wide, narrowing to ca. 1.2 mm wide at isthmus below the lateral lobes. Column winged at apex. Ovary rugulose; capsule with pedicel up to 11 mm long; capsules smooth to minutely verrucose, ellipsoid, 20–22 mm long; pedicels slender, 10–12 mm long.

LITTLE CAYMAN: *Kings* LC117A (type, BM). Reported to occur also on Cayman Brac and Grand Cayman, though no specimens have been collected.

— Endemic; perhaps related to *Encyclia gravida* of the Greater Antilles, Mexico and C. America based on the paniculate inflorescence and rugulose ovaries.

Encyclia phoenicia (Lind.) Neum., *Rev. Hort.* ser. 2, 4: 137 (1845–1846). PLATE 8.

Epidendrum phoeniceum Lind. (1841).

Epiphytic, caespitose, herbs with pyriform to cylindrical-ovoid pseudobulbs 1.5–2.5 mm diam. Leaves 1–3 from the apex of pseudobulb, coriaceous, rigid when small, conduplicate at base, linear to elliptic-lanceolate, acute to obtuse, 3.5–38 cm long. Inflorescence erect on a smooth peduncle 11–93 cm long, racemose or a sparsely branched panicle, smooth to verrucose, 2–55 cm long, 2–25(–38) flowered; floral bracts broadly ovate, 1.5–6 mm long. Flowers showy, odorless to strongly fragrant of vanilla, resupinate; sepals and petals dark red to purplish brown, spreading, margins undulate to flat; dorsal sepal oblong-elliptic to elliptic-spatulate, acute to obtuse, 18–28 mm long, lateral sepals elliptic-oblanceolate to oblanceolate, acute, 18.5–28 mm long; petals spatulate to oblanceolate, acute to rounded, 17–28 mm long; lip white to lilac with purple lines, free to basally adnate to column, trilobed, 18.5–30 mm long, lateral lobes erect, flanking the column, narrowly oblong to ovate, subfalcate, acute to rounded, apical margins inrolled making the lobes appear lanceolate, 10.5–17 mm long, isthmus 1.5–4.5 mm long, 1.5–3.5 mm wide, midlobe truncate to cordate, suborbicular, flabellate-crisped, deeply emarginate at apex, 10–19 mm long, 8–23 mm wide, callus canaliculate, oblanceolate, 5–13 mm long, lateral ridges thickest at isthmus, terminating with 2 free palps at the tip of the column and up to 3 mm beyond the base of the midlobe. Column white with purple spots at base, semiterete, ventrally canaliculate, clavate, stout, straight, 9.5–13 mm long, 2.5–5 mm diam. near base, auricles subquadrate; clinandrium apex rounded, exceeded by bifid lateral teeth. Anther prominently bilobed. Capsules green to purple green with three verrucose ribs, ellipsoid to obovoid, 2.5–4.5 cm long, on a curved pedicel 9–10 mm long.

CAYMAN BRAC: *Mrs. Rena Reid*, photographs; also a specimen without specific origin other than from the Cayman Islands: *Dr. John E. Davies s.n.* (K). The latter specimen was determined as *Encyclia phoenicia* by Phillip Cribb at Kew. It is not clear from the literature whether these specimens were wild collected on the island. LITTLE CAYMAN: *Proctor 47841* (IJ).

— Cuba, a variable and widespread species. The description above is based on Cuban material. In the first edition of this flora, this plant was identified as '*Epidendrum* aff. *plicatum* Lindl.', a species also found in Cuba and very similar in size to *Encyclia phoenicia*.

Prosthechea Knowles & Westc.

Epiphytic or epilithic plants with pseudobulbous, fusiform, often flattened stems. Leaves 1–5, articulate, conduplicate, thin. Inflorescence scapose or sessile racemes, often subtended by a prominent spathe. Flowers commonly non-resupinate; lip adnate to basal half of the column, callus usually a thickened pad. Column usually gibbous, wingless, mid-tooth usually large, erect at apex of column, often covered by a fleshy knob-like, obtuse or truncate appendage, ligulate (connected to the anther cap by a thin flap of tissue), deltoid, subquadrate, or subflabellate, and sometimes fimbriate, lateral teeth stout, separated from mid-tooth by deep narrow sinuses. Capsules, three-winged or sharply three-angled, the suture covered by a strap of tissue that lifts upon dehiscence.

A neotropical genus of about 90 species variously treated by authors as *Anacheilium*, *Encyclia*, *Epidendrum*, *Epithecia* or *Hormidium*. Molecular and anatomical evidence support the morphological data that suggest that the genus should be recognized as a monophyletic taxon.

KEY TO SPECIES

1. Pseudobulbs flat, round to ovate; flowers resupinate; sepals 12–14 mm long, yellow with brown markings, lip white, 6–7 mm long: *P. boothiana*
1. Pseudobulbs laterally compressed, fusiform, flowers non-resupinate; sepals 26–41 mm long, pale green; lip dark purple with green lines, 10–19 mm long: *P. cochleata*

Prosthechea boothiana (Lind.) W. E. Higgins, *Phytologia* 82: 376 (1997). PLATE 9.

> *Epidendrum boothianum* Lindl. (1838).
> *Encyclia boothiana* (Lind.) Dressler (1961).

Caespitose epiphytes with flat, round to ovate pseudobulbs, 17–40 mm long. Leaves 2–3 from apex of pseudobulb, coriaceous, oblong to elliptic, or oblanceolate, obtuse to acute, 3.8–12 cm long. Inflorescence an erect, scapose raceme, laxly 1–11-flowered, to 13.5 cm

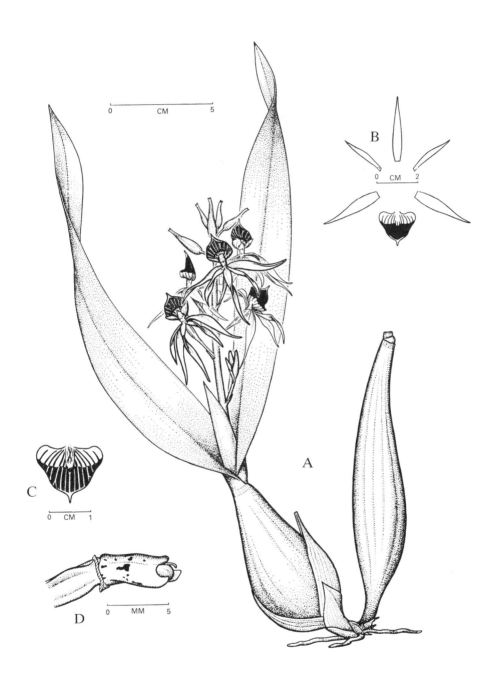

FIG. 73B **Prosthechea cochleata.** A, habit. B, perianth segments. C, lip. D, apex of ovary and column. (H.U.)

long; floral bracts triangular-ovate, 1.2–1.5 mm long. Flowers resupinate; sepals and petals yellow to yellow-green with numerous brown spots or bars, spreading; dorsal sepal oblong-elliptic, acute, 13 mm long, lateral sepals oblanceolate, cuspidate, apiculate, 12–13.5 mm long; petals linear-oblanceolate to spatulate, obtuse, 11–12.5 mm long; lip white to creamy-yellow, sometimes with a green apex, trilobed, adnate to the base of the column for 3 mm, overall length 6–7 mm, 6 mm wide across lateral lobes, lateral lobes triangular, rounded, midlobe subquadrate-triangular, rounded. Column white with a green base, straight, clavate, 6–7 mm long. Capsules on slender pedicels 5–7 mm long, 3-winged, broadly elliptic, 20–31 mm long.

GRAND CAYMAN: *Brunt* 1872 (BM); *Kings* GC 281.

— Florida, Bahamas, Cuba and Haiti.

Prosthechea cochleata (L.) W. E. Higgins, *Phytologia* 82: 377 (1997). **FIG. 73B**. PLATE 9.

Epidendrum cochleatum L. (1763).
Encyclia cochleata (L.) Dressler (1961).

Epiphytic, epilithic or terrestrial, caespitose herbs with pseudobulbous stems 2–20 cm long. Leaves 1–3, produced from the apex of the pseudobulb, narrowly oblong, acute to acuminate, 15–38 cm long. Inflorescence terminal, erect, scapose raceme 13–62 cm long with 3–29-flowers, produced sequentially, few open at one time. Flowers non-resupinate; sepals and petals similar, pale green, linear, reflexed, often twisted, 26–41 mm long; lip dark purple-maroon with green veins (rarely cream), erect, simple, adnate to the basal half of the column for 2–3 mm, free portion cochleate, 10–19 mm long. Column erect, stout, 6–8 mm long. Anthers 1–3 (if more than 1 then lateral anthers aberrant and often poorly developed, and flowers autogamous), clinandrium bluntly trilobed; pollinia 4, hard, waxy. Capsules pendent, pedicel 1.5–2 cm long, capsule 3-winged, triangular in cross-section, 20–50 mm long.

CAYMAN BRAC: *Mrs. Rena Reid s.n.* (IJ). The specimen consists of but a single flower which leaves no doubt as to its identity.

— Florida, Bahamas, Greater and Lesser Antilles, Mexico south to Colombia, Venezuela, Guyana and French Guiana.

Myrmecophila Rolfe

Plants robust, usually caespitose, epiphytic, terrestrial or lithophytic herbs with tough, usually ridged, usually conical, hollow pseudobulbs. Leaves 2–4, leathery or fleshy, stiff, sessile, oblong, elliptic to elliptic-ovate. Inflorescences terminal, erect; peduncle elongate, bracteate; racemes, or rarely panicles. Flowers generally large, showy, resupinate; sepals free, often longer and wider than petals, otherwise similar, margins often undulate; lip free or slightly connate with the base of the column, trilobed, rarely simple, disc with several undulate lamellae or raised veins. Column often parallel with the labellum, stigma entire, pollinia hard, ovoid or laterally compressed, 8 in 2 equal sets. Capsules pendent, large.

A genus of approximately 8 species of the Greater Antilles, Mexico, C. America and Venezuela. *Myrmecophila* is a segregate genus of *Schomburgkia*. Throughout its nomenclatural history, *Myrmecophila* has been bounced into and out of *Schomburgkia*. The latest trend in opinion among systematists is that *Myrmecophila* is a distinct taxon. In fact, recent phylogenetic studies using molecular sequence data indicate that *Myrmecophila* is quite distinct and sister to *Cattleya*, *Brassavola* and Brazilian *Laelia*, whereas *Schomburgkia* sensu stricto is in a different clade altogether and allied with the Mexican *Laelia* (van den Berg, C., Higgins, W. E., Dressler, R. L., Whitten, W. M., Soto Arenas, M. A., Culham, A. & Chase, M. W. (2000). A phylogenetic analysis of Laeliinae (Orchidaceae) based on sequence data from internal transcribed spacers (ITS) of nuclear ribosomal DNA. *Lindleyana* 15: 96–114.).

Myrmecophila thomsoniana (Rchb. f.) Rolfe, *Orchid Rev.* 25: 51 (1917). PLATE 9.

Schomburgkia thomsoniana Rchb. f. (1887).
Schomburgkia brysiana var. *thomsoniana* (Rchb. f.) H. G. Jones (1963).
Schomburgkia thomsoniana var. *albopurpurea* Strachan ex Fawc. (1894).
Schomburgkia thomsoniana var. *atropurpurea* [sic] Hook.f. (1902).
Schomburgkia brysiana var. *atropurpurea* [sic] (Hook. f.) H. G. Jones (1963).
Schomburgkia albopurpurea (Strachan ex Fawc.) Withner (1993).
Myrmecophila albopurpurea (Strachan ex Fawc.) Nir (2000).

Epiphytes with short, stout, rhizomes and conical to fat, fusiform, straight to curved-ascending, ribbed, hollow, pseudobulbs, 5.3–20 cm long. Leaves 3–5, sessile, entire, oblong, elliptic-lanceolate, ovate-elliptic to broadly oblong-elliptic, 3.8–16.5 cm long, 2.3–6 cm wide. Inflorescences peduncle erect, 4.2–10(20) cm long, bracts; raceme 2–5 cm long, or a loosely branched panicle, laxly flowered, 1–25 cm long, 2–20 flowers produced in succession. Flowers large and showy; sepals and petals white to creamy with slightly wavy margins; dorsal sepal 24–33 mm long, lateral sepals slightly oblique, oblong-oblanceolate, acute, 29–37 mm long; petals oblong, rounded, margins undulate, 30–40 mm long; lip dark purple to blackish violet at apex, trilobed, base rounded, overall length 27–31 mm, 25–39 mm wide across lateral lobes when spread, lateral lobes erect and flanking the column, triangular with a rounded apex, 20–22 mm long, midlobe 13–14 mm long, 8–13 mm wide across base, 9–10 mm wide across flabellate-crispate apex, retuse, reflexed, disc lamellae 3 reaching near the lip apex, more prominent and undulate on midlobe. Column slightly arched, semiterete, clavate, 14–15 mm long. Capsule ellipsoid, beaked, 2.5–3.5 cm long on pedicel 22–25 mm long.

The nomenclatural history of *Myrmecophila thomsoniana* must be one of the most complex and confused of any species in the Antilles. Although most plants on Grand Cayman have paler sepals and petals than those on Cayman Brac, lip color is consistent and is actually the characteristic used by locals to distinguish the two varieties (Dressler personal communication, 2003). Most recent authors were under the assumption that the type had come from Little Cayman or Cayman Brac where populations are rather uniform with flowers that are yellow to yellow-orange with a reddish- or pinkish-purple lip. Thus, the varietal names *albopurpurea* or *atropurpurea* [sic], applied to the common variety on Grand Cayman, are superfluous because they represent the type, var. *thomsoniana* (Dresssler, R. L. & Carnevali, G. (2000). The wild banana orchid of the Cayman Islands. *Orchid Digest* 64: 81–83.)

Variety *thomsoniana* shows considerable variation on Grand Cayman; some forms are intermediate in flower color and size between var. *thomsoniana* and var. *minor*. White flowers are the norm, but pale yellow flowers occur just about everywhere and are more frequent on the eastern end of the island, the side nearest to Cayman Brac and Little Cayman (Dressler & Carnevali, 2000). Introgressive hybridization may be involved, although there are alternative explanations.

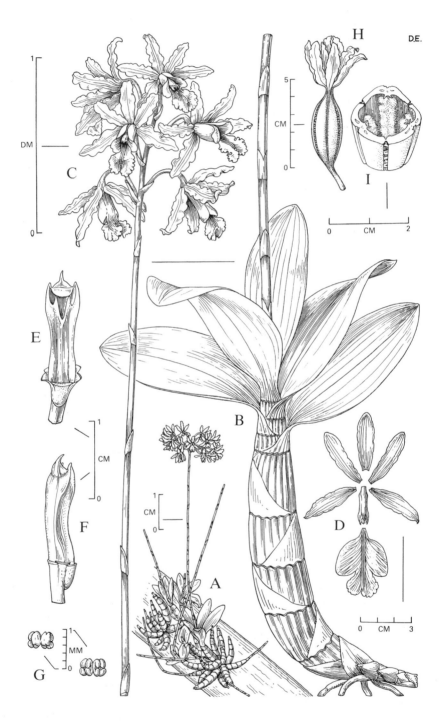

FIG. 74 **Myrmecophila thomsoniana** var. **minor.** A, habit. B. pseudobulb, leaves, and base of peduncle. C, inflorescence. D, flower parts. E, column, ventral view. F, column, lateral view. G, pollinarium, ventral view showing viscidium and dorsal view showing pollinia. H, capsule with withered flower attached. I, fruit, cross-sectioned. (D.E.)

Myrmecophila thomsoniana has two varieties that may be distinguished according to the following key:

1. Lip dark purple to blackish-violet; petals 30–36 mm long, white to
   yellow-cream adaxially   **M. thomsoniana** var. **thomsoniana**
1. Lip reddish- or pinkish-purple; petals 23–28 mm long, yellow to
   yellow-orange adaxially: **M. thomsoniana** var. **minor**

Myrmecophila thomsoniana var. *thomsoniana*

GRAND CAYMAN: *Fawcett s.n.* (K); *Brunt* 1663, 1692 (BM); *Kings* GC 15, GC 200, GC 292 (BM).

— Endemic.

Myrmecophila thomsoniana var. *minor* (Strachan ex Fawc.) Dressler, *Orquideología* 22: 230 (2003). **FIG. 74**. PLATE 10.

Schomburgkia thomsoniana var. *minor* Strachan ex Fawc. (1894).
Schomburgkia brysiana var. *minor* (Hook. f.) H. G. Jones (1966).

LITTLE CAYMAN: *Kings* LC 4, LC7 (BM). CAYMAN BRAC: *Kings* CB18a, CB18b (BM); *Proctor* 15319 (BM); *Fawcett s.n.* (K, NY).

— Cuba(?). The Cuban records are reported by A. D. Hawkes (1951) (Studies in Antillean botany — 3. A preliminary check-list of Cuban orchids. *Brittonia* 7: 173–183.). He questioned one locality (Matanzas) but not another (Sancti Spíritus). We have not yet seen any specimens from outside the Cayman Islands.

Brassavola R. Br.

Epiphytes or lithophytes with short, stout rhizomes and slender to pseudobulbous stems. Leaves mostly solitary, articulate, fleshy, coriaceous, conduplicate, subterete to flattened. Inflorescences terminal, pedunculate, racemose, 1–several-flowered. Flowers large, white and/or green, showy, resupinate with a strong nocturnal fragrance; sepals and petals subsimilar, spreading, linear to narrowly lanceolate; lip clawed, tubular around column, apically spreading. Column erect, short, winged. Anther terminal, operculate, pollinia 8 or 12 in unequal groups of 4, hard, waxy. Stigma entire. Capsules pedicellate and long-beaked.

A neotropical genus of about 15 species. The localities of all West Indian species are confused. *Brassavola nodosa* (L.) Lindl. has been reported from Puerto Rico and Jamaica, but it has not been seen on those islands outside of cultivation since the initial reports. The types of *B. cordata* Lindl., *B. subulifolia* Lindl. and *B. cucullata* (L.) R. Br. are all attributed to localities where they are not known today. Could these errors have been intentional and designed to mislead collectors and competitors?

Brassavola nodosa (L.) Lindl., *Gen. Sp. Orch. Pl.* 114 (1831).

Epiphytes or lithophytes, somewhat caespitose. Stems erect, cylindrical, slender to 15 cm tall, unifoliate. Leaves erect, subapical, very fleshy and coriaceous, linear to linear-oblong, 6–32 cm long, 3–25 mm wide. Racemes 1–several-flowered, 6–20 cm long. Flowers showy, nodding, 6–8 cm across, nocturnally fragrant; pedicellate ovary 3.5–5 cm long; sepals and petals similar, pale green to yellowish, spreading, linear, acuminate, 4–10 cm long; lip white, sometimes with a few purple spots, tubular at its base for ca. 3 cm, enfolding column, broadening and flattening abruptly to become cordate to nearly orbicular, apex acuminate, 3.5–9 cm long, 1.1–5 cm wide. Column short, stout, 7–8 mm long. Capsule ellipsoidal, 3–4 cm long; pedicel ca. 10 mm long; beak ca. 10 mm.

GRAND CAYMAN: Rocky woodland N of Cottage Point: found and cultivated by E. McLaughlin. No voucher specimen has been collected (Proctor, G. R. (1996). Additions and corrections to "Flora of the Cayman Islands". *Kew Bull.* 51: 483–507.).

— Jamaica(?), Puerto Rico(?), Mexico through C. America to Colombia and Venezuela.

Dendrophylax Rchb. f.

Monopodial epiphytes reduced to a mass of velamentous, terete to broad and flat roots. Stems very short, inconspicuous. Leaves absent or minute and scale-like. Inflorescences axillary, with filiform scapes, often sparsely branched, few- to several-flowered. Flowers minute, resupinate or not; sepals and petals similar, free; lip simple to deeply lobed, with a spur, short saccate to long and slender. Column short. Anther terminal, incumbent, operculate, pollinia 2, hard, waxy, each attached to a separate stipe or both attached to a common stipe, viscidium elongate. Ovary subsessile. Capsules dehiscence by 3 broad valves alternating with 3 linear ribs, all separating from the apex and reflexing.

A Caribbean genus of 14 species. The circumscription of the genus has been in dispute for more than a century. Different species of leafless orchids have bounced from one genus to the next: *Campylocentrum*, *Dendrophylax*, *Harrisella*, *Polyrrhiza* and *Polyradicion*. Recent molecular evidence has added non-morphological evidence to the data pool from which we can make taxonomic judgements. For example, in this treatment, the monotypic *Harrisella* is included in *Dendrophylax*. Although there are a number of unique features (autapomorphies) of the genus, the molecular phylogeny indicates that *Harrisella* is so closely related to *Dendrophylax* that segregation of the genus cannot be justified without creating other monotypic genera and thereby obscuring relationships (Carlsward, B. S., Whitten, W. M. & Williams, N. H. (2003). Molecular phylogenetics of neotropical leafless Angraecinae (Orchidaceae): re-evaluation of generic concepts. *Intl. J. Plant Sci.* 164: 43–51.).

KEY TO SPECIES

1. Inflorescences 1-flowered; flowers showy; lip white, 24–42 mm long with a pendent, curved spur 11–15 cm long:	D. fawcettii
1. Inflorescences racemose or paniculate; flowers inconspicuous; lip pale yellow, 2.7–4 mm long with a saccate spur 1 mm long:	D. porrectus

Dendrophylax fawcettii Rolfe, *Gard. Chron.*, ser. 3, 4: 533 (1888). **FIG. 75**. PLATE 10.

Polyrrhiza fawcetii (Rolfe) Cogn. (1910).

Epiphytic, leafless herbs with numerous, grey-green roots 1.5–3 mm diam. Stems to 13 mm long, partially concealed by roots, occasionally producing very long stolons. Leaves absent. Inflorescences 1-flowered, erect or ascendent; peduncles stiff, 5–28 cm long, floral bracts tubular, truncate, 4 mm long. Flowers resupinate, showy, fragrant; sepals and petals cream; dorsal sepal lanceolate, acuminate, 20–28 mm long, lateral sepals subfalcate, lanceolate,

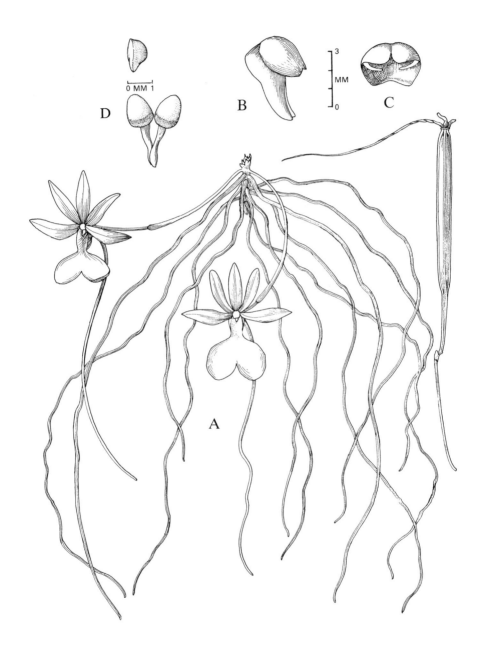

FIG. 75 **Dendrophylax fawcettii.** A, habit. B, column with terminal anther, side view. C, column, front view with anther removed. D, pollinarium. Unlabelled, fruit. (G.)

acuminate, 20–26 mm long; petals reflexed, narrowly elliptic to oblanceolate, acuminate to obtuse, 20–25 mm long; lip white, bilobed, obcordate, rounded, emarginate, 24–42 mm long, 21–30 mm wide, base expanded below column and extended into a pendent, slender, curved spur, 11–15 cm long. Column short, stout, 3–3.5 mm long, wings triangular, flanking the stigma, 2 mm long. Ovary pedicellate, 3–3.7 cm long; capsules cylindric, 6-ribbed, 8–8.5 cm long, 6–6.5 mm diam., pedicel 25–33 mm long.

GRAND CAYMAN: *Dressler* 2904, *Fawcett* (type, K); *Kings* GC 19, GC 19a.

— Endemic: grows on trees and rocks in sheltered situations. This species is rare and should be protected, otherwise it may become extinct in the wild.

Dendrophylax fawcettii and *D. funalis* of Jamaica have the unusual habit of vegetatively spreading by producing elongate stolons from which new main shoot axes develop.

Dendrophylax porrectus (Rchb. f.) Carlsward & Whitten, *Intl. J. Plant Sci.* 164: 51 (2003).

Harrisella porrecta (Rchb. f.) Fawc. & Rendl. (1909).

Plants twig epiphytes with silvery gray-green roots, 0.25–1 mm diam. Stems erect with a few reddish-brown, scarious scales, covered by roots, to 15 mm long. Inflorescences few to many, filiform, racemes or sparsely branched panicles arising from the stem through the mass of roots, several-flowered, 2–5 cm long. Flowers small, pale yellow, non-resupinate; dorsal sepal elliptic-obovate, rounded to obtuse, 1.7–2.5 mm long, lateral sepals obliquely elliptic-ovate, rounded to obtuse, 2–2.5 mm long; petals oblong-elliptic to ovate, obtuse, 1.7–2.3 mm long; lip broadly ovate, rounded to obtuse, concave, margins involute, base with a short, apically expanded, scrotiform spur, ca. 1 mm wide, disc with an erect tubercle, total length of lip 2.7–4 mm, 2 mm wide. Column stout, 0.3–0.4 mm long, 0.6 mm diam. Ovary pedicellate ca. 2 mm long. Capsules reddish brown when mature, ovoid, sparsely verruculose, 4.5–6 mm long, the valves reflexed at dehiscence, pedicels 1.5–2 mm long.

GRAND CAYMAN: *Proctor* 48246.

— Florida, Greater Antilles, Mexico and El Salvador. This species was discovered by Frank Roulstone III in a low, rocky woodland.

Oeceoclades Lindl.

Caespitose or rhizomatous herbs, terrestrial or rarely epiphytes with aggregated or approximate pseudobulbs, ovoid to cylindrical or globose to pear-shaped, of one elongate internode. Leaves 1–3, articulate, commonly petiolate, conduplicate, coriaceous, linear, narrowly ovate, narrowly elliptic, broadly ovate or elliptical, acute, obtuse or acuminate. Inflorescence lateral, arising from the base of the pseudobulb, erect, racemose or paniculate, few-flowered. Flowers membranous, resupinate, often showy; sepals and petals membranous, variously spreading; sepals spatulate, narrowly ovate, narrowly obovate or obovate, obtuse, acute or subacuminate; petals elliptic to subcircular, acute or obtuse; lip sessile, basally produced into a conspicuous spur, blade 3-lobed, the midlobe 2-lobed or shallowly emarginate, disk with a pair of variously shaped calli or with 3 axial, variously thickened, sparsely papillose or hirsute ridges. Column erect, short, slightly arched, semiterete. Anther terminal, operculate, incumbent, 2-locular; pollinia 2, yellow, cartilaginous, subtriangular. Fruits capsular.

A genus of about 30 species from tropical America, tropical Africa, Madagascar, the Mascarene Islands and the Seychelles.

Oeceoclades maculata (Lindl.) Lindl., *Gen. Sp. Orch. Pl.* 237 (1833). Plate 10.

Plants terrestrial, caespitose herbs with fleshy, white roots and slightly compressed, unifoliate, ovoid pseudobulbs to 4.0 cm tall. Leaves subpetiolate, blade green with darker mottling, conduplicate, coriaceous, subsucculent, oblong to elliptical, 8–25 cm long, 1.5–5.0 cm wide. Inflorescence lateral, erect, sparsely bracteate, peduncle to 30 cm, racemose or rarely paniculate, 5–15-flowered, produced sucessively. Flowers with straw-colored sepals and petals; sepals 8–14 mm long, 2–3 mm wide, dorsal sepal lanceolate to linear-elliptic, acute, lateral sepals linear-elliptic, falcate, acute; petals connivent with the dorsal sepal, elliptic-oblanceolate, 8–14 mm long, 3–4 mm wide; lip white with two lateral pink blotches extending to the base of the midlobe, trilobed, concave, basally produced into a conspicuous, recurved spur, continuous with a short column foot, lateral lobes erect, the median lobe subquadrate, emarginate, disc with two subtriangular lamellae at the base, the spur ca. 4 mm long, slightly bilobed at the apex. Column semiterete, 4–5 mm long. Pollinia yellow, attached to a short stipe with a broad viscidium. Capsules 2–3 cm long.

GRAND CAYMAN: *Davies s.n.* (pickled flower identified by P. Cribb at Kew). Seen near the center of the island and at the Botanic Park.

— Florida, Bahamas, Greater and Lesser Antilles, Mexico south through C. America and tropical S. America, also widespread in tropical Africa. Plants show considerable ecological amplitude growing in wet to dry forests. In the past forty years, *Oeceoclades maculata* has been rapidly spreading throughout the neotropics.

Cyrtopodium R. Br.

Mostly terrestrial or lithophytic, epiphytic in the flora area, caespitose herbs with stout rhizomes and large ovoid, fusiform to cylindrical pseudobulbs of several elongate internodes. Leaves several, distichous, coriaceous, acuminate, plicate, articulate, deciduous at the end of the growing season. Inflorescences lateral, arising from the base of the pseudobulbs, erect, racemose to paniculate, many-flowered, usually much longer than leaves. Flowers resupinate, often showy; floral bracts linear-ovate to elliptic, often showy, the margins undulate; sepals and petals membranous, spreading, often conspicuously undulate, lateral sepals laterally adnate to the base of the column foot, sometimes wider than the dorsal one and/or slightly oblique; petals usually wider than the sepals; lip subsessile to clawed, adnate to the apex of the column foot, conspicuously 3-lobed, the lateral lobes erect to spreading, sometimes spatulate, the margins entire, undulate and/or verrucose, disk with a keeled, tuberculate, cristate or verrucose callus. Column short to somewhat elongate, slightly arched, clavate to subclavate, semiterete, with a conspicuous foot. Anther terminal, incumbent, operculate, imperfectly 2-locular, pollinia 2, yellow, cartilaginous, sulcate, subtriangular in outline, semiterete in cross-section, attached to a short, trullate viscidium. Capsule massive.

A neotropical genus of about 30 species.

Cyrtopodium punctatum (L.) Lindl., *Gen. Sp. Orch. Pl.* 188 (1833).

Epiphytic plants 50–150 cm tall with fusiform pseudobulbs 30–50 cm long. Leaves numerous, linear-lanceolate, linear-elliptic or oblanceolate, 10–80 cm long, 2–5 cm wide. Inflorescences erect, paniculate, 10–90 cm long, many-flowered; floral bracts yellow spotted brown, lanceolate, acute to acuminate, margins undulate, 2–5 cm long, 0.5–1.5 cm wide; bracts at the base of the inflorescence branches similar to the floral bracts but much more conspicuous, 5–7 cm long. Flowers bright yellow or yellowish-green with maroon and green markings, showy, 4–5 cm across; sepals oblong-lanceolate to elliptic-lanceolate, acute, margins strongly undulate, 11–20 mm long; petals more or less elliptic, truncate to rounded, margins slightly undulate, 10–21 mm long, 7–12 mm wide; lip trilobed, narrowly clawed, lateral lobes spotted or solid dark orange or maroon, erect, curving upward around the column, obovate to rotund, 5–11 mm wide, midlobe transversely elliptic, 4–5 mm long, 6–7 mm wide, truncate, margin conspicuously verruculose-tuberculate, disc with a conspicuous, cristate to verruculose callus. Column green, clavate, 5–7 mm long, with a prominent foot, pollinia 2, yellow, juxtaposed, sulcate. Capsules pendent, ellipsoidal-obovoid, 8–10 cm long.

GRAND CAYMAN: *Ebanks* (photo).

— Florida, Cuba, Hispaniola, Puerto Rico, Venezuela and Colombia.

Ionopsis Kunth

Small, epiphytic herbs with rhizomes bearing pseudobulbous stems, often concealed by leaf-sheaths. Leaves articulate, apical and/or lateral, conduplicate, fleshy, coriaceous, sometimes semiterete. Inflorescence 1–3, lateral, long-pedunculate, racemose or paniculate, several–many-flowered. Flowers pedicellate, resupinate; dorsal sepal free, lateral sepals united at the base forming a short spur-like chin with the column foot; lip clawed, attached to the column foot. Column slender, semiterete, foot short. Anther terminal, operculate, incumbent, pollinia 2, hard, waxy, on a slender stipe attached to a viscidium. Stigma entire. Capsule ovoid, ellipsoid to cylindrical.

A neotropical genus of about 3 species.

Ionopsis utricularioides (Sw.) Lindl., *Coll. Bot.* t. 39A (1826).

Small twig epiphytes with cylindrical pseudobulbs, 4–15 mm long, generally hidden by leaf-sheaths. Leaves 1–4, apical and lateral, thick, fleshy, narrowly oblong, 4–25 cm long, 3–18 mm wide. Inflorescences to 80 cm long; peduncles to 55 cm long, bracts few, distant; racemes in small plants, panicles in larger plants, branches spreading, few to many, 2–70-flowered; floral bracts lanceolate, acuminate, 1–2 mm long. Flowers pale pink to rose or violet, sepals and petals connivent and parallel with the column; sepals oblong to ovate-lanceolate, acute, 3–5 mm long, 1–1.5 mm wide, lateral sepals basally united, gibbous, forming a short spur-like mentum with the column foot; petals broadly oblong, obtuse, 4–6 mm long, 1.5–2 mm wide; lip simple, claw with paired oblong callosities and pubescent above, lamina pubescent at junction with claw, obovate, deeply notched, 7–12 mm long, 8–15 mm wide. Column, stout, broadest at stigma, 1.5–2 mm long. Pollinia yellow, globose. Ovary filiform, 5–11 mm long. Capsules cylindrical, 1.5–2 cm long, beak 3–6 mm long, pedicel slender, 6–8 mm long.

GRAND CAYMAN: *Kings* GC 239.

— Florida, Greater and Lesser Antilles, Mexico through C. America and tropical S. America, and the Galapagos Islands.

Tolumnia Raf.

Mostly small twig epiphytes, rarely lithophytic or terrestrial, with short or long and wiry rhizomes. Stems small, usually completely hidden by leaf bases, sometimes pseudobulbous. Leaves articulate, fan-like arrangement, fleshy coriaceous, conduplicate, laterally compressed or subterete, psygmoid (incorrectly referred to as "equitant"). Inflorescences lateral from between leaf bases, pedunculate, racemose or paniculate, few–many-flowered. Flowers generally colorful, showy, resupinate; dorsal sepal free, erect, lateral sepals free or variously connate; petals free, spreading, usually larger than the dorsal sepal; lip attached to the base of the column, nectary absent, trilobed, midlobe usually emarginate, disc callus absent to tuberculate, glabrous to pubescent. Column erect, semiterete, footless, subapical lateral wings flanking the stigma footless, tabula infrastigmatica present. Anther terminal, operculate, incumbent, pollinia 2, hard, waxy, attached to an elongate stipe with a small viscidium. Capsules ellipsoidal.

A genus of 25 species, some not well-defined. Most species occur in the Greater Antilles, but some are known from the United States (southern Florida), the Bahamas and the Lesser Antilles.

Popularly known as the "variegata" or "equitant" group of oncidiums, *Tolumnia* is vegetatively distinct from *Oncidium* but the flowers are very similar, probably owing to evolutionary convergence. Vegetative morphology, chromosome numbers and molecular sequence data are all congruent and support the segregation of *Tolumnia* from *Oncidium*.

KEY TO SPECIES

1. Leaves subterete; flowers yellow; lip midlobe broadly ovate, dentate-lacinate:	T. calochila
1. Leaves triquetrate, flat, folded (psygmoid); flowers multi-colored, mostly pink or white and marked with brown; lip midlobe transversely reniform:	T. variegata

Tolumnia calochila (Cogn.) Braem, *Orchidee* 37: 58 (1986).

Oncidium calochilum Cogn. (1910).

Epiphytes with very short rhizomes concealed by bracts and roots. Leaves 2–3, aciculate, straight to falcate, laterally compressed, subterete, 2.5–10 cm long, 1–1.5 mm wide. Inflorescences lateral; peduncle erect to spreading, slender, 2.5–12.5 cm long; raceme slender, 1–3 distant flowers. Flowers yellow, sometimes with greenish midveins in the sepals, fragrant; dorsal sepal erect, concave, lanceolate, acuminate, 8–10 mm long, 1.7–2 mm wide, lateral sepals connate, concave, elliptic-lanceolate, acuminate, synsepal bifid, 9–10.5 mm long, 2.5 mm wide; petals spreading, apices recurved, ovate to triangular-lanceolate, acuminate, 9–12 mm long, 3.5 mm wide; lip clawed, trilobed, lateral lobes inconspicuous, ovate to lanceolate, 2–2.2 mm long, separated from the midlobe by an isthmus, midlobe broadly ovate, rounded, dentate-lacinate, 10–12 mm long, 11–12 mm wide, disc callus with raised ridges becoming 5-pronged, arching, 3.5–4 mm long above the base of the midlobe. Column straight, somewhat perpendicular to the lip, 6–7 mm long, broadly winged below with smaller wings above. Pollinia yellow, obovoid, 0.8–0.9 mm long. Ovary pedicellate, slender with upper half nearly twice as thick as the lower half, 7–10 mm long. Capsules not seen.

GRAND CAYMAN: *Brunt* 2169; *Proctor* 28000.

— Cuba and Hispaniola. Generally rare. On Grand Cayman, this species occurs in dense woodlands near the center of the island.

Ecology. Dry forests from coastal and montane regions. Epiphytic on thorny trees and shrubs. Elevation sea level to 650 m. Locally common. Self-incompatible (Charanasri & Kamemoto, 1977). Flowering: Apr.–Jul. Fruiting: unknown.

Tolumnia variegata (Sw.) Braem, *Die Orchidee* 37: 59 (1986).

Oncidium variegatum (Sw.) Sw. (1800).

Oncidium caymanense Moir (1968), emend (1969).

Epiphytes with short to long, wiry rhizomes. Leaves 3–7, triquetrous, laterally compressed and concave, short, coriaceous, and rigid to long, falcate, succulent and lax, acute to acuminate, 1.5–17.2 cm long, 3–11 mm wide. Inflorescences 1–few, lateral from leaf axils, horizontal to erect, total length 2–65 cm; racemes or sparsely branched panicles 1–16-flowered. Flowers mostly white to rose-purple, 10–24 mm long; sepals and petals marked reddish brown; dorsal sepal arched to geniculate, concave, oblanceolate to spatulate, rounded to acute, 2–7 mm long, lateral sepals connate to near apex, apically bicarinate, bicuspidate, concave, slightly shorter, 2–6.5 mm long; petals spreading, pandurate to obovate, truncate, rounded to acute, apiculate, 4.5–10 mm long; lip trilobed, lateral lobes auriculate, spreading to ascending, entire, oblong to obovate, 2–7 mm long, isthmus 1.5–5 mm wide, margin entire to minutely fimbriate, midlobe reniform, broader than long, 2.5–14 mm long, 7–17.5 mm wide, disk marked reddish-brown, callus yellow with reddish-brown markings, glabrous to velvety, composed of 2 basal spreading or drooping horns and 3 erect apical horns connected by a median ridge. Column white, green or rose-purple, erect, glabrous or partially velvety, 2.5–4.5 mm long, tabula infrastigmatica yellow, wings spreading, dolabriform, rounded to ovate with a rounded base and a falcate apex, 2–3 mm long, 1–1.5 mm wide. Ovary pedicellate, slender, 8–26 mm long. Capsules green, pendent, ellipsoid to cylindrical, ca. 1.5 cm long.

GRAND CAYMAN: *Brunt* 2168; *Proctor* 27999; *Moir s.n.* (type AMES)

— Cuba, Hispaniola, Puerto Rico, United States Virgin Islands and British Virgin Islands. This species was common in a very limited area and intermixed with *Tolumnia calochila*. The Grand Cayman population may be endangered by real estate development and perhaps by orchid fanciers.

The Cayman Island plants have been described as a distinct species, *Oncidium caymanense*. However, Norris Williams (personal communication 2003) obtained material of *O. caymanense* for DNA sequencing and the results unequivocally showed that *O. caymanense* is no more different from *T. variegata* than other *T. variegata* populations are among themselves.

In the first edition of this flora, the author gave a history of the species describing how he had originally collected *Oncidium caymanense*, pressed specimens and then sent some live plants to Gauntlett in Jamaica. Gauntlett then sent the plants to the Bahamas, where they were passed to another grower before they ended up in the hands of Moir, a grower and amateur taxonomist in Hawaii. When Moir described *O. caymanense*, he failed to designate a type, but emended his protologue in the following year.

DICOTYLEDONES

The dicotyledonous families are arranged in the following sequence, in accordance with the taxonomic system of Cronquist:

MAGNOLIALES	Annonaceae, Canellaceae, Lauraceae
PIPERALES	Piperaceae
NYMPHAEALES	Nymphaeaceae
RANUNCULALES	Menispermaceae
PAPAVERALES	Papaveraceae
URTICALES	Ulmaceae, Moraceae, Urticaceae
MYRICALES	Myricaceae
CASUARINALES	Casuarinaceae
CARYOPHYLLALES	Phytolaccaceae, Nyctaginaceae, Cactaceae, Aizoaceae, Portulacaceae, Basellaceae, Chenopodiaceae, Amaranthaceae
BATALES	Bataceae
POLYGONALES	Polygonaceae
PLUMBAGINALES	Plumbaginaceae
THEALES	Clusiaceae
MALVALES	Tiliaceae, Sterculiaceae, Malvaceae
VIOLALES	Salicaceae, Turneraceae, Passifloraceae, Caricaceae, Cucurbitaceae
CAPPARALES	Capparaceae, Cruciferae, Moringaceae
EBENALES	Sapotaceae
PRIMULALES	Theophrastaceae, Myrsinaceae
ROSALES	Crassulaceae, Chrysobalanaceae, Leguminosae
MYRTALES	Lythraceae, Thymelaeaceae, Myrtaceae, Onagraceae, Combretaceae
CORNALES	Rhizophoraceae
SANTALALES	Olacaceae, Loranthaceae, Viscaceae
RAFFLESIALES	Apodanthaceae
CELASTRALES	Celastraceae
EUPHORBIALES	Buxaceae, Euphorbiaceae
RHAMNALES	Rhamnaceae, Vitaceae
SAPINDALES	Surianaceae, Sapindaceae, Burseraceae, Anacardiaceae, Simaroubaceae, Rutaceae, Meliaceae, Zygophyllaceae
GERANIALES	Oxalidaceae
LINALES	Erythroxylaceae
POLYGALALES	Malpighiaceae, Polygalaceae
UMBELLALES	Araliaceae, Umbelliferae
GENTIANALES	Loganiaceae, Gentianaceae, Apocynaceae, Asclepiadaceae
POLEMONIALES	Solanaceae, Convolvulaceae, Menyanthaceae, Hydrophyllaceae
LAMIALES	Boraginaceae, Verbenaceae, Avicenniaceae, Labiatae

SCROPHULARIALES	Oleaceae, Scrophulariaceae, Myoporaceae, Bignoniaceae, Acanthaceae
CAMPANULALES	Goodeniaceae
RUBIALES	Rubiaceae
ASTERALES	Compositae

KEY TO THE DICOTYLEDONOUS FAMILIES

1. Plants evidently parasitic:

2. Plants without stems or leaves, consisting of single parasitic flowers (or fruits) on stems of *Bauhinia*:
Apodanthaceae

2. Plants with obvious stems:
  3. Stems twining and leafless: **Lauraceae**
  3. Stems shrubby and bearing green leaves:
    4. Flowers bisexual, borne in racemes: **Loranthaceae**
    4. Flowers unisexual, borne in rows partly embedded in internodes of articulate spikes:
Viscaceae

1. Plants not, or not evidently, parasitic:

5. Trees with leaves reduced to minute scales forming whorls at joints of thin, green, ribbed twigs; fruits in small, hard, globose heads: **[Casuarinaceae]**
5. Not as above:
  6. Aquatic herbs with roundish, usually floating leaves:
    7. Leaves reticulate-ribbed beneath; flowers with numerous separate glabrous petals arranged spirally, the inner ones passing gradually into the many stamens: **Nymphaeaceae**
    7. Leaves smooth beneath; flowers with 5-lobed, fringed-hairy corolla and 5 distinct stamens:
Menyanthaceae
  6. Terrestrial plants, or if aquatic the leaves aerial:
    8. Perianth absent; flowers often small or minute:
      9. Flowers bisexual, in dense, slender but fleshy spikes; ovary 1-locular: **Piperaceae**
      9. Flowers unisexual; inflorescence various:
        10. Leaves gland-dotted:
          11. Leaves glandular on both sides: **Myricaceae**
          11. Leaves glandular beneath only: **Oleaceae**
        10. Leaves not gland-dotted; ovary 2–4-locular:
          12. Leaves opposite, fleshy, subterete; inflorescence cone-like; ovary 4-locular, each cell with a solitary basal ovule: **Bataceae**
          12. Leaves alternate (or if opposite, the plants with latex); inflorescence not cone-like; ovary usually 2–3-locular, each locule with 1 or 2 pendulous ovules: **Euphorbiaceae**
    8. Perianth present:
      13. Plants sometimes with latex; flowers unisexual, plants monoecious or dioecious; fruit a capsule, splitting into as many cocci as there are ovary-locules, or else a drupe: **Euphorbiaceae**
      13. Plants not with combined characters as given above:
        14. Perianth in 1 series, or apparently so: **Plumbaginaceae**
          15. Leaves not in basal rosettes; flowers not in one-sided spikes:
            16. Leaves opposite:
              17. Anthers united into a ring or tube, the filaments free; flowers in heads, subtended by an involucre of bracts: **Compositae**
              17. Anthers free:

18. Stipules present: **Urticaceae**

18. Stipules absent:

  19. Ovary 1-locular:

    20. Perianth petaloid, more or less tubular; fruit an anthocarp: **Nyctaginaceae**

    20. Perianth dry, scarious, neither petaloid nor tubular; fruit an utricle:

        Amaranthaceae

  19. Ovary 2- or more-locular:

    21. Shrubs; flowers unisexual:

      22. Monoecious; stems erect, bearing flat leathery leaves; fruit a 3-horned capsule:

        Buxaceae

      22. Dioecious; stems arching, bearing fleshy subterete leaves; flowers minute, in small

        cone-like spikes: **Bataceae**

    21. Herbs; flowers bisexual; capsules not horned:

      23. Trailing succulent herbs with terete stems: **Aizoaceae**

      23. Erect herbs with 4-angled stems: **Lythraceae**

16. Leaves alternate:

  24. Anthers united into a ring or tube, the filaments free; flowers in heads, subtended by an

    involucre of bracts: **Compositae**

  24. Anthers free:

    25. Fruits united in a head-like or hollow syncarp; plants with latex: **Moraceae**

    25. Fruits not syncarpous; latex absent:

      26. Ovary inferior or nearly so:

        27. Flowers in loose umbels:

          28. Trees: **Araliaceae**

          28. Herbs: **Umbelliferae**

        27. Flowers in spikes, panicles or dense heads: **Combretaceae**

      26. Ovary superior or nearly so:

        29. Stipules present:

          30. Stipules united to form a membranous or papery sheath closely surrounding the

            stem above each node: **Polygonaceae**

          30. Stipules free, often minute and inconspicuous:

            31. Flowers bisexual:

              32. Leaves trifoliate; creeping herb with yellow flowers: **Oxalidaceae**

              32. Leaves simple; plants herbaceous or woody; flowers not yellow:

                Phytolaccaceae

            31. Flowers unisexual or polygamous:

              33. Stamens 8 or more: **Salicaceae**

              33. Stamens fewer than 8: **Ulmaceae**

        29. Stipules absent or apparently so:

          34. Ovary 3–6-locular; viscid shrub; fruit a papery-winged capsule: **Sapindaceae**

          34. Ovary 1-locular; fruit not a papery-winged capsule:

            35. Shrubs or small trees:

              36. Flowers bisexual; perianth 6-parted; stamens 9, the anthers 4-locular; fruit a black

                or green drupe: **Lauraceae**

              36. Flowers unisexual, plants monoecious or dioecious; perianth 4-parted; stamens

                8, the anthers 2-celled; fruit a white drupe: **Thymelaeaceae**

            35. Herbs (rarely somewhat shrubby) or vines:

              37. Stamens same in number as the perianth segments:

                38. Slender twining herbaceous vine without tendrils: **Basellaceae**

              38. Plants erect or prostrate, not twining:

39. Inflorescence a fleshy spike with minute flowers embedded, or else a cluster with bracts becoming hard and cristate-toothed in fruit: **Chenopodiaceae**

39. Inflorescence neither fleshy nor the bracts becoming cristate-toothed: **Amaranthaceae**

37. Stamens more numerous than the perianth segments; vine with tendrils and pink flowers: **Polygonaceae**

14. Perianth in 2 series, or apparently so:

40. Segments of the inner perianth series (petals) separate, or united only at the extreme base:

41. Petals more than 10; plants succulent, leafless or with rudimentary leaves; plant-body usually (but not always) more or less spiny: **Cactaceae**

41. Petals less than 10; plants obviously leafy:

42. Fruit a legume (i.e. a pod-like capsule formed from a single free carpel that splits along both ventral and dorsal edges into two halves), or an indehiscent or fragmenting modification of this type: **Leguminosae**

42. Fruit not a legume:

43. Stamens more than twice as many as the petals:

44. Leaves opposite:

45. Ovary superior: **Clusiaceae**

45. Ovary more or less inferior: **Myrtaceae**

44. Leaves alternate:

46. Flowers 3-parted (or with 6 petals); leaves simple; ovary with carpels not completely united (becoming more or less fused in aggregate fruits): **Annonaceae**

46. Flowers 4–5-parted; ovary syncarpous or of 1 carpel:

47. Leaves compound, with numerous glandular-pellucid dots: **Rutaceae**

47. Leaves simple:

48. Stipules, stipular glands or stipular hairs present:

49. Ovary at least partly inferior:

50. Succulent herbs: **Portulacaceae**

50. Shrub or small tree: **Chrysobalanaceae**

49. Ovary superior:

51. Ovary 1-locular, exserted on a stalk (gynophore): **Capparaceae**

51. Ovary 2- or more-locular, sessile:

52. Filaments free except at base: **Tiliaceae**

52. Filaments more or less united into a tube or sheath:

53. Petals 5, equal and regular; hairs often stellate: **Malvaceae**

53. Petals 3, unequal (2 upper ones and a boat-shaped keel); hairs simple: **Polygalaceae**

48. Stipules absent:

54. Prickly herb with yellow sap: **Papaveraceae**

54. Shrub or small tree without prickles; sap colorless; foliage aromatic: **Canellaceae**

43. Stamens twice as many as the petals or fewer:

55. Ovary more or less inferior:

56. Plants herbaceous:

57. Vines with tendrils: **Cucurbitaceae**

57. Not vines; tendrils absent:

58. Flowers in umbels; ovary 2-locular with 1 ovule in each cavity; leaves sheathing at the base: **Umbelliferae**

58. Not as above:

59. Plants succulent, stems prostrate or ascending; ovary 1-locular; capsule splitting transversely: **Portulacaceae**
59. Plants not succulent, stems erect; ovary 4–6-locular; capsule splitting lengthwise: **Onagraceae**
56. Plants woody (shrubs or trees):
60. Leaves opposite:
61. Stilt-roots present; inflorescence 1–3-flowered; petals hairy: **Rhizophoraceae**
61. Stilt-roots absent; inflorescence many-flowered; petals glabrous: **Combretaceae**
60. Leaves alternate: **Rhamnaceae**
55. Ovary more or less superior, at least in flower:
62. Ovary half-immersed in the disk, which increases in size as the ovary ripens until it almost completely envelops the fruit; shrub or small tree: **Olacaceae**
62. Ovary completely superior, or if partly immersed in a disk, this not enlarging in fruit:
63. Leaves simple:
64. Stipules present (these small, or sometimes conspicuous):
65. Vines with tendrils:
66. Ovary stalked; flowers with persistent petals and conspicuous filamentous corona: **Passifloraceae**
66. Ovary sessile; flowers with minute petals, these soon falling; corona absent: **Vitaceae**
65. Not vines:
67. Leaves opposite:
68. Calyx bearing prominent sessile glands: **Malpighiaceae**
68. Calyx not glandular:
69. Flowers in terminal panicles; stamens 8: **Lythraceae**
69. Flowers in axillary cymes; stamens 4 or 5: **Celastraceae**
67. Leaves alternate:
70. At least some flowers unisexual; fruit a 2–3-seeded leathery capsule or dry drupe: **Celastraceae**
70. All flowers bisexual:
71. Leaves without glands; flowers not yellow; ovary 3-or 5-locular: **Turneraceae**
71. Leaves without glands; flowers not yellow; ovary 3-or 5-locular:
72. Ovary 5-locular; hairs often stellate: **Sterculiaceae**
72. Ovary 3-locular; hairs never stellate:
73. Pedicels terete; stamens free to the base; fruit a 3-seeded capsule: **Rhamnaceae**
73. Pedicels angled; stamens united below; fruit a 1-seeded drupe: **Erythroxylaceae**
64. Stipules absent:
74. Ovary lobed or the carpels more or less distinct:
75. Twining vine; flowers unisexual; plants dioecious: **Menispermaceae**
75. Erect shrub; flowers bisexual with yellow petals: **Surianaceae**
74. Ovary entire, with fused carpels:
76. Tree-like, usually unbranched giant herb, with thick hollow stem and large palmately lobed leaves: **[Caricaceae]**
76. Not as above:
77. Small seaside herb with 4-parted white flowers and 6 stamens (4 long and 2 short); fruit a spindle-shaped indehiscent capsule: **Cruciferae**
77. Plants woody (trees):

78. Stamens 1–5; fruit a large 1-seeded drupe; leaves not gland-dotted:
Anacardiaceae

78. Stamens 20 or more; fruit a large juicy berry with aromatic rind; leaves with pellucid glandular dots: **Rutaceae**

63. Leaves compound:

79. Leaves 1-pinnate (or 1-palmate):

80. Leaves opposite; stipules present; flowers yellow or blue: **Zygophyllaceae**

80. Leaves alternate; stipules absent; flowers neither yellow nor blue:

81. Plants herbaceous; leaves palmately divided: **Capparaceae**

81. Plants woody (shrubs or trees); leaves pinnate or 3-foliolate:

82. Leaflets more than 18 per leaf (up to 51), small, whitish beneath; flowers unisexual, plants dioecious: **Simaroubaceae**

82. Leaflets 3–17 per leaf; flowers bisexual or plants polygamo-dioecious:

83. Stamens united to form a tube; fruit a dehiscent capsule:
Meliaceae

83. Stamens free or united only at the base; fruit a capsule or drupe:

84. Ovary 1-locular or the carpels free; fruit a drupe or of drupelets:

85. Foliage pellucid-dotted, or if not, the stems spiny: **Rutaceae**

85. Foliage not pellucid-dotted; stems not spiny; sap irritating to the skin: **Anacardiaceae**

84. Ovary 3–5-locular:

86. Fruit a capsule:

87. Capsule 3-valved; stamens 8–10:

88. Capsule usually more than 6 cm long; stamens 8: **Sapindaceae**

88. Capsule less than 1.3 cm long; stamens 10: **Burseraceae**

87. Capsule 5-valved; stamens 4–6: **Meliaceae**

86. Fruit a drupe:

89. Drupe thin-fleshed, less than 3 cm in diameter, with dry leathery rind; flowers white or greenish: **Sapindaceae**

89. Drupe thick-fleshed, more than 3 cm in diameter, with thin epidermis; flowers red-purple: **Anacardiaceae**

79. Leaves 2-pinnate or more divided:

90. Ovary 1-locular; small tree with long triangular capsules containing many 3-winged seeds: **[Moringaceae]**

90. Ovary 3–6-locular:

91. Vines with tendrils; stamens 8, free; fruit a membranous inflated capsule:
Sapindaceae

91. Small tree; stamens 10–12, united into a tube; fruit a small drupe:
Meliaceae

40. Segments of the inner perianth series united into one more or less tubular structure (corolla) at more than the extreme base:

92. Ovary inferior, at least during flowering:

93. Stipules present and often conspicuous, interpetiolar between opposite leaves:
Rubiaceae

93. Stipules absent; leaves alternate or opposite:

94. Herbaceous vines with tendrils: **Cucurbitaceae**

94. Not vines; tendrils absent:

95. Flowers in heads subtended by an involucre of bracts; calyx highly modified in the form of scales, bristle or awns: **Compositae**

95. Flowers few and separate in stalked axillary dichasia; small fleshy seaside shrub; fruit a small black drupe: **Goodeniaceae**

92. Ovary not inferior at time of flowering:

  96. Ovary apparently inferior in fruit:

    97. Plants herbaceous with opposite leaves and funnel-shaped corolla-like perianth:
        Nyctaginaceae

    97. Plants woody with alternate leaves; ovary half-immersed in the disk, which enlarges as the ovary ripens until it almost completely envelopes the fruit: **Olacaceae**

  96. Ovary superior in flower and fruit:

    98. Stamens twice as many as the corolla-lobes:

      99. Plants woody (branched shrub or small tree); all leaves simple and entire; corolla bearded within: **Olacaceae**

      99. Plants herbaceous (if tree-like in stature, the stem hollow and usually unbranched, the leaves palmately lobed or compound):

        100. Corolla 5-lobed; tree-like unbranched giant herb with thick hollow stem and large palmate leaves: **[Caricaceae]**

        100. Corolla 4-lobed; succulent herbs with simple or pinnately divided leaves:
        Crassulaceae

    98. Stamens as many as or fewer than the corolla-lobes:

      101. Stamens opposite to and equal in number with the corolla-lobes:

        102. Plants with latex; ovary 4–14-locular: **Sapotaceae**

        102. Plants without latex; ovary 1-celled:

          103. Flowers in terminal racemes, the parts in 5s; leaves of stiff hard texture:
          Theophrastaceae

          103. Flowers in small lateral clusters among or below the leaves (often at leafless nodes), the parts in 4s; leaves of thin, herbaceous texture: **Myrsinaceae**

      101. Stamens alternating with the corolla-lobes and equal in number or fewer:

        104. Corolla regular (actinomorphic), the divisions of equal size and similar in shape:

          105. Leaves all opposite, or rarely whorled:

            106. Stipules present: **Loganiaceae**

            106. Stipules absent:

              107. Stamens connected with the stigma to form a central column of intricate structure; plants with latex: **Asclepiadaceae**

              107. Stamens separate:

                108. Stamens 2; fruit a drupe or berry: **Oleaceae**

                108. Stamens 4 or 5; fruit a capsule or follicle:

                  109. Corolla strongly twisted in bud:

                    110. Ovary 1-locular; fruit a capsule with numerous naked seeds; plants without latex: **Gentianaceae**

                    110. Ovary of 2 separate carpels initially connected by a common style; fruit of 2 follicles, the seeds with or without hairs or other appendages; plants usually (but not always) with latex: **Apocynaceae**

                  109. Corolla not twisted in bud; latex absent:

                    111. Ovary 2-locular with numerous ovules in each cavity; flowers solitary or paired: **Scrophulariaceae**

                    111. Ovary 1–4-locular with 1–4 ovules in each cavity; flowers in cymes, spikes or heads:

                      112. Ovary 1-locular, the ovules pendulous from a free basal placenta; shrub or tree of saline habitats, the roots bearing upright aerial pneumatophores: **Avicenniaceae**

                      112. Ovary 2–4-locular; herbs, shrubs or trees not of saline habitats; pneumatophores lacking: **Verbenaceae**

105. Leaves all or mostly alternate (rarely some of them opposite or whorled):
  113. Ovules 1 or 2 in each cavity:
    114. Ovary 2- or 4-locular, entire; corolla usually trumpet- or salver-shaped, strongly twisted in bud; mostly trailing or twining herbaceous vines, sometimes with latex, rarely a small erect shrub: **Convolvulaceae**
    114. Ovary 4-locular, usually 4-lobed; flowers in cymes, the corolla not or but slightly twisted in bud; herbs, shrubs, woody vines or trees, never with latex: **Boraginaceae**
  113. Ovules several to many in each cavity:
    115. Stamens 4, the fifth a short staminode or absent; fruit a 2-locular, 2- or 4-valved capsule; ovules numerous, borne on swollen axile placentas: **Scrophulariaceae**
    115. Stamens 5, all functional:
      116. Ovary of 2 separate carpels united initially by the style; fruit of 2 follicles containing numerous winged seeds; latex copious: **Apocynaceae**
      116. Ovary simple with fused carpels; seeds not winged; latex absent:
        117. Ovary 1-locular with 2 parietal placentas; prostrate or decumbent annual herb with solitary or paired axillary flowers; fruit a small capsule: **Hydrophyllaceae**
        117. Ovary 2- or 4-locular with thick axile placentas; erect herbs or shrubs, or woody vines, the flowers solitary or in cymose clusters or racemes: **Solanaceae**
104. Corolla irregular or oblique (zygomorphic), at least one of the divisions differing in size or shape from the others:
  118. Leaves compound: **Bignoniaceae**
  118. Leaves simple:
    119. Flowers solitary or rarely several in a cluster:
      120. Corolla 4 cm long or more, broadly bell-shaped, glabrous; ovary 1-locular with numerous ovules; fruit a large globose berry with a hard shell: **Bignoniaceae**
      120. Corolla 2 cm long, narrowly cylindric with 2-lipped limb, the middle lobe of the lower lip densely bearded; ovary 2-locular with 4 ovules in each cavity; fruit a small drupe: **Myoporaceae**
    119. Flowers in distinct inflorescences:
      121. Ovules solitary in each cavity; fruit a drupe, or else a schizocarp of nutlets; plants often aromatic:
        122. Style arising from the base of the ovary; ovary 4-locular, 4-lobed: **Labiatae**
        122. Style terminal on the ovary; ovary 1–4-locular, entire: **Verbenaceae**
      121. Ovules 2 or more in each cavity; fruit a capsule; plants not aromatic:
        123. Fruit a beaked, 2-valved capsule, splitting elastically to the very base, flinging the seeds out by a sling action: **Acanthaceae**
        123. Fruit a 2–4-valved capsule opening or splitting at the top but never all the way to the base, the segments not elastic: **Scrophulariaceae**

ANNONACEAE

Trees or shrubs with alternate, entire leaves; stipules lacking. Flowers mostly perfect and 3-parted; sepals usually 3, valvate or imbricate; petals commonly 6 in two series, valvate or imbricate, the inner ones often rudimentary or absent. Stamens numerous, the anther-cells adnate; carpels numerous (rarely few), generally free, with 1 or more ovules in each cell, in fruit free or united to form a fleshy multiple fruit. Seeds with or without an aril, with copious ruminate endosperm and a minute embryo.

A pantropical family of about 75 genera or more, and more than 2,000 species, the majority in the Old World.

Annona L.

Trees or shrubs, glabrous or with simple or stellate pubescence. Flowers solitary or in few-flowered clusters, these terminal, opposite the leaves, or apparently internodal; sepals 3, small, valvate; petals 6, free or connate at base, biseriate, the inner ones small or lacking, the outer valvate, fleshy and usually concave. Stamens extrorse, the connective produced above the cells into a disk. Carpels often connate, containing solitary, basal, erect ovules. Fruit a fleshy aggregate, often edible. A chiefly tropical American genus of about 120 species. In addition to those occurring in the Cayman Islands, a number of others are important for their edible fruits.

KEY TO SPECIES

1. Flowers globose or broadly pyramidal in bud; fruits smooth or covered with soft spines:	
2. Leaves more or less rounded at base and widest at or below the middle; pedicels glabrous; fruits smooth:	A. glabra
2. Leaves more or less acute at base and widest above the middle; pedicels sericeous; fruits soft-spiny:	A. muricata
1. Flowers oblong or narrowly oblong in bud; fruits covered with rounded tubercles:	A. squamosa

Annona glabra L., *Sp. Pl.* 1: 537 (1753).

POND-APPLE

A shrub or small tree to 6 m tall or more, the trunk often enlarged or buttressed at base. Leaves of stiff texture, petiolate, ovate-elliptic to oblong-elliptic, 7–14 cm long, 3–8 cm broad, acute at apex and rounded or obtuse at base. Flowers solitary, internodal, on pedicels 1.5–2 cm long; sepals 3–5 mm long, rounded and apiculate; petals 6, the outer ones ovate, 2.5–3 cm long, the inner ones smaller. Fruit globose-ovoid, 5–12 cm long, smooth, yellowish at maturity, edible but insipid; seeds brown.

GRAND CAYMAN: *Brunt* 1837, 2046.

— Common at or near sea-level in suitable habitats throughout tropical America; also occurs in western Africa. This species grows in swamps or wet thickets, or often near mangroves.

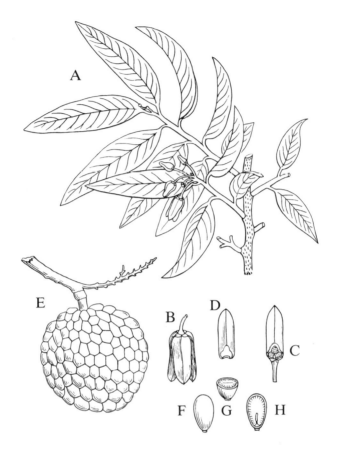

FIG. 76 **Annona squamosa**. A, habit, × ¹/₃. B, flower, × ²/₃. C, receptacle and petal, × ²/₃. D, petal, × ²/₃. E, fruit, × ²/₃. F, seed, × ²/₃. G, H, cross- and long. sections of seed. (F. & R.)

Annona muricata L., *Sp. Pl.* 1: 536 (1753).

SOURSOP

A small tree up to 8 m tall, the young branchlets clothed with minute reddish-brown hairs, soon becoming glabrate. Leaves lustrous, petiolate, oblong-obovate, 8–15 cm long, 3–6 cm broad, minutely sericeous beneath when young and with persistent domatia in the nerve-axils, abruptly acute at apex and less abruptly so at base. Flowers solitary, terminal or opposite leaves, on pedicels 1.5–2 cm long; sepals ca. 3 mm long, broadly deltate; petals 6, the outer ones ovate-acuminate, thick, 2.5–3.5 cm long, the inner ones smaller. Fruit asymmetrically oblong-ovoid, up to 20 cm long or more, green at maturity and covered with curved, flexible spines, edible; seeds black.

CAYMAN BRAC: *Proctor 29107*. Probably occurs on Grand Cayman, but no specimens were seen during the preparation of the present work.

— Apparently indigenous to the West Indies, but widely cultivated and becoming naturalized in many tropical countries.

Annona squamosa L., *Sp. Pl.* 1: 537 (1753). **FIG. 76.**

SWEETSOP

A shrub or small tree up to 6 m tall with a rounded or spreading crown, the young branchlets grayish-sericeous. Leaves petiolate, elliptic or lance-elliptic, 5–11 cm long, 2–5 cm broad, acute at both ends, usually paler beneath and bearing small, scattered, soft hairs. Flowers greenish, solitary or several in small clusters opposite leaves, on pedicels 1–2 cm long; sepals 1.5–2 mm long, rounded-deltate; petals 6, the outer linear-oblong and 1.5–3 cm long, the inner ones rudimentary. Fruit globose or ovoid, usually 7–9 cm in diameter, glaucous, the carpels incompletely fused and projecting as smooth rounded tubercles, edible; seeds brown.

GRAND CAYMAN: *Brunt* 2196; *Kings* GC 144; *Proctor* 15250. CAYMAN BRAC: *Proctor* 28964.

— Apparently indigenous to the West Indies, but widely cultivated in tropical America and easily becoming naturalized.

The custard-apple, *Annona reticulata* L., has been reported in Grand Cayman but no specimens have been seen by the present writer.

CANELLACEAE

Small trees or shrubs with aromatic bark and alternate, entire leaves; stipules lacking. Flowers perfect, in axillary or terminal cymes or corymbs, rarely solitary; sepals 3, imbricate; petals 4 or 5, usually free, imbricate, sometimes alternating with petaloid scales; stamens rather numerous, the filaments united into a tube, the anthers attached closely together outside the tube and opening longitudinally; ovary free, 1-celled, with 2 or more ovules; style short, thick, with a 2–6-lobed stigma. Fruit a berry with 2 or more seeds, these smooth and hard; endosperm copious, fleshy, the embryo short.

A small family of 5 genera and about 16 species widely scattered in tropical America, eastern Africa and Madagascar.

Canella P. Br.

Evergreen glabrous trees or shrubs, the bark and leaves pleasantly aromatic. Flowers purple, red or violet in terminal, bracteolate corymbs; petals 5, petaloid scales lacking; stamens 10–20, the anthers contiguous on the filament-tube; ovary with 2 or 3 parietal placentas, each bearing 2 ovules. Berry globose, the gelatinous pulp enclosing few, obovoid to globose seeds.

A genus of 2 species, one occurring in Florida and the West Indies, the other in Colombia.

Canella winterana (L.) Gaertn., *Fruct. & Sem. Pl.* 1: 373 (1788). **FIG. 77.** PLATE 10.

PEPPER CINNAMON

A shrub or small tree rarely to 15 m tall, leaves oblanceolate or spatulate, 3–12 cm long, dark green and rather glossy above, the tissue minutely and closely gland-dotted, on petioles ca. 1 cm long. Inflorescence several- to many-flowered, the flowers on slender

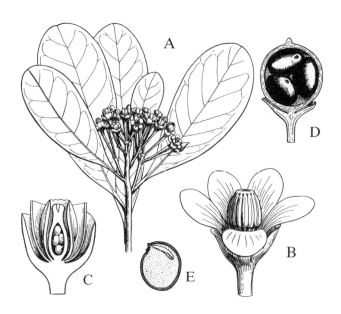

FIG. 77 **Canella winterana**. A, branch with leaves and flowers, × ²/₃. B, flower, × 4. C, flower cut lengthwise, × 4. D, fruit cut vertically, × 2. E, seed cut lengthwise, × 2. (F. & R.)

pedicels; sepals rounded, ca. 3 mm long, ciliolate; petals blood-red, obovate, nearly twice as long as the sepals. Staminal tube 3–4 mm long, bearing yellow anthers. Berries deep crimson, 8–10 mm in diameter; seeds black.

GRAND CAYMAN: *Brunt* 1990, 2143, 2151; *Correll & Correll* 51041; *Hitchcock*; *Proctor* 15186. LITTLE CAYMAN: *Kings* LC 15; *Proctor* 28117, 28177. CAYMAN BRAC: *Kings* CB 48; *Matley*; *Proctor* 28955, 29052.

— Florida and the West Indies south to Barbados. Common in rocky woodlands and thickets, chiefly on limestone and at lower to middle elevations. *Canella* bark was formerly used in medicine as an aromatic stimulant and tonic. The berries, when eaten by pigeons, impart to the flesh a distinct and pleasing flavour. This species has horticultural value as an ornamental, though apparently not often planted.

LAURACEAE

Trees, shrubs, or leafless parasitic herbs; leaves alternate and simple; stipules lacking. Flowers perfect or dioecious, axillary panicles or (in *Cassytha*) in spikes; perianth segments usually 6, in 2 whorls, fused below into a short tube that develops into a cupule at the base of the fruit. Stamens in a double ring, an outer ring of 6 perfect stamens and an inner ring of which 3 are perfect, all opening by valves, alternating with 3 staminodes that are often very small or apparently lacking. Ovary superior, free from the perianth tube, 1-celled with 1 pendulous ovule. Fruit a 1-seeded berry (drupe), with the persistent perianth tube usually forming a cupule at the base, or else enveloped by the fleshy receptacle; seed relatively large, without endosperm, the cotyledons thick and fleshy.

A chiefly pantropical family of about 32 genera and more than 1,100 species, especially common in S. America and southeast Asia. Many species are important as timber trees, whereas others contain aromatic spices of commercial value. Examples of the latter are those producing cinnamon (*Cinnamomum zeylanicum*) and camphor (*C. camphora*).

KEY TO GENERA

1. Trees with green leaves; fruit not enclosed by a succulent perianth:

    2. Perfect stamens 3, the anthers 2-locular; fruiting calyx a double-margined cupule: **Licaria**

    2. Perfect stamens 9, the anthers 4-locular; fruiting cupule single margined or absent:

        3. Wild tree; staminodes minute or lacking; fruiting calyx a single-margined cupule; fruit small, not edible: **Ocotea**

        3. Cultivated tree; staminodes prominent, sagittate; perianth segments pertsistent in fruit, a cupule absent; fruit large, edible: **[Persea]**

1. Twining parasitic herb lacking chlorophyll, the stems filiform and apparently leafless; fruit enclosed by the enlarged succulent perianth: **Cassytha**

Licaria Aubl.

Trees. Leaves alternate or rarely opposite, coriaceous, pinnately veined. Inflorescences paniculate, axillary or seemingly terminal. Flowers perfect; tepals 6, equal or nearly so, caduceus; 6 stamens of outer 2 rows modified into foliaceous scale-like staminodes, third row of 3 stamens fertile, with basal glands, anthers 2-locular, extrorse or rarely introrse, fourth row wanting; ovary included in perianth tube. Fruiting cupules clearly or obscurely double margined; berry ellipsoid to subglobose, largely to completely included in cupule.

A genus of about 45 species primarily of Central and tropical S. America. For more information, see A. J. G. H. Kostermans, *Recueil Trav. Bot. Néerl.* 34: 500–609 (1937).

Licaria triandra (Sw.) Kosterm., *Recueil Trav. Bot. Néerl.* 34: 588 (1937). PLATE 11.

Tree to 20 m tall. Leaves with petioles 0.5–1 cm long; blades oblong-elliptic, elliptic or oval, 5–13 × 1.8–5.5 cm, broad, coriaceous, shiny and glabrous both surfaces, base acute or rounded and slightly cuneate-decurrent, apex often abruptly acuminate with an obtuse point. Perianth lobes 1.8 mm, obtuse. Cupules 9–12 mm high, 12–18 mm in diameter, red when mature; berry oblong-ovoid, 2–2.5 cm long, 1.2–1.4 cm in diameter.

GRAND CAYMAN: *Burton* 194, 195; *Proctor* 52115. A rare component of primary tropical forest; occurs in a relictual fragment of such forest along Walkers Road, S. of George Town.

Florida, Greater and Lesser Antilles.

Ocotea Aubl.

Evergreen trees or rarely shrubs with coriaceous leaves, glabrous or variously pubescent. Flowers small, perfect or polygamo-monoecious, in axillary or terminal panicles; perianth 6-parted, its lobes nearly equal. Stamens 9, shorter than the perianth; anthers 4-celled, on very short filaments; staminodes none or small. Berry globose or ellipsoid, seated in a cupule (enlarged calyx-base).

A chiefly tropical American genus of more than 300 species. The genus *Nectandra* is here not considered separable from *Ocotea*.

Ocotea coriacea (Sw.) Britton in Britton & Millsp., *Fl. Baham.* 143 (1920). PLATE 11.

Nectandra coriacea (Sw.) Griseb. (1860).

SWEETWOOD

A small tree to 10 m tall or more; leaves elliptic, lance-elliptic, or lance-oblong, chiefly 6–13 cm long, 2–4.5 cm broad, acuminate, glabrous or nearly so, glossy dark green and of hard texture, with finely reticulate veins. Panicles axillary, shorter than the leaves; flowers creamy-white, sweet-scented, on red pedicels. Berry black, ovoid or subglobose, 10–13 mm long, each seated in a 1-margined, bright red cupule.

GRAND CAYMAN: *Brunt* 2160; *Lewis* 3862. CAYMAN BRAC: *Proctor* 29015, 29341.

Florida, West Indies and the Yucatan Peninsula, in woodlands at low elevations. This tree has some use in carpentry; it is reported that the flowers yield good honey, and cattle occasionally eat the fruits. It is quite attractive when planted as a shade tree.

[*Persea* Mill.

The 'avocado' or 'pear', *Persea americana* Mill., is often cultivated.

GRAND CAYMAN: *Millspaugh* 1317.

— Apparently indigenous to the region of Honduras, Guatemala and southern Mexico. This species is a highly variable complex of cultivars whose genetic relationships are not well understood.]

Cassytha L.

Twining parasitic herbs, adhering to the host by means of suckers (haustoria); leaves reduced to small scales. Flowers perfect, in spikes or racemes; perianth segments 6, fused at base, arranged in 2 series, the 3 segments of the outer series much the smaller. Stamens 9, the staminodes 3. Ovary 1-celled with 1 ovule. Fruit globose and fleshy, enclosed by the succulent perianth tube and crowned by the persistent perianth segments.

A pantropical genus of about 20 species, the majority Australian.

Cassytha filiformis L., *Sp. Pl.* 1: 35 (1753). FIG. 78.

OLD MAN BERRY

Twining parasitic herb with slender, elongate yellow or greenish stems, freely branched, up to 4 m long or more, sometimes matted. Scales (reduced leaves) few, 1–2 mm long. Spikes few- to several-flowered, 1–2 cm long; flowers white, ca. 2 mm long, the inner perianth segments ovate, larger than the outer. Fruit globose, white, 6 mm in diameter.

GRAND CAYMAN: *Brunt* 1909; *Hitchcock*; *Kings* GC 165, GC 265; *Proctor* 15226. LITTLE CAYMAN: *Kings* LC 29, LC 82, LC 85; *Proctor* 28080. CAYMAN BRAC: *Kings* CB94; *Millspaugh* 1168; *Proctor* 29096.

— Pantropical; indiscriminately parasitic on other vegetation.

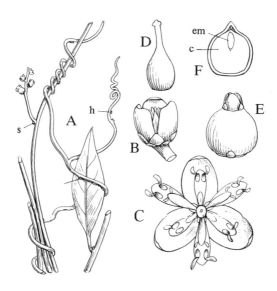

FIG. 78 **Cassytha filiformis**. A, habit, showing haustoria (h) and rudimentary leaf (s), × ²/₃. B, flower as normally seen, × 6. C, flower flattened out, × 7. D, pistil, × 18. E, fruit, × 2. F, long. section of fruit, showing inner face of cotyledon (c) and axis of embryo (em). (F. & R.)

PIPERACEAE

Herbs, shrubs, or rarely small trees; leaves alternate, opposite or whorled, nearly always simple and entire, palmately or pinnately veined, often succulent. Flowers minute, perfect, without a perianth, whorled or spirally arranged in tail-like spikes that may be terminal, opposite the leaves or rarely axillary, sometimes several together on a common peduncle. Stamens 2–6, rarely more; ovary superior, sessile or rarely stalked, 1-celled, with 1 basal ovule. Fruit a small 1-seeded berry (drupe); seed with both perisperm and endosperm, usually pungently oily.

A pantropical family of about 9 genera and perhaps 2,000 species or more, the majority in the Western Hemisphere.

KEY TO GENERA

1. Plants herbaceous; stigma 1:	**Peperomia**
1. Plants more or less woody shrubs; stigmas 2–4:	**Piper**

Peperomia Ruiz & Pav.

Terrestrial or epiphytic herbs, creeping to erect, with more or less succulent stems. Leaves simple, variously arranged. Spikes terminal or axillary, simple or in branched inflorescences. Flowers numerous, minute, borne in the axils of round, peltate bracts; stamens 2; ovary sessile (in Cayman species), the apex obtuse, acute or beaked; stigma terminal or lateral below the beak. Fruit very small, seed-like.

A large genus of nearly 1,000 species, of no economic importance except for a few species that are used as ornamental pot plants.

KEY TO SPECIES

1. Leaves less than 5 cm long; fruits not beaked:

  2. Leaves opposite (Cayman Brac):   **P. pseudopereskiifolia**

  2. Leaves alternate (Grand Cayman):

    3. Stems branching and trailing, rooting at nodes; leaves somewhat fleshy, finely black-dotted; petioles ciliate: **P. glabella**

    3. Stems ascending or erect, rooting at base; leaves not fleshy, finely pellucid-dotted; petioles glabrous: **P. simplex**

1. Leaves up to 10 cm long or more; fruits beaked:

    4. Peduncle glabrous; fruit-beak subulate, curved but not abruptly hooked at apex: **P. magnoliifolia**

    4. Peduncle microscopically hairy; fruit-beak abruptly hooked at apex: **P. obtusifolia**

Peperomia pseudopereskiifolia C. DC., *Prodr.* 16(1): 448 (1869). PLATE 11.

Ascending glabrous herb with quadrangular stems. Leaves opposite or sometimes a few ternate, short-petiolate (petioles 3–8 mm), the blades lanceolate to ovate, abruptly but bluntly acuminate at apex, cuneate and slightly decurrent at base, 3–4.5 cm long, 1.3–2 cm broad, the tissue subcoriaceous and opaque, 5-nerved, the mid-vein prominulous. Spikes terminal and from upper axils, densely flowered, the bracts round-peltate; ovary deeply immersed; stigma penicillate. Fruits not seen.

CAYMAN BRAC: *Proctor* 47323, 52159.

Cuba; reported once in Guatemala.

Peperomia glabella (Sw.) A. Dietr. in L., *Sp. Pl.* ed. 6, 1: 156 (1831).

A creeping or sometimes pendent herb, often rooting at the lower nodes, with ascending flowering branches. Leaves light green, petiolate, densely black-dotted, elliptic or lance-elliptic, chiefly 2–4 cm long, more or less acuminate at apex and narrowed at base, and palmately 3-nerved from near the base. Spikes slender, 1 mm thick, black-dotted, not densely flowered, 6–12 cm long, often a cluster of several spikes at the apex of the stem and solitary spikes in the leaf-axils. Fruit ovoid, 0.3–1 mm long, the apex oblique with stigma below the apex.

GRAND CAYMAN: *Hitchcock*; *Kings GC 183*; *Proctor* 47799, 49351.

West Indies and continental tropical America, epiphytic or on rocks. The Cayman plants were collected on shaded limestone rocks near Jackson Point, and in the forest behind the University College of the Cayman Islands.

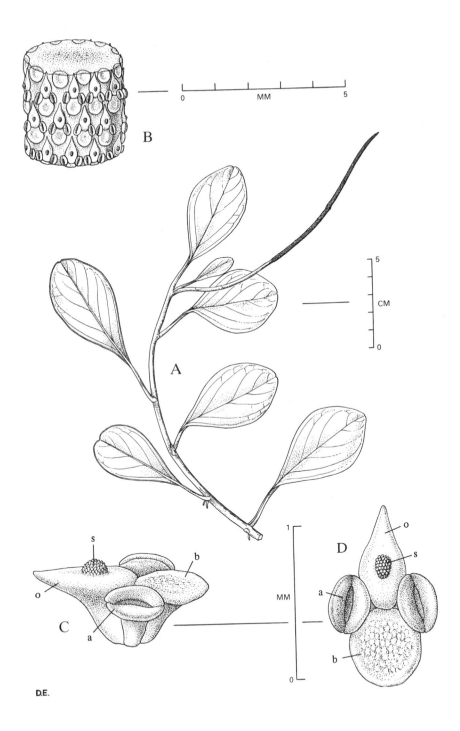

FIG. 79 **Peperomia magnoliifolia**. A, habit. B, portion of spike. C, D, two views of a flower showing bract (b), anther (a), ovary (o), and stigma (s). (D.E.)

Peperomia simplex Hamilt., *Prodr. Pl. Ind. Occ.*: 2 (1825). PLATE 11.

Ascending to erect (rarely trailing) glabrous herb up to 20 cm or more tall; leaves very short-petiolate, elliptic to elliptic-lanceolate, acute at apex and base, sparsely ciliolate at apex, mostly 2.5–4.5 cm long and 1–2(–3) cm wide, subsucculent, paler beneath. Spikes solitary or paired, terminal or axillary, up to 8 cm long; fruit globose, ca. 0.8 mm long.

GRAND CAYMAN: *Proctor 49216.*

Jamaica. Cayman plants have been found only in the area of Spots, in rocky woodland on a ridge where most of the plants have been destroyed by housing development.

Peperomia magnoliifolia (Jacq.) A. Dietr. In L.. *Sp. Pl.* ed. 6, 1: 153 (1831). FIG. 79.

WILD BALSAM, VINE BALSAM

A creeping-ascending glabrous herb with succulent stems 3–5 mm thick, rooting at many of the nodes. Leaves thick and fleshy, rather dark green, mostly obovate on long, narrowly winged petioles, 5–13 long, 3–6 cm broad above the middle, with 3–4 indistinct pinnate veins on each side. Spikes solitary in leaf-axils or up to 3 terminal ones, the longest as much as 15 cm long, 3 mm thick, densely flowered, on glabrous peduncles 2–3 cm long. Fruit ellipsoid, ca. 1 mm long, with a curved beak not obviously hooked.

GRAND CAYMAN: *Brunt 2157; Kings GC 368; Proctor 27986.*

West Indies and continental tropical America. This species is apparently always terrestrial, rooting in accumulations of humus on rocks and boulders.

Peperomia obtusifolia (L.) A. Dietr. in L., *Sp. Pl.* ed. 6, 1: 154 (1831).

Trailing and rooting herb with ascending stem terminations to ca. 20 cm tall; leaves long-petiolate, the blades elliptic-obovate, 5–15 cm long, 3.5–6.5 cm wide, cuneate-decurrent at the base, thick and leathery. Spikes usually terminal, solitary or paired, 5–15 cm long with peduncles 2–4 cm long; fruit ellipsoid, ca. 1.3 mm long including the hooked beak.

GRAND CAYMAN: *Proctor 47348.* Known only from the Forest Glen area.

Florida, Mexico, C. America, Greater and Lesser Antilles, and S. America, apparently uncommon throughout its range.

Piper L.

Shrubs or small trees (rarely vines), with the branches often jointed at the nodes. Leaves alternate, often unequal-sided at the base, sometimes with pellucid dots, the venation pinnate or palmate. Spikes opposite the leaves, typically solitary and simple. Flowers usually perfect, numerous; stamens 2–6; ovary sessile; stigmas 2–4, sessile or on a short and thick style. Fruit variable, usually small.

A vast genus of more than 1,000 species, often numerically important in the understory of tropical vegetation, a few weedy species ubiquitous in clearings of many areas, chiefly where rainfall is high. The chief economic species is the black pepper, *Piper nigrum*, a subwoody vine.

Piper amalago L., *Sp. Pl.* 1: 29 (1753). PLATE 11.

JOINTER, PEPPER ELDER

A somewhat aromatic shrub usually 2 or 3 m tall, nearly or quite glabrous, the slender branches with swollen joint-like nodes. Leaves chiefly ovate on distinct petioles and with a long-acuminate apex, mostly 6–12 cm long and up to 5 cm broad below the middle, palmately 5-veined. Spikes 5–7 cm long and ca. 1.5 mm thick, rather loosely flowered, minutely grayish-scurfy between the flowers. Fruit ovoid-conical, separated on the spike.

GRAND CAYMAN: *Brunt* 2150; *Hitchcock*; *Kings* GC 233; *Millspaugh* 1299; *Proctor* 15010.

— West Indies and continental tropical America; a common, variable and widespread species.

NYMPHAEACEAE

Aquatic herbs with thick, starchy, perennial rhizomes. Leaves usually floating, with very long petioles and peltate blades. Flowers solitary, large, floating on the surface of water or held a few inches above it; sepals 4, free; petals numerous, free, inserted in a close spiral on the receptacle. Stamens numerous, often with petaloid filaments. Carpels numerous, free or more or less united; ovules numerous, pendulous, parietal. Fruit a spongy berry, dehiscing by swelling of the mucilage surrounding the seeds; seeds with some endosperm and abundant perisperm.

A small, world-wide family of about 5 genera and 80 species.

Nymphaea L.

Rhizome fleshy, creeping, edible; leaves with roundish floating blades cleft nearly to the middle on one side. Flowers showy and often fragrant; inner stamens with narrow filaments and yellow anthers, passing into outer ones with broad petaloid filaments and small anthers. Carpels united into a many-celled, half-inferior ovary. Fruit ripening under water and breaking irregularly to free the seeds.

A widely distributed genus of about 50 species, known generally as water-lilies. The seeds of many species are both edible and nutritious, and are commonly eaten in several parts of the world.

KEY TO SPECIES

1. Petioles glabrous; leaf-blades usually toothed; flowers diurnal:	N. ampla
1. Petioles hairy at apex (just below attachment to leaf-blade); leaf-blades entire; flowers nocturnal:	N. amazonum

Nymphaea ampla (Salisb.) DC., *Syst.* 2: 54 (1821).

WATER LILY

Leaves with blades 10–45 cm broad, the margin irregularly wavy-toothed, the upper surface green or yellowish-green, the lower surface purple and with conspicuous raised reticulate veins; petiole peltately attached almost centrally. Flowers 8–15 cm broad, opening during the day and closing at night, fragrant; sepals oblong-lanceolate, green with brown speckles on the outside; petals and petaloid filaments white; anthers of the outer stamens with connective prolonged into a appendage.

GRAND CAYMAN: *Hitchcock*; *Kings GC* 190; *Proctor* 31025, 50234.

West Indies and C. America, in ponds and sluggish streams, apparently tolerant of brackish water. The Cayman plants were mostly found in swampy ponds southeast of George Town.

Nymphaea amazonum C. Mart. & Zucc. in *Abh. Math.-Phys. Cl. Koenig. Bayer. Akad. Wiss.* 1: 363 (1832).

Rhyzome subcylindric, to 10 cm long and 3 cm thick. Leaves with floating blades with petioles mostly 3–5 mm in diameter, glabrous except for an apical ring of simple hairs, these 3–6 mm long; blades broadly elliptic to suborbicular, mostly 8–24 cm in diameter with entire margin, the base subpeltate with sinus edges parallel and tapering into obtuse lobes, dark green above, reddish-purple beneath. Flowers floating, nocturnal, opening 2 nights, 8–14 cm in diameter, fragrant, white, with glabrous peduncle; sepals 4, ovate-attenuate, light green outside; petals 16–20, ovate-oblong, creamy-white, obtuse to rounded apex, diminishing in size toward the center; stamens 100–200, the outer with petaloid filaments; carpels 18–40, fused marginally, the stigmatic disks bright yellow. Fruit depressed obovoid-subspherical, to 3.2 cm in diameter, enclosed by persistent perianth. Seeds ovoid, 0.9–1.3 mm, brownish-red, with scattered long silvery hairs.

GRAND CAYMAN: *Proctor & Roulstone* 50233. Found in a natural temporary pond NNW of the Queen Elizabeth II Botanic Park.

Mexico and tropical America to eastern Brazil, and in the Greater Antilles; often not observed because of its nocturnal flowering.

MENISPERMACEAE

Dioecious trees, shrubs, or vines with alternate entire leaves; stipules lacking. Flowers minute and inconspicuous, in axillary bracteate panicles. Sepals and petals various in number in dimerous or trimerous whorls, or sometimes solitary. Stamens of male flowers 4 or 6; staminodes in female flowers 6 or lacking; filaments free or united into a column. Carpels 3 or 1, free, each with a single ovule. Fruit a drupelet, often succulent; seeds horseshoe-shaped, with large embryo and scanty endosperm.

A chiefly tropical family with about 65 genera and 400 species. Members of this family often contain toxic alkaloids, and a number of S. American species provide ingredients for arrow-poisons used by Amerindian tribes. These complex poisons, which vary in composition according to the species used, are collectively known as curare.

Cissampelos L.

Slender climbing shrubs or vines; leaves roundish, more or less cordate, sometimes peltate, palmately veined. Panicles cymose, the staminate many-flowered with branches much exceeding the bracts, the pistillate few-flowered with branches enclosed by the bracts. Staminate flowers with 4 sepals and petals united into a short cup. Pistillate flowers with 1 obovate sepal and 1 smaller petal opposite to the sepal. Carpel 1, with 3-lobed style; drupelet globose, more or less hairy.

About 30 species, mostly tropical American.

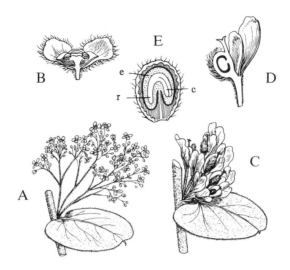

FIG. 80 **Cissampelos pareira**. A, staminate inflorescence, × $^2/_3$. B, male flower in section, × 10. C, pistillate inflorescence, × 6. D, female flower in section. E, drupe cut lengthwise, × 4, showing endosperm (e), cotyledons (c), and radicle (r). (F. & R.)

Cissampelos pareira L. *Sp. Pl.* 2: 1031 (1753). **FIG. 80. Plate 12.**

QUACORI

A slender, elongate vine, often high-climbing; leaves mostly 3–9 cm long and broad, varying from glabrous to softly velvety-hairy; petioles usually attached marginally but sometimes peltately. Flowers creamy-white, the staminate scarcely 1 mm across. Fruit scarlet, 3–5 mm in diameter.

GRAND CAYMAN: *Brunt* 2041; *Correll & Correll* 51039; *Kings* GC 120.

— Pantropical, frequent in dryish thickets and woodlands. The roots contain alkaloids similar to those used in the preparation of curare. The only Cayman specimens seen are staminate.

PAPAVERACEAE

Mostly herbs (rarely shrubs) with yellow or orange sap; leaves alternate, simple or variously lobed; stipules lacking. Flowers often rather large, solitary or rarely paniculate; sepals 2 or 3, deciduous; petals 4–6, free (rarely lacking). Stamens numerous, free. Ovary superior, 1-celled, with parietal placentas; ovules numerous; stigmas sessile or nearly so. Fruit a capsule, opening by valves or pores.

A chiefly Northern Hemisphere family most common in temperate regions, with about 26 genera and 200 species. Relatively few species occur in the tropics or south of the equator. Many species contain characteristic alkaloids, the best-known being those of the opium poppy, *Papaver somniferum*.

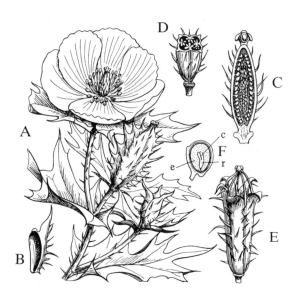

FIG. 81 **Argemone mexicana.** A, branch with bud, fruit and flower, × ²/₃. B, sepal, × ²/₃. C, pistil cut lengthwise, × ²/₃. D, cross-section of ovary. E, ripe capsule, × ²/₃. F, seed cut lengthwise; e, endosperm; c, cotyledons; r, radicle. (F. & R.)

Argemone L.

Glaucous spiny herbs with yellow sap. Leaves sessile, pinnately lobed, with spiny teeth. Flowers solitary, terminal on branches, yellow or white; ovary with 4–7 placentas and numerous ovules. Capsule very spiny, opening by 4–7 short valves at the apex.

A genus of about 10 species, mostly of the southwestern United States and Mexico.

Argemone mexicana L., *Sp. Pl.* 1: 508 (1753). **FIG. 81.** PLATE 12.

THISTLE, THOM THISTLE

Weedy, glabrous herb to 1 m tall, usually shorter. Leaves very sharply spiny and noticeably glaucous, the lower ones 15 cm long or more, the upper ones much shorter. Flowers sessile at the apex of axillary or terminal branches, subtended by 1–3 leafy bracts; sepals ca. 2 cm long, acuminate with spiny apex; petals yellow, 2–3 cm long, in 2 series. Capsule 3–4 cm long, ellipsoid; seeds black, numerous, ca. 2 mm in diameter.

GRAND CAYMAN: *Brunt* 1958; *Hitchcock*; *Kings GC* 129, *GC* 161; *Proctor* 27993. CAYMAN BRAC: *Kings CB* 40; *Proctor* 29100.

— West Indies and continental tropical America; widely naturalized throughout the tropics as a weed of open waste ground. The yellow sap is said to be slightly caustic and useful for removing warts.

ULMACEAE

Trees or shrubs, unarmed or sometimes spiny, with alternate, simple leaves 3-nerved and often asymmetric at base; stipules small, soon falling. Flowers usually in cymose clusters, unisexual or rarely perfect, regular, small; if unisexual, the plants monoecious; perianth sepaloid, the 4 or 5 segments free or united. Staminate flowers with erect stamens equalling the perianth lobes in number and opposite them; rudimentary ovary (pistillode) present. Pistillate flowers with superior, 1-celled ovary containing 1 pendulous ovule; styles 2, simple or forked, stigmatic on the inner sides. Fruit (of Cayman species) a small hard drupe; seeds with straight embryo; endosperm none or scanty.

A world-wide family of about 15 genera and 200 species.

KEY TO GENERA

1. Leaves nearly smooth on upper (adaxial) surface; staminate perianth segments imbricate; pistillate perianth deciduous: **Celtis**

1. Leaves very rough on upper surface (like sandpaper); staminate perianth segments valvate; pistillate perianth persistent: **Trema**

Celtis L.

Small trees, shrubs, or scramblers, unarmed or spiny; leaves entire or serrate. Staminate and pistillate flowers mixed in small axillary cymose clusters appearing with young foliage, the staminate several or numerous, the pistillate solitary or few. Perianth usually 5-partite. Drupe subglobose, with a somewhat flattened, reticulate or tuberculate stone-like seed.

A chiefly Northern Hemisphere genus of about 80 species.

KEY TO SPECIES

1. Leaves serrate, very oblique at base; erect unarmed tree; styles simple; fruit pedicel longer than petioles: **C. trinervia**

1. Leaves entire, not oblique at base; scrambling shrub with spines; styles forked; fruit pedicel shorter than petioles: **C. iguanaea**

Celtis trinervia Lam., *Encycl. Méth. Bot.* 4: 140 (1797). **FIG. 82.** PLATE 12.

Small unarmed tree to 10 m tall; leaves obliquely ovate, mostly 4–10(–13) cm long, long-acuminate, the margins coarsely serrate except toward the base and apex, sparsely pubescent on both sides and finely black-dotted beneath. Staminate flowers greenish, in fascicles of 3–5; pistillate flowers usually solitary. Drupe globose-ovoid, 7–10 mm long, purple-black.

GRAND CAYMAN: *Proctor* 15127.

— Greater Antilles, in thickets and forests at low to medium elevations.

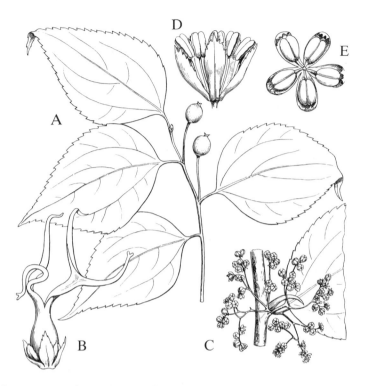

FIG. 82 **Celtis trinervia**. A, fruiting branch, × ¹/₂. B, **C. iguanaea**, pistillate flower × 4; C, staminate inflorescence, × ¹/₂; D, E, staminate flower, × 8. (F. & R.)

Celtis iguanaea (Jacq.) Sarg., *Silv. N. Amer.* 7: 64 (1895). **FIG. 82.** PLATE 12.

A scrambling or often high-climbing shrub armed with sharp recurved spines; leaves equilaterally ovate, elliptic or oblong (rarely slightly oblique at base), 5–12 cm long, blunt to subacuminate, the margins nearly entire, almost glabrous on both sides except for domatia in the main nerve-axils beneath. Flowers in short axillary cymes, the staminate in fascicles of 2–4, the pistillate solitary. Drupe globose-ellipsoid, 12–14 mm long, yellow.

LITTLE CAYMAN: *Proctor* 28027. CAYMAN BRAC: *Proctor* 29344. GRAND CAYMAN *P. Ann van B. Stafford et al.*, photos, found on the Old Man Bay Caves Trail, Canaan Land and Jasmin Lane.

— West Indies and continental tropical America, often growing in thickets on rocky limestone escarpments.

Trema Lour.

Unarmed trees or shrubs; leaves serrate and often with scabrid upper surface. Staminate and pistillate flowers mixed in small axillary cymose clusters, the perianth mostly 5-partite; pistillate flowers sometimes with functional stamens. Drupe surrounded by the persistent perianth; seed with scant but fleshy endosperm.

A pantropical genus of about 30 species, characteristic of clearings, thickets and forest-margins.

Trema lamarckiana (Roem. & Schult.) Blume, *Mus. Bot. Lugd.-Bat.* 2: 58 (1856).

A shrub or small tree to 7 m tall or more, with slender, rough-pubescent twigs. Leaves more or less lance-oblong, 2–6 cm long, acute at apex, very rough on upper side, reticulate-veined and finely tomentose beneath. Flowers greenish, ca. 2 mm across when open. Drupe ovoid, 2.8–3 mm long, red, glabrous.

GRAND CAYMAN: *Brunt* 1911, 2040; *Correll & Correll* 51018; *Kings* GC 134; *Proctor* 27930; *Sachet* 421. LITTLE CAYMAN: *Kings* LC 79.

— Florida and the West Indies south to St. Vincent, often common in thickets and on the borders of woodlands. The Cayman plants grow in sandy soils or in fragmented coral limestone.

MORACEAE

Trees or shrubs, rarely herbs, often with milky sap (latex); leaves alternate, entire, toothed or lobed either pinnately or palmately; stipules present. Flowers unisexual, in spikes or heads, or on a flat entire or lobed receptacle, or on the inside of a closed receptacle; perianth sepaloid, of 2–5 (usually 4) free or fused segments. Male flowers with stamens generally as many as the perianth lobes and opposite them, or only 1. Female flowers with 1-celled ovary, superior to inferior, bearing a simple, 2-toothed or 2-partite style; ovule 1. Fruit 1-seeded, free or more commonly united with fruits of other flowers to form a more or less fleshy pseudocarp, in whose formation the receptacle may also be involved.

A mostly tropical family of about 53 genera and more than 1,000 species.

KEY TO GENERA

1. Flowers borne on the inner surface of a more or less globose, hollow receptacle, this having a small opening at the apex that is concealed by scales:	**Ficus**
1. Flowers variously arranged but never inside a closed receptacle:	
2. Fruit a small globose head less than 1.5 cm in diameter; wild dioecious tree with spines:	**Maclura**
2. Fruit a large, head-like syncarp more than 10 cm in diameter; cultivated monoecious trees without spines:	**[Artocarpus]**

Ficus L.

Trees or shrubs, rarely vine-like, often epiphytic at least when young, with milky or rarely watery sap. Leaves entire (in American species); stipules small, usually soon falling. Staminate and pistillate flowers mingled on the inner surface of a hollow, usually globose, fleshy receptacle, this with a small opening (ostiole) at the apex, covered or concealed by several small scales; receptacles axillary, each subtended by a lobate involucre of bracts. Staminate perianth of usually 3 small segments; stamens usually 1 or 2. Pistillate perianth of 4–6 small segments; ovary with 1 pendulous ovule. Fruit a small, seed-like achene, these numerous within a single receptacle, which as a whole is known as a 'fig'. All figs are edible, but most are rather insipid.

A pantropical genus of more than 600 species. The true edible fig, *Ficus carica* L. of the Mediterranean region, is widely cultivated.

KEY TO SPECIES

1. Figs distinctly stalked:	F. citrifolia
1. Figs sessile in leaf-axils:	
2. Leaves obtuse; wild shrub or tree:	F. aurea
2. Leaves sharply acuminate; cultivated tree:	[F. benjamina]

Ficus citrifolia Mill., *Gard. Diet.* ed. 8 (1768). PLATE 13.

Ficus laevigata Vahl (1805).
Ficus populnea Willd. (1806).*Ficus brevifolia* Nutt. (1846)

BARREN FIG, WILD FIG

A much-branched glabrous tree up to 15 m tall, or shrub-like in exposed situations, often with dense masses of aerial roots hanging from the branches; when these roots reach the ground they thicken to form supplementary trunks or rib-like extensions of the main trunk. Leaves ovate to elliptic, long-petiolate, 5–10 cm long, 2–5 cm broad, more or less acute at apex, truncate or subcordate (rarely acutish) at base of blade, the upper surface dotted with numerous minute elevated papillae (cystoliths). Ripe figs yellow, sometimes red-spotted, globose, 7–12 mm in diameter, on slender peduncles usually 5–10 mm long, rarely longer.

GRAND CAYMAN: *Hitchcock*; *Kings* GC 109, GC 130; *Proctor* 15124. LITTLE CAYMAN: *Kings* LC 88A. CAYMAN BRAC: *Proctor* 28982, 29085.

— Florida, the West Indies and parts of C. America. Common, most frequent in rocky woodlands, coastal thickets and along the base of cliffs; occasionally cultivated. The island of Barbados is supposed to have been named for this tree, whose abundant aerial roots often give it a 'bearded' appearance.

Ficus aurea Nutt., *Sylva* 2: 4 (1846). **FIG. 83.** PLATE 13.

Ficus dimidiata Griseb. (1859).

WILD FIG

A much-branched glabrous tree up to 20 m tall, or diffuse and shrub-like in exposed situations, in habit like *Ficus citrifolia*. Leaves elliptic to obovate, long-petiolate, 6–16 cm long, 2–7 cm broad, rounded or subacute at apex, narrowed or truncate at the base of the blade; minute punctate cystoliths present or absent on upper surface. Figs sessile and subtended by broad bracts, when ripe rose-red to crimson, depressed-globose, 4–7 mm in diameter.

GRAND CAYMAN: *Brunt* 1929, 1992a; *Hitchcock*; *Proctor* 15029. LITTLE CAYMAN: *Kings* LC 87, LC 88; *Proctor* 28042. CAYMAN BRAC: *Kings* CB 96; *Proctor* 28937.

— Florida, the Bahamas and the Greater Antilles except Puerto Rico, also in the Swan Islands, frequent in coastal thickets and inland at low to medium elevations; in favourable situations, it can become an enormous tree with a trunk as much as 2 m in diameter.

[**Ficus benjamina** L., a large Malayan species with glossy, ovate, abruptly acuminate leaves, 4–10 cm long. It is sometimes planted as a shade tree. GRAND CAYMAN: *Proctor* 15149.]

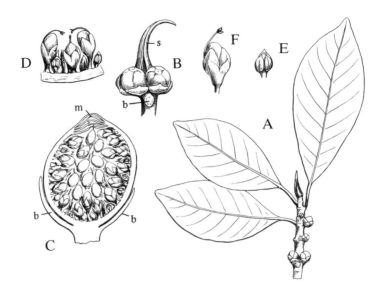

FIG. 83 **Ficus aurea**. A, flowering branch, × ½. B, apex of shoot with two figs; s, stipule; b, leaf-scar. C, vertical section of fig, × 4; b, basal bracts; m, ostiole. D, individual flowers within a fig, × 5. E, staminate flower, × 8. F, pistillate flower, × 8. (F. & R.)

Maclura Nutt.

Dioecious trees with yellowish latex, often armed with spines; leaves entire or toothed, pinnately veined; stipules small, soon falling. Staminate flowers in axillary catkin-like spikes; perianth 4-parted; stamens 4, inflexed in bud; ovary rudimentary. Pistillate flowers in dense axillary heads, mingled with bracts similar to the perianth segments; perianth 4-parted, the oblique ovary included, the style filiform, exserted; ovule laterally attached. In fruit, the perianths become somewhat fleshy and are tightly crowded together to form a globose syncarp; achenes compressed, oblique at apex.

A genus of perhaps a dozen species, one temperate American, one Mexican, one tropical American, and the rest occurring in Africa and Madagascar.

Maclura tinctoria (L.) D. Don ex Steud., *Nom. ed.* 2, 2: 87 (1841). **FIG. 84.**

Chlorophora tinctoria (L.) Gaudich. ex Benth. in Benth. & Hook. (1880).

FUSTIC

A deciduous tree up to 10 m tall or more, with horizontal, often wide-spreading branches frequently armed with spines. Leaves lanceolate to oblong-elliptic or elliptic, short-petiolate, 3–8 cm long or more, mostly 1.5–3 cm broad, blunt to acuminate at apex, the margins entire or toothed, and chiefly glabrous except for minute hairs along the midrib beneath. Staminate catkins pendulous, 3–5 cm long or more, on short peduncles. Pistillate heads green, globose, 5–8 mm in diameter when flowering, increasing to 12 mm or more when in fruit.

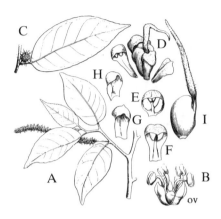

FIG. 84 **Maclura tinctoria**. A, branch with Staminate inflorescence, × ¹/₄. B, Staminate flower, × 4; **ov**, rudimentary ovary. C, leaf with pistillate inflorescence, × ¹/₄. D, portion of pistillate inflorescence dissected, showing flower with enveloping bracts, × 4. E, F, gland-bearing bracts, × 4. G, H, outer and inner perianth segments, × 4. I, pistil, × 4. (F. & R.)

GRAND CAYMAN: *Brunt* 2014; *Kings* GC 408; *Proctor* 27982. CAYMAN BRAC: *Proctor* 52170.

— West Indies and continental tropical America, common in woodlands at low to medium elevations. The fruits are edible (though scarcely palatable!); the wood is very tough and close-grained, and has many uses; also, a yellowish dye known as fustic or khaki is extracted from the wood and commonly used for dyeing the cloth used for military uniforms.

[*Artocarpus* J. R. & G. Forst.

Monoecious trees with white latex; leaves medium-sized to large, entire or lobed, pinnately veined; stipules paired, often large and papery. Inflorescences unisexual, axillary or cauliflorous on old wood. Staminate flowers in stout, cylindric spikes; perianth 2–4-lobed, with 1 exserted stamen. Pistillate flowers in fleshy more or less globose heads in which they are mixed with peltate bracts, all being fused together; style entire or 2-lobed. Mature syncarp (compound fruit) formed by enlargement of the entire pistillate head, when fertile thus enclosing numerous achenes each representing an individual fruit; seeds large, lacking endosperm but with fleshy cotyledons.

A genus of 47 species indigenous to the Indo-Malayan region eastward to the Solomon Islands. Two species, *Artocarpus altilis* (Parkinson ex F. A. Zorn) Fosberg (breadfruit) and *Artocarpus heterophyllus* Lam. (jackfruit), are cultivated throughout the tropics for their edible fruits. Both of these occur in the Cayman Islands, but no voucher specimens have been noted, except one for *A. altilis*, GRAND CAYMAN: *Sachet* 446. The ordinary breadfruit is a seedless cultivar propagated by suckers; seeded breadfruit trees are also grown on many West Indian islands.]

URTICACEAE

Herbs, shrubs or rarely small trees, without latex, sometimes armed with stinging hairs, and often with epidermal inclusions (cystoliths). Leaves opposite or alternate; stipules present, sometimes conspicuous. Flowers small and usually wind-pollinated, regular, unisexual, monoecious or dioecious, usually in cymose clusters or panicles; perianth segments 2–5, free or united, sepaloid. Staminate flowers with stamens as many as and opposite to the perianth segments, incurved in bud, straightening to an exserted position and dehiscing explosively when mature. Pistillate flowers with superior ovary, 1-celled with 1 basal ovule. Fruit an achene, sometimes enclosed in the persistent perianth.

A family of about 45 genera and 900 species, widely distributed in tropical and temperate regions.

Pilea Lindl.

Annual or perennial herbs, sometimes shrubby at base. Leaves opposite, those of a pair usually unequal in size, usually 3-veined, with conspicuous epidermal cystoliths; stipules connate, intrapetiolar. Flowers minute, monoecious or dioecious, the clusters axillary. Staminate perianth usually 4-partite; pistillate perianth 3-partite, the segments usually unequal, each subtending a scale-like staminode; stigma sessile, shortly penicillate. Achene compressed, more or less enclosed by the persistent perianth, the median segment of which is usually larger than the lateral ones. The achenes in at least some species are expelled by the catapult action of the enlarged inflexed staminodes, similar in mechanism to the explosive ejection of the pollen from staminate flowers.

A very large pantropical genus of more than 500 species, notable for their local endemism. More than two-thirds of the species occur in the American tropics, about 180 of them in the West Indies. A few species are commonly grown as ornamental pot plants or as foliage plants in gardens.

KEY TO SPECIES

1. Stems erect, spreading or tufted; leaf-blades oblanceolate to obovate, contracted gradually into the petiole, in strongly unequal pairs, the larger leaf of a pair 3–12 mm long: **P. microphylla**

1. Stems mostly prostrate, forming a delicate mat with short erect flowering branches; leaf-blades suborbicular, abruptly contracted into the petiole, in subequal or moderately unequal pairs, the larger leaf of a pair up to ca. 4 mm long including the filiform petiole: **P. herniarioides**

Pilea microphylla (L.) Liebm. in *Danske Vid. Selsk. Skr.* V 2: 296 (1851).

A glabrous, pale green herb, annual or perennial in various forms, with succulent, freely branched stems, these usually more or less erect, rarely prostrate or pendent. Leaves small, entire, 1-nerved, varying in shape from broadly obovate to oblanceolate or elliptic, the larger 2–12 mm long, the upper surface beset with linear cystoliths transversely arranged.

Flowers in very small sessile or stalked cymules, these monoecious or unisexual. Achenes 0.5 mm long, slightly exceeding the persistent perianth segment.

243

This taxon is a pantropical complex of differing forms, some of them weeds of disturbed soil, others occurring in various natural habitats. Several horticultural variants are known, often called lace-plant or artillery-plant. One of the best-known is var. *trianthemoides* (Sw.) Griseb., a low shrubby herb with arching branches, which is commonly cultivated as an edging for flower-beds. Two varieties occur as wild plants in the Cayman Islands; these can be characterized as follows:

1. Annual; stems 0.5–1 mm thick; leaves thin; plants of damp
   shady ground near paths or habitations: var. **microphylla**
1. Perennial; stems to 2.5 mm thick; leaves thick, succulent;
   plants of exposed limestone crevices: var. **succulenta**

Pilea microphylla var. *microphylla*

Characters as given in the key; upper leaf-surface with conspicuous parallel elongate-linear cystoliths.

GRAND CAYMAN: *Kings GC 352.*
— Pantropical.

Pilea microphylla var. *succulenta* Griseb., *Fl. Br. W.I.* 155 (1860).

Characters as given in the key; cystoliths much shorter, less conspicuous and less regularly arranged. The Cayman plants have much more narrow leaves than the typical form in Jamaica, with the underside lacking the minutely foveolate (pitted) texture found in the latter.

GRAND CAYMAN: *Kings GC 393; Proctor 15209.* CAYMAN BRAC: *Kings CB 68, CB 100; Proctor 29043, 29118.*
— Jamaica, Hispaniola, Puerto Rico and the Virgin Islands.

Pilea herniarioides (Sw.) Wedd., *Ann. Sci. Nat. sér.* 3, Bot. 18: 207 (1852).

Stems filiform, 0.2–0.5 mm thick, few-branched, often rooting at the nodes. Leaves thin and delicate, with blades 1–3 mm long and broad, usually glabrous; cystoliths short-linear, more or less transversely scattered on the upper surface; petioles filiform, 1–2 mm long. Cymules in upper or terminal axils, sessile or stalked; perianth of staminate flowers 1.2–1.5 mm long, the pistillate flower smaller. Achenes ovoid-ellipsoid, 0.6–0.7 mm long.

GRAND CAYMAN: *Hitchcock,* Jan. 18 1891.
— Widely distributed in the Caribbean region from Florida to Curacao; also in Costa Rica. This species is quite rare throughout its range, usually occurring in hollows of honeycombed limestone near the sea. Some authors have treated it as a variety of *Pilea microphylla.*

MYRICACEAE

Aromatic shrubs or small trees; leaves alternate, simple, entire or toothed (rarely lobed) resin-dotted; stipules usually absent. Flowers small, unisexual, monoecious or dioecious, sessile under scale-like bracts and grouped in short axillary spikes; bracteoles often present within the bract. Perianth lacking; staminate flower with usually 4–8 stamens inserted on a receptacle, the filaments short, distinct or slightly united; anthers 2-celled, dehiscing longitudinally. Pistillate flower a solitary 1-celled ovary subtended by 2–8

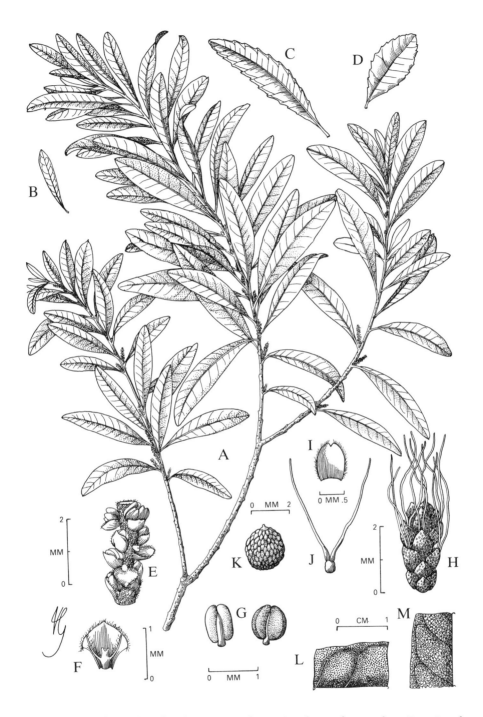

FIG. 85 **Myrica cerifera.** A, branch with staminate spikes. B, C, D, leaves of various form. E, portion of staminate inflorescence. F, bract of staminate flower with filaments. G, two views of anther. H, pistillate inflorescence. I, bracteole of pistillate flower. J, pistillate flower. K, fruit. L, upper surface of leaf. M, lower surface of leaf. (G.)

bracteoles; ovule 1, erect; style short and bearing 2 linear stigmas. Fruit a small oblong or globose drupe covered with resinous or waxy papillae; seed without endosperm, the embryo straight with flat cotyledons.

The family consists of a single genus with about 40 species, widely distributed in tropical, temperate and arctic regions.

Myrica L.

With the characters of the family.

Myrica cerifera L., *Sp. Pl.* 2: 1024 (1753). **FIG. 85.**

BAYBERRY

A shrub up to 4 m tall or more, the twigs glabrate or sparsely pubescent; leaves short-petiolate, narrowly oblanceolate, 5–10 cm long, with acute apex and attenuate base, the margins entire or with a few teeth, the surfaces with numerous minute golden resin-dots, those of the upper side sunken in small pits. Staminate spikes sessile, mostly 1 cm long or less; pistillate spikes longer. Drupes globose, 2–3 mm in diameter, densely covered with whitish wax.

GRAND CAYMAN: *Brunt* 1950; *Proctor* 31042.

— Eastern United States, the West Indies, and Mexico to Costa Rica, in widely various habitats. The Cayman plants grow in swampy woodlands toward the interior of the island. The wax of the fruits is often extracted in the United States and C. America for the making of candles; these burn with a pleasing aroma.

[CASUARINACEAE

Trees of pine-like aspect, with green, jointed, angular, striate branchlets that perform the function of leaves; leaves reduced to minute scales, these borne in whorls at the nodes and often short-connate into a sheath. Flowers unisexual, the plants monoecious; staminate flowers borne in narrow cylindric spikes on the ends of short lateral branchlets, several bracts combining at each node of the spike to form a serrate cup over the edge of which hang several stamens; each stamen represents a male flower with a concealed 1- or 2-segmented perianth and 2 bracteoles. Pistillate flowers in lateral, dense spherical heads, each female flower in the axil of a bract, without a perianth but subtended by 2 bracteoles; ovary small, 1-celled, with 2 long-exserted stigmas; ovules 1 or 2. After fertilization the pistillate flower-head becomes hard and cone-like, the woody bracts and bracteoles subtending winged achenes. Seed solitary, without endosperm.

A taxonomically isolated family, with a single genus of more than 40 species, these mostly Australian.

Casuarina Adans.

With the characters of the family.

Casuarina equisetifolia L., *Amoen. Acad.* 4: 123, 143 (1759); J. R. & G. Forst. (1776). **FIG. 86.**

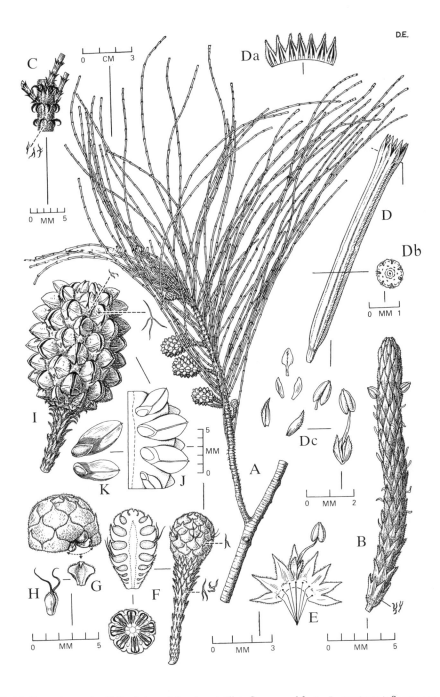

FIG. 86 **Casuarina equisetifolia.** A, branch bearing pistillate flowers and fruits. B, staminate inflorescence. C, portion of staminate spike with scales removed. D, perianth-like cup of united bracts; a, apex spread out; b, cross-section; c, single staminate flower and separated parts. E, staminate flower in position on spread-out apex of bract cup. F, pistillate inflorescence in various aspects. G, pistillate bract. H, pistillate flower. I, fruiting inflorescence. J, long. section through portion of fruiting inflorescence. K, achenes. (D.E.)

247

Casuarina litorea L. in Stickman (1754) (illegit.).

AUSTRALIAN PINE, WEEPING WILLOW, BEEFWOOD

A fast-growing tree up to 20 m tall or more, assuming a tall, narrow but open shape resembling a conifer; trunk to 1 m in diameter; branchlets pale or rather dark green, slender (resembling pine-needles), 0.6–0.8 mm thick; scales in whorls of 6–8, 1–3 mm long, acute, ciliate. Staminate spikes 1–4 cm long. Fruiting heads globose, ca. 1.5 cm in diameter.

GRAND CAYMAN: *Kings GC 83; Proctor* 15144, 31051. CAYMAN BRAC: *Proctor* (sight record).

— Native of Australia, but widely planted and naturalized in most warm countries; common in the West Indies, especially in sandy or gravelly soils near the sea, but also flourishing inland. It is not definitely known when this species was introduced to the Cayman Islands, but this probably occurred during the second half of the nineteenth century. It is now well-established and naturalized in many localities. It is often planted and trimmed as a hedge, and is very useful as a windbreak near seashores because of its high tolerance of salt. The wood is very hard, heavy and fine-grained, but unfortunately very susceptible to attack by termites.]

PHYTOLACCACEAE

Herbs, shrubs or trees, mostly glabrous; leaves alternate, simple, entire; stipules minute or lacking. Flowers regular, perfect, in terminal and axillary racemes or spikes; perianth of 4 or 5 segments, persistent in fruit. Stamens 4, 5, or more, variously inserted. Ovary superior, of 1–several 1-ovuled carpels, the ovules basal; style short or lacking. Fruit an achene or berry of 1–several carpels; seeds with mealy or fleshy endosperm.

A widespread, chiefly tropical family of up to 22 genera and about 120 species. Species of *Phytolacca* provide an edible herb used in cooking in some countries.

KEY TO GENERA

1. Erect herb; stamens 4; stigma capitate:	**Rivina**
1. Climbing or scrambling shrub; stamens 8–16; stigma penicillate:	**Trichostigma**

Rivina L.

Erect herbs, somewhat shrubby at base; leaves alternate or subopposite. Flowers in slender terminal and axillary racemes, the bracts and bracteoles minute and deciduous. Perianth 4-partite, the segments not enlarging in fruit; stamens 4; ovary of 1 carpel, the stigma capitate on a very short style. Fruit a small globose berry with fleshy pericarp; seed with annular embryo and mealy endosperm.

A small tropical American genus of about 3 species.

Rivina humilis L., *Sp. Pl.* 1: 121 (1753). PLATE 13.
Rivina humilis var. *glabra* L. (1753).
Rivina humilis var. *laevis* (L.) Millsp. (1900).

FOWL BERRY, BLOOD BERRY

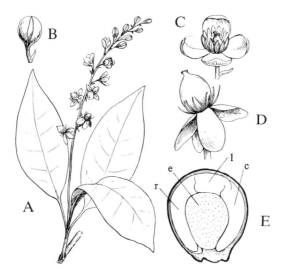

FIG. 87 **Trichostigma octandrum.** A, portion of branch with inflorescence, × ²/₃. B, flower-bud, × 2.
C, the same, opened, × 2. D, fruit, × 3. E, section of seed, × 5; **e**, endosperm; **c**, cotyledons; **1**, lobed base
of cotyledons; **r**, radicle. (F. & R.)

A perennial herb up to 1.5 m tall (often much smaller), glabrous varying to finely
pubescent. Leaves thin, long-petiolate, 2–14 cm long, the blades ovate to lanceolate, acute
to acuminate at apex, obtuse or rounded at base. Racemes lax, 3–10 cm long; perianth
white or creamy, 2–3 mm long, the stamens shorter. Fruit a crimson berry, at length
becoming dry, ca. 3 mm in diameter when dry.

GRAND CAYMAN: *Brunt* 1891, 2030; *Hitchcock*; *Kings* GC 71; *Lewis* GC 53, 3832;
Millspaugh 1343, 1352; *Proctor* 15102. LITTLE CAYMAN: *Proctor* 28036. CAYMAN
BRAC: *Kings* CB 47, CB 75, CB 101; *Proctor* 29075.

— Florida, Texas, West Indies and continental tropical America, frequent in shady
thickets and glades, also sometimes a weed of gravelly waste ground.

Trichostigma A. Rich.

Scrambling or vine-like shrubs; leaves alternate or subopposite, with minute, deciduous
stipules. Flowers in terminal and axillary racemes, with deciduous bracts and persistent
bracteoles. Perianth 4-partite, the segments somewhat enlarging and becoming reflexed in
fruit; stamens 8–16; ovary of 1 carpel, the stigma sessile, brush-like. Fruit a globose, 1-
seeded berry, the leathery pericarp adhering to the seed.

A tropical American genus of 3 species.

Trichostigma octandrum (L.) H. Walt, in *Engler. Pflanzenr.* 4(83): 109 (1909).
FIG. 87. PLATE 13.

A glabrous climbing or sprawling shrub, the stems scrambling over other vegetation often to
a height of 6 or 7 m with long, slender branches. Leaves membranous, long-petiolate, blades
elliptic or lanceolate, in all 3–10 cm long or more, 1.5–4 cm broad, acuminate at apex, mostly

249

cuneate at base. Racemes equal in length to the leaves or longer; perianth white or greenish, turning red or purple in fruit. Berry purple to nearly black, ca. 5 mm in diameter.

CAYMAN BRAC: *Proctor* 15323, 52160.

— Florida, West Indies, and continental tropical America, in rocky thickets and woodlands. The berries are often eaten by birds.

NYCTAGINACEAE

Herbs, shrubs or trees; leaves opposite or sometimes partly alternate, simple, entire or nearly so; stipules lacking. Flowers regular, unisexual or perfect, usually in terminal or axillary cymes, the bracts minute or (in *Mirabilis*) forming a calyx-like involucre. Perianth segments usually 5, fused to form a tube, the base of which persists on the ripe fruit. Stamens 2–10, typically 5, the filaments often unequal in length. Ovary superior, enclosed by the base of the perianth tube, 1-celled; style elongate, with capitate stigma. Fruit (called an anthocarp) consisting of an indehiscent utricle enclosed by the enlarged adhering base of the perianth; seed with large cotyledons enclosing the endosperm.

A pantropical family most abundant in the Western Hemishpere, with about 25 genera and more than 250 species in all. The ornamental woody vines *Bougainvillea spectabilis* Willd., *B. glabra* Choisy and various hybrids of these, are widely cultivated.

KEY TO GENERA

1. Plants herbaceous:	
2. Perianth very small (less than 3 mm long), subtended by minute bracts:	**Boerhavia**
2. Perianth large and showy (up to 5 cm long), subtended by a calyx-like involucre:	**Mirabilis**
1. Plants woody:	
3. Fruit dry, with 5 rows of glands:	**Pisonia**
3. Fruit drupe-like, fleshy, without glands:	**Guapira**

Boerhavia L.

Annual or perennial herbs often growing from a woody taproot, the main stem bifurcate near the base, otherwise erect, prostrate or diffuse. Leaves opposite. Flowers very small, perfect, sessile or subsessile in small clusters on panicles; bracts minute. Perianth 5-lobed, bell- or funnel-shaped, with tubular base surrounding the ovary. Stamens 2 (rarely 3), the filaments fused at base. Ovary narrowed towards the base; style with peltate stigma. Anthocarp small, angled or ribbed.

A pantropical genus of about 40 species, often weeds of open waste ground.

KEY TO SPECIES

1. Anthocarps glabrous, lacking glandular hairs; flowers white or pale rose:	**B. erecta**
1. Anthocarps with glandular hairs; flowers crimson:	
2. Anthocarps oblong-clavate, 4–6 mm long; leaves usually roundish at apex:	**B. diffusa**
2. Anthocarps obovoid-clavate, 2.5–4 mm long; leaves usually acutish at apex:	**B. coccinea**

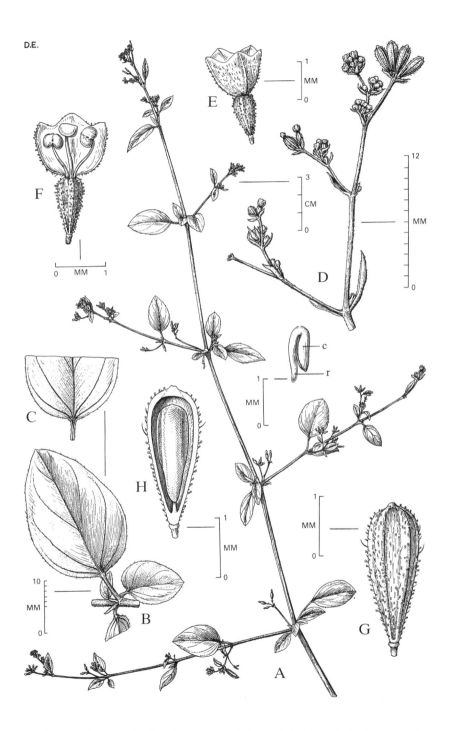

FIG. 88 **Boerhavia coccinea.** A, habit. B, node with leaves. C, base of leaf. D, branch with flowers and fruits. E, single flower. F, same with portion of perianth removed. G, anthocarp. H, same in long. section. (D.E.)

Boerhavia erecta L., *Sp. Pl.* 1: 3 (1753).

BROOMWEED

Annual herb, erect or ascending, 15–100 cm tall, the stems glabrate or sparsely puberulous. Leaves deltate-ovate or lanceolate, petiolate, 2–6 cm long, the apex sharply acute, the margins sinuate-repand, the lower surface whitish. Flowers white or pale rose, 2–6 in a cluster, many of these on a diffuse panicle; perianth limb 1.5–2 mm long. Anthocarp club-shaped, 5-angled, 3–4 mm long, glabrous.

GRAND CAYMAN: *Brunt* 1711; *Correll & Correll* 51025A; *Kings* GC 301; *Millspaugh* 1277. LITTLE CAYMAN: *Proctor* 28167. CAYMAN BRAC: *Proctor* 29114.

— Southern United States, West Indies and continental tropical America, a frequent weed of sandy paths and clearings at low elevations.

Boerhavia diffusa L., *Sp. Pl.* 1: 3 (1753).

Boerhavia paniculata Rich. (1792).

CHICKWEED

Perennial herb with woody taproot, the main branches prostrate or decumbent, the flowering branches ascending; glabrate or minutely puberulous, often glutinous. Leaves elliptic to roundish, petiolate, 2–6 cm long, the apex obtuse or roundish, the margins somewhat irregular and minutely ciliate, the lower surface pale or whitish. Flowers crimson, 2–6 in a cluster, the perianth limb ca. 1 mm long. Anthocarp oblong-clavate with truncate apex, glandular-hairy, 4–6 mm long.

LITTLE CAYMAN: *Kings* LC 59; *Proctor* 28166. CAYMAN BRAC: *Millspaugh* 1189.

— West Indies and continental tropical America, a weed of sandy clearings.

Boerhavia coccinea Mill., *Gard. Dict.* ed. 8 (1768). **FIG. 88**.

Boerhavia hirsuta Willd. (1794).

CHICKWEED

A loosely branched herb with long, somewhat sprawling stems, these dark and minutely puberulous. Leaves broadly ovate or subrhombic, petiolate, mostly 1–5 cm long, the apex obtuse to acute, sometimes apiculate, the margins somewhat irregular and minutely ciliate. Flowers crimson, 2–5 in a cluster, the perianth limb 1.5–2 mm long. Anthocarp narrowly obovoid-clavate with rounded apex, glandular-hairy, 2.5–4 mm long.

GRAND CAYMAN: *Correll & Correll* 51025B; *Proctor* 15056. CAYMAN BRAC: *Kings* CB 21; *Proctor* 29113.

— West Indies and continental tropical America, a weed of sandy waste ground.

Mirabilis L.

Perennial herbs with opposite leaves. Flowers perfect, in cymes; bracts forming a 5-lobed, calyx-like involucre subtending the perianth. Perianth segments 5, fused to form a long, funnel-shaped tube, deciduous after flowering from a constriction level with the top of the ovary. Stamens 5 or 6, unequal in length, exserted; filaments fused at base to form a fleshy cup. Stigma capitate, hairy, at the end of a long-exserted style. Anthocarp ribbed; testa of seed adhering to the pericarp.

A tropical American genus of about 24 species.

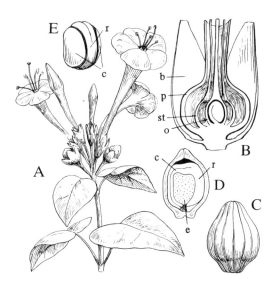

FIG. 89 **Mirabilis jalapa**. A, portion of plant, × ²/₃. B, lower part of flower cut lengthwise, × 4; b, bract; p, perianth; st, stamen; o, ovary. C, anthocarp, × 2. D, same cut lengthwise, × 2; c, cotyledons; r, radicle; e, endosperm. E, embryo, × 2; c, cotyledons; r, radicle. (F. & R.)

Mirabilis jalapa L., *Sp. Pl.* 1: 177 (1753). **FIG. 89**.

FOUR-O'CLOCK

An erect herb up to 60 cm tall or more, freely branched. Leaves long-petiolate, 2–16 cm long, the blades ovate and long-acuminate, sometimes unequal-sided at base, glabrous or minutely pubescent. Flowers in small terminal clusters, with calyx-like involucres 7–8 mm long; perianth opening about 4 p.m. (hence the common name), usually red, purple or white, ca. 5 cm long. Anthocarp black, ovoid, ca. 1 cm long, ribbed and tuberculate.

GRAND CAYMAN: *Millspaugh* 1402. CAYMAN BRAC: *Proctor* (sight record).

— Florida and Texas, West Indies and continental tropical America, often cultivated and naturalized so that its true natural range is uncertain. It also occurs as an escape from cultivation in the Old World tropics.

Pisonia L.

Trees, shrubs or woody vines, some species armed with spines. Leaves mostly opposite. Flowers small, unisexual, in cymose panicles; plants dioecious. Staminate perianth funnel-shaped, 5-toothed, with 6–10 exserted stamens, the filaments slightly fused at the base. Pistillate perianth tubular, 5-toothed, with short usually sterile stamens; ovary with exserted style and capitate stigma. Anthocarps dry and rather hard, with 5 rows of extremely viscid stalked glands (rarely the glands restricted to the distal end).

A pantropical genus of about 30 species. The viscid-glandular anthocarps are capable of ensnaring and disabling birds.

KEY TO SPECIES

1. Scrambling woody vine armed with recurved spines; leaves glabrous, acute or acuminate: P. aculeata

1. Erect bushy shrub or very small tree without spines; leaves puberulous, broadly rounded or blunt at apex: P. margaretiae

Pisonia aculeata L., *Sp. Pl.* 2: 1028 (1753). PLATES 13 & 14.

A deciduous scrambling shrub often climbing high in trees, armed with stout recurved spines; young twigs minutely puberulous. Leaves rhombic-elliptic to ovate, long-petiolate, 3–15 cm long, the blades acute to acuminate at apex (rarely blunt or rounded), attenuate at base, glabrate or sparsely puberulous beneath. Cymes axillary and terminal, compact, puberulous; flowers yellowish-green, fragrant. Staminate perianth ca. 3.5 mm long, the pistillate shorter. Anthocarps ellipsoid or clavate, 8–15 mm long, 5-angled.

LITTLE CAYMAN: *Proctor* 35078. CAYMAN BRAC: *Proctor* 28960.

Pantropical, variable, uncommon in the Cayman Islands. In Little Cayman, it occurs in dense rocky virgin forest near the center of the island; in Cayman Brac, it grows in dense thickets along the base of cliffs.

Pisonia margaretiae Proctor, sp. nov. PLATE 14.

Unarmed shrub or small tree to 5 m tall with smooth bark; ultimate twigs and terminal bud densely and minutely puberulous. Leaves broadly ovate to rotund, rounded to blunt and minutely apiculate at apex, shallowly cordate at base, long-petiolate, the puberulous petioles 2.5–8 cm long, the blades 6–15 cm long, 6–13 cm broad below the middle; venation strongly prominulous beneath, the midvein densely puberulous especially beneath, the secondary and tertiary veins puberulous mainly on the lower side. Plants mostly dioecious but occasionally sequentially monoecious. Staminate inflorescence pale green, densely subglobular to hemispheric, mostly 2–2.5 cm in diameter, the individual flowers 7–9 mm long. Pistillate inflorescence densely cymose, becoming diffuse at maturity; anthocarps 8–10 mm long, 3–5 mm thick, 5-ribbed, the ribs glandular throughout from apex to base.

Differs from *Pisonia acueata* by lacking spines and having leaves puberulous rather than glabrous and leaves rounded to obtuse at the apex rather than acute or acuminate.

Pisonia margaretiae Proctor, sp. nov. a *P.acueata* spinis carentibus, foliis puberulis (non glabris) ad apicem rotundatis vel obtusis (non acutis neque acuminatis) differt.

GRAND CAYMAN: *Margaret Barwick* s. n., Apr. 1996; *Proctor* 48679, 49213, 49239, 49343 (type, IJ; isotype SJ), 52071, 52113, 52142.

Endemic. This little tree, originally discovered by Margaret Barwick (to whom the species is dedicated) on the estate of Arthur Hunter near Bats Cave, is chiefly known from the Spots area in the vicinity of Jasmin Lane. It has been taken into cultivation at the Queen Elizabeth II Botanic Park.

Guapira Aubl.

Unarmed trees or shrubs with opposite, often somewhat fleshy leaves. Flowers small, unisexual, in cymose panicles similar to those of *Pisonia*; plants dioecious. Stamens usually 10, exserted; stigma multifid. Anthocarp more or less fleshy, drupe-like, obovoid to ellipsoid or subglobose, lacking viscid glands.

A tropical American genus of 65 species or more. There is considerable doubt that this taxon can be maintained as sufficiently distinct from *Pisionia*, and some authors combine them under the latter name, which has priority.

Guapira discolor (Sprengel) Little, *Phytologia* 17: 368 (1968). PLATE 14.

Pisonia discolor Sprengel (1825).
Torrubia discolor (Sprengel) Britton (1904).

CABBAGE TREE

A shrub or small tree to about 6 m tall, with drooping branches; leaves thin, glabrous, long-petiolate, variable in size and shape, 2–8 cm long, blades oblong to narrowly or broadly elliptic, rarely obovate, obtuse at both ends or cuneate at base. Cymes lax, minutely puberulous, rather few-flowered, the flowers sessile or nearly so; staminate perianth ca. 3.5 mm long, the pistillate a little shorter. Fruit crimson, ellipsoid, 5–8 mm long.

GRAND CAYMAN: *Kings* GC 106; *Proctor* 15082. LITTLE CAYMAN: *Kings* LC 37; *Proctor* 28037, 28061. CAYMAN BRAC: *Proctor* 28958.

— Bahamas and Greater Antilles, frequent in rather dry, rocky coastal woodlands. This species appears to be the chief Cayman host of the parasitic mistletoe *Phoradendron rubrum*, known locally as scorn-the-ground.

CACTACEAE

Succulent perennials, often very woody, of various habit and usually spiny, the stems rounded, ribbed, angular or tuberculate, the surface dotted with small hairy cushions (areoles) from which variously arise branches, spines, flowers, hairs, glands, and leaves (if any). Leaves absent, or small and scale-like if present (foliaceous in *Pereskia*), soon falling; stipules absent. Spines of various types, usually clustered, sometimes lacking. Flowers regular, mostly bisexual, usually solitary or sometimes in clusters; receptacle united with the inferior ovary (collectively called the pericarpel), sometimes prolonged above it as a perianth tube. Perianth of few to numerous segments, often intergrading from sepaloid to petaloid or else sharply differentiated. Stamens usually numerous, the filaments usually borne on the throat of the perianth. Ovary 1-celled, sessile or partly immersed in a branch; ovules numerous on several parietal placentas; style long and simple, terminated by a several lobed stigma. Fruit a many-seeded berry, usually fleshy.

A large family of 124 genera and 1,816 taxa, confined in the natural state to the Western Hemisphere except for the genus *Rhipsalis*. A number of species of other genera (especially *Opuntia*) are now naturalized in the Old World and Australia. The curious spiny stems of cacti, which are adapted to endure extreme drought, are valued for their ornamental or bizarre ap pearance; in addition, many species bear beautiful flowers. Some species are planted to form "living fences". The large nocturnal flowers of various species, known as night-blooming cereus, are prized in horticulture. The fruits of many species are edible.

The nomenclature and taxonomy of the Cactaceae of the Cayman Islands have been updated on the basis of The New Cactus Lexicon and follow the lead of the world's authorities on the Cactaceae. Thanks to Nigel Taylor of the Royal Botanic Gardens, Kew for his advise and assistance.

KEY TO GENERA

1. Ultimate branches more or less flattened:

    2. Plants scrambling and vine-like, the lower stems thin and woody; spines lacking (except on seedlings); flowers with long slender perianth tube: **Epiphyllum**

    2. Plants erect, not vine-like; all parts thick and succulent; spines and glochids usually present, barbed; an obvious perianth tube absent: **Opuntia**

1. All parts of stems cylindric and ribbed:

    3. Stems erect, never vine-like:

        4. Flowering areoles bearing conspicuous tufts of white hairs; ovary naked: **Pilosocereus**

        4. Flowering areoles lacking conspicuous tufts of white hairs; ovary scaly or woolly: **Harrisia**

    3. Stems elongate and vine-like, climbing on rocks and trees and bearing aerial roots: **Selenicereus**

Epiphyllum Haw.

Plants mostly epiphytic, the main stems terete and woody, the ultimate and penultimate branches much flattened, often thin and leaf-like, sometimes 3-winged; areoles small, borne along the margins of flattened branches; spines usually absent on the mature plants but often present on seedlings; true leaves lacking but cotyledons may be long-persistent on young plants. Flowers usually of medium or large size, mostly nocturnal, odorless or fragrant; perianth tube elongate, longer than the perianth segments, in some species greatly elongated. Filaments usually long, borne at the top of the perianth tube or within the throat. Style long-exserted, white or colored; stigma-lobes several, linear. Perianth soon dropping. Fruit globular or oblong, often with low ridges or sometimes tuberculate, red or purple, edible or insipid; seeds glossy black.

    A tropical American genus with 18 taxa.

Epiphyllum phyllanthus (L.) Haw., *Syn. Pl. Succ*, 197 (1812) var. **plattsii** Proctor, var. nov.

Differs from var. *phyllanthus* in habit (terrestrial rather than epiphytic); in its shorter, dark red perianth tube (up to 15 cm long rather than 25–30 cm long and green); and its red style with orange stigma-lobes (rather than a pinkish or white style and white stigma-lobes). TYPE: *Proctor* 48205 (SJ, photo IJ), collected 25 Jul. 1992.

Epiphyllum phyllanthus (L.) Haw. var. **plattsii** Proctor, var. nov. a var. *phyllantho* habitu terrestri non epiphytico, tubo perianthii fusco-rubro (non viridi) breviore usque 15 cm (non 25–30 cm) longis, stylo rubro (non subroseo neque albo) et lobis stigmatis aurantiis (non albis) differt. PLATE 14.

Stems elongated and much branched, the main branches woody, terete or angled, less than 2 cm thick, the further or ultimate divisions flat, thin, bright green, up to 5 cm broad with sinuate margins, tapering to an obtuse apex. Flowers nocturnal, fragrant; perianth tube 25–30 cm long, ca. 5 mm in diameter, green; perianth segments white, reflexed at anthesis, narrowly oblong, ca. 2.5 cm long, sharply acute. Filaments numerous, delicately filiform, exserted ca. 1.5 cm. Style dark red, exserted ca. 2 cm, the stigma lobes orange, closely cohering. Fruit ovoid-oblong, 6–7 cm long, red.

CAYMAN BRAC: *Proctor* 47322, 48205 (type), 52153. Endemic to Cayman Brac; this is the only record of the genus *Epiphyllum* growing naturally in the West Indies. Found in natural rocky tropical dry limestone forest on the summit ridge of the bluff ca. 0.6 km due south of the coastal site called "The Rock". The plants are terrestrial, sprawling over fractured and eroded dolomitic rocks, in contrast to S. American forms of this species, which are always epiphytic.

— var. *phyllanthus* is confined to Panama and tropical S. America.

Opuntia Mill.

Succulent spiny perennials, branched from the base or developing a trunk, the branches conspicuously jointed, usually more or less flattened; leaves very small, terete, soon falling, the areoles (nodes) otherwise bearing spines, short barbed hairs (glochids), simple hairs, glands and sometimes flowers. Flowers sessile, solitary; perianth bell-shaped, with numerous lobes. Ovary glabrous but with glochidiate areoles; style scarcely longer than the stamens, terminating in 2–7 erect stigmatic rays. Berry pear-shaped, depressed at the apex, with or without spines.

A complex American genus, occurring in both tropical and temperate areas. The fruits are edible.

The New Catcus Lexicon has assigned the three species of Cayman cacti formerly recognized as *Opuntia* to three different genera which are recognised here, but retaining the original keys. The New cactus Lexicon reognises 75 species of *Opuntia*, 4 species of *Nopalea* and 3 species of *Consolea*.

KEY TO SPECIES

1. Plants noticeably spiny; stamens shorter than the perianth; wild species:

  2. Stems spreading or bushy, branching at or near the base and not developing a trunk; flowers yellow, the perianth 3 cm long or more: **O. dillenii**

  2. Stems developing a distinct trunk, becoming tree-like in habit; flowers orange, the perianth 1 cm long or less: **Consolea macracantha** (syn. **O. millspaughii** subsp. **caymanensis**)

1. Plants mostly lacking spines (except as seedlings); stamens much longer than the perianth; cultivated species: **[Nopalea cochenillifera** (syn. **O. cochenillifera)]**

Opuntia dillenii (Ker-Gawl.) Haw., *Suppl. Pl. Succ.* 79 (1819). **FIG. 90.** PLATE 14.

Opuntia tuna of some authors, not Mill. (1768).

PRICKLY PEAR

Plant usually 1–2 m tall, branched, sprawling and bushy; joints obovate or elliptic, 10–20 cm long; leaves 4–6 mm long; areoles large, with numerous glochids and 2–6 stout, slightly flattened, yellowish spines up to 4 cm long (but often much shorter),

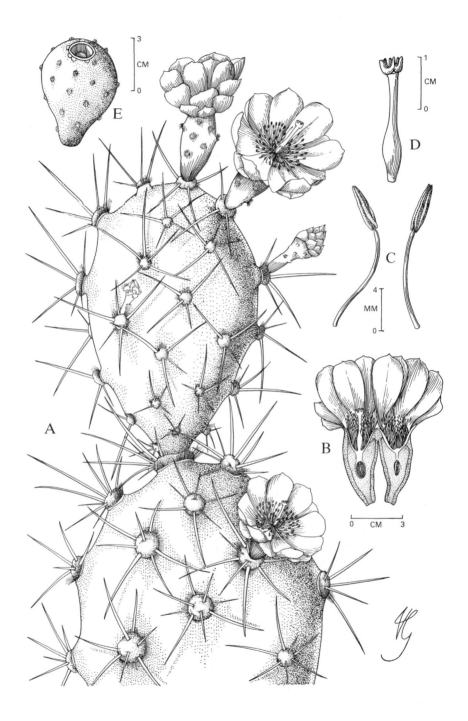

FIG. 90 **Opuntia dillenii**. A, habit. B, flower split lengthwise. C, two stamens. D, style terminated by stigmatic rays. E, fruit. (G.)

these straight or slightly curved-spreading. Perianth pale yellow, concolorous or sometimes shading to reddish at the base within, 3–5 cm long. Fruit obovoid, purplish-red, 5–7 cm long.

GRAND CAYMAN: *Kings GC 226; Proctor 15197.* LITTLE CAYMAN: *Proctor 35097.* CAYMAN BRAC: *Kings CB 29, CB 78; Proctor 29069.*

— Florida, the West Indies and northern S. America, chiefly coastal but also occurring inland in clearings and on dry stony plains. Cayman plants vary widely in length of spines, and also somewhat in the size and color of the flowers. In some West Indian islands, the fleshy joints are peeled, diced and cooked with rice as a vegetable.

Consolea macracantha (Griseb.) A. Berger, *Entwickl. Kakteen* 94 (1926). PLATE 15.

Opuntia millspaughii Britton subsp. *caymanensis* (Areces) Proctor, comb. nov.
Consolea millspaughii (Britton) A. Berger subsp. *caymanensis* Areces, *Brittonia* 53: 100, fig. 6 (2001).
Opuntia spinosissima of this flora, first ed., not Mill. (1768).

Low to medium-sized erect shrub 1.5–2.5 m tall, branching at or near the top of a woody trunk, this at base elliptic in cross-section, at the summit 5–10 cm × 5–8 cm in diameter, with smooth dull brown bark, throughout loosely covered with spiny areoles. Ultimate segments determinate in growth, flattened, smooth, slightly asymmetric, oblong-ovate to narrowly ovate, mostly 20–40 cm long, 6–10 cm wide, and 0.8–1 cm thick, dull green, with spines along the margins, usually rounded at the apex. Areoles on the trunk nearly round, ca. 5 mm in diameter, separated mostly by 1–1.5 cm, the upper border of each bearing a crown of glochidia 2–4 mm long and also with 6–15 spines, these straight, stiff, pungent and very unequal, the longest one deflexed and to 10 cm long. Areoles on old lateral and ultimate segments usually pitted. Flowers orange-red changing to vermillion with age, 1.5–2 cm broad when expanded, commonly 1–3 along the margins of ultimate stem segments; perianth short-campanulate with 16–19 segments arranged in 3 series, the outermost segments short and succulent. Stamens up to 10 mm long, spreading; anthers oblong, 0.7–1 mm long; style usually not exserted beyond the anthers. Fruit not described.

LITTLE CAYMAN: *Proctor 47284* (SJ). CAYMAN BRAC: *Areces 2402* (type, K; isotype NY); *Kings CB27* (BM); *Proctor 29134, 47802, 48674.*

Formerly considered an endemic subspecies occurring in the Bahamas and islands off northern Cuba. Now considered to be the more widespread species occurring throughout the West Indies.

[*Nopalea cochenillifera* (L) Salm-Dyck, *Cact. Hort. Dyck.* 64 (1850)] FIG. 91. PLATE 15.

Opuntia cochenillifera (L.) Mill., *Gard. Dict.* ed. 8 (1768).

COCHINEAL

Erect plant of somewhat tree-like habit at maturity, then up to 4 m tall or more with trunk 10–20 cm in diameter at the base, usually unarmed or nearly so; joints bright green, oblong or oblanceolate, 15–30 cm long or more; areoles with deciduous glochids. Flowers red or scarlet with pink, long-exserted stamens; perianth 1.5–2 cm long; stigma 6–7-lobed. Fruit red, ca. 5 cm long.

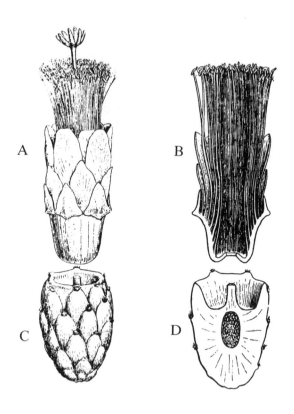

FIG. 91 **Nopalea cochenillifera**. A, flower removed from ovary, × 1. B, same cut lengthwise and style removed, × 1. C, ovary with base of style, × 1. D, same cut lengthwise, × 1 (F. & R.)

GRAND CAYMAN: *Hitchcock*; *Kings* GC 340. LITTLE CAYMAN: *Proctor* 35195. CAYMAN BRAC: *Kings* CB 33.

— Widely planted and locally spontaneous throughout the tropics, its natural origin probably Mexico or C. America. This is the chief species of cactus on which the cochineal insect was formerly grown for the production of a highly prized red dye, an industry the Spanish invaders found already established when they conquered Mexico in 1518. The fruits are edible.]

Pilosocereus Byl. & Rowl.

Erect columnar cacti, the stems simple or branched, the elongate joints leafless, ribbed and grooved, the upper areoles often densely woolly or long-bristly. Flowers solitary at the upper areoles, opening at night; perianth short-funnel-form or campanulate; ovary subglobose, naked and spineless, rarely bearing a few small scales; style usually short-exserted. Fruit a more or less globose smooth berry with numerous small black or brown seeds.

A tropical and subtropical American genus of 49 taxa.

Pilosocereus royenii (L) Byles & Rowley, *Cact. Succ. J. Gt Brit.* 19: 67 (1957). PLATE 15.

Pilosocereus aff. *swartzii* (Griseb) Byles & Rowley

DILDO

Stems simple or branched, to 3 m tall or more, slightly glaucous; ribs about 10, obtuse, indented between the areoles; spines 8–20 from an areole, variable in length from 0.5 to 3 cm or more. Flowers greenish or pinkish, subtended by masses of woolly hairs; perianth 3–5 cm long. Fruit depressed-globose, 2–3.5 cm in diameter.

GRAND CAYMAN: *Proctor* 15178. LITTLE CAYMAN: *Proctor* 35074. CAYMAN BRAC: *Kings* CB 30; *Proctor* 15312.

Harrisia Britton

Arborescent, arching or rarely prostrate cacti with slender, branched stems, the elongate branches fluted or angled; areoles borne on ribs or prominences, spiny. Flowers nocturnal, solitary at areoles near branch ends; perianth funnel-form, large, with a cylindric scaly tube as long as the limb or longer, the scales subtending areoles that bear tufts of woolly hairs. Stamens numerous, shorter than the perianth. Ovary and young fruit tuberculate. Fruit more or less globose, usually bearing sharp deciduous scales and small tufts of woolly hairs; seeds numerous, black.

A genus of 9 species, widely distributed from Florida to Argentina, of which one, *Harrisia gracilis*, is West Indian.

Harrisia gracilis (Mill.) Britton, *Bull. Torr. Bot. Club* 35: 563 (1908). PLATE 16.

Erect cactus of columnar or tree-like habit, up to 7 m tall, with a short trunk and few to numerous elongate arching branches, these 9–11-ribbed and 1.5–2.5 cm thick; ribs rounded, the grooves shallow; areoles with 6–10 or more needle-like spines of varying length from 0.5 to 3 cm. Flower-buds enveloped in tawny wool. Perianth 15–20 cm long, the inner segments white. Fruit globose, yellow, 4.5–6 cm in diameter, the white edible pulp containing numerous black seeds.

GRAND CAYMAN: *Kings* GC 56. LITTLE CAYMAN: *Proctor* 35118. CAYMAN BRAC: *Proctor* 29005, 29068.

— Otherwise known only from the southern side of Jamaica, where it is frequent in dry thickets; a similar plant, perhaps conspecific, occurs in the Swan Islands. The Cayman plants occur on exposed limestone rocks and cliffs.

Selenicereus (Berg.) Britton & Rose

Plants slender and elongate, the ribbed or angled stems trailing, clambering or climbing over rocks and trees, producing aerial roots at irregular intervals; areoles small, sometimes elevated on small knobs, usually spiny. Flowers large, nocturnal; perianth funnel-shaped with a slender elongate tube and flaring limb, the outer segments brownish or reddish, the inner ones white; ovary and perianth tube furnished with scales, bristles, and hairs at the areoles. Stamens numerous, the weak filaments in two distinctly separate series, one from a circle at the top of the perianth tube, the other from scattered points within the tube. Style exceeding the stamens; stigma-lobes slender, numerous. Fruit rather large, red or reddish, covered with tufts of deciduous spines, bristles, and hairs.

— A genus of 15 taxa, ranging from Texas and the West Indies to S. America. The flowers resemble some of those of the true night-blooming cereus species of the genus *Epiphyllum*, and are fully as ornamental, though apparently not as often or as prolifically produced.

KEY TO SPECIES

1. Spines acicular, to 9 mm long, the mature vegetative areoles usually lacking hairs; flowers not over 21 cm long, the inner perianth segments gradually narrowed to a pointed apex: **S. grandiflorus**

1. Spines conical, less than 2 mm long, usually accompanied by numerous much longer rather stiff whitish hairs; flowers 24–39 cm long, the inner perianth segments widest near the apex, abruptly terminated in a short-acuminate point: **S. pteranthus**

Selenicereus grandiflorus (L.) Brit. & Rose, *Contr. U.S. Nat. Herb.* 12: 430 (1909). PLATE 16.

VINE PEAR

Stems trailing or high-climbing, up to 10 m long or more, sometimes forming tangles. In Cayman, plants seldom more than 1 cm thick, with an average of 8 ribs; spines 5–10 per areole, needle-like, rarely accompanied by a few shorter brown hairs; juvenile plants have the spines both fewer and shorter. Flower-buds and perianth tube densely covered with brown hairs and sharp bristles; perianth 19–21 cm long, the outer segments light brown, the inner white and less than 1 cm broad. Fruit ovoid, ca. 8 cm long.

GRAND CAYMAN: *Brunt* 1697, 1919; *Kings GC* 56, *GC* 237; *Proctor* 15179, 31032.

— Cuba and Jamaica; also said to be widely cultivated and escaping in various parts of tropical America. The indigenous Grand Cayman population differs from most plants of other areas particularly in its more slender stems. However, the extreme variability of this species, especially in Jamaica, mitigates against recognising the Cayman plants as a distinct variety.

Selenicereus pteranthus (Dietr) Brit. & Rose, *Contr. U.S. Nat. Herb.* 12: 431 (1909).

Selenicereus boeckmannii (Otto) Brit. & Rose, *Contr. U.S. Nat. Herb.* 12: 429 (1909).

Stems light green, extensively trailing or climbing, often 10 m long or more, sometimes forming tangles, ca. 1 cm thick and with up to 8 ribs; spines 3–6 per areole, most of them scarcely over 1 mm long, accompanied by 10–20 setaceous hairs, these up to 5 mm long or more. Perianth similar in colouration to that of *S. grandiflorus* but larger, the tube densely clothed with pale tawny hairs, the oblanceolate inner segments up to 3 cm broad near the apex. Fruit globular, 5–6 cm in diameter.

CAYMAN BRAC: *Proctor* 35147.

— Cuba, Hispaniola, and eastern Mexico; introduced into Florida and the Bahamas. Both this and the preceding species occur naturally in dense rocky woodlands.

AIZOACEAE

Succulent unarmed herbs, the stems usually prostrate or semi-prostrate, the nodes often jointed. Leaves alternate, opposite or whorled, simple, entire. Flowers regular, perfect, terminal or axillary, solitary to several together or in branched inflorescences; perianth with 5 segments; stamens 5–many. Ovary free, 1–many-locular; styles as many as the locules. Fruit a membranous dehiscent capsule, included in the persistent perianth; seeds usually many, rarely only 1.

— A chiefly tropical and subtropical family of about 22 genera and 600 species, most abundant in Africa.

Sesuvium L.

Perennial, mostly creeping herbs with branched fleshy stems; leaves opposite, the bases often dilated and connate; stipules absent. Flowers axillary, mostly solitary; perianth lobes united toward base, coloured and petaloid within; stamens 5 or numerous, inserted on the perianth tube. Ovary 3–5-celled, with 3–5 styles. Capsule 3–5-celled, splitting transversely (circumscissile); seeds numerous, with annular embryo.

— A pantropical genus of about 5 species.

KEY TO SPECIES

1. Flowers solitary, with numerous stamens; perianth more than 4 mm long:
  2. Leaves 2–6 cm long; capsule 9–11 mm long: — S. portulacastrum
  2. Leaves 1–2 cm long; capsule 4–5 mm long: — S. microphyllum
1. Flowers 1 or several together, each with 5 stamens; perianth less than 4 mm long: — S. maritimum

Sesuvium portulacastrum L., *Syst. Nat.* ed. 10, 2: 1058 (1759). **FIG. 92.** PLATE 16.

SEA-PUSLEY

Prostrate succulent herb, often widely and extensively creeping, occasionally with short suberect branches. Leaves linear-oblong to oblanceolate, mostly 2–6 cm long (rarely less or more), 2–10 mm broad, the bases clasping. Flowers solitary on peduncles 5–14 mm long; perianth cup ca. 4 mm long, the lobes 5 mm long, usually pink within. Capsule oblong-conical, 9–11 mm long.

GRAND CAYMAN: *Brunt* 2104; *Hitchcock*; *Kings GC 34, GC 219, GC 282*; *Proctor* 15201; *Sachet* 394. LITTLE CAYMAN: *Proctor* 28066. CAYMAN BRAC: *Kings CB 55*; *Millspaugh* 1221; *Proctor* 29128.

— Pantropical, chiefly on sandy or rocky sea-shores or on saline flats near mangroves.

Sesuvium microphyllum Willd., *Enum. Hort. Berol.* 521 (1809).

Creeping succulent herb, forming mats, with numerous short tangled branches. Leaves oblanceolate or spatulate, chiefly 1–2 cm long. Flowers solitary on peduncles 3–7 mm long; perianth cup 1.5–2 mm long, the lobes 2–3.5 mm long, white within. Capsule ovoid-conical, 4–5 mm long.

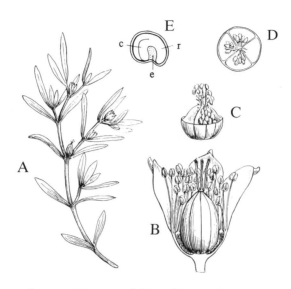

FIG. 92 **Sesuvium portulacastrum**. A, portion of plant, × ²/₃, B, flower with portion of perianth and some stamens removed, × 3. C, capsule with upper part fallen, × 3. D, lower part of C cut across, × 3. E, section of seed, × 10; e, endosperm; c, cotyledons; r, radicle. (F. & R.)

LITTLE CAYMAN: *Proctor* 28056.

— Cuba and a few other West Indian islands, not common; often grows in sandy hollows among seaside rocks.

Sesuvium maritimum (Walt.) B.S.P., *Prem. Cat. N.Y.* 20 (1888).

Creeping succulent herb, forming mats or rosettes, with numerous short branches. Leaves oblanceolate or spatulate, chiefly 0.5–1.5 cm long. Flowers solitary or 2–3 together in leaf-axils, nearly sessile; perianth cup ca. 1 mm long, the lobes 1.5–2 mm long; stamens 5. Capsule ovoid, ca. 3 mm long.

GRAND CAYMAN: *Proctor* 15115, 52242.

— Atlantic coast of N. America from Long Island to Florida; also in the Bahamas, Cuba and Puerto Rico, often occurring on salinas and along the edge of mangrove swamps.

PORTULACACEAE

Usually succulent unarmed herbs, prostrate or erect, rarely somewhat shrubby at the base. Leaves alternate, subopposite or opposite, simple, entire. Flowers regular, perfect, solitary, clustered at the tips of branches, or else in terminal racemes or panicles; sepals 2; petals usually 4–6, free. Stamens equal in number to the petals, or sometimes less or more. Ovary superior or half-inferior, 1-celled, with free-central placentation; style with 3–7 stigmatic branches; ovules 2 to many. Fruit a capsule, circumscissile or splitting by 3 valves; seeds 2–numerous, with curved embryo.

— A family of about 20 genera and 500 species, mostly American.

KEY TO GENERA

1. Flowers solitary or cymose-capitate; ovary half-inferior; capsule circumscissile; stipules scarious or represented by hairs:	**Portulaca**
1. Flowers in racemes or panicles; ovary superior; capsule 3-valved; stipules absent:	**Talinum**

Portulaca L.

Fleshy herbs prostrate or erect; leaves alternate, subopposite or opposite, usually whorled around the flowers; stipules minute or reduced to hairs. Flowers solitary or clustered at the tips of branches; sepals united at base; stamens 8–20. Ovary with numerous ovules; style 3–9-cleft or -parted. Capsule many-seeded.

— A genus of about 20 species, the majority American.

KEY TO SPECIES

1. Plants glabrous; leaves flat:	
2. Leaves 6–30 mm long; sepals 3–5 mm long; petals 4–6 mm long; capsule ca. 6 mm long:	P. oleracea
2. Leaves 1–3.5 mm long sepals 2–3 mm long; petals ca. 3.5 mm long; capsule 2–3.5 mm long:	P. tuberculata
1. Plants hairy; leaves more or less cylindrical:	
3. Flowers yellow:	
4. Annual; petals ca. 3 mm long; capsule circumscissile below the middle; seeds black, 0.4 mm broad:	P. halimoides
4. Perennial; petals more than 6 mm long; capsule circumscissile above the middle; seeds brown, 0.5–0.6 mm broad:	P. rubricaulis
3. Flowers purple or crimson:	P. pilosa

Portulaca oleracea L., *Sp. Pl.* 1: 445 (1753). **FIG. 93.**

PUSLEY, WILD PARSLEY

Annual fleshy glabrous herb, prostrate and radiating from a taproot, or sometimes ascending or erect, the stems often reddish. Leaves subopposite or sometimes alternate, subsessile, obovate or spatulate, up to about 30 mm long but usually less, the apex rounded or retuse. Flowers sessile; sepals broadly ovate, keeled, acute; petals 5, yellow, deeply notched at the apex, soon withering; stamens 6–12; stigmas 3–6. Capsule circumscissile at about the middle; seeds black, 0.7–0.8 mm in diameter, granulate.

GRAND CAYMAN: *Brunt* 1859, 1959; *Hitchcock*; *Kings* GC 383; *Proctor* 15048. LITTLE CAYMAN: *Kings* LC 24; *Proctor* 28169, 28170. CAYMAN BRAC: *Kings* CB 23, CB 46; *Proctor* 29078.

— World-wide in distribution, occurring commonly in tropical and temperate regions of both hemispheres, a weed of sandy soil and open waste ground. This species is widely eaten as a vegetable, and is also used in salads and as an ingredient of soup.

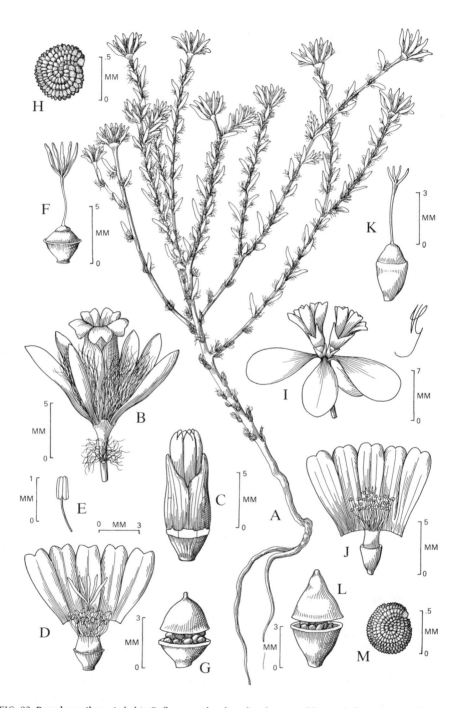

FIG. 93 **Portulaca pilosa**. A, habit. B, flower with subtending leaves and hairs. C, flower-bud. D, flower opened out to show stamens and pistil. E, stamen. F, ovary, style and stigmas. G, fruit showing dehiscence. H, seed. **Portulaca oleracea**. I, flowers with subtending leaves. J, flower opened out to show stamens & pistil. K, ovary, style and stigmas. L, fruit showing dehiscence. M, seed. (G.)

Portulaca tuberculata Leon, *Contr. Ocas. Mus. Hist. Nat. Col. "De La Salle"*, no. 9: 3 (1950).

Small fleshy herb of flaccid texture, with prostrate branches 2–10 cm long, forming rosettes from a taproot. Leaves opposite, flat, obovate to orbicular, 1–3.5 mm long. Flowers sessile, solitary or 2 together; sepals deltate-ovate; petals 4 or 5, yellow, scarcely exceeding the sepals; stamens 5–8; stigmas 4–5. Capsule terminated by a rounded-conical tubercle, and circumscissile at or slightly below the middle; seeds black or nearly so, 0.5–0.6 mm in diameter, granulate.

LITTLE CAYMAN: *Proctor* 28191, 35089. CAYMAN BRAC: *Proctor* 29012, 29041, 29351.

— Otherwise known only from eastern Cuba. The Cayman plants grow in small soil-filled pockets of limestone rocks and cliffs, or rarely in sandy clearings.

Portulaca halimoides L., *Sp. Pl.* ed. 2, 1: 639 (1762).

An erect or diffuse annual herb usually less than 10 cm tall, the stems bearing tufts of white hairs in the leaf-axils; leaves subcylindrical, 6–10 mm long, deciduous. Flowers sessile, embedded in a tuft of white hairs; sepals 2–2.5 mm long; petals yellow or white with yellow centre, ovate, 3 mm long; stamens 8–20. Capsule usually hidden in white hairs, circumscissile below the middle; seeds black, 0.4 mm in diameter, minutely tuberculate.

GRAND CAYMAN: *Kings* GC 224.

— West Indies and Mexico, in dry or occasionally moist clearings near the sea.

Portulaca rubricaulis Kunth in H.B.K., *Nov. Gen. & Sp.* 6: 73 (1820).

Portulaca phaeosperma Urban (1905).

A fleshy prostrate or ascending herb usually less than 15 cm high, often with red stems, and bearing small tufts of white axillary hairs. Leaves subcylindrical, linear or lance-linear, up to 10 mm long or a little more. Flowers sessile, often ca. 1 cm across when open; sepals 4–5 mm long; petals yellow, usually 6–7 mm long, occasionally less; stamens 12–16; stigmas 5–7. Capsule subglobose, 2.5–3 mm in diameter, circumscissile above the middle; seeds brown, 0.5–0.6 mm in diameter, rugulose.

GRAND CAYMAN: *Brunt* 2006, 2122, 2149; *Kings* GC 329; *Proctor* 15221, 27943. LITTLE CAYMAN: *Proctor* 28176, 35088, 35140. CAYMAN BRAC: *Proctor* 28978, 28979.

— Florida and the West Indies to northern S. America, often in pockets of limestone rocks, occasionally in seasonally wet rocky pastures.

Portulaca pilosa L., *Sp. Pl.* 1: 445 (1753). **FIG. 93.**

TEN-O'CLOCK

Annual or semiperennial herb, the stems prostrate or ascending, less than 15 cm long, densely white- or brownish-pilose from the leaf-axils. Leaves linear-subcylindrical, 5–15 mm long. Flowers sessile, surrounded by a tuft of whitish or pale brownish hairs and a whorl of 6–10 leaves; sepals ovate, 2–3 mm long, not keeled; petals purple or crimson, obovate-retuse, 3–5.5 mm long; stamens 15–22, with crimson filaments; stigmas 4–6. Capsule subglobose, 3–4 mm in diameter, circumscissile at about the middle; seeds black, 0.5 mm in diameter, minutely tuberculate.

GRAND CAYMAN: *Proctor* 15204. CAYMAN BRAC: *Proctor* 29070.

— Widespread from southern U.S.A. to S. America, growing on open sandy ground or in soil-filled pockets of limestone rock.

The large-flowered *Portulaca grandiflora* Hook., native of Argentina, is sometimes cultivated. Its pink, red, yellow, orange or white flowers, which open only in the morning, have petals 15–25 mm long; 'double' forms are also grown.

Talinum Adans.

Fleshy glabrous erect or ascending herbs. Leaves alternate, without stipules. Petals 5; stamens 10–30, adherent to the base of the petals. Ovary free, with many ovules; style 3-lobed or 3-cleft. Capsule 3-valved; seeds numerous, borne on a central globose placenta.
— A genus of about 15 species, chiefly American.

Talinum triangulare (Jacq.) Willd. in L., *Sp. Pl.* 2: 862 (1800).

Annual herb with taproot, the erect stems to 50 cm tall or more; leaves oblanceolate to obovate, 2–6 cm long, 0.5–2 cm broad or more, rounded or acute at the apex. Racemes simple or branched, few–many-flowered; pedicels 3-angled. Sepals persistent, ca. 5 mm long; petals pink in Cayman plants, elsewhere often yellow or sometimes white, 6–9 mm long and 6 mm broad; stamens numerous. Capsule ca. 5 mm long; seeds black, 0.8 mm long.

LITTLE CAYMAN: *Proctor* 35142.
— Widely distributed from the Bahamas and Greater Antilles to S. America and in W. Africa. The Little Cayman plants were found along sandy roadsides and may have escaped from cultivation.

BASELLACEAE

Herbaceous glabrous vines, often rather succulent, with tuberous roots. Leaves alternate, simple, entire; stipules absent. Flowers small, regular, perfect, in axillary and terminal racemes or panicles; each pedicel with 1 bract at base and 2 bracteoles at top subtending the flower. Sepals 2, sometimes winged in fruit; petals 5; stamens 5, opposite the petals and inserted on a hypogynous disc adnate to the base of the corolla. Ovary superior, 1-celled, with 1 basal ovule; styles usually 3, free or united; stigmas entire or cleft. Fruit an indehiscent 1-seeded utricle, included in the perianth; seeds with endosperm.
— A small family of 5 genera and about 22 species, mostly in tropical America.

Anredera Juss.

Climbing herbs with perennial tuberous roots and somewhat fleshy leaves. Flowers in axillary and terminal spike-like racemes, these simple or branched; bracteoles usually adnate to the base of the perianth; sepals petaloid, shorter than the petals; petals white or greenish-white, sometimes changing to purple with age, the stamens inserted at their base. Ovary ovoid, the styles separate or somewhat united. Seed with semiannular embryo.
— A tropical American genus of 10 species.

Anredera vesicaria (Lam.) C. F. Gaertn., *Suppl. Carp.* 3: 176 (1807). **FIG. 94.**

Anredera leptostachys (Moq.) van Steenis, *Fl. Males.* (ser. 1) 5 (3): 302 (1957).
Boussingaultia leptostachya Moq. (1849).

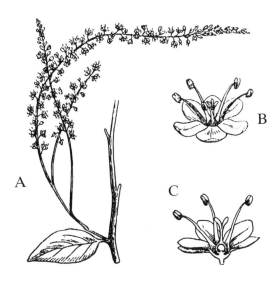

FIG. 94 **Anredera vesicaria**. A, portion of flowering branch, × ²/₃. B, flower of a related species (Boussingaultia baselloides), × 3. C, the same cut lengthwise, × 3 (F. & R.)

Slender twining vine sometimes forming tangles; leaves short-petiolate, the blades ovate to elliptic, 2–6 cm long, 1.5–4 cm broad, acute to acuminate at both apex and base. Racemes lax, simple, slender, longer than the leaves; flowers pale greenish-yellow to white, on pedicels 1 mm long; petals ca. 2 mm long. Fruits not seen.

GRAND CAYMAN: *Proctor* 15190. LITTLE CAYMAN: *Proctor* 28147.

— Florida, West Indies, and continental tropical America at low elevations, the Cayman plants found growing in sandy coastal thickets.

CHENOPODIACEAE

Annual or perennial herbs, rarely shrubs; leaves alternate or rarely opposite, simple, sometimes reduced to scales; stipules lacking. Flowers small, regular, perfect or unisexual, variously clustered, usually greenish and wind-pollinated. Perianth absent or of 2–5 segments, these more or less united at the base, persistent after flowering; stamens as many as or fewer than the perianth segments and opposite them, the anthers 2–4-celled. Ovary superior, 1-celled, with 1 basal ovule; styles 1–3, the stigmas capitate or 2–3-lobed and elongate. Fruit a 1-seeded, usually indehiscent utricle with a thin or hard pericarp, usually included in the persistent perianth; seed with annular embryo enclosing the endosperm, or endosperm sometimes lacking.

— A large family of wide distribution, with over 100 genera and 1,400 species, best represented in Asia, and especially characteristic of deserts and saline soils. *Beta vulgaris* L., the European beet and its variants, and *Spinacia oleracea* L., spinach, are well-known vegetables of this family.

KEY TO GENERA

1. Leaves evident, alternate, more or less toothed or lobed; stem not jointed; stamens 3–5:	Atriplex
1. Leaves reduced to opposite pairs of scales; stem jointed; stamens 2:	Salicornia

Atriplex L.

Herbs or low shrubs, more or less scurfy-canescent or silvery; leaves mostly alternate. Flowers dioecious or monoecious, small, green, in axillary capitate clusters or panicled spikes. Staminate flowers bractless, consisting of a 3–5-parted perianth and an equal number of stamens; filaments free or united by their bases; a pistillode sometimes present. Pistillate flowers subtended by 2 bracts which enlarge in fruit and are more or less united; perianth none; stigmas 2. Utricle completely or partly enclosed by the fruiting bracts; seeds with mealy endosperm.

— A widely distributed genus of about 150 species, often found in saline or arid habitats.

Atriplex pentandra (Jacq.) Standley, N. Amer. Fl. 21: 54 (1916).

Annual bushy or suffrutescent herb, the branches procumbent or ascending, up to 8 cm long. Leaves silvery-scurfy especially on the under side, sessile or nearly so, oblong to rhombic or obovate, 1–2 cm long, with more or less repand-dentate margins. Flowers monoecious, the staminate in dense short terminal spikes, the pistillate clustered in the axils of leaves. Fruiting bracts united at base only, sharply toothed on the margins and crested or tuberculate on the sides.

LITTLE CAYMAN: *Sauer* 3315.

— Florida to Texas, Bermuda, the West Indies, and northern S. America, on sandy seashores.

Salicorna L.

Succulent glabrous herbs with opposite jointed branches, the leaves reduced to mere opposite scales. Flowers perfect or the lateral ones staminate, sunken 3–7 together in the axils of the upper scales, thus forming narrow terminal spikes. Perianth obpyramidal or rhomboid, fleshy, 3–4-toothed or truncate, becoming spongy in fruit and falling with it. Stamens 2 or rarely solitary, exserted. Ovary ovoid, with 2 stigmas. Fruit enclosed by the adherent perianth; seed often hairy, without endosperm.

— A cosmopolitan halophytic genus of about 35 species.

KEY TO SPECIES

1. Perennial; main stems prostrate and subwoody, with several or many erect branches:	S. perennis
1. Annual; main stem erect with lateral branches:	S. bigelovii

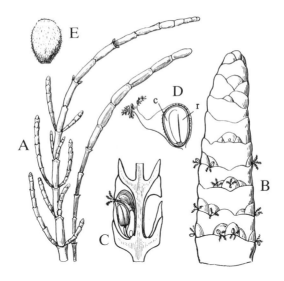

FIG. 95 **Salicornia perennis**. A, portion of stem and branch, × ²/₃. B, portion of stem in flower, × 5. C, small portion of B cut lengthwise, showing a flower enclosed in the perianth and another perianth empty, × 7. D, fruit cut lengthwise, × 11; **c**, cotyledons; **r**, radicle. E, seed, × 11. (F. & R.)

Salicorna perennis Mill., *Gard. Dict.* ed. 8 (1768). **FIG. 95.**

Salicornia ambigua Michx. (1803).

Perennial by a woody rootstock; main stems trailing or decumbent, up to 60 cm long or more, rooting at nodes, the branches ascending or erect; ultimate branchlets terete, ca. 1 mm thick between the nodes. Scales connate to form a shallow cup surrounding each node of the branchlets. Fertile spikes 1–4 cm long, scarcely distinguishable from the sterile branchlets, their joints about as long as thick, the flowers about equalling the joints in length.

GRAND CAYMAN: *Brunt* 1736; *Kings* GC 175, GC 176B; *Proctor* 27971, 48260.

— N. American coasts of the Atlantic Ocean, also occurring in the northwestern half of the West Indies and along the Gulf of Mexico, chiefly on saline shores and salt flats.

Salicornia bigelovii Ton., *Bot Mex. Bound. Surv.* 184 (1859).

Annual with an erect fleshy stem, few–many-branched upwardly, the nearly terete ascending branches 1.5–2 mm thick between the nodes. Scales triangular-ovate, sharply pointed, 2–3 mm long, connate at the sides. Fertile spikes 1–12 cm long, obviously thicker than the sterile branches, their joints about as long as thick, the flowers a little shorter.

LITTLE CAYMAN: *Kings* LC 117; *Proctor* 35072, 48238.

— Atlantic and Gulf coasts of N. America and the northwestern half of the West Indies (but not reaching Jamaica), also in California; not common, occurring on sand-spits and salinas.

[*Chenopodium ambrosioides* L., sometimes called 'mexican tea', is an aromatic herb often cultivated in the West Indies as a vermifuge. There is no record of it in the Cayman Islands, but it may occur in a few old-time gardens. It tends to persist after cultivation, and sometimes becomes naturalised in sandy or gravelly waste places.]

AMARANTHACEAE

Herbs or rarely small shrubs, occasionally succulent or spiny; leaves opposite or alternate, simple, usually entire; stipules lacking. Flowers small, perfect or unisexual, bracteolate, variously clustered, usually in terminal or axillary inflorescences; perianth of 2–5 segments, scarious, persistent after flowering. Stamens 2–5, opposite the perianth segments; filaments free or commonly more or less fused, sometimes forming a corolla-like tube. Ovary superior, 1-celled; ovule 1 in Cayman genera, basal. Fruit a membranous or fleshy utricle, indehiscent, irregularly rupturing, or circumscissile, enclosed in or resting on the persistent perianth. Seed naked or arillate, usually lustrous and smooth or nearly so, with copious endosperm and annular or hippocrepiform embryo.

— A world-wide family of about 65 genera and 850 species, of little or no economic importance. A few are edible ('calalu'), and one species of *Amaranthus* produces seeds that are used as a cereal in Asia.

KEY TO GENERA

1. Leaves alternate:	**Amaranthus**
1. Leaves opposite:	
2. Flowers in very elongate unbranched spikes:	**Achyranthes**
2. Flowers not in elongate spikes:	
3. Flowers in loose panicles; anthers 2-locular:	**Iresine**
3. Flowers in dense heads:	
4. Stems erect; heads 20–25 mm thick; anthers 2-locular:	**[Gomphrena]**
4. Stems prostrate to decumbent-ascending; heads 3–10 mm thick; anthers 1-locular:	
5. Mat-like herb with basal leaf-rosette; stamens 2; staminodes present:	**Lithophila**
5. Trailing or decumbent-ascending herbs lacking a basal leaf-rosette; stamens 5; staminodes absent:	
6. Leaves succulent; flowering heads pedunculate:	**Blutaparon**
6. Leaves thin, not succulent; flowering heads sessile:	**Alternanthera**

Amaranthus L.

Annual erect or prostrate herbs, glabrous or pubescent; leaves entire, long-petiolate. Flowers unisexual, monoecious, in axillary clusters or terminal panicles; perianths of 3–5 equal segments. Stamens 2–5; filaments free; staminodes none; anthers 2-celled. Ovary with 2–3 stigmas. Fruit compressed, indehiscent or circumscissile, often with 2 or 3 beaks. Seed compressed, with annular embryo.

— A world-wide genus of about 60 tropical and temperate species.

KEY TO SPECIES

1. Flowers in axillary clusters or very short spikes; leaf-blades oblong or obovate and less than 3.5 cm long:	A. crassipes
1. Flowers in terminal panicles as well as axillary spikes; leaf-blades ovate or rhombic-ovate, mucronate, up to 10 cm long:	
2. Flowers with 3 stamens and perianth segments; utricle indehiscent:	A. viridis
2. Flowers with 5 stamens and perianth segments; utricle circumscissile:	A. dubius

Amaranthus crassipes Schltdl., *Linnaea* 6: 757 (1831).

Glabrous herb with somewhat fleshy decumbent to ascending stems mostly less than 20 cm long or sometimes longer; leaves broadly oblong to obovate, 1–3.5 cm long, the blades with apex rounded and notched, the base attenuate, the tissue prominently whitish-veined beneath. Flowers straw-coloured, in small dense axillary clusters, the pistillate flowers with short, much-thickened peduncles that detach with the fruit, the staminate flowers in separate clusters in axils of the upper leaves. Perianth of the pistillate flowers with 4 or 5 segments, 1–1.5 mm long. Utricle finely wrinkled, indehiscent.

CAYMAN BRAC: *Proctor* 29071.

— Florida, West Indies, and on the continent from Mexico to Peru, usually a weed of dryish waste places. The Cayman Brac plants were growing in pockets of limestone rock near a path on The Bluff overlooking Spot Bay.

Amaranthus viridis L., *Sp. Pl.* ed. 2, 2: 1405 (1763). **FIG. 96.**

CALALU

Erect herb up to 1 m tall, sparingly branched, nearly glabrous; leaves ovate or rhombic-ovate on long petioles, up to 8 cm long or sometimes much more, the apex blunt and

FIG. 96 **Amaranthus viridis**. A, portion of plant in flower, × ²/₃. B, staminate flower, × 10. C, pistillate flower with one perianth-segment removed, × 10. D, fruit with persistent perianth, × 10. E, seed cut lengthwise, × 10; **c**, cotyledons; **r**, radicle; **e**, endosperm. (F. & R.)

notched, with a small mucro in the notch. Flowers green, in terminal panicles and axillary spikes or clusters; perianth of 3 segments; stamens 3. Utricle globose, wrinkled, indehiscent, about as long as the perianth.

GRAND CAYMAN: *Brunt* 2097; *Hitchcock*; *Kings* GC 367; *Millspaugh* 1347; *Proctor* 15101, 27946.

— Pantropical, often cultivated as a vegetable.

Amaranthus dubius Mart. ex Thell., *Mem. Soc. Sci. Nat. Cherbourg* 38: 203 (1912).

Amaranthus tristis Griseb. not L. (1753).

Erect herb up to 1 m tall, more or less branched, glabrous or pubescent above. Leaves rhombic-ovate on long, often red petioles, mostly 2–10 cm long, the apex usually acute and mucronate. Flowers greenish, in terminal panicles and dense axillary clusters; perianth of 5 segments; stamens 5. Utricle with circumscissile dehiscence.

GRAND CAYMAN: *Hitchcock*; *Millspaugh* 1390. CAYMAN BRAC: *Kings* CB 66; *Millspaugh* 1191.

— West Indies and continental tropical America, also in tropical Africa, a weed of waste ground.

The weedy, pantropical 'spiny salalu' or 'macca calalu', *Amaranthus spinosus* L., has not been recorded from the Cayman Islands, but is to be expected.

Achyranthes L.

Annual or perennial herbs, glabrous or pubescent, sometimes shrubby at base; leaves opposite, petiolate, entire. Flowers perfect, in long slender spikes, deflexed in fruit, the bracts spine-tipped. Perianth of 5 subequal segments (rarely 2 or 4), glabrous or pubescent, becoming hard and ribbed. Stamens 5 (rarely 2 or 4), alternating with laciniate staminodes; filaments united into a short cup at the base; anthers 4-celled. Ovary with filiform style and capitate stigma; ovule 1, suspended from an elongate funicle. Utricle indehiscent, included in the persistent perianth; seed oblong, with annular embryo.

— A pantropical genus of about 10 species.

Achyranthes aspera L. var. *aspera*, *Sp. Pl.* 1: 204 (1753). **FIG. 97.**

Achyranthes aspera var. *obtusifolia* (Lam.) Griseb. (1859).
Achyranthes indica (L.) Mill. (1768).

DEVIL'S HORSEWHIP

An erect herb to 1 m tall, the stems whitish-pubescent, with a few low branches; leaves obovate or rotund, 2–7 cm long and wide, rounded at the apex and sometimes with a small abrupt point, the surfaces pilose-sericeous especially beneath. Flowers green, in terminal whip-like spikes up to 40 cm long, more densely flowered toward the end; bracts broadly ovate, 3 mm long, tipped by a rigid spine; bracteoles aristate, shorter than the perianth. Perianth segments 4 mm long, acuminate; stamens 1 mm long, twice as long as the staminodes. Utricle enclosed in the perianth and by the spiny bracteoles, readily detached and adhering to animals and human passers-by.

GRAND CAYMAN: *Brunt* 2111a; *Hitchcock*; *Millspaugh* 1272; *Proctor* 15247. CAYMAN BRAC: *Proctor* (sight record).

— A pantropical weed of pastures and open waste ground.

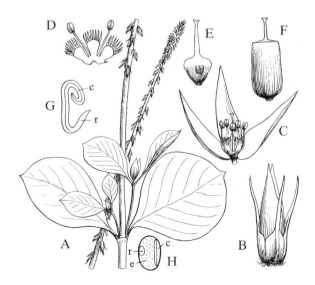

FIG. 97. **Achyranthes aspera** var. **aspera**. A, portion of plant, × ²/₃. B, flower with bract and bracteoles; × 7. C, flower with two perianth-segments removed, × 7. D, portion of staminal tube spread out, with 2 stamens and 3 staminodes, × 7. E, ovary with ovule, × 7. F, utricle, × 7. G, embryo, × 7; **c**, cotyledons; **r**, radicle. H, section through seed, × 7; **e**, endosperm; **r**, radicle; **c**, cotyledons. (F. & R.)

Iresine P. Browne

Erect to decumbent or scandent herbs, sometimes shrubs or small trees, glabrous or pubescent; leaves opposite, entire, petiolate. Flowers perfect or unisexual, monoecious or dioecious, with bracts and bracteoles, mostly in large panicles or panicled spikes; perianth of 5 segments, often hairy. Stamens 5, the subulate filaments united at base to form a short tube; anthers 2-celled. Style short or none; stigmas 2 or 3, subulate or filiform, or in the staminate flowers sometimes capitate but non-functional. Utricle more or less compressed, indehiscent; seed with annular embryo.

A chiefly tropical American genus of about 45 species, a few also in tropical Africa.

Iresine diffusa Humb. & Bonpl. ex Willd. in L., *Sp. Pl.* ed. 4, 4: 765 (1806).

Iresine celosia L., 1759, illegit.

Annual or sometimes perennial, the stems erect, procumbent, or clambering, sometimes elongate and up to 3 m long, glabrous or nearly so. Leaves thin, broadly ovate to lanceolate, 2–14 cm long, acute to acuminate at apex. Flowers very numerous in panicled spikes, the pistillate with copious long wool at the base; bracts and perianth silvery or pale greenish, the perianth segments 1–1.5 mm long. Seeds shining dark red, 0.5 mm in diameter.

GRAND CAYMAN: *Hitchcock*; *Kings GC 70*.

— Southeastern U.S.A., West Indies, and Central and S. America, often in thickets or else a weed of waste places, common except at high elevations. The Cayman specimens are rather small for the species.

[*Gomphrena* L.

Pubescent annual or perennial herbs. Flowers perfect, in dense heads or spikes, white, yellow, or red; bracteoles keeled, the keel often crested. Perianth 5-lobed or 5-parted. Stamens 5, the filaments united into a lobed tube. Stigma 2-lobed, the lobes recurved. Utricle flattened; seed smooth.

A pantropical genus of about 100 species.

Gomphrena globosa L., *Sp. Pl.* 1: 224 (1753).

Annual, usually branched or bushy herb 30–80 cm tall, the stems appressed-pilose, the nodes swollen. Leaves oblong-elliptic, 2–10 cm long, up to 4 cm broad, acute at the apex; petioles 5–20 mm long. Flower-heads subglobose, magenta, yellowish or white, usually subtended by 2 or 3 small leaves; perianth woolly, shorter than the acute to acuminate bracts and bracteoles.

GRAND CAYMAN: *Correll & Correll* 51019.

— Although originally described from India, this species probably originated in tropical America. It is widely cultivated in warm countries, and often escapes along roadsides and in open waste land.]

Lithophila Sw.

Perennial herbs with stout woody taproot; leaves of two types: tufted elongate ones forming a basal rosette, and much smaller opposite ones along the flowering scapes. Flowers bracteolate, perfect, in small heads or short, dense spikes; perianth of 5 segments. Stamens 2, the filaments united at base into a short tube; anthers 1-celled; staminodes 3. Ovary with short style and 2 slender stigmas; ovule 1. Utricle compressed; seed lenticular.

A small genus of about 6 species, occurring in the West Indies, Venezuela, and the Galapagos Islands.

Lithophila muscoides Sw., *Nov. Gen. & Sp. Pl.* 14 (1788).

A prostrate, mat-like herb, the branched stems usually less than 15 cm long; leaves whitish-villous near the base but otherwise glabrous, those of the basal rosette linear-oblanceolate and up to 5 cm long, those of the stems oblong or oblanceolate and mostly less than 1 cm long. Flower-heads sessile, more or less globose to subcylindrical, 3–13 mm long; bracts whitish-membranous, ovate; perianth segments whitish with a median black spot, 1.5–2.5 mm long. Seed shining brown, 0.5 mm in diameter.

GRAND CAYMAN: *Brunt* 1797; *Kings* GC 258, GC 331; *Proctor* 27972.

— Widespread in the West Indies but apparently lacking from Jamaica, rather variable; several varieties have been described. The Cayman plants grow in black humus in hollows of limestone pavements, in sheltered situations toward the interior of the island.

Blutaparon Raf.

Creeping perennial herbs, branched and rather succulent, glabrous or pubescent. Leaves opposite, narrow, entire. Flowers perfect, with bracts and bracteoles, in small dense heads or spikes, these axillary and terminal. Perianth of 5 subequal segments, thickened at the base and short-stipitate. Stamens 5, fused at the base; anthers 1- or 2-celled; staminodes

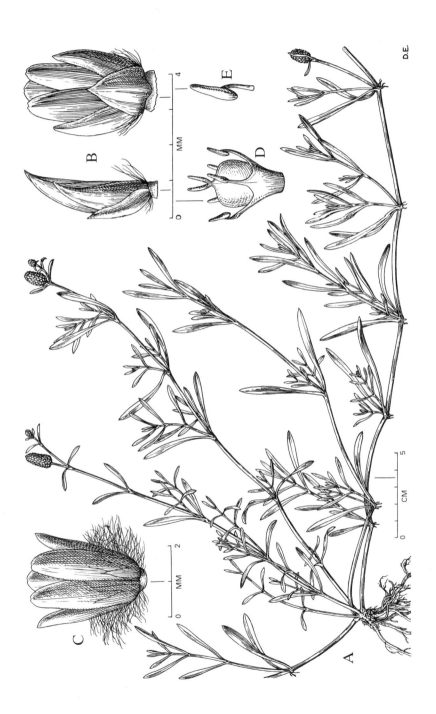

FIG. 98 **Blutaparon vermiculare.** A, habit. B, two aspects of flower, with bracts. C, flower with bracts removed. D, flower with perianth removed. E, stamen. (D.E.)

absent. Ovary with a short style and 2 stigmas. Utricle compressed, indehiscent; seed lenticular, with annular embryo.

— A small genus of 10 species, chiefly on seashores of tropical America and W. Africa.

Blutaparon vermiculare (L.) Mears in *Taxon* 31: 118 (1982). **FIG. 98.**

Philoxerus vermicularis (L.) R. Br. ex Smith in Rees (1814); Adams (1972).
Caraxeron vermicularis Raf. (1837); Proctor (1984).

Fleshy creeping herb with main stems up to 60 cm long or more, rooting at the nodes, with short ascending flowering branches. Leaves sessile, linear to oblanceolate, mostly 1–3 cm long. Flower-heads sessile, globose to cylindrical, 5–20 mm long; bracts and perianth whitish, the perianth segments 3–5 mm long, the inner ones woolly near the base. Seed orbicular, shining dark brown, 1 mm in diameter.

GRAND CAYMAN: *Brunt* 1626, 1771, 1860; *Hitchcock; Proctor* 15298, 52125; *Sachet* 456. LITTLE CAYMAN: *Kings* LC 27; *Proctor* 28181. CAYMAN BRAC: *Millspaugh* 1220; *Proctor* 29362.

— Florida, West Indies, continental tropical America, and the west coast of tropical Africa, frequent on seashores and in brackish situations.

Alternanthera Forsskål

Perennial or annual weeds, herbaceous to suffruticose; young stems and leaves often pubescent, but glabrous with age. Leaves opposite, entire. Inflorescences axillary or terminal, sessile or pedunculate. Flowers perfect; tepals 5, generally white, equilong or outer 3 clearly longer than and enclosing inner 2; stamens 5 (3), united below; anthers oblong or ovate, 2-locular with 1 line of dehiscence; pseudostaminodia shorter or longer than stamens, subulate or ligulate, dentate to fimbriate distally; style obscure to slender; stigma capitate and globose to more or less punctuate; ovule 1. Utricles indehiscent.

— Perhaps 200 species, all tropical or subtropical.

Alternanthera sessilis (L.) R. Br. ex DC., *Cat. Pl. Hort. Monsp.* 77 (1813).

Trailing or scandent herb of open places or shallow water; stems ribbed with lines of pubescence between ribs; horizontal rows of longer trichomes at nodes. Leaves with indistinct petioles; blades ovate to elliptic, to 5 cm long, sparsely pubescent below, apex acute to obtuse, base tapering. Inflorescences congested, axillary, bracts often persistent after flowers have fallen; bracts and bracteoles ovate 0.3–0.8 mm long, mucronate, hyaline, 1-nerved, apiculate, white, glabrous; anthers 3, ovate; pseudostaminodia subulate or dentate, equal to or shorter than filaments; style <0.2 mm long; stigma punctate. Utricles compressed, obcordate, at maturity exceeding tepals; seed lenticular, ca. 1.2 mm in diameter, red-brown.

GRAND CAYMAN: *Proctor* 50738.
— Pantropical.

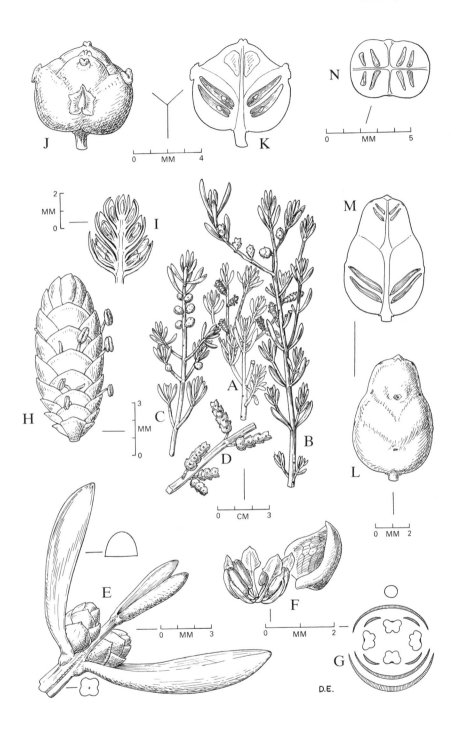

FIG. 99 **Batis maritima.** A, portion of staminate plant. B, C, portions of pistillate plant. D, fruiting branch. E, node with two staminate spikes (immature). F, staminate flower. G, floral diagram, staminate flower. H, staminate spike anthesis. I, long. section through portion of staminate spike. J, pistillate spike. K, long. section of same. L, fruiting spike. M, N, sections through same. (D.E.)

BATACEAE

Decumbent shrub with numerous opposite ascending branches; leaves opposite, sessile, succulent and semiterete, entire; stipules lacking. Flowers unisexual, dioecious, in fleshy axillary spikes. Staminate spikes sessile, with many persistent imbricated scales, each subtending a flower, the perianth cup-shaped and transversely 2-lobed above the middle. Stamens usually 4, inserted at base of the perianth, alternating with 4 staminodes; anthers 2-celled, opening inwardly. Pistillate spikes stalked, 4–12-flowered, with small roundish deciduous scales in alternating pairs; perianth lacking. Ovary sessile, 4-celled, those of a spike united to form eventually a fleshy compound fruit; ovule 1 in each cavity, erect; stigma sessile, 2-lobed. Seeds with no endosperm and large cotyledons.

— A single genus with one species.

Batis L.

The genus has the characters of the family.

Batis maritima L., *Syst. Nat.* ed. 10, 2: 1289 (1759). **FIG. 99.** PLATE 17.

A glabrous shrub with characteristic sweetish odour, less than 1 m high with stout spreading or prostrate main stems; branches angular. Leaves mostly 1–1.5 cm long, acutish or blunt. Spikes 5–14 mm long; stamens exserted. Fruit 1–2 cm long, yellowish-green.

GRAND CAYMAN: *Brunt* 1634; *Kings* GC 280, GC 281A; *Proctor* 15136. LITTLE CAYMAN: *Kings* LC 118, LC 119.

— Southeastern U.S.A., West Indies, continental tropical America, California, and the Galapagos and Hawaiian Islands, frequent or common on salinas and coastal marshes.

POLYGONACEAE

Herbs, shrubs or trees of various habit; leaves alternate, simple, mostly entire, the petiole often dilated and clasping; stipules of characteristic tubular form (ocreae) ensheathing the stem above each leaf-base (except in *Antigonon*). Flowers small, regular, perfect or unisexual, in terminal and axillary racemose inflorescences; perianth of 3–6 segments, persistent after flowering, sometimes becoming fleshy. Stamens 3–9 from a central disc, the filaments free or fused at the base; anthers 2-celled, dehiscent by longitudinal slits. Ovary superior, 1-celled, with usually 3 styles; ovule 1, basal, sessile or erect at the apex of an elongate funicle. Fruit an achene, trigonous or compressed, usually surrounded by the persistent perianth; seed often grooved or lobate, with abundant endosperm, the embryo usually lateral and either curved or straight.

A family of about 30 genera and 800 species, of wide geographic distribution but the majority in temperate regions. Few of them are economically important, but buckwheat (*Fagopyrum sagittatum* Gilib.) is a widely grown Asiatic species whose seeds produce a kind of flour.

KEY TO GENERA

1. Plants woody, often tree-like:	**Coccoloba**
1. Plants herbaceous:	
2. Plants vine-like, climbing by means of tendrils terminating the inflorescences; ocreae absent:	**[Antigonon]**
2. Plants not vine-like and without tendrils; ocreae present:	**Polygonum**

Coccoloba L.

Trees or shrubs, glabrous or sometimes pubescent, with alternate, simple, entire leaves; leaves deciduous or persistent, the petioles not sheathing; ocreae deciduous or persistent, cylindrical, truncate and not ciliate. Flowers small, unisexual or functionally so, dioecious, in spike-like, axillary or subterminal, simple or rarely branched racemes, and subtended by minute bracts (ocreolae); perianth of 5 subequal segments united at the base. Staminate flowers in small clusters along the raceme; stamens 8, the filaments fused at the base; ovary usually abortive. Pistillate flowers solitary (not in clusters) along the raceme, with abortive or rarely functional stamens; ovary 3-angled, with 3 styles. Fruit more or less 3-angled, enclosed by the thickened and succulent perianth, thus appearing drupe-like.

— A chiefly tropical American genus of more than 150 species.

Coccoloba uvifera (L.) L., *Syst. Nat.* ed. 10, 2: 1007 (1759). **FIG. 100.**

SEA-GRAPE

Diffuse shrub or tree to 15 m tall, the branchlets finely pubescent when young, soon becoming glabrous. Leaves orbicular to reniform, mostly 8–15 cm long, sometimes wider than long, glabrous and minutely punctate on both sides, commonly bearded in the axils of the basal veins beneath; ocreae rigid, deciduous, 3–8 mm long, puberulous to pilose. Flowers creamy-white, the staminate in clusters of up to 7, with pedicels 1–2 mm long, the pistillate on pedicels 3–4 mm long. Fruit obpyriform, 1.2–2 cm long, in drooping clusters resembling bunches of grapes, edible; mature fruiting perianth rose-purple, the concealed achene black.

GRAND CAYMAN: *Brunt* 1760; *Hitchcock*; *Kings* GC 298; *Proctor* 15119. LITTLE CAYMAN: *Proctor* 28051, 28099. CAYMAN BRAC: *Kings* CB 99; *Millspaugh* 1225.

— From Florida throughout the Caribbean area to northern S. America, a common plant of sandy coastal thickets, sometimes occurring inland. The fruits have an acidulous flavour, but are sometimes used to make jellies and preserves; they are also occasionally fermented with sugar to make an alcoholic beverage. The wood is hard, heavy, compact, and of fine texture; it takes a high polish and is sometimes used in cabinet-work. When cut, the bark yields an astringent red gum known as West Indian Kino; this was formerly used in medicine. The trees are quite ornamental in appearance and have considerable horticultural value in plantings near the sea, as they are quite tolerant of salt spray.

It seems quite remarkable that no other species of Coccoloba is known to occur in the Cayman Islands, as members of this genus are otherwise common throughout the West Indies, and several species are found very widely in habitats like the Cayman woodlands.

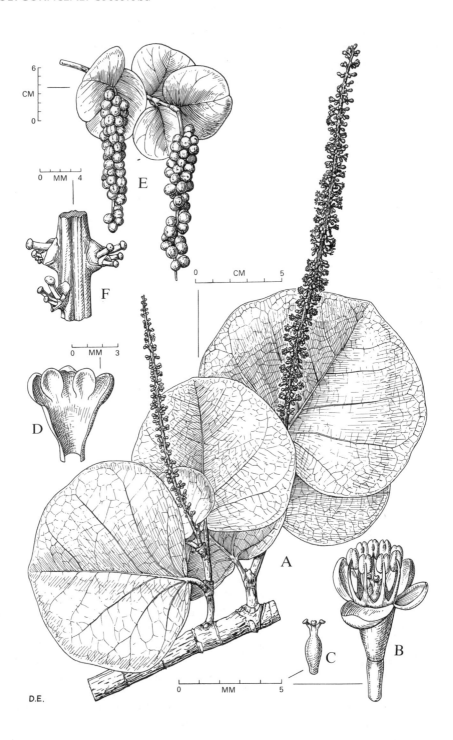

FIG. 100 **Coccoloba uvifera**. A, habit, with staminate inflorescences. B, staminate flower. C, sterile ovary. D, perianth. E, portion of fruiting branch. F, portion of staminate inflorescence after flower have fallen. (D.E.)

282

[*Antigonon* Endl.

Herbaceous vines, sometimes suffrutescent below; leaves petiolate, the blades cordate or deltate, entire; ocreae lacking or represented by a transverse line. Flowers usually pink, in racemes opposite the leaves or terminal, each raceme usually ending in a branched tendril. Perianth of 5 segments, the 3 outer ones larger and broadly cordate, the 2 inner ones narrower and oblong. Stamens 7–8, the filaments fused at the base. Ovary 3-angled, with 3 short styles, the stigmas capitate or peltate; ovule attached to a long funicle and at first pendulous. Achene 3-angled, hidden by the enlarged perianth; seed subglobose, 3–6-lobed, with ruminate endosperm.

— A Mexican and C. American genus of 5 species, often cultivated for their showy pink flowers.

Antigonon leptopus Hook. & Arn., *Bot. Capt. Beechey Voy.* 308, t. 69 (1839).

CORALILLA

Stems often 4 m long or more, climbing by means of axillary tendrils, finely pubescent. Leaves with deltate-ovate blades up to 9 cm long, the apex acute to acuminate, the base cordate and non-decurrent, densely puberulous on both sides in Cayman specimens; petioles 1–4 cm long, somewhat clasping at the base. Racemes 3–8 cm long; pedicels slender, up to 8 mm long; larger perianth segments ca. 1 cm long at anthesis, bright pink, rarely white; filaments with glandular hairs.

GRAND CAYMAN: *Kings* GC 125; *Proctor* 15150. CAYMAN BRAC: *Proctor* (sight record).

— A native of Mexico, widely planted elsewhere as an ornamental, and frequently becoming naturalised in roadside thickets and waste places. The roots bear tubers that are said to be edible.]

Polygonum L.

Annual or perennial herbs, glabrous or pubescent, often glandular, the stems usually enlarged at the nodes. Leaves entire and rather thin, often glandular-punctate; petioles enlarged and sheathing at base; ocreae cylindrical or funnel-shaped, usually membranous or hyaline, often ciliate or fringed with bristles. Flowers perfect, in spikes, racemes or narrow panicles, rarely capitate in corymbs; perianth of 4–6 lobes, these subequal or the outer ones larger, in fruit closely adhering to the achene. Stamens 3–9, inserted on the perianth. Ovary with 2 or 3 styles united below, with capitate stigmas; ovule usually stalked. Achene flattened or 3-angled; seed with embryo excentric.

— A cosmopolitan, chiefly temperate genus of about 150 species, commonly occurring in marshy or wet places.

KEY TO SPECIES

1. Ochreae (stipular sheaths) fringed with bristly hairs:
  2. Flower clusters separated; sepals glandular-punctate: **P. punctatum**
  2. Flower clusters contiguous; sepals not glandular-punctate: **P. hydropiperoides**
1. Ochreae glabrous or nearly so: **P. densiflorum**

FIG. 101 **Polygonum punctatum.** A, leaf and flower-spikes, × ²/₃. B, portion of flower-spike, × 5. C, perianth cut open, × 5. D, stamen, × 13. E, achene, × 6. F, same in transverse section, × 6; e, endosperm; c, embryo. (F. & R.)

Polygonum punctatum Ell., *Bot. S.C. & Ga.* 1: 455 (1817). **FIG. 102.**

SMARTWEED

A slender annual or perennial, often forming large colonies, the stems erect or the lower part creeping and rooting from nodes, simple or branched. Leaves lanceolate, mostly 3–8 cm long, shortly acuminate at each end, densely punctate; ocreae cylindrical, 1–1.5 cm long, bristly hairy and ciliate, persistent. Racemes linear, slender, often interrupted, 1–6 cm long; ocreolae ciliate, 2.5–3 mm long. Perianth white or greenish-white, 2 mm long, glandular-punctate. Achene usually 3-angled, 2 mm long, brown or black.

GRAND CAYMAN: *Kings* GC 178; *Proctor* 15291.

— Common throughout temperate and tropical America, occurring in wet thickets, ditches and swales, along the border of streams and lakes, and in moist waste ground. In some countries (e.g. Guatemala), poultices of the leaves are applied to dogs suffering from mange.

Polygonum hydropiperoides Michx., *Fl. Bor. Am.* 1: 239 (1803).

Plants perennial, slender, glabrous or sparsely strigillose, erect or decumbent at base and rooting at nodes, simple or branched. Ochreae more or less cylindrical, 1–2 cm long, strigose and fringed with long bristles. Leaves lanceolate to oblong-lanceolate or linear-lanceolate, subsessile, mostly 5–10 cm long and 5–15 mm broad, attenuate at each end, not punctuate, the margins ciliate. Racemes narrowly cylindrical to almost linear, erect, 3–6 cm long. Flowers white, greenish or pinkish, the perianth 2 mm long; style 3-parted to below middle. Achenes ovoid, triquetrous, pointed, 3 mm long, lustrous black.

GRAND CAYMAN: *Proctor* 50225. Found in a seasonal pond after heavy rains ca. 1 km NNW of the Queen Elizabeth II Botanic Park.

U.S.A., Mexico, Central and S. America, rare in the West Indies.

Polygonum densiflorum Meisn. in Mart., *Fl. Bras.* 5 (1): 13 (1855).

Polygonum glabrum of Adams, 1972; Proctor, 1984; not Willd.

A stout herb erect from a decumbent and rooting base, glabrous throughout. Leaves lanceolate, mostly 5–11 cm long, acuminate at both ends, densely glandular-punctate; ocreae loose, truncate, 1.3–2 cm long, the upper part deciduous from a cup-shaped base. Racemes continuous, 3–7 cm long, loosely but densely flowered, arranged in terminal panicles; perianth pink in bud, opening white, ca. 3 mm long, glandular-punctate. Achene orbicular-biconvex, apiculate, ca. 2 mm long, lustrous black.

GRAND CAYMAN: *Brunt* 1798, 1823, 2100; *Hitchcock*; *Kings* GC 178, GC 179. *Proctor* 27934.

— Pantropical, in moist situations, the Cayman plants often in *Typha* swamps and in seasonally flooded pastures.

[PLUMBAGINACEAE

Perennial herbs or shrubs with basal or alternate entire simple leaves and perfect and regular flowers. Calyx inferior, 4- or 5-toothed, sometimes plaited at the sinuses, the tube 5–15-ribbed. Corolla hypogynous, of 4 or 5 clawed segments connate at the base or united into a tube. Stamens 4 or 5, opposite the corolla-segments, hypogynous; anthers 2-locular, dorsally attached to the filaments, the locules longitudinally dehiscent. Disc none. Ovary superior, 1-locular; ovule solitary, anatropous, pendulous; styles 5. Fruit a utricle, achene, or capsule, enclosed by the calyx. Seed solitary.

A family of wide distribution with about 10 genera and 500 species, many of them occurring in saline or arid habitats. Only the genus *Limonium* has been found growing wild in the Cayman Islands, but *Plumbago capensis*, a small shrub with blue flowers, is sometimes cultivated.

Limonium Mill., nom. cons.

Perennial (rarely annual) herbs or dwarf shrubs. Leaves simple, usually in a basal rosette but sometimes with densely leafy branches; leaves often absent at anthesis. Inflorescence a corymbose panicle with terminal secund spikes or sometimes with terminal non-flowering branches; spikes of bracteate 1–5-flowered spikelets. Calyx infundibuliform, the limb scarious, sometimes shortly dentate between lobes. Corolla with a short tube or the petals connate only at the base. Stamens inserted at the base of the corolla. Styles 5, glabrous, free or connate at the base; stigmas filiform. Fruit a capsule with circumscissile or irregular dehiscence.

— A cosmopolitan genus of 150 species.

Limonium companyonis (Grenier & Billot) Kuntze, *Rev. Gen. Pl.* 1: 395 (1891). PLATE. 17.

Perennial herb with a woody taproot; leaves subsucculent, entire, spatulate, tapering downward to a narrow pseudopetiolar base, mostly 1.5–3 cm long and up to 1 cm broad near the emarginate apex, forming a dense basal rosette. Inflorescence leafless but with minute bracts, mostly dichotomously branching, up to ca. 25 cm tall with numerous ultimate flowering branches, these somewhat loosely flowered one-sided spikes, the flowers singly attached and separated by 1 mm or more at the points of attachment. Calyx-tube dark brown, ca. 4 mm long, subtended by a broadly ovate bract ca. 1 mm long. Petals light violet, exceeding the calyx by ca. 2 mm. Fruits and seeds not observed.

GRAND CAYMAN: *Joanne Ross s. n.*, from George Town Barcadere; *Proctor 52237*, from vicinity of Safe Haven, north of George Town. All plants were growing on sterile white saline marl and were numerous at the Safe Haven locality.

Mediterranean coast of France and the Balearic Islands, in saline soil among coastal rocks. Its mode of introduction to the Cayman Islands is unknown.]

CLUSIACEAE

Trees, shrubs, or sometimes herbs, terrestrial or epiphytic, often with white or yellowish latex. Leaves mostly opposite and decussate, simple, entire, sometimes with black or transparent dots or lines; stipules lacking. Flowers regular, perfect, or unisexual and dioecious or monoecious, in terminal or axillary inflorescences or sometimes solitary, generally white or yellow, sometimes pink; sepals 2–6, persistent; petals 2–6. Staminate flowers with numerous stamens of indefinite number, these free or united at the base, an ovary lacking or rudimentary. Pistillate or perfect flowers with fewer stamens or staminodes, these often definite in number; ovary superior, with 1 to many cells; stigmas sessile or terminating separate styles; ovules 1 to numerous in each cell. Fruit a capsule, berry or drupe; seeds often enveloped in an aril, without endosperm.

A chiefly tropical family of about 45 genera and 1,000 species, related to the Theaceae but differing especially in the presence of oil glands or tubes in the tissues. Many species yield useful timber, and one of the most delicious tropical fruits, the mangosteen (*Garcinia mangostana* L.), belongs to this family.

KEY TO GENERA

1. Flowers unisexual; fruit a resinous dehiscent capsule:	Clusia
1. Flowers bisexual; fruit a large fleshy drupe:	[Mammea]

Clusia L.

Glabrous trees or shrubs, rarely woody vines, with gummy latex, often epiphytic or growing on rocks or cliffs. Leaves usually thick and more or less leathery. Flowers unisexual or rarely perfect, mostly dioecious; inflorescences terminal, usually several-flowered and alternately branched, rarely 1-flowered. Bracteoles subtending the flowers in 1 to many decussate pairs, often resembling sepals. Sepals 4–6, roundish. Petals 4–10, free or somewhat connate at the base. Pistillate flowers with 5 to numerous staminodes, free or united, with or without anthers. Ovary 4–15-celled, with an equal number of sessile, radiating stigmas; ovules numerous in each cell, arillate. Fruit a leathery or fleshy resinous capsule, septicidally dehiscent.

A mostly tropical American genus of about 145 species, with 2 isolated species in New Caledonia and 1 in Madagascar. The aboriginal inhabitants of the West Indies made use of the waterproof latex of *Clusia* species to seal cracks in their water vessels and canoes, and even today similar use is occasionally made of this substance, by people who are far from being aboriginal.

KEY TO SPECIES

1. Leaves thick and rigid; bracteoles 2; petals 6, rosy-white; capsule 5–8 cm in diameter: **C. rosea**

1. Leaves thinner and more flexible; bracteoles 6–14; petals 4, thick and pale yellow; capsule 2–3 cm in diameter   **C. flava**

Clusia rosea Jacq., *Enum. Syst. Pl. Carib.* 34 (1760). **FIG. 102.** PLATE 17.

BALSAM

A shrub or more commonly a tree up to 10 m tall or more; leaves extremely thick and leathery, on very short, broad, winged petioles, rounded-obovate, mostly 12–16 cm long or sometimes more, with numerous lateral prominulous veins ascending at an angle of 45°. Inflorescences few-flowered; staminate flowers with stamens in several series, united at base to form a cup or ring. Pistillate flowers with staminodes wholly fused to form a cup surrounding the ovary; sepals 4, the inner ones up to 2 cm long; petals 3–4 cm long, of waxy texture. Capsule 6–8-celled, subglobose.

GRAND CAYMAN: *Proctor* 32486. CAYMAN BRAC: *Proctor* 15311, 29050.

— West Indies and southern Mexico south to northern S. America. The Cayman Brac trees grow in rocky woodlands on the central plateau.

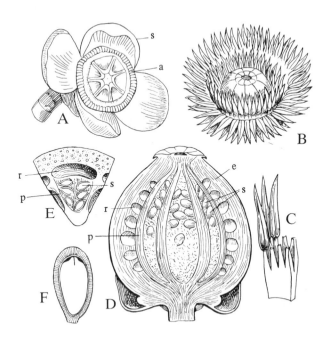

FIG. 102 **Clusia rosea.** A, pistillate flower with petals removed, × 1; **s**, sepal; **a**, staminodes. B, double ring of stamens surrounding pistil of perfect flower, × 1. C, portion of staminal ring, enlarged. D, fruit cut lengthwise, × 2/3; **e**, exocarp; **r**, resin duct; **p**, placenta; **s**, seeds. E, portion of same cut across. F, seed cut lengthwise, × 3. (F. & R.)

Clusia flava Jacq., *Enum. Syst. Pl. Carib.* 34 (1760).

BALSAM

A diffuse shrub or small tree up to 9 m tall, terrestrial or epiphytic. Leaves leathery, on short petioles, cuneate-obovate, 6–16 cm long, with numerous lateral veins ascending at an angle of 45° or narrower. Staminate inflorescence recurved, with usually 2–7 flowers; stamens inserted on a receptacle, crowded, free, the sterile ovary with 4 3-rayed stigmas. Pistillate flowers usually solitary on a recurved pedicel; staminodes 8–12 in 4 bundles, free, bearing anthers; stigmas 12. Bracteoles sepal-like, decreasing in size gradually downward. Sepals 4, 9–11 mm long; petals 4, opposite the sepals, pale yellow and very thick, 2–2.5 cm long, one pair larger than the other. Capsule about 12-locular, subglobose.

GRAND CAYMAN: *Brunt* 1644; *Hitchcock*; *Kings GC* 229; *Millspaugh* 1379; *Proctor* 15141, 31047; *Sachet* 431. CAYMAN BRAC: *Proctor* 29009.

— Jamaica and C. America, in rocky woodlands. The Cayman Brac trees have leaves noticeably smaller than those of Grand Cayman.

[*Mammea* L.

Trees with resinous sap; leaves hard and leathery, dark green and with numerous pellucid dots, the lateral veins prominulous and reticulate with cross-veins. Inflorescences axillary, sessile, 1–3-flowered, or sometimes clustered along the branches below the leaves. Flowers perfect; calyx closed in bud, splitting into 2 segments when the flower opens; petals 4–6. Stamens numerous, free or united at base, the anthers linear and longitudinally dehiscent; ovary 2-celled, with 2 ovules in each cell, or 4-celled, the ovules solitary; ovules basal, erect. Style thick, capped by a large, shield-like, 2-lobed stigma. Fruit a large, fleshy drupe with 1–4 seeds; seeds large, rough-surfaced, the embryo with thick, fleshy cotyledons.

— A genus of 4 species, one tropical American, the other 3 in tropical Africa.

Mammea americana L., *Sp. Pl.* 1: 512 (1753).

MAMMEE

A glabrous tree up to 15 m tall or more; leaves oblong-elliptic or narrowly obovate, 10–25 cm long, short-petiolate, densely pellucid-dotted. Flowers white, fragrant; sepals 1–1.7 cm long; petals 1.5–2 cm long or more. Fruit subglobose, apiculate, 10–15 cm long, with rough, russet-coloured skin and yellowish edible flesh, rather firm and juicy but sometimes bitter; seeds about two-thirds as long as the fruit, reddish and rough.

GRAND CAYMAN: *Brunt* 1992, 2112; *Kings GC* 198.

— West Indies, apparently native on some islands, but introduced in Grand Cayman. The tree is widely planted in tropical countries for its handsome evergreen foliage, fragrant flowers, and edible fruits.]

TILIACEAE

Trees, shrubs, or sometimes herbs, the pubescence often of branched hairs; leaves alternate, rarely opposite, simple; stipules usually present. Flowers perfect, solitary or in few-flowered cymose inflorescences, axillary or terminal; sepals usually 5, free or united; petals usually 5 or else none. Stamens rather numerous, free or united at the base in fascicles; anthers 2-celled, opening by longitudinal slits or apical pores. Ovary superior, sessile, 2–10-celled; style usually simple, with as many stigmas as ovary-cells; ovules 1 to many in each cell, on axial placentae. Fruit a capsule, drupe or berry; seeds 1 to many, with endosperm and straight embryo.

— A widely distributed family of about 50 genera and 400 species.

KEY TO GENERA

1. Fruit a more or less elongate capsule, lacking spines or bristles:	Corchorus
1. Fruit globose, indehiscent, covered all over with stiff, often hooked, spines:	Triumfetta

Corchorus L.

Herbs or low shrubs with pubescence of simple or stellate hairs; leaves mostly thin, with serrate or crenate margins; stipules present. Flowers yellow, small, solitary or in few-flowered clusters, axillary or opposite the leaves; sepals and petals 5, rarely 4. Stamens numerous or else twice as many as the petals, free, inserted on a flat torus. Ovary 2–5-celled; ovules numerous in each cell. Fruit a silique-like capsule, splitting lengthwise into 2–5 valves, many-seeded, sometimes with transverse partitions between the seeds.

A pantropical genus of about 30 species. Two Asiatic species (*Corchorus olitorius* L. and *C. capsularis* L.) are the source of jute, a fibre of great economic importance. In Jamaica and the Cayman Islands, bags made of jute are called crocus bags, an obvious corruption of *Corchorus*.

KEY TO SPECIES

1. Leaves and capsules nearly or quite glabrous:	
2. Calyx 6–7 mm long; capsule 2-locular, not winged, 3–7 cm long, with 4 minute teeth at the apex:	C. siliquosus
2. Calyx 3–4 mm long; capsule 3-locular, wing-angled, 1.5–2.5 cm long, with 3 horizontal beaks at the apex:	C. aestuans
1. Leaves densely covered with stellate pubescence; capsules woolly, 4-locular:	C. hirsutus

Corchorus siliquosus L., *Sp. Pl.* 1: 529 (1753). **FIG. 103.** PLATE 17.

A small shrub or shrubby herb up to 1 m tall or more, the tough stems usually bearing a line of short simple hairs. Leaves small, glabrous except for simple hairs on the petiole, lance-oblong to ovate, 0.4–2 cm long, the apex acute, the margins finely serrate; stipules hair-like. Flowers solitary or in pairs on short axillary peduncles; sepals linear, 6–7 mm long; petals obovate, shorter than the sepals. Capsule 3–7 cm long, linear, compressed, 2–3 mm broad, minutely puberulous along the join between the valves; seeds 3-angled, ca. 1 mm long.

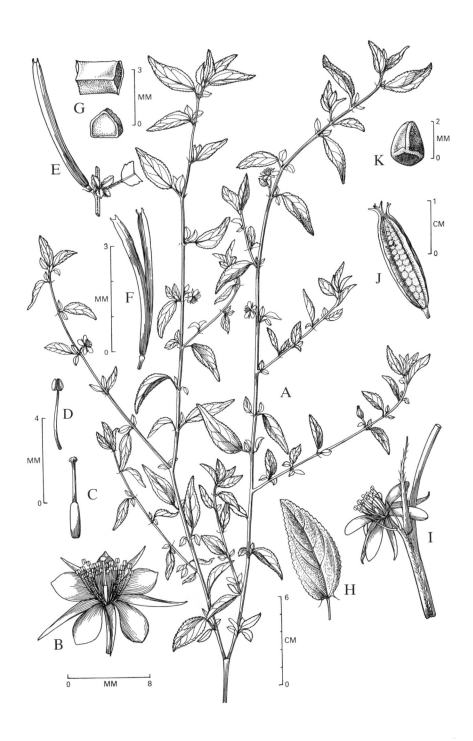

FIG. 104 **Corchorus siliquosus**. A, habit. B, flower. C, pistil. D, stamen. E, fruit. F, fruit after dehiscence. G, seed, 2 views. **Corchorus aestuans**. H, leaf. I, flower habit. J, fruit. K, seed. (G.)

GRAND CAYMAN: *Brunt* 1648; *Hitchcock*; *Millspaugh* 1344; *Proctor* 15095. LITTLE CAYMAN: *Proctor* 28160. CAYMAN BRAC: *Proctor* 28973.

— West Indies and on the continent from Florida and Texas south to northern S. America, a weedy persistent plant of open waste places, sandy roadsides, and pastures.

Corchorus aestuans L., *Syst. Nat.* ed. 10, 2: 1079 (1759). **FIG. 103.**

An annual subwoody herb, the stems low-spreading or ascending, puberulous or short-pilose with simple hairs. Leaves glabrous or with scattered hairs, slender-petiolate, 1–6 cm long, the blades ovate to orbicular with obtuse or subacute apex, the margins finely serrate with the lowermost serration on one or both sides often elongated into a hair-like bristle; stipules subulate, 5–7 mm long. Flowers solitary or paired in the leaf-axils, almost sessile; sepals hooded, ca. 4 mm long; petals obovate, 3–4 mm long. Capsule 1.5–2.5 cm long, narrowly oblong, 4–5 mm thick, triangular in cross-section, narrowly winged on the angles, glabrous; seeds discoid, dark brown, less than 1 mm broad.

CAYMAN BRAC: *Proctor* 28938.

— Pantropical but not usually very common. The Cayman Brac plants grow in moist hollows, thickets, and along the borders of pastures.

Corchorus hirsutus L., *Sp. Pl.* 1: 530 (1753).

Shrub up to 2 m tall, clothed throughout with dense, soft, stellate pubescence. Leaves petiolate, the blades lance-oblong to ovate or elliptic, 1.5–6 cm long, the apex blunt to acute, the margins irregularly serrate or crenulate-serrate; stipules subulate, 4–5 mm long, soon deciduous. Flowers pedicellate, in axillary umbels of 2 to 8; sepals 5–6 mm long; petals obovate, about equal in length to the sepals. Capsule oblong or nearly globose, 7–11 mm long, densely woolly, with a short erect terminal beak; seeds irregularly ellipsoidal, black, 1.5–2 mm long.

GRAND CAYMAN: *Correll & Correll* 51024.

— West Indies and tropical Africa, mostly in rather dry thickets.

Triumfetta L.

Herbs or shrubs, usually bearing stellate hairs. Leaves thin, variable, often 3–5-angled or -lobed, irregularly serrate. Flowers yellow or red, few or densely clustered in cymes or panicles, axillary or opposite the leaves, or sometimes in terminal racemes; sepals 5, somewhat hooded at the apex; petals 5 or sometimes lacking. Stamens numerous or sometimes 10, free, borne on the elevated 5-glandular receptacle. Ovary 2–5-celled with 2 ovules in each cell; stigma 2–5-toothed. Fruit a subglobose prickly capsule, indehiscent or separating into cocci; prickles hooked at the apex, by this means often clinging to animals or to clothing; seeds 1 or 2 in each cell, with endosperm.

— A pantropical weedy genus of about 50 species.

KEY TO SPECIES

1. Flowers with petals; body of the fruit glabrous:	T. semitriloba
1. Flowers lacking petals; body of the fruit stellate-pubescent:	T. lappula

Triumfetta semitriloba Jacq., *Enum. Syst. Pl. Carib.* 22 (1760).

BUR-WEED

A short-lived shrub or woody herb up to 1.5 m tall or more; leaves long-petiolate, the petioles 1–9 cm long; blades ovate to broadly ovate below, 2–10 cm long, or the upper lance-oblong and much smaller, acute at apex and rounded or truncate to cordate at base, the margins unequally dentate and often angled or shallowly 3-lobed, stellate-pubescent on both sides; stipules linear, 5–6 mm long. Inflorescences rather few-flowered; sepals 5–8 mm long, green; petals slightly shorter than the sepals. Ovary 3-celled. Fruit 3–5 mm in diameter (excluding prickles), the body glabrous or nearly so, the prickles with scattered retrorse hairs.

GRAND CAYMAN: *Hitchcock*; *Millspaugh* 1297; *Proctor* 15019.

— West Indies and continental tropical America, a common weed of waste ground, roadside thickets, and old fields.

Triumfetta lappula L., *Sp. Pl.* 1: 444 (1753).

BUR-WEED

A shrub or woody herb 1–2 m tall; leaves long-petiolate, the petioles 1–8 cm long; blades roundish-angulate or very broadly ovate below, up to 13 cm long, the upper ones ovate or elliptic and much smaller, acuminate at apex and usually truncate at base, the margins very unequally dentate and often angulate or 3–5-lobed, densely stellate-pubescent on both sides. Inflorescences many-flowered, dense; sepals 3–4 mm long, densely pubescent; petals absent. Ovary 2-celled (but in fruit often apparently 3- or 4-celled because of the development of both ovules in one or both cells). Fruit 2–3.5 mm in diameter (excluding the prickles), the body densely stellate-pubescent, the prickles with scattered or numerous retrorse hairs.

GRAND CAYMAN: *Hitchcock*.

— West Indies and continental tropical America; adventive at scattered localities in the Old World tropics, a weed of secondary thickets, waste ground and roadsides. The bark is said to contain a tough fibre suitable for making rope.

STERCULIACEAE

Herbs, shrubs or trees, the pubescence often at least partly of stellate hairs. Leaves alternate, simple, and entire, dentate or lobate, rarely digitately compound; petioles often pulvinate at the apex; stipules usually present, but soon deciduous. Flowers perfect or rarely unisexual, in axillary or terminal racemes or cymose panicles, rarely solitary. Calyx 5-lobed, persistent. Petals 5, often persistent after withering, or sometimes lacking. Stamens 5 or more, united at the base or beyond the middle, then forming a tubular column on which the anthers often alternate with staminodes. Ovary superior, sometimes stalked, 2–5-locular (reduced to 1 carpel in *Waltheria*); styles as many as the carpels, free or more or less united; ovules 2 to many in each locule. Fruit with the carpels partly free, or else united to form a capsule, dehiscent or indehiscent; seeds with or without endosperm, the pericarp never woolly.

— A pantropical family of about 50 genera and 750 species. Chocolate (from *Theobroma cacao* L.) and cola (from *Cola* spp.) are economically important products of this family, widely used in the foods and beverage industries. They both contain a stimulating drug known as caffeine.

KEY TO GENERA

1. Ovary long-stalked, in age longer than the fruit; carpels spirally twisted in fruit: — **Helicteres**

1. Ovary nearly or quite sessile; carpels not twisted in fruit:

  2. Herbs or low shrubs; petals flat, withering and persisting in fruit; capsules thin, splitting lengthwise:

    3. Ovary 1-locular; petals yellow: — **Waltheria**

    3. Ovary 5-locular; petals pink or purple:

      4. Flowers with a long-exserted gynophore: — **Neoregnellia**

      4. Flowers lacking a long-exserted gynophore: — **Melochia**

  2. Trees:

    5. Leaves entire; flowers with petals; fruits subglobose, black when ripe, nearly indehiscent: — **Guazuma**

    5. Leaves palmately lobed; flowers without petals; fruits with red carpels, dehiscent: — **[Sterculia]**

Helicteres L.

Shrubs or small trees with stellate pubescence; leaves serrate or entire, petiolate. Flowers axillary, solitary or in clusters; calyx 2-lobed or else tubular with a 5-lobed apex; petals 5, flat, equal or unequal, clawed, with auriculate appendages on the claws. Stamens 6, 8, 10, or indefinite, the filaments slightly united in pairs at the base; anthers 2-celled, the cells sometimes confluent; staminodes present. Ovary 5-lobate and 5-celled, borne on a very long, pedicel-like gynophore, this curved and noose-like at first, finally straightening and long-exserted beyond the calyx and petals; ovules many. Fruit composed of 5 hard, cohering, tube-like carpels, these spirally twisted or sometimes straight, splitting open along the inner seams; seeds small, flattened-ovoid, with scanty endosperm and straight embryo.

— A tropical American genus of about 30 species; the hard, screw-like fruits are unique.

Helicteres jamaicensis Jacq., *Enum. Syst. Pl. Carib.* 30 (1760). **FIG. 104.** PLATE 17.

WILD COW ITCH, SCREW-BUSH

Shrub, sometimes arborescent and up to 5 m tall, the twigs densely stellate-pubescent. Leaves long-petiolate, mostly 4–15 cm long, the blade ovate, acuminate at apex and cordate at base, the margins irregularly crenate-toothed, and both sides densely stellate-pubescent. Peduncles opposite the leaves, 1.5–3 cm long, 1–3-flowered. Calyx bell-shaped, 1.5–2 cm long, stellate-pubescent, 2-cleft and unequally 5-toothed. Petals oblong, white or creamy, longer than the calyx. Stamens 10, with short filaments; staminodes 5. Gonophore 5–8 cm long. Fruit more or less cylindrical or ellipsoid, 2–3.5 cm long, woolly pubescent with stellate hairs, the carpels twisted around about twice, free at the apex.

GRAND CAYMAN: *Brunt* 1993; *Correll & Correll* 51013; *Fawcett*; *Hitchcock*; *Kings* GC 371; *Millspaugh* 1370; *Proctor* 15015. LITTLE CAYMAN: *Kings* LC 32; *Proctor* 28034. CAYMAN BRAC: *Millspaugh* 1183; *Proctor* 29026.

— West Indies and C. America, usually in rocky thickets and woodlands.

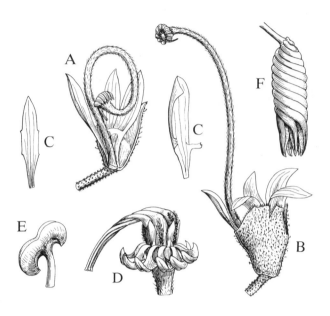

FIG. 104 **Helicteres jamaicensis**. A, bud of flower just opening cut lengthwise, × 1. B, flower, × 1. C, petals, × 1. D, apex of gonophore with stamens and pistil, × 4. E, stamen, × 8. F, fruit, × ²/₃. (F. & R.)

Waltheria L.

Herbs or shrubs with stellate hairs; leaves petiolate, irregularly serrate; stipules small and narrow. Flowers yellow, small, in dense axillary or terminal cymose heads, each flower subtended by 3 linear bracteoles. Calyx 5-lobed. Petals 5, spatulate, withering-persistent. Stamens 5, united at base; staminodia absent. Ovary sessile, 1-celled; ovules 2; style slightly lateral and the stigma club-shaped or fringed. Capsule 2-valved, 1-seeded; seed with endosperm and straight embryo.

— A chiefly tropical American genus of more than 30 species.

Waltheria indica L., *Sp. Pl.* 2: 673 (1753). **FIG. 106**. PLATE 18.

Waltheria americana L. (1753).

Erect or decumbent shrub or shrubby herb up to 1 m tall or a little more, densely stellate-pubescent throughout, or rarely glabrate. Leaves oblong to ovate or oblong-lanceolate, mostly 2–6 cm long, with apex obtuse, the base obtuse to subcordate, the petioles up to 2 cm long. Inflorescence with bracteoles 3–4 mm long; calyx 3.5–5 mm long, with linear-lanceolate lobes; petals 5–6 mm long. Staminal tube 2 mm long. Capsule 2–3 mm long. GRAND CAYMAN: *Brunt* 1637, 1898a; *Hitchcock*; *Millspaugh* 1325, 1327, 1336, 1382; *Proctor* 15084; *Sachet* 415. LITTLE CAYMAN: *Proctor* 28084.

— West Indies and continental tropical America; naturalised in the Old World tropics. In the Cayman Islands, this species grows in dry sandy thickets, rough pastures, and in rocky scrublands.

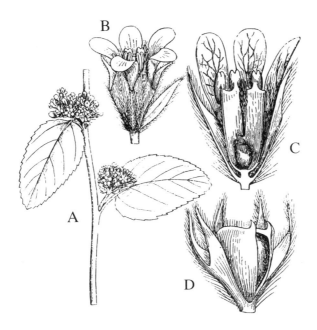

FIG. 105 **Waltheria indica**. A, portion of branch with leaves and flowers, × ²/₃. B, flower with bracteole, × 4. C, flower cut lengthwise, × 7. D, fruit with persistent calyx, showing seed, × 7. (F. & R.)

Neoregnellia Urban.

Shrubs with stellate pubescence. Leaves alternate, distichous, with dentate margins. Flower axillary, solitary, polygamous; calyx 5-lobed; petals dilated at the apex, clawed and auriculate at the base, marcescent. Stamens 10, the anthers 2-locular; staminodes 5, membranous. Gynophore filiform, long-exerted; ovary 5-carpellate, each locule with 1 or 2 ovules; styles free. Fruit a 5-lobed capsule, each lobe a 1–2-seeded carpel.

— A monotypic genus endemic to the Greater Antilles.

Neoregnellia cubensis Urban, *Repert. Spec. Nov. Regni. Veg.* 20: 306 (1924). Plate 18.

Small branched shrub seldom more than 1 m tall; leaves subsessile or very short petiolate, lanceolate, oblong-lanceolate, or oblong, mostly 1–3 cm long and 0.3–0.9 cm broad at or below the middle, blunt at the apex and subcordate at the base, the margins dentate chiefly on the apical half, the venation pinnate, prominulous on the lower surface, and beset with numerous minute stellate hairs on both sides. Flowers pedicellate (pedicels ca. 2 mm long); calyx dark brown, 5.5–6.5 mm long; petals crimson, 11–12 mm long, emarginate; gynophore ca. 2 cm long. Fruit 5-lobed, 2–4 mm in diameter, densely stellate-tomentose.

GRAND CAYMAN: *Proctor* 47353, 47827.

— Otherwise known only from Cuba and Hispaniola.

Melochia L.

Herbs or shrubs, rarely trees, with pubescence chiefly of stellate hairs; leaves petiolate, narrow to broad, with serrate margins. Flowers mostly small, perfect but heterostylous (i.e. with either long or short styles), more or less densely clustered in the leaf-axils or opposite the leaves, or else in terminal spikes, cymes, or panicles. Calyx 5-lobed or 5-toothed, bell-shaped; petals 5, usually pink, purple or violet, sometimes white, spatulate to oblong, withering-persistent. Stamens 5, opposite the petals, more or less united into a tube at the base, rarely alternating with 5 tooth-like staminodes. Ovary 5-celled, sessile or short-stalked; ovules 2 in each cell; styles 5, free or more or less united. Fruit a 5-valved capsule, longitudinally dehiscent, 5–10-seeded; seeds with endosperm and straight embryo.

— A pantropical genus of about 60 species, the majority in tropical America.

KEY TO SPECIES

1. Leaves and stems densely whitish-pubescent with stellate hairs:	**M. tomentosa**
1. Leaves and stems glabrous or minutely puberulous with mostly simple hairs:	
2. Inflorescence stalked, opposite the leaves; capsule pyramidal:	**M. pyramidata**
2. Inflorescence sessile, axillary; capsule depressed-globose:	**M. nodiflora**

Melochia tomentosa L., *Syst. Nat.*, ed. 10, 2: 1140 (1759). PLATE 18.

VELVET-LEAF

Shrub up to 2 m tall or sometimes more, densely whitish stellate-pubescent; leaves lance-oblong to ovate 1–7 cm long, rounded to acute at apex, the base rounded or subcordate, the veins channelled on the upper side and prominently raised beneath. Flowers in small axillary and terminal pedunculate cymes; calyx 6–8 mm long, deeply and narrowly lobed, and densely stellate-pubescent on the outside; petals pink, mauve, or rosy-purple, 10–12 mm long or more. Stamens 6–8 mm long, united for about half their length. Ovary short-stalked; styles united for about half their length. Capsule broadly pyramidal, long-beaked, ca. 9 mm long; seeds 1 or 2 in each cell.

GRAND CAYMAN: *Brunt* 1810, 1999; *Fawcett*; *Kings* GC 423; *Millspaugh* 1288, 1326; *Proctor* 15054; *Sachet* 405. LITTLE CAYMAN: *Kings* LC 95; *Proctor* 28141. CAYMAN BRAC: *Kings* CB 15; *Millspaugh* 1184; *Proctor* 28961.

— West Indies and continental tropical and subtropical America, in sandy or rocky thickets and scrublands.

Melochia pyramidata L., *Sp. Pl.* 2: 674 (1753).

A decumbent to erect shrubby herb to 1 m tall but often lower, the stems usually minutely puberulous on one side or nearly glabrous. Leaves thin, lanceolate to ovate, 1–6 cm long, acute or obtuse at the apex, rounded at the base, glabrous or minutely puberulous with simple hairs mixed with a few branched or stellate ones. Flowers in pedunculate umbels of 2–5, opposite the leaves; calyx 3.5–4 mm long, with linear lobes; petals pink or purplish, obovate, 6–7 mm long, clawed. Capsule pyramidal, 5-angled, 5–8 mm long, bearing a few minute stellate hairs.

GRAND CAYMAN: *Kings* GC 324; *Millspaugh* 1345; *Proctor* 15252, 27995.

— West Indies and continental tropical America, chiefly along roadsides and in weedy fields.

Melochia nodiflora Sw., *Nov. Gen. & Sp. Pl.* 97 (1788).

Slender low shrub up to 1.5 m tall, often suffrutescent, the stems dark red, the younger parts usually puberulous with mostly simple hairs. Leaves on slender petioles, the blades oblong to very broadly ovate, mostly 1.5–7 cm long and up to 5.5 cm broad, the apex acute to acuminate, the base rounded or subcordate, conspicuously veined beneath. Flowers mingled with thin brown bracteoles in dense sessile axillary clusters. Calyx 3–4 mm long, with lanceolate lobes; petals pink, often striped, ca. 5 mm long, short-clawed. Stamens completely united, forming a tube ca. 2 mm long. Ovary sessile, 5-lobed; styles free. Capsule 5-lobed, pubescent, the carpels separating when ripe.

GRAND CAYMAN: *Correll & Correll* 51011; *Hitchcock*; *Lewis* 3834; *Proctor* 15106. CAYMAN BRAC: *Proctor* 28948.

— West Indies and continental tropical America, a weed of roadsides, pastures and old fields.

Guazuma Adans.

Trees with stellate pubescence; leaves short-petiolate, sometimes inequilateral at the base, and irregularly toothed; stipules present. Flowers small, in axillary short-pedunculate cymes; calyx 2–3-parted; petals 5, clawed and hooded-concave, the apex 2-cleft and bearing a terminal linear 2-cleft appendage. Stamens united to form a 5-lobed tube, the acuminate lobes (equivalent to staminodes) alternate with the petals, the short-stalked anthers in groups of 2 or 3 in the sinuses, 2-celled. Ovary sessile or short-stalked, 5-lobed, 5-celled; ovules few to numerous in each cell, on axile placentae; styles more or less united. Fruit a subglobose, woody capsule, covered with short, hard tubercles or (in one species) plumose-setose, indehiscent or imperfectly 5-valvate at the apex; seeds with endosperm and slightly curved embryo, the cotyledons leaf-like and inflexed-folded.

— A tropical American genus of 4 species (Freytag in *Ceiba* 1 (4): 193–225 (1951)).

Guazuma ulmifolia Lam., *Encycl. Méth. Bot.* 3: 52 (1789). **FIG. 106.**

BA`CEDAR

A small tree up to 10 m tall or more; leaves ovate to ovate-oblong, 3–15 cm long, acute to acuminate at the apex, usually cordate or subcordate and inequilateral at base, pubescent with stellate hairs on both sides. Inflorescences 2–3 cm long; sepals 3–4 mm long, reflexed; petals yellowish or cream, 3–4 mm long, minutely stellate-pubescent on the outside, the appendage 4–6 mm long and 0.2–0.5 mm wide. Fruit globose or oblongoid, 17–37 mm long, indehiscent, the blackish tubercles separating deeply and irregularly at maturity; seeds numerous, obovoid, 2–3.8 mm long.

GRAND CAYMAN: *Hitchcock*. Apparently very rare in the Cayman Islands.

— West Indies and continental tropical America, usually in dry open woodlands or on the borders of pastures. It is a rather useful tree: in addition to being resistant to drought, rope and twine can be made from the tough fibrous bark, the flowers are a good source of honey, and both the foliage and the immature fruits make nutritious fodder for cattle and other stock.

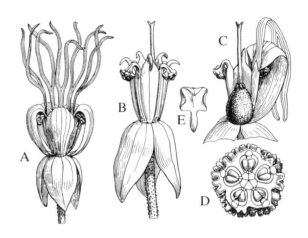

FIG. 106 **Guazuma ulmifolia**. A, flower, × 4. B, same with petals removed, showing staminal tube with stamens and staminodes, × 7. C, portion of flower showing the apex of a petal with appendage lying on a stamen, also ovary and style, × 7. D, x-section of fruit, × 1. E, embryo, × 11. (F. & R.)

[*Sterculia* L.

Mostly large trees, usually buttressed, evergreen or deciduous, andromonoecious. Leaves alternate, petiolate, palmately compound or lobed, or simple and entire. Stipules modified into stipulaceous bracts. Flowers staminate or bisexual, with parts borne on elongate androgynophores. Sepals with lobes sometimes appendaged; petals absent. Stamens 10, 12, 15 or many, the anther loculi irregularly arranged. Gynoecium tomentose or pilose; ovules 2–20 per carpel; style recurved, the stigma capitate or 5-lobed. Fruit of 1–5 woody follicles, bright red or brown and densely minutely velvety externally. Seeds 2–20 per follicle, pendant from the follicular suture, ellipsoid, with glistening, blue-black, papery tegument.

— A pantropical genus of about 300 species.

Sterculia apetala (Jacq.) Karst., *Fl. Columb.* 2: 35 (1869). PLATE 18.

A large tree with buttressed trunks and a broad, dense, spreading crown. Leaves with stout petioles mostly 12–20 cm long or more, the blades coriaceous and deeply 5-lobed with central lobe 15 cm long or more, the upper surface glabrous and lower surface densely and minutely stellate-puberulous. Flowers in dense panicles, bell-shaped calyx, 5-lobed or less spreading brown woody follicles, these red within, dehiscing to release the black ellipsoid seeds ca. 2 cm long; interior of follicles covered with numerous stiff needle-like hairs, these very irritating to the skin.

GRAND CAYMAN: *Proctor* 48290; found in Lower Valley.

— Mexico to Peru and Brazil; widely planted elsewhere.]

MALVACEAE

Herbs, shrubs or trees, often with stellate hairs; leaves alternate, simple but often lobed, usually palmately veined, at least at the base; stipules present. Flowers regular, usually perfect, axillary and solitary, or else in axillary or terminal racemes, fascicles, or panicles; each flowers subtended by 3 or more bracteoles, these often large and forming an 'epicalyx' outside the true calyx. Sepals usually 5, more or less united, the lobes valvate; petals 5, free but adnate to the base of the staminal column (thus appearing gamopetalous), twisted and overlapping in bud. Stamens numerous, rarely 5 or 10, the filaments united below into a tube (column); anthers 1-celled. Ovary superior, 2- to many-celled; style single at the base, dividing above into as many branches as there are ovary-cells; ovules 1 or more in each cell, attached along its inner angle. Fruit usually a dry schizocarp or capsule; seeds with scant endosperm and curved embryo, the cotyledons folded and foliaceous.

— A nearly world-wide family of more than 45 genera (up to 75 recognised by some authorities) and about 1,500 species. Many are of economic value for food (e.g. okra, *Hibiscus esculentus* L., and sorrel, *Hibiscus sabdariffa* L.); textile fibres (e.g. cotton, *Gossypium* spp.); timber (e.g. blue mahoe, *Hibiscus elatus* Sw.); and ornament (e.g. *Hibiscus* spp., *Malvaviscus* spp.).

KEY TO GENERA

1. Petals convolute, never opening; style-branches 10; carpels fleshy, united into a berry:	**Malvaviscus**
1. Petals spreading; style-branches 5, separate or united; carpels dry, not berry-like:	
2. Fruit a schizocarp, the carpels more or less separating at maturity into separate cocci:	
3. Flower subtended by large, cordate-ovate, leaf-like bracts and almost concealed by them; staminal column bearing anthers on the side; plants with harsh stinging hairs:	**Malachra**
3. Flowers not subtended by leaf-like bracts (but with or without an epicalyx); staminal column bearing anthers at the apex:	
4. Epicalyx absent:	
5. Carpels 1-seeded:	
6. Carpels with a transverse partition or ring inside; leaves whitish-tomentose beneath:	**Wissadula**
6. Carpels not partitioned or ringed inside; leaves glabrous or pubescent, but not whitish-tomentose beneath:	**Sida**
5. Carpels 2- or 3-seeded:	
7. Carpels leathery, beaked:	**Abutilon**
7. Carpels membranous, inflated and bladder-like, rounded at apex and not beaked:	**Herissantia**
4. Epicalyx present:	
8. Carpels of fruit covered with numerous short, barbed spines:	**Urena**
8. Carpels of fruit unarmed or with 1–3 long, simple spines:	**Malvastrum**
2. Fruit a capsule splitting at maturity, the carpels not separating:	
9. Epicalyx absent:	**Bastardia**
9. Epicalyx present:	
10. Epicalyx of 3 bracteoles larger than the sepals:	**Gossypium**
10. Epicalyx of 5 or more bracteoles smaller than the sepals:	

Wissadula Medic.

Shrubs or shrubby herbs, usually densely clothed in whitish or yellowish tomentum composed of minute stellate hairs; leaves often long-petiolate, cordate, and long-acuminate. Inflorescences axillary and terminal, loosely paniculate; flowers on long slender pedicels, lacking an epicalyx; calyx broadly 5-lobed; petals yellow or orange. Staminal column divided at apex into an indefinite number of filaments. Ovary 5-celled; ovules 1 or 3 in each cell; style-branches with capitate stigmas. Fruits somewhat top-shaped; ripe carpels beaked, the beaks erect or pointing outward, partially divided inside by an incomplete transverse partition sometimes represented by a ring, and opening by two valves; seeds 1–3 in each carpel.

— A genus of about 40 species, all but a single pantropical one occurring in tropical America.

Wissadula fadyenii Planch, ex. R. E. Fries in *Svensk. Vet. Akad. Handl.* 43 (4): 30, t. 1, f. 1–2; t. 6, f. 2–4 (1908).

Wissadula divergens Griseb. (1859), not Benth.

A shrubby herb to 1 m tall or more; leaves 3–17 cm long, the blades triangular-ovate and long-acuminate, cordate or subcordate at base, the margins entire, densely soft stellate-pubescent beneath, and 5–7-nerved from the base; petioles varying from 0.5 cm on the upper leaves to 6 cm on the lower ones. Pedicels to 5 cm long. Calyx 3–3.5 mm long, divided about halfway into triangular lobes; petals 4–5 mm long. Fruit ca. 6 mm broad at maturity, the dark brown carpels minutely puberulous, their beaks ca. 0.5 mm long; seed one in each carpel, ca. 2 mm long, minutely hairy.

GRAND CAYMAN: *Hitchcock*, collected 19 January 1891. No subsequent collector has found this species in the Cayman Islands.

— Jamaica, Trinidad, and northern S. America, a weedy plant of clearings and old fields.

Abutilon Mill.

Herbs or shrubs, rarely trees, usually velvety-pubescent with stellate hairs; leaves mostly cordate, often angulate or lobed. Flowers small or rather large, axillary and solitary or in small cymes, or else terminal and paniculate. Calyx 5-lobed; petals commonly white, yellow or red. Staminal column divided at the apex into numerous slender filaments. Ovary 5–many-celled; ovules 3–9 in each cell; style-branches as many as the cells, filiform or club-shaped. Mature carpels coalescent at the base or completely separating, 2-valved; seeds subreniform.

A genus of more than 100 species, occurring widely in tropical and subtropical regions.

Abutilon permolle (Willd.) Sweet, *Hort. Brit.* 1: 53 (1826).

A shrubby herb with an erect stem up to 1.5 m tall, all parts softly stellate-pubescent; leaves mostly 4–10 cm long, the blades broadly ovate and deeply cordate, the apex acuminate, the margins irregularly crenate; petioles up to 5 cm long. Flowers axillary, solitary, long-pedunculate; calyx 8–10 mm long, densely stellate-pubescent; petals 1.2–1.7 cm long, yellow. Carpels 7–10, when ripe 9–10 mm long, dark brown, villous, 3-seeded, beaked at the apex; seeds 2 mm in diameter.

GRAND CAYMAN: *Hitchcock*; *Millspaugh* 1799.

— Florida, West Indies, southern Mexico and northern Guatemala, often in coastal thickets and dry fields.

Herissantia Medic.

Annual or sometimes short-lived perennial herb, the slender branches trailing, spreading, or weakly erect, more or less finely stellate-tomentose; leaves cordate. Flowers mostly solitary on axillary peduncles, the peduncles filiform. Calyx deeply 5-cleft; petals 5, whitish. Carpels about 12, the styles slender with terminal stigmas; ovules 2–6 in each carpel. Seeds glabrous or thinly setulose.

A monotypic genus, native of tropical and subtropical America, now pantropical in distribution.

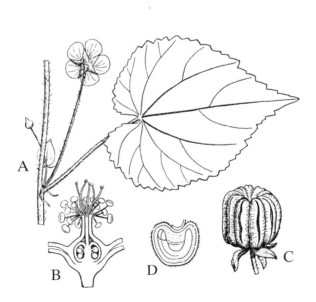

FIG. 107 **Herissantia crispa**. A, leaf and flower, × ²/₃. B, flower cut lengthwise, with calyx and petals removed, enlarged. C, capsule, × 1. D, section of seed, × 6. (F. & R.)

Herissantia crispa (L.) Brizicky, *Jour. Arnold Arb.* 49 (2): 278–9 (1968). **FIG. 107.**

Abutilon crispum (L.) Medik. (1787).

Stems up to 1.2 m long, frequently much shorter in arid situations; leaves 2–8 cm long, the petioles equal in length to the blades; blades acute to shortly acuminate, the margins crenate-dentate, stellate-tomentose on both sides. Peduncles 1.5–5 cm long. Calyx 4–8 mm long; petals as long as to twice as long as the calyx, obovate. Carpels when ripe aggregated in a head 10–15 mm in diameter, thinly pubescent; seeds black, ca. 2 mm in diameter.

GRAND CAYMAN: *Hitchcock*; *Proctor* 15220. CAYMAN BRAC: *Proctor* 29059.

— Florida and Texas southward throughout the West Indies and continental tropical America; also said to occur in southeast Asia. This species chiefly grows in sandy waste ground, dry fields, and gravelly clearings in woodlands.

Malvastrum A. Gray

Annual or perennial herbs, sometimes subwoody, with pubescence of simple, branched and stellate hairs. Leaves entire, serrate, lobed or cleft; linear stipules present. Flowers axillary or in terminal clusters or spikes, stalked or sessile; epicalyx of 3 linear to lanceolate bracteoles about as long as the calyx, rarely lacking. Calyx 5-lobed; petals 5, usually yellow or white. Ovary of 5–many carpels; ovules 1 in each cell; style-branches as many as the carpels; stigmas linear, club-shaped or capitate. Ripe carpels separating, somewhat bivalvate or else indehiscent; seeds reniform.

— A genus of 50 or more species, the majority in tropical America and a few in S. Africa.

KEY TO SPECIES

1. Flowers in dense terminal and axillary spikes; stellate hairs with 5 or more radiating branches: M. americanum

1. Flowers mostly solitary in the axils and in small terminal heads; hairs 4-branched, with paired branches directed forward and backward:

    2. Carpels without spines: M. corchorifolium

    2. Carpels with 3 spines on the back: M. coromandelianum

Malvastrum americanum (L.) Torr. in Emory, *Rep. U.S. & Mex. Bound. Surv.* 2 (1): 38 (1859).

Malvastrum spicatum (L.) A. Gray (1849).

Perennial shrubby herb up to 2 m tall but usually lower, stellate-pubescent throughout; leaves 3–20 cm long, the blades ovate or triangular-ovate, acute or subacuminate at apex, truncate at base, the margins unequally serrate; petioles 1–9 cm long. Flowers sessile in dense terminal and axillary spikes intermingled with leaf-like bracts, the axillary spikes sometimes of only 2 or 3 flowers; bracteoles 5–7 mm long. Calyx ca. 5 mm long; petals yellow, 5–6 mm long. Carpels up to 15, beak-like at the apex but without spines, hispid on the upper side.

CAYMAN BRAC: *Millspaugh* 1167-bis, 1188.

— Pantropical, often a weed of roadsides and waste places. The stems contain a strong, hemp-like fibre.

Malvastrum corchorifolium (Desr.) Britton in Small, *Fl. Miami* 119 (1913).

A woody herb up to 2 m tall but usually much lower, the stems and leaves clothed with stiff adpressed 4-branched hairs; leaves 1–8 cm long, the blades oblong to ovate, acutish, with sharply serrate margins; petioles 0.5–3.5 cm long. Flowers subsessile, solitary in leaf-axils and clustered in small terminal heads; bracteoles about equal in length to the calyx, 4–5 mm long, clothed with long simple and minute stellate hairs; petals orange or yellow, ca. 5 mm long. Carpels up to 15, without spines, but bearing long simple hairs on the upper part of the back.

GRAND CAYMAN: *Brunt* 1706; *Proctor* 15100. CAYMAN BRAC: *Proctor* 28934.

— Florida and the West Indies, often in dry sandy fields and along roadsides.

Malvastrum coromandelianum (L.) Garcke, *Bonplandia* 5: 295 (1857). **FIG. 108.**

Malvastrum tricuspidatum (Ait.) A. Gray (1852).

M. *americanum* of some authors, not Torr. (1859).

A woody annual herb, decumbent or up to 1 m tall, the stems clothed with stiff adpressed 4-branched hairs, the leaves with similar hairs and also simple ones; leaves 1–10 cm long, the blades oblong to ovate, obtuse, with coarsely serrate margins; petioles 0.5–3 cm long. Flowers short-stalked, solitary in the leaf-axils or occasionally several together at apex of the stem; bracteoles about equal in length to the calyx and inserted on its base. Calyx 5–8 mm long, hairy; petals pale yellow or dull orange, 7–9 mm long. Carpels 10 or more, with 2 short spines at the middle of the back and a longer one at the top, surrounded by hairs.

GRAND CAYMAN: *Hitchcock*.

— Pantropical, said to be introduced from America to the Old World, a weed of roadsides and open waste ground.

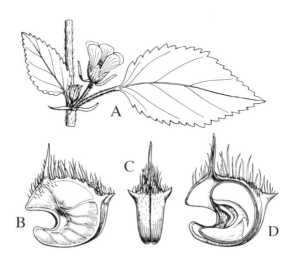

FIG. 108 **Malvastrum coromandelianum.** A, portion of flowering branch, × 1. B, ripe carpel, × 7. C, back of same beginning to split open, × 7. D, ripe carpel with seed, cut through, × 7. (F. & R.)

Sida L.

Herbs or shrubs with stellate or simple hairs; leaves usually more or less serrate; narrow stipules present. Flowers axillary or terminal, solitary or in heads, spikes, racemes, cymes or panicles; epicalyx lacking. Calyx 5-lobed or 5-toothed; petals mostly yellow, orange, or white. Staminal column bearing anthers at the apex. Carpels 5 or more; ovules 1 in each cell. Ripe carpels separating, 2-valvate above and often bearing spines at the apex; seeds solitary in each carpel.

— A pantropical genus of about 125 species.

KEY TO SPECIES

1. Peduncles adnate to the petioles of leaf-like bracts; flowers orange; stems prostrate from a woody taproot:	**S. ciliaris**
1. Peduncles free; flowers yellow or whitish; stems erect or ascending:	
2. Carpels 5:	
3. Flowers mostly solitary in the leaf-axils; leaves lance-oblong, minutely stellate-pubescent, not over 3 cm long:	**S. spinosa**
3. Flowers in clusters, axillary and terminal; leaves ovate-cordate; up to 10 cm long:	
4. Stems and leaves with long simple hairs and not viscid-glandular; flowers in small dense axillary and terminal heads:	**S. urens**
4. Stems viscid-glandular and leaves with small soft stellate hairs; flowers in loose cymes or cymose panicles:	**S. glutinosa**
2. Carpels more than 5:	
5. Leaves densely and softly stellate-pubescent, the blades ovate; calyx densely pubescent; carpel-spines with simple hairs:	**S. cordifolia**
5. Leaves nearly or quite glabrous, the blades lance-oblong; calyx glabrate; carpel-spines with minute stellate hairs:	**S. stipularis**

Sida urens L., *Syst. Nat.* ed. 10, 2: 1145 (1759).

Annual or sometimes persisting herb, much-branched, the stems trailing or ascending, clothed throughout with long, stiff, simple hairs; leaves long-petiolate, 3–10 cm long, the blades ovate or lance-ovate, long-acuminate, the base cordate, the margins unequally serrate. Flowers several together in dense globose clusters, these axillary and terminal, subsessile or pedunculate. Calyx 6–8 mm long, 5-angulate, the lobes long-acuminate; petals yellow with red spot at the base, ca. 7 mm long. Carpels 5, glabrous, spineless or with 2 short teeth at the apex.

GRAND CAYMAN: *Millspaugh 1284*.

— West Indies, continental tropical America and tropical Africa, a weed of fields, roadsides, pastures and secondary thickets.

Sida spinosa L., *Sp. Pl.* 2: 683 (1753).

Annual or perennial herb with erect or spreading stems, all the younger parts clothed with very minute stellate pubescence, the stems often with 1 or 2 small spine-like tubercles just below the attachment of each leaf. Leaves petiolate, 1–3 cm long, the blades linear or narrowly lance-oblong to ovate, obtuse or acute at the apex, the base truncate or subcordate, the margins serrate. Flowers mostly solitary in the leaf-axils on short peduncles. Calyx densely stellate-pubescent, 5–7 mm long, with 5 triangular lobes; petals yellow, equal in length to the calyx. Carpels 5, each bearing 2 short spines at the apex.

GRAND CAYMAN: *Brunt* 2114; *Kings* GC 382; *Proctor* 27935.

— Pantropical, also adventive in temperate N. America, a weed of roadsides and pastures.

Sida glutinosa Commers. ex Cav., *Monad. Diss.* 1: 16, t. 2, f. 8 (1785).

Wissadula divergens of Millspaugh (1900), non Benth.

A somewhat shrubby erect herb up to 1 m tall or more, the stems densely puberulous with simple, often glandular-viscid hairs; leaves long-petiolate, 3–12 cm long, the blades ovate, acuminate, cordate at the base, and with crenate-serrate margins, both surfaces clothed with fine stellate pubescence. Flowers initially solitary from the leaf-axils, but almost immediately becoming accompanied by elongate pedunculate cymes of several flowers from the same axils; these, together with a terminal cyme, form a large paniculate inflorescence with numerous leaf-like bracts. Calyx angulate with acuminate lobes, 5–6 mm long, clothed with numerous simple hairs; petals salmon-yellow, pale orange, or buff, ca. 8 mm long. Carpels 5, puberulous on the inner part and with 2 short beaks at the apex.

GRAND CAYMAN: *Millspaugh* 1346, 1350.

— West Indies and continental tropical America, in clearings, along roadsides, and in dry waste ground.

Sida ciliaris L., *Syst. Nat.* ed. 10, 2: 1145 (1759).

Prostrate trailing herb with perennial woody taproot, the stems to 30 cm long but usually shorter, clothed with adpressed stellate hairs. Leaves petiolate, mostly 6–15 mm long, the blades oblong or oblanceolate, the margins serrate toward the apex, glabrous on the upper side and stellate-pubescent beneath. Flowers solitary or several together, terminal, the short peduncles adherent to the petiole of a leaf-like bract. Calyx 4–5 mm long, with long simple hairs and very minute stellate ones; petals orange or salmon, often purplish-red at the base, ca. 8 mm long. Carpels 7–8, covered with short tubercles toward the apex.

GRAND CAYMAN: *Brunt* 2119; *Kings* GC 381-A; *Proctor* 15099. CAYMAN BRAC: *Proctor* 28993.

— Florida and Texas, West Indies, and continental tropical America, in sandy clearings, open fields, and in pockets of limestone rocks.

Sida cordifolia L., *Sp. Pl.* 2: 684 (1753).

Annual or persisting erect herb up to 1.5 m tall but usually less, all parts densely and softly stellate-pubescent. Leaves petiolate, 3–8 cm long or more, the blades ovate or lance-oblong, obtuse at the apex, truncate or slightly cordate at the base, the margins serrate. Flowers pedunculate, axillary and terminal in dense racemes or corymbs. Calyx angulate, 6–7 mm long; petals orange or salmon, or yellow with an orange spot at the base, ca. 10 mm long. Carpels 7–12, 2-beaked at the apex, the beaks divergent, hairy.

CAYMAN BRAC: *Proctor* 29058.

— Pantropical, in clearings, sandy fields, and along paths.

Sida stipulata Cav., *Monad. Diss.* 1: 22, t. 3, f. 10 (1785).

Sida acuta of many authors, non Burm. f. (1768).
Sida carpinifolia var. *antillana* Millsp. (1900).

BROOM-WEED

A shrubby annual or persisting herb with erect (rarely decumbent), tough stems up to 1 m tall, the young parts very minutely stellate-pubescent, becoming mostly glabrate with age. Leaves short-petiolate, mostly 1.5–6 cm long, the blades lance-oblong to elliptic, blunt or acute at both ends, the margins finely to coarsely serrate, essentially glabrous and often reddish on the upper side, but with scattered very minute stellate hairs beneath. Flowers axillary, mostly solitary but sometimes 2 or 3 together, short-stalked or subsessile. Calyx 6–8 mm long, glabrate or with a few very minute stellate hairs on the outside and a few simple hairs on the margins of the lobes; petals pale yellow, buff, or whitish, ca. 10 mm long. Carpels 7–12, with 2 spines at the apex, these bearing very minute stellate hairs.

GRAND CAYMAN: *Brunt* 2078; *Kings* GC 361; *Millspaugh* 1303; *Proctor* 15099; *Sachet* 448. LITTLE CAYMAN: *Proctor* 28164. CAYMAN BRAC: *Proctor* 28974.

— Pantropical, also in warm-temperate N. America, an often abundant weed of pastures, roadside banks, and waste ground.

Bastardia Kunth

Herbs or shrubs with stellate pubescence and often also glandular-viscid; leaves petiolate, cordate, and entire, crenate or dentate; stipules filiform, deciduous. Flowers axillary, solitary or 2 to 3 together; epicalyx lacking. Calyx 5-lobed; petals yellow. Anthers borne at the apex of the staminal column. Ovary 5–8-locular; ovules 1 in each cell, pendent, attached above at the inner angle; style-branches as many as the carpels. Fruit a dehiscent 5–8-valved capsule, the more or less hairy seeds solitary in the locules.

— A tropical American genus of 6 species, differing from *Sida* in the capsular fruit.

Bastardia viscosa (L.) Kunth in H.B.K., *Nov. Gen.* 5: 256 (1822).

An erect shrubby herb up to 1 m tall or more, viscid-glandular throughout and with a strong unpleasant odour. Leaves 2–5 cm long, acuminate, the margins glandular-dentate, and both surfaces densely pubescent with minute stellate hairs. Peduncles slender, 1–3 cm long; calyx 3.5–4 mm long, deeply lobed; petals yellowish, ca. 5 mm long. Capsule 5–8-celled, 3–4 mm long, minutely stellate-pubescent, not beaked; seeds black, puberulous with white hairs.

GRAND CAYMAN: *Brunt* 1964, 2145; *Proctor* 15264. The Cayman specimens represent the small-leafed variety known as var. *parvifolia* (Kunth) Griseb.

— West Indies and continental tropical America, a foetid weed of waste places.

Malachra L.

Annual to somewhat persisting herbs, often bristly with stiff, harsh, simple or branched hairs; leaves long-petiolate, frequently palmately angled or -lobed; stipules linear. Flowers in dense axillary or terminal heads subtended by large foliaceous bracts; bracts (rarely more), prominently veined, folded down the middle, shortly stalked or sessile, each bract with 2 or 4 stipule-like outgrowths at the base; epicalyx lacking. Calyx 5-lobed; petals 5, yellow or white (rarely purple). Staminal column short, truncate or 5-toothed, with anthers borne on the outside. Ovary 5-celled, each cell with 1 ovule; style 10-branched. Fruit a schizocarp, the mature obovoid carpels separating from the central axis; seeds reniform.

— A small genus of 9 species indigenous to tropical and subtropical America, but some of them widely naturalised as weeds in the Old World tropics.

Malachra alceifolia Jacq., *Collect.* 2: 350 (1789).

WILD OKRA

A coarse weedy herb up to 1 m tall or more, the stems erect and often unbranched, clothed with numerous small soft stellate hairs mixed with much longer, stiff, yellowish ones, these simple or stellate. Leaves 5–18 cm long, the blades roundish or broadly ovate, somewhat angled to shallowly 3–5-lobed, the margins dentate, the surfaces stellate-pubescent, especially beneath. Flower-heads axillary, sessile or stalked; bracts triangular-ovate, acute to acuminate at the apex, 1.5–2.5 cm long, often white-spotted between the veins; calyx 4–6 mm long, the lobes lanceolate; petals yellow, 1.5 cm long. Carpels of the ripe fruit 3–3.5 mm long, puberulous with simple hairs.

GRAND CAYMAN: *Brunt* 2095, 2141; *Correll & Correll* 51048; *Hitchcock*; *Proctor* 27952.
— West Indies and continental tropical America. The Cayman plants often occur in seasonally wet hollows in pastures.

Urena L.

Herbs or shrubs, the pubescence all or chiefly of stellate hairs; leaves long-petiolate, usually palmately angled or lobed, the 3 central nerves each bearing a slit-like gland near its base; stipules linear. Flowers axillary, solitary or in small clusters, sometimes forming long terminal interrupted spikes; epicalyx of 5 united bracteoles, adherent to the calyx. Calyx 5-lobed, the lobes alternating with the lobes of the epicalyx; petals 5, pink, obovate or obcordate. Staminal tube about as long as the petals, bearing anthers on the outside. Ovary 5-celled, each cell with 1 ovule; style 10-branched. Fruit a schizocarp, each carpel covered with barbed spines.

— A pantropical genus of 6 species.

Urena sinuata L., *Sp. Pl.* 2: 692 (1753).

An erect, branched herb up to 1 m tall; leaves 2–9 cm long, the blade rounded- or ovate-angulate in outline, more or less deeply lobed, in larger leaves the lobes narrowed downward forming rounded bays between them. Epicalyx 4–4.5 mm long in flower, slightly lengthening in fruit, stellate-pubescent; calyx nearly as long as the epicalyx; petals ca. 1.5 cm long. Ripe carpels ca. 5 mm long, puberulous and covered with spines, the spines each bearing 2–4 retrorse barbs at the tip.

GRAND CAYMAN: *Hitchcock*; *Millspaugh* 1321.
— A pantropical weed, readily spread by its adherent spiny carpels. This species intergrades in many regions, presumably by hybridisation with *Urena lobata* L., but the latter has not been found in the Cayman Islands. Both species yield a strong fibre that can be used as a substitute for flax.

Malvaviscus Adans.

Shrubs, sometimes arborescent, the pubescence chiefly of stellate hairs; leaves long-petiolate, the blades usually dentate and often angled or lobed, the tissue with numerous minute pellucid dots; stipules linear or subulate. Flowers mostly solitary in the upper leaf-axils, or sometimes forming short terminal corymbs or racemes; epicalyx of 5–many, more or less linear bracteoles. Calyx 5-lobed, bell-shaped, somestimes with 2 or 3 of the lobes united; petals 5, usually bright red or pink, convolute into a loose tube and not opening. Staminal column spirally grooved, long-exserted beyond the petals, the anthers borne on the outside toward the apex. Ovary 5-celled, each cell with 1 ovule; style 10-branched, with capitate stigmas. Fruit fleshy, berry-like at first, the carpels ultimately separating, indehiscent.

— A tropical American genus of 3 or more species, two of them rare, the other extremely variable and difficult to classify.

Malvaviscus arboreus Cav. var. *cubensis* Schlecht., *Linnaea* 11: 360 (1837). PLATE 18.

Malvaviscus jordan-mottii Millspaugh (1900).

MAHOE

A somewhat straggling shrub, sometimes arborescent and up to 5 m tall, the upper branches, petioles and pedicels subglabrous or pubescent with fine stellate hairs. Leaves 5–17 cm long, the blades lanceolate to very broadly ovate, the apex acutish to short-acuminate, the base cordate, the margins sinuate-dentate, the surfaces glabrate to densely stellate-pubescent, especially beneath. Epicalyx and calyx ca. 1 cm long, stellate-pubescent; petals about twice as long as the calyx. Staminal column projecting about 1 cm beyond the petals.

GRAND CAYMAN: *Brunt* 2159; *Correll & Correll* 51017; *Hitchcock*; *Kings* GC 404; *Millspaugh* 1313; *Proctor* 11980, 15211; *Rothrock* 180, 237. LITTLE CAYMAN: *Proctor* 35103. CAYMAN BRAC: *Millspaugh* 1166 (type of *M. jordan-mottii*); *Proctor* 29008.

— Bahamas, Cuba, and perhaps the Yucatan area of C. America, usually in rocky limestone woodlands at low elevations. Several other varieties occur in various regions. The Grand Cayman and Cayman Brac populations differ markedly in pubescence. In the former island, the leaves are covered beneath with a fine, soft stellate pubescence in which also occur numerous much larger, stiff and harsh, 3-branched hairs. Cayman Brac specimens, on the other hand, lack the fine pubescence, but the larger hairs are usually even more abundant, making the foliage very unpleasant to handle. Sometimes, however, almost glabrous plants occur (including those of Little Cayman), and one such was the basis of the name *M. jordan-mottii*. Millspaugh (1900) reported that the nettle-like quality of mahoe has caused it to be used in the Cayman Islands as a flagellant for rheumatic patients, but I have heard no recent report of such a drastic treatment.

Malvaviscus arboreus var. *penduliflorus* (DC.) Schery, sometimes called sleeping hibiscus or pepper hibiscus, is often cultivated in the West Indies as an ornamental; it differs from var. *cubensis* in having very much larger flowers (more than 4.2 cm long) and in always lacking nettle-like stinging hairs.

Hibiscus L.

Herbs, shrubs or trees, variously pubescent or almost glabrous; leaves various, often lobed or toothed, sometimes simple and entire. Flowers usually solitary in the axils of the upper leaves, often large and showy; epicalyx of usually many bracteoles, rarely as few as 5, free or united, sometimes more or less adherent to the calyx. Calyx 5-lobed; petals 5, variously coloured, often large and showy. Ovary 5-locular, each cell with 3–many ovules; style 5-branched, the stigmas more or less capitate; seeds reniform or subglobose, glabrous or pubescent, often numerous.

— A pantropical genus of about 200 species, a few also occurring in temperate regions.

KEY TO GENERA

1. Leaves serrate, sharply toothed or lobed; stipules linear-subulate; bracteoles free:

   2. Wild shrub or shrubby herb; leaves softly pubescent:

      3. Shrubby herb rarely more than 1 m tall; leaves 3–9 cm long, not lobed; flowers pink with whitish veins: **H. lavateroides**

      3. Shrub up to 6 m tall; leaves up to 15 cm long and broad, angularly 3-lobed; flowers dull tawny yellow, becoming reddish with age: **H. clypeatus**

   2. Cultivated shrub; leaves coarsely toothed but not lobed, glabrate; calyx-lobes lance-acuminate, about one-fourth as long as the petals or less: **[H. rosa-sinensis]**

1. Leaves entire or crenulate; stipules oblong, large; bracteoles united to form a cup, free from the calyx: **H. pernambucensis**

Hibiscus lavateroides Moric., *Mém. Soc. Phys. & Hist. Nat. Genève* 7: 263, t. 16 (1836).

Shrubby herb mostly 0.6–1.2 m tall; leaves deltate-ovate, cordate or truncate at base, acuminate or obtuse at apex, the margins irregularly serrate. Flowers solitary in upper axils; calyx 1.7–2.2 cm long; corolla 2.5–4.5 cm long, the petals with numerous large stellate hairs on outside. Capsule shorter than the calyx.

CAYMAN BRAC: *Proctor* 47336 (wrongly cited as from Grand Cayman in *Kew Bull.* 51 (3): 491 (1996).)

— Jamaica, Mexico and Honduras.

Hibiscus clypeatus L., *Syst. Nat.* ed. 10, 2: 1144 (1759).

A shrub up to 4 m tall or more, softly stellate-pubescent throughout. Leaves long-petiolate, 7–25 cm long or more, the blades very broadly angular-ovate, usually shallowly but rather sharply 3-lobed toward the apex, cordate at base, the margins glandular-dentate; stipules linear-subulate, ca. 1.5 cm long. Flowers on stout peduncles 5–11 cm long; epicalyx of 9–11 linear bracteoles, these unequal in length and much shorter than the calyx. Calyx 3.5–4 cm long, with foliaceous 5-nerved lobes. Petals 4.5–6 cm long, pale yellow, velvety pubescent on the outside. Capsule shorter than the calyx, yellowish-hairy; seeds glabrous, dark brown, ca. 4 mm long.

CAYMAN BRAC: *Fawcett* 10. Not collected in the Cayman Islands since 1888.

— Greater Antilles, Yucatan and Peten, in thickets and dry rocky woodlands.

[*Hibiscus rosa-sinensis* L., *Sp. Pl.* 2: 694 (1753).

The commonly cultivated ornamental hibiscus or 'shoe-black' occurs in many colour-forms. It is a native of tropical Asia.]

Hibiscus pernambucensis Arruda, *Diss. Pl. Brazil* 44 (1810). PLATES 18 & 19.

Hibiscus tiliaceus of Proctor (1984), not L.
Pariti tiliaceum (L.) Button (1918).

SEASIDE MAHOE

An arborescent shrub or small diffuse tree to 6 m tall, all parts finely stellate-puberulous except the upper side of the leaves; leaves long-petiolate, 5–28 cm long, the blades roundish-reniform or very broadly ovate, abruptly acuminate, the base cordate, the margins entire or finely crenulate, the upper surface green and glabrous or nearly so, beneath finely whitish-tomentose, the median 1 or 3 nerves bearing a slit-like gland near the base; stipules oblong, 2–4 cm long, deciduous. Flowers solitary or occasionally 2 or 3 together, axillary and terminal, on short peduncles mostly 0.5–1.5 cm long; epicalyx cup-shaped, 8–11-lobed, shorter than the calyx. Calyx 2–3 cm long, 5-lobed to the middle; petals yellow or orange, 5–7 cm long. Staminal column nearly as long as the petals, bearing anthers along its entire length. Capsule ovoid, hairy, 1.5–2 cm long; seeds numerous, 4 mm long, glabrous and papillose.

GRAND CAYMAN: *Brunt* 1625, 1791; *Fawcett* 9. LITTLE CAYMAN: *Kings* LC 71; *Proctor* 35110.

— Neotropical, on protected seashores and borders of mangrove swamps. The wood is very light and can be used as a substitute for cork. The bark contains strong fibres which have the quality of becoming stronger when wet; in many tropical regions this material is used for making rope, mats, and coarse cloth.

Kosteletzkya Presl

Herbs or shrubs, usually with harsh simple and stellate pubescence; leaves mostly sagittate or often angular-lobed; stipules linear or subulate. Flowers axillary and terminal, solitary or in few-flowered open cymes; epicalyx of 5–10 bracteoles, these sometimes minute or lacking. Calyx 5-lobed or 5-toothed; petals white, pink, purple or yellowish, either spreading or erect and convolute. Staminal column with anthers on the outside. Ovary 5-celled, each cell with 1 ovule; style 5-branched, with capitate or dilated stigmas. Fruit a depressed, 5-angled, dehiscent capsule; seeds reniform, glabrous or puberulous.

— A genus of 12 species occurring in America, Africa, and the Mediterranean region.

Kosteletzkya pentasperma (Bert.) Griseb., *Fl. Brit. W. Ind.* 83 (1859). FIG. 109. PLATE 19.

A slender, branched herb up to 1 m tall or more, the stems green, finely puberulous in a line on one side, otherwise plentifully clothed with long, stiff, nettle-like simple hairs each from a pustulate base; leaves petiolate, 2–9 cm long, the blades varying from narrowly lanceolate or hastate to ovate, the apex narrowly acute, the base cordate, the margins crenate-serrate, clothed on both sides with harsh oppressed hairs, those on the upper side mostly simple, those beneath 3-branched. Flowers solitary on long slender peduncles from the leaf-axils, often accompanied by a secondary cymose branchlet bearing several flowers; bracteoles linear or subulate, ciliate, shorter than the calyx. Calyx 4.5–5 mm long,

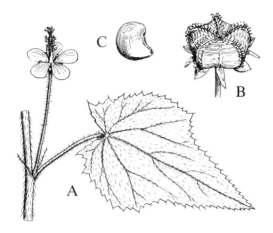

FIG. 109 **Kosteletzkya pentasperma**. A, leaf and flower, × ²/₃. B, fruit, × 2. C, seed, × 4. (F. & R.)

bearing 3-branched hairs; petals white, ca. 10 mm long. Capsule ca. 9 mm broad and 5 mm high, deeply 5-lobed, recurved-hairy on the angles; seeds minutely puberulous.

GRAND CAYMAN: *Brunt* 1964a, 2091, 2108; *Kings* GC 287, GC 327; *Proctor* 27951.

— West Indies and continental tropical America, often occurring in swamps, wet ditches and swales. The harsh, nettle-like hairs covering this plant make it very unpleasant to handle.

Thespesia Soland. ex. Correa

Trees or sometimes shrubs; leaves long-petiolate, entire or sometimes lobed; stipules linear, deciduous. Flowers solitary in the leaf-axils, usually large; epicalyx of 3–5 narrow deciduous bracteoles. Calyx truncate and 5-toothed or rarely 5-lobed; petals yellow. Ovary 5-celled, each cell with several ovules; style club-shaped at the apex, with 5 grooves or short branches. Capsule hard or leathery, indehiscent or sometimes splitting by 5 valves; seeds usually 2 or 3 in each cell, glabrous or hairy, the cotyledons much-folded and black-dotted.

— A small pantropical genus of about 17 species.

Thespesia populnea (L.) Soland. ex. Correa in *Ann. Mus. Hist. Nat. Paris* 9: 290, t. 25, f. 1 (1807). **FIG. 110**. Plate 19.

POPNUT, PLOPNUT

A shrub or small tree up to 10 m tall, the younger parts clothed with numerous very minute peltate scales, the whole plant otherwise glabrous. Leaves mostly 7–20 cm long, the petioles often as long as the blades; blades broadly ovate, acuminate at apex, cordate at base, the margins entire. Flowers on stout peduncles shorter than the petioles; bracteoles 1–2.5 cm long. Calyx cup-shaped, 7–9 mm long; petals 5–6 cm long, pale yellow with red spot at the base, the whole turning reddish with age, the margins ciliate toward the base. Staminal column much shorter than the petals. Capsule subglobose, ca. 3 cm in diameter; seeds 8–10 mm long, finely hairy on the angles.

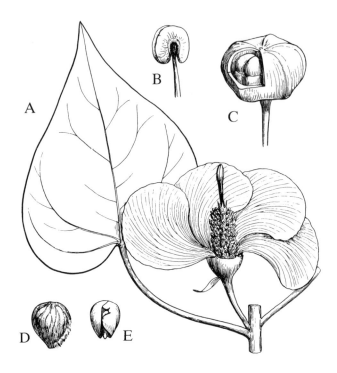

FIG. 110 **Thespesia populnea.** A, leaf and flower with a petal removed, × ²/₃. B, stamen, × 7. C, fruit partly cut open, × ²/₃. D, seed, × 1. E, embryo, × 1. (F. & R.)

GRAND CAYMAN: *Brunt* 1664, 1673; *Hitchcock*; *Kings* GC 62; *Millspaugh* 1238; *Proctor* 15098. LITTLE CAYMAN: *Kings* LC 70; *Proctor* 28076; *Sauer* 4174. CAYMAN BRAC: *Proctor* 29104.

— Pantropical, chiefly near seashores and along the borders of mangrove swamps. Sometimes planted for ornament.

Gossypium L.

Subwoody herbs or shrubs, sometimes arborescent, generally marked all over with numerous minute black dots, often pubescent with simple or stellate hairs. Leaves usually long-petiolate and 3–7-lobed, or sometimes entire. Flowers solitary in the leaf-axils, large, on stout peduncles; epicalyx of 3 large cordate bracteoles, these more or less deeply incised or rarely entire. Calyx truncate or 5-toothed; petals yellow or red. Ovary 5-celled, the cells with many ovules; style club-shaped at the apex, grooved and bearing 3–5 stigmas. Fruit a dehiscent capsule; seeds densely woolly with very long or short hairs, or both, or sometimes nearly glabrous; endosperm scant or none.

A widespread tropical and subtropical genus of more than 20 species (or twice as many recognised by some authors). Several species have been cultivated (as cotton) for a very long time, and as they hybridise freely their classification is often very difficult.

KEY TO SPECIES

1. Seeds with long, loose, easily detachable hairs and without a covering of 'fuzz'
   or short hairs; staminal column relatively long, with anthers compactly arranged
   on short filaments which are all about the same length: G. barbadense

1. Seeds with a double coat, consisting partly of long, firmly adherent hairs and
   also a dense covering of short 'fuzz'; staminal column short, with anthers loosely
   arranged on filaments longer above than below: G. hirsutum

Gossypium barbadense L., *Sp. Pl.* 2: 693 (1753). **FIG. 111.**

SEA-ISLAND COTTON, LONG-STAPLE COTTON

A coarse herb or a shrub up to 4 m tall; leaves 5–15 cm long, the blades 3–5-lobed or
nearly entire, cordate at base, usually glabrous or nearly so. Bracteoles broadly cordate,
4–7 cm long, with 3–7 lacerations. Petals 4.5–8 cm long, pale yellow turning dull reddish.
Capsule ovoid, acuminate, mostly 3-celled; seeds with long, white, easily detachable lint
but no short fuzz.

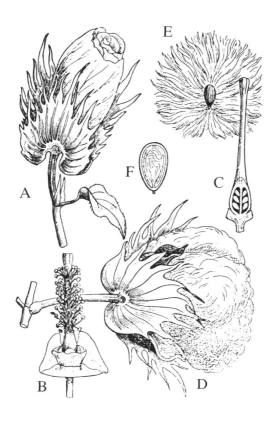

FIG. 111 **Gossypium barbadense.** A, flower about to open, × ²/₃. B, flower with calyx and corolla cut
away, showing staminal tube enclosing pistil, × ²/₃. C, pistil with ovary cut lengthwise, × 1. D, capsule
open, showing mass of cotton, × ²/₃. E, seed with cotton attached, × ²/₃. F, seed cut lengthwise, showing
twisted embryo, × 1¹/₂ (F. & R.)

313

GRAND CAYMAN: *Brunt* 1945; *Hitchcock*; *Millspaugh* 1367; *Proctor* 15053.

— West Indies and tropical S. America; widely cultivated and spontaneous in tropical and subtropical regions.

Gossypium hirsutum L., *Sp. Pl.* ed. 2, 2: 975 (1763). var. ***punctatum*** (Schum.) J. B. Hutch., *Sil & Steph. Evol. Gossyp.* 40 (1947).

SHORT-STAPLE COTTON

A coarse herb or a shrub up to 4 m tall, often clothed with chiefly simple hairs; leaves 5–15 cm long, the blades mostly 3-lobed and cordate. Bracteoles broadly cordate, 3–6 cm long, with 9–13 lacerations. Petals 3.5–5 cm long, pale lemon-yellow turning pink. Capsule ovoid-ellipsoid, acuminate, rough, mostly 3-celled; seeds with short dense fuzz and long white adherent lint.

GRAND CAYMAN: *Sauer* 4102. LITTLE CAYMAN: *Kings* LC 10, LC 11; *Proctor* 28150. CAYMAN BRAC: *Kings* CB 43, CB 85; *Proctor* 28980; *Sauer* 4140.

— Coasts and islands of the Gulf of Mexico, Florida, Bahamas, and Hispaniola; in various forms cultivated in many regions and frequently becoming naturalised. Cultivars of this species are commonly and abundantly grown as cotton in the U.S.A., and are among the world's most important economic plants.

[LECYTHIDACEAE

Barringtonia asiatica (L.) Kurz, a large tree with evergreen foliage, large flowers with innumerable long, pink stamens, and curious 4-cornered green fruits up to 15 cm in diameter or more, is sometimes planted for shade or ornament.

GRAND CAYMAN: *Kings* GC 232.]

SALICACEAE

(Formerly Flacourtiaceae)

Trees or shrubs, sometimes dioecious. Leaves simple, alternate to subopposite, rarely opposite, with or without stipules; margins often with salicoid dentation. Inflorescence terminal or axillary, very diverse, commonly in the form of racemes, spikes, catkins, panicles, corymbs, short cymes, glomerules, or reduced to a solitary flower. Flowers unisexual or bisexual, often small. Sepals (0–)3–8(–15), valvate or imbricate, free or partly connate toward base; calyx occasionally obconical or turbinate and connate with lower part of ovary. Petals either absent or equal in number and similar to the sepals, free. Disc usually present and cup-like and more or less adnate to adaxial surface of calyx, or annular and extrastaminal, or divided into separate glands. Stamens (1–)2-numerous, inserted singly or in groups on the receptacle, often between the disc lobes or glands, occasionally united into a column; anthers usually small, ovoid or elliptic, rarely linear. Ovary superior or less often semi-inferior, unilocular; styles 1–8, entire or branched. Fruit a berry or capsule, or less often a drupe or samara; seeds often arillate.

A world-wide family of 52 genera and more than 1,200 species occurring from the tropics to arctic regions.

KEY TO GENERA

1. Flowers in racemose corymbs; sepals valvate; petals present; leaves strongly reticulate-veined on upper side: **Banara**

1. Flowers solitary, cymose, or in compact clusters; sepals imbricate; petals absent; leaves at most only slightly reticulate-veined on upper side:

   2. Flowers perfect (bisexual); staminodes present, alternating with the stamens; spines absent or if present simple and spur-like:

     3. Style present; stamens 6–15:   **Casearia**

     3. Style absent; stamens 20–40:   **Zuelania**

   2. Flowers unisexual, the plants dioecious; staminodes absent; at least the lower trunk armed with branched spines.   **Xylosma**

Banara Aubl.

Shrubs or small trees; leaves simple, alternate, somewhat inequilateral, the major glandular-serrate, the tissue punctuate; stipules very small. Flowers small, bisexual; calyx persistent, 3–5-lobulate, the lobules valvate; petals opposite the calyx-lobules, imbricate, persistent. Stamens numerous, with capillary filaments; staminodes absent. Ovary ovoid, superior, 1-locular or semi-3–8-locular by intrusion of the placentas; style slender, stigma capitate; ovules numerous. Fruit a leathery many-seeded berry.

— A widely distributed neotropical genus of about 37 species, absent from Jamaica and the Lesser Antilles.

Banara caymanensis Proctor, *Kew Bull.* 51 (3): 491 (1996). PLATE 19.

Shrub 1.5–2.5 m tall, the vegetative parts glabrous throughout, the ultimate branchlets slender and somewhat zigzag. Leaves with slender petioles 4–9 mm long; blades penninerved, ovate to broadly elliptic or subrotund, 2.5–5.5 cm long, 15–4 cm broad, obtuse or very shortly and bluntly acuminate at the apex, cordulate and without glands at base, the margins obscurely glandular-serrate, the surface glossy and densely prominulous-reticulate, the texture thin-coriaceous. Stipules lance-subulate, 1.5–2 mm long, caducous. Flowers 4–5, pale green, in small axillary or subterminal corymbs, the peduncles 3–4 mm long, the pedicels ca. 2 mm long. Sepals and petals minutely puberulous on the outside and about 2 mm long, the sepals lance-acuminate, the petals ovate-orbicular. Stamens, style, ovary and fruits not seen.

CAYMAN BRAC: *Proctor* 47816 (type).
— Endemic. Top of The Bluff just ENE of Hawksbill Bay.

Casearia Jacq.

Trees or shrubs, usually unarmed but sometimes the trunk spiny or the branches with spine-like spurs. Leaves petiolate, entire or toothed, often with pellucid dots and lines in the tissue; stipules small and soon falling. Flowers perfect, clustered or umbellate in the leaf-axils, rarely solitary or racemose, with jointed pedicels bracteate at the base. Calyx of 4–6 overlapping lobes; petals none. Stamens 6–15, inserted on the calyx near its base, alternating with an equal number of staminodes; filaments free or united with the staminodes to form a ring. Ovary superior, free, ovoid or oblong, narrowed above into a short style, the stigma capitate or divided; ovules numerous, produced on 3 parietal placentas. Fruit a fleshy or dry capsule, 3–4-valvate, with numerous seeds; seeds with fleshy aril.

— A pantropical genus alleged to have about 200 species.

KEY TO SPECIES

1. Mature leaves less than 2 cm long:	**C. staffordiae**
1. Mature leaves mostly over 2 cm long, up to 18 cm long:	
2. Leaves velvety-pubescent beneath:	**C. hirsuta**
2. Leaves glabrous or nearly so:	
3. Stigma 2–3-lobed; calyx 1.5–2.5 mm long:	**C. sylvestris**
3. Stigma entire, capitate; calyx 4–5 mm long:	
4. Leaves mostly 9–18 cm long; pedicels jointed close to the base:	**C. guianensis**
4. Leaves 2–7 cm long; pedicels jointed below the middle:	
5. Sepals united for ¼ to ⅓ their length; branches often with stout, naked or leafy, spine-like spurs, these bearing flower-clusters:	**C. aculeata**
5. Sepals essentially free, united only at the base; branches always unarmed:	**C. odorata**

Casearia staffordiae Proctor, sp. nov. PLATES 19 & 20.

Unarmed shrub to ca. 2 m tall with lax, arching or drooping elongate branches; bark light gray-corticate. Leaves alternate, solitary, or sometimes 2–3-fasciculate at nodes, glabrous petioles ca. 1 mm long, the blades obovate, sometimes elliptic-obovate or often subrotund, mostly 0.8–2 cm long and 0.5–1.3 cm broad, the apex rounded or slightly emarginate, the margins entire except often with 2 obscure teeth or minute crenations one on each side near the apex, the base cuneate, the thin tissue containing numerous very minute pellucid dots and lines (seen best when dry). Flowers white, solitary, nearly sessile in leaf axils; calyx-lobes 5, spreading, acute or subacuminate, ca. 1.5 mm long. Stamens 10, the pubescent filaments ca. 1.5 mm long. Fruit a globose capsule 8 mm in diameter.

GRAND CAYMAN: first found by Mrs. P. Ann van B. Stafford along the Chisholm Trail near North Side, vouchered by photographs and *Proctor* 52106; also collected along the Mastic Trail, *Proctor* 52138, 52139; and *van B. Stafford* s. n., 9 January 2005 (type IJ). The type specimen bears fruits. Endemic.

This species is closely related to *Casearia emarginata* Wr. ex Griseb. of Cuba, Haiti, and Yucatan but differs in habit, in the complete absence of spines or spinescent branchlets, in having solitary subsessile white flowers with pointed calyx lobes, and in having much larger fruits (globose, 8 mm in diameter, rather thab ovoid, 4 mm long).

Casearia staffordiae Proctor, sp. nov. *C. emarginatae* Wr. ex Griseb. valde affinis sed spinis atque ramulis spinescentibus omnino carentibus, floribus solitariis subsessilibus, lobis calycis acutis et fructibus globosis (non ovoideis) multo majoribus 8 mm diametro (non 4 mm longis) differt.

Casearia sylvestris Sw., *Fl. Ind. Occ.* 2: 752 (1798).

An unarmed shrub or small tree, often with long, slender, puberulous branches; leaves lance-oblong, 5–10 cm long, long-acuminate at apex, glossy green and densely pellucid-dotted; stipules roundish, 1–1.5 mm long. Flowers in dense, sessile, axillary clusters, yellow-green or whitish; pedicels mostly less than 3 mm long or often nearly obsolete. Calyx minutely puberulous and ciliolate. Stamens 10, free; staminodes hairy. Fruit subglobose, 3–5 mm in diameter, red, 3-valved, 2–6-seeded; seeds flattened-ellipsoidal, 2 mm long.

GRAND CAYMAN: *Proctor* 15011.

— West Indies and continental tropical America. The Grand Cayman plants were found to be rare in dry rocky woodland between North Side and Forest Glen.

Casearia guianensis (Aubl.) Urban, *Symb. Ant.* 3: 322 (1902). PLATE 20.

An unarmed shrub, sometimes arborescent; leaves obovate or elliptic, up to 18 cm long, short-acuminate at the apex, the margins distantly and minutely glandular-denticulate, the tissue with numerous pellucid dots and lines; stipules linear-subulate, 2–5 mm long. Flowers in loose or rather dense axillary clusters, greenish-white or cream; pedicels 5–7 mm long, finely appressed brownish-hairy; calyx 4–5 mm long, hairy like the pedicel. Stamens 8; staminodes hairy. Stigma capitate. Fruit subglobose or ellipsoid, 6–12 mm in diameter, white or greenish, sometimes flushed with red, obtusely 6-angled, 3–10-seeded; seeds subovoid, 3–3.5 mm long.

CAYMAN BRAC: *Proctor* 29116.

— West Indies and continental tropical America; the Cayman Brac specimen was collected in rocky woodland near the base of cliffs at Spot Bay.

Casearia hirsuta Sw., *Fl. Ind. Occ.* 2: 755 (1798). PLATE 20.

An unarmed shrub, the young branches velvety brownish-pubescent; leaves narrowly obovate to oblong-elliptic or elliptic, 5–12 cm long, the apex blunt to very shortly acuminate, the margins distantly and minutely glandular-denticulate, the tissue with numerous pellucid dots but scarcely any lines, finely puberulous on the upper side, beneath velvety brownish-pubescent; stipules linear-lanceolate, 3 mm long. Flowers in axillary clusters or at leafless nodes, greenish-white, fragrant; pedicels 4–6 mm long, spreading or bristly puberulous; calyx 4–5 mm long, hairy like the pedicel. Stamens 8 or 10; staminodes hairy. Fruit ovoid, 10–15 in diameter, green or yellow-green, 3-angled; seeds oblong-conical, 2.5–3 mm long, minutely pitted.

GRAND CAYMAN: *Proctor* 15012, 15083. CAYMAN BRAC: *Proctor* 15318, 28966.

— Cuba, Hispaniola, Jamaica, and northern S. America, in rocky thickets and woodlands. This and the preceding species are sometimes called wild coffee.

Casearia aculeata Jacq., *Enum. Syst. Pl. Carib.* 21 (1760). PLATE 20.

THOM PRICKLE

A shrub up to 3 m tall or more, or frequently wand-like and few-branched, or semi-scrambling, often beset with sharp, spine-like, woody spurs 1.5–3 cm long, these representing modified branchlets and usually bearing 1 or 2 leaves and a cluster of flowers, or even developing tiny secondary branchlets, the youngest parts very minutely puberulous. Leaves narrowly or broadly elliptic or obovate, mostly 1.5–5 cm long, blunt to acute at the apex, the tissue with numerous pellucid dots and lines; stipules triangular-acuminate, 1 mm long. Flowers cream, in clusters often on spur-like branchlets; pedicels puberulous, jointed about the middle or below; calyx 4–5 mm long. Stamens 8; staminodes hairy. Fruit subglobose, obtusely 3-angled, 6–12 mm long.

GRAND CAYMAN: *Brunt* 1898, 2087, 2163; *Proctor* 27990.

— West Indies and continental tropical America, in dry thickets and along fencerows.

Casearia odorata Macf., Fl. Jam. 1: 215 (1837). Plates 20 & 21.

An unarmed shrub up to 4 m tall, sometimes with straggling branches, rarely tree-like; leaves elliptic or obovate, 2–9 cm long, the apex obtuse or acutish, the tissue with numerous pellucid dots but scarcely any lines; stipules lance-subulate, 0.5–1 mm long. Flowers cream, very fragrant, in axillary clusters; pedicels appressed-puberulous, jointed about the middle or below; calyx 4–5 mm long. Stamens 8; staminodes hairy. Fruit globose, obscurely angled, 8–10 mm long.

GRAND CAYMAN: *Kings* GC 359; *Proctor* 27931.

— Jamaica. It is possible that this and the preceding are really only forms of a single variable species.

Zuelania A. Rich.

Unarmed deciduous trees or shrubs, usually flowering when leafless or just as the new leaves appear; young parts softly brownish-puberulous. Leaves petiolate, usually inequilateral at the base, with crenulate or crenate margins and numerous pellucid dots. Flowers perfect, in dense terminal clusters, the flowering shoots prolonged by new growth after flowering so that the fruits are lateral on the stems below the new leaves; pedicels jointed, with bracts at the base; calyx of 4–5 overlapping lobes; petals lacking. Stamens 20–40, alternating with an equal number of staminodes. Ovary superior, free; stigma sessile or subsessile, peltate; ovules numerous on 3 placentas. Fruit a large fleshy berry-like capsule, at length splitting open by 3 valves; seeds numerous, arillate.

— A chiefly West Indian genus of 2 or 3 species, closely related to *Casearia*.

Zuelania guidonia (Sw.) Britton & Millsp., Bahama Fl. 285 (1920). **FIG. 112.** Plate 21.

Jeremiah-bush

A shrub or small tree to 10 m tall or more; leaves lance-oblong or elliptic, 5–20 cm long, subacuminate at the apex, obtuse to subcordate at base, the margins subentire to crenate, mostly glabrescent on the upper surface, densely soft-puberulous beneath. Flowers with puberulous pedicels 6–10 mm long; calyx whitish or greenish, 6–7 mm long. Ovary tomentose. Fruit subglobose, green and juicy, 3–5 cm in diameter; seeds 5 mm long.

GRAND CAYMAN: *Kings* GC 143; *Proctor* 15251. LITTLE CAYMAN: *Proctor* 35213. CAYMAN BRAC: *Proctor* 29342.

— Bahamas, Cuba, Hispaniola, Jamaica and C. America, in rocky woodlands. The wood is sometimes used for construction.

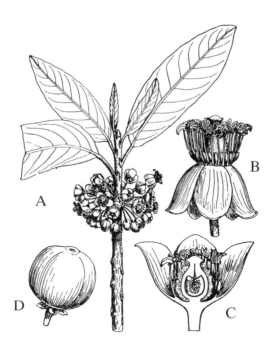

FIG. 112 **Zuelania guidonia**. A, branchlet with flowers and young leaves, × $^2/_3$. B, flower, × 3. C, opening flower cut lengthwise, × 3. D, fruit, × $^1/_2$. (F. & R.)

Xylosma Forster

Shrubs or trees, often armed with simple, straight spines in the leaf-axils or branched ones on the trunk. Leaves toothed or sometimes entire; stipules lacking. Flowers unisexual and dioecious, in small axillary clusters or short racemes. Sepals 4 or 5, usually small, scale-like, and ciliate; petals lacking. Male flowers with several to many stamens, these often surrounded by a glandular disc or mixed with glandular staminodes. Female flowers with ovary inserted on a glandular disc; ovules 2 on each of 2 (rarely 3–6) parietal placentas; style short, entire or divided, or rarely the stigmas subsessile and peltate-lobed. Fruit a small berry with 2–8 seeds obovoid, smooth and hard.

— A pantropical genus of perhaps 65 species, many of them variable and difficult to distinguish.

Xylosma bahamense (Britton) Standl., *Tropical Woods* 34: 41 (1933). PLATE 21.

SHAKE HAND

An intricately branched shrub or small tree to 6 m tall, the trunk and often the larger limbs densely armed with branched spines up to 5 cm long, the leafy twigs with or without simple spines. Leaves mostly ovate or broadly elliptic, 6–15(–30) mm long, entire or with 1–4 blunt teeth, the apex acute, sometimes mucronate; petiole ca. 1 mm long. Fruit obovoid-oblong, 6 mm long.

GRAND CAYMAN: *Kings* GC 332; *Proctor* 15138, 15183.

— Otherwise known only from the northern Bahamas.

TURNERACEAE

Herbs or shrubs, rarely trees, glabrous or pubescent, the hairs simple or branched; leaves alternate, simple, usually serrate, often 2-glandular at the base; stipules small or lacking. Flowers perfect, regular, axillary, solitary or few, sessile or stalked, rarely in racemes; peduncles free or united with the petiole, often jointed and 2-bracteolate. Calyx tubular, 5-lobed, soon deciduous. Petals 5, often yellow, clawed and inserted in the throat of the calyx, sometimes bearing a fimbriate scale at the apex of the claw. Stamens 5, inserted at the base, middle, or throat of the calyx tube, rarely hypogynous; filaments free; anthers oblong, opening inwardly. Ovary superior, free, 1-celled; styles 3, terminal, thread-like, simple or divided; stigmas brush-like or rarely merely dilated; ovules numerous, 2-seriate on 3 parietal placentas. Fruit a capsule splitting partially or completely by 3 valves; seeds oblong-cylindrical, slightly curved, arillate, with hard pitted coat; endosperm abundant; cotyledons plano-convex.

— A pantropical family of 8 genera and about 150 species, the majority American.

Turnera L.

Herbs or low shrubs, glabrous or pubescent with simple hairs; leaves serrate or rarely entire, often 2-glandular at the base. Flowers solitary in the axils, small or large, the peduncles usually adnate to the petioles, each with 2 bracteoles. Calyx with short tube and oblong, linear or lanceolate lobes; petals usually yellow, obovate or spatulate. Stamens inserted below the petals, sometimes hypogynous. Ovary sessile; styles simple, the stigmas brush-like or fan-like. Capsule splitting to the base, usually many-seeded; aril unilateral.

— A chiefly tropical American genus of more than 60 species.

KEY TO SPECIES

1. Leaves more than 2 cm long (up to 8 cm or more), with protuberant glands at the base of the blade; calyx 1 cm long or more:

    2. Leaves broadly lanceolate or ovate, mostly 2.8 cm long, appressed-puberulous, the margins crenate-serrate or coarsely dentate; apex of petiole 2-glandular: **T. ulmifolia**

    2. Leaves narrowly lanceolate, with long attenuate apex, up to 12 cm long, glabrous except for a few midvein hairs, the margins subentire but faintly notched; apex of petiole often 3-glandular: **T. triglandulosa**

1. Leaves less than 1.5 cm long, without protuberant glands at the base of the blade; calyx 0.4–0.7 cm long: **T. diffusa**

Turnera ulmifolia L., *Sp. Pl.* 1: 271 (1753). **FIG. 113.**

Turnera ulmifolia var. *angustifolia* (Mill.) Willd. (1979).

CAT-BUSH

An erect bushy herb or short-lived shrub up to ca. 1 m tall, the younger parts and leaves finely appressed-puberulous. Leaves narrowly to broadly lanceolate, mostly 2–8 cm long, acute or acuminate at the apex, the margins irregularly crenate-serrate; basal glands 0.7–0.8 mm in diameter. Peduncles almost completely united to the petioles; bracteoles

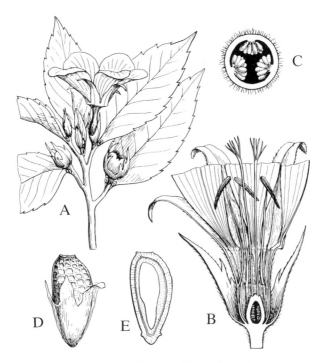

FIG. 113 **Turnera ulmifolia**. A, branchlet with flower and fruit, × ²/₃. B, flower cut lengthwise (petals cut), × 2. C, cross-section of ovary, × 6. D, seed with aril, × 10. E, seed cut lengthwise, × 10. (F. & R.)

leaf-like, 1–3 cm long. Calyx-tube ca. 1 cm long; petals bright yellow, 2–3 cm long or sometimes more. Capsule 6–9 mm long; seeds ca. 2.5 mm long, densely pitted.

GRAND CAYMAN: *Brunt* 1849, 2079; *Hitchcock*; *Kings* GC 391; *Lewis* 3847; *Millspaugh* 1260; *Proctor* 15086. LITTLE CAYMAN: *Kings* LC 102; *Proctor* 28049. CAYMAN BRAC: *Kings* CB 25; *Proctor* 29031.

— West Indies and continental tropical America; introduced in the Old World tropics. A somewhat weedy, variable species, but quite handsome in flower.

Turnera triglandulosa Millsp., *Field Mus. Bot.* 2: 77 (1900). PLATE 21.

Erect, nearly glabrous shrub up to 1.5 m tall or more, with long, virgate, wide-spreading branches. Leaves linear-lanceolate with long-tapering apex, narrowly cuneate base, and subentire but distantly notched margins, widely varying in length up to 12 cm long but usually shorter, distinctly paler on the underside; petioles 1–1.4 cm long, often 3-glandular at junction with the blade. Peduncles entirely adherent to the petiole, the flower appearing to be sessile at base of leaf-blade; flower-parts similar to those of *T. ulmifolia* except the bracteoles entire, the calyx lobes equalling the corolla, and the petals deep yellow.

GRAND CAYMAN: *Proctor* 47783 (cult. at Botanic Park). LITTLE CAYMAN: *Proctor* 48229, 49335, 49342, 51176. CAYMAN BRAC: *Millspaugh* 1152 (type), 1195, 1209.

— Endemic. This entirely seems very distinct when grown side by side with *T. ulmifolia*, but the latter species is so variable over its whole range as to cast some doubt on the validity of closely related forms such as *T. triglandulosa*, which was placed in synonymy in the first edition of this book. No doubt further investigation of this complex is needed.

321

Turnera diffusa Willd. ex Schult. in L., *Syst. Veg.* 6: 679 (1820).

A low shrub, often densely branched, less than 1 m tall; younger parts and underside of leaves densely puberulous. Leaves oblanceolate or obovate, 6–15 mm long, coarsely crenate-dentate, the nerves grooved on the upper side and prominent beneath, the underside minutely and densely glandular all over the surface as well as puberulous. Peduncles very short or obsolete; bracteoles linear-subulate, 2–4 mm long. Calyx 5-toothed, 3–4 mm long; petals bright yellow, up to 8 mm long. Capsule 3–4 mm long; seeds 1.5–2 mm long, reticulate-striate.

LITTLE CAYMAN: *Proctor* 28113, 35127.

— West Indies and continental tropical America, usually in arid rocky scrublands. In some countries, the leaves are used to make an aromatic tea, which is reputed to be an aphrodisiac.

PASSIFLORACEAE

Vines, trailing or climbing by means of axillary unbranched tendrils, herbaceous or woody, or rarely erect shrubs or trees. Leaves alternate, simple, entire or lobed, or very rarely compound, the petiole usually bearing glands; stipules 2, thread-like or large and ovate. Flowers axillary, perfect or rarely unisexual, solitary or variously cymose-racemose or paniculate; peduncles usually jointed; bract and bracteoles 3, small and distant from the flower, or large, foliaceous and close to the flower, forming an involucre. Calyx-tube (receptacle) almost flat to saucer-like or bell-shaped, giving rise in the middle to a gynophore (ovary-stalk), and nearly always bearing toward the margin or within its throat a ring of one or several series of erect or radiate filamentous or membranous outgrowths (called the corona); sepals usually 5, inserted in the throat of the receptacle, overlapping, often with a small horn-like process on the back near the apex, and usually coloured inside. Petals either lacking or usually 5, alternate with the sepals, free, overlapping, and withering-persistent. Stamens usually 5, attached below the ovary on the gynophore. Ovary superior and free, located at the apex of the gynophore, 1-celled; style simple or 3–4-branched, or else 3 separate styles present; ovules numerous, attached to 3 or 5 parietal placentas. Fruit berry-like or sometimes a capsule opening by 3 apical valves; seeds numerous, usually compressed-ovoid, covered with a fleshy aril or surrounded by pulp; endosperm fleshy; cotyledons often foliaceous.

A widespread tropical and subtropical family of about 12 genera and 600 species, especially concentrated in S. America. Most of the species belong to the single genus *Passiflora*; several of these produce edible fruits known by such names as passion-fruit, granadilla, sweet-cup, and golden apple.

Passiflora L.

Vines with characters as given for the family; receptacle shorter than the rest of the flower; corona prominent and often brightly coloured, its filaments distinct or more or less united. Sepals and petals 4 or 5, or petals sometimes lacking. Stamens 4 or 5, the filaments joined to the gynophore below, free above. Styles 3; stigmas capitate. Fruit a dry or pulpy berry.

— A largely American genus of more than 500 species. These are easily recognised by their complicated and rather curious flowers, which to the early Spanish explorers suggested some of the emblems associated with the Crucifixion, hence the Spanish name 'pasionaria', the English name 'passion-flower', and the Latin term *Passiflora*.

KEY TO SPECIES

1. Flowers large (6–10 cm across) and intensely fragrant, subtended by an involucre of 3 large segments; fruit ellipsoidal, 7–8 cm long: **[P. laurifolia]**

1. Flowers smaller, not noticeably fragrant, and lacking an involucre; fruit globose, less than 2 cm in diameter:

  2. Leaves entire or commonly 3-lobed, the apex more or less acute; flowers greenish or cream; petals lacking: **P. suberosa**

  2. Leaves always entire, the apex usually notched and bearing a small bristle in the notch; flowers deep red-purple; petals present: **P. cupraea**

[*Passiflora laurifolia* L., *Sp. Pl.* 2: 956 (1753).

GOLDEN APPLE, WATER LEMON

High-climbing vine, the stems grooved-striate; leaves glabrous, elliptic or broadly oblong, 6–14 cm long, 1-nerved, the apex abruptly pointed, the base rounded or subcordate; petioles with 2 glands near the apex; stipules linear, 6–9 mm long. Flowers solitary on peduncles longer than the petioles; segments of the involucre 3.5–4 cm long, leafy, not united, with margins usually glandular-dentate. Sepals pale green, horned below the apex; petals shorter and narrower than the sepals, cream with minute red speckles; corona in several unlike series of various lengths, bright purple with white transverse bands toward the base. Fruit narrowly ellipsoid, 7–8 cm long, orange-yellow, edible.

GRAND CAYMAN: *Kings* GC 293. This was probably gathered from a cultivated plant. — West Indies and northern S. America, indigenous and cultivated.]

Passiflora suberosa L., *Sp. Pl.* 2: 958 (1753). **FIG. 114.** PLATE 22.

Passiflora pallida L. (1753).
Passiflora minima L. (1753).
Passiflora angustifolia Sw. (1788).

WILD PUMPKIN

A small, chiefly herbaceous vine, creeping or climbing, glabrous nearly throughout or sometimes pubescent (at least on the petioles); leaves usually membranous, extremely variable in outline even on the same plant, frequently deeply 3-lobed, but also often entire and narrowly to broadly elliptic, 2–13 cm long, the apex acutish and bristle-tipped, and with 2 short-stalked glands near apex of the petiole. Flowers 1.5–3 cm across, 1 or 2 in the leaf-axils on filamentous peduncles longer than the petioles; involucre lacking. Sepals greenish or cream, 5–8 mm long; petals absent; corona half as long as the sepals, 2-seriate, the outer ring with filaments recurved, yellow at apex, white in the middle, and purple or violet toward base. Berry globose, usually black, 6–15 mm in diameter; seeds 3–4 mm long, abruptly acuminate, coarsely reticulate.

FIG. 114 **Passiflora suberosa**. A, habit. B, C, D, E, F, various forms of leaf. G, flower. H, stamen. I, pistil. J, fruit. **Passiflora cupraea**. K, flower. L, two stamens. M, fruit. (G.)

GRAND CAYMAN: *Brunt* 2144, 2164; *Hitchcock*; *Kings* GC 238; *J. Popenoe* in 1968 (MO); *Proctor* 1194. LITTLE CAYMAN: *Proctor* 28057. CAYMAN BRAC: *Proctor* 29010.

— Widespread and common in the warmer parts of the Western Hemisphere, often in thickets, rocky woodlands, and pastures.

Passiflora cupraea L., *Sp. Pl.* 2: 955 (1753). **FIG. 114.** PLATE 22.

A subwoody, glabrous vine, trailing or high-climbing; leaves entire, ovate to very broadly elliptic or rotund, 3–7 cm long, shallowly notched and bristle-tipped at apex, subcordate at base, reticulate-veined on both sides, and with a few distant, flat, circular glands beneath. Flowers solitary in the axils, usually red-purple, on peduncles longer than the petioles; involucre lacking. Sepals narrowly oblong, ca. 2.5 cm long; petals narrower and shorter; corona 1-seriate, yellowish, the filaments linear-oblong, 2–4 mm long. Gynophore 2–3 cm long. Berry globose, dark blue or purple, 1.5–2 cm in diameter; seeds ca. 3 mm long, blunt or acute, transversely rugose.

GRAND CAYMAN: *Correll & Correll* 51026. LITTLE CAYMAN: *Proctor* 35113. CAYMAN BRAC: *Proctor* 29053.

— Bahamas, eastern Cuba, and Tortue Island (Haiti), in rocky woodlands, scrublands, and coastal thickets.

[CARICACEAE

Coarse or giant unbranched or few-branched herbs, often tree-like or shrubby, with terminal crowns of leaves and milky sap; leaves alternate, usually very large and long-petiolate, palmately lobed or -foliolate; stipules lacking. Flowers yellow, whitish or greenish, in axillary inflorescences, unisexual or perfect, monoecious, dioecious or polygamous; staminate and polygamous inflorescences pendulous and paniculate, cymose or racemose; pistillate inflorescences short, few-flowered and cymose. Calyx 5-lobed; corolla of 5 petals, in staminate and perfect flowers united below into a long tube, in pistillate flowers almost free. Stamens usually 10, inserted at the mouth of the corolla-tube; filaments with adnate 2-celled anthers. Pistillate flowers without staminodes; ovary free, sessile, 1-celled or imperfectly 5-celled; stigmas sessile or nearly so, 3–5-lobed; ovules numerous, in 2 or more series on 5 placentas. Fruit a large, fleshy berry; seeds numerous, compressed or globose, with fleshy endosperm.

— A small family of about 4 genera and 40 species; one genus occurs in tropical Africa, the others all being tropical American.

Carica L.

With the characters of the family, the stems unarmed. Flowers dioecious; stamens inserted in 2 series, free, the outer ones with elongate filaments, the inner ones short, the connective often produced into a ligule beyond the anther-cells; stigmas 5, linear or variously cleft. Fruit filled with pulp, or a large cavity often present; seeds rugose-tuberculate, covered with a succulent membrane.

— About 30 species, the majority in S. America.

Carica papaya L., *Sp. Pl.* 2: 1036 (1753). **FIG. 115.**

PAWPAW

A giant tree-like herb up to 6 m tall or more, the mostly simple, columnar trunk soft and hollow, and marked with the scars of fallen leaves; leaves up to 1 m long or more, with long hollow petioles; blade mostly 30–60 cm broad, palmately 5–7-lobed, the segments pinnately lobed, more or less glaucous beneath. Flowers with corolla twisted in the bud; staminate flowers 2–3.5 cm long, the pistillate ones 4–5 cm long, creamy or pale yellow. Fruit pendulous, varying greatly in size, shape and colour, up to 50 cm long, edible; seeds black, 6–7 mm long.

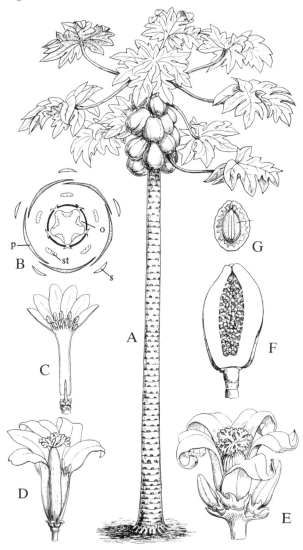

FIG. 115 **Carica papaya**. A, pistillate (fruiting) plant, much reduced. B, floral diagram, perfect flower; **s**, sepal; **p**, petal; **st**, stamen; **o**, ovary. C, staminate flower cut open, × 1½. D, perfect flower cut open, × 1. E, pistillate flower cut open, × 1. F, young fruit cut lengthwise, much reduced. G, seed cut lengthwise, × 2; **e**, endosperm. (F. & R.)

GRAND CAYMAN: *Crosby, Hespenheide & Anderson* 28 (GH); *Proctor* (sight record). LITTLE CAYMAN: *Kings* LC 80. CAYMAN BRAC: *Proctor* 29340.

— Native of tropical America, but the precise extent of its original range is unknown; cultivated and becoming naturalised in almost all tropical countries. The milky sap of this and related species contains a substance called papain, which is often used to tenderise meat. The ripe Pawpaw or Papaya is one of the most delicious of all tropical fruits.]

CUCURBITACEAE

Climbing or creeping herbs of rapid growth and with abundant watery sap, annual or perennial, usually with lateral tendrils which are interpreted as modified stipules; leaves alternate, simple or palmately lobed or divided, often cordate and of membranous texture, and usually with long petioles. Flowers regular, axillary, solitary or in racemes, cymes or panicles, unisexual, monoecious or rarely dioecious. Calyx of 5 (rarely of 3 or 6) sepals united below into a bell-shaped tube. Petals 5 (rarely 3 or 6), free or united. Stamens 3 (rarely 5) of which one has a 1-celled anther, the others 2-celled; filaments short; anthers free or cohering into a head, often contorted. Ovary inferior (or in *Sechium* with the apex free), 3-celled, each cell representing a carpel, the placentas meeting in the axis; style terminal, simple or lobed; ovules few to numerous. Fruit usually a fleshy or corky berry (rarely a capsule), usually indehiscent (but in *Momordica* opening by valves); seeds 1 or few to many, compressed, without endosperm, the cotyledons foliaceous.

— A largely pantropical family of more than 100 genera and perhaps 850 species. Many produce edible fruits, but a few are poisonous.

KEY TO GENERA

1. Flowers in panicles, small; plants dioecious; stamens 5 with free anthers; fruit large, woody, with discoid seeds:	Fevillea
1. Flowers solitary or in clusters or racemes; plants monoecious with unisexual flowers; stamens 3 with coherent anthers:	
2. Leaves nearly entire or else shallowly 3–5-angled, rarely 3-lobed:	
3. Tendrils simple:	
4. Corolla ca. 0.5 cm long; fruit ca. 1 cm in diameter:	Melothria
4. Corolla ca. 4 cm long; fruit up to 7 cm in diameter:	Cionosicyos
3. Tendrils branched:	
5. Perennial with high-climbing stems; flowers clustered, greenish or white; fruits 1-seeded:	[Sechium]
5. Annual with trailing stems; flowers solitary, yellow; fruits many-seeded:	[Cucurbita]
2. Leaves deeply 5 to many-lobed:	
6. Tendrils simple; fruit prickly or tuberculate:	
7. Fruits indehiscent, prickly, with greenish or yellowish pulp, not bitter:	[Cucumis]
7. Fruits splitting open by 3 valves, tuberculate, with crimson pulp, very bitter:	[Momordica]
6. Tendrils branched; fruits smooth:	[Citrullus]

Fevillea L.

Woody vine; leaves cordate, entire or more commonly palmately lobed or angled, membranous. Tendrils lateral, 2-branched at apex. Peduncles collateral with tendrils and leaves. Flowers in hanging panicles, small, yellow or greenish, dioecious. Staminate flowers with short bell-shaped or cup-like tube; sepals 5; petals 5, clawed, spreading with an erect longitudinal tongue along the midrib. Staminodes 5, in the centre of the receptacle; anthers 1-locular, opening by a longitudinal slit. Pistillate flowers with sepals similar to those of staminate flowers; petals 5, clawed, spreading with 5 or more staminodes; 20 minute glands occurring at base of petals. Ovary 3-locular; styles 3; stigmas 2- or 3-lobed; ovules 6 or fewer in each locule. Fruit large, ringed above or below the middle with the scar of the fallen calyx-limb, with 3 cavities; seeds usually 6 in each cavity, large, compressed, roundish in outline, overlapping and hanging from the margins of the placentas in each cavity; cotyledons large.

— A genus of 10 neotropical species.

Fevillea cordifolia L., *Sp. Pl.* 2: 1013 (1753).

ANTIDOTE CACCOON (Jamaica).

High climbing perennial vine; early leaves ovate, entire, the latter ones 5-angled or -lobed, membranous, glabrous, up to 17 cm long, the margins with glandular teeth at vein-ends and with petioles 3–7 cm long. Panicles up to 60 cm long, pendulous. Calyx shallowly campanulate, greenish-yellow, puberulous, 2–3 mm long, 5-lobed, the lobes broadly lanceolate and acute; corolla subrotate, greenish-yellow or dusky orange, 6–10 mm in diameter, 5-lobed, the lobes orbicular and rounded. Staminate flowers with 5 free stamens ca. 1 mm long, the anthers 1-locular. Pistillate flowers with 5 staminodes; ovary globose, 3–4 mm in diameter, puverelous; styles ca. 1 mm long, connate to near the middle; stigmas fused, capitate and 3-lobed. Fruit globose, hard, green, 7–12 cm in diameter with a circumfrential scar below the middle; seeds flattened-discoidal, 4–6 cm in diameter and ca. 1 cm thick, smooth, yellowish-grey.

GRAND CAYMAN: *Proctor 47279*. Found near the middle of the island in rocky woodland ca. 0.6 km west of the Frank Sound Road.

— Greater Antilles, Guadeloupe, Martinique, C. America and northern S. America.

Melothria L.

Slender herbaceous vines with simple tendrils; leaves lobed, angular, or nearly entire. Plants monoecious in the Cayman species; flowers small, yellow or white. Staminate flowers racemose (rarely solitary); calyx 5-toothed; corolla deeply 5-parted; stamens 3. Pistillate flowers solitary or clustered, on long slender peduncles, the calyx and corolla as in the staminate flowers; staminodes 3 or none; ovary constricted below the calyx-tube; placentas 3; style short, surrounded at the base by a ring-like disc; stigmas 3. Berry small, globose or ovoid.

— A genus of 10 tropical American species.

Melothria pendula L., *Sp. Pl.* 1: 35 (1753). **FIG. 116.**

Melothria guadalupensis (Spreng.) Cogn. in DC. (1881); Adams (1972).

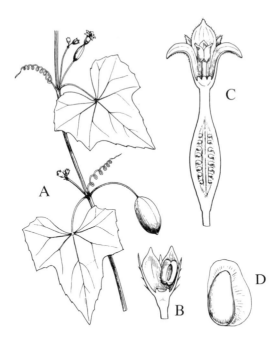

FIG. 116 **Melothria pendula.** A, portion of branch with leaves, flowers and fruit, × ²/₃. B, staminate flower of a related species (*M. cucumis*) cut lengthwise, one stamen removed, × 5. C, pistillate flower of same, cut lengthwise, × 3. D, seed, × 4. (F. & R.)

A tender glabrous climber; leaves broadly ovate-cordate, 2–5 cm long or more, 3–5-angled or shallowly 3–5-lobed, the margins distantly denticulate, the upper (adaxial) surface very rough, the lower less so. Staminate flowers several toward the end of a peduncle, on pedicels 1–2 mm long; calyx-tube 2 mm long; corolla yellow, 2 mm long. Pistillate flowers solitary, sometimes in the same axil as the staminate ones but on much longer peduncles; calyx and corolla similar to the staminate ones; ovary 4–5 mm long. Fruit purplish-black, ovoid, 1–1.5 cm long; seeds flat, 4 mm long.

GRAND CAYMAN: *Proctor* 27984.

— West Indies and continental tropical America, variable, occurring in rocky thickets and woodlands.

Cionosicyos Benth. & Hook.f.

High-climbing glabrous vines; leaves membranous, broadly ovate, cordate or subcordate at base, acuminate at apex, entire or 3-lobed, the margins with small glandular points at major vein-ends. Plants monoecious; flowers rather large, solitary in leaf axils, long-stalked with a joint just below the flower. Calyx tube top-shaped or funnel-shaped; sepals 5, leathery, ovate- or oblong-lanceolate. Corolla funnel-shaped-rotate, 5-parted; segments obovate, ribbed, cream colour; stamens 3–5; filaments free, hirsute, inserted into the base of the corolla; anthers all united in a cylindrical column which is shortly exceeded by 5 connectival lobes; locules conduplicate. Pistillate flowers with cup-shaped calyx tube, otherwise similar to the staminate flowers. Ovary globose with 6 locules; stigmas 3, bent back, leaf-like; ovules numerous. Fruit a globose fleshy berry, glabrous, with numerous seeds.

— A genus with perhaps 3 poorly differentiated species of Cuba, Jamaica, Cayman Islands and Mexico.

Cionosicyos pomiformis Griseb., *Fl. Brit. W. Ind.* 288 (1860). PLATE 22.

WILD PUMPKIN, DUPPY PUMPKIN

Stems slender, angled, glabrous. Leaves 5–10 cm long, 3-nerved with the lateral veins 2-branched, the margins minutely and remotely toothed. Peduncles of staminate flowers 4–5 cm long, those of pistillate flowers ca. 1.5 cm long. Flowers greenish-white or cream; corolla 3.5–4 cm long, tomentose or the pistillate corolla papillose on the outside. Fruit a globose berry ca. 4 cm in diameter, green with whitish bands when immature, turning yellow, then bright orange, when ripe; seeds numerous, embedded in pulp.

GRAND CAYMAN: *Kings GC 237; Proctor 47355, 47784, 484686.*

— Otherwise typically confined to Jamaica.

[*Momordica* L.

Slender herbaceous vines, annual or perennial, with simple tendrils; leaves lobed or in some species entire. Peduncles often with a small leafy bract. Flowers yellow, monoecious or dioecious. Staminate flowers solitary or in panicles; calyx-tube short, closed by 2 or 3 oblong, incurved scales; petals 5, free or nearly so; stamens 3, the anthers at first coherent, at length free, contorted. Pistillate flowers solitary, with calyx and corolla like those of the staminate flowers; staminodes none or represented by 3 glands at the base of the style; stigmas 3; ovules numerous. Berry opening by 3 valves or indehiscent.

— A genus of 45 species indigenous to the Old World tropics; one or two are naturalised in nearly all warm countries.

Momordica charantia L., *Sp. Pl.* 2: 1009 (1753). PLATES 22 & 23.

SERASEE, sometimes spelled CERASEE

Tender climber with stems 1–3 mm thick; leaves 2–7 cm long and wide, deeply palmately lobed into 5–7 obovate, irregularly toothed lobes, glabrous or nearly so except on the veins. Staminate flowers solitary on long filiform peduncles 5–15 cm long, with a roundish green bract near the middle. Pistillate flowers on short peduncles 2–4 cm long, with a bract near the base. Fruit bright orange, 4–6 cm long (var. *charantia*) or 1.5–3 cm long (var. *abbreviata* Ser. ex DC., 1828), ovoid or spindle-shaped, tuberculate, when ripe splitting by 3 valves from the apex; seeds embedded in crimson pulp.

GRAND CAYMAN: *Brunt 1717, 1897a; Hitchcock; Kings GC 223; Maggs II 63; Millspaugh 1329; Proctor 15078; Sachet 381.* CAYMAN BRAC: *Proctor 29122.* All the Cayman specimens appear to represent var. *abbreviata.*

— Now occurring in all warm countries, cultivated and naturalised. Both the leaves and the fruits are believed by many people to be of medicinal value, perhaps because they are so bitter.]

[*Citrullus* Schrad.

Citrullus lanatus (Thumb.) Matsumura & Nakai (*C. vulgaris* Schrad.) the watermelon, is cultivated and often found growing as an escape in open waste ground.

GRAND CAYMAN: *Brunt* 2129; *Proctor* 27960.

— Native of tropical and southern Africa.]

[*Sechium* Juss.

Sechium edule (Jacq.) Sw., the chocho, is sometimes cultivated. It is native to continental tropical America, where it is usually called 'chayote', and is cultivated and semi-naturalised in many tropical countries. Both the fruits and the root are edible, the latter resembling a yam when cooked.]

[*Cucurbita* L.

Cucurbita pepo L., the pumpkin, is cultivated and has been observed as an escape, but scarcely persisting.

GRAND CAYMAN: *Proctor* (sight record). CAYMAN BRAC: *Proctor* 29347.

— Native of continental tropical America, and almost world-wide in cultivation. Many varieties occur.]

[*Cucumis* L.

Annual or perennial herbs, usually trailing, or sometimes climbing by means of simple tendrils; leaves deeply lobed. Plants usually monoecious, with yellow flowers. Staminate flowers clustered or rarely solitary, the calyx 5-lobed, the corolla more or less campanulate, 5-parted; stamens 3, with linear or curved anther-cells, the connective prolonged above; ovary represented by a gland. Pistillate flowers solitary, the calyx and corolla as in the staminate flowers; staminodes 3; ovary with 3–5 placentas; stigmas 3–5; ovules numerous. Fruit fleshy and often juicy, of various shapes, indehiscent, many-seeded.

A chiefly African genus of 25 species, the following usually considered native to tropical America, but actually probably introduced at an early date.

Cucumis anguria L., *Sp. Pl.* 2: 1011 (1753). **FIG. 117.**

WILD CUCUMBER

An annual trailing herb, the angled stems and leaves rough with stiff white hairs; leaves 5–9 cm long, cordate and deeply 5-lobed, with toothed and somewhat wavy margins; petioles equal in length to the blades or longer. Flowers small, the staminate on peduncles 1–2 cm long and with calyx ca. 6 mm long; anthers with a 2-lobed appendage. Pistillate flowers solitary on peduncles 5–10 cm long. Fruit pale yellow, ovoid, 4–7 cm long, more or less prickly, edible; seeds 4–4.5 mm long.

GRAND CAYMAN: *Brunt* 2115; *Kings* GC 417; *Proctor* 27959.

— Tropical Africa; West Indies and continental tropical and subtropical America, often a weed of open fields and waste ground. The cucumber (*Cucumis sativa* L.) and muskmelon or cantaloupe (*Cucumis melo* L.) are related species of African origin].

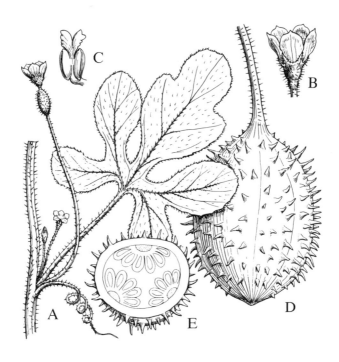

FIG. 117 **Cucumis anguria**. A, portion of branch with leaf, tendril, and staminate and pistillate flowers, × $^2/_3$. B, staminate flower, × 3. C, stamen much enlarged. D, fruit, × $^2/_3$. E, cross-section of same, × $^2/_3$ (F. & R.)

CAPPARACEAE

Herbs, shrubs or trees, glabrous or pubescent, sometimes glandular or beset with closely adhering scales; leaves usually alternate, simple or palmately compound; stipules present or absent, sometimes represented by spines. Flowers regular or irregular, perfect, solitary in the leaf-axils or in axillary or terminal racemes or corymbs; sepals 4, free or partly united; petals 4 or rarely none, sessile or clawed. Stamens 4 to numerous, the filaments usually free or sometimes united to the stalk of the ovary. Ovary borne on a stalk (gynophore) or rarely sessile, 1-celled or sometimes apparently several-celled by false septa; style usually short or none, the stigma orbicular; ovules numerous on parietal placentas in 1 to many series, rarely solitary. Fruit a silique-like capsule or a berry, rarely a drupe; seeds with little or no endosperm and curved embryo.

— A widely distributed, chiefly tropical family of about 40 genera and 700 species. Few are of any economic value, but capers, the pickled flower-buds of *Capparis spinosa* L. of the Mediterranean region, are used as a condiment.

KEY TO GENERA

1. Shrubs or trees; fruit a globose to elongate berry, the seeds embedded in pulp:	Capparis
1. Herbs; fruit an elongate capsule, the seeds not embedded in pulp:	Cleome

Capparis L.

Shrubs or trees, unarmed or sometimes spiny, glabrous, pubescent, or lepidote. Leaves simple, often of hard or leathery texture; stipules present or absent. Flowers usually several together in axillary and terminal racemes or corymbs; sepals free or partly united, or rarely united in bud and irregularly rupturing, and often with a gland at the base. Petals imbricate, usually white. Stamens numerous, with filiform free filaments. Ovary on a long gynophore, 1–4-celled, with 2–6 placentas; ovules numerous; stigmas sessile. Fruit a stalked globose to cylindrical berry, indehiscent or rarely irregularly dehiscent; seeds numerous, with convolute embryo.

— A widespread, chiefly tropical genus of about 150 species.

KEY TO SPECIES

1. Fruits subglobose; younger parts and underside of leaves covered with minute stellate hairs: — **C. ferruginea**

1. Fruits elongate; stellate hairs absent:

2. Younger parts and underside of leaves covered with minute peltate scales; mature fruit dry within, rupturing irregularly: — **C. cynophallophora**

2. All parts glabrous and devoid of scales; mature fruits with seeds embedded in red pulp, and rupturing by two valves (like a legume pod): — **C. flexuosa**

Capparis ferruginea L., *Syst. Nat.* ed. 10, 2: 1071 (1759). PLATE 23.

DEVIL HEAD

Shrub or small tree to 5 m tall or more, the young branches densely clothed with minute rusty stellate hairs; leaves oblanceolate or narrowly elliptic, mostly 3–8 cm long, sharply acute or subacuminate at the apex, glabrous above, beneath densely clothed with minute whitish stellate hairs. Flowers in pedunculate axillary corymbs; calyx 2–2.5 mm long, deeply cleft into narrow segments; petals ca. 5 mm long; stamens usually 8, about as long as the petals. Fruit subglobose, ca. 1.5 cm long, covered with minute rusty stellate hairs, and rupturing irregularly at maturity.

GRAND CAYMAN: *Brunt* 1786, 2134; *Kings* GC 240; *Proctor* 15049.
— Cuba, Hispaniola and Jamaica, in dry rocky woodlands and coastal thickets.

Capparis cynophallophora L., *Sp. Pl.* 1: 504 (1753). **FIG. 118.** PLATE 23.

Capparis longifolia Sw., 1788 (juvenile form).

HEADACHE BUSH

Shrub or small tree to 6 m tall or more, the young branches covered with minute, closely adhering peltate scales. Leaves oblong, lance-oblong, or elliptic (or in some juvenile forms narrowly linear), mostly 3–8 (–12) cm long, the apex shallowly notched, rounded, or sometimes acutish, the upper surface naked and dark shining green, the underside pale and densely covered with minute, closely adhering, peltate scales. Flowers in loose, few-flowered subterminal corymbs, the peduncles arising from axils of the upper leaves; sepals free, 8–11 mm long; petals 10–14 mm long, scaly outside, at first white, turning purplish. Stamens numbering 20–30, 2–3 times as long as the petals, the filaments purple. Gynophore as long as the filaments. Fruit torulose, 5–30 cm long.

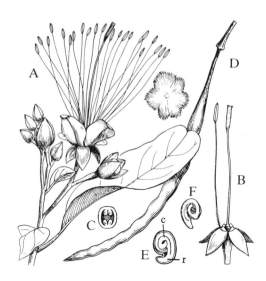

FIG. 118 **Capparis cynophallophora**. A, portion of flowering branch, × ²⁄₃. B, flower with petals and all but one stamen removed, × ²⁄₃. C, cross-section of ovary, × 5. D, fruit, × ²⁄₃. E, seed cut lengthwise, slightly enlarged; c, cotyledon; r, radicle. F, embryo slightly enlarged. G, scale of leaf, very much enlarged. (F. & R.)

GRAND CAYMAN: *Brunt* 1757, 1784, 2002; *Kings* GC 142; *Proctor* 15125. LITTLE CAYMAN: *Proctor* 35214, 35218. CAYMAN BRAC: *Proctor* 29138.

— Florida, West Indies, and continental tropical America, in rocky thickets and woodlands.

Capparis flexuosa (L.) L., *Sp. Pl.* ed. 2, 1: 722 (1762). PLATES 23 & 24.

RAW BONES, RAW HEAD, BLOODY HEAD

A shrub or small tree, sometimes scrambling or with arching or flexuous branches, entirely glabrous and devoid of scales. Leaves oblong, lance-oblong, or elliptic, 4–9 cm long, blunt or often notched at the apex, an oblong gland in each axil. Flowers nocturnal, large; sepals 6–10 mm long; petals white or pale greenish, 1.5–3 cm long; stamens numerous, the white filaments 4–6 cm long. Ovary 0.6–1 cm long, on an elongate gynophore. Fruit red, 10–20 cm long, cylindrical or torulose, with scarlet pulp and several to numerous large white seeds.

GRAND CAYMAN: *Brunt* 1667, 1700, 2044; *Correll & Correll* 51004; *Kings* GC 289; *Proctor* 15126. LITTLE CAYMAN: *Kings* LC 109; *Proctor* 28044. CAYMAN BRAC: *Kings* CB 16; *Proctor* 29006.

— Florida, West Indies and continental tropical America, in rocky thickets and woodlands.

Cleome L.

Herbs, sometimes somewhat shrubby, rarely scandent, glabrous or pubescent; leaves simple or palmately 3–7-foliolate, the leaflets entire or serrulate. Flowers solitary in the upper axils, forming a more or less leafy raceme; calyx 4-toothed or 4-parted, persistent or deciduous; petals subequal, sessile or clawed. Stamens usually 6, rarely 4 or 12–20, all or only 2 bearing anthers; filaments usually unequal and declinate. Ovary sessile or at the apex of a gynophore; style very short or the stigma sessile; ovules numerous. Fruit a more or less elongate 1-celled capsule with membranous valves; seeds more or less reniform, usually rough or pubescent.

— A pantropical genus of about 75 species.

KEY TO SPECIES

1. Leaves simple, linear-lanceolate; plant perennial, decumbent from the apex of a woody taproot:	**C. procumbens**
1. Leaves palmately compound; plants annual, erect:	
2. Stems armed with stipular spines; flowers white or pale rose; stamens 4–6:	**C. spinosa**
2. Stems unarmed; flowers yellow; stamens 12–20:	**C. viscosa**

Cleome procumbens Jacq., *Sel. Stirp. Amer.* 189, t. 120 (1763).

A low, more or less decumbent herb arising from a woody perennial taproot, glabrous throughout; leaves simple, linear-lanceolate, 0.5–2 cm long. Flowers on slender pedicels less than 1 cm long; sepals free, unequal, up to 3 mm long; petals yellow, ca. 5 mm long. Capsule narrowly cylindrical, usually 15–20 mm long, the apex apiculate with the persistent 1.5–2 mm style.

GRAND CAYMAN: *Brunt 2120; Correll & Correll 51036; Proctor 27937; Sachet 411.*
— Cuba, Hispaniola, and Jamaica, in seasonally wet savannas and pastures.

Cleome spinosa Jacq., *Enum. Syst. Pl. Carib.* 26 (1760). **FIG. 119.**

CAT'S WHISKERS

A coarse, erect herb up to 1 m tall or more, usually glandular-puberulous throughout, the stems armed with pairs of sharp, yellowish stipular spines. Leaves palmately (3–)5–7-foliolate, the leaflets 2–9 cm long; petioles 2–11 cm long, usually armed with prickles. Flowers in terminal racemes, each pedicel subtended by a large sessile oval bract; sepals linear, 4–5 mm long; petals white or light purple, long-clawed. Stamens 6, crimson, long-exserted. Gynophore much longer than the pedicel, the ovary glandular, 7–8 mm long, with sessile stigma. Capsule linear-cylindrical, 5–12 cm long, glabrous or puberulous; seeds nearly 2 mm in diameter, almost smooth.

CAYMAN BRAC: *Proctor 29093.*
— West Indies and continental tropical and subtropical America, usually a roadside or pasture weed, sometimes cultivated for ornament.

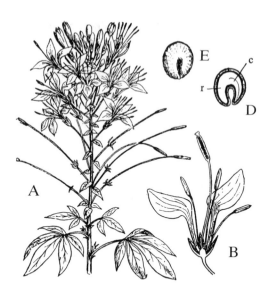

FIG. 119 **Cleome spinosa.** A, portion of flowering branch. × ½. B, flower cut lengthwise, slightly enlarged. C, seed × 5. D, seed cut lengthwise, × 5; **c**, cotyledons; **r**, radicle. (F. & R.)

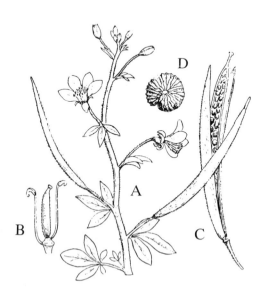

FIG. 120 **Cleome viscosa.** A, portion of flowering branch, × 1. B, flower with sepals, petals, and all but two stamens removed, × 2. C, ripe fruit, × 1. D, seed, much enlarged. (F. & R.)

Cleome viscosa L., *Sp. Pl.* 2: 672 (1753). **FIG. 120.**

   Cleome icosandra L. (1753).

   Polanisia viscosa (L.) DC. (1824).

A coarse erect viscid herb to about 1 m tall, glandular-pubescent throughout, the stems unarmed. Leaves 3–5-foliolate, the obovate or elliptic leaflets 1–4 cm long. Flowers solitary in the upper leaf-axils; sepals narrowly oblong, 5–7 mm long, deciduous; petals yellow, ca. 10 mm long. Stamens 12–20, usually shorter than the petals. Ovary sessile; style short, elongating in fruit, with capitate stigma. Capsule linear-cylindrical, 6–8 cm long, densely glandular-pubescent, the apex terminated by the persistent style; seeds transversely crested.

   CAYMAN BRAC: *Proctor* 29040.

   — A pantropical weed of open waste ground.

CRUCIFERAE

Annual or perennial herbs with watery acrid sap, the pubescence of simple or branched hairs; leaves alternate, simple or compound, the basal ones often forming a rosette; stipules absent. Flowers regular, perfect, in terminal or axillary racemes; sepals 4, free, the inner two sometimes pouch-like at the base; petals 4, rarely absent. Stamens usually 6, with 4 long and 2 short; anthers mostly 2-celled, longitudinally dehiscent. Ovary superior, sessile or rarely stalked, 2-carpellate and usually 2-celled; style simple with 2 stigmas; ovules usually numerous. Fruit a silique (a 2-valved capsule whose 2 parietal placentas persist as a 'replum' from which the valves separate) or (in *Cakile*) an indehiscent capsule that is transversely 2-jointed; seeds small, often mucilaginous when wet, frequently winged or marginate, mostly lacking endosperm, or sometimes an oily endosperm present.

   A large, world-wide, but chiefly temperate-climate family of about 225 genera and more than 2,500 species. There are no poisonous plants in this family, and a number have found wide use as food. Among these are the turnip (*Brassica campestris* L.), the cabbage and its several races, such as cauliflower, broccoli, and brussels sprouts (all forms of *Brassica oleracea* L.), and mustard (*Brassica nigra* (L.) K. Koch).

KEY TO GENERA

1. Fruit a lomentum, transversely 2-jointed, indehiscent; leaves succulent; plants of sandy sea-beaches:	**Cakile**
1. Fruit an orbicular, compressed, dehiscent, 1-seeded silicula; leaves thin; plants of inland waste ground and cultivated areas, weedy:	**Lepidium**

Cakile Mill.

Succulent, glabrous, annual or biennial herbs, the branched stems often decumbent; leaves entire to pinnatifid. Flowers in rather long racemes, the pedicels short, becoming thickened in fruit; sepals erect, the outer ones obtuse and somewhat hooded at the apex; petals clawed and white, pink or purple. Ovary cylindrical, sessile, articulate near the middle, each joint containing 1 ovule; stigma depressed-capitate, narrower than the style.

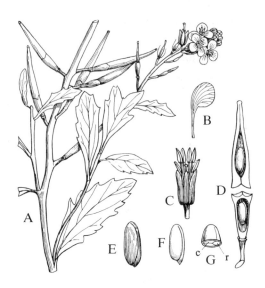

FIG. 121 **Cakile lanceolata**. A, portion of plant, × ²/₃ B, petal, × 2. C, flower without petals, × 2. D, fruit with the joints separated, slightly enlarged. E, seed, × 2. F, embryo, × 2. G, seed cut across, × 2; c, cotyledons; r, radicle. (F. & R.)

Capsule indehiscent, consisting of two 1-seeded joints, the upper terminating in a beak and easily separating from the lower; seeds rather large, somewhat rugulose.

— A widely distributed genus of perhaps 4 species, chiefly on seashores.

Cakile lanceolata (Willd.) O. E. Schulz in Urban, *Symb. Ant.* 3: 504 (1903). **FIG. 121**. PLATE 24.

Cakile maritima of some authors, not Scop.

With characters of the genus; leaves petiolate, linear-oblanceolate to oblong-elliptic, 3–7 cm long, entire, variably toothed or undulate. Petals white, 6–8 mm long. Fruit narrowly spindle-shaped, mostly 1.8–2 (–3) cm long, 2–4 mm thick, the lower joint cylindrical, the upper much longer and narrowly conical-subacuminate.

GRAND CAYMAN: *Kings GC 271; Millspaugh 1308; Proctor 15052*. LITTLE CAYMAN: *Kings LC 91*. CAYMAN BRAC: *Kings CB 56; Millspaugh 1159; Proctor 28922*.

— Florida, West Indies, and the Caribbean coasts of Central and S. America, on sandy seashores.

Lepidium L.

Annual, biennial, or perennial herbs, rarely suffutescent. lower leaves petiolate, pinnately incised to tripinnate, rarely entire. Flowers in dense, non-bracteate racemes that often elongate in fruit; sepals not saccate; petals small, white (rarely pink or yellow), rarely absent; stamens 6, 4, or 2, the filaments neither appendaged nor toothed; ovary 2-ovuled; style short or absent, the stigma capitate. Siliculas orbicular, ovate, elliptic or oblong, strongly compressed, dehiscent. Seeds solitary or rarely 2 per locule, pendulous, mucilaginous.

— A cosmopolitan genus of about 175 species, distributed mainly in temperate regions.

Lepidium virginicum L., *Sp. Pl.* 2: 645 (1753).

Annual or biennial herb usually 20–90 cm tall; stems erect, often many-branched above the spathulate in outline, often deeply pinnatifid; lobes incised to dentate; uppermost leaves lanceolate to linear, incised to dentate or entire. Sepals ca. 1 mm long, glabrous or sparsely pubescent. Petals white, spathulate, 1–3 mm long, rarely absent. Stamens 2, rarely 4; nectar glands 4. Fruiting pedicels slender, 4–10 mm long, spreading to horizontal, glabrous or pubescent. Siliculas orbicular to broadly elliptic-ovate, 2.5–4 mm long, shallowly notched at apex; seeds 1.5 mm long.

GRAND CAYMAN: *Penny Clifford* s.n., Jan. 1994 (CAYM). Found near iguana-breeding cays at MRCU, George Town.

— Native to N. America; introduced to all other regions except Australia.

[MORINGACEAE

Trees with pungent roots, the bark exuding gum; leaves alternate, 3–4-pinnately compound, the divisions opposite, the ultimate leaflets entire; stipules none or reduced to glands. Flowers perfect, irregular, in axillary panicles; sepals 5, unequal, united at base into a short, cup-like tube; petals 5, the upper two smaller. Stamens 10, 5 perfect ones alternating with 5 sterile, all inserted on the edge of a disc lining the calyx-tube; filaments free; anthers 1-celled. Ovary superior, stalked, 1-celled with 3 parietal placentas; style tubular, open at the apex; ovules numerous. Fruit a many-seeded elongate capsule opening by 3 valves; seeds large, 3-winged or wingless; embryo without endosperm.

— A small family of 1 genus and 10 species, indigenous to Africa, Arabia and India.

Moringa Adans.

With characters of the family.

Moringa oleifera Lam., *Encycl. Meth. Bot.* 1: 398 (1785).

MARONGA, HORSERADISH TREE

A small tree to 6 m tall or more, with whitish bark, the branchlets and leaves usually puberulous. Leaves to 40 cm long, with numerous leaflets, these 1–2 cm long. Flowers numerous, fragrant, the petaloid sepals 9–13 mm long, white tinged with crimson; petals slightly larger and white to yellowish, tinged crimson near the base outside. Capsule obtusely 3-angled, pendent, 20–45 cm long, 1–2 cm thick; seeds broadly 3-winged, 2.5–3 cm long.

GRAND CAYMAN: *Hitchcock*; *Kings* GC 231.

— Native of eastern Africa but widely planted and naturalised in tropical America. The roots have the flavour of horseradish, for which they can be used as a substitute. The young leaves, pods, and flowers can be cooked as an excellent vegetable. The seeds yield 'ben oil', used for lubricating watches and other delicate machinery.]

SAPOTACEAE

Trees or shrubs, often with milky sap (latex); pubescence often of 2-branched hairs; leaves mostly alternate, simple, entire and without stipules. Flowers rather small, usually perfect, solitary or clustered in the leaf-axils; sepals 4–12, imbricate; corolla 4–8-lobed but usually the lobes 5, often externally appendaged between the lobes. Stamens inserted on the corolla-tube opposite the lobes and equal to them in number, frequently alternating with more or less petaloid staminodes; filaments distinct; anthers 2-celled. Ovary superior, 4–14-locular; style simple; ovules solitary in each locule. Fruit usually a berry or drupe, sometimes of large size; seeds large, smooth and shining, with a rather conspicuous basal or lateral scar (hilum); endosperm present or lacking, usually scanty; embryo large, with broad foliaceous cotyledons.

— A chiefly pantropical family of between 35 and 75 ill-defined genera and about 800 species. Many are important timber trees, while several are valuable for their edible fruits. The latex of several C. American species is an important ingredient of chewing-gum.

KEY TO GENERA

1. Sepals imbricate or spiralled in 1 series; appendages of the corolla-lobes lateral or absent; fruits smooth, not mealy roughened:

  2. Staminodes present; leaves not persistently silky-hairy beneath:

    3. Fruit less than 3 cm long; seed-scar basal or nearly so: Sideroxylon

    3. Fruit over 4 cm long; seed-scar lateral, extending nearly the whole length of the seed: [Pouteria]

  2. Staminodes absent; leaves persistently silky-hairy beneath: [Chrysophyllum]

1. Sepals in two series (usually 3+3); corolla-lobes often with paired dorsal appendages; fruit mealy roughened, the seeds with lateral scar: Manilkara

Sideroxylon Jacq.

Unarmed or spiny trees or shrubs. Stipules absent. Leaves spirally arranged or rarely opposite, often fascicled on short lateral shoots. Flowers axillary or at leafless nodes. Flowers solitary of fasciculate, sessile or rarely pedunculate, bisexual or rarely unisexual; calyx a single whorl of 5(–8) free sepals; corolla cyathiform, glabrous, the tube nearly always shorter than the lobes; lobes usually 5, more or less imbricate, spreading, entire or divided into a larger median segment and two smaller lateral segments. Stamens usually 5 (rarely 4 or 6) in a single whorl attached to the top of the corolla tube, exserted, with well-developed filaments; anthers extrorse, usually glabrous. Staminodes usually 5 (rarely 4 or 6), usually well-developed, alternating with the stamens. Ovary (1–)5(–8) locular, hairy or glabrous; style exserted or included; usually only 1 ovule maturing to form a seed in each fruit. Fruit somewhat fleshy; seed with nearly basal scar.

— A tropical genus of 49 neotropical species and 20 others in Africa, Madagascar, and the Mascarene Islands.

KEY TO SPECIES

1. Unarmed trees; leaves up to 11 cm long:

    2. Leaves with minute inrolled pocket at base of blade on upper side; fruits more than 10 mm in diameter: — **S. foetidissimum**

    2. Leaves flat on both sides at base; fruits 5–6 mm in diameter: — **S salicifolium**

1. Spiny shrub; leaves usually less than 2 cm long: — **S. horridum**

Sideroxylon foetidissimum Jacq., *Enum. Syst. Pl. Carib.* 15 (1760). **FIG. 122.** PLATE 24.

MASTIC

A glabrous, medium-sized tree often 10–15 m tall or more; leaves mostly 6–11 cm long, long-petiolate, the blades broadly elliptic or rotund. Flowers strongly scented, cream or yellowish, the sepals 1.3–2.1 mm long, broadly rounded; corolla 3.5–5 mm long, the filaments and lacerate or toothed staminodes somewhat shorter. Fruit yellow, ellipsoid or subglobose, usually 1.5–2.5 cm long (rarely more), with thin but fleshy pulp covering the single large seed.

GRAND CAYMAN: *Proctor* 15002.

— Florida and the West Indies as to typical variety; a related form occurs in Mexico and C. America. This is a species of dry rocky woodlands; its wood is hard, heavy, strong and durable, and is used for construction, boats and furniture. It has become uncommon or rare through much of its range because of over-cutting and the total neglect of replanting.

Sideroxylon salicifolium (L.) Lam, *Tab. Encycl.* 2: 42 (1794). **FIG.123.** PLATE 24.

Bumelia salicifolia (L.) Sw. (1788).
Dipholis salicifolia (L.) A. DC. (1844).

WILD SAPODILLA, WHITE BULLET

A small unarmed tree to 8 m tall or more, the youngest parts clothed with fine reddish-brown hairs; leaves oblanceolate to narrowly or rather broadly elliptic, 5–11 cm long, with petioles mostly 1 cm long or less. Flowers numerous in clusters at defoliated nodes or sometimes in leaf-axils, on pedicels 1–4 mm long; sepals 1.4–3 mm long, finely hairy; corolla 3.3–4.5 mm long, the lobes narrowed to a claw-like base. Staminodes 1.5–2 mm long, erose-lacerate. Ovary glabrous or slightly hairy. Fruit black, subglobose, 6–10 mm in diameter, 1-seeded or rarely with 2 or 3 seeds.

GRAND CAYMAN: *Brunt* 2161; *Proctor* 27988. LITTLE CAYMAN: *Proctor* 28115, 35173, 35217. CAYMAN BRAC: *Proctor* 15322.

— Florida, West Indies and C. America, frequent in rocky woodlands. The wood is said to be hard, very heavy, strong, tough, and moderately durable.

Sideroxylon horridum (Griseb.) Pennington, *Fl. Neotrop.* 52, figs. 21, 22c (1990). **FIG. 124.** PLATE 25.

Bumelia glomerata Griseb. (1862).

SHAKE HAND, WHITE THORN

J.C.W.

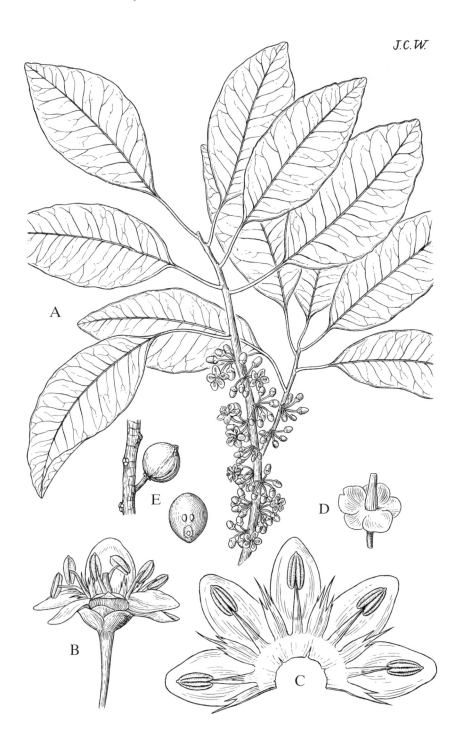

FIG. 122 **Sideroxylon foetidissimum.** A, branch with leaves and flowers, × ⅓. B, flower, × 4. C, corolla spread out, × 6, showing anthers and staminodes. D, calyx and pistil, × 4. E, two views of fruit, × ⅔, × 1. (J.C.W., ex St.)

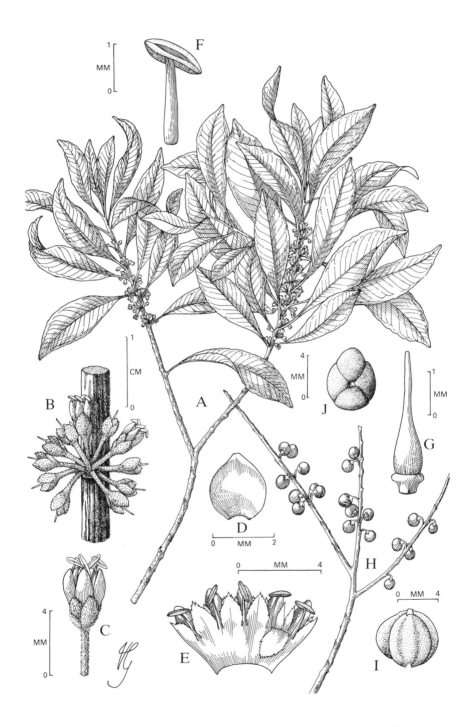

FIG. 123 **Sideroxylon salicifolium**. A, branch with leaves and flowers. B, single cluster of flowers. C, flower. D, sepal (inner side). E, corolla spread out to show stamens and staminodes; one staminode bent down. F, stamen. G, pistil. H, fruiting branch I, lateral view of fruit. J, basal view of fruit. (G.)

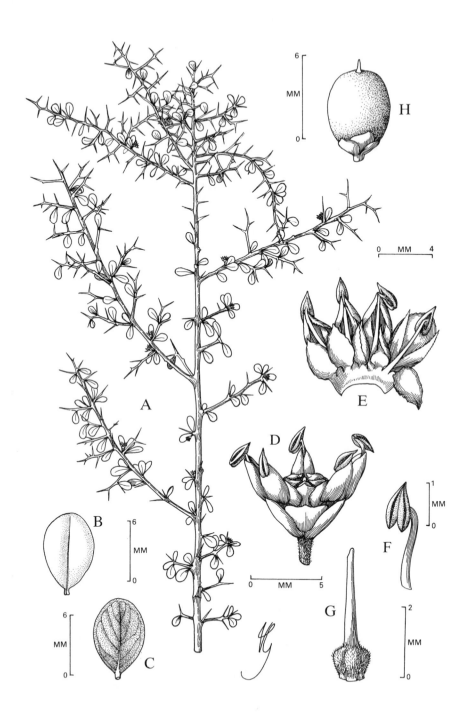

FIG. 124 **Sideroxylon horridum**. A, branch with leaves and flowers. B, leaf, dorsal (upper) surface. C, leaf, ventral (lower) surface. D, flower. E, corolla spread out to show stamens and staminodes; one staminode bent down. F, stamen. G, pistil. H, fruit. (G.)

A spiny shrub, nearly prostrate or erect and up to 3 m tall or more, the youngest parts minutely reddish-strigillose, otherwise glabrous. Leaves oblanceolate to broadly obovate or suborbicular, rounded at the apex, the veins (except the midrib) generally not visible. Flowers 1–4 together, subsessile; sepals 1.1–2.2 mm long; corolla 3.1–3.8 mm long; stamens included, with filaments ca. 1 mm long. Staminodes ovate to reniform, 1–1.3 mm long. Fruit subglobose, green or dark red, mostly 5–9 mm in diameter.

GRAND CAYMAN: *Brunt* 1781, 1782; *Kings* GC 326. LITTLE CAYMAN: *Proctor* 28089. CAYMAN BRAC: *Kings* GC CB 38.

— Cuba and Haiti, in dry rocky thickets and woodlands. This is perhaps the most obtrusive of several unrelated spiny shrubs known as 'shake hand' in the Cayman Islands, the others being *Xylosma bahamense* (Salicaceae) and *Zanthoxylum coriaceum* (Rutaceae). The three can usually be distinguished easily by differences in the leaves, those of *Sideroxylon* being simple and entire, and minutely reddish-hairy beneath when young; those of the *Xylosma* being simple and often with 1–4 blunt teeth, and always glabrous; while those of *Zanthoxylum* are pinnately compound, the leaflets being very shiny and with prominulous venation.

[*Pouteria* Aubl.

Trees with alternate or occasionally subopposite leaves, these with primary lateral veins widely apart, strongly developed, and distinctly upcurved toward the margins. Flowers solitary or few together in the leaf-axils, or sometimes at defoliated nodes; sepals 4–12, paired when 4, otherwise imbricate or spiralled; corolla usually 4–6-lobed. Stamens attached at base of corolla or else to near the level of the corolla-tube sinuses; staminodes petaloid or not, rarely absent. Ovary more or less hairy, 1–10-celled; ovules laterally attached. Fruit usually fleshy, rarely hard, containing 1–6 or more seeds, these with a long lateral scar; endosperm absent.

— A tropical American genus of 50 or more species.

Pouteria campechiana (Kunth) Baehni, *Candollea* 9: 398 (1942).

EGG FRUIT

Tree to 15 m tall or more (often smaller), nearly glabrous except for minute reddish hairs on the growing tips; leaves oblanceolate to narrowly obovate, mostly 5–18 cm long, acuminate at apex. Flowers solitary or several together on pedicels averaging 10 mm long; sepals 4–6, minutely pubescent outside, the margins ciliolate; corolla greenish-cream, 7–12 mm long, with 4–7 lobes. Stamens attached near level of sinuses, the filaments 1.3–2 mm long; staminodes linear or lanceolate, 2.4–2.9 mm long. ovary 4–10-celled. Fruit more or less pear-shaped, 4–7 cm long, with pointed apex and yellow edible flesh; seeds 4–7 (rarely less), 2–4 cm long, lustrous brown with elongate lateral scar.

GRAND CAYMAN: *Brunt* 2047, 2048; *Lewis* 2868; *Proctor* 15114, 15235.

— C. America; occurs as an escape from cultivation on the Florida Keys, the Bahamas, and Cuba. There is a tradition on Grand Cayman that this tree was introduced from C. America and cultivated for its edible fruits. However, it now occurs as a component of some of the natural coastal woodlands, and in such areas cannot be distinguished from the indigenous flora.]

[*Chrysophyllum* L.

Shrubs or small to medium-sized trees; leaves alternate, with secondary lateral veins nearly parallel to the primary ones. Flowers solitary to numerous in the leaf-axils; sepals usually 5, nearly free; corolla usually 5-lobed. Stamens arising from near the level of the sinuses, often connected by a slightly thickened ring in the corolla-throat; staminodia nearly always absent. Ovary 4–12-celled; ovules attached laterally or basilaterally. Fruit fleshy, with 1–5 or more seeds; seeds with large lateral or basilateral scar; endosperm copious.

— A tropical, chiefly American genus of about 150 species.

Chrysophyllum cainito L., *Sp. Pl.* 1: 192 (1753).

STAR APPLE

A small to rather large tree, the leaves densely reddish-sericeous beneath, glabrous and shining green on the upper side. Fruit green or purple, commonly globose or nearly so, 3–10 cm in diameter, several-seeded, edible; seeds with lateral scar.

GRAND CAYMAN: *Kings* GC 358, collected from a cultivated plant in George Town. The species does not occur wild in the Cayman Islands.

— West Indies; widely cultivated and naturalised both in the West Indies and continental tropical America, so that its original natural range would be difficult to determine.]

Manilkara Adans.

Trees with alternate leaves often of hard, leathery texture, the primary lateral veins parallel, straight, and often obscure. Flowers perfect or rarely unisexual, solitary or few together in the leaf-axils; sepals biseriate, commonly 6 (3, 3), occasionally 8 or 4; corolla with lobes as many as the sepals each with a pair of dorsal more or less petaloid appendages at the base, sometimes partly or wholly fused to the lobes, or rarely vestigial. Stamens as many as the corolla-lobes and opposite them, usually alternating with an equal number of staminodes. Ovary 6–14-celled; ovules attached laterally. Fruit often mealy roughened, several-seeded; seeds with long lateral scar; endosperm copious.

— A pantropical genus of about 35 species.

Manilkara zapota (L.) van Royen, *Blumea* 7: 410 (1953).

Achras zapota L. (1759), not L. (1753).
Sapota achras Mill. (1768).
Manilkara zapotilla (Jacq.) Gilly (1943).

NASEBERRY, a corruption of the Spanish 'nispero', a name originally applied to the European Medlar-tree (*Mespilus germanica* L.), whose fruits those of *Manilkara* resemble.

A medium-sized to large tree with copious white latex; leaves clustered toward the ends of the branchlets, elliptic or nearly so, mostly 5–12 cm long, loosely reddish-tomentose beneath when young, soon glabrate; petioles mostly 1–2 cm long. Flowers solitary in the leaf-axils, the pedicels about equal in length to the petioles; sepals 6–10 mm long, minutely pubescent; corolla 5–13 mm long, the lobes entire or toothed at the apex. Staminodes petaloid and similar to the corolla-lobes. Ovary 9–12-celled, densely short-hairy. Fruit brown, mealy roughened, and varying from ellipsoid to ovoid or subglobose, up to 10 cm in diameter, edible; seeds strongly flattened, 16–23 mm long.

GRAND CAYMAN: *Kings* GC 347; *Proctor* 15088. LITTLE CAYMAN: *Proctor* 28148. CAYMAN BRAC: *Proctor* 28967.

— Florida, West Indies and C. America, often planted and naturalised, so that its true natural range would be difficult or impossible to ascertain. It is doubtfully native in the Cayman Islands, although it can be found growing in woodlands remote from habitations. The fruit is one of the most important in regions where it grows. The very heavy light red wood is hard and durable, and is sometimes used to make furniture. This is one of the species used as a source of 'chiclé' in C. America.

THEOPHRASTACEAE

Trees or shrubs; leaves evergreen and mostly of stiff, hard texture, alternate, opposite or whorled, and often gland-dotted; stipules absent. Flowers regular, perfect or unisexual, solitary or in axillary or terminal clusters or racemes. Calyx 5-parted, with obtuse, overlapping lobes. Corolla 5-lobed, the tube bell-shaped or cylindrical. Stamens 5, borne near the base of the corolla-tube and opposite the lobes, alternating with 5 staminodes. Ovary superior, 1-celled, consisting of 5 fused carpels; ovules numerous on a basal placenta; style rather short, with a capitate or discoid stigma. Fruit a hard or fleshy indehiscent berry, few- or several-seeded; seeds more or less flattened; endosperm cartilaginous.

— A tropical American family of 5 genera and about 60 species. Some of them are extremely ornamental, but they are not often cultivated.

Jacquinia L.

Shrubs or small trees, often with pale or whitish bark; leaves scattered or whorled, short-petioled, usually gland-dotted. Flowers perfect, mostly in terminal or axillary racemes, rarely umbellate or solitary; pedicels often thickened upwardly; sepals united at the base; corolla with spreading lobes alternating with petaloid staminodes. Stamens attached to a fleshy ring at base of the corolla-tube; staminodes inserted near top of the corolla-tube. Fruit ovoid or globose, of rather hard texture and relatively few-seeded.

— A genus of about 30 species, widely distributed in the West Indies and continental tropical America.

KEY TO SPECIES

1. Petioles glabrous:

    2. Leaves mostly less than 4 cm long; corolla-lobes ca. 2.5 mm long; corolla light-yellow: **J. proctorii**

    2. Leaves usually more than 5 cm long; corolla-lobes 3–4 mm long; corolla white: **J. arborea**

1. Petioles glandular-puberulous: **J. keyensis**

Jacquinia proctorii Stearn, *Nord. J. Bot.* 12: 234–236 (1992). PLATE 25.

WASH-WOOD

A much-branched, dense-crowned glabrous shrub 1.5–3 m tall, the young twigs covered with minute granulate scales. Leaves narrowly to broadly obovate 1.5–4 cm long, flat or somewhat bullate, the margins narrowly revolute, the apex acutish or retuse, sometimes mucronate. Inflorescence usually shorter than the leaves, 1.5–2.5 cm long; bracteoles

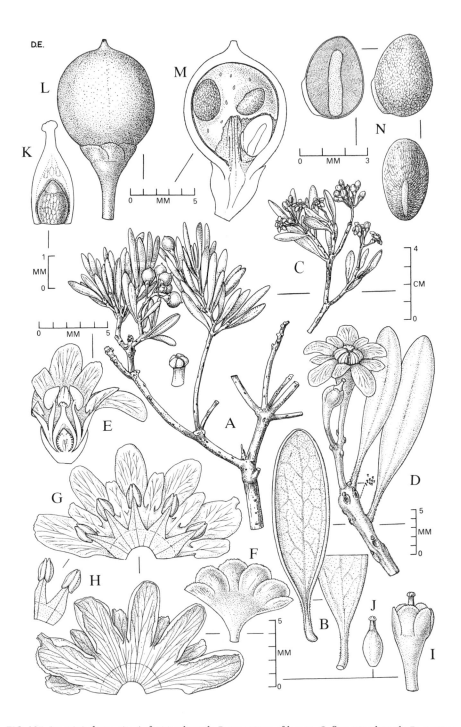

FIG. 125 **Jacquinia keyensis**. A, fruiting branch. B, two views of leaves. C, flowering branch. D, portion of same, enlarged. E, long. section of flower. F, calyx spread out. G, two views of corolla spread out. H, two stamens. I, calyx with pistil enclosed. I, pistil K, long. section of pistil. L, fruit. M, same, long. section. N, three views of seed. (D.E.)

appressed to bases of the pedicels, ca. 1 mm long, brown-tipped; pedicels 3–6 mm long. Flowers odorous; calyx ca. 2 mm long, with orbicular lobes; corolla light yellow, with lobes reflexed and ca. 2.5 mm long. Fruit yellowish-green, 4–5 mm in diameter.

GRAND CAYMAN: *Kings GC 234*; *Proctor 49359, 49364*. LITTLE CAYMAN: *Proctor 28105, 28184, 28189, 35077*. CAYMAN BRAC: *Proctor 15328, 52161, 52162*.

— Jamaica.

Jacquinia arborea Vahl, *Eclog. Amer.* 1: 26 (1796).

Jacquinia barbasco (Loefl.) Mez (1903), in part.

WASH-WOOD

A shrub or small tree to 5 m tall or more, the young branchlets minutely whitish-scurfy; leaves mostly whorled at the upper nodes, narrowly to broadly obovate, 4–10(–13) cm long, the very short glabrous petioles usually orange. Flowers fragrant, in terminal racemes usually a little shorter than the leaves, the axils 2.5–4 cm long, minutely glandular-puberulous in the axils of the pedicels; bracteoles whitish, ca. 2 mm long; pedicels 7–12 mm long. Calyx 3–4 mm long; corolla white, the lobes ca. 3.5 mm long. Fruit bright red, 9–10 mm in diameter; seeds 4.5 mm long.

GRAND CAYMAN: *Kings GC 334*.
— West Indies, frequent in coastal thickets.

Jacquinia keyensis Mez in Urban, *Symb. Ant.* 2: 444 (1901). **FIG. 125.** PLATE 25.

WASH-WOOD

A shrub or small tree up to 5 m tall or more, with smooth grey bark, the young branchlets minutely whitish-scurfy. Leaves alternate or crowded, oblong to obovate, mostly 2–5 cm long, often bullate-revolute, the apex obtuse or retuse, the very short petioles densely glandular-puberulous. Inflorescence equal in length to or usually longer than the leaves, the axis 2–8 cm long; bracteoles 0.5 mm long, blackish and bullate with a pale triangular apex; pedicels 5–8 mm long. Flowers fragrant; calyx ca. 2 mm long; corolla cream, the lobes 3–4 mm long. Fruit orange, 7–9 mm in diameter; seeds 2.5 mm long.

GRAND CAYMAN: *Brunt 1885, 2102*; *Kings GC 172*; *Proctor 27965*.
— Florida, the Bahamas, Cuba, and Jamaica, in coastal scrub-lands and thickets.

MYRSINACEAE

Trees or shrubs, glabrous or pubescent; leaves alternate, simple, entire or with crenulate or serrate margins, the surfaces often gland-dotted; stipules none. Flowers perfect, or unisexual and dioecious, regular, with parts in 4s or 5s in terminal or axillary clusters, racemes, or panicles. Calyx segments free or somewhat connate, often ciliate and gland-dotted; corolla with segments united at least toward the base. Stamens as many as the corolla-lobes, shorter than these and opposite them, almost free or inserted on and somewhat connate with the corolla-tube; anthers opening by slits or apical pores; staminodes none. Ovary superior, sessile, 1-celled, with a free central placenta; ovules numerous or few; style short or long; stigma simple and variously capitate, lobed, or fimbriate. Fruit a 1-seeded drupe; seed with copious endosperm.

— A pantropical family of about 30 or 40 rather weakly differentiated genera and 900 species, few if any of economic value.

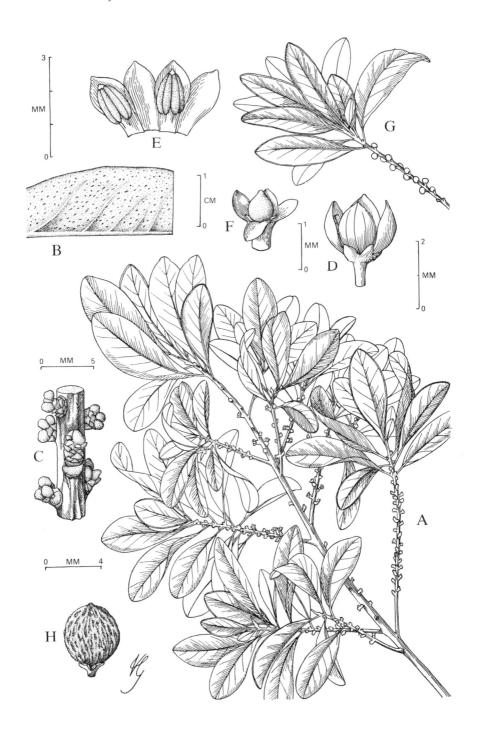

FIG. 126 **Myrsine acrantha**. A, branch with leaves and flowers. B, portion of leaf showing lower (ventral) surface. C, portion of stem showing clusters of buds in glomerules. D, staminate flower. E, corolla of staminate flower, spread out, showing stamens. F, pistillate flower. G, fruiting branch. H, fruit. (G.)

Myrsine L.

Shrubs or trees, glabrous or pubescent; leaves petiolate, entire or nearly so. Flowers small, dioecious, 4–5-parted, densely clustered on short, spur-like racemes or glomerules in the leaf-axils or at defoliated nodes. Sepals connate at base; corolla-lobes usually united below, rarely free, usually dark-lined, often papillose on the margins. Stamens inserted on throat of the corolla, the filaments obsolete, the anthers sessile, opening by slits, producing no pollen in pistillate flowers. Ovary globose or ellipsoid, in pistillate flowers the sessile stigma relatively large and subcapitate or variously lobed; ovules few, uniseriate. Fruit dry or fleshy, with a hard endocarp; seed globose and smooth; embryo curved, transverse.

— A pantropical genus of more than 200 species.

Myrsine acrantha Krug & Urban, *Notizbl. Bot. Gart. Berlin* 1: 79 (1895). **FIG. 127.** PLATE 26.

Rapanea acrantha (Krug & Urban) Mez (1901).

A glabrous shrub or small tree to 7 m tall or more; leaves oblong or narrowly obovate to obovate, mostly 5–8(–12) cm long and usually over 2.5 cm broad, blunt or minutely notched at the apex, the lower surface minutely gland-dotted. Flowers 4-parted, in dense clusters of 3–7 on lateral spurs shorter than the petioles, mostly along naked branches below the foliage; corolla-lobes elliptic, free to the base. Fruits ca. 5 mm in diameter, greenish with longitudinal glandular markings.

GRAND CAYMAN: *Brunt* 1977; *Proctor* 11997.

— Jamaica and perhaps Cuba, at medium to rather high elevations. The Cayman plants grow at or near sea-level in dry, rocky woodlands.

[CRASSULACEAE

Succulent annual or perennial herbs, sometimes suffrutescent; leaves opposite, simple or sometimes compound, usually very thick; stipules absent. Flowers regular, perfect, usually in cymes or panicles; sepals 4 or 5, united below; corolla 4–5-lobed, the lobes spreading. Stamens as many as or twice as many as the corolla-lobes, the anthers dehiscent by longitudinal slits. Ovary superior, with as many carpels as sepals, distinct or partly united, with a scale at the base of each; ovules numerous, in 2 series along the ventral suture of the carpel. Fruit consisting of 4 or 5 1-celled follicles containing numerous small or minute seeds; seeds with fleshy endosperm; embryo terete, with short, obtuse cotyledons.

— A widespread family of at least 33 genera and more than 1,000 species, in this hemisphere best represented in Mexico.

Bryophyllum Salisb.

Erect, often large perennial herbs; leaves simple or pinnate with a terminal leaflet. Flowers nodding, in paniculate, many-flowered cymes; calyx bell-shaped, inflated, 4-toothed; corolla more or less urn-shaped, with 4 short lobes. Stamens 8, inserted on the corolla-tube at or below the middle; anthers borne at or near the mouth of the corolla-tube. Carpels 4, free or partly united; fruit of 4 follicles.

— A genus of about 20 species, all native to Madagascar, one of them (*B. pinnatum*) often cultivated and widely naturalised in tropical countries.

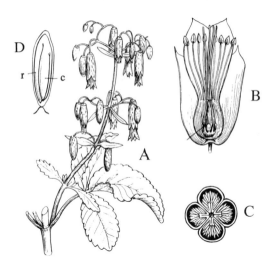

FIG. 127 **Bryophyllum pinnatum**. A, inflorescence and leaf. × ¹⁄₆. B, flower cut lengthwise, × 1; sc, scale. C, cross-section of ovary, × 3. D, seed, × 40; c, cotyledon; r, radicle. (F. & R.)

Bryophyllum pinnatum (Lam.) Oken in *Allgem. Naturgesch.* 3 (3): 1966 (1841). **FIG. 127**.

LEAF-OF-LIFE, CURIOSITY PLANT

Fleshy, glabrous herb to 1 m tall; lower leaves simple, the upper pinnate, the leaves or leaflets oblong to elliptic and up to 14 cm long or more, with coarsely crenate margins, often bearing plantlets in the notches. Calyx greenish-yellow, the tube 2.5–3.5 cm long; corolla reddish, 2.5–4.5 cm long, the tube constricted below the middle, the lobes acute and 1–1.5 cm long. Stamens attached to the constriction of the corolla-tube. Carpels 1.2–1.5 cm long.

GRAND CAYMAN: *Hitchcock*; *Kings* GC 35a, GC 97a, GC 117, GC 146; *Millspaugh* 1311. LITTLE CAYMAN: *Proctor* 28178. CAYMAN BRAC: *Kings* CB 69; *Proctor* 29029.

— Naturalised throughout the tropics. The leaves are sometimes juiced and with added salt taken as a cold remedy; it has a very unpleasant taste. An allegedly medicinal tea is also made from the leaves.]

CHRYSOBALANACEAE

Trees or shrubs with alternate, simple, entire leaves; stipules present. Flowers perfect or rarely unisexual, usually more or less zygomorphic, in simple or compound racemes or cymes. Calyx 5-lobed from a turbinate or bell-shaped tube, more or less unequal or spurred at the base. Petals 5 or lacking, often unequal, short-clawed, inserted in the mouth of the calyx-tube. Stamens 2–many, inserted with the petals, often with larger fertile ones opposite the larger calyx-lobes; filaments slender, exserted. Ovary superior, sessile or more often stalked, 1-celled with 2 basal erect ovules, or rarely 2-celled, each cell with 1 ovule; style simple, lateral or almost gynobasic. Fruit a sessile or stalked drupe, with bony almost 2-valved endocarp, or sometimes a crustaceous berry; seed lacking endosperm; embryo with thick, fleshy cotyledons.

— A widespread tropical and subtropical family of 10 genera and about 400 species, the majority in tropical America. This family is sometimes included with the Rosaceae.

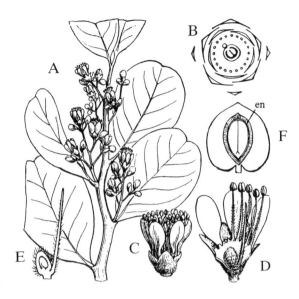

FIG. 128 **Chrysobalanus icaco.** A, portion of flowering branch, × ²/₃. B, floral diagram. C, flower, × 2. D, flower cut lengthwise and four stamens removed, × 3. E, pistil with ovary cut lengthwise, × 4. F, fruit cut lengthwise, × ²/₃; **en**, endocarp. (F. & R.)

Chrysobalanus L.

Shrubs or small trees; leaves glabrous, the stipules small and deciduous. Flowers small, white or cream, in terminal and axillary pubescent cymes; stamens 10–numerous, a few of them sterile. Ovary of 1 carpel, sessile at base of the calyx-tube; style attached laterally near base of the ovary; ovules 2. Drupe fleshy, 1-seeded.

— A genus of about 3 species, 2 of them American, the other in Africa.

Chrysobalanus icaco L., *Sp. Pl.* 1: 513 (1753). **FIG. 128**. PLATE 26.

COCO-PLUM

Usually a shrub or sometimes a small tree with glabrous branches; leaves roundish-elliptic or very broadly obovate, mostly 3–7 cm long, the apex blunt, rounded or retuse. Cymes paniculate, mostly shorter than the leaves; flowers finely woolly, the calyx-lobes 2.5 mm long, the petals ca. 5 mm long; stamens longer than the petals, with hairy filaments. Drupe purplish-red or blackish, more or less ellipsoid, 1.3–3 cm long, the pericarp fleshy and edible.

GRAND CAYMAN: *Brunt* 2053; *Hitchcock*; *Kings* GC 355; *Lewis* 3824; *Proctor* 15118, 15223. LITTLE CAYMAN: *Proctor* 28144. CAYMAN BRAC: *Proctor* 29004.

— Florida, West Indies, and on the continent from Mexico to northern S. America, common especially in sandy thickets near the sea. The fruit is edible but rather insipid. The seeds are said to be rich in oil, and the Carib Indians are alleged to have strung them on sticks and burned them like candles.

LEGUMINOSAE

Trees, shrubs or herbs of various habit, sometimes climbing, often armed with spines or prickles. Leaves alternate or rarely opposite, sometimes simple but mostly pinnately compound or often bipinnate, the leaflets few or numerous, small or large, entire, lobed, or rarely toothed; stipules always present, stipels sometimes so. Flowers mostly irregular and perfect, or regular and polygamous, the peduncles axillary or terminal, 1-many-flowered; pedicels solitary, paired, or clustered, usually in the axil of a bract; bracteoles usually 2 at the base of the calyx. Sepals free or more or less united, usually 5 or sometimes 4 in irregular flowers, 3–6 but most often 5 in regular flowers. Petals as many as the sepals, in irregular flowers one borne higher than the rest, either outside and enclosing the others in bud, or inside and enclosed by the others in bud. Stamens usually twice as many as the petals, rarely as many or fewer, or (in Mimosoideae) often more numerous, inserted on the margin of the receptacle, free or partly connate; anthers 2-celled, mostly opening by longitudinal slits. Ovary superior and usually elongate, sessile, 1-celled; ovules numerous or rarely 1, attached along the inner angle of the ovary in 1 or 2 series. Fruit a pod, usually dry but sometimes more or less pulpy within, rarely drupe-like, most often 2-valved but sometimes indehiscent, continuous within or septate; seeds 1–many, with a hilum or scar on one edge, attached along the upper valve of the pod; endosperm usually none or scanty, rarely copious; cotyledons flat and leaf-like or thick and fleshy.

One of the largest families of flowering plants and of world-wide distribution, with about 500 genera and nearly 13,000 species. Included are many of the most valuable members of the plant kingdom, variously useful for food or other purposes. Especially important are those with edible seeds, such as the peanut (*Arachis hypogaea* L.), congo or pigeon pea (*Cajanus cajan* (L.) Millsp.), soy bean (*Glycine max* (L.) Men.), kidney bean or red peas (*Phaseolus vulgaris* L.), and the pea or green peas (*Pisum sativum* L.); there are many others, too numerous to mention here. Numerous other species are useful for fodder, timber, fibre, dyes, gums and resins, and medicinal products.

The roots of most leguminous plants possess small nodules which contain minute bacterial organisms that are able to "fix", or chemically combine, atmospheric nitrogen with other elements. This process enriches soils in which leguminous plants grow.

Some botanists divide the group into three families, but their close relationship is so evident that it seems more logical to treat these taxa as subfamilies.

KEY TO SUBFAMILIES

1. Flowers more or less irregular (zygomorphic), often conspicuously so; stamens 10 or less; mostly shorter than the longest petals:

  2. Flowers pea-like, the uppermost petal outside the others and covering them in bud: SUBFAM. 1. FABOIDEAE

  2. Flowers not pea-like, the uppermost petal attached inside the others and covered by them in bud: SUBFAM. 2. CAESALPINIOIDEAE

1. Flowers regular (actinomorphic), the petals all equal; stamens 10 or usually more numerous, usually much more conspicuous than the petals: SUBFAM. 3. MIMOSOIDEAE

SUBFAMILY 1. FABOIDEAE

Herbs, shrubs or trees; leaves mostly pinnately or rarely palmately compound, sometimes simple or trifoliate, the leaves and leaflets mostly entire. Flowers irregular; sepals usually 5, partly united below into a cup; petals 5, overlapping, the uppermost (the standard) outside and enclosing the others in bud, the two lateral (the wings) more or less parallel to each other, the two lower innermost, generally parallel and united by their lower margin to form a keel. Stamens usually 10, the filaments partly united and sheath-like below, sometimes 9 of them so united and the uppermost free or absent.

— Among the Leguminosae, this subfamily is especially characteristic of temperate regions, though many species also occur in tropical countries.

KEY TO GENERA

1. Leaves simple or with one leaflet:

  2. Pods continuous, not jointed; plants erect or climbing, herbaceous or woody:

    3. Pod enveloped in a large membranous bract; shrub: **[Moghania]**

    3. Pod not enveloped in a bract:

      4. Erect herbs; pods inflated: **Crotalaria**

      4. Woody scrambler or an arching shrub; pods flat: **Dalbergia**

  2. Pods jointed, the joints separating:

    5. Leaflet longer than broad, usually less than 15 mm long and 10 mm broad (rarely larger); stems glabrous; calyx not inflated in fruit: **Alysicarpus**

    5. Leaflet much broader than long, 3–8 cm broad; stems with numerous minute hooked hairs; calyx inflated in fruit: **Christia**

1. Leaves compound:

  6. Leaves abruptly pinnate, without a terminal leaflet:

    7. Slender twining woody vine; seeds red with a black spot: **Abrus**

    7. Erect shrubs or trees; seeds brown: **Sesbania**

  6. Leaves pinnate with a terminal leaflet:

    8. Leaflets 3:

      9. Pods jointed, the joints separating:

        10. Stipules united to the petiole; leaflets without stipels; flowers yellow: **Stylosanthes**

        10. Stipules free from the petiole; leaflets subtended by stipels; flowers not yellow: **Desmodium**

      9. Pods not jointed (but sometimes constricted between the seeds):

        11. Erect trees or shrubs:

          12. Tree with orange or scarlet flowers; seeds not edible: **Erythrina**

          12. Shrub with yellow flowers (sometimes reddish-tinged); seeds commonly eaten: **[Cajanus]**

        11. Twining vines or suberect herbs:

          13. Standard more then 1 cm long:

            14. Standard different in size from the other petals:

              15. Standard flat, much larger than the other petals: **Centrosema**

              15. Standard folded, shorter than the other petals: **Mucuna**

            14. Standard about the same size as the other petals:

              16. Suberect herb; flower deep red; keel incurved, forming a complete spiral: **Phaseolus**

16. Twining vines; flower not deep red; keel not forming a complete spiral:
17. Style hairy:
18. Flowers yellow; stigma oblique or lateral; pod smooth: **Vigna**
18. Flowers white, purplish or violet; stigma terminal; pod with warty projections along the margins: **[Lablab]**
17. Style glabrous:
19. Calyx 4-lobed, the lobes equal and acuminate; pod not ridged: **Galactia**
19. Calyx 2-lipped, the lips more or less rounded, the upper larger than the lower; pod ridged longitudinally on each side near the upper margin: **Canavalia**
13. Standard less than 1 cm long:
20. Flowers white; pod more than 3 cm long: **Teramnus**
20. Flowers yellow; pod less than 2 cm long: **Rhynchosia**
8. Leaflets 5 or more:
21. Trees:
22. Flowers white or nearly so; all 10 stamens united into a sheath; pod broadly 4-winged: **Piscidia**
22. Flowers pink; 9 stamens united, 1 free; pod smooth and unwinged: **[Gliricidia]**
21. Herbs or shrubs:
23. Flowers yellow or whitish:
24. Pods terete but constricted between the seeds; filaments all free: **Sophora**
24. Pods flat, more or less indented on one side; filaments united: **Aeschynomene**
23. Flowers not yellow; stamens all united except the uppermost one; pod not constricted or indented between the seeds:
25. Twining vine; flowers blue: **[Clitoria]**
25. Erect herbs or subshrubs; flowers not blue:
26. Leaflets subtended by minute stipels; racemes axillary; pods subcylindrical: **Indigofera**
26. Leaflets without stipels; racemes terminal or opposite the leaves; pods flat: **Tephrosia**

Crotalaria L.

Herbs or shrubs; leaves simple or digitately 3–5-foliolate; stipules free from the petiole, sometimes decurrent on the stem, often small or none. Flowers yellow or sometimes blue or purplish, in racemes terminal or opposite the leaves, rarely solitary and axillary. Calyx with free lobes or the calyx rarely 2-lipped. Standard roundish, short-clawed; wings obovate or oblong, shorter than the standard; keel incurved, beaked. Stamens 10, united below into a 2-cleft sheath; anthers alternately small, versatile, and large, basifixed. Ovary with 2–many ovules; style incurved or abruptly inflexed above the ovary, minutely barbed above along the inner side. Pod globose or oblong, 2-valved, more or less inflated, continuous within.

— A widespread, chiefly tropical genus of more than 200 species. *C. juncea* L. or 'sunn hemp' is cultivated in some countries for its valuable fibre, usable as a substitute for flax and true hemp (*Cannabis*).

KEY TO SPECIES

1. Leaves simple:	
2. Stipules roundish and leaf-like; flowers lavender-blue; pods hairy:	**C. verrucosa**
2. Stipules minute or lacking; flowers yellow; pods glabrous:	**C. retusa**
1. Leaves 3-foliolate:	**C. incana**

Crotalaria verrucosa L., *Sp. Pl.* 2: 715 (1753). PLATE 26.

SWEET PEA

A low bushy annual less than 1 m tall, with 4-angled stems. Leaves simple, somewhat glaucous, ovate or roundish from an abruptly tapered base, 2–7 cm long, glabrous above, minutely pubescent beneath; stipules roundish-lunate and deflexed, up to 10 mm long. Racemes densely several–many-flowered; calyx 2-lipped, 7–9 mm long, about half as long as the corolla, the lobes narrowly triangular and acuminate; standard whitish or pale lavender-blue; keel violet or purple. Pod obovoid-oblong, beaked, 3–3.5 cm long, finely appressed-puberulous.

GRAND CAYMAN: *Brunt* 2017; *Kings* GC 307; *Proctor* 15080. LITTLE CAYMAN: *Proctor* 28064. CAYMAN BRAC: *Kings* CB 5, CB 20a; *Proctor* 28986.

— Pantropical, in pastures and along sandy roadsides.

Crotalaria retusa L., *Sp. Pl.* 2: 715 (1753). **FIG. 129.**

Tough-stemmed annual, the stems striate and appressed-puberulous; leaves simple, oblanceolate, 3–10 cm long, the apex rounded or retuse, glabrous above, minutely puberulous beneath, the tissue with numerous minute pellucid dots; stipules awl-shaped, ca. 1 mm long. Racemes elongate, many-flowered; calyx 2-lipped, ca. 12 mm long, about half as long as the corolla, the lobes triangular-acuminate; petals yellow, the standard up to 2.4 cm broad, tinged reddish on the back and with fine purplish lines within. Pod oblong, blackish, 3–4 cm long, glabrous.

GRAND CAYMAN: *Brunt* 1878; *Kings* GC 209.

— Pantropical, a common weed of sandy roadsides, pastures, and open waste ground. The stems are said to contain a useful fibre.

Crotalaria incana L., *Sp. Pl.* 2: 716 (1753).

Erect, shrubby annual up to 1 m tall, the stems, petioles and racemes densely brownish-hirsute; leaves long-petiolate, trifoliolate, the leaflets broadly obovate or roundish, 1–5 cm long, of thin texture, glabrous above, pale beneath and appressed-pilose especially along the midvein; stipules awl-shaped, small, deciduous. Racemes rather few-flowered; calyx 5-lobed, 7–9 mm long, with lance-acuminate lobes; corolla yellow, 10–13 mm long. Pod pendulous, oblong, 2.5–3 cm long, soft-hairy.

GRAND CAYMAN: *Hitchcock*; *Millspaugh* 1387.

— Pantropical, a weed of roadsides and open waste ground.

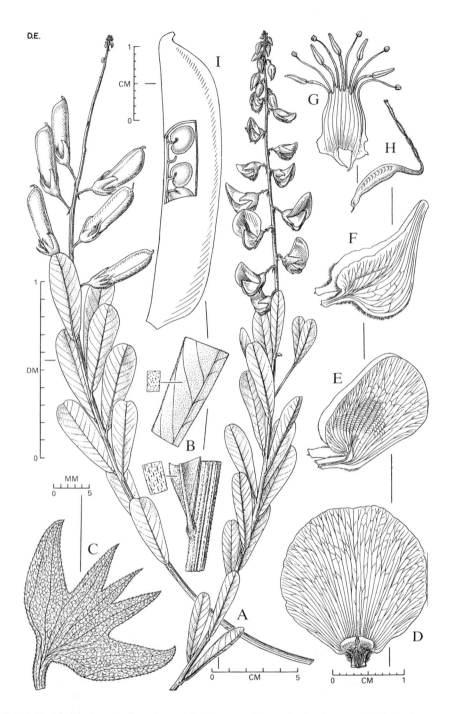

FIG. 129 **Crotalaria retusa**. A, flowering and fruiting stems. B, two details of stem and leaf. C, calyx spread out. D, standard. E, wings. F, keel. G, stamens, H, pistil. I, pod with portion cut away to show seeds. (D.E.)

Indigofera L.

Herbs or shrubs, often with strigose pubescence, the hairs attached by the middle; leaves pinnate with a terminal leaflet, rarely 3-foliolate or 1-foliolate; stipules small, subulate, slightly adnate to the petioles. Racemes axillary; calyx 5-lobed, the lobes sometimes unequal; standard broad or roundish, sessile or short-clawed, strigillose outside; keel usually spurred on each side near the base. Stamens 10, the uppermost free, the others united into a sheath. Ovary sessile; style bent upward, with capitate stigma; ovules 1–many. Pods narrow, subcylindrical, 2-valved but usually opening along the upper suture only, partitioned inside.

— A pantropical genus of more than 300 species, most abundant in Africa.

KEY TO SPECIES

1. Leaflets alternate:	[I. spicata]
1. Leaflets opposite:	
2. Pod sickle-shaped, less than 2 cm long, with 3–6 seeds:	I. suffruticosa
2. Pod straight or only slightly curved, 2–4 cm long, with 8–15 seeds:	I. tinctoria

[*Indigofera spicata* Forssk., *Fl. Aegypt.–Arab.* 138 (1775).

Prostrate to ascending herb from a thick rootstock, the stems to 60 cm long, flattened, pubescent. Stipules lanceolate, 6–8 mm long; leaflets 7–11, obovate or oblong-oblanceolate, obtuse or rounded at the apex, cuneate at base, 1.3–3 cm long, with rough appressed pubescence. Inflorescence up to 15 cm long in fruit; flowers crimson-red, 5 mm long in dense many-flowered clusters. Fruits reflexed, linear, straight, 1.1–2.5 cm long with thick margins, appressed-pubescent, 5–8-seeded; seeds 1.5 mm long, 1 mm thick.
GRAND CAYMAN: *P. Ann van B. Stafford* s.n.; found beside rugby pitch, Mary Read Crescent, South Sound area, 29 Aug. 2003.

— Widespread in Africa and Asia; introduced and naturalised in scattered areas of the neotropics.]

Indigofera suffruticosa Mill., *Gard. Dict.*, ed. 8, no. 2 (1768).

Indigofera anil L. (1771).

WILD INDIGO

A stiff, shrubby herb up to 1.5 m tall but usually lower, finely white-strigillose throughout; leaves with 9–15 leaflets, these elliptic or oval, 1.5–3 cm long, mucronate at the apex. Racemes shorter than the leaves; calyx ca. 1.5 mm long; corolla salmon-red, 5–6 mm long. Pods numerous and densely crowded, 1.5–2 cm long, strongly curved, 2 mm thick.

GRAND CAYMAN: *Hitchcock*; *Kings* GC 273.

— West Indies and continental tropical America, often a weed of sandy waste ground or along roadsides; naturalised in the Old World tropics. This plant was formerly much cultivated, especially in C. America, as the chief source of the blue dye known as indigo, a colour supplanted in modern times by synthetic substitutes.

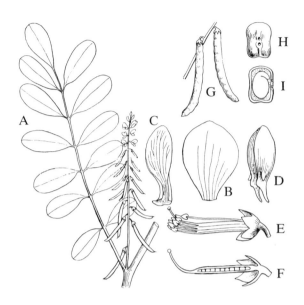

FIG. 130 **Indigofera tinctoria**. A, leaf and raceme, × ²/₃. B, standard, × 6. C, wing, × 6. D, keel, × 6. E, flower with petals removed, × 7. F, ovary and calyx cut lengthwise, × 7. G, ripe pods, × ²/₃. H, seed, × 3. I, seed cut lengthwise, × 3. (F. & R.)

[*Indigofera tinctoria* L., *Sp. Pl.* 2: 751 (1753). **FIG. 130**.

INDIGO

A shrubby herb up to 1 m tall, the stems sparsely white-strigillose, or densely so near the ends; leaves with 9–15 leaflets, these broadly obovate or oval, 1–2.5 cm long, mucronate at the apex, finely strigillose. Racemes shorter than the leaves; calyx ca. 1.5 mm long; corolla salmon-red, 5–6 mm long. Pods numerous, 2–4 cm long, slightly curved or straight, 1.5–2 mm thick, beaked at the apex.

GRAND CAYMAN: *Proctor* 27996. LITTLE CAYMAN: *Proctor* 35196.

— Native of southern India, but now naturalised in most warm countries; formerly cultivated as a source of indigo.]

Tephrosia Pers.

Annual or perennial herbs or sometimes shrubs; leaves pinnate with a terminal leaflet; leaflets few or often numerous, with many conspicuous lateral nerves beneath, these parallel and very oblique; stipules setaceous or broader, stipels absent. Flowers in rather few-flowered racemes, these terminal or opposite the leaves, rarely axillary; bracteoles absent. Calyx 5-lobed, the lobes subequal or the lower one longer, the upper 2 usually more or less joined. Petals clawed, the standard roundish and more or less sericeous outside. Stamens 10, the uppermost more or less free. Ovary sessile; style inflexed or incurved, usually glabrous; ovules numerous. Pods flat, 2-valved, beaked, many-seeded.

— A widely distributed genus of more than 125 species, commonest in tropical regions. Several species have been used as fish poisons, especially in South and C. America.

KEY TO SPECIES

1. Leaflets 9–15, narrowly oblanceolate, obtuse or acute; racemes with short peduncles; corolla 10–15 mm long: **T. cinerea**

1. Leaflets 5–9, more or less obovate, rounded or retuse at the apex; racemes with long peduncles; corolla 7–10 mm long: **T. senna**

Tephrosia cinerea (L.) Pers., *Syn. Pl.* 2: 328 (1807). PLATE 27.

Perennial herb, trailing, decumbent or ascending from a woody taproot, often much-branched, finely strigose; leaves with 9–15 leaflets, these mostly linear-oblanceolate or oblong-oblanceolate, 1–3 cm long, 2.5–5 mm wide, usually acutish and mucronate, whitish-strigose beneath; stipules lance-acuminate, 3–8 mm long. Racemes opposite the leaves, on peduncles mostly 1–3 cm long. Calyx ca. 5 mm long; corolla purple, 10–15 mm long. Pods 4–5 cm long, 3–5 mm broad, finely strigose, 6–12-seeded.

GRAND CAYMAN: *Hitchcock*; *Proctor* 15238, 52095.

— West Indies and continental tropical America, especially in sandy soils along roadsides and in open ground near the sea.

Tephrosia senna Kunth in H. B. K., *Nov. Gen.* 6: 458 (1824).

Tephrosia cathartica (Sesse & Moç.) Urban (1905).

An erect perennial herb, much-branched and finely strigose; leaves with 4–9 leaflets, these obovate or cuneate-oblong, 1–3.5 cm long, 4–16 mm wide, usually rounded or retuse with a small mucro in the notch, whitish-strigillose on both surfaces; stipules subulate, 7–10 mm long. Racemes opposite the leaves, on peduncles 3.5–7 cm long; flowers widely spaced. Calyx ca. 5 mm long; corolla rose-red or pale mauve, 7–10 mm long. Pods 2.5–4 cm long, 3–4 mm broad, finely strigose, 6–8-seeded.

CAYMAN BRAC: *Kings* CB 82; *Millspaugh* 1158; *Proctor* 29067.

— West Indies except Jamaica, also in the Yucatan area and Colombia, in sandy or loamy clearings and along borders of thickets.

[*Gliricidia* Kunth ex Endl.

Deciduous shrubs or small trees; leaves pinnate with a terminal leaflet; stipules small, stipels lacking. Flowers in axillary racemes, often appearing during the dry season when the plant is more or less leafless; bracteoles absent. Calyx shortly bell-shaped, truncate or 5-toothed. Corolla usually pink or whitish; standard roundish, reflexed, short-clawed, often with 2 callosities at the base of the blade within; wings oblong, free; keel curved, the petals clawed and free at the base, united at the apex. Stamens 10, 9 united and the uppermost free. Ovary stalked; style glabrous, bent at almost a right angle, the stigma capitate and papillose; ovules 7–12. Pod elongate, short-stalked, flat, 2-valved; seeds roundish, compressed.

— A tropical American genus of about 5 species.

Gliricidia sepium (Jacq.) Kunth ex Griseb., *Abh. Ges. Wiss. Götting. 7; Phys. Cl.* 52 (1857).

A small tree up to about 10 m tall, or sometimes shrub-like, branching from near the base; leaves with 7–17 leaflets, these lance-oblong, elliptic or ovate, 3–6 cm long, acutish at both ends, glabrate above and sparsely fine-hairy beneath; stipules ovate or lanceolate, 2 mm long. Racemes 5–10 cm long, often densely flowered; pedicels 5 mm long; calyx 4–5 mm long; corolla 1.5–2 cm long, usually light pink. Pods 10–12 cm long, 1–1.5 cm broad, glabrous; seeds dark brown, ca. 1 cm in diameter.

GRAND CAYMAN: *Proctor* 15234. CAYMAN BRAC: *Proctor* 29139.

— Central and northern S. America, commonly planted and becoming naturalised in the West Indies. Uncommon in the Cayman Islands. The wood is very hard, heavy, strong and durable, and takes a high polish. The leaves are apparently safely eaten by cattle, but all parts of the plant are poisonous to rats, mice and also dogs. Poultices of the fresh crushed leaves are said to have a healing effect on skin ulcers and sores.]

[*Sesbania* Scop.

Herbs, shrubs, or small trees; stipules soon falling. Leaves alternate, even-pinnate, the rhachis ending in a setaceous point; leaflets numerous. Inflorescence axillary, racemose, usually few-flowered; flowers yellow, purplish or variegated and of small to medium size, or else white or red and of large size; calyx broadly campanulate, 5-lobed; standard ovate to orbicular, spreading or reflexed; wings oblong, clawed, with a basal auricle; keel sharply incurved, clawed. Stamens geniculate near the base; ovary short-stipulate; ovules many; style incurved, glabrous, the stigma small, capitate. Fruit linear to subterete, beaked, transversely septate, dehiscent by 2 valves; seeds cylindrical-oblong, brown, smooth.

— A genus of 50 species widespread in the tropics and subtropics.

[*Sesbania sericea* (Willd.) Link, *Enum. Hort. Berol.* 2: 244 (1822).

Annual shrub 2–5 m tall with slender branches; plant silky-pubescent on stems and under surfaces of the leaves. Leaflets 12–25 pairs, linear-oblong, glabrous on upper side, 12–25 mm long, 4–7 mm broad. Racemes slender, usually 5–8-flowered; calyx 3–4 mm long; corolla ca. 8 mm long, yellow, the standard petal spotted blackish-purple; pod 10–20 cm long, ca. 3 mm broad.

GRAND CAYMAN: *Proctor* 52123, found in a brackish wetland along the road to Patrick Island.

— Native of Ceylon; naturalised in the neotropical region from the Bahamas to northern S. America.]

[*Sesbania grandiflora* (L.) Poir., a small tree probably native either to India or Australia, is sometimes cultivated for ornament; it is locally called picashia or Spanish Armada. It has short, 2-flowered racemes, curved buds, and large white to red flowers 7–8 cm long.

GRAND CAYMAN: *Kings GC* 147, GC 288.]

Stylosanthes Sw.

Mostly low herbs, annual or perennial, or sometimes small shrubs, often with viscid pubescence; leaves pinnately 3-foliolate, stipules united to the base of the petiole, stipels absent. Flowers small, yellow, in chiefly terminal spikes or heads, each flower nearly sessile in the axil of a 2-dentate or divided bract, the pedicel very short and adnate to the bract, sometimes accompanied by a bristle-like sterile flower. Calyx with a long, stalk-like tube and with the 4 upper lobes joined, the lower one narrower and distinct. Petals and stamens inserted at top of the calyx-tube; standard roundish; stamens 10, all united into a closed sheath, the anthers alternately long and short. Ovary subsessile; style hair-like; ovules 2 or 3. Pods compressed, 1–2-jointed, the joints reticulate or muricate, the apical one with a hooked beak.

— A genus of about 30 species, chiefly in tropical America, a few in tropical Africa and Asia.

Stylosanthes hamata (L.) Taub. in *Verh. Bot. Ver. Brandenb.* 32: 22 (1889). **FIG. 131.**

Stylosanthes procumbens Sw. (1788).

LUCY JULIA, PENCIL FLOWER

Perennial subwoody herb with prostrate or ascending slender stems reaching about 45 cm in length; leaflets lanceolate to elliptic, 5–15 mm long, 2–5 mm broad, glabrate but with finely ciliate margins, the nerves prominent beneath; stipules 6–8 mm long. Flowers subtended by leaf-like bracts; calyx-tube 3–4 mm long; corolla 4–5 mm long. Pods 7–10 mm long, with an apical beak about as long as one joint.

GRAND CAYMAN: *Brunt* 1875, 2086, 2121; *Hitchcock*; *Kings* GC 69; *Millspaugh* 1335; *Proctor* 15061. LITTLE CAYMAN: *Kings* LC 75; *Proctor* 28062. CAYMAN BRAC: *Proctor* 29095.

— Florida, West Indies, and C. America south to Colombia, often a weed of pastures and lawns, and to be considered beneficial in such situations.

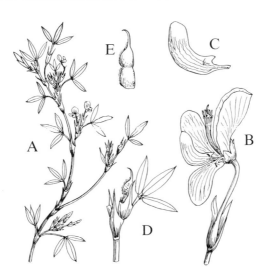

FIG. 131 **Stylosanthes hamata**. A, branch with leaves and flowers, × ²/₃. B, flower with one wing turned down, × 5. C, wing, inner face, × 5. D, portion of branch with pod, × 2. E, pod, × 2. (F. & R.)

Desmodium Desv.

Annual or perennial herbs, procumbent to erect, sometimes scandent or vine-like, often somewhat woody; pubescence often of hooked hairs. Leaves trifoliolate or rarely 1-foliolate, stipulate; leaflets with stipels. Flowers in terminal or axillary racemes or panicles, rarely solitary or in small clusters, usually pink, purple, or sometimes white. Calyx with short tube, 2-lipped, the upper lip with 2 teeth, the lower with 3 acute or attenuate lobes. Standard obovate or roundish; wings obliquely oblong, adhering to the keel; keel nearly straight, obtuse. Stamens 10, the upper sometimes more or less free; anthers all alike. Ovules 2 to many. Pod of 2–many flat or twisted 1-seeded joints, usually with hooked hairs, indehiscent and separating easily from each other.

— A cosmopolitan genus of more than 500 species. The fruits adhere tenaciously to clothing, feathers, or fur by means of their minute hooked hairs, and by this means are often widely dispersed.

KEY TO SPECIES

1. Leaves mostly 1-foliate:	[D. gangeticum]
1. Leaves mostly 3-foliate:	
2. Flowers solitary or in small clusters of 2–4; plants small, creeping, with leaflets less than 1 cm long:	D. triflorum
2. Flowers in more or less elongate racemes; stems ascending or erect; leaflets more than 1 cm long:	
3. Upper margin of pod continuous, not notched:	D. incanum
3. Upper and lower margins of pod equally notched:	D. tortuosum

[*Desmodium gangeticum* (L.) DC., *Prodr.* 2: 327 (1825).

Trailing or lax herb sometimes ascending to nearly 1 m tall. Leaflet up to 10 cm long or more and to 6 cm broad, often much smaller, ovate-elliptic, broadly acute at apex, rounded at base. Corolla 3–4 mm long, light yellow or white tinged rose; inflorescence becoming elongated in fruit. Fruit 5–10-jointed, curved, the segments 2–2.5 mm long and ca. 2 mm broad.

GRAND CAYMAN: *Proctor 52114*, a weed in the Queen Elizabeth II Botanic Park.

— Native of the Old World tropics; naturalised in Jamaica and Trinidad; attributed to St. Lucia by Adams (1972) but not by Howard (1988).]

Desmodium triflorum (L.) DC., *Prodr.* 2: 334 (1825).

A prostrate creeping herb with pubescent stems, often rooting at the nodes. Leaves with obovate or obcordate leaflets mostly 3–8 mm long. Flowers axillary or opposite the leaves; calyx 2–3 mm long, with lanceolate lobes; corolla pink or purplish, 3–4 mm long. Pods flat, curved, with continuous upper margin; joints 2–6, each ca. 3 mm long, the sides prominently net-veined.

GRAND CAYMAN: *Correll & Correll 51053; Kings GC 400; Millspaugh 1368, 1800*.

— Pantropical, a frequent weed of lawns and pastures.

Desmodium incanum DC., *Prodr.* 2: 332 (1825); Nicolson, *Taxon* 27(4): 365–370 (1978).

Desmodium supinum (Sw.) DC. (1825).

D. canum (J. F. Gmel.) Schinz & Thell. (1914); Adams (1972).

CHICK WEED

Decumbent woody herb with ascending branches, more or less pubescent throughout. Leaves petiolate, the leaflets chiefly oblong-elliptic or elliptic, 1–6 cm long or more, the terminal the largest, usually pale beneath. Racemes terminal; calyx red, ca. 3 mm long, with triangular teeth; corolla pink, fading to bluish; standard ca. 6 mm long and broad. Uppermost stamen free. Pods flat, deeply notched on the lower side, the upper side continuous; joints 3–8, each ca. 4 mm long.

GRAND CAYMAN: *Brunt* 1899, 2127; *Hitchcock*; *Kings* GC 249, GC 277; *Millspaugh* 1330, 1384; *Proctor* 15262. CAYMAN BRAC: *Kings* CB 22; *Proctor* 28970.

— Florida, West Indies, continental tropical America, and tropical Africa, common weed of roadsides, pastures, clearings and thickets.

Desmodium tortuosum (Sw.) DC., *Prodr.* 2: 332 (1825).

An erect subwoody herb to 1 m tall or more, the stem finely pubescent with hooked hairs. Leaves petiolate, the leaflets lanceolate, mostly ovate or elliptic, 1–7 cm long. Racemes terminal and in the upper axils, together forming a large, open, paniculate inflorescence; flowers often 2 or 3 together, on slender pedicels 1–1.5 cm long. Calyx 2.5–3 mm long; corolla pink or purple, ca. 4 mm long. Pods more or less twisted, equally indented on both margins; joints 3–7, each ca. 4 mm long.

GRAND CAYMAN: *Brunt* 2074; *Kings* GC 93, GC 360; *Proctor* 27926, 52094; *Sachet* 371.

— West Indies, continental tropical America, and tropical Africa, a weed of open waste ground and roadside banks.

[*Alysicarpus* Desv.

Perennial herbs with 1-foliolate leaves; stipules dry, papery, enclosing 2 stipels. Flowers small, in short terminal or axillary racemes; calyx deeply 5-lobed, the lobes elongate and rigid; corolla equal in length to or shorter than the calyx and included in it; standard roundish or obovate, clawed; wings obliquely oblong, adhering to the keel; keel obtuse, incurved. Stamens 10, 9 of them united, the uppermost free. Pods cylindrical with several or numerous joints, the joints separating, indehiscent.

— A genus of about 16 species indigenous to the tropics of Africa, Asia and Australia.

Alysicarpus vaginalis (L.) DC., *Prodr.* 2: 353 (1825).

An herb with prostrate or trailing stems growing from a woody taproot, nearly glabrous throughout; leaves with leaflet stalked and variously lanceolate, oblong, obovate or roundish, 0.4–4 cm long, the apex rounded but with a minute mucro, the base subcordate; stipules lanceolate, ca. 7 mm long, whitish. Racemes mostly 1–3 cm long; calyx 4–5 mm long; corolla deep salmon-red or purple, about as long as the calyx. Pods up to 2 cm long, 2–8-jointed, minutely puberulous.

GRAND CAYMAN: *Correll & Correll* 51052; *Proctor* 15279. CAYMAN BRAC: *Proctor* 28985, 29358.

— Native of tropical Asia and Africa; widely naturalised in the West Indies, the Cayman plants occurring in second-growth thickets, damp pastures, and in soil-pockets of exposed limestone pavements.]

[*Christia* Moench

Herbs, usually erect with few branches; stipules ovate to lanceolate, free, striate. Leaves usually with 1 leaflet wider than long (occasionally also with a pair of much smaller obovate leaflets). Flowers in lax terminal and axillary racemes; bracts acuminate, soon falling. Calyx broadly campanulate, enlarged after flowering, the lobes equal; corolla with standard obovate to obcordate, clawed, the wings obliquely oblong, adherent to the keel; keel slightly incurved; stamens 10; ovary with 2–8 ovules; style subulate, curved, glabrous, the stigma broadly capitate. Fruits subsessile or stipitate, constricted between the seeds, the joints compressed or turgid, folded within the enlarged calyx; seeds orbicular or subglobose.

— A genus of 12 species native to southeast Asia and Australia.

Christia vespertilionis (L.f.) Backh.f., *Reinwardtia* 6: 90 (1961).

Slender sparingly branched herb to 60 cm tall, the stems puberulous with minute hooked hairs; terminal (or only) leaflet usually less than 1.5 cm long but with extended lateral lobes producing a total width of 3–8 cm, these lobes horizontal or somewhat curved-ascending or -descending, on upper side tinged purplish with lighter streaks marking veins; lateral leaflets if present very small. Flowers simple or paired in racemes up to 12 cm long, the pedicels 2–3 mm long; calyx inflated in fruit to 11 mm long, standard 6 mm long, yellow to creamy. Fruit 4–6-jointed, each segment ca. 3 mm long; seeds oblong ca. 2 mm long, brown.

GRAND CAYMAN: *Proctor* 47826. Found in a North Side agricultural area on 20 April 1992.

— Native of southeast Asia; naturalised in Jamaica, Hispaniola, on several Lesser Antillean islands and in Trinidad.]

Abrus Adans.

Slender woody vines; leaves even-pinnate, lacking a terminal leaflet, the rhachis terminating in a bristle; leaflets numerous; stipels absent. Flowers in short racemes, terminal or axillary on short branches, the pedicels clustered at nodes of the raceme. Calyx truncate or very shortly 5-toothed; corolla usually pinkish or whitish; standard ovate, the short claw adherent to the stamen-tube; wings oblong-falcate; keel curved, longer than the wings. Stamens 9, united into a sheath split above; anthers all alike. Ovary subsessile; style short, incurved, with capitate stigma; ovules numerous. Pod somewhat compressed, 2-valved, with partitions between the seeds.

— A small pantropical genus of about 5 species.

Abrus precatorius L., *Syst. Nat.* ed. 12, 2: 472 (1767). **FIG. 132.** PLATE 27.

LICORICE, JOHN CROW BEAD

An elongate, sometimes high-climbing vine, the younger green stems sparsely pubescent. Leaves 4–10 cm long; leaflets in 10–20 pairs, oblong or obovate-oblong, 5–18 mm long, mucronulate at the apex, glabrous above and sparsely appressed-pubescent beneath.

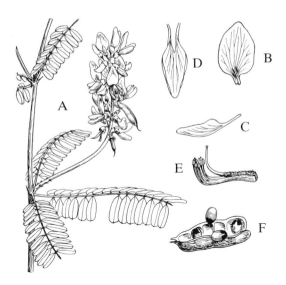

FIG. 132 **Abrus precatorius**. A, leaves and raceme, × $^2/_3$. B, standard, × $1^1/_2$. C, wing, × $1^1/_2$. D, keel, × $1^1/_2$. E, stamens and pistil × $1^1/_2$. F, pod and seeds, × $^2/_3$. (F. & R.)

Racemes 4–8 cm long, many-flowered; calyx 2–3 mm long; corolla pinkish or rose with whitish keel, 8–11 mm long. Pods oblong, beaked, 2–2.5 cm long, finely pubescent, 3–5-seeded; seeds ca. 5 mm long, bright scarlet with a black spot at the hilum.

GRAND CAYMAN: *Brunt* 2172; *Kings* GC 182a; *Proctor* 15231; *Sachet* 460a. LITTLE CAYMAN: *Proctor* 28065. CAYMAN BRAC: *Kings* CB 87; *Proctor* 28951, 29357.

— Pantropical, in thickets and woodlands, usually near sea-level. The roots allegedly contain glycerrhizin, the same substance as that found in commercial licorice. There are conflicting reports in the literature about the use of this material as a substitute for true licorice. The seeds, on the other hand, are known to be extremely poisonous if eaten; however, they are often used to make bracelets, necklaces, and other ornaments, and have been much used in some Asian countries as weights by jewel merchants. The unit of weight called the "carat" (one twenty-fourth of an ounce) is said to have been associated with the weight of *Abrus* seeds, 2 seeds allegedly weighing 1 carat.

Centrosema (DC.) Benth.

Herbaceous twining vines, sometimes woody below; leaves pinnately 3-foliolate, with stipels; stipules persistent, striate. Flowers rather large, on 1–several-flowered axillary peduncles, these solitary or paired; bracteoles relatively large, striate, appressed to the calyx. Calyx bell-shaped, unequally 5-lobed; standard large, roundish, flattened, spurred or slightly pouched at the base; wings and keel much shorter than the standard, the keel incurved. Stamens 10, all united or the uppermost free; anthers all alike. Ovary subsessile; style incurved, somewhat dilated at the apex; ovules numerous. Pod thin and flat, 2-valved, often with a longitudinal flange near each margin, and terminated by a long straight beak, partially septate within between the seeds; seeds transversely oblong.
— A genus of about 30 species in tropical and warm-temperate America.

Centrosema virginianum (L.) Benth., *Comm. Legum. Gen.* 56 (1837). PLATE 27.

A slender twining herbaceous vine, the stems nearly glabrous; leaflets glabrous, narrowly ovate or narrowly elliptic, 1.5–4 cm long or more. Flowers 1–3 on each peduncle; bracteoles ovate, ca. 5 mm long, minutely puberulous; calyx-lobes lance-linear, 8–12 mm long; corolla lavender-pink sometimes bluish or nearly white; standard 2.5–3 cm broad, minutely puberulous on the back. Pods 7–10 cm long, 3 mm wide, the terminal beak up to 10 mm long.

GRAND CAYMAN: *Sachet* 369. CAYMAN BRAC: *Proctor* 28950.

— Widespread in warm-temperate and tropical America, consisting of many varying races. The Cayman plants grow in secondary woodland thickets and are quite rare.

[*Clitoria* L.

Erect, trailing, or climbing herbs or shrubs, or sometimes large trees; leaves simple or more usually pinnately 3–7-foliolate, rarely 9–11-foliolate; stipules striate, persistent; stipels usually present. Flowers rather large, solitary or clustered in the leaf-axils, or in short racemes; bracts paired, persistent; bracteoles mostly larger than the bracts, striate. Calyx tubular, 5-lobed; standard large, rounded, retuse at the apex, narrowed and without appendages at the base; wings falcate-oblong, adherent to middle of keel; keel shorter than the wings, acute, incurved. Stamens 10, the uppermost free, the rest united; anthers all alike. Ovary stalked; style incurved, dilated at the apex, hairy on inner side. Pod stalked, elongate, compressed, 2-valved, interrupted or continuous within; seeds subglobose or compressed.

— A pantropical genus of about 30 species.

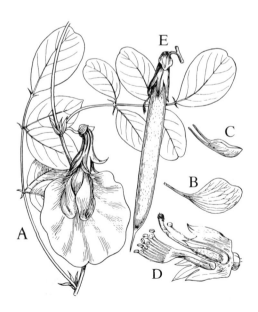

FIG. 133 **Clitoria ternatea**. A, portion of branch with leaves and flower, × ²/₃. B, wing, × ²/₃. C, keel, × ²/₃. D, calyx, stamens and pistil, × 1. E, pod, × ²/₃. (F. & R.)

Clitoria ternatea L., *Sp. Pl.* 2: 753 (1753). **FIG. 133.** PLATE 27.

BLUEBELL

A twining herbaceous vine, more or less woody at the base, finely pubescent more or less throughout, but often sparsely so. Leaves petiolate, 5–7-foliolate; leaflets elliptic or ovate, 2–5 cm long; stipules linear, 5–7 mm long. Flowers solitary on axillary peduncles 3–8 mm long; bracteoles roundish, 6–7 mm long; calyx 1.5–2 cm long; standard intense blue with white blotch in the middle, rarely all white, 4–4.5 cm long; wings and keel much shorter. Pods 6–10 cm long, 8–9 mm broad, beaked at the apex; seeds 5–6 mm long, mottled dark brown.

GRAND CAYMAN: *Brunt* 1645, 1727; *Hitchcock*; *Kings* GC 43, GC 186, GC 227; *Maggs* II 54; *Millspaugh* 1318, 1883; *Proctor* 15077, 15182; *Sachet* 377. CAYMAN BRAC: *Kings* CB 63; *Proctor* 28949, 29353 (flowers white).

— Native of tropical Africa, but widely cultivated as an ornamental, and now somewhat naturalised in nearly all warm countries.]

Teramnus P. Br.

Slender twining herbaceous vines, more or less pubescent; leaves pinnately 3-foliolate, with stipels; stipules small. Flowers very small, in axillary racemes or clusters; bracteoles linear or lanceolate; calyx-lobes 5, or 4 with the upper one 2-toothed; standard obovate, narrowed below into a long claw, not appendaged; wings narrow, adherent to the keel; keel shorter than the wings, almost straight, obtuse. Stamens 10, all united into a closed sheath, the filaments alternately long and short, the anthers on short filaments sterile. Ovary sessile, the style thick and glabrous, the stigma capitate; ovules many. Pod linear-elongate, compressed, 2-valvate, straight or curved, septate within, at apex often beaked.

— A small pantropical genus of about 4 species.

KEY TO SPECIES

1. Leaflets mostly 1.5–4 cm long; pods 2–4 cm long, sparsely pubescent:	**T. labialis**
1. Leaflets mostly 5–8 cm long; pods 5 cm long or more, densely brown-pubescent:	**T. uncinatus**

Teramnus labialis (L.f.) Spreng. in L., *Syst. Veg.* ed. 16, 3: 235 (1826).

A slender twiner, finely pubescent throughout; leaflets ovate or elliptic, 1.5–4 cm long (rarely longer), acutish and mucronate at the apex; stipules lance-linear, 2–3 mm long. Racemes slender, about as long as the leaves, the flowers scattered; calyx 5-lobed, 3–4 mm long, the lobes mostly shorter than the tube; corolla white, slightly longer than the calyx. Pods 2–4 cm long, rarely more, ca. 3 mm wide, slightly curved, with an upturned beak 1.5–2 mm long; seeds dark brown, ca. 2 mm long.

GRAND CAYMAN: *Correll & Correll* 51008; *Proctor* 31037. CAYMAN BRAC: *Proctor* 29369.

— Pantropical, a frequent weed of thickets and overgrown waste ground.

Teramnus uncinatus (L.) Sw, *Nov. Gen. & Sp. Pl.* 105 (1788).

An elongate slender twiner, the stems densely pilose with reflexed hairs; leaflets oblong to lance-linear, mostly 5–8 cm long, obtuse and mucronate at the apex, appressed hairy on both-sides, densely so beneath; stipules lanceolate, 3–5 mm long. Racemes usually much longer than the leaves, the flowers remotely scattered; occasionally some flowers solitary in the leaf-axils; calyx 5-lobed, pilose, 5–6 mm long, the lobes longer than the tube; corolla purplish or whitish, scarcely longer than the calyx. Pods mostly 5–7 cm long, 3–4 mm wide, densely brown-pilose, nearly straight, with right-angled beak 3–5 mm long; seeds lustrous orange-brown, ca. 4 mm long.

GRAND CAYMAN: *Correll & Correll 51007; Hitchcock.*

— West Indies and continental tropical America, usually in rather moist second-growth thickets.

Erythrina L.

Trees or shrubs, the trunks, branches, petioles, and sometimes even the leaf midribs and veins often armed with more or less woody thorns or spines. Stipules minute, soon falling. Leaves alternate, pinnately 3-foliate; stipels small, usually fleshy and glandular. Inflorescence axillary or terminal, pseudoracemose (flowers clustered on short axis); flowers large, brightly coloured; calyx inequilaterally campanulate or tubular, the margin truncate, oblique or lobed, often split down one side; standard large, clawed or narrowed at the base, usually greatly exceeding the keel; wings often small; keel petals often coherent, rarely free, usually much shorter than the standard. Stamens 10, alternately long and short. Ovary stipitate, fusisorm, curved, 2-several ovuled; style elongate, incurved, glabrous, the stigma capitate. Fruit stipitate, linear-oblong, constricted between the seeds, often curved, dehiscent, the valves papery, leathery, or woody; seeds several, ellipsoid, of brightly coloured or bicolourous; hilum lateral, linear.

— A genus of more than 100 species widely distributed in tropical and subtropical areas of the world.

Erythrina velutina Willd., *Ges. Naturf. Freunde, Neue Schr.* 3: 426 (1801). PLATE 27.

Tree to 20 m tall with spines on trunk and branches. Leaflets rhomboid-ovate to subrotundate, tomentose beneath, mostly 5–15 cm long and up to 10 cm broad. Inflorescence 12–25 cm long; standard up to 6 cm long, vermillion; wing and keel petals olive brown with crimson margins. Pods mostly 8–13 cm long; seeds red, ca. 1.5 cm long. GRAND CAYMAN: *Hitchcock; Proctor 49927.*

— Native of northern S. America; widely planted and naturalised elsewhere.]

Mucuna Adans.

Herbaceous or woody vines; leaves pinnately 3-foliolate, the lateral leaflets usually unequal-sided; stipules deciduous; stipels present. Flowers rather large, in axillary long-stalked racemes or clusters, the peduncles often greatly elongate and pendent; calyx shortly bell-shaped, 4-lobed; corolla mostly dark purple or yellowish; standard shorter than the wings, with inflexed auricles at the base; wings incurved, often adherent to the keel; keel equalling the wings or longer, incurved at the acute or beak-like apex.

Stamens 10, 9 of them united, the uppermost free; anthers alternately long and short, the short ones hairy. Ovary sessile, hairy; style glabrous; ovules few. Pod more or less oblong, thick, leathery, hairy (the hairs often stinging), septate within; seeds few, large, rounded-oblong.

— A pantropical genus of about 50 species. Several, such as the velvet bean (*Mucuna deeringiana* (Bort) Merrill), have been widely grown as cattle-fodder. The seeds of that species are also sometimes parched and ground as a substitute for coffee.

Mucuna pruriens (L.) DC., *Prodr.* 2: 405 (1825). PLATE 28.

COW-ITCH

An elongate herbaceous vine, growing over shrubs and small trees; leaflets rhombic-ovate, the lateral very asymmetric, 7–14 cm long, of thin texture, appressed-pilose especially beneath. Racemes mostly short-pedunculate, few-flowered; calyx broader than long, the lowest lobe up to 10 mm long; corolla dull dark purple; standard 2 cm long, half as long as the wings; keel 4.5 cm long, yellowish-beaked. Pods curved-oblong, mostly 5–7.5 cm long, 1–1.5 cm thick, covered with abundant easily detached irritant hairs; seeds 2–6, transverse-oblong, 1–1.5 cm long, black with a white hilum.

GRAND CAYMAN: *Hitchcock; Kings GC 228*.

— Pantropical. The stinging hairs of the pods can cause almost unbearable itching and burning of the skin, and are very dangerous to the eyes.

Galactia P. Br.

Slender herbaceous or woody vines, rarely erect shrubs; leaves pinnately 3-foliolate, with stipels; stipules small, often deciduous. Flowers paired or clustered at raised nodes on axillary racemes, rarely solitary in the leaf-axils; bracteoles minute. Calyx 4-lobed, the lobes acuminate, unequal; corolla pink or white; standard roundish, the margins slightly inflexed at the base; wings adherent to the keel and equalling it in length. Stamens 10, the lower 9 united, the uppermost free or partly so; anthers all alike. Ovary nearly sessile; style very slender, glabrous, with a minute stigma. Pod elongate, straight or slightly curved, usually compressed, 2-valved, partly septate within.

— A pantropical genus of more than 80 species.

Galactia striata (Jacq.) Urban, *Symb. Ant.* 2: 320 (1900).

A slender and often elongate herbaceous vine, the stems finely pubescent; leaflets elliptic or ovate-elliptic, 2.5–8 cm long, mucronate at the apex, the upper side glabrate and with finely prominulous venation, the underside rather densely appressed-pubescent. Inflorescence up to 15 cm long but often less, short- to long-pedunculate; flowers paired at the nodes; calyx appressed-pubescent, 5–7 mm long, with narrowly lance-acuminate lobes; corolla 8–9 mm long, light rose-violet, the standard, wings and keel of about equal length. Pods 4–5 cm long, ca. 5 mm broad, pubescent.

GRAND CAYMAN: *Kings GC 42, GC 311; Proctor 31040*.

— Florida, West Indies and continental tropical America, in thickets and woodlands. Over its wide range this is an extremely variable species; the above description is based chiefly on Cayman material.

Canavalia Adans.

Climbing or trailing herbs; leaves pinnately 3-foliolate, with stipels; stipules small, deciduous. Flowers rather large, borne in clusters or solitary at swollen nodes of axillary racemes; calyx 2-lipped, the upper lip large and truncate or 2-lobed, the lower much smaller and entire or 3-cleft; standard roundish, reflexed; wings narrow, subfalcate or somewhat twisted; keel broader than the wings, incurved, obtuse or with an inflexed or spiral beak. Stamens 10, all united above, the uppermost one free at the base; anthers all alike. Ovary subsessile; style incurved or spiral with the keel, glabrous, the stigma small; ovules few. Pod oblong or broadly linear, compressed or convex, winged or ribbed longitudinally, 2-valved; seeds rounded or bean-like, with a linear hilum.

— A pantropical genus of about 50 species, a few sometimes cultivated as forage or for the edible seeds and young pods.

KEY TO SPECIES

1. Nodes of inflorescence each with 3 or more flowers; pods with sutural ribs only; seeds concolourous, with hilum nearly as long as the seed: **C. nitida**

1. Nodes of inflorescence each with 1 or 2 flowers; pods with an additional rib on either side 3–5 mm from the ventral suture; seeds marbled orange and brown, with hilum less than half as long as the seed: **C. rosea**

Canavalia nitida (Cav.) Piper, *Contr. U.S. Nat. Herb.* 20: 559, 562 (1925).

Canavalia ekmanii Urban (1918).

HORSE BEAN

Herbaceous vine, climbing on rocks or over bushes, the stems glabrous or nearly so; leaflets glabrous, oblong-elliptic or elliptic, mostly 5–8 cm long, 2.5–3.5 cm broad, bluntly subacuminate and minutely refuse at the apex, the fine reticulate venation slightly raised and evident on both sides. Racemes longer than the subtending leaves; calyx with the upper lip deeply 2-cleft, the lower lip 3-lobed; corolla purple or dark violet with a pink keel, twice as long as the calyx, up to 1.6 cm long. Pods oblong, 10–15 cm long and ca. 3 cm broad, strongly convex and of thick texture; seeds lustrous wine-red or brick-red, the linear hilum blackish.

GRAND CAYMAN: *Kings GC 104* growing on "the cliff at East End".

— Originally thought to be endemic to Cuba, this species has more recently been ascribed a broader range including the Bahamas, the Greater Antilles (except Jamaica), the Virgin Islands and Mexico.

Canavalia rosea (Sw.) DC., *Prodr.* 2: 404 (1825). PLATE 28.

Canavalia maritima Thouars (1813), illegit.
C. obtusifolia (Lam.) DC. (1825), in part.

SEA BEAN

A long-trailing herb, the stems becoming rope-like; puberulous when young. Leaflets sparsely strigillose, roundish or very broadly obovate, mostly 5–12 cm long, 2.5–10 cm broad, rounded to broadly refuse at the apex, the fine venation scarcely evident. Racemes often much longer than the leaves; calyx helmet-like, the upper lip broad and marginate, the lower much shorter and 3-toothed; corolla pink or rose, fading bluish-purple, ca. 2 cm long. Pods oblong, 7–13 cm long and 2.5–3 cm broad, somewhat compressed and with a distinct flange on either side parallel to the ribs of the ventral suture and 3–5 mm from them; seeds marbled orange and brown, 12–16 mm long (usually less than 15 mm), the whitish hilum less than half as long as the length of the seed.

GRAND CAYMAN: *Brunt 2052; Hitchcock; Kings GC 68; Millspaugh 1307; Proctor 15060; Sachet 404; Sauer 3320.* LITTLE CAYMAN: *Kings LC 23; Proctor 28085; Sauer 4164.* CAYMAN BRAC: *Kings CB 13; Millspaugh 1170; Proctor 29074.*

— Pantropical, especially common in sandy clearings near the sea.

Phaseolus L.

Annual or perennial herbs, twining or erect, rarely prostrate, sometimes with woody or tuberous roots; leaves pinnately 3-foliolate with stipules; stipules striate, persistent. Flowers in clusters at enlarged nodes along the axes of axillary racemes, the nodes bracteolate. Calyx 4- or 5-toothed or -lobed, the upper two teeth or lobes often more or less united; corolla white, yellow, red, or purple; standard roundish, recurved-spreading or somewhat twisted; wings usually obovate, equal in length to or longer than the standard; keel linear to obovate, with a low spirally twisted or coiled beak. Stamens 10, the uppermost free, the rest united; anthers all alike. Ovary subsessile; style twisted with the keel, usually hairy on one side; stigma oblique; ovules few or many. Pod linear or falcate, compressed or subcylindrical, 2-valved, with tissue between the seeds; seeds rounded-oblong or flattened, the hilum small or linear.

— A world-wide genus of more than 100 species, some often cultivated. As a source of food, this genus is of major importance, as it includes the majority of the various kinds of beans. The only wild Cayman species is often placed in a segregate genus *Macroptilium* distinguished especially on the basis of having 5 (instead of usually 4) calyx lobes.

Phaseolus lathyroides L., *Sp. Pl.* ed. 2, 2: 1018 (1763).

Phaseolus semierectus L. (1767).
Macroptilium lathyroides (L.) Urban (1928); Adams (1972).

An erect or straggling, annual or persistent herb with subwoody rhizomes; leaflets mostly lanceolate to elliptic, 1.5–5 cm long or more, acutish at the apex, glabrous or nearly so except for the densely pubescent petiolules. Racemes few-flowered, on peduncles much longer than the leaves; flowers in pairs remotely scattered on the puberulous rachis; bracts and bracteoles subulate. Calyx 4–6 mm long, 5-toothed; corolla deep red-purple; standard 1.5 cm long; keel forming a single spiral. Pods narrowly linear, 8–10 cm long, ca. 2 mm broad, subcylindrical, strigillose; seeds oval, ca. 3 mm long, brownish-grey speckled with black.

GRAND CAYMAN: *Brunt 1902, 1968; Hitchcock; Kings GC 323; Millspaugh 1324; Proctor 15035; Sachet 454.* CAYMAN BRAC: *Kings PCB 2; Proctor 29089.*

— West Indies and continental tropical America; naturalised in the East Indies and Polynesia. This is commonly a weed of open fields and roadsides.

Vigna Savi

Twining herbaceous vines, sometimes prostrate or erect; leaves pinnately 3-foliolate, with stipels; stipules sessile or sometimes extended at the base below the point of attachment. Flowers in short, crowded racemes at the ends of long axillary peduncles; bracts and bracteoles small, soon falling. Calyx bell-shaped, 5-lobed or the upper 2 lobes united; corolla yellow or rarely pale purple; standard roundish, with inflexed basal auricles; wings falcate-obovate, shorter than the standard; keel equal in length to the wings, incurved but not forming a complete loop or spiral. Stamens 10, the uppermost free, the rest united; anthers all alike. Ovary sessile; style thread-like or thickened above, often bearded along the inner side, the stigma oblique or lateral; ovules many. Pod linear, straight or nearly so, subcylindrical, 2-valved, interrupted within between the seeds; seeds reniform or subquadrate with a short lateral hilum.

— A pantropical genus of about 40 species.

Vigna luteola (Jacq.) Benth. in Mart., *Fl. Bras.* 15 (1): 194 (1859).

Vigna repens (L.) Kuntze (1891), not Baker (1876).

A trailing or twining herbaceous vine, glabrous or somewhat pubescent, sometimes rampant and forming dense tangles; leaflets lanceolate to ovate, 3–9 cm long, blunt or acutish at the apex, glabrous or nearly so except for the pubescent petiolules. Peduncles longer than the leaves; calyx 4-lobed, 4–5 mm long, the upper lobe rounded with 2 small teeth; corolla yellow; standard ca. 1.5 cm long, retuse at the apex; keel obtuse. Pods straight or slightly curved, 4–7 cm long, ca. 5 mm wide, sparsely pilose; seeds 4–5 mm long, lustrous black with white hilum.

GRAND CAYMAN: *Brunt* 1748, 1914, 2126; *Hitchcock*; *Kings* GC 79, GC 212; *Lewis* 3828; *Maggs* II 52; *Millspaugh* 1241. LITTLE CAYMAN: *Kings* LC 73. CAYMAN BRAC: *Proctor* 29372.

— Pantropical, a weedy species of damp thickets and more or less open waste ground.

[*Lablab* Adans.

Twining herbaceous vines, perennial from a subwoody root; leaves pinnately 3-foliolate, with stipels; stipules small, striate. Peduncles axillary, elongate; flowers clustered at nodes of a raceme; bracts and bracteoles small, soon falling. Calyx campanulate, 4-toothed; corolla white or purple; standard roundish, clawed, with 2 auricles at the base; wings falcate-obovate; keel incurved at a right-angle. Stamens 10, the uppermost free, the rest united; anthers all alike. Ovary sessile; style flattened toward apex and bearded along the inner side; stigma terminal. Ovules few. Pod falcate-oblong, slightly curved, strongly compressed, 2-valved with thickened margins, beaked at the apex, and 2–4-seeded; seeds somewhat compressed, with linear hilum.

A genus of 1 species, probably native of Africa but cultivated widely both for its edible seeds and young pods, and as an ornamental.

Lablab purpureus (L.) Sweet, *Hort. Brit.* 481 (1827).

Dolichos lablab L. (1753).

D. purpureus L. (1763).

BONAVIST

With the characters of the genus. Leaflets ovate-deltate or ovate-rhombic, 4–13 cm long, the lateral ones unequal-sided. Racemes many-flowered; standard 1.5 cm long and somewhat broader. Pods ca. 7.5 cm long and 2.5 cm broad, the lower suture curved and finely warty-roughened; seeds ca. 1 cm long.

GRAND CAYMAN: *Millspaugh* 1396.

— Now pantropical in distribution, cultivated or often spontaneous in roadside thickets and open waste ground.]

[*Cajanus* DC.

Cajanus cajan (L.) Millsp., the congo (gungo) or pigeon pea, is sometimes cultivated for its edible seeds. It shows little or no tendency to naturalise.

GRAND CAYMAN: *Hitchcock*.]

Rhynchosia Lour.

Twining herbaceous or subwoody vines, rarely erect; leaves pinnately 3-foliolate, with or without stipels; leaflets minutely gland-dotted beneath; stipules lanceolate to ovate. Flowers borne singly or in pairs along the axillary racemes; calyx 4–5-lobed, the 2 upper lobes more or less united; corolla yellow, often tinged or striped with dark red; standard roundish, spreading or reflexed, with inflexed auricles at the base; wings narrow; keel falcate or incurved at the apex. Stamens 10, the uppermost free, the rest united. Ovary subsessile; style thread-like, incurved above, the stigma terminal; ovules 2 or rarely 1. Pod oblong or falcate, compressed, beaked at the apex, 2-valved, continuous within or rarely septate; seeds usually 2, roundish-compressed, often red, with short lateral hilum.

— A large pantropical genus of about 150 species, a few extending into temperate N. America.

Rhynchosia minima (L.) DC., *Prodr.* 2: 385 (1825). **FIG. 135.**

A small twining or trailing vine, the stems angled or grooved, puberulous when young; leaflets rhombic or subrhombic, 1–2.5 cm long, minutely puberulous on both sides; stipules subulate, ca. 2 mm long. Racemes slender, 5–10 cm long, often exceeding the leaves; flowers remote, not paired, reflexed in age; calyx 3–4 mm long, puberulous; corolla pale yellow, 5–6 mm long; standard puberulous and gland-dotted. Pods 1–1.7 cm long, 3–5 mm broad, puberulous and gland-dotted, slightly constricted between the seeds; seeds 3–3.5 mm long, mottled with light and dark brown.

GRAND CAYMAN: *Brunt* 1969; *Hitchcock*; *Kings* GC 74; *Millspaugh* 1283, 1349; *Proctor* 15113; *Rothrock* 167. LITTLE CAYMAN: *Proctor* 28161. CAYMAN BRAC: *Kings* CB 4; *Proctor* 28996.

— A pantropical weed of open waste ground and roadsides, or sometimes in pastures and thickets.

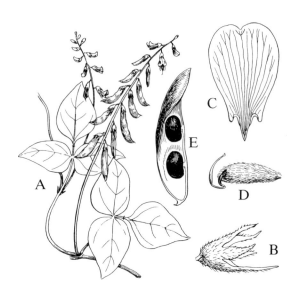

FIG. 134 **Rhynchosia minima.** A, leaves, inflorescence, and pods, × ²/₃. B, calyx, × 5. C, standard, × 5. D, pistil, × 5. E, one valve of pod with seeds, × 2. (F. & R.)

[*Moghania* J. St. Hil.

Erect shrubs or woody herbs; leaves 1-foliolate or digitately 3-foliolate; leaflets prominently veined and with numerous minute glandular dots beneath. Flowers in axillary and terminal spikes or racemes, or in small cymes along a raceme-like axis; bracts small and soon falling, or large, inflated and persistent. Calyx 5-lobed, the lowest lobe longer than the others; standard roundish, auricled at the base. Stamens 10, the uppermost free, the rest united. Ovary with 2 ovules; stigma small, terminal. Pod oblong, 2-valved, inflated; seeds 2 or sometimes 1.

— A genus of about 35 species occurring in tropical Africa, Asia, and Australia.

Moghania strobilifera (L.) J. St. Hil. ex Ktze., *Revis. Gen. Pl.* 1: 199 (1891).

Flemingia strobilifera (L.) Ait. F. (1812).

WILD HOPS

A weedy shrub up to 1.5 m tall or more, the young stems and petioles appressed-pubescent; leaflet 1, elliptic to ovate, mostly 5–15 cm long, 3–10 cm broad, acutish at the apex. Inflorescences axillary and terminal, paniculate, consisting of 2–5 raceme-like branches, these densely beset with large, distichously arranged, inflated pale green bracts, each enclosing a small cyme of flowers; racemes 5–15 cm long; bracts cordate, 1–2.5 cm long, broader than long, pubescent, turning light brown with age. Calyx ca. 5 mm long; corolla whitish, 5–6 mm long. Pods ca. 1 cm long, concealed by the persistent bracts.

GRAND CAYMAN: *Kings GC 294.*

— Native of the East Indies, but long naturalised almost throughout the West Indies and common on many islands. It appears, however, to be rare in the Cayman Islands.]

Dalbergia L.

Trees or shrubs, the branches often scandent or trailing; leaves pinnate with a terminal leaflet or (in Cayman species) 1-foliolate. Flowers white or purplish, in axillary or terminal racemes, cymes or panicles; bracts and bracteoles minute. Calyx 5-lobed, the lobes unequal, the upper two broader and the lowest one longer than the others; standard ovate or roundish; wings oblong; keel obtuse, its petals united dorsally at the apex. Stamens 10, the uppermost one more or less free, the sheath split on the upper side, sometimes also on the lower side; anthers small, erect, and paired. Ovary stalked; style short, incurved, with small terminal stigma; ovules few. Pod roundish to oblong or linear, very compressed and flat, indehiscent, 1–4-seeded; seeds kidney-shaped, compressed.

— A pantropical genus of more than 100 species.

KEY TO SPECIES

1. Leaves glabrous on both sides and more or less cordate at the base; pods oblong, 3–4-seeded: **D. brownei**

1. Leaves puberulous beneath and truncate or rounded at the base; pods roundish, 1-seeded: **D. ecastaphyllum**

Dalbergia brownei (Jacq.) Urban, *Symb. Ant.* 4: 295 (1905). **FIG. 135**. PLATE 28.

COCOON

A suberect, tangled or trailing shrub, the glabrous branches often ending in a woody tendril; leaflet 1, glabrous, ovate, 4–8.5 cm long, acutish at the apex, subcordate or cordate at the base, very lustrous on the upper side. Flowers white, fragrant, in small dense panicles scarcely exceeding the petioles; panicle-branches and petioles minutely

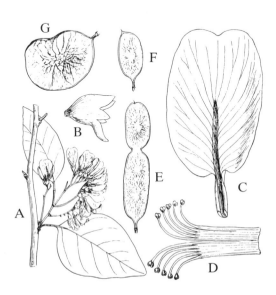

FIG. 135 **Dalbergia brownei**. A, portion of flowering branch, × ²/₃. B, calyx, × 3. C, standard, × 5. D, stamens, × 5. E, F, pods, × ²/₃. G, pod of D. ecastaphyllum × ²/₃. (F. & R.)

puberulous. Calyx ca. 4 mm long; standard ca. 8 mm long; keel much shorter than the wings. Pods more or less oblong from a stalk-like base, mostly 2–5 cm long, 3–4-seeded or occasionally 1-seeded, glabrous and covered with fine raised reticulations.

GRAND CAYMAN: *Brunt* 1629, 1912; *Hitchcock*; *Proctor* 15192.

— Florida, the Greater Antilles, and continental tropical America, chiefly along the borders of mangrove swamps.

Dalbergia ecastaphyllum (L.) Taub. in Engl. & Prantl, *Nat. Pflanzenfam.* 3 (3): 335 (1894). PLATE 28.

Ecastaphyllum brownei Pers. (1807).

A rambling shrub, sending up long vertical shoots from the more or less tangled main branches, reaching 3 m high or more, the young leafy branches finely pubescent; leaflet 1, oblong-ovate or elliptic, 5–12 cm long, shortly acuminate at the apex and usually rounded or truncate at the base, glabrous above, paler beneath and finely puberulous. Flowers white, faintly scented, in small dense panicles about 2 cm long, about twice as long as the petioles. Calyx ca. 3 mm long, pubescent; standard 7–8 mm long. Ovary with stalk longer than the calyx. Pods roundish or kidney-shaped, 2–3 cm across, finely pubescent, on a short stalk, always 1-seeded.

GRAND CAYMAN: *Brunt* 2045. LITTLE CAYMAN: *Kings* LC 89, LC 90; *Proctor* 35208.

— West Indies, continental tropical America, and west tropical Africa, along borders of mangrove swamps and borders of streams near the coast.

Piscidia L.

Deciduous trees or large shrubs, the dark twigs marked with numerous small whitish lenticels; leaves alternate, pinnate with a terminal leaflet; leaflets opposite, in 3–5 pairs; stipels absent. Flowers in small lateral panicles, appearing while the tree is leafless; calyx 5-lobed, the lobes short, triangular; corolla whitish or pink; standard rounded, retuse; wings and keel clawed at the base. Stamens 10, all united above, the uppermost one free toward the base. Ovary sessile, with many ovules; style thread-like, incurved, the stigma small, terminal. Pod linear, compressed, indehiscent, 3–7-seeded, with 4 broad longitudinal wings; seeds oval, compressed.

— A small tropical American genus of about 6 species.

Piscidia piscipula (L.) Sarg., *Gard. & For.* 4: 436 (1891). PLATE 29.

Ichthyomethia piscipula (L.) Hitchc. (1891).

DOGWOOD

A small tree usually 6–10 m tall; leaflets elliptic-oblong or obovate, 5–10 cm long, shortly acuminate at the apex, glabrous above, finely pubescent beneath. Panicles 8–20 cm long, somewhat dense with numerous flowers; pedicels 2–7 mm long; calyx 6–7 mm long, greyish-strigillose; corolla whitish or pale pink; standard ca. 1.5 cm long, about equalled by the wings and keel. Pods 2–8 cm long, the body 4–5 mm wide, much exceeded in width by the wings; wings thin, glabrate, pale green, undulate or ruffled.

GRAND CAYMAN: *Kings* GC 318.

— Florida, West Indies, and continental tropical America, occasional or frequent in dry woodlands. The wood of this species is hard, heavy, strong and durable; it is difficult to work but takes a high polish. The bark, especially of the roots, is well-known for its narcotic and poisonous properties. In some places, people apply this material locally to relieve toothache, and in Jamaica it is said to be used occasionally for curing mange in dogs. If the bark and leaves are crushed and thrown into water, most nearby fish will soon become stupefied and will float on the surface.

Sophora L.

Herbs, shrubs or trees; leaves pinnate with a terminal leaflet; leaflets opposite, subopposite, or irregularly alternate; stipels absent. Flowers in terminal racemes or panicles; calyx bell-shaped, obliquely truncate or with 5 short teeth; corolla yellow; standard elliptic or rounded; wings and keel about the same length, clawed. Stamens 10, all free. Ovary short-stalked; style slightly incurved, with a minute terminal stigma; ovules rather few. Pod stalked, elongate, constricted between the seeds, indehiscent; seeds rounded-oblong or somewhat compressed.

— A tropical and warm-temperate genus of about 50 species, some of the arboreous ones noted for their extremely hard wood.

Sophora tomentosa L., *Sp. Pl.* 1: 373 (1753). **FIG. 136.**

MICAR

A shrub usually 1–2.5 m tall, variably more or less tomentose throughout; leaves with 11–17 subopposite or irregularly alternate leaflets, these ovate or elliptic, 2–5 cm long, blunt or slightly refuse at the apex, densely pubescent beneath or sometimes glabrate. Racemes mostly 10–30 cm long, often many-flowered; calyx truncate or obscurely

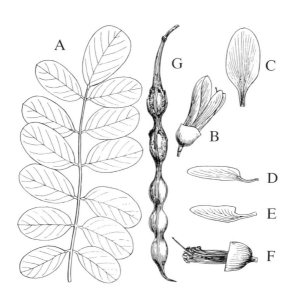

FIG. 136 **Sophora tomentosa**. A, leaf, × 1/2. B, flower, × 2/3. C, standard, × 2/3. D, wing, × 2/3. E. keel, × 2/3. F, calyx, stamens and pistil, × 2/3. G, pod partially decayed above, showing seeds, × 2/3. (F. & R.)

toothed, ca. 7 mm long; standard elliptic, folded, 2–3 cm long. Pods 5–15 cm long, 5–9-seeded, tomentose; seeds rounded-oblong, 7–8 mm long, orange-brown.

GRAND CAYMAN: *Kings GC 387; Proctor 15147.*

— Pantropical in coastal thickets.

Aeschynomene L.

Trailing or erect herbs or shrubs. Stipules either peltate or else attached at the base, often with appendages extending below the point of attachment. Leaves pinnately compound, a terminal leaflet present or absent; leaflets few to many. Inflorescence of terminal or axillary racemes; calyx campanulate or bilabiate; petals white or yellow to red or purple; stamens 10, the filaments united; ovary stipitate, the style glabrous, the stigma capitate. Fruits stipitate, 2–many-jointed, the segments flat or convex, indehiscent. Seeds reniform, lustrous brown or black; hilum circular.

— A genus of 150 species, the majority occurring in the Western Hemisphere and Africa.

KEY TO SPECIES

1. Leaflets 2–several-costate:	A. americana
1. Leaflets 1-costate:	A. sensitiva

Aeschynomene americana L., *Sp. Pl.* 2: 713 (1753).

Low slender-branched to robust bushy annual up to more than 1 m tall; stems glabrous to thinly hispid, often with glandular hairs. Stipules peltate, appendaged below the point of attachment. Leaflets up to ca. 30 pairs, 5–15 mm long, 1–2 mm broad. Inflorescence few-flowered; calyx 3–6 mm long; corolla 5–10 mm long, usually white tinged pale violet or yellowish. Fruit up to about 8-jointed, the upper margin entire or slightly crenate, the lower margin deeply indented, the segments ca. 4 mm long, glabrous to puberulous. Seeds 2–3 mm long, dark brown.

GRAND CAYMAN: *Proctor 48299, 48659,* recorded only from the Forest Glen area, North Side.

— Western Hemisphere from Florida and Mexico to Argentina; common in the West Indies; reported also from Sierra Leone in W. Africa.

Aeschynomene sensitiva Sw., *Prodr.* 107 (1788).

Erect somewhat woody annual to 40 cm tall or more, mostly glabrous or glabrate. Stipules basally attached, up to 20 mm long, 1.5–5 mm wide acuminate at apex, truncate and erose at base. Leaves 2–10 cm long, with 5–20 pairs of leaflets, the petiole and rachis hispidulous; leaflets oblong, 4–15 mm long, 1.5–3 mm broad. Calyx 4–8 mm long, ciliate; corolla 5–9 mm long, yellowish with dark speckles. Fruit 4–12-jointed, the segments 5–7 mm long, glabrate, smooth to veruccose, the upper margin entire, the lower margin crenate. Seeds 3–4 mm long, brown.

GRAND CAYMAN: *Proctor 48318,* recorded only from open marshy ground at The Common, near West Bay.

— Mexico, C. America, Greater and Lesser Antilles (except Jamaica), Trinidad, and S. America.

SUBFAMILY 2. CAESALPINIOIDEAE

Herbs, shrubs or trees, with or without prickles, erect or sometimes trailing, scandent or climbing; leaves usually pinnate or bipinnate, or (in *Bauhinia*) with 2 leaflets partly united; stipules usually present. Flowers irregular or rarely regular, terminal or axillary, solitary or in racemes or panicles; sepals usually 5, imbricate or partly united; petals usually 5, imbricate, the uppermost surrounded by the rest in bud. Stamens 10, rarely fewer, sometimes some of them sterile; filaments usually free or sometimes partly united. Pods dehiscent or indehiscent.

— The members of this subfamily are chiefly found in the tropics.

KEY TO GENERA

1. Leaves pinnate, or rarely bipinnate at the base only:

    2. Leaflets 2, free or united at base:

        3. Leaflets united at base; pods flat, dehiscent: — **Bauhinia**

        3. Leaflets free; pods turgid, woody, indehiscent: — **Hymenaea**

    2. Leaflets more than 2:

        4. Petals 3; stamens 3, united into a sheath; pods not strongly flattened, indehiscent, containing edible pulp: — **[Tamarindus]**

        4. Petals 5; stamens 6–10; pods strongly flattened:

            5. Pods splitting at the middle of the valves, not at the margins; leaflets obovate; flowers numerous in racemes: — **[Haematoxylum]**

            5. Pods splitting at one or both margins, or not splitting; leaflets various, but if obovate, the flowers solitary: — **Cassia**

1. Leaves amply bipinnate:

    6. Sprawling or erect shrubs, with or without prickles; pods less than 12 cm long: — **Caesalpinia**

    6. Tree, without prickles; pods 20–50 cm long: — **[Delonix]**

Caesalpinia L.

Trees or shrubs, erect or sometimes sprawling or scandent, variously prickly or else unarmed; leaves bipinnate without a terminal pinna; pinnae with an odd number of leaflets, these small and numerous or few and relatively large; stipels sometimes present, or sometimes represented by spines; stipules large or minute, or rarely absent. Flowers usually yellow or red, often showy, in axillary racemes or terminal panicles; bracteoles absent. Calyx with short tube and 5 imbricate lobes; petals roundish to oblong, strongly imbricate, slightly unequal, sometimes clawed. Stamens all free, usually hairy or glandular at the base; anthers all alike. Ovary sessile, free from the calyx; style thread-like with small terminal stigma; ovules few. Pods compressed or somewhat turgid, 2-valved or sometimes indehiscent; seeds roundish or globose, sometimes flattened; endosperm lacking.

— A pantropical genus of 70 or more species.

KEY TO SPECIES

1. Sprawling or scrambling wild shrubs, often prickly; seeds oblong to subglobose, not flattened:

  2. Pods nearly or quite without prickles: **C. caymanensis**

  2. Pods densely beset with prickles:

    3. Stipules large and leaf-like; seeds grey: **C. bonduc**

    3. Stipules minute or lacking; seeds yellow or olive:

      4. Leaflets mostly 2–3.5 cm long; pods less than 3.5 cm broad, often narrowed toward base: **C. wrightiana**

      4. Leaflets mostly 4–6 cm long; pods ca. 4 cm broad, abruptly rounded or subtruncate at base: **C. intermedia**

1. Erect cultivated shrub; pods compressed, without prickles, the seeds strongly flattened: **[C. pulcherrima]**

Caesalpinia caymanensis Millsp., *Field Mus. Bot. Ser.* 2: 49, t. 60 (1900).

Sprawling shrub, the young branches white-ciliate and with few or no prickles, soon becoming glabrous; leaves golden-tomentose throughout, 20–25 cm long; pinnae about 6 pairs; leaflets about 15 per pinna, oblong-elliptic, 2–2.7 cm long, blunt and mucronate, each subtended on the rhachilla by a recurved thorn. Flowers undescribed. Pods dark brown, broadly oblong, 5.5–8 cm long, spineless or nearly so; seeds usually 2, grey or greenish-grey, subglobose, ca. 2 cm long.

GRAND CAYMAN: *Kings GC 96, GC 97; Millspaugh 1263* (type).

— Endemic. This species used to occur in extensive thickets just north of George Town, where in recent years it appears to have been exterminated by growth of the town and the construction of hotels. The validity of this species, based chiefly on the absence of prickles, is now considered doubtful.

Caesalpinia bonduc (L.) Roxb., *Fl. Ind.* 2: 362 (1832). **FIG. 137.** PLATE 29.

Caesalpinia bonducella (L.) Fleming (1810).

Caesalpinia crista L. (1753), in part but not as to present interpretation.

COCKSPUR, GREY NICKEL

Sprawling or scrambling thorny shrub, the young branches finely pubescent and often bearing numerous prickles of variable length and stoutness; leaves finely golden-tomentose more or less throughout when young, 25–60 cm long, the rhachis and rhachillae with numerous recurved thorns. Pinnae in 5–8 pairs; leaflets about 15 per pinna, ovate or oblong-elliptic, 2–4 cm long, mucronate; stipules leafy, of 2 or 3 broadly rounded segments to 3 cm long. Racemes simple or branched, 10–20 cm long, with numerous crowded flowers; calyx 6–7 mm long, reddish-woolly; petals yellow, 9–12 mm long. Stamens shorter than the petals. Pods dark brown, broadly elliptic-oblong, 5–8 cm long, covered with numerous straight prickles; seeds usually 2, grey or greenish-grey, subglobose, ca. 2 cm long.

GRAND CAYMAN: *Brunt 1762, 1803; Hitchcock; Kings GC 98, GC 99; Milspaugh 1250; Proctor 15120, 31043; Sachet 425.* LITTLE CAYMAN: *Kings LC 46; Proctor 28032.* CAYMAN BRAC: *Kings CB 77; Proctor 29119.*

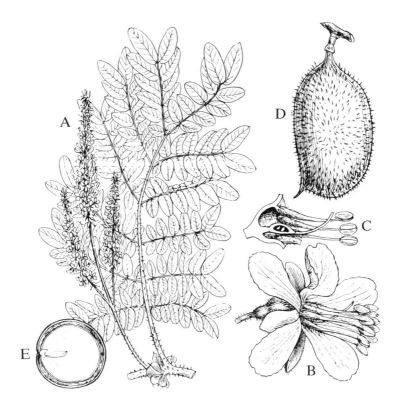

FIG. 137 **Caesalpinia bonduc.** A, leaf with inflorescence, much reduced. B, flower, × 2. C, flower cut lengthwise, with sepals and petals removed, × 2. D, pod, × ¹/₂. E, seed cut to show cotyledon and plumule, × ²/₃. (F. & R.)

— Pantropical, on sandy seashores and in coastal thickets or open waste land. The seeds, when roasted, ground, and prepared as a beverage like coffee, are alleged to have medicinal value in controlling oedema, but unroasted are said to be poisonous.

Caesalpinia wrightiana Urban, *Symb. Ant.* 2: 274 (1900).

COCKSPUR, YELLOW NICKEL

Trailing or scrambling thorny shrub, the young branches minutely puberulous or glabrate, with rather few uniform prickles; leaves lustrous, puberulous when very young, soon becoming glabrate, 20–50 cm long or more, the rhachis and rhachillae armed with strongly recurved thorns. Pinnae mostly in 3–6 pairs; leaflets mostly 9–13 per pinna, ovate, rather coriaceous, short-acuminate, mucronate, glossy on the upper surface; stipules minute and subulate or usually lacking. Racemes simple or branched, mostly 8–12 cm long; pedicels slender, 2–3(–6) mm long; calyx 7 mm long; petals yellow, ca. 5 mm long. Stamens about equal in length to the petals. Pods light brown, often slightly glaucous, obovate-oblong, 5–6 cm long, covered with numerous prickles up to 1 cm long; seeds usually 2, yellow or olive, oblong, mostly 1.7–1.9 cm long.

LITTLE CAYMAN: *Proctor* 35109, 35124.

— Cuba, Jamaica, and perhaps the Swan Islands, occurring in coastal thickets and dry woodlands over limestone.

Caesalpinia intermedia Urban, *Symb. Ant.* 2: 274 (1900).

Caesalpinia major of Adams in *Fl. Pl. of Jamaica* (1972), not Dandy & Exell (1938).

COCKSPUR, YELLOW NICKEL

Scrambling shrub like the last, but with larger, often more elliptic leaflets of a thinner texture, the tips more strongly acuminate. Racemes simple or branched, 15–30 cm long; pedicels rather stout, 6–8 mm long; calyx 8–9 mm long; petals yellow, 10–11 mm long. Stamens shorter than the petals. Pods dark brown at maturity, ovate-oblong, 5–7 cm long, prickly more or less like the last. Seeds oblong-rounded to subglobose, ca. 2 cm long or sometimes slightly less.

GRAND CAYMAN: *Kings* GC 127; *Proctor* 15143. LITTLE CAYMAN: *Proctor* 28063.

— Cuba and Jamaica. This species and the previous one (*C. wrightiana*) have long been confused and have usually been listed under one or another incorrect name. Type material of both species has been examined in order to establish their correct identity; for access to these authentic specimens the author is indebted to the late Dr. William T. Gillis of Michigan State University.

[*Caesalpinia pulcherrima* (L.) Sw., *Obs. Bot.* 166 (1791).

BARBADOS PRIDE

An erect, glabrous shrub to 3 m tall, unarmed or with a few scattered prickles; leaves 10–40 cm long; pinnae in 5–20 pairs; leaflets 17–25 per pinna, oblong, 1–2 cm long. Racemes usually terminal, pyramidal, the lower pedicels much longer than the upper. Sepals unequal, 10–16 mm long; petals yellow or red, long-clawed, 2–2.5 cm long. Stamens 6–8 cm long, with yellow or red filaments, these long-exserted. Pods flat, irregularly oblong, 8–10 cm long or sometimes a little longer, 6–8-seeded.

GRAND CAYMAN: *Hitchcock*; *Kings* GC 397. CAYMAN BRAC: *Proctor* (sight record, cultivated).

— Origin unknown; widely cultivated in most warm countries, and sometimes becoming naturalised.]

[*Haematoxylum* L.

Glabrous trees, often armed with spines; leaves even-pinnate or occasionally bipinnate by elongation of the lowest pair of pinnae; pinnae 2–4 pairs. Flowers small, yellow, in axillary racemes; bracteoles absent; calyx deeply 5-lobed, the lobes imbricate and slightly unequal; petals oblong, spreading, not clawed. Stamens free, hairy at the base; anthers all alike. Ovary short-stalked; style thread-like with small terminal stigma; ovules 2 or 3. Pods flat and wing-like, opening by a longitudinal slit along the middle of each valve; seeds oblong, compressed, without endosperm, the cotyledons 2-lobed.

— A tropical American genus of 2 species.

Haematoxylum campechianum L., *Sp. Pl.* 1: 384 (1753). **FIG. 138.** PLATE 29.

LOGWOOD

A shrub or small tree with deeply fluted trunk, usually less than 8 m tall, often armed with stout spines; pinnae cuneate-obovate, 0.7–2.5 cm long, rounded, truncate, or emarginate at the apex, finely many-nerved; stipules sometimes spine-like. Racemes mostly dense,

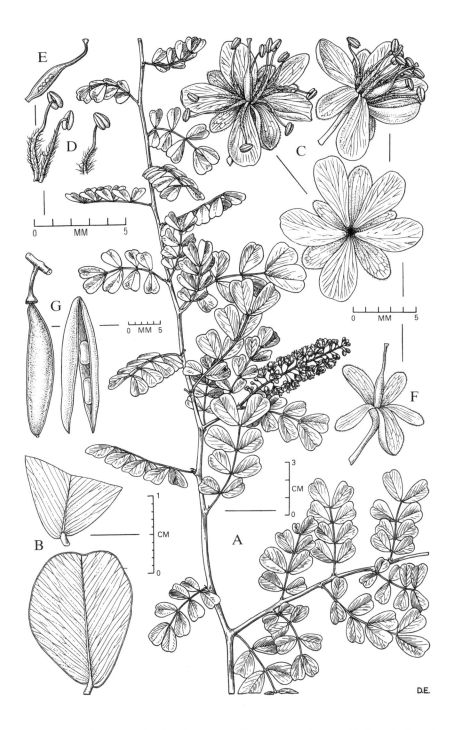

FIG. 138 **Haematoxylum campechianum.** A, portion of flowering branch. B, details of two leaflets. C, details of three flowers. D, stamens. E, pistil. F, pistil and calyx. G, two views of fruit. (D.E.)

2–8 cm long; calyx 3–4 mm long; petals 5–6 mm long. Pods oblanceolate or narrowly oblong-elliptic, 3–5 cm long, 1–3-seeded.

GRAND CAYMAN: *Brunt* 1800, 1808; *Hitchcock*; *Kings* GC 416; *Millspaugh* 1369, 1731; *Sachet* 409.

— Native to C. America, chiefly in the Yucatan Peninsula; widely naturalised in the West Indies, where it was introduced early in the 18th century. The heartwood, which is bright red when freshly cut, is the source of a dye formerly much used for textiles, and which is still highly valued as a bacteriological and cytological stain. It is extracted by boiling the wood chips in water, and must be used with a chemical 'mordant' in order to be permanent. The usual final dye-colour obtained is black. The wood itself is very hard and heavy, and normally will not float; although brittle, it is very durable and takes a high polish.]

[*Delonix* Raf.

Delonix regia (Bojer) Raf., the poinciana, is often planted for ornament. This tree, whose flowers occur in several colour-forms, is a native of Madagascar.

GRAND CAYMAN: *Brunt* 2027; *Kings* GC 77. CAYMAN BRAC: *Proctor* (sight record).]

Cassia L., *Senna* Mill., *Chamaechrista* Moench

Herbs, shrubs or trees of various habit; leaves even-pinnate, often with glands on the rhachis and petiole. Flowers solitary, or in axillary or terminal racemes, or in terminal panicles; calyx of 5 imbricate sepals, these deciduous; petals 5, subequal or the lower ones smaller. Stamens 10, all perfect, or with 6–7 perfect and the remainder staminodial, or else 3 lacking; anthers all alike or those of the lower stamens larger, all opening by an apical pore or slit. Ovary sessile, stalked, free from the calyx; ovules few or many. Pods cylindrical or flattened, indehiscent or 2-valved, rarely with longitudinal wings, continuous or septate within, or often filled with pulp; seeds more or less compressed, with endosperm.

These 3 genera comprise approx. 660 species (*Senna* approx. 295–300 species; *Cassia* approx. 30 species; *Chamaechrista* approx. 330 species — see Lewis *et al.* 2005, *Legumes of the World*), most plentiful in tropical regions. The laxative drug 'senna' consists of the dried leaves of several Asiatic species of *Senna*; otherwise, few are of any economic value other than as showy garden and street trees and shrubs (e.g. *Cassia fistula, Cassia javanica* var. *indochinensis*) are often planted as ornamentals.

Author note: the treatment here repeats that of the first edition, but the editors have brought the taxonomy and nomenclature in line with the monograph of Irwin & Barneby ('*The American Cassiinae*', *Mem. N.Y. Bot. Gard.* 35: 1–917 (1982)). According to this publication, the genus *Cassia* in the strict sense does not occur in the Cayman Islands except for two or three cultivated tree species. All the native or naturalised Cayman species belong to the genera *Senna* or *Chamaecrista* and can be summarised as follows: *Senna uniflora* (Miller) H. S. Irwin & Barneby; *Senna obtusifolia* (L.) H. S. Irwin & Barneby; *Senna occidentalis* (L.) Link; *Senna ligustrina* (L.) H. S. Irwin & Barneby; *Senna pallida* (Vahl) H. S. Irwin & Barneby (*Cassia biflora* of ed. 1); *Chamaecrista nictitans* (L.) Moench var. *aspera*; *Chamaecrista lineata* (Sw.) Greene var. *lineata* (*Cassia clarensis* (Britton & Rose) R. A. Howard of ed. 1).

KEY TO NATIVE SPECIES OF SENNA AND CHAMAECHRISTA

1. **Leaflets 2–4 pairs, pubescent all over:**
  2. Weedy herbs with leaflets mostly 2–5 cm long:
    3. Petals 5–7 mm long; pods straight, 2.5–5 cm long, transversely grooved between the seeds: **S. uniflora**
    3. Petals ca. 10 mm long; pods curved, 15–20 cm long, not transversely grooved: **S. obtusifolia**
  2. Small bushy shrub with leaflets less than 1.5 cm long: **C. lineata var. lineata**
1. **Leaflets mostly 5 pairs or more, glabrous or nearly so:**
  4. Petals 1.2 cm long or more; pods glabrous or puberulous, more than 5 cm long:
    5. Leaves glandular at the base of the petiole; leaflets acute or acuminate:
      6. Petiolar glands globose-tuberculate; seeds dark olive: **S. occidentalis**
      6. Petiolar glands cylindrical-pointed; seeds black with grey margins: **S. ligustrina**
    5. Leaves glandular between the lowest pair of leaflets; leaflets obtuse or rounded at the apex: **S. pallida**
  4. Petals less than 1 cm long; pods hairy, 1.5–3 cm long: **C. nictitans var. aspera**

Senna uniflora (Miller) H. S. Irwin & Barneby *Mem. N.Y. Bot. Gard.* 35: 258 (1982). **FIG. 139.**

Cassia uniflora Miller (1768).

An erect annual herb to 1 m tall, clothed throughout with more or less appressed, brownish hairs. Leaves petiolate, the petiole not glandular; rhachis with 1 or more cylindrical or club-shaped glands between the leaflets; leaflets 2–4 pairs, oblong to broadly obovate, 2–5 cm long, rounded and mucronate at the apex. Racemes short, axillary; sepals rounded, 4–6 mm long; petals yellow, about twice as long as the sepals. Perfect stamens 7. Pods linear, 2.5–5 cm long, 3–4 mm broad, deeply grooved between the seeds and ultimately separating into joints; seeds 5–10, trapezoidal, ca. 4 mm long, dark brown.

GRAND CAYMAN: *Kings GC 48, GC 49, GC 90.*

— Bahamas, Greater Antilles, Mexico to northern S. America, also in the Galapagos Islands, a weed of grassy roadsides, clearings, and thickets.

Senna obtusifolia (L.) H. S. Irwin & Barneby, *Mem. N.Y. Bot. Gard.* 35: 252 (1982).

Cassia obtusifolia L. (1753).

Annual shrubby herb to 1 m tall or more; leaflets 2–3 pairs, obovate or broadly oblong-obovate, puberulous on both sides, 1.5–4 cm long, 1–2.5 cm broad, broadly obtuse and shallowly emarginate, with a gland on the leaf-rhachis between the lowest pair of leaflets only. Flowers mostly paired on short peduncles in the leaf-axils; petals light yellow or orange-yellow. Pods narrowly linear, slightly 4-angled, smooth; seeds 20–24 per pod.

GRAND CAYMAN: *Sachet 383.*

— Pantropical except for Australia, a weed of open waste ground, roadsides and clearings.

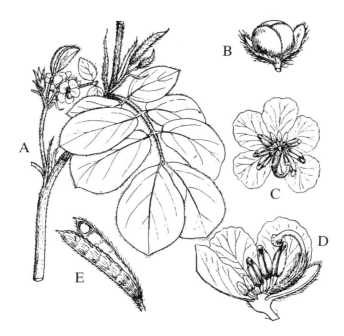

FIG. 139 **Senna uniflora**. A, portion of branch showing leaf, flowers, and young pods, × ¹/₂. B, calyx in bud, × 2. C, flower from above, × 2. D, vertical section through flower, × 3. E, pod, × 1. (F. & R.)

Senna occidentalis (L.) Link, *Handb.* 2: 140 (1831).

Cassia occidentalis L. (1753).

DANDELION

An erect annual herb up to about 1 m tall, often subwoody near the base, glabrous or nearly so. Leaves long-petioled, the petiole bearing a sessile globose gland near the base; leaflets mostly 4–6 pairs, ovate or lance-ovate, 3–7 cm long, acuminate at the apex, the margins finely ciliate. Racemes few-flowered, in the upper axils; sepals greenish, 6–9 mm long; petals yellow, twice as long as the sepals. Perfect stamens 6, two of them longer than the other four, and 4 staminodes. Ovary pubescent. Pods oblong-linear, slightly curved, 6–12 cm long, 6–9 mm broad, slightly grooved transversely between the seeds, minutely puberulous or glabrate; seeds flattened-obovoid, brown, 3–4 mm long.

GRAND CAYMAN: *Hitchcock*; *Kings* GC 220; *Millspaugh* 1393; *Proctor* 15112. CAYMAN BRAC: *Proctor* 28971.

— Southern U.S.A., West Indies, and continental tropical America; naturalised in the Old World tropics. In some countries, the roasted and pulverised seeds are used as a substitute for coffee.

Senna ligustrina (L.) H. S. Irwin & Barneby, *Mem. N.Y. Bot. Gard.* 35: 409 (1982). PLATE 29.

Cassia ligustrina L. (1753).

A small erect shrub up to 1.5 m tall or more, the young branches puberulous. Leaves long-petioled, the petiole bearing a pointed-cylindrical or conical gland below the middle; leaflets 5–7 pairs, lanceolate and unequal-sided, 1.5–4 cm long or more, narrowly acute, glabrous except for the finely ciliate margins. Racemes few-flowered, in the crowded upper axils, forming a panicle-like leafy inflorescence; sepals 6–8 mm long; petals yellow, dark-veined, 1.2–1.5 cm long. Stamens as in *S. occidentalis*. Pods oblong-linear, somewhat curved, 7–10 cm long, 6–8 mm broad, glabrous or sparsely puberulous; seeds numerous, oblong-ellipsoid, 4–4.5 mm long, black with a grey rim.

GRAND CAYMAN: *Fawcett*; *Proctor* 15266.

— Florida, Bermuda, Bahamas, and the Greater Antilles, in thickets and clearings.

Senna pallida (Vahl) H. S. Irwin & Barneby, *Mem. N.Y. Bot. Gard.* 35: 531 (1982).

Cassia biflora L. (1753).
Cassia pallida Vahl (1807)

A shrub usually 2–4 m tall, the young twigs, petioles, and inflorescence glabrous or puberulous. Leaves petiolate, the petiole not glandular, but a cylindrical and often acuminate gland occurs between the lowest pair of leaflets; leaflets 4–11 pairs, oblong-elliptic or narrowly obovate, 1–3.5 cm long, rounded and mucronulate at the apex. Inflorescence a subumbellate axillary raceme with 2–4 rather large flowers, the pedicels bearing cylindrical glands at the base; calyx 5.5–8 mm long, the lobes unequal; petals yellow, very unequal, the largest 2–2.3 cm long, subsessile, the smaller ones clawed. Perfect stamens 7, three of them larger than the others and beaked. Ovary sessile. Pods linear, compressed, 2-valved, 7–15 cm long, 5–8 mm broad; seeds 14–20, oblong with notch at one end, brown, 5–6 mm long.

CAYMAN BRAC: *Proctor* 28956.

— West Indies, C. America and northern S. America, in thickets and woodlands.

Chamaecrista nictitans (L.) Moench var. aspera (Muhl. ex Elliott) H. S. Irwin & Barneby, *Mem. N.Y. Bot. Gard.* 35: 838 (1982). **FIG. 140.**

Cassia nictitans (L.) Greene var. *aspera* (Muhl. ex Elliott) Torr. & Gray (1840).
Chamaecrista confusa Britton (1930), not *Cassia confusa* Phil. (1894).
Cassia caymanensis C. D. Adams (1970).

WILD SHAME-FACE

A small woody herb or subshrub, the villous-pubescent stems ascending or rigidly erect, branched or unbranched. Leaves much longer than wide, 2.5–6.5 cm long, short-petiolate, the pubescent petioles with 1 or 2 stalked glands; rhachis pubescent; leaflets glabrous, in 10–25 pairs, linear, unequal-sided, 6–12 mm long, aristate at the apex. Flowers solitary on axillary peduncles shorter than the petioles; calyx ca. 5 mm long, hairy; petals yellow, the larger 7–8 mm long. Perfect stamens 10. Pods compressed, hairy, linear-oblong, 15–25 mm long, 3–4 mm wide, 2-valved and elastically dehiscent; seeds 4–7, flattened obovoid, dark brown, ca. 3 mm long.

GRAND CAYMAN: *Brunt* 1904; *Kings* GC 44, GC 92; *Millspaugh* 1305; *Proctor* 15072, 27929, 31048; *Sachet* 370.

— Southeastern U.S.A., Bahamas and Jamaica, in pastures and sandy clearings.

Chamaecrista lineata (Sw.) Greene, *Pittonia* 4: 31 (1899) var. lineata. PLATE 30.

Cassia lineata Sw. (1788).
Cassia clarensis (Britton) Howard (1947).

A bushy, much-branched shrub up to 1.5 m tall, densely appressed-puberulous throughout. Leaves often broader than long, the short petiole bearing a round sessile gland above the

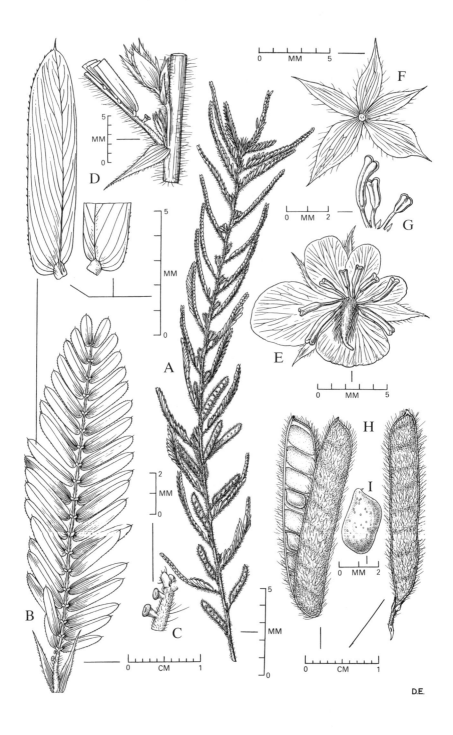

FIG. 140 **Chamaecrista nictitans var. aspera**. A, branch with flowers and fruits. B, leaf. C, glands of petiole. D, details of stem and leaflets. E, complete flower. F, calyx. G, stamens. H, fruits. I, seed. (D.E.)

middle; leaflets 2–4 pairs, oblanceolate or oblong-obovate, unequal-sided, 8–15 mm long, rounded or truncate and mucronate at the apex, the lateral veins prominent. Flowers axillary, solitary or 2 together, on slender pedicels ca. 1 cm long, much longer than the petioles; calyx ca. 8 mm long, puberulous in a longitudinal line on the back; petals yellow, the larger 9–10 mm long. Perfect stamens 10. Pods compressed, appressed-puberulous, linear and slightly curved, 20–35 mm long, 3–4 mm wide, apiculate, 2-valved and elastically dehiscent; seeds 8–11, oblong-stipitate, light brown, ca. 3 mm long.

GRAND CAYMAN: *Brunt* 1907.

— Bahamas and Greater Antilles except Puerto Rico, in sand or pockets of exposed limestone, usually near the sea.

Bauhinia L.

Trees or shrubs, erect or climbing, with or without spines or tendrils; leaves simple, of 2 united leaflets more or less parted at the apex, or rarely completely 2-foliolate; stipules small and soon falling. Flowers solitary or in racemes, these simple and terminal or axillary, or paniculate; calyx with a short elongate tube, the limb more or less spathe-like, before anthesis either closed and entire, or else contracted at the apex and 5-toothed, after anthesis remaining entire or variously splitting. Petals 5, slightly unequal. Perfect stamens 10 or less, some or most often being reduced to staminodia or lacking; anthers attached at the middle and opening longitudinally. Ovary stalked or subsessile, with 2–many ovules; style various, the stigma small or often dilated and peltate. Pods oblong or linear, compressed, 2-valved with elastic valves, or indehiscent; seeds roundish, compressed, and with endosperm; cotyledons flat, more or less fleshy.

— A pantropical genus of more than 200 species, several often cultivated for ornament.

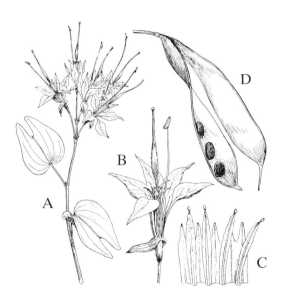

FIG. 141 **Bauhinia divaricata**. A, inflorescence and two leaves. × ⅓. B, flower, × ⅔. C, sterile stamens opened flat, × 2. D, pod, × ⅔. (F. & R.)

Bauhinia divaricata L., *Sp. Pl.* 1: 374 (1753). **FIG. 141.** PLATE 30.

Bauhinia porrecta Sw. (1788).

BULL-HOOF

An erect to arching shrub or small tree to 4 m tall or more, the branches glabrous or nearly so; leaves petioled, the bifurcate blades mostly 2–6 cm long, often wider than long, the lobes variously united from 1/4 to 7/8 of their length, the nerves puberulous beneath. Racemes short and dense, with up to about 10 flowers, all perfect or often several with abortive ovary and functionally staminate only; calyx 12–15 mm long, terminated by short, bristle-like teeth; petals at first white, then turning pink, 2–2.5 cm long, slender-clawed. Fertile stamen 1, twice as long as the petals, the 9 sterile ones much shorter and united into a tube for most of their length. Ovary long-stalked. Pods linear- or oblanceolate-falcate, 5–12 cm long, 9–15 mm wide, long-stalked, finely puberulous or glabrate; seeds 3–10, flattened-ellipsoidal, 6–8 mm long, dark brown.

GRAND CAYMAN: *Brunt* 1787, 1887; *Correll & Correll* 50998; *Hitchcock*; *Lewis* GC 32; *Millspaugh* 1289; *Proctor* 15024. LITTLE CAYMAN: *Kings* LC 106; *Proctor* 28040. CAYMAN BRAC: *Kings* CB 14, CB 91a; *Millspaugh* 1160, 1207, 1208; *Proctor* 29047.

— West Indies and on the continent from southwestern Texas south to Honduras, often frequent in rocky woodlands. Such ornamental species as *Bauhinia galpinii* N. E. Br. and *Bauhinia monandra* Kurz are sometimes cultivated.

Cayman plants of *Bauhinia divaricata* on all three islands are sometimes infected with the parasitic flower *Pilostyles globosa* var. *caymanensis* (Apodanthaceae: see page 428).

Hymenaea L.

Large, unarmed, resinous trees; stipules soon falling. Leaves petiolate, 2-foliolate, the leaflets asymmetric, subsessile, coriaceous. Inflorescence a terminal corymbose panicle. Flowers large; calyx narrowly campanulate with 4 or 5 imbricate lobes; petals 5, white, sessile or clawed. Ovary short-stipitate, the stipe adnate to the calyx tube; style filiform with small stigma. Fruit woody, thick, indehiscent; seeds bony.

— A genus of 14 species, 13 in the neotropics (mostly in S. America) and 1 in eastern tropical Africa.

Hymenaea courbaril L., *Sp. Pl.* 2: 1192 (1753).

LOCUST, STINKING TOE

Tree up to 30 m tall and d.b.h. 2 m (under favourable conditions); bark smooth, greyish, the young branches densely pale-puberulous at first, becoming glabrescent with age; stipules sheathing, ca. 1 cm long, densely puberulous, soon falling. Leaves with petioles up to 14 mm long or more; leaflets subsessile, ovate to elliptic, 4.4–12.8 cm long, coriaceous, strongly asymmetric, acute to acuminate at apex, cuneate or truncate at base, with scattered abaxial dark gland dots. Inflorescence axillary, paniculate, up to 10 cm long, densely puberulous; pedicels short and stout; calyx-tube 11–21 mm long, exuding resin, the lobes 14–18 mm long, slightly hooded at apex; petals 5, white,

13–20 mm long, densely gland-dotted; stamens 10, free, exserted; ovary glabrous, the style exserted, the stigma capitate. Fruit oblong, 7.5–10.3 cm long; seeds 1–8, embedded in dryish edible pulp.

GRAND CAYMAN: *Proctor* 52236, found in wooded pasture near the East End Farm Road.

— Mexico, C. America, Greater and Lesser Antilles, and S. America.

[*Tamarindus* L.

Unarmed trees; leaves even-pinnate, with small, numerous leaflets, the petiole and rhachis not glandular; stipules minute, soon falling. Flowers in short terminal racemes; bracts and bracteoles ovate-oblong, soon falling. Calyx 4-lobed; petals 5, three of them subequal and evident, the other two minute and scale-like. Perfect stamens 3, united into a sheath; anthers oblong, opening longitudinally. Ovary stalked, the stalk adnate to the calyx-tube; ovules numerous; style elongate. Pods more or less oblong, thick and scarcely compressed, indehiscent, the outer covering (epicarp) crust-like and fragile, enclosing an edible pulp; seeds roundish, compressed, separated by hard partitions; endosperm lacking, the cotyledons thick.

— The genus consists of a single species, native of tropical Asia, but now cultivated in most tropical countries and often becoming naturalised.

Tamarindus indica L., *Sp. Pl.* 1: 34 (1753).

TAMARIND

A handsome evergreen tree 10 m tall or more; leaves glabrous or nearly so; leaflets 10–18 pairs, oblong, 12–25 mm long, the venation finely prominulous. Racemes mostly shorter than the leaves; calyx 8–10 mm long; petals pale yellow with red veins, the larger petals slightly longer than the calyx. Pods 5–10 cm long or more, 2 cm thick, brown and finely scaly; seeds 1 cm in diameter, lustrous brown.

GRAND CAYMAN: *Brunt* 2171; *Kings* GC 81. CAYMAN BRAC: *Kings* CB 6.

— The juicy, acidulous pulp of the fruit is used to make a refreshing drink, and as an ingredient of candies and condiments. The plant itself makes a fine shade tree, resistant to drought, and there are also numerous uses for the leaves and wood.]

SUBFAMILY 3. MIMOSOIDEAE

Herbs, shrubs, or trees, with or without prickles; leaves usually bipinnate. Flowers regular, perfect or rarely polygamous, in axillary globose heads, cylindrical spikes or racemes, or in terminal panicles of globose heads; calyx usually 4–5-lobed or -toothed, the lobes valvate; petals 4 or 5, free or united below, valvate. Stamens as many or twice as many as the petals, or more numerous up to 100 (always 10 or more in Cayman genera except *Mimosa*); filaments thread-like, usually elongate, free or united below, the tube so formed often exserted and showy.

— A subfamily chiefly occurring in the tropics.

KEY TO GENERA

1. Leaflets alternate; flowers in racemes; stamens scarcely longer than the petals; seeds bright red: **[Adenanthera]**

1. Leaflets opposite; flowers in small round heads; stamens long-exserted; seeds not red:

    2. Pods with continuous persistent margins, the valves separating from them and breaking into joints; flowers pink; stamens 4 per flower; leaves sensitive to the touch: **Mimosa**

    2. Pods with valves not separating from the margins and not breaking into joints; flowers white or yellow; stamens 10 or more per flower; leaves not sensitive to the touch:

        3. Flowers yellow; pods swollen, marked with lines: **Acacia**

        3. Flowers white; pods flat, not marked with lines:

            4. Pods more than 1 cm broad: **Leucaena**

            4. Pods less than 1 cm broad:

                5. Green-stemmed wiry herb; leaflets less than 1 cm long; stamens 10, free: **Desmanthus**

                5. Large shrub or small tree; leaflets more than 1 cm long; stamens more than 10, united at the base: **Calliandra**

[*Adenanthera* L.

Unarmed trees; leaves bipinnate, the pinnae subopposite, the leaflets alternate. Flowers very small, in long slender racemes, perfect or polygamous; calyx bell-shaped, 5-toothed; petals 5, united below the middle or nearly free. Stamens 10, free, the filaments about as long as the petals; anthers bearing a deciduous gland. Ovary sessile, with many ovules. Pods flat, linear, 2-valved, the valves becoming twisted and curled after dehiscence; seeds roundish, thick and hard.

— A small genus of 3 or 4 species, native of Africa, Asia and Australia.

Adenanthera pavonina L., *Sp. Pl.* 1: 384 (1753). PLATE 30.

CURLY BEAN

A tree to 10 m tall or more, nearly glabrous; leaves with 2–5 pairs of subopposite pinnae, these 10–20 cm long; leaflets alternate, elliptic or oblong-elliptic, 2–4 cm long, obtuse. Racemes 5–15 cm long, simple from the leaf-axils, panicled at the ends of branches. Flowers cream, turning yellow or pale orange, on slender pedicels; calyx ca. 1 mm long; petals ca. 3 mm long. Pods 15–25 cm long, 12–16 mm broad, swollen over the seeds; seeds somewhat compressed, ca. 8 mm in diameter, bright red.

GRAND CAYMAN: *Hitchcock*; *Kings* GC 76.

— Native of tropical Asia, but now planted and becoming naturalised in most warm countries. The wood is hard, close-grained, strong and durable. The seeds are often used to make necklaces; in India, they are also used as a standard measure of weight, 1 seed weighing about 4 grains.]

Desmanthus Willd.

Unarmed woody herbs or slender shrubs with angulate-striate branches; leaves bipinnate, with a sessile gland on the petiole just below the lowest pair of pinnae; leaflets very small; stipules bristle-like, persistent. Flowers in small, few-flowered heads, all perfect or the lowest sterile or staminate only, on solitary axillary peduncles. Calyx shortly 5-toothed; petals 5, free or slightly coherent, valvate. Stamens usually 10, free, exserted; anthers without glands. Ovary subsessile, with many ovules. Pods linear, straight or falcate, flat, 2-valved, continuous or septate within; seeds flattened-ovoid.

— A tropical American genus of perhaps more than 20 species, the exact number uncertain due to great variability within the group.

Desmanthus virgatus (L.) Willd. in L., *Sp. Pl.* 4: 1047 (1806). **FIG. 142.**

Desmanthus depressus Humb. & Bonpl. ex Willd. in L. (1806).

A somewhat shrubby herb or slender shrub, variable in habit from decumbent to erect, to 1 m tall or more, glabrous or nearly so; leaves with pinnae in mostly 3–5 pairs; leaflets 8–20 pairs, linear or linear-oblong, 3–9 mm long, of thin texture. Peduncles 1–5 cm long; flowers whitish, in small brush-like heads 3–5 mm across. Pods 3–8 cm long, 3–5 mm broad, acute or acuminate; seeds oblique, brown, 2 mm long.

CAYMAN BRAC: *Proctor* 28972.

— Florida and Texas southward to S. America, and also naturalised in tropical Asia, a weed of secondary thickets and open waste ground. Some writers separate the smaller and larger forms into two species, but there appears to be complete intergradation between the extremes. The Cayman specimen combines the leaf and fruit characters of D. '*depressus*' with the erect, shrubby habit of typical D. *virgatus*.

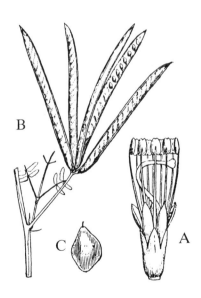

FIG. 142 **Desmanthus virgatus.** A, flower, × 5. B, peduncle and ripe pods, × ²/₃. C, seed much enlarged. (F. & R.)

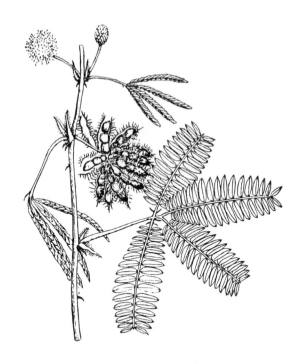

FIG. 143 **Mimosa pudica**. Portion of branch with flowers in bud and open and cluster of ripe pods. (F. & R.)

Mimosa L.

Herbs, shrubs or trees, or sometimes woody vines, usually armed with prickles; leaves bipinnate, often sensitive to the touch, and the petioles mostly lacking glands. Flowers perfect or polygamous, in stalked heads, mostly 4–5-parted, or sometimes the parts in 3s or 6s; peduncles axillary and solitary or clustered, or sometimes terminal and paniculate. Calyx usually minute; petals more or less united. Stamens as many or twice as many as the petals, free, exserted. Ovary sessile or rarely stalked, and with 2–many ovules. Pods oblong or linear, usually flat, the valves jointed and separating from the persistent continuous margins, the joints 1-seeded; seeds compressed, roundish.

— A pantropical genus of more than 400 species, best represented in the warmer parts of America.

Mimosa pudica L., *Sp. Pl.* 1: 518 (1753). **FIG. 143**. PLATE 30.

SHAME-FACE, SHAME-LADY, SENSITIVE PLANT

A subwoody annual or persistent herb, prostrate or erect, the stems up to 60 cm long, bearing long spreading hairs or glabrate, also armed with sharp recurved prickles below the stipules and sometimes elsewhere. Leaves collapsing when touched; pinnae 2 pairs, 2.5–8 cm long; leaflets in 10–20 pairs, oblong-linear, 5–10 mm long, very inequilateral at the base, and sparsely hairy. Flowers pink, in globose heads 1–1.5 cm in diameter, the peduncles 1–3 in the leaf-axils; petals ca. 2 mm long; stamens 4, pink, 7–8 mm long. Pods linear-oblong, 1–1.5 cm long, 3–4 mm broad, with 2–5 joints, glabrous, the margins densely armed with bristles.

GRAND CAYMAN: *Kings* GC 421.

— West Indies and continental tropical America, also naturalised in the Old World tropics. This is chiefly a weed of pastures and open waste ground, often in moist places and along paths.

Leucaena Benth.

Unarmed shrubs or small trees; leaves bipinnate, usually with a gland on the petiole; stipules small. Flowers white, in globose, stalked, many-flowered heads, the peduncles solitary or clustered in the leaf-axils, or sometimes arranged in terminal naked racemes, each peduncle bearing 2 bracts at or below the apex. Calyx tubular, 5-toothed; petals 5, free, valvate. Stamens 10, free, exserted, the anthers often hairy. Ovary stalked, with filiform style; ovules many. Pods oblong-linear, flat, 2-valved, continuous within, short-stalked; seeds transverse, compressed-ovoid.

— An American genus of perhaps as many as 50 species, but probably fewer.

Leucaena leucocephala (Lam.) DeWit in *Taxon* 10: 54 (1961). PLATE 31.

Leucaena glauca Benth. (1842), in part, not *Mimosa glauca* L. (1753).

A shrub or small tree to about 6 m tall (in our area), the younger parts and leaflet-margins puberulous; pinnae in 3–8 pairs, 6–9 cm long; leaflets in 8–14 pairs, linear-oblong and inequilateral, 7–16 mm long, 2–4 mm broad, often pale or glaucous beneath. Flower-heads ca. 2 cm in diameter, on axillary peduncles 2–4 cm long; calyx 2–3 mm long; petals 4–5 mm long, puberulous. Stamens 8–10 mm long. Pods dark brown, pointed, 10–18 cm long, 1.3–1.8 cm broad, usually 3–10 in a cluster.

GRAND CAYMAN: *Brunt* 1891, 2083; *Hitchcock*; *Kings* GC 78; *Millspaugh* 1391; *Proctor* 15205. CAYMAN BRAC: *Kings* CB 83; *Proctor* 28933.

— Florida, West Indies, and continental Caribbean countries; naturalised in the Old World tropics; characteristic of second-growth thickets, old fields, and roadsides. It is alleged that if horses or mules eat any part of this plant, their hair will fall out, but that cattle and goats are not affected.

Acacia L.

Shrubs or trees, armed with prickles or unarmed; leaves bipinnate, usually with a petiolar gland; leaflets mostly small and numerous; stipules often spine-like. Flowers in globose heads or cylindrical spikes, perfect or polygamous, usually yellow, on solitary or clustered peduncles from the leaf-axils, or in terminal panicles. Sepals 4–5-toothed or -lobed; petals 4 or 5, more or less united. Stamens numerous (sometimes 50 or more), free or nearly so, exserted. Ovary sessile or stalked, with hair-like style; ovules 2–many. Pods various in form, cylindrical or compressed, linear to ovate, straight, curved or contorted, membranous to woody, 2-valved or indehiscent; seeds transverse or longitudinal, usually compressed-ovoid.

— A large, widespread genus of about 500 species, most plentiful in tropical America, Africa and Australia. It is sometimes subdivided into a number of smaller genera, but these are mostly not very distinct.

Acacia farnesiana (L.) Willd. in L., *Sp. Pl.* ed. 4, 4: 1083 (1806).

A small bushy tree or thicket-forming shrub, the branches glabrous or nearly so, with prominent lenticels, and armed with stipular spines; leaves with 1–3 pairs of pinnae, rarely more, 1–4 cm long; petiole puberulous and with a small gland; leaflets in 10–20 pairs, oblong-linear, 2–6 mm long, ca. 1 mm broad; stipular spines 4–30 mm long. Flowers yellow, fragrant, in globose heads 7–15 mm in diameter; peduncles solitary or clustered, 2–4 cm long, puberulous. Calyx 1–1.5 mm long; corolla 2–3 mm long. Pods dark brown and marked with fine longitudinal lines, glabrous, linear-oblong, curved, subcylindrical, 4–7 cm long, 1–1.5 cm thick, filled with sweet pulp.

GRAND CAYMAN: *Brunt* 1890. CAYMAN BRAC: *Proctor* 29087.

— Southern U.S.A., West Indies, and continental tropical America, often planted and naturalised, so that its true natural range is obscure; introduced in the Old World tropics and subtropics. Commonly occurs in thickets along the borders of pastures; the leaves and pods are much eaten by livestock. The wood is hard and close-grained, but used chiefly as fuel. The bark and pods are rich in tannin. The viscid juice of the pods can be used to mend china, while the gum exuding from the trunk is similar to gum arabic and is suitable for making mucilage. The flowers yield a delicious, high-priced perfume by petroleum ether extraction further purified by alcoholic extraction; commercial production at present is mostly confined to Lebanon.

Calliandra Benth.

Mostly unarmed shrubs or small trees, rarely herbs; leaves bipinnate, the leaflets small and numerous or large and few; stipules usually persistent, often crowded at the base of young shoots, rarely spine-like. Peduncles solitary or clustered in the leaf-axils or in terminal racemes; flowers in globose heads, polygamous; calyx 5-toothed or -lobed; corolla funnel-shaped, 5-lobed to about the middle, the lobes valvate. Stamens numerous (10–100), united toward the base, long-exserted; anthers usually minute and either glandular-pubescent or glabrous. Ovary sessile, with hair-like style; ovules many. Pods linear, straight or nearly so, narrowed at the base, flat with thickened margins, 2-valved, the valves elastically dehiscent, continuous within; seeds roundish, compressed.

— A genus of more than 150 species, the majority in tropical America. Several with showy red stamens are commonly cultivated for ornament.

Calliandra cubensis (Macbr.) Léon in *Contr. Ocas. Mus. Hist. Nat. Col. "De La Salle"*, no. 9: 7 (1950). PLATE 31.

Calliandra gracilis of Hitchc (1893), not Griseb. (1861).
Calliandra formosa var. *cubensis* Macbr. (1919).

A small glabrous tree to 5 m tall or more; leaves with 2 or 3 pairs of pinnae; leaflets 6–12 pairs, oblong or oblong-obovate, mostly 10–24 mm long with rounded apex and very unequal-sided base. Peduncles 3–5 cm long; calyx 2.5–3 mm long; corolla 4–6 mm long. Stamens with white filaments 1–1.5 cm long. Pods 5–9 cm long, usually 7–8 mm broad, rounded-truncate and obliquely short-apiculate at the apex, long-attenuate to a stalk-like base, 8–10-seeded.

GRAND CAYMAN: *Hitchcock*. LITTLE CAYMAN: *Proctor* 35189. CAYMAN BRAC: *Proctor* 29034, 29355.

— Eastern Cuba and the Bahamas, in rocky limestone woodlands.

LYTHRACEAE

Herbs, shrubs or trees, the young stems often 4-angled; leaves usually opposite, simple and entire, or rarely alternate; stipules absent or sometimes present. Flowers regular or zygomorphic, usually perfect, 3–16-parted, axillary or extra-axillary and solitary or in cymes or clusters, or rarely in terminal panicles. Calyx tubular, with valvate primary teeth or lobes, sometimes also with an 'epicalyx' of as many accessory teeth; petals as many as the calyx-lobes, or fewer, or none, inserted in the throat of the calyx between the lobes, often crumpled in bud. Stamens as many as, or fewer or more than, the calyx-lobes, inserted at different levels on the inside of the calyx-tube. Ovary superior, sessile or stalked, completely or incompletely 2–6-locular, the style simple or none, the stigma small and capitate or rarely 2-lobed; ovules numerous on axile placentas, or rarely only 2. Fruit a dry dehiscent or indehiscent capsule; seeds usually small and many, lacking endosperm.

— A world-wide but chiefly tropical and subtropical family of about 22 genera and 475 species. Few are of any commercial value, but several, especially species of *Lagerstroemia* (with alternate leaves and 15–many stamens) are widely cultivated for ornament.

KEY TO GENERA

1. Herbs; flowers sessile and solitary or few together in the leaf-axils:	Ammannia
1. Shrubs; flowers stalked and numerous in terminal panicles:	[Lawsonia]

Ammannia L.

Annual herbs, glabrous or nearly so, with stems more or less 4-angled; leaves narrow, opposite, sessile. Flowers small, sessile, solitary or clustered in the leaf-axils; calyx bell-shaped to globose or ovoid, 4-toothed, with 4 accessory teeth in the sinuses; petals 4, deciduous, or else none. Stamens 4 or 8. Ovary sessile, 1–5-locular, with very short or longer and exserted style. Capsule subglobose, thinly membranous, more or less enclosed by the persistent calyx, rupturing irregularly.

— A mostly tropical or subtropical genus of about 20 species, mostly occurring in swamps or moist places.

KEY TO SPECIES

1. Style more than 1.5 mm long, conspicuous; petals present; capsule 2–3 mm in diameter:	A. coccinea
1. Style very short or obsolete; petals absent; capsule usually more than 3 mm in diameter:	A. latifolia

Ammannia coccinea Rottb., *Pl. Hort. Rar. Progr.* 7 (1773).

Erect annual herb; stem glabrous, simple or commonly branched above, often spongy at base, up to 60 cm tall or more. Leaves decussate, linear-lanceolate, up to 6 or more cm long and 0.8 cm broad, acute at apex, glabrous but the margins minutely scabridulous. Calyx ca. 2 mm long in flower, much longer in fruit; petals 4(–5), 1.5–2 mm long, bright rose-pink to deep magenta; stamens 4(–5). Capsules 3.5–5 mm in diameter.

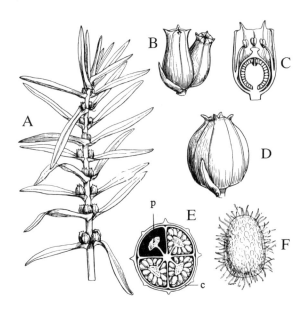

FIG. 144 **Ammannia latifolia.** A, portion of branch with flowers and fruit, × ²/₃. B, two flowers from leaf-axil, × 4. C, flower cut lengthwise and spread out, × 4. D, fruit enclosed in the globose calyx, × 4. E, cross-section of fruit, × 4; c, calyx; p, placenta with seeds removed. F, seed, × 30. (F. & R.)

GRAND CAYMAN: *Proctor* 47366, 48319.

— Central and Southeastern U.S.A. from South Dakota and Ohio to Texas, the Carolinas and Florida, the Bahamas, Grater and Lesser Antilles, and Mexico to northwestern coastal S. America; introduced in California, various Pacific islands, Japan, Philippine Islands, and southwestern Europe.

Ammannia latifolia L., *Sp. Pl.* 1: 119 (1753). **FIG. 144.**

A sprawling or erect herb to 1 m tall or more, simple or much-branched; leaves linear or linear-lanceolate, 3–8 cm long, 3–10 mm broad, clasping at the base. Flowers solitary or 2–3 together in the leaf-axils; calyx ca. 4 mm long, the teeth indistinct or nearly obsolete, the accessory teeth much larger and spreading; petals none. Stamens and style enclosed by the calyx. Capsule 4–5 mm in diameter.

GRAND CAYMAN: *Brunt* 1826, 1879, 1880, 1881, 2106; *Hitchcock*; *Kings* GC 253; *Proctor* 15246, 15302, 27948.

— West Indies and continental tropical America, commonly in swampy places.

[*Lawsonia* L.

A glabrous shrub or small tree, the 4-angled branchlets sometimes with a spine-like tip; leaves rather small, opposite, with short petioles; stipules minute, conical, whitish. Flowers small, 4-parted, fragrant, in terminal panicles; calyx broadly top-shaped, with lobes slightly longer than the tube, lacking accessory teeth; petals short-clawed. Stamens 8, the filaments thick, exserted. Ovary sessile, 2–4-locular, with a stout style. Capsule globose, indehiscent or rupturing irregularly; seeds thick, trigonous-pyramidal, spongy at the apex.

— The genus consists of a single species.

Lawsonia inermis L., *Sp. Pl.* 1: 349 (1753).

HENNA

A bushy shrub 2 m tall or more; leaves elliptic to obovate, 1–3.5 cm long, sharply acuminate. Panicles 5–20 cm long; calyx 3–5 mm long; petals dull creamy or tawny, 4–6 mm long. Capsules 4–7 mm in diameter.

GRAND CAYMAN: *Brunt* 1856; *Howard & Wagenknecht* 15027-A. LITTLE CAYMAN: *Proctor* (seen in cultivation). CAYMAN BRAC: *Proctor* 28939.

— Probably native of eastern Africa and tropical Asia, but now planted and often naturalised in most warm countries. In the Cayman Islands, it now grows wild in thickets and along the borders of pastures. In some countries, chiefly in Asia, the leaves are used to make a yellow dye for staining the nails, hands and feet. If a paste of the leaves is applied to the hair, it soon produces a bright red colour. The plant also yields a dull red dye for cloth, and an excellent perfume can be extracted from the flowers. However, the yield of essential oil from the flowers is relatively so small and its extraction commercially unprofitable.]

THYMELEACEAE

Unarmed trees or shrubs, rarely herbs, the inner bark made up of very strong meshed fibres; leaves opposite or alternate, entire, mostly pinnate-nerved; stipules lacking. Flowers regular, perfect or by abortion polygamous or unisexual, rarely solitary, usually in heads, umbels, racemes or spikes, these stalked or sessile, terminal or axillary. Calyx tubular or urn-shaped, 4–5-lobed, the lobes equal or unequal, imbricate; petals as many or twice as many as the calyx-lobes, or often absent. Stamens as many or twice as many as the calyx-lobes, attached within the tube above the middle; anthers 2-celled, opening by longitudinal slits. Ovary superior, sessile or short-stalked, 1- or 2-celled, subtended by a ring-like or cup-like disc, sometimes represented by 4 or 5 scales, or lacking; stigma terminal, more or less capitate; ovules solitary in each cell, laterally attached near the apex of the cell, anatropous. Fruit nut-like or drupe-like, indehiscent; seeds with or without endosperm, the embryo straight and with fleshy cotyledons.

— A widespread family of about 40 genera and more than 500 species, the majority in Australia and South Africa. The strong bark-fibres of many species have been used for making rope.

Daphnopsis Mart. & Zucc.

Mostly dioecious shrubs or small trees; leaves alternate or sometimes apparently whorled. Flowers unisexual, axillary or terminal in heads, umbels or racemes, the latter simple or branched. Calyx 4-lobed; petals 8, 4, or absent. Stamens in staminate flowers 8, inserted in two series in the calyx-tube, the anthers sessile or nearly so. Pistillate flowers usually smaller than the staminate, with 8, 4, or no staminodes; ovary sessile, 1-celled. Fruit a small drupe; seeds without endosperm.

— A tropical American genus of 46 species.

KEY TO SPECIES

1. Peduncles simple; perianth lobes acute:	**D. occidentalis**
1. Peduncles dichotomously branched; perianth lobes obtuse:	**D. americana**

Daphnopsis occidentalis (Sw.) Krug & Urban, *Engl. Bot. Jahrb.* 15: 349 (1892). FIG. 145. Plate 31.

A shrub or small tree to 7 m tall or more, all the younger parts lightly clothed with fine appressed hairs; leaves oblanceolate or narrowly obovate, mostly 4–11 cm long, acuminate or blunt to rounded at the apex, usually pale beneath. Flowers apparently monoecious in small heads on peduncles 1–5 cm long, greenish- or yellowish-white, the tube appressed-hairy; petals absent; staminate calyx-tube 8 mm long, the lobes 4 mm long; pistillate calyx-tube 4 mm long, the lobes 2–2.5 mm long. Drupe ellipsoidal, 1.4–1.8 cm long, whitish when ripe.

GRAND CAYMAN: *Proctor* 31041. CAYMAN BRAC: *Proctor* 29020.

— Jamaica; usually occurs in rocky limestone woodlands.

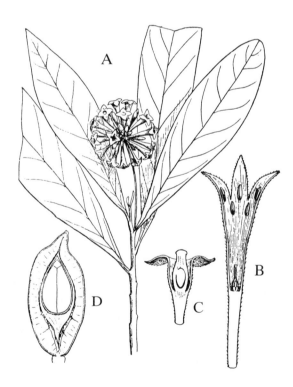

FIG. 145 **Daphnopsis occidentalis.** A, branchlet with leaves and inflorescence, × ²/₃. B, staminate flower cut lengthwise, × 4. C, pistillate flower cut lengthwise, × 4. D, fruit cut lengthwise, × 3. (F. & R.)

Daphnopsis americana (Mill.) J. R. Johnston in *Contrib. Gray Herb. n.s.* 37: 242 (1909).
PLATE 31.

Large shrub or tree up to 20 m tall; bark smooth, dull brown. Leaves often deciduous during dry periods, oblong-lanceolate to elliptic, 6–12 cm long, acute or acuminate at the apex, cuneate at base. Inflorescences terminal mostly 3-branched; flowers greenish- to yellowish-white, fragrant; staminate perianth tube ca. 9 mm long, the lobes recurved and 2.5 mm long; anthers orange; pistillate perianth tube 3.5–5 mm long, the lobes spreading and 1 mm long; stigma exserted. Ripe fruit ovoid, 9–13 mm long, crowned by persistent style base.

GRAND CAYMAN: *Burton* 056, 058, 192; *Ebanks* 001, as reported by Guala *et al.* in *Kew Bull.* **57**: 237 (2002). It has not been determined to which of Nevling's subspecies this material belongs, but the Jamaican population of this species represents subspecies **cumingii** (Meisn.) Nevling in *J. Arnold Arb.* 41: 413 (1960); the description supplied above applies to this subspecies.

MYRTACEAE

Unarmed shrubs or trees with simple, entire, opposite or alternate leaves, these pinnate-veined and usually punctate with resinous or pellucid glands; principal lateral veins usually united toward the margins by an often obscure submarginal vein extending the length of the blade; stipules absent. Flowers axillary or terminal, solitary or in bracteate inflorescences with opposite branching, usually modified in various ways, e.g. (a) elongation of the axis and reduction of the lateral branches to one flower each, forming a raceme; (b) suppression of the axis and reduction of the lateral branches to one flower, forming 'glomerules' or umbel-like clusters; (c) reduction of the lateral branches to one pair, these arising just below the flower terminating the main axis, forming a 'dichasium'; (d) elongation of both central axis and lateral branches, resulting in a panicle; and (e) transitional forms in which a panicle terminates in triads or dichasia. Flowers usually perfect, regular or essentially so. Ovary inferior, the calyx-tube ('hypanthium') adnate to the ovary its whole length or sometimes prolonged beyond it; calyx-lobes usually 4 or 5, either distinct and subequal, or the calyx more or less closed and rupturing irregularly at anthesis, or else the calyx closed and circumscissile, the top lifting off like a lid. Petals usually 4 or 5, or sometimes absent. Stamens indefinitely many, in one to many series originating around the margin of the thickened calyx-disc, usually inflexed in bud; anthers mostly 2-celled, usually opening by longitudinal slits. Ovary 2–many-celled, bearing a simple, elongate style with a small capitate stigma; ovules 2 or more to each cell, borne on axial or parietal placentas. Fruit a berry, drupe or capsule; seeds usually without endosperm.

— A family of perhaps 60 genera and nearly 3,000 species, chiefly in the tropics, but well represented in subtropical and temperate areas of the Southern Hemisphere.

The wood of this family is usually hard, tough and close-grained, and many species yield useful timber. Many also produce edible fruit, while the aromatic oils of numerous *Eucalyptus* species have found a wide use in the pharmaceutical, soap, and perfume industries. The clove tree (*Syzygium aromaticum*) is well-known as a source of culinary spice and an aromatic oil used in various food-products, dental preparations and perfumes. In the West Indies, the pimento or allspice tree (*Pimenta dioica*) and the bay rum tree (*Pimenta racemosa*) likewise produce commercially important spice or oil.

KEY TO GENERA

1. Calyx without lobes, closed in bud, the top falling off like a lid at anthesis; petals absent or minute and adherent to the inside of the calyx-lid; vegetative branching dichotomous: — **Calyptranthes**

1. Calyx with 4 or 5 evident lobes, these usually persistent; petals distinct; vegetative branching not dichotomous:

    2. Inflorescence compound, paniculate, foliage strongly aromatic: — **Pimenta**

    2. Inflorescence with flowers solitary, clustered, racemose, or dichasial, never in panicles; foliage aromatic to varying degrees:

        3. Calyx closed in bud or nearly so, at anthers splitting into 4 or 5 irregular or unequal lobes; stigma capitate; flowers usually solitary in leaf-axils: — **[Psidium]**

        3. Calyx open in bud, the lobes 4, subequal; stigma not or but slightly thicker than the style:

            4. Flowers 3 or 7 in a dichasium; cotyledons 2, distinct in the seed: — **Myrcianthes**

            4. Flowers solitary, clustered, or in racemes; cotyledons not distinguishable in the seed:

                5. Hypanthium abruptly contracted at base, the apex scarcely prolonged beyond the ovary; inflorescences always axillary: — **Eugenia**

                5. Hypanthium attenuate downward at base, the apex much prolonged beyond the ovary and forming a flaring lip; inflorescences terminal or lateral: — **[Syzygium]**

Calyptranthes Sw.

Shrubs or small trees with more or less dichotomous branching, the young branchlets often flattened, keeled or winged, glabrous or pubescent; if pubescent, the hairs usually attached near the middle (dibrachiate). Flowers usually in axillary or terminal cymose panicles, rarely solitary or few in a simple cluster. Calyx completely closed in bud, circumscissile at anthesis, the cap-like top falling away or sometimes remaining attached at one edge. Petals none, or 2–5 and very small and inconspicuous, often falling with the calyx-cap. Stamens numerous in several series. Ovary 2-celled, with 2 ovules in each cell. Fruit a 1–4-seeded berry, crowned by the basal part of the calyx; seeds subglobose; cotyledons relatively large, thin and contorted.

— A tropical American genus of more than 100 species.

KEY TO SPECIES

1. Branchlets, leaves and inflorescence entirely glabrous; individual flowers distinctly stalked: — **C. zuzygium**

1. Branchlets, leaves and inflorescence more or less densely sericeous; flowers sessile in small clusters at the ends of panicle-branches: — **C. pallens**

Calyptranthes zuzygium (L.) Sw., *Nov. Gen. & Sp. Pl.* 79 (1788).

Shrubs or small tree to 10 m tall, glabrous throughout; leaves stiff, short-petioled, elliptic or narrowly obovate, 2.5–7 cm long, with obtuse or bluntly acuminate apex, the midrib prominent on the upper side toward the base. Panicles trichotomous, rather few-flowered, the flowers fragrant; pedicels to 8 mm long; mature unopened buds ca. 4 mm long, glabrous; hypanthium 2.5 mm long. Berries globose, 7–9 mm in diameter, red turning blackish-glaucous, edible.

LITTLE CAYMAN: *Kings* LC 68?

— Florida, Bahamas, the Greater Antilles, and the Swan Islands, chiefly in woodlands over limestone. The presence of this species in the Cayman Islands needs further confirmation.

Calyptranthes pallens Griseb. in *Gött. Abh.* 7: 67 (1857). PLATE 32.

BASTARD STRAWBERRY, STRAWBERRY TREE, RED STRAWBERRY

Shrub or small tree to 7 m tall or more, the young branchlets reddish-sericeous and somewhat 2-edged. Leaves stiff, petiolate, lanceolate to elliptic, 2.5–7 cm long, the apex often long-acuminate, the midrib grooved on the upper side toward the base, the tissue beneath (abaxial side) sericeous and often distinctly paler. Panicles many-flowered, the flowers in small sessile clusters; mature unopened buds 2–3 mm long, sericeous; hypanthium 1–2 mm long. Berries globose, 4–5 mm in diameter, dark red, edible.

GRAND CAYMAN: *Brunt* 1806, 1894, 2050; *Kings* GC 284, GC 413; *Proctor* 15301. LITTLE CAYMAN: *Kings* LC 16; *Proctor* 28127. CAYMAN BRAC: *Kings* CB 58a; *Proctor* 15329, 28911, 29057.

— Florida, West Indies, and Mexico, in rocky or sandy woodlands and in thickets bordering pastures.

Pimenta Lindley

Trees, pubescent to nearly glabrous, with foliage strongly aromatic when crushed; young branches often more or less acutely angled. Flowers in axillary panicles with deciduous bracteoles; individual flowers sometimes unisexual or apparently so, 4- or 5-merous; calyx lobes persistent; petals free, white; stamens numerous, the filaments filiform; anthers short, longitudinally dehiscent; ovary usually bilocular with 1–7 ovules per locule, attached to subapical placenta; style subulate with stigma subcapitate or scarcely enlarged. Fruit a more or less globose black berry; seeds usually 1 or 2, subglobose or reniform.

— A neotropical genus of 15 species, mainly occurring in the West Indies and C. America.

Pimenta dioica (L.) Merril in *Contr. Gray Herb.* 165: 37 (1947).

PIMENTO, ALLSPICE

Tree to 20 m tall with trunk up to 30 cm in diameter; vigorous young branches flattened and 4-angled, the angles terminating distally in the position of stipules; pubescence of sordid yellowish-white hairs. Leaves with petioles 1.5–2 cm long or more; blades ovate or elliptic, 9–20 cm long, 3–9 cm broad, the apex obtuse to acute or subacuminate, the base rounded to acute with decurrent margins, the midvein deeply grooved on upper side.

Panicle 6–12 cm long, 3–4-times compound, many-flowered. Calyx lobes 4, subequal, rounded, 1.5 × 2 mm, canescent-tomentulose within; petals ca. 5 mm long. Stamens ca. 150; ovules 1 or 2 per locule; style 4–5 mm long with stigma twice as thick. Fruits mostly 6–8 mm in diameter with many small convex oil-bearing glands. Seeds usually 2, suborbicular, compressed; embryo forming a double spiral.

GRAND CAYMAN: *Proctor 49238*; collected in a yard in George Town. This species was at first thought to be solely an introduced cultivated plant until the dead remains of several very large old trees were found in a George Town building site. Later, documents were found in the Cayman Archives that recorded the export of significant amounts of Allspice from the Cayman Islands in the early to mid nineteenth century. It appears that groves of pimento trees formerly grew in the part of Grand Cayman that is now urban George Town, but all original trees have now disappeared. Meanwhile, a few young trees are now developing from seeds or seedlings brought from Jamaica. However, the evidence suggests that the Allspice tree should be considered indigenous to Grand Cayman.

[*Psidium* L.

Shrubs or trees, often with strongly pinnate-veined leaves. Flowers axillary, solitary or in 3-flowered dichasia, or otherwise clustered; calyx closed or somewhat open in bud, usually splitting irregularly down to the ovary at anthesis, producing 4 or 5 distinct teeth or lobes; hypanthium usually prolonged beyond the summit of the ovary, the free portion splitting with the calyx-lobes. Petals 4 or 5, usually white. Stamens numerous in several series from a broad disc. Ovary usually 3–4-celled (rarely 2- or up to 7-celled), with numerous ovules on bilamellate placentas of parietal origin. Fruit a fleshy berry, sometimes large and edible; seeds more or less kidney-shaped, hard and bony; embryo curved or C-shaped, with small cotyledons and a long radicle.

— A tropical American genus of more than 100 species.

[*Psidium guajava* L., *Sp. Pl.* 1: 470 (1753). **FIG. 146.** PLATE 32.

GUAVA

Shrub or small tree rarely over 7 m tall, with 4-angled branchlets; leaves elliptic or oblong, mostly 7–14 cm long, prominently veined, the apex obtuse or acute, beneath clothed with soft, greyish, mostly appressed hairs; tissue with numerous pellucid dots. Flowers usually solitary in the leaf-axils, the peduncles 1–2 cm long; calyx completely closed in bud, the mature buds 1.3–1.6 cm long, puberulous toward the base. Petals white, 1–1.2 cm long. Style about as long as the petals, with flat, peltate stigma. Berry globose or pear-shaped, 2–6 cm long, with pinkish or yellowish edible flesh and numerous small seeds.

GRAND CAYMAN: *Brunt 1827, 1980; Hitchcock; Kings GC 385; Millspaugh 1378; Proctor 11985.* CAYMAN BRAC: *Proctor 28981.*

— Widely distributed in the American tropics, often planted and naturalised, its true natural range not known; introduced in the Old World tropics.]

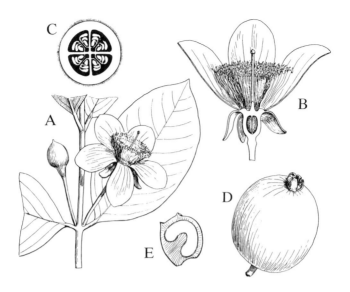

FIG. 146 **Psidium guajava**. A, portion of branch with leaf, bud and flower, × ²/₃. B, flower cut lengthwise, × 1¹/₃. C, cross-section of ovary, × 6. D, fruit, × ²/₃. E, seed cut to show embryo, × 7. (F. & R.)

Myrcianthes Berg

Small or medium-sized trees; leaves opposite or sometimes in 3s. Flowers solitary or usually in 3- or 7-flowered dichasia (rarely more), the peduncles 1–several in the upper leaf-axils, or sometimes the inflorescence modified into a terminal panicle bearing the flowers in small dichasia at the tips; central flower of a dichasium usually sessile. Calyx mostly 4-lobed; petals 4 or in some species 5. Ovary usually 2-celled, with numerous (usually 8–20) ovules radiating from a centrally attached placenta. Fruit a 1–4-seeded berry; seeds with fleshy, plano-convex cotyledons, short terete radicle, and evident plumule.

— A widespread American genus of perhaps 25 species, the majority in S. America.

Myrcianthes fragrans (Sw.) McVaugh, *Fieldiana Bot.* 29: 485 (1963). PLATE 32.

Eugenia fragrans (Sw.) Willd. (1800).
Anamomis fragrans (Sw.) Griseb. (1860).
Anamomis lucayana Britton (1920).

CHERRY

Aromatic shrub or small tree to 7 m tall or more, the bark smooth and whitish, the young branchlets and petioles puberulous; leaves narrowly obovate or elliptic, mostly 1.5–4 cm long, blunt or slightly notched at the apex, the margins more or less revolute, densely gland-dotted on both sides, glabrous except for the short petiole. Peduncles usually solitary, 1.5–3.5 cm long, usually 3-flowered, but occasionally the flowers 1 or 7. Hypanthium puberulous, 2–2.5 mm long; calyx-lobes in two unequal pairs; petals white, 4–6 mm long, gland-dotted. Berry black, up to 1 cm in diameter, usually with a single bean-shaped seed maturing.

GRAND CAYMAN: *Brunt* 1778, 2189, 2193; *Kings* GC 379; *Proctor* 15202, 27975. LITTLE CAYMAN: *Kings* LC 65; *Proctor* 28092, 35076. CAYMAN BRAC: *Proctor* 29056, 35157.

— Florida, West Indies, Mexico and C. America, and northern Venezuela, varying considerably from one region to another. The Cayman population described above closely resembles that of Cuba and the Bahamas, and is quite unlike that typical of Jamaica. It grows in rocky thickets and woodlands.

Eugenia L.

Shrubs or trees, glabrous or pubescent; leaves opposite. Flowers axillary, usually at leafy nodes, solitary or usually in racemes, the axis elongate or variously shortened, or sometimes lacking so that the flowers appear to be in umbels or glomerules. Calyx 4-lobed, the lobes distinct in bud and usually persistent on the fruit; hypanthium closely subtended by a pair of bracteoles, at the apex little or not at all prolonged beyond the summit of the ovary. Petals 4, usually white and conspicuous. Stamens borne on a flat disc surrounding the base of the style. Ovary usually 2-celled, with numerous ovules in each cell (rarely only 2). Fruit a small drupe or berry, usually with thin pulp over a single massive seed, or occasionally the fruit 2-seeded; embryo with short, thick cotyledons usually not distinguishable from each other.

— A very large tropical and subtropical genus of at least 1,000 species.

KEY TO SPECIES

1. Leaves widest above the middle, the apex rounded, triplinerved from the base; flowers very small, often clustered at leafless nodes: **E. foetida**

1. Leaves widest at or below the middle, the apex more or less acuminate (sometimes bluntly so), pinnately nerved (the nerves often obscure); flowers or inflorescences usually axillary among leaves:

  2. Leaves with midrib grooved on the upper side; flowers racemose; ripe fruits black, smooth:

    3. Racemes very short; pedicels not over 1.5 mm long; sepals less than 1 mm long, minutely ciliate: **E. axillaris**

    3. Racemes up to 2 cm long or more; pedicels 3–8 mm long or more; sepals 1.5–3 mm long, silky pubescent: **E. biflora**

  2. Leaves with midrib flat on upper side; flowers solitary or in umbel-like clusters with pedicels 10–18 mm long; ripe fruits red, ribbed: **[E. uniflora]**

Eugenia foetida Pers., *Syn. Pl.* 2: 29 (1806).

    E. buxifolia (Sw.) Willd. (1799), not Lam. (1789).

    E. maleolens of Proctor in Adams (1972) (= *E. monticola* (Sw.) DC. as to name).

Shrub or small tree to 8 m tall, more or less puberulous. Leaves oblanceolate to obovate (rarely elliptic), mostly 2–5 cm long, usually rounded at apex and cuneate at base, with petioles 1.5–3 mm long; veins prominulous. Inflorescence very shortly racemose, with rachis not over 2 mm long; flowers crowded, 2–5 pairs, the pedicels 1–2 mm long; calyx lobes rounded, subequal, 0.5–0.7 mm long. Ripe fruits globose, black, ca. 5 mm in diameter.

D.E.

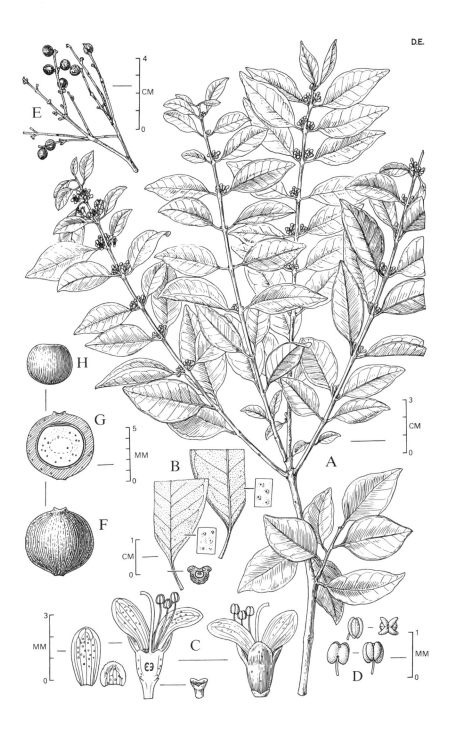

FIG. 147 **Eugenia axillaris**. A, flowering branch. B, details of leaf. C, details of flower. D, four views of anther. E, fruiting branchlets. F, fruit, G, same, long. section through fruit. H, seed. (D.E.)

CAYMAN BRAC: *Proctor* 47328; found in rocky woodland on top of The Bluff northeast of Hawksbill Bay on Nov. 14, 1991. Subsequent attempts to relocate this small tree have not been successful.

— Florida, Yucatan, Belize, Bahamas, Greater Antilles, Virgin Islands and northern Lesser Antilles.

Eugenia axillaris (Sw.) Willd. in L., *Sp. Pl.* ed. 4, 2: 960 (1800). **FIG. 147.** PLATE 32.

Eugenia baruensis Jacq. (1789).
Eugenia monticola of Griseb. (1860), not DC.

STRAWBERRY

Shrub or small tree to 8 m tall, glabrous throughout; leaves rather leathery, elliptic or ovate, 3–7 cm long, obtusely acuminate, the glandular dots but faintly pellucid. Flowers 4–10 on very short racemes, 2–4 mm long; pedicels 1–1.5 mm long; sepals mostly 0.8 mm long. Fruits black, more or less globose, 7–10 mm in diameter, alleged to be edible.

GRAND CAYMAN: *Brunt* 2088, 2107, 2166, 2198; *Correll & Correll* 51009; *Hitchcock*; *Kings* GC 285; *Proctor* 15228, 27932; *Sachet* 424. LITTLE CAYMAN: *Proctor* 35099, 35167, 35204. CAYMAN BRAC: *Millspaugh* 1157; *Proctor* 28912.

— Florida, West Indies, Mexico and northern C. America, in sandy or rocky thickets and woodlands.

Eugenia biflora (L.) DC., *Prodr.* 3: 276 (1828).

Shrub or small tree to 5 m tall (rarely 10 m or more), puberulous. Leaves variable, ovate to elliptic, rarely roundish, the blades mostly 3–7 cm long and not over 3 cm broad on petioles 3–8 mm long; apex more or less acuminate, the base shortly cuneate, the midrib grooved on upper side. Flowers in racemes often 2 cm long or more; pedicels 5–15 mm long; sepals 1.5–3 mm long silky-pubescent. Fruits obliquely globose or obovoid, 6–8(–10) mm in diameter, black, glandular-veruccose.

GRAND CAYMAN: *Proctor* 47861; found in the North Side area called "The Mountain" (probably near what is now known as The Mastic Trail).

— Southern Mexico, C. America, Greater Antilles except Cuba, Lesser Antilles and northern S. America.

[*Eugenia uniflora* L., the Surinam cherry, is sometimes cultivated in the George Town area. It is a shrub with ovate leaves, the white flowers on very long slender pedicels. The red, ribbed fruits are edible.

GRAND CAYMAN: *Proctor* 15308.

— Cultivated sparingly in most tropical countries, and often becoming naturalised; probably native of S. America.]

[*Syzygium* P. Browne ex Gaertn. (nom. cons.)

Trees or shrubs, mostly glabrous. Leaves pinnately nerved, usually with an intramarginal vein on each side. Inflorescences usually terminal, or else terminal and lateral, or the flowers solitary. Flowers 4- (seldom 5-)merous; hypanthium variously or not at all prolonged beyond the ovary, the base usually prolonged to a pseudostalk above the bractioles, the upper part remaining entire or rarely splitting between calyx lobes after anthesis, calyx lobes various, persistent or deciduous; petals free or often more or less united into a calyptra; stamens usually numerous in several series, free or collected into bundles, the filaments filiform, the anthers longitudinally dehiscent. Ovary bilocular (rarely 3- or 4-locular); ovules few or numerous per locule; style filiform with minute stigma. Fruit a fleshy or else dry and leathery berry with 1 or 2 (rarely more) seeds; seeds usually large with fleshy cotyledons, these completely free or partially fused.

— A large palaeotropical genus of between 700 and 800 species, included in *Eugenia* by some authors. Several species are often planted in the neotropics, sometimes becoming naturalised.

Syzygium jambos (L.) Alston in Trimen, *Handb. Fl. Ceylon* 6: Suppl 115 (1931).

ROSE APPLE

Glabrous shrub or small tree. Foliage dense, dark green; leaves lanceolate or very narrowly elliptic, 12–20 cm long, the apex narrowly acuminate. Filaments cream, to 40 mm long. Fruit 1-seeded, depressed-globose, 3–6 cm in diameter, light yellow, rose scented, edible.

GRAND CAYMAN: *Proctor* 48291; from Lower Valley district.
— Nature of the Indo-Malayan region, but now pantropical.]

[PUNICACEAE

Punica granatum L., the pomegranate, is often cultivated around houses, and may persist after cultivation.

GRAND CAYMAN: *Kings* GC 295.]

ONAGRACEAE

Annual or perennial herbs, sometimes aquatic, rarely shrubs or trees; leaves alternate, opposite, or whorled, entire to dentate or pinnatifid; stipules absent. Flowers usually perfect and regular, mostly axillary and solitary but sometimes in spikes, racemes or panicles; calyx often prolonged above the ovary into a slender tube, cleft at apex into 2–6 (often 4) valvate lobes; petals 2–5 (often 4), rarely none, often inserted at base of a disc, soon falling. Stamens 1–8, inserted with the petals, the anthers 2-celled and attached dorsally to slender filaments. Ovary inferior, 1–6- (often 4-)celled; stigma capitate and entire or 4-lobed; ovules numerous or rarely solitary. Fruit a capsule, or sometimes nut-like or berry-like, often elongate and splitting into 4 valves, the valves separating from the seed-bearing axis; seeds usually numerous and small, with little or no endosperm.

— A primarily temperate-climate family of about 20 genera and 650 species, widely distributed in both hemispheres.

Ludwigia L.

Annual or perennial herbs, often aquatic or growing in wet places, rarely shrubs or small trees. Leaves opposite or alternate, entire in our species. Flowers axillary, solitary; calyx not prolonged above the ovary, the lobes usually 4, rarely more; petals usually 4, sometimes lacking, yellow or rarely white. Stamens mostly 4 or 8. Ovary 4–6-celled; ovules numerous on prominent placentas. Fruit a terete, ribbed or angled capsule crowned by the calyx-lobes and disc; seeds numerous, minute.

— A cosmopolitan genus of about 75 species, the majority in the American tropics.

KEY TO SPECIES

1. Sepals and petals usually 4; seeds pluriseriate in each cavity of the fruit:
    2. Capsule 4-angled, ca. 1.5 cm long; calyx 4–5 mm long, scarcely exceeded by the petals: **L. erecta**
    2. Capsule cylindrical, 3–4.5 cm long; calyx 5–12 mm long, much exceeded by the petals: **L. octovalvis**
1. Sepals and petals usually 5; seeds uniseriate in the fruit cavities: **L. affinis**

Ludwigia erecta (L.) H. Hara in *Jour. Jap. Bot.* 28: 292 (1953).

    Jussiaea erecta L. (1753).

Erect, glabrous herb to 1 m tall, much-branched, the branches obscurely angled by the decurrent petioles. Leaves almost linear to narrowly or broadly lance-acuminate, 5–8 cm

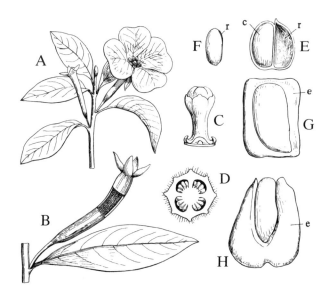

FIG. 148 **Ludwigia octovalvis.** A, branchlet with leaves, flower and young capsule, × 2/3. B, leaf and ripe capsule with part of pericarp removed, × 2/3. C, style, much enlarged. D, cross-section of ovary enlarged. E, seed, × 20; c, embryo; r, the hollow enlarged raphe. F, L. erecta, seed with inconspicuous raphe (r), × 20. G, H seeds of two other species (L. repens, L. leptocarpa), × 20; e, endocarp. (F. & R.)

long, mostly 0.8–1.5 cm broad, slightly rough on the margins, the petioles 2–5 mm long. Flowers small, sessile or subsessile, the yellow petals 4–5 mm long. Capsule 2–3 mm in diameter; seeds about 0.4 mm long.

GRAND CAYMAN: *Brunt* 1893; *Proctor* 15303, 27954, 48314.

— Nearly pantropical in distribution, chiefly in wet ditches and moist fields.

Ludwigia octovalvis (Jacq.) Raven in *Kew Bull* 15: 476 (1962). **FIG. 148.**

Jussiaea suffruticosa L. (1753).

Erect herb often subwoody at base, up to 1 m tall or more, almost glabrous to more or less villous-pubescent, the branches angled. Leaves usually lanceolate to oblong or lance-ovate, 3–11 cm long, up to 2.5 cm broad, narrowed at both ends. Flowers petiolate, with showy yellow petals up to 2 cm long. Capsule 2.5–3.5 mm in diameter; seeds about 0.6 mm long.

GRAND CAYMAN: *Brunt* 1928; *Proctor* 15278, 48274.

— Pantropical, frequently a weed of low moist ground.

Ludwigia affinis (DC.) Hara, *Jour. Jap. Bot.* 28: 291 (1953).

More or less erect woody herb or subshrub sometimes up to 2 m tall or more, with unicellular brown-tipped hairs on the stems. Leaves ovate to elliptic, up to 8 cm long or more and 3(–6) cm broad, acute or obtuse at the apex, pilose on both surfaces and minutely pellucid-dotted. Sepals 3–6 mm long; petals yellow, 6–8 mm long. Capsule 2–3.5 cm long, subsessile.

GRAND CAYMAN: *Proctor* 48275; found in wet depression in old pasture just south of the Salina Reserve on August 8, 1992.

— Jamaica, Guatemala to Peru, Brazil and the Guianas; also reported from Trinidad. Reports from the Lesser Antilles were tentatively rejected by Howard (*Fl. Lesser Ant.* 6: 2 (1989)).

COMBRETACEAE

Trees or shrubs, often climbing, unarmed or spiny; leaves alternate, opposite, or rarely whorled, petiolate, simple, entire, and without stipules. Flowers usually perfect, sometimes polygamo-dioecious or unisexual, usually in spikes, racemes or heads, rarely paniculate. Calyx with tube adnate to the ovary, divided above into usually 4 or 5 lobes, these valvate in bud, persistent or deciduous. Petals none or 4–5. Stamens 4–5, or 8–10, inserted on the limb or base of the calyx, the filaments inflexed in bud; anthers attached at the middle, opening by longitudinal slits. Ovary wholly adnate to the calyx, thus appearing inferior, 1-celled, the style simple or filiform, the stigma simple or lobed; ovules 2–6, hanging from apex of the cell by slender stalks (stalks lacking in *Laguncularia*). Fruit leathery or drupe-like, often angled or winged, 1-seeded, usually not opening; seed usually elongate and grooved, without endosperm, the cotyledons often fleshy and oily.

— A pantropical family of about 15 genera and more than 500 species.

KEY TO GENERA

1. Leaves alternate; petals lacking:	
2. Flowers in spikes:	**Terminalia**
2. Flowers in dense globose heads:	**Conocarpus**
1. Leaves opposite; petals present:	**Laguncularia**

Terminalia L.

Erect shrubs or trees without spines; leaves alternate or apparently subopposite, often crowded at ends of the branches, and often bearing glands at the base beneath. Flowers perfect or staminate, small, usually green or white, borne in spikes; calyx with limb bell-shaped, cut to the middle with 4 or 5 lobes, soon falling; petals lacking. Stamens usually 10 in 2 equal series, the lower opposite the calyx-lobes, the upper alternating; filaments exserted. Stigma simple; ovules usually 2. Fruit dry or drupe-like, often winged.

— A widely distributed tropical genus of about 200 species, many yielding valuable timber.

KEY TO SPECIES

1. Leaves mostly 5–10(–12) cm long, cuneate at base; fruit with flat, papery wings, not edible:	**T. eriostachya var. margaretiae**
1. Leaves mostly 10–30 cm long, minutely cordulate at base; fruit a bony ellipsoid drupe 4–7 cm long with thin edible flesh:	**[T. catappa]**

Terminalia eriostachya A. Rich in Sagra, *Hist. Phys. Cuba Pl. Vasc.* 524 (1846) var. *margaretiae* Proctor, var. nov. PLATE 33.

Tree up to 20 m tall with glabrous ultimate branches. Leaves obovate or elliptic-obovate, rarely oblanceolate, mostly 5–10 cm long, 1.5–4 cm broad, blunt or rounded at apex with a minute point, cuneate at base, minutely prominulous on both sides with lateral veins more prominent beneath, sometimes with small domatia in the axils, otherwise the leaves glabrous. Spikes densely ferruginous-tomentose, up to 12 cm long or more, usually longer than the leaves. Calyx 5 mm long; stamens exserted. Fruits 2-winged, the wings broadly rounded, ca. 2 cm wide.

GRAND CAYMAN: *Margaret Barwick et al.* s.n., 7 April 1991; *Proctor* 47360, 47785 (type), 48300.

— The variety is endemic; the species is otherwise confined to Cuba where it is widespread but not common. Var. *margaretiae* differs from typical *T. eriostachya* in its thinner leaves with minutely prominulous (instead of deeply channelled) veins on the upper surface and the lower surface with only the main side veins (instead of all the veins) prominent. Also, the leaf apices in var. *margaretiae* are minutely pointed (instead of rounded or emarginate). The twigs of var. *margaretiae* are glabrous but finely ferruginous-tomentose in typical *T. eriostachya*. The fruits of var. *margaretiae* have broadly rounded (instead of falcate) wings.

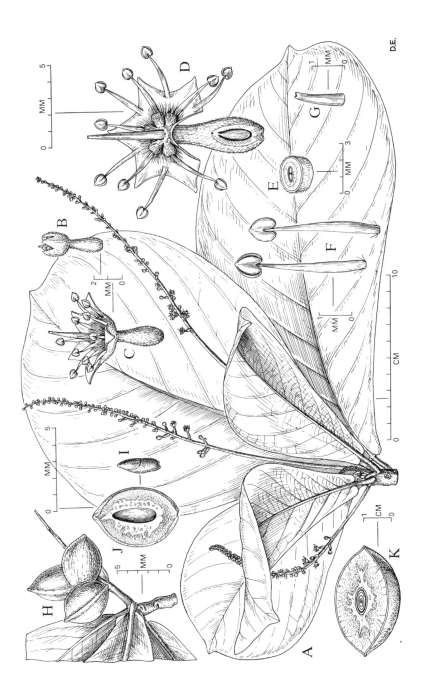

FIG. 149 **Terminalia catappa**. A, flowering branch. B, C, two views of staminate flowers. D, perfect flower. E, cross-section of ovary. F, stamens. G, apical portion of style. H, fruiting inflorescence. I, J, seed and long. section of drupe. K, cross-section of drupe and seed. (D.E.)

Terminalia eriostachya A. Rich. var. *margaretiae* Proctor, var. nov. a var. *eriostachya* foliis tenuioribus, pagina superiore folii venas minute prominulas (non profunde canaliculatas) ferenti, pagina inferiore venas laterales principales tantum (non omnes) prominentes ferenti, apicibus foliorum minute acutis (non rotundatis neque emarginatis), ramunculis glabris (non subtiliter ferrugineo-tomentosis) et fructibus alas late rotundatas (non falcatas) ferentibus differt.

[*Terminalia catappa* L., *Mant. Pl.* 128 (1767). **FIG. 149.** Plate 33.

ALMOND, INDIAN ALMOND

A fast-growing tree with conspicuously whorled, horizontal branches; leaves clustered near ends of the branches, obovate, 10–30 cm long, rounded and abruptly pointed at the apex, tapered downward to a minutely subcordate base, almost glabrous. Flower-spikes 5–15 cm long, the perfect (fruit-producing) flowers toward the base, the distal flowers all being staminate only. Fruit a bony, flattened-ellipsoid, 2-edged drupe, 4–7 cm long, with thin edible flesh.

GRAND CAYMAN: *Brunt* 1759; *Hitchcock*; *Kings GC* 30; *Millspaugh* 1316; *Proctor* 15116. LITTLE CAYMAN: *Kings LC* 55; *Proctor* 35094. CAYMAN BRAC: *Proctor* (sight).

— Native of tropical Asia, commonly planted for ornament and shade throughout the tropics, especially in localities near the sea, and often becoming naturalised. The wood is hard and close-grained; the bark and leaves are astringent and contain tannin, and also together with the fruits yield a black dye used in some countries for making ink or dyeing textiles. The seeds are edible; they are very rich in oil and have an almond-like flavour.]

Conocarpus L.

Shrubs or trees, the leaves alternate, of firm texture, and often biglandular at the base. Flowers perfect and staminate, minute, in dense globular heads, these rather numerous in terminal panicles. Calyx-limb 5-parted, soon falling; petals none. Stamens 5, with exserted filaments and small, cordate anthers. Ovary compressed, 1-celled, with short subulate style and simple stigma; ovules 2, pendulous from apex of the cell. Fruits scale-like, overlapping, 1-seeded, aggregated in a cone-like head.

— A small genus of 2 species, one occurring in tropical America, the other in W. Africa.

Conocarpus erectus L., *Sp. Pl.* 1: 176 (1753) (as 'erecta'). **FIG. 150.** Plate 33.

BUTTONWOOD, BUTTON MANGROVE

A low, trailing shrub in exposed situations, or an erect shrub or tree in more favourable localities. Leaves alternate, lanceolate to elliptic, obtuse or acute at both ends, glabrous (or in var. *sericeus* DC. covered with silky whitish tomentum (**FIG. 150C**)), 2–10 cm long, the petiole with 2 glands on the upper (adaxial) surface at the base of the blade. Flower-heads ca. 1 cm in diameter or less; calyx-limb ca. 1 mm long.

GRAND CAYMAN: *Brunt* 1628, 1971, 1972 (var. *sericeus*); *Hitchcock*; *Kings GC* 27, GC 101, GC 137 (var. *sericeus*), GC 138; *Lewis GC* 27; *Millspaugh* 1306; *Proctor* 15224; *Sachet* 390. LITTLE CAYMAN: *Kings LC* 21, LC 49 (var. *sericeus*); *Proctor* 28103, 28174 (var. *sericeus*). CAYMAN BRAC: *Brunt* 1677, 1678, 1679; *Kings CB* 11 (var. *sericeus*); *Millspaugh* 1212 (var. *sericeus*); *Proctor* 28925, 29048 (var. *sericeus*).

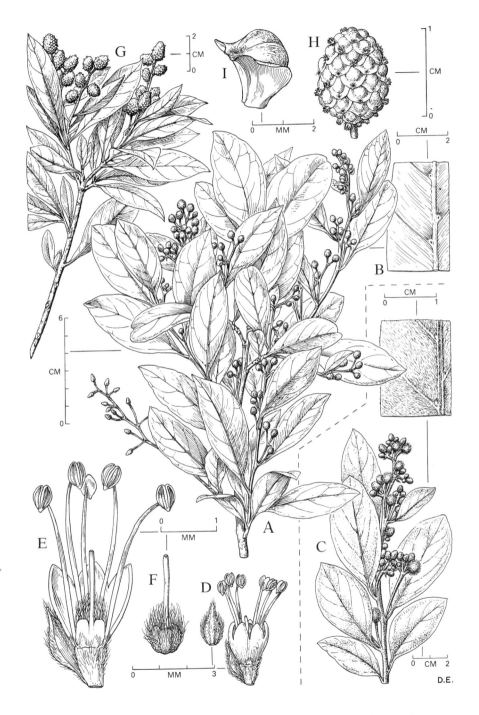

FIG. 150 **Conocarpus erectus**. A, flowering branch. B, detail of leaf. (C, flowering branch of var. *sericeus* and detail of leaf.) D, flower and bract. E, long. section of flower. F, pistil. G, fruiting branch. H, fruiting spike. I, single fruit. (D.E.)

417

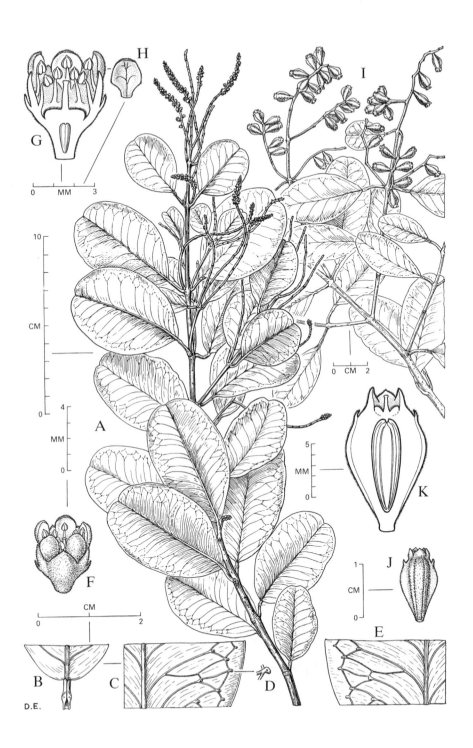

FIG. 151 **Laguncularia racemosa**. A, flowering branch. B, C, D, E, details of leaf. F, flower. G, long. section of flower. H, petal. I, fruiting branch. J, single fruit. K, long. section of fruit. (D.E.)

— Widespread in tropical America, a common element on the landward side of mangrove swamps but also occurring along rocky seashores. The wood is very fine-textured, hard and durable; it is highly valued as a fuel and for making charcoal. The bark is widely used for tanning leather. Var. *sericeus* is often planted as an ornamental shrub.

Laguncularia Gaertn. f.

A shrub or trees; leaves opposite, rather thick and leathery, the veins obscure, with minute glands in the tissue on both sides near the margins, and with 2 glands at apex of the petiole. Flowers mostly perfect with a few staminate ones intermixed, in pubescent axillary spikes and in a terminal panicle of usually 3 spikes, the central one often 3-branched. Bracteoles 2, scale-like, below the calyx. Calyx-limb cup-like, 5-cleft to the middle, persistent; petals 5, small and roundish, soon falling. Stamens 10 in 2 series, the filaments included, the anthers cordate. Ovary 1-celled, crowned by a disc; stigma 2-lobed; ovules 2, pendulous from apex of cell. Fruit leathery, angled and unequally ribbed, the angles narrowly winged or marginate.

— A small genus of 2 species occurring in tropical American and W. Africa.

Laguncularia racemosa (L.) Gaertn. in Gaertn. f., *Fruct.* 3: 209, t. 217 (1805). FIG. 151. PLATE 34.

WHITE MANGROVE

A bushy small tree sometimes 15 m tall or more; leaves oblong to oval, 3–11 cm long, obtuse or rounded at both ends, often notched at the apex, glabrous, with petioles, 0.5–2 cm long. Calyx finely pubescent, 2–3 mm long; petals about equal in length to the calyx, fruit 1.5–2 cm long.

GRAND CAYMAN: *Kings* GC 66, GC 171; *Sachet* 458. LITTLE CAYMAN: *Kings* LC 29, LC 30; *Proctor* 28075, 28173. CAYMAN BRAC: *Brunt* 1685; *Kings* CB 7; *Proctor* 29127.

— Widespread in saline swamps of tropical America and W. Africa, usually associated with other mangrove species. The wood is hard, heavy, strong and dense, but is little used except for fuel. The bark is sometimes used for tanning leather.

RHIZOPHORACEAE

Trees or shrubs, usually glabrous; leaves mostly opposite and with stipules, rarely alternate and lacking stipules, simple and more or less leathery, entire or sometimes serrulate; stipules interpetiolar, united in pairs, soon falling. Flowers perfect, usually subtended by bracteoles, and arranged in small axillary cymes, panicles, spikes or racemes, rarely solitary. Calyx-tube more or less adnate to the ovary or rarely free, the limb extending beyond the ovary and cleft into 3–14 lobes, these valvate and persistent. Petals as many as the sepals and often shorter, mostly concave and embracing the stamens, sessile or clawed, usually more or less lacerate or fringed. Stamens usually 2–4-times as many as the petals, or rarely the same number, inserted on the margin of a perigynous or epigynous disc, the lobes of the disc sometimes elongated into staminodes; filaments short or elongate; anthers 2-celled, opening by longitudinal slits. Ovary usually inferior, mostly 2–5-celled, or the septa disappearing and 1-celled; style with simple or lobed stigma; ovules usually 2 in each cell, hanging side by side from the axis above the middle. Fruit leathery or fleshy, crowned by the persistent calyx-limb, mostly indehiscent or tardily splitting, 1-seeded or the cells 1-seeded; seeds pendulous, with or without an aril; endosperm fleshy or lacking.

— A pantropical family of about 16 genera and 120 species.

Rhizophora L.

Glabrous shrubs or trees supported by curved prop-roots and aerial roots, the branchlets thick and marked by leaf-scars; leaves opposite, petiolate, leathery, elliptic and entire. Peduncles 2- or 3-forked, bearing few flowers; flowers leathery; calyx mostly 4-parted; petals 4 or rarely 5. Stamens 8–12, inserted with the petals on a thick, cup-like disc; filaments very short; anthers elongate, acuminate, with numerous round pollen-sacs but eventually 2-valved. Ovary half-inferior, 2-celled, prolonged above the calyx as a fleshy cone; style awl-shaped, with a 2-toothed stigma; ovules 2 in each cell. Fruit leathery, surrounded above the base by the reflexed sepals, 1-celled, 1-seeded; seed without endosperm; radicle becoming elongate and perforating the apex of the fruit while still on the tree, eventually falling upright into the mud or water beneath.

— Three or more species, widely distributed along tropical seashores.

Rhizophora mangle L., *Sp. Pl.* 1: 443 (1753). **FIG. 152.** PLATE 34.

RED MANGROVE

A tree, sometimes low, shrubby and diffuse, otherwise up to 18 m tall or more; leaves 5–15 cm long, obtuse, entire, deep green, the nerves obscure; stipules 2.5–4 cm long. Peduncles mostly 2–3-flowered; calyx 1 cm long, becoming reflexed; petals creamy yellow, 7–8 mm long, villous inside, chiefly below the apex. Stamens 8, ca. 5 mm long. Fruit 2.5–3.5 cm long but soon appearing much longer because of the developing radicle.

GRAND CAYMAN: *Hitchcock*; Kings GC 176A; *Sachet* 457. LITTLE CAYMAN: *Kings* LC 31, LC 31A. CAYMAN BRAC: *Kings* CB 8; *Millspaugh* 1211; *Proctor* 29126.

— Muddy or sandy seashores of tropical America and islands of the Pacific. The red mangrove with associated species forms an important plant community of tropical coasts, and is abundant in the Cayman Islands. *Rhizophora* is especially adapted to saline aquatic habits partly by virtue of its numerous prop-roots; these roots catch and hold mud and silt so that coast-lines are not only protected from erosion but are actually extended into the sea, the mangrove thickets gradually filling with soil and eventually becoming more or less solid land. Mangrove seedlings also drift about and take root on sand-bars to form small islands. The prop-roots form a tangle that becomes the habitat for numerous marine and semi-marine organisms. *Rhizophora* has hard, heavy, strong, durable wood, used in some regions for many purposes, such as miscellaneous construction and for posts, pilings, and railway ties. It produces a superior kind of charcoal, and the bark has been much used for tanning leather. The young shoots are used to produce various dyes, the colours depending on the kind of salts used in preparing them. In spite of these various potential uses, mangroves should not be indiscriminately cut for any purpose without due consideration of their very real value in protecting shorelines from sea-erosion.

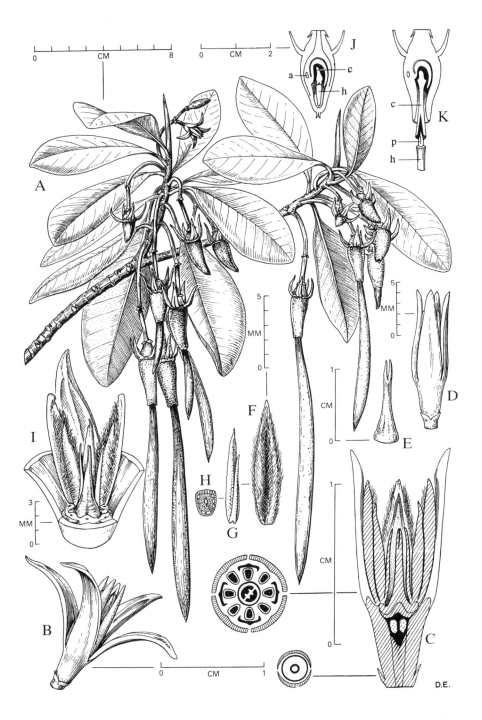

FIG. 152 **Rhizophora mangle**. A, branch with flower, fruits, and developing radicles. B, flower. C, long. section of flower, and cross-sections at two levels. D, flower with sepals removed. E, pistil. F, inner face of petal. G, stamen. H, cross-section of anther. I, dissection of flower. J, long. section of fruit; a, abortive cell; c, cotyledon; h, hypocotyl. K, long. section of fruit after germination of seed. (D.E.)

OLACACEAE

Shrubs or trees, the leaves usually alternate, entire and without stipules. Flowers small, perfect or unisexual, solitary or in cymes or racemes; calyx cup-shaped, persistent, with 4–6 teeth or lobes, sometimes greatly enlarged in fruit. Petals 4–6, free or united below into a bell-shaped corolla, inserted on the receptacle or at the margin of a disc, valvate. Stamens 4–12, more or less adnate to the petals, all fertile or some of them sterile; anthers 2-celled. Ovary free, 1-celled or imperfectly 3–5-celled below, with a central placenta at the apex; ovules usually 3, pendulous. Fruit a drupe, 1-seeded; embryo minute, at the apex of the fleshy endosperm.

— A widely dispersed tropical family of 25 genera and about 250 species.

KEY TO GENERA

1. Corolla-lobes densely bearded within; plants armed with spines:	Ximenia
1. Corolla-lobes not bearded; plants unarmed:	Schoepfia

Ximenia L.

Shrubs or trees, often armed with spines, these formed from abortive branchlets; leaves often clustered on short spurs. Flowers in short axillary cymes or sometimes solitary; calyx 4–5-toothed or -lobed; petals 4 or 5, valvate, densely white-bearded within. Stamens free, twice as many as the petals; anthers linear. Ovary partly 3-celled; style simple, with subcapitate stigma; ovules 3, linear. Fruit a drupe with abundant pulp and a hard stone.

— A genus of at least 9 species, 5 of them in Mexico, 1 confined to Hispaniola, 1 in South Africa, 1 in certain Pacific islands, and another of pantropical distribution.

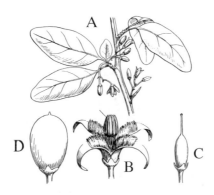

FIG. 153 **Ximenia americana.** A, portion of branch with leaves and flowers. B, flower; a, anthers. C, pistil. D, drupe. (F. & R.)

Ximenia americana L., *Sp. Pl.* 2: 1193 (1753). **FIG. 153**.

HOG PLUM

A densely branched shrub or small tree to 6 m tall or more, the branches usually spiny; leaves glabrous, oblong to elliptic, 3–7 cm long, rounded or obtuse at both ends, deciduous. Flowers fragrant; corolla yellowish-white, 4-lobed, the lobes reflexed, 6–9 mm long. Fruit yellow or reddish, 1.4–1.7 cm long, the flesh edible.

GRAND CAYMAN: *Brunt* 2142; *Kings GC* 353; *Proctor* 15161. LITTLE CAYMAN: *Proctor* 28086, 35091.

— Pantropical, chiefly in coastal thickets but in mountainous countries often with a very wide altitudinal distribution. This species is occasionally cultivated for its edible fruits, which can be eaten raw or cooked. The wood is fragrant, reddish-yellow, fine-textured, very hard and heavy, and has been used in India as a substitute for sandalwood. The bark is occasionally used for tanning leather. Although *Ximenia* is called hog plum in the Cayman Islands, this name is usually applied elsewhere (e.g. in Jamaica) to a species of *Spondias* (Anacardiaceae), while the general common name for *Ximenia* elsewhere is 'tallow plum'.

Schoepfia Schreber

Glabrous shrubs or small trees with coriaceous leaves. Flowers small and inconspicuous, in short racemes which are solitary or clustered in the leaf-axils; calyx cup-shaped, obscurely toothed; disc entire, adnate to the ovary, enlarging as the ovary ripens, at length almost enveloping the fruit, forming an evident ring around it near the apex. Petals 4–6, inserted at margin of the disc, cohering to form a bell-shaped corolla with free lobes at the apex. Stamens equal in number to the petals, adnate to the corolla-tube. Ovary partly immersed in the disc, imperfectly 3-celled, with short or long style and 3-lobed stigma; ovules 3, pendulous from apex of the placenta. Fruit a drupe with thin flesh and hard seed.

— A small genus of about 12 species in the American and Asiatic tropics. Some have been shown to have parasitic root connections to other plants.

Schoepfia chrysophylloides (A. Rich.) Planch, *Ann. Sci. Nat.* ser. 4, 2: 261 (1854). PLATE 34.

A shrub or small tree up to 7 m tall but usually much lower; leaves petiolate, the blades ovate or elliptic, 2–7 cm long, blunt at both ends. Flowers dimorphic (long-or short-styled), paired on short peduncles, the minute calyces of each pair coherent; corolla cream, 3–3.5 mm long. Fruit red, narrowly obovoid, 6.5–7.5 mm long.

LITTLE CAYMAN: *Proctor* 28121. CAYMAN BRAC: *Proctor* 29051. Both of the cited specimens are sterile; flower and fruit details have been taken from material collected elsewhere, and from literature.

— Bahamas, Cuba, Jamaica, and Hispaniola, in dry rocky woodlands, apparently nowhere very common.

LORANTHACEAE

Parasitic shrubs, vines, or sometimes small trees containing chlorophyll, attached to host plants by specialised roots called haustoria. Branches terete or angled, glabrous or pubescent; leaves usually present and well-developed, opposite, simple, often somewhat fleshy. Flowers small or more usually large, often showy, perfect or if unisexual then the plants dioecious, usually borne in groups of 3 (dichasia) on axillary or terminal racemes or panicles, rarely solitary. Calyx represented in rudimentary form by a circular rim or calyculus at apex of ovary. Petals 4–8, valvate, free or united into a tube, regular or (in some African species) zygomorphic. Stamens as many as the petals, adnate to their base or middle and usually shorter; anthers usually 4-locular (or 2-locular by coalescence), opening longitudinally. Disc present or lacking. Ovary inferior, 1-locular, capped by the calyculus; style simple (sometimes elongate) with minute terminal stigma; ovules in the strict sense lacking, the equivalent structure undeveloped until after pollination. Fruit berry-like, with small solitary seed surrounded by viscid pulp; endosperm compound and without chlorophyll.

— A chiefly tropical family of about 57 genera and more than 200 species, a few occurring in temperate regions. Widely known as mistletoes, a term also (and primarily) applied to members of the related family Viscaceae.

Dendropemon (Blume) Reichb.

Small parasitic shrubs with coriaceous or somewhat fleshy, mostly flat leaves, and small bracteolate flowers in simple axillary racemes, the bracteoles connate, cupulate. Receptacle-rim produced into a truncate or 4–6-toothed calyculus. Corolla of usually 5 or 6 free petals, these usually small. Stamens short, borne at the base of the petals; anthers dorsifixed. Style relatively short, with a terminal stigma which in some species is the same diameter as the style, in others more or less capitate.

— A primarily West Indian genus of about 15 species.

Dendropemon caymanensis Proctor, *Phytologia* 35: 403 (1977). PLATE 34.

Stems ca. 2 mm thick near the base, up to ca. 20 cm long, terete, cinereous, branched; branches more or less compressed, smooth and glabrous, with internodes 0.7–1.5 cm long. Leaves with petioles 1–1.5 mm long, the blades oblong to oblanceolate-oblong, 0.5–2.5 cm long, 3–10 mm broad, subacute and apiculate at the apex, narrowed toward the base, the median nerve prominent especially beneath, the lateral nerves few, obscurely prominulous or obsolete, the margins flat and entire, the texture subcoriaceous. Inflorescences much shorter than most leaves, with peduncles 1.5–4.5 mm long and rhachis up to 1 cm long, rather stiff, 4–8-flowered, smooth and glabrous throughout, the rhachis more or less compressed. Pedicels 0.5–1 mm long; bracteoles with shortly triangular free apex. Flower-buds oblong, obtuse; calyculus cup-shaped and truncate; petals of open flowers ligulate, ca. 1.8 mm long and 0.5–0.6 mm broad. Style 0.9–1.1 mm long, with subcapitate stigma. Berries obovoid, blackish-purple, up to 5.5 mm long.

LITTLE CAYMAN: *Proctor* 35215 (type), 47309.

— Endemic. This species is known only from the type locality, where it grows as a parasite on *Capparis cynophallophora* and other trees. It appears to be related to *D. purpureus* (L.) Krug & Urban of the Bahamas, Cuba, Hispaniola and Puerto Rico, and to *D. rigidus* Urban & Ekman of the Dominican Republic, but is smaller than either of these.

VISCACEAE

Parasitic shrubs (some taxa reduced to minute size) containing chlorophyll, attached to host plants by specialised roots called haustoria. Branches articulate and terete, angled or flattened, glabrous or pubescent. Leaves opposite, simple, often rather fleshy, or else reduced to scales and apparently lacking. Plants either monoecious or dioecious, the flowers very small and always unisexual, rarely solitary, usually in rows or series on the internodes of articulated axillary or terminal spikes, which in turn may be arranged in symmetric compound inflorescences. Perianth segments usually 3 (rarely 4), valvate, or in very reduced genera sometimes lacking or indistinguishable from the apical part of the ovary, united at base. Stamens as many as the perianth segments, adherent to them and much shorter, the anthers nearly sessile; anthers 1–4-locular, the cells splitting transversely. Ovary inferior, 1-locular, at apex with minute, sessile, knob-like stigma; calyculus absent. True ovules lacking, but embryo development different in many details from that of Loranthaceae. Fruit berry-like, crowned by the persistent perianth segments, with solitary seed surrounded by viscid pulp. endosperm simple and green.

— A chiefly tropical family of about 7 genera and more than 350 species, some occurring in North Temperate regions. The great majority of the species belong to the genus *Phoradendron*, which is confined to the Western Hemisphere. The European *Viscum album* L. was the original mistletoe, allegedly with magical properties, but all members of the family are now known by this name.

Phoradendron Nutt. PLATE 35.

Small shrubs mostly parasitic on broad-leaved woody plants, rarely on conifers or on other species of *Phoradendron*. Stems terete, 4-angled, or flattened and 2-edged, rarely winged. Leaves usually evident or conspicuous, rarely apparently absent. Inflorescence of axillary jointed spikes with a pair of small fleshy or scarious bracts subtending each joint; flowers arranged in 2–6 (rarely 8) rows on each internode, each flower sessile and more or less immersed in a small pit; plants monoecious or dioecious. Staminate perianth bearing an almost sessile 2-locular anther at the base of each lobe. Ovary with capitate stigma. Berries white, yellow, orange, or red.

— A widespread Western Hemisphere genus of 234 species, the majority occurring in tropical areas (J. Kuijt: Monograph of *Phoradendron* (Viscaceae), *Syst. Bot. Mon.* 66: 1–643 (2003)).

KEY TO SPECIES

1. Stems somewhat lax, more or less quadrangular in cross-section, not expanded and flattened at distal end of internodes; flowers not red:

    2. Leaves faintly 3-nerved at base, lanceolate, narrowly oblanceolate, or narrowly elliptic, usually not more than 1 cm broad; flowers 3–7 per fertile bract: **P. quadrangulare**

    2. Leaves 3–5 nerved at base, obovate to nearly elliptic, up to 2.8 cm broad; flowers 1 or sometimes 3 per fertile bract: **P. trinervium**

1. Stems stout, the internodes expanded and flattened distally; flowers often red: **P. rubrum**

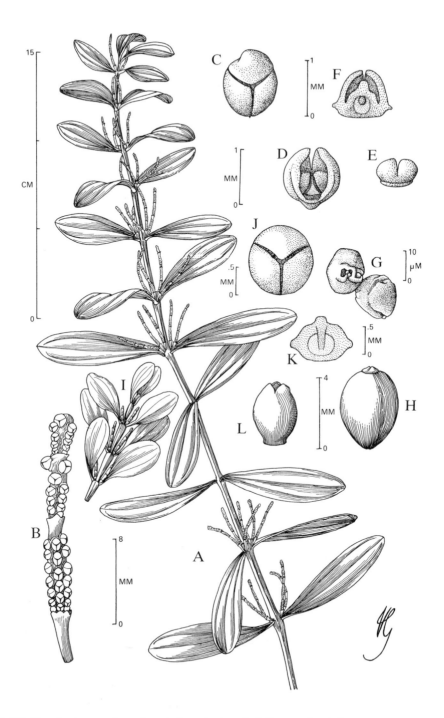

FIG. 154 **Phoradendron quadrangulare**. A, flowering branch. B, flower-spike. C, flower. D, staminate flower, opened. E, single stamen. F, long. section of pistillate flower. G, pollen-grains as shown by scanning electron microscope. H, fruit. **Phoradendron rubrum**. I, habit of small branch. J, flower. K, pistillate flower, cross-section. L, fruit. (G.)

Phoradendron quadrangulare (Kunth) Krug & Urban, *Engl. Bot. Jahrb.* 24: 35 (1897). **FIG. 154.**

P. *quadrangulare* var. *gracile* Krug & Urban (1897).

P. *gracile* (Krug & Urban) Trelease (1916).

Stems up to 40 cm long, and to 5 mm thick near the base; branches with the younger internodes acutely and subequally 4-angled; cataphylls (sheathing scales) occurring only at the base of the lowest internode of a branch. Leaves narrowly oblanceolate, flaccid, mostly 1.5–4 cm long, 3–9 mm broad, obtuse at the apex, attenuate at base, very obscurely 3-nerved from the base. Spikes usually solitary in the leaf-axils, 1–2 cm long, 2–3-jointed; ripe fruits subglobose, smooth, 2.5–3 mm in diameter, pale orange, yellow, or nearly white, somewhat translucent.

GRAND CAYMAN: *Correll & Correll* 51033; *Kings* GC 14, GC 150, GC 365, GC 388. LITTLE CAYMAN: *Proctor* 35194. CAYMAN BRAC: *Kings* CB 91.

— In the broad sense widely distributed in tropical America, but many of the variants have been described as separate species. The Cayman collections resemble the population occurring in Jamaica, where it has been called *P. gracile.* The Grand Cayman plants grew as parasites on *Guapira discolor* (Nyctaginaceae), while that of Little Cayman was on *Croton linearis* (Euphorbiaceae). The Cayman Brac specimen was collected on *Bauhinia divaricata* (Leguminosae: Caesalpinioideae, *Kings* CB 91 A).

Phoradendron trinervium (Lam.) Griseb., *Fl. Brit. W.I.* 314 (1860).

Stems much branched, the internodes somewhat quadrangular but often obscurely so, not expanded or obviously flattened distally; basal cataphylls 1 pair, on lateral branches only, deeply bifid, axillary or nearly so, sometimes with a second pair borne ca. 5 mm above the base. Leaves narrowly obovate to nearly elliptic, up to 4.5 cm long, of thin texture, the apex rounded, notched, or sometimes subacute. Plants monoecious but male flowers few. Inflorescences 1–1.5 cm long, lengthening in fruit; fertile internodes 3–6. Fruits ovoid, 4 mm long, 2.5 mm in diameter, yellowish to orange.

GRAND CAYMAN: *Proctor* 15111. LITTLE CAYMAN: *Proctor* 28122.

— Mexico, Central and northern S. America, Bahamas, Greater and Lesser Antilles.

Phoradendron rubrum (L.) Griseb, *Fl. Brit. W. I.* 314 (1860). **FIG. 154.**

SCORN-THE-GROUND

Stems up to 50 cm long (but often less), and to 6 mm thick near the base, the oldest terete, the intermediate internodes more or less 4-angled, and those nearest the apex flattened and rather sharply 2-angled, especially just below a node; cataphylls occurring only at the base of the lowest internode of a branch. Leaves oblanceolate to obovate or sometimes elliptic, rather thick and leathery, 2–8 cm long, (5–)10–28 mm broad, rounded at the apex, more or less cuneate at the base, with a very short petiole, obscurely 3–5-nerved from the base. Spikes 1–3 in the leaf-axils, 1–3 cm long, 3–5-jointed; ripe fruits subglobose, smooth, ca. 4 mm in diameter, reddish or orange.

GRAND CAYMAN: *Brunt* 1652; *Kings* GC 165, GC 388; *Lewis* GC 14; *Proctor* 15111, 15259. LITTLE CAYMAN: *Kings* LC 36, LC 43; *Proctor* 28185, 35075, 35100. CAYMAN BRAC: *Proctor* 28959, 35148.

— Bahamas and Cuba. This common species is most often found as a parasite on *Guapira discolor* (Nyctaginaceae), but has been recorded twice on *Swietenia mahagoni* (Meliaceae)

(*Brunt* 1652, *Proctor* 15259) and once on *Croton linearis* (Euphorbiaceae) (*Proctor* 35148). *Phoradendron* species are usually not host-specific, but nevertheless seem to show definite preferences that may vary from one area to another. As these plants are adapted to dispersal by birds, it is possible that what seem to be host-preferences of the *Phorandendron* species may actually reflect the perching preferences of the birds that eat the berries.

APODANTHACEAE

Dioecious endoparasites without chlorophyll. Plant body filamentous, resembling a fungal mycelium inhabiting the stem tissues of the host plant. The aerial portions of the parasite consist of flowering shoots, each producing a single flower, which bursts out of the cortical tissues of the host. These flowers are less than 1 cm in diameter, often only a few millimetres; a flower consists of a fleshy column bearing reproductive structures, subtended by 4–10 perianth-like scales (bracts?), these arranged in whorls; true petals are apparently absent. Staminate flowers produce sessile anthers in 1–3 whorls attached to the fleshy central column; these anthers are 2-locular and may open by slits or pores; the pollen is often viscous. Pistillate flowers have an ovary with a single cavity containing numerous ovules; style absent, the 1–several stigmas sessile on apex of the ovary. Fruit a small berry with many seeds, indehiscent or rupturing irregularly; seeds minute.

— A small family of 3 genera with a total of about 23 species of wide distribution in tropical, subtropical, and warm temperate areas.

Pilostyles Guillamin

With characters of the family; plants always parasitic on Leguminosae. Flowers arising from small depressed cupules on stems of the host. Perianth scales (bracts?) attached by a broad base; anthers transversely dehiscent; stigma annular; fruit surrounded by the dry perianth scales.

— A genus of about 20 species of scattered distribution from the neotropics and subtropics to the Mediterranean area, southwestern Asia, and subtropical southwestern Australia.

Pilostyles globosa (S. Wats.) Hemsley, *J. Linn. Soc. Bot.* 31: 311 (1896) var. *caymanensis* Proctor, var. nov. PLATE 35.

With characters as given for family and genus, the Cayman population differing from typical *P. globosa* of Mexico and Jamaica in the smaller size of its flowers and fruits (2–2.5 mm in diameter vs. 3–3.5 mm in diameter) and in having entire or nearly entire bract-margins (rather than margins minutely erose).

Pilostyles globosa (S. Wats.) Hemsley var. *caymanensis* Proctor, var. nov. a var. *globosa* floribus fructibusque minoribus 2–2.5 mm (non 3–3.5 mm) diametro et marginibus bractearum integris vel subintegris (non minute erosis) differt.

GRAND CAYMAN: *Proctor* 47281, 47781, 48666, 52140 (type, IJ). LITTLE CAYMAN: *Proctor* 47311. CAYMAN BRAC: *Proctor* 47319, 47327, 47337, 52157. All specimens were found parasitic on branches of *Bauhinia divaricata* L. The earliest collection of *Pilostyles* in the Cayman Islands was made on 10 November 1991. The type collection of *P. globosa* var. *caymanensis* was gathered in natural woodland within the Queen Elizabeth II Botanic Park on 9 April 2003.

CELASTRACEAE

Shrubs or trees, erect or sometimes climbing; leaves opposite or alternate, rarely whorled, simple and entire or toothed but never lobed; stipules absent or small and soon falling. Flowers usually in axillary cymes, greenish or white, perfect or else unisexual and monoecious or dioecious; pedicels frequently jointed. Calyx small, with 4–5 imbricate lobes, persistent. Petals 4–5, short, spreading, sessile below the margin of the disc, imbricate in bud. Stamens 4–5, inserted on or near the margin of the disc, with awl-shaped filaments; anthers dehiscing introrsely. Ovary 3–5-celled, the short thick style entire or sometimes 3–5-lobed; ovules 2 or sometimes 1 in each cell, erect or rarely pendulous. Fruit a drupe or capsule; if capsular, the carpels fused; seeds with or without an aril; endosperm usually present and fleshy; embryo usually rather large, with flat, foliaceous cotyledons.

— A cosmopolitan family of more than 40 genera and 850 species.

KEY TO GENERA

1. Fruit a capsule; seed with an aril; flower-parts in 5s:	**Maytenus**
1. Fruit a drupe; seed without an aril:	
2. Flower-parts in 5s:	**Elaeodendron**
2. Flower-parts in 4s:	
3. Flowers bisexual; some leaves in whorls:	**Crossopetalum**
3. Flowers unisexual; plants mostly dioecious:	
4. Leaves opposite:	**Gyminda**
4. Leaves alternate:	**Schaefferia**

Maytenus Molina

Shrubs or small trees, mostly glabrous; leaves alternate, coriaceous, entire or serrate; stipules minute and soon falling. Flowers small, polygamous or rarely functionally unisexual and the plants dioecious, axillary and solitary, clustered, or in small cymes. Stamens inserted outside and below the disc, with ovate-cordate anthers; disc round with wavy margin. Ovary immersed in the disc and confluent with it, 2–4-locular; style absent or very short; ovules 1 or 2 in each cell, erect. Fruit a hard, 1–3-locular capsule, 2–3-valved; seeds more or less enclosed by a thin aril; endosperm fleshy or apparently none.

— Tropical American genus of perhaps 70 species, the majority in S. America.

KEY TO SPECIES

1. Leaves more than 4 cm long; calyx-lobes 1.2–1.5 mm long; fruit globose, orange, 15 mm in diameter or more:	M. jamaicensis
1. Leaves less than 3 cm long; calyx-lobes 0.5 mm long; fruit globose-obovoid, red or orange, 5–9 mm long:	M. buxifolia

Maytenus jamaicensis Krug & Urban in *Notizbl. Bot. Gart. Berlin* 1: 78 (1895).

Shrub or densely leafy slender tree up to 10 m tall. Leaves (in Cayman Islands) obovate, 4–4.5 cm long, 2–2.5 cm broad above the middle, rounded at the apex, tapering downward to a cuneate base and short petiole up to 3 mm long, of somewhat thick texture, the midrib prominent. Flowers short-stipitate; seeds completely enclosed by white aril.

CAYMAN BRAC: *Mrs. Joanne Ross* s.n. (photo only, no specimens collected).
— Hispaniola, Jamaica, and Panama. Specimens are needed in order to substantiate the Cayman record.

Maytenus buxifolia (A. Rich.) Griseb., *Cat. Pl. Cub.* 53 (1866).

BASTARD CHELAMELLA

A shrub or rarely a small tree with grey twigs; leaves oblong to obovate or spatulate, mostly 1–3 cm long, of smooth hard texture, obtuse and slightly notched at the apex. Flowers few in clusters, on pedicels 1–4 mm long; calyx-lobes roundish, 0.5 mm long; petals greenish-yellow, ovate, nearly 2 mm long. Fruits red or orange, globose-obovoid, 5–9 mm long, apiculate.

LITTLE CAYMAN: *Kings* LC 67; *Proctor* 28091.
— Bahamas, Cuba, and Hispaniola, in dry rocky thickets and scrublands.

Elaeodendron Jacq. f. ex Jacq.

Shrubs or small trees; leaves opposite and alternate, entire or crenate; stipules minute, soon falling. Flowers perfect, or often unisexual and dioecious, in cymes on axillary peduncles, the parts in 5s. Stamens inserted under and outside the disc, with globose anthers; petaloid staminodes present in pistillate flowers. Ovary confluent with the disc, 2–5-celled; style very short, the stigma 2–5-lobed; ovules 2 in each cell, erect. Fruit a dry or pulpy drupe, the stone 1–3-celled with 1 or 2 seeds in each cell; seeds without an aril.
— A pantropical genus of nearly 60 species.

Elaeodendron xylocarpum (Vent.) P. DC., *Prodr.* 2: 11 (1825) var. *attenuatum* (A. Rich.) Urban, *Symb. Ant.* 5: 88 (1904). PLATE 35.

A small tree up to 10 m tall, sometimes lower and shrubby; leaves opposite or some of them alternate, oblong to elliptic or obovate, mostly 4–10 cm long, the apex obtuse to acute, the margins nearly entire or sparingly crenulate. Inflorescence shorter than the leaves; flowers pale green on pedicels 1–3 mm long; sepals ca. 1 mm long, petals 2–3 mm long. Fruits yellowish, globular to ellipsoid, 1.5–3 cm long, rounded and smooth or minutely apiculate.

GRAND CAYMAN: *Brunt* 1698, 1829, 1979; *Hitchcock*; *Proctor* 15033, 15110.
— As interpreted here, one of several intergrading and variable varieties of a species occurring throughout the West Indies. Var. *attenuatum* is best represented in the Bahamas, Cuba and Hispaniola but does not occur in Jamaica. It grows in dry thickets and rocky woodlands.

This taxon is placed in the genus *Cassine* by Adams (1972) and several other modern writers. However, *Cassine* interpreted strictly is confined to Africa, and interpreted broadly causes confusion in delimiting other genera in the Celastraceae.

Crossopetalum P. Br.

Shrubs or low trees; leaves opposite or in whorls of 3 (rarely alternate), short-petiolate, the margins entire, crenate, or spiny; stipules minute. Flowers small, perfect, in cymes or subsolitary at the ends of short or long peduncles, the parts mostly in 4s. Calyx lobed; petals reflexed. Stamens inserted between lobes of the disc; anthers subglobose. Ovary usually 4-celled, with short style and 4-lobed stigma; ovules solitary in each cell, erect. Fruit a small dry or fleshy drupe; seed with or without an aril and with fleshy endosperm.

— A tropical American genus of about 20 species.

KEY TO SPECIES

1. Stems and underside of leaves glabrous; flowering pedicels 1–2.5 mm long:	C. rhacoma
1. Stems and underside of leaves densely pubescent; flowering pedicels 4–6 mm long:	C. caymanense

Crossopetalum rhacoma Crantz, *Inst. Rei Herb.* 2: 321 (1766). **FIG. 155**. PLATE 35.

Rhacoma crossopetalum L. (1759).

SNAKE BERRY, TOBACCO BERRY, WILD TOBACCO

Usually a shrub 1–3 m tall, the glabrous young twigs with 4 raised longitudinal lines; leaves opposite or sometimes in whorls of 3, in Cayman plants oblanceolate, oblong, or obovate, 1–3 cm long, the apex obtuse or rounded and often slightly notched, the margins crenulate. Inflorescence a 1–3-times-forked cyme with glabrous peduncle and bracts; pedicels 1–2.5 mm long; calyx 0.7 mm long, glabrous; petals obovate-elliptic, 1–1.2 mm long, green tinged with red. Style erect with 4 recurved linear stigmas at the apex. Fruit obliquely obovoid, 5–7 mm long, scarlet, crowned by the persistent style.

GRAND CAYMAN: *Brunt* 2054; *Hitchcock*; *Proctor* 15091, p.p. (IJ). LITTLE CAYMAN: *Proctor* 28098. CAYMAN BRAC: *Proctor* 29141.

— Florida and throughout the West Indies to St. Lucia, also in the Belize cays, Bonaire and Colombia, mostly in coastal scrublands.

Crossopetalum caymanense Proctor in *Sloanea* 1: 2 (1977). **FIG. 155**. PLATES 35 & 36.

A shrub 1–2 m tall, the densely pubescent young branches with 4 prominent longitudinal ridges; leaves opposite or sometimes in whorls of 3, lanceolate to ovate or sometimes oblong, 1–5 cm long, the apex acute and sharply mucronate, the margins distantly serrulate. Inflorescence a 2–5-times-forked cyme with densely pubescent peduncle and bracts; pedicels glabrous, at anthesis and in fruit flexuous and 4–6 mm long; calyx 0.8–0.9 mm long, pubescent; petals roundish, 1.1–1.5 mm long, cream or pale salmon. Style nearly obsolete, the stigma minutely 4-lobed. Fruit globose or ellipsoid, 5–6 mm long, crimson.

GRAND CAYMAN: *Proctor* 15026, 15091 (BM), 15184 (type), 27968. CAYMAN BRAC: *Kings* CB 12.

— Endemic, occurring in rocky thickets and woodlands. The Cayman Brac specimens bear larger and more ovate leaves than those of Grand Cayman.

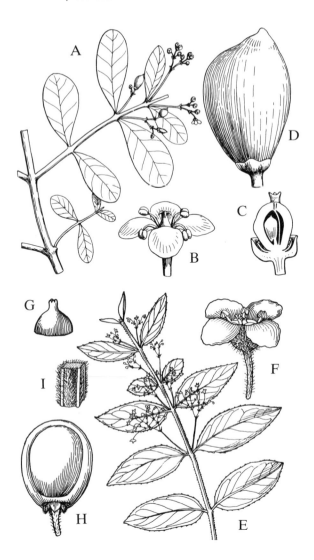

FIG. 155 **Crossopetalum rhacoma**. A, twig with leaves and flowers, × 1. B, flower, × 10. C, ripening ovary cut lengthwise, × 10. D, fruit, × 5. (F. & R.). **Crossopetalum caymanense**. E, twig with leaves and flowers, × 1. F, flower, × 10. G, ovary × 20. H, fruit, × 4. I, section of stem, × 8. (G.)

Gyminda Sarg.

Shrubs or small trees; leaves opposite, short-petiolate, the margins entire or crenulate-serrate above the middle; small stipules present. Flowers small, unisexual and dioecious, in small, stalked, axillary cymes, the parts in 4s. Calyx deeply lobed; petals reflexed. Staminate flower with stamens inserted in lobes of the disc; anthers oblong. Pistillate flower with 3-celled ovary and a 2-lobed stigma; ovule solitary and pendulous in each cell. Fruit a small drupe, 1- or sometimes 2-seeded; seeds with thin, fleshy endosperm.

— A pan-Caribbean genus of 3 species.

Gyminda latifolia (Sw.) Urban, *Symb. Ant.* 5: 80 (1904). **FIG. 156**. Plate 36.

A glabrous shrub or small tree to 7 m tall, the young twigs marked with 4 raised lines. Leaves oblanceolate to obovate, 1–5 cm long, the apex rounded or obtuse. Flowers fragrant, whitish; sepals 0.6–0.8 mm long, petals 1.6–2.2 mm long. Fruit black, more or less ellipsoid, 4–8 mm long.

GRAND CAYMAN: *Brunt* 2051; *Correll & Correll* 51028; *Proctor* 15066, 15196, 15230. LITTLE CAYMAN: *Proctor* 28093, 35135. CAYMAN BRAC: *Proctor* 29140.

— Florida and the West Indies, in rocky thickets and woodlands.

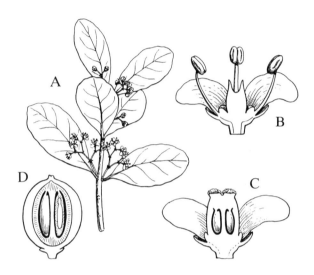

FIG. 156 **Gyminda latifolia**. A, branchlet with leaves and flowers, × ²/₃. B, staminate flower cut lengthwise, × 8. C, pistillate flower cut lengthwise, × 8. D, drupe cut lengthwise, × 6. (F. & R.)

Schaefferia Jacq.

Glabrous shrubs or small trees; leaves alternate or clustered on short, spur-like branches, the margins entire; stipules minute. Flowers unisexual and usually dioecious, solitary or clustered in the leaf-axils, with parts in 4s. Calyx deeply lobed; petals hypogynous. Stamens hypogynous or inserted below the margin of a small, inconspicuous disc. Ovary 2-celled; ovules solitary in each cell, erect; style short with prominent 2-lobed stigma. Fruit a small dry drupe, 2-seeded; seeds without aril and with scant endosperm.

— A tropical American genus of 16 species.

Schaefferia frutescens Jacq., *Enum. Syst. Pl. Carib.* 33 (1760). **FIG. 157**. Plate 36.

Shrub or small tree to 6 m tall, the young twigs striate with raised lines; leaves narrowly to broadly elliptic, 2–5 cm long, sharply acute or subacuminate at apex, the veins prominulous on both sides. Flowers 1–several in a cluster, pale green, on pedicels 1–3 mm long; calyx 0.7–1 mm long; petals 3–4 mm long. Fruit spherical to ovoid, 4–6 mm long, orange or scarlet, crowned by the persistent stigmas.

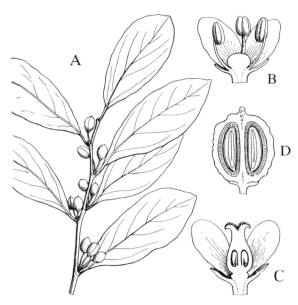

FIG. 157 **Schaefferia frutescens.** A, fruiting branch, × 2/3. B, staminate flower cut lengthwise, × 7. C, pistillate flower cut lengthwise, × 7. D, drupe cut lengthwise, × 4. (F. & R.)

GRAND CAYMAN: *Correll & Correll* 51031; *Proctor* 11996, 15164. LITTLE CAYMAN: *Proctor* 28182. CAYMAN BRAC: *Proctor* 15330, 29028.

— Florida and the West Indies; attributed also to Mexico and Ecuador. This species grows in dry, rocky woodlands.

BUXACEAE

Shrubs or small trees, rarely herbs; leaves opposite or alternate, usually entire and of hard texture; stipules lacking. Flowers unisexual, monoecious or rarely dioecious, in axillary or supra-axillary, lax or dense spikes or racemes, the terminal flowers often pistillate and the others staminate, each one subtended by a bract. Perianth of 4–6 imbricate sepals, or lacking; petals none. In staminate flowers, the stamens free and opposite the sepals, or indefinite; anthers 2-celled. In pistillate flowers, the ovary usually 3-celled, terminating in 3 simple styles; ovules 2 or rarely 1 in each cell, pendulous. Fruit a dehiscent capsule or sometimes drupe-like, usually crowned by the 3 persistent styles; seeds with more or less fleshy endosperm, or rarely none.

— A small, scattered, cosmopolitan family of about 7 genera and 100 species.

Buxus L.

Shrubs or small trees, usually densely branched; leaves opposite, sessile or short-petiolate. Bracts numerous, several often without flowers. Staminate flowers usually stalked, with 6 sepals in 2 series; stamens 4. Pistillate flowers sessile, with 6 sepals in 2 series. Fruit a 3-horned capsule, splitting loculicidally through the horns; seeds oblong, 3-cornered, with a small strophiole.

— A widely distributed genus of about 45 species, the majority West Indian.

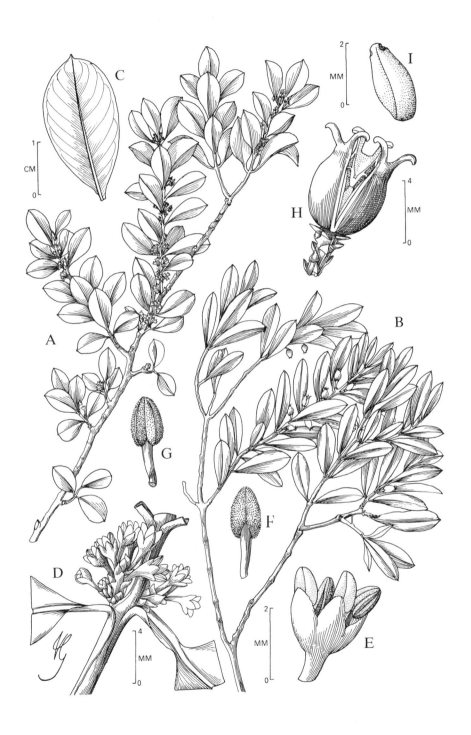

FIG. 158 **Buxus bahamensis**. A, flowering branch. B, fruiting branch. C, leaf. D, flower-clusters. E, staminate flower. F, G, two views of stamen. H, fruit. I, seed. (G.)

435

Buxus bahamensis Baker in Hook., *Ic. Pl.* t. 1806 (1889). **FIG. 158.** PLATE 36.

Usually a shrub 1–2 m tall, glabrous throughout; leaves elliptic or sometimes oblong, rigid, 1.5–3 cm long, sharply acute and mucronulate at the apex, the venation obscure. Flowers greenish-white, the staminate on pedicels 2–3 mm long; sepals 1–2 mm long. Capsule 5–7 mm long.

GRAND CAYMAN: *Proctor* 47367, 48241. LITTLE CAYMAN: *Proctor* 28138, 35136.

— Bahamas, Cuba and Jamaica, in dry rocky scrublands. The identity of the Grand Cayman population needs to be verified.

EUPHORBIACEAE

Herbs, shrubs or trees, sometimes scandent or twining, often with milky sap; leaves usually alternate, sometimes opposite or whorled, simple or rarely digitately compound, palmate-lobed, or absent; stipules often present. Inflorescence various in form, the flowers usually small, unisexual, monoecious or dioecious, generally regular. Perianth sometimes lacking, often dissimilar in staminate and pistillate flowers, with either a calyx only, or a calyx and corolla, the segments free or united, imbricate or valvate. Staminate flowers with receptacle sometimes expanded into a disc within the stamens, or the glands or lobes of the disc alternating with stamens of the outer series; stamens sometimes indefinite, or often as many as the sepals or fewer, sometimes only 1; filaments free or united; rudimentary ovary present or absent. Pistillate flowers with disc ring-like or cup-like, or of separate glands, or absent; ovary usually 3-locular (rarely otherwise), styles usually 3, free or united, entire, cleft or fringed; ovules 1 in each locule, or 2 collateral ones, pendulous, anatropous, attached at inner angle of the cavity, the raphe ventral; micropyle often covered with a cushion-like caruncle (obturator), this persisting on the seed. Fruit usually a 'schizocarp' capsule splitting 3 ways into 2-valved cocci, these separating from the persistent axis, or sometimes drupe-like and indehiscent. Seeds as many as the ovules; endosperm usually copious and fleshy; cotyledons mostly broad and flat or folded.

— One of the largest and most diverse plant families, with more than 300 genera and 7,500 species. Many are easily recognisable by the combination of milky sap and dry 3-locular fruit, but there are numerous exceptions to these characters. The family includes a number of plants of great economic importance, of which *Hevea*, the source of most natural rubber, is the most valuable.

SYNOPSIS OF SUBFAMILIES

Ovules 2 in each loculus of the 3-locular ovary; fruit a capsule; sap never milky; leaves entire, unlobed, and non-glandular: *SUBFAMILY 1. PHYLLANTHOIDEAE*

Ovules 2 in each loculus of the 2-locular ovary; fruit a drupe; sap never milky; leaves 3-foliolate, non-glandular: *SUBFAMILY 2. OLDFIELDIOIDEAE*

Ovules 1 in each loculus of the ovary; fruit a capsule or drupe; sap often coloured or milky; leaves often lobed and/or with petiolar glands or marginal teeth:

SUBFAMILY 3. EUPHORBIOIDEAE

SUBFAMILY 1. PHYLLANTHOIDEAE

KEY TO GENERA

1. Flowers with petals; leaves pinnate-veined, usually deciduous at flowering time:

    2. Staminate flowers with free filaments; pistillate flowers subsessile, the fruits on stalks not longer than diameter of the fruit: **Savia**

    2. Staminate flowers with filaments united into a column, this expanded at apex into a disc bearing the anthers; pistillate flowers and fruits very long-stalked: **Astrocasia**

1. Flowers without petals; leaves variously veined or absent, but if present not deciduous at flowering time:

    3. Spiny shrubs; leaves very small (5–15 mm long), often clustered: **Securinega**

    3. Unarmed shrubs or small trees; leaves (if present) larger and obviously alternate:

        4. Plants dioecious; leaves always present but may be deciduous at flowering time:

            5. Small tree up to 8 m tall; leaves pinnate-veined, green on both sides; calyx 4-lobed: **Margaritaria**

            5. Shrubs rarely more than 2 m tall; leaves 3–5 veined from base, pale or whitish on underside; calyx 5-lobed: **Chascotheca**

        4. Plants monoecious; leaves present or absent, but if present always pinnate-veined: **Phyllanthus**

Savia Willd.

Dioecious shrubs or small trees; leaves alternate, entire; stipules present. Flowers axillary, the parts in 5s; rudimentary petals present. Staminate flowers densely clustered, subsessile; disc ring-like, outside the stamens; filaments free, alternate with the petals; a rudimentary ovary present, with 3 short styles. Pistillate flowers with 3-celled ovary and separate styles, each with 2 awl-shaped acuminate branches; ovules 2 in each cell. Seeds ovoid or 3-edged, without a caruncle.

    — A West Indian genus of probably less than 10 species.

Savia erythroxyloides Griseb. in *Mem. Amer. Acad. n.s.* 8: 157 (1860). Plate 36.

WILD COCO-PLUM

A shrub or small tree to 5 m tall; leaves glabrous, leathery, elliptic to obovate, 2–7 cm long, the apex obtuse or rounded, sometimes shallowly notched; veins prominulous especially on the upper side. Capsules glabrous, depressed-globose, ca. 8 mm in diameter, slightly and obtusely 3-lobed.

    GRAND CAYMAN: *Brunt* 1874, 1895, 2003, 2188; *Proctor* 15023, 15030, 15167, 15177, 52080. LITTLE CAYMAN: *Kings* LC 64; *Proctor* 28094, 35108. CAYMAN BRAC: *Proctor* 15326, 28943.

    — Florida, Bahamas, Cuba, Jamaica, Hispaniola, and the Swan Islands, in dry rocky thickets and woodlands. As here interpreted, not specifically distinct from *Savia bahamensis* Britton.

Astrocasia Robins. & Millsp.

Glabrous shrubs or small trees; leaves alternate, membranous, on long slender petioles, the margins entire. Flowers dioecious, clustered in the leaf-axils; petals present. Staminate flowers on short pedicels; sepals 5, imbricate; petals 5, erect; stamens 10, the filaments connate into a slender column, this expanded at the apex into a disc; anthers ellipsoid, sessile, horizontally dehiscent; ovary rudiment lacking. Pistillate flowers on very long slender pedicels; ovary 3-celled. Fruit an elastically dehiscent, 3-grooved capsule; seeds irregularly globose, without a caruncle.

— A chiefly C. American genus of about 4 species.

Astrocasia tremula (Griseb.) Webster in *Jour. Arnold Arb.* 39: 208 (1958). PLATE 36.

Phyllanthus glabellus Fawc. & Rendle (1919), not *Croton glabellus* L. (1759).

An arborescent shrub or small tree to 5 m tall; leaves with petioles 3–5 cm long or more, the blades elliptic or rhombic-ovate, mostly 4–8 cm long, blunt at apex, whitish-glaucous on the under surface, often deciduous before flowering; stipules lance-linear, soon falling. Staminate flowers on pedicels mostly 5–10 mm long; sepals 1.5 mm long, petals to 2.5 mm long; anthers 4, attached around margin of the peltate connective, opening transversely. Pistillate flowers on pedicels lengthening in fruit to as much as 4.5 cm; sepals 5, to 2 mm long; petals 5, to 3.5 mm long; style-branches 2-lobed, fleshy. Capsules depressed-globose, 8–10 mm in diameter; seeds 3 or 6, often all abortive, flattish-ellipsoidal, 4.5 mm long.

GRAND CAYMAN: *Proctor* 11975 (?), 15140 (?).

— Jamaica, in rocky woodlands.

Securinega Commers. Ex Juss.

Shrubs or small trees, monoecious or dioecious, with alternate or more often fasciculate leaves. Flowers apetalous, axillary, solitary or clustered. Staminate flowers very small, numerous; sepals 5, imbricate; glands 5, alternating with the 5 stamens; filaments free, inserted opposite the sepals; anthers dehiscing longitudinally. Pistillate flowers few or solitary, with calyx like that of staminate flowers; ovary 3-locular, the styles free, bifid; ovules 2 in each locule. Capsules dehiscing in 3 bivalved segments.

— A genus of about 20 species widely scattered in tropical and subtropical regions.

Securinega acidoton (L.) Fawc. & Rendle in *J. Bot.* 57: 68 (1919). PLATE 37.

GREEN EBONY

Shrub or small tree to 6 m tall, much-branched with zigzag spiny twigs. Leaves clustered, obovate or oblanceolate with prominulous veins, 5–15 mm long, 3–9 mm broad, short-petiolate. Sepals roundish, greenish-yellow, up to 2 mm long. Capsules 4–5 mm in diameter; seeds light brown, ca. 2 mm long.

GRAND CAYMAN: *P. Ann van B. Stafford* s.n.; *Proctor* 49241, 52185.

— Greater Antilles and Virgin Islands. Cayman specimens are all sterile; flowering and/or fruiting specimens are needed.

Sideroxylon horridum resembles both *Securinega acidoton* and *Rochefortia acanthophora* and needs to be considered when making an identification.

Margaritaria L.f.

Dioecious shrubs or small trees, the branches conspicuously lenticellate. Leaves simple, alternate, deciduous, with entire margins; stipules present. Inflorescence axillary, of apetalous flowers clustered on short peduncles. Staminate flowers long pedicellate; calyx lobes 4, fused toward base; disc annular, entire; stamens 4, a rudimentary pistil lacking. Pistillate flowers with shorter pedicels; calyx lobes 4, fused toward base; ovary of 2–5 carpels with 2 ovules per locule; styles free, flattened and bifid at apex. Capsules long-pedicellate, irregularly dehiscent; seeds 2 in each locule, carunculate.

— A pantropical genus of 14 species.

Margaritaria nobilis L.f., *Suppl. Pl.* 428 (1782).

Shrub or small tree to 8 m tall. Leaves short-petiolate, the blades ovate-lanceolate, elliptic, or oblong, 4.5–12 cm long, 2–4.5 cm broad, acuminate at apex, cuneate at base, glabrous. Staminate flowers 5–20 per cluster with pedicels up to 5 mm long; pistillate flowers 1–3 per cluster, the pedicels elongating to 1.5 cm in fruit, the calyx lobes becoming reflexed. Capsules depressed-globose, 9–12 mm in diameter. Seeds oblong, 3–4 mm long, blue-black.

GRAND CAYMAN: *Proctor* 49316, 49371; found only in the area of Queen Elizabeth II Botanic Park, in natural forest.

— Mexico, Central and S. America, and the Greater and Lesser Antilles.

Chascotheca Urban

Dioecious, glabrous shrubs with terete branchlets; leaves alternate, distichous, entire, 3–5-nerved from the base; stipules small, biauriculate. Flowers in small clusters from minute, cushion-like spurs in the leaf-axils or often at defoliated nodes; sepals 5; petals lacking. Stamens 5, with filaments connate at the base, the rudimentary ovary columnar and terminating in 3 recurved styles. Pistillate flowers with ovary 3-celled and 2 ovules in each cell; styles 3, free, forked at the apex. Capsule of 3 two-valved cocci; seeds bullate-rugose, with abaxial hilum; caruncle absent; endosperm fleshy.

— A small genus of 3 species confined otherwise to Cuba and Hispaniola.

KEY TO SPECIES

1. Leaves elliptic, with midrib whitish beneath and petiole 0.5–0.7 mm in diameter at the top; fruiting pedicels 4–10 mm long; capsules 2–3 mm in diameter:	C. neopeltandra
1. Leaves broadly ovate to orbicular, with midrib dark beneath and petiole 0.3–0.4 mm in diameter at the top; fruiting pedicels 12–25 mm long; capsules 4–5 mm in diameter:	C. domingensis

Chascotheca neopeltandra (Griseb.) Urb., *Symb. Ant.* 5: 14 (1904). PLATE 37.

A dense-crowned shrub to 3 m tall or more, with numerous wide-spreading branches; leaves with petioles 2–5 mm long, the blades 2–4 cm long, obtuse to acutish at the apex, pale or glaucous beneath with the fine venation faintly or scarcely evident. Sepals membranous, 1.5 mm long, the margins minutely ciliate. Staminate flowers subsessile. Pistillate flowers not seen.

GRAND CAYMAN: *Brunt* 1817 ?; *Proctor* 52069, 52105.

— Cuba and Hispaniola, in thickets on limestone rock.

Chascotheca domingensis (Urb.) Urb., *Symb. Ant.* 5: 14 (1904).

Shrub 1.5–2 m tall, the branchlets with horizontal frond-like habit; leaves with delicate, filiform petioles 2–6 mm long, the blades light green 1–2 cm long, obtuse or rounded at the apex, the fine, densely reticulate venation evident on the under side. Sepals membranous, 1.5 mm long, with margins minutely ciliate. Staminate flowers subsessile or on pedicels to 2 or 3 mm long. Pistillate flowers not seen, apparently undescribed.

GRAND CAYMAN: *Brunt* 2187 (sterile); *Proctor* 15165 (mixture, ? flowers and fruits). CAYMAN BRAC: *Proctor* 35150 (sterile).

— Hispaniola, apparently rare in dry rocky woodlands.

Phyllanthus L.

Herbs, shrubs, or trees of very diverse habit, the branching either unspecialised and either spiral or distichous, or else 'phyllanthoid', i.e. the spiralled leaves on the main axes reduced to stipule-like 'cataphylls' which subtend deciduous branchlets, these either bearing distichous leaves or else leafless and more or less broadly flattened (then called 'phylloclades'). Leaves present or absent, always simple and entire, usually glabrous, the petiole always shorter than the blades; stipules deciduous or persistent. Plants usually monoecious, rarely dioecious. Inflorescences axillary; calyx 4–6-lobed, the lobes imbricate in bud; petals lacking; disc nearly always present, commonly divided into segments alternating with the calyx-lobes. Staminate flowers stalked; stamens 2–15 (mostly 3–6), the filaments free or connate, the anthers free or connate; rudimentary ovary absent. Pistillate flowers stalked or subsessile; staminodes absent (except in *P. acidus*); ovary usually 3-celled with 2 ovules in each cell; styles erect or spreading, free or united into a column, variously divided or else dilated into an entire or lacerate stigma. Fruit mostly a dehiscent capsule, rarely berry-like or drupe-like; seeds usually 2 in each cell, rarely only 1 developing; endosperm cartilaginous.

— A nearly cosmopolitan genus of about 750 species, the majority of them Old World in distribution. *Phyllanthus acidus* (L.) Skeels, the 'jimbling', originally from Brazil, is commonly cultivated for its abundant edible (but sour) fruits.

KEY TO SPECIES

1. Plants clothed with green leaves:

  2. Small annual herbs:

    3. Stamens 5, free; capsules pendulous, on pedicels 3–7 mm long; seeds papillose: **[P. tenellus]**

    3. Stamens usually 3, united; pedicels less than 3 mm long; seeds longitudinally ribbed: **P. amarus**

  2. Shrubs usually 1–3 m tall; leaf-blades more than 1 cm long (up to 8 cm); capsules more than 4 mm in diameter (up to 10 mm):

    4. Leaf-blades not over 2.5 cm long; pedicels of staminate flowers 5–7 mm long, the flowers themselves ca. 1.5–2 mm in diameter; capsules 4–4.5 mm across: **P. caymanensis**

    4. Leaf-blades up to 8 cm long; pedicels of staminate flowers 8–15 cm long or more, the flowers themselves ca. 2.5–4 mm in diameter; capsules up to 10 mm across: **P. nutans**

1. Plants (except seedlings) leafless, the ultimate branches being flattened, green and leaf-like with nodes represented by marginal notches whence are produced the flowers and fruits: **P. angustifolius**

[*Phyllanthus tenellus* Roxb., *Fl. Ind.* 3: 668 (1832).

Annual or short-lived perennial herb 20–60 cm tall often subwoody at base. Leaves thin, elliptic to obovate, mostly 10–20 mm long, acute or obtuse at the apex, somewhat pale beneath. Calyx lobes whitish except for narrow green midrib. Capsule 1.7–1.9 mm in diameter, obscurely reticulate-veined. Seeds trigonous, light brown.

GRAND CAYMAN: *Proctor* 49350; a weed in the George Town area.

— Native of the Mascarene Islands, introduced and naturalised widely in the neotropics and subtropics.]

Phyllanthus amarus Schum. & Thonn. in *Kongl. Danske Vidensk. Selsk. Skr.* 4: 195–196 (1829).

    Phyllanthus niruri of many authors, not L. (1753).

Erect annual herb mostly 10–30 cm tall, with phyllanthoid branching, the main stem simple or branched; deciduous branchlets 4–12 cm long, with 15–30 leaves; leaves membranous, usually elliptic-oblong, mostly 5–11 mm long, obtuse or rounded and often apiculate, more or less glaucous beneath. Flowers paired beneath the leaf-axils, all but the lowest axils having one staminate and one pistillate flower, each minutely stalked. Capsules oblate, obtusely 3-angled, mostly 1.9–2.1 mm in diameter; seeds sharply 3-angled.

GRAND CAYMAN: *Hitchcock*; *Millspaugh* 1339, 1363; *Proctor* 15103, 15244. LITTLE CAYMAN: *Proctor* 35108. CAYMAN BRAC: *Proctor* 29092.

— A pantropical weed of open waste ground.

Phyllanthus caymanensis Webster & Proctor in *Rhodora* 86 (846): 121 (1984). PLATE 37.

Glabrous shrub up to 2.5 m tall with simple (pinnatiform) deciduous branchlets; leaves membranous, short-petiolate, the blades ovate or rhombic-ovate, mostly 1.5–2.5 cm long, obtusely to acutely pointed at the apex, somewhat pale or subglaucous beneath; stipules 1.2–1.5 mm long. Flowers in axillary cymules, each cluster with one central female flower and several lateral male flowers; calyx-lobes of male flowers 1.3–1.7 mm long; stamens 3, the filaments connate into a column 0.4–0.5 mm high; female flowers with filiform pedicels becoming 8–12 mm long in fruit, the calyx-lobes 1.4–1.7 mm long; styles free, spreading, bifid. Capsules oblate, prominently veiny, greenish; seeds angled, 1.9–2 mm long, nearly smooth.

LITTLE CAYMAN: *Proctor* 35145. CAYMAN BRAC: *Proctor* 35151 (type).

— Endemic; related to *P. mocinianus* and *P. mcvaughii* of Mexico and C. America.

Phyllanthus nutans Sw., *Prodr. Veg. Ind. Occ.* 27 (1788). PLATE 37.

An irregularly branched glabrous to hirsutulous shrub 1–3 m tall, without deciduous branchlets; leaves short-petiolate, the blades usually elliptic or ovate, mostly 2–8 cm long, obtuse or rarely acute at the apex, often somewhat glaucous or purple-tinged beneath; stipules 3–5 mm long, yellowish. Inflorescences axillary or pseudoterminal, often racemiform; male flowers in small clusters (cymules) at intervals on the inflorescence-axis; filaments united into a column 0.7–1.1 mm high; female flowers solitary or 2–3 in small clusters mostly on pseudoterminal inflorescences, with pedicels mostly 10–27 mm long. Capsules oblate-spheroidal, obscurely 6-ribbed, ca. 6 mm high and up to 10 mm broad; seeds 3-angled, 4.2–7 mm long, smooth.

The Cayman representatives of this species have been assigned to two subspecies, as follows:

1. Leaves ovate, flat, obtuse or rounded at base; flowers in axillary and pseudoterminal inflorescences; calyx-lobes subentire; stylar column 0.5–2.3 mm high: ssp. **nutans**
1. Leaves elliptic, with narrowly revolute margins, acute at both ends; flowers all axillary, solitary or the male and female paired at each axil; calyx-lobes denticulate; stylar column 0.5–0.7 mm high: ssp. **grisebachianus**

Phyllanthus nutans ssp. *nutans* FIG. 159.

GRAND CAYMAN: *Brunt* 1994, 2139, 2194; *Proctor* 11977, 15245. LITTLE CAYMAN: *Proctor* 28039, 49333.

— Jamaica and the Swan Islands, in various habitats; the Cayman plants occur in dry rocky woodlands.

Phyllanthus nutans ssp. *grisebachianus* (Muell. Arg.) Webster in *Jour. Arnold Arb.* 39: 61 (1958).

LITTLE CAYMAN: *Kings* LC 42.

— Eastern Cuba; in habitats similar to those of ssp. *nutans*. It has been pointed out that the Cayman populations of *P. nutans* show transitional features between the two subspecies. Several collections from Little Cayman (*Proctor* 35111, 35209, 35212) have not been determined to subspecies.

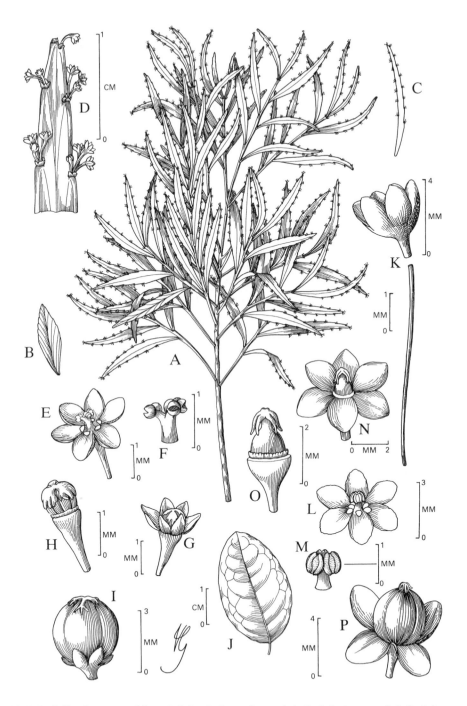

FIG. 159 **Phyllanthus angustifolius**. A, habit. B, C, two forms of phylloclade. D, apex of phylloclade with clusters of flowers. E, staminate flower. F, stamens (androecium). G, pistillate flower. H, gynoecium. I, fruit. **Phyllanthus nutans** ssp. **nutans**. J, leaf. K, flower showing long pedicel. L, staminate flower. M, stamens (androecium). N, pistillate flower. O, gynoecium. P, fruit. (G.)

Phyllanthus angustifolius (Sw.) Sw., *Fl. Ind. Occ.* 2: 1111 (1800). **FIG. 159.** PLATES 37 & 38.

Phyllanthus linearis of Hitchcock (1893), not Swartz (1788).

DUPPY BUSH, DUPPY BASIL

A shrub mostly 1–3 m tall, or occasionally tree-like and 6 m tall, with phyllanthoid branching, the axes glabrous; cataphylls of main stems clustered at apex in a scaly cone rounded in outline and 3–6 mm thick, light to dark brown. Leaves absent except on seedlings. Branchlets bipinnatiform, the ultimate axes broadened into phylloclades, these lance-linear to elliptic or obovate-lanceolate and mostly 3–8 cm long and 2–10 mm broad. Flowers in small staminate or bisexual clusters at notches of the phylloclades; pedicels mostly 2–6 mm long; calyx greenish-cream, pale buff, or shades of red. Capsules oblate, 3–4 mm in diameter; seeds 3-angled, 1.4–2.6 mm long.

GRAND CAYMAN: *Brunt* 1758, 1828; *Correll & Correll* 51002; *Hitchcock*; *Howard & Wagenknecht* 15030; *Kings* GC 116, GC 202; *Proctor* 11974, 11976; *Sauer* 4111. LITTLE CAYMAN: *Proctor* 28052, 28188, 35085, 35086, 35191; *Sauer* 4171. CAYMAN BRAC: *Fawcett*; *Kings* CB 70; *Matley*; *Proctor* 28915, 28916.

— Jamaica and the Swan Islands, in rather dry rocky thickets and woodlands. The Little Cayman population is notably variable in phylloclade shape and flower colour.

[*Breynia disticha* J. R. & G. Forst., a *Phyllanthus*-like shrub with green and white variegated leaves, is cultivated in Grand Cayman (*Kings* GC 308) and Cayman Brac (*Proctor*, sight record). It is said to originate from islands of the Pacific.]

SUBFAMILY 2. OLDFIELDIOIDEAE

Picrodendron Planch.

Small bushy trees; leaves deciduous, alternate, digitately 3-foliolate; stipules small, soon falling. Flowers dioecious, without petals. Staminate flowers lacking a calyx, the naked stamens in clusters of 3–54 on a convex receptacle subtended by usually several imbricate bracts, the clusters on stalked axillary spikes crowded near the ends of the branchlets and appearing with the young leaves; anthers 2-celled, opening lengthwise, rudiment of ovary lacking. Pistillate flowers solitary in leaf-axils, long-stalked, the pedicel expanded at apex into a concave receptacle; calyx of 4–5 unequal valvate sepals; ovary superior, 2-celled, with slender terminal style and 2 large spreading stigmas; ovules 2 in each cell, pendulous and inserted below an obturator. Fruit a thin-fleshed 1- or 2-seeded drupe; seeds without endosperm.

— An Antillean endemic genus once thought to contain 3 species, but here considered probably to consist of but a single variable species. *Picrodendron* has been placed by some authors in a family of its own, of dubious affinity, but more recent consensus places it in the Euphorbiaceae.

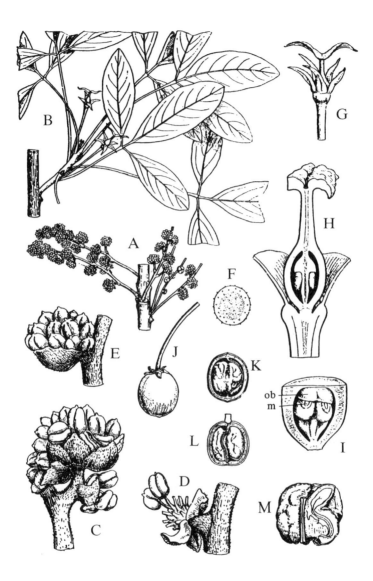

FIG. 160 **Picrodendron baccatum.** A, part of twig with staminate inflorescence, × ²/₃. B, same with pistillate flowers, × ²/₃. C, D, E, clusters of stamens with subtending bracts, × 7. F, pollen grain, × 400. G, pistillate flower, × 2. H, ovary and style cut lengthwise, × 5. I, ovary cut lengthwise through one cavity, × 10; **ob**, obturator; **m**, micropyle. J, fruit, × ²/₃. K, same cut open, with one seed, × ²/₃. L, same cut open, with two seeds, × ²/₃. M, seed cut to show radicle and cotyledons × 1¹/₂. (F. & R.)

Picrodendron baccatum (L.) Krug & Urban in *Engl. Bot. Jahrb.* 15: 308 (1892). FIG. 160. PLATE 38.

WILD PLUM, CHERRY, JAMAICA WALNUT, BLACK IRONWOOD

A tree to 12 m tall, or in exposed situations low and shrubby; leaflets mostly narrowly to very broadly elliptic, rarely obovate, mostly 2.5–9 cm long, acutish to blunt or rounded at the apex, glabrous to rather densely puberulous chiefly on veins beneath; ultimate venation finely reticulate. Staminate inflorescences minutely puberulous; anthers ca. 1 mm long. Pistillate flowers with denticulate sepals mostly 3–5 mm long, thickening in fruit. Fruits subglobose or broadly ellipsoid, ca. 2 cm in diameter, greenish to orange-yellow, with bitter flesh; seeds alleged to be edible, "but should be eaten with caution" (Fawcett & Rendle (1920), p. 275).

GRAND CAYMAN: *Brunt* 1763; *Kings* GC 131; *Proctor* 15248, 27976. LITTLE CAYMAN: *Kings* LC 77; *Proctor* 28038. CAYMAN BRAC: *Proctor* 29125.

— Bahamas, Cuba, Hispaniola, Jamaica and the Swan Islands, in rocky woodlands and thickets. The fruits are a favourite food of iguanoid lizards (genus *Cyclura*).

SUBFAMILY 3. EUPHORBIOIDEAE

KEY TO GENERA

1. Flowers solitary or in spikes, cymes, or panicles, not fused and aggregated into bisexual cyathia:

   2. Leaves palmately veined, lobed, or divided:

      3. Shrubs or trees; flowers in panicle-like inflorescences:

         4. Stamens branched and fasciculate, anthers many (up to 1,000) per flower; leaves peltate: **Ricinus**

         4. Stamens relatively few and unbranched; leaves not peltate:

            5. Flowers with petals; sap not milky; roots not tuberous:

               6. Calyx-lobes free, imbricate; fruit a capsule; seeds with a caruncle: **Jatropha**

               6. Calyx-lobes valvate, fused into a spathe; fruit indehiscent; seeds without a caruncle: **[Aleurities, p. 449]**

            5. Flowers without petals; sap milky; roots with starchy tubers: **[Manihot, p. 450]**

      3. Herb; flowers in terminal spikes: **Croton lobatus,** (see **Croton** p. 450)

   2. Leaves simple and pinnate-veined (rarely 3-nerved at base):

      7. Petals present, at least in staminate flowers:

         8. Stamens more or less inflexed in bud; plants bearing stellate hairs or scales: **Croton**

         8. Stamens erect in bud; plants clothed with malpighiaceous hairs: **Argythamnia**

      7. Petals absent in both staminate and pistillate flowers:

         9. Fruit a dry capsule; sap not milky:

            10. Twining vine with stinging hairs: **Tragia**

            10. Stems upright, not twining, woody or herbaceous, without stinging hairs:

11. Flowers clustered on small cushions in the leaf-axils; leaves without glands: **Adelia**
11. Flowers in spikes (the distal staminate portion often deciduous in fruit):
    12. Leaves with conspicuous glands at top of petiole; pubescence stellate: **Bernardia**
    12. Leaves without glands; pubescence if present not stellate:
        13. Pistillate flowers sessile and subtended by conspicuous bracts; pubescent small herbs (or cultivated shrubs): **Acalypha**
        13. Pistillate flowers with long-stalked ovary and fruit, the bracts if present minute; glabrous shrubs or small trees: **Gymnanthes**
9. Fruit a fleshy drupe; sap milky, poisonous: **Hippomane**

1. Flowers fused into a usually bisexual cyathium, usually with one central female flower surrounded by 4 or 5 staminate flowers each represented by a single stamen; cyathium bordered by 1–several usually conspicuous glands, these often with petaloid appendages, the whole cyathium rather flower-like in general appearance:
    14. Cyathium spurred, somewhat slipper-shaped; glands hidden within spur: **[Pedilanthus p. 469]**
    14. Cyathium not spurred, the glands on its rim:
        15. Gland of cyathium usually 1 or rarely more, never with petaloid appendage; leaves subtending inflorescence often coloured or pale, at least at base: **Poinsettia**
        15. Glands of cyathium usually 4 or 5, with or without petaloid appendages:
            16. Leaves alternate or apparently absent; stipules absent; stem-axis persisting: **Euphorbia**
            16. Leaves opposite; stipules present and persistent; main stem-axis aborting within 2 nodes of cotyledons, the stem always forking at this point: **Chamaesyce**

Jatropha L.

Herbs, shrubs, or small trees; leaves alternate with blades entire, toothed, or palmately lobed; stipules present, often glandular. Plants monoecious, rarely dioecious; flowers in terminal, often long-stalked dichasia, the lower flowers pistillate, the distal ones staminate; calyx 5-lobed; petals 5, imbricate to contorted, free or coherent. Staminate flowers with mostly 8–12 connate stamens; rudimentary ovary absent. Pistillate flowers with 2–3-celled ovary; styles 2 or 3, connate at base, the stigmas 2-forked; ovules solitary. Fruit a capsule with 1 seed in each locule; seeds with caruncle; endosperm copious.

— A large tropical genus of perhaps 150 species, the majority in America and Africa.

[*Jatropha multifida* L., and doubtless other ornamental members of this genus, is cultivated in Grand Cayman (*Proctor*, sight record).]

KEY TO SPECIES

1. Leaves palmately lobed to below the middle; leafy nodes with conspicuous glandular stipules: **J. gossypiifolia**

1. Leaves palmately veined but not or only slightly and broadly lobed; leafy nodes without glandular stipules:

    2. Inflorescence repeatedly 2-forked, the branches diverging; leaf-blades rounded at base; capsules ca. 2 cm long: **J. divaricata**

    2. Inflorescence of corymbose cymes; leaf-blades openly cordate at base; capsules 2.5–4 cm long: **J. curcas**

Jatropha gossypiifolia L., *Sp. Pl.* 2: 1006 (1753).

BITTER CASSAVA, WILD CASSAVA

A low shrub often less than 1 m tall; leaves long-petiolate, the petioles with glandular hairs, the blades mostly 3–12 cm across, deeply divided into 3–5 lobes and narrowly cordate at base, more or less pubescent; stipules cut into thread-like glandular hairs 3–5 mm long. Flowers deep crimson or purple; sepals glandular-ciliate, 3–5 mm long; petals glabrous, ca. 4 mm long. Stamens 10–12, the filaments much longer than the anthers and united above the middle. Capsule ca. 1 cm in diameter, glabrous; seeds 7–8 mm long.

CAYMAN BRAC: *Proctor 28917*.

— Pantropical, often a weed of pastures and open waste ground, occasionally planted for ornament.

Jatropha divaricata Sw., *Prodr.* 98 (1788). PLATE 38.

Shrub or small tree 2.5–8 m tall with resinous aromatic sap. Leaves ovate to elliptic, softly leathery, 7–16 cm long, 3–9 cm broad, acute to acuminate at apex, rounded or rarely narrowly cordate at base, the margins entire, glabrous except often with small patches of hairs at base of main lateral veins beneath (abaxially). Peduncles 3–10 cm long; male petals ca. 5 mm long; pistillate flowers larger, solitary at junctions of panicle branches. Seeds purplish, 10–12(–14) mm long.

GRAND CAYMAN: *Proctor 47860, 48663*. Found near Forest Glen and along the Mastic Trail.

— Jamaica.

Jatropha curcas L., *Sp. Pl.* 2: 1006 (1753).

PHYSIC NUT

A shrub or small tree to 7 m tall, with milky sap; leaves long-petiolate, without glandular hairs, the blades roundish-ovate, mostly 4–12 cm across, widely cordate at the base, and entire, slightly angulate, or broadly lobed, glabrous. Flowers yellowish-green, in small corymbose cymes; sepals 3.5–4.5 mm long; petals ca. 6.5 mm long, pubescent within. Stamens 9, the 5 inner filaments united halfway, the 4 outer united at base. Capsule fleshy on the outside at first, 2–3-celled, 2.5–4 cm long; seeds ca. 2 cm long.

GRAND CAYMAN: *Hitchcock*.

— Widespread in tropical America, sometimes planted as a 'quickstick' fence. There is some doubt whether it still occurs in the Cayman Islands. The seeds of this species are valued for their purgative properties, and for their oil, which can be used in the manufacture of soap.

[*Aleurites moluccana* (L.) Willd., the candlenut, is cultivated in George Town, GRAND CAYMAN (*Brunt* 1989; *Hitchcock*; *Millspaugh* 1337). It is a monoecious tree to 12 m tall or more, the young twigs and other parts densely clothed with minute stellate hairs, the leaves long-petioled, with blades 3–5-nerved at the base, entire or lobed, and up to 20 cm long or more. The small white flowers are borne in terminal panicle-like cymes. The fruits are 5–6 cm in diameter and contain 1 or 2 large, nut-like seeds. These seeds are

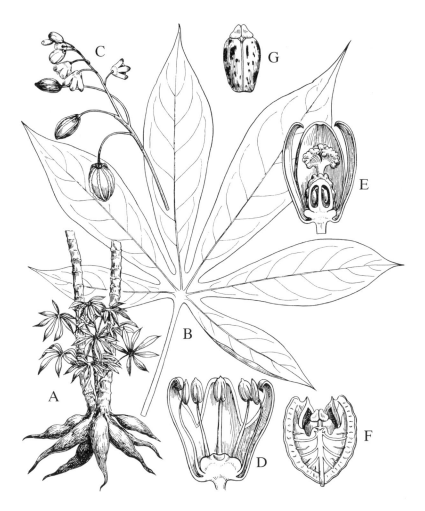

FIG. 161 **Manihot esculenta.** A, lower part of stem showing young shoot and tubers, much reduced. B, small leaf, × 1. C, portion of inflorescence, × 1/3. D, male flower cut lengthwise, × 5. E, female flower cut lengthwise, × 2. F, one carpel of ripe fruit showing seed, × 1. G, seed, × 1. (F. & R.)

edible "but should be eaten with caution" (Fawcett & Rendle, 1920); pickling in alcohol or roasting are said to reduce their purgative effect. Although locally called 'walnut', this species is not, of course, related to the true walnut. Another species of *Aleurites*, *A. fordii*, is widely cultivated in southeastern U.S.A. and elsewhere as a source of tung oil, used in the manufacture of varnish.]

[*Manihot esculenta* Crantz, the cassava plant (**FIG. 161**), is cultivated rather widely in Grand Cayman (*Kings* GC 309, *Millspaugh* 1293), and to a lesser extent on Little Cayman and Cayman Brac, and may become naturalised in the long run. It is a shrubby plant with tuberous, starchy roots and alternate, palmately lobed leaves. The unisexual flowers are monoecious in terminal panicles. This species is a native of Brazil, but is cultivated in many tropical countries for its starchy tubers. Cassava starch is low in protein, lacking in vitamins, and nutritionally is an inferior food. On the other hand, it has the advantage of being drought-resistant and easy to grow, and can be harvested at various seasons. The leaves can be eaten as a cooked vegetable.]

Croton L.

Herbs, shrubs or trees, the stems often containing coloured or resinous sap but not milky latex; indumentum at least in part of branched or scale-like trichomes. Leaves alternate, pinnately or palmately veined or sometimes lobed; petioles sometimes bearing glands at the junction of the leaf-blades. Plants monoecious or sometimes dioecious; flowers usually in spike-like, often terminal racemes, the staminate several on the apical portion, the pistillate often solitary near the base; petals usually present on staminate flowers and absent from pistillate ones. Staminate calyx 5-lobed; stamens mostly 8–20, free, the filaments indexed in bud; rudimentary ovary absent. Pistillate calyx usually 5–7-lobes; ovary mostly 3-celled, each cell with 1 ovule; styles free or nearly so, 1–several-times-forked. Fruit a 3-seeded capsule; seeds with copious endosperm.

   — A very large, widely distributed, but chiefly tropical American genus of between 700 and 1,000 species (authorities disagree on the number), well represented in the West Indies. Very few have any economic importance, although the aromatic oils and alkaloids of a few have medicinal or flavouring value. The horticultural crotons are properly called *Codiaeum* (see below, p. 454).

KEY TO SPECIES

1. **Leaves simple and entire; shrubs:**

   2. Leaves narrowly linear (less than 2 mm wide) and less than 2 cm long: — **C. rosmarinoides**

   2. Leaves broader and longer:

      3. Leaves clothed with stellate hairs or non-silvery scales beneath, or rarely glabrous:

         4. Leaves linear-oblong, white or yellowish beneath; plants dioecious: — **C. linearis**

         4. Leaves elliptic or ovate, green beneath; plants monoecious: — **C. lucidus**

      3. Leaves clothed with minute silvery scales: — **C. nitens**

1. **Leaves 3–5-lobed, the margins crenate-serrate; an annual herb:** — **C. lobatus**

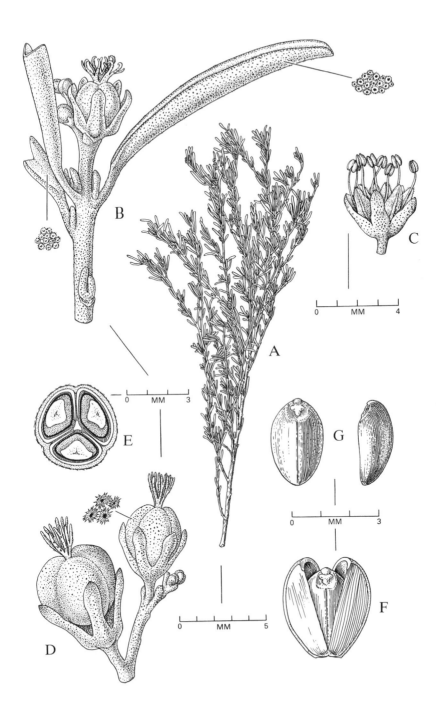

FIG. 162 **Croton rosmarinoides**. A, branch, habit. B, branchlet, much enlarged, with young raceme. C, staminate flower. D, pistillate flower and young fruit. E, cross-section of fruit. F, portion of capsule, seed within. G, two views of a seed. (D.E.)

Croton rosmarinoides Millsp. in Britton & Millsp., *Bahama Fl.* 222 (1920). **FIG. 162**

A densely branched monoecious shrub about 1 m tall or less; leaves subsessile, mostly 6–19 mm long and 0.6–2 mm broad, the margins strongly incurved beneath, the lower (abaxial) surface densely clothed with minute whitish, brown-centreed stellate scales, the upper by very minute greenish punctate scales. Racemes terminal, short and densely flowered, densely scaly throughout. Petals of staminate flowers woolly ciliate; stamens 6, the filaments woolly at the base. Pistillate flowers 1–4 per inflorescence; styles 4-branched, stellate-scaly. Capsules ellipsoid-globose, densely stellate-scaly.

GRAND CAYMAN: *Proctor 47859*. LITTLE CAYMAN: *Proctor 28137, 35073, 47315, 48218*. CAYMAN BRAC: *Proctor 29062, 52164*.

— Bahamas and Cuba, in dry rocky scrublands.

Croton linearis Jacq., *Enum. Syst. Pl. Carib.* 32 (1760); *Sel. Stirp. Amer.* 256, t. 162, f. 4 (1763). **FIG. 163.** PLATE 38.

Croton cascarilla L. (1753), in part, not as to type.

ROSEMARY

An aromatic shrub mostly 1–2 m tall, the branchlets and undersides of leaves densely clothed with white or yellowish stellate-hairy scales; leaves short-petiolate, linear-oblong, up to 5 cm long and mostly 2–9 mm broad, glabrous above; petioles with 2 cylindrical glands at apex on the upper (adaxial) side. Plants dioecious; racemes usually longer than the leaves, with scattered subsessile flowers. Staminate calyx 1.5 mm long; stamens 13–15, with glabrous filaments. Pistillate calyx 1.8 mm long; ovary clothed with minute stellate scales; styles 2-branched. Capsules globose, ca. 5 mm in diameter, 3-grooved, minutely stellate-scaly.

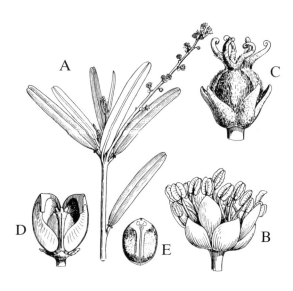

FIG. 163 **Croton linearis**. A, branch with staminate flower, × 7. C, pistillate flower, × 10. D, coccus with seed, × 3. E, seed, × 4. (F. & R.)

GRAND CAYMAN: *Brunt* 1655; *Correll & Correll* 50995; *Hitchcock*; *Kings* GC 52, GC 392; *Millspaugh* 1312; *Proctor* 15058; *Sachet* 365. LITTLE CAYMAN: *Kings* LC 9; *Proctor* 28029. CAYMAN BRAC: *Kings* CB 49; *Millspaugh* 1217; *Proctor* 28969.

— Florida, Bahamas, and the Greater Antilles except Puerto Rico, in rocky thickets and old pastures.

Croton lucidus L., *Syst. Nat.* ed. 10, 2: 1275 (1759). PLATES 38 & 39.

A monoecious shrub 1–2 m tall; leaves usually long-petiolate, glabrate or stellate-pubescent, the blades elliptic or ovate, 2–12 cm long, acute to acuminate at the apex, the tissue minutely pellucid-dotted. Racemes 3–8 cm long; staminate calyx and petals ca. 2 mm long, densely pubescent; stamens 11–12; pistillate calyx 5–6.5 mm long, densely pubescent; styles short, deeply 4-branched. Capsules oblong-ellipsoidal, glabrate, 10–12 mm long.

GRAND CAYMAN: *Brunt* 2135, 2148; *Hitchcock*; *Proctor* 15025, 27981; *Rothrock* 77, 172, 177. LITTLE CAYMAN: *Kings* LC 13, LC 53; *Proctor* 28139. CAYMAN BRAC: *Millspaugh* 1199, 1204, 1206, 1216; *Proctor* 29061.

— Bahamas and Greater Antilles; frequent in rocky thickets and woodlands.

Croton nitens Sw., *Nov. Gen. & Sp. Pl.* 100 (1788). PLATE 39.

Croton glabellus of Fawcett & Rendle (1920), not L. (1763).

Croton eluteria of Adams (1972), not (L.) Sw. (1788).

WILD CINNAMON

An aromatic shrub or small tree to 10 m tall or more, the branchlets brownish-scaly; leaves oblong or elliptic to very broadly ovate, 2–10 cm long, blunt to acutish or acuminate at the apex, densely clothed beneath with minute pale silvery scales having brown-punctate centres, scattered similar scales also on the upper (adaxial) side; tissue also minutely pellucid-dotted. Plants monoecious. Racemes simple or branched, axillary and terminal, much shorter than the leaves; all flowers stellate-scaly and with woolly margined petals. Staminate calyx 2–2.4 mm long; stamens 10–13, the long-exserted filaments glabrous except at the base. Pistillate flowers like the staminate; styles 4–6-branched. Capsules obovoid-globose, tuberculate and stellate-scaly, 7–9 mm long.

GRAND CAYMAN: *Brunt* 2116, 2137; *Hitchcock*; *Proctor* 15239. LITTLE CAYMAN: *Proctor* 28030, 28110. CAYMAN BRAC: *Proctor* 28968.

— Jamaica, Swan Islands, Mexico and C. America; related forms in Colombia and Ecuador. In the Cayman Islands, this species is frequent in dry rocky woodlands. *Croton nitens* is closely related to *C. eluteria* of the Bahamas, Cuba, and Hispaniola, and the two are united under the latter name by some authors. However, the two populations are distinctively different in appearance, and specimens placed side by side can be distinguished at a glance. Submerging them under one name would conceal the fact that the Cayman plants are allied to those of Jamaica and not to those of Cuba and the Bahamas. *C. eluteria* as it occurs in the Bahamas is the source of high-priced aromatic cascarilla oil, used in medicines, alcoholic beverages and perfumes. Similar oils occur in *C. nitens*, but have not been exploited.

Croton lobatus L., *Sp. Pl.* 2: 1005 (1753).

A weedy monoecious herb up to 1 m tall, the branches pilose with simple hairs; leaves thin and glabrous to sparingly pilose, usually 3-lobed, rarely 5-lobed, the lobes mostly 2–7 cm long, constricted toward the base, the tips acuminate, the margins crenate-serrate; petioles often as long as the blades. Racemes terminal or axillary, up to 10 cm long. Staminate flowers glabrous; stamens 10–13; pistillate flowers with glandular-ciliate sepals; ovary stellate-pubescent and pilose; styles 3–8-branched. Capsule globose-ellipsoid, ca. 8 mm long; seeds 5 mm long.

GRAND CAYMAN: *Hitchcock.*

— Florida, West Indies (except Jamaica) continental tropical America, and tropical Africa. This plant is usually a weed of fields and open waste places; its continued presence in the Cayman Islands requires confirmation.

[*Codiaeum variegatum* (L.) Blume is the correct name for the ornamental shrubs commonly called croton. This species is, in fact, not closely related to the botanical genus of that name. Although the leaves of *Codiaeum* are exceedingly variable in size, shape and colouration, all the forms belong to a single botanical species, this native to the region of Malaya and islands of the Pacific; it is much cultivated throughout the tropics.]

Argythamnia Sw.

Monoecious or very rarely dioecious perennial herbs, shrubs or rarely small trees, always bearing 'malpighiaceous' hairs at least when young, and often containing purplish pigment. Leaves alternate, petiolate, entire to serrulate; stipules present. Inflorescence an axillary raceme, with pistillate flowers at the base. Staminate flowers with 4 or 5 pubescent valvate sepals and 4 or 5 petals alternate with the sepals; glands 4 or 5; stamens 4–15 in 1 or 2 series, the filaments joined at the base; rudimentary ovary absent. Pistillate flowers with 5 pubescent valvate sepals, the petals 5, rudimentary or absent; glands 5, opposite the sepals, inserted on disc of the ovary; ovary superior, subglobose, 3-celled, each cell with 1 pendulous ovule; styles 3, 1–2-forked. Capsule splitting into 3 1-seeded cocci; seeds rough-coated, and with endosperm.

— An American genus of about 50 tropical or warm-temperate species.

Argythamnia proctorii Ingram, *Gentes Herb.* 10: 25 (1966). PLATE 39.

A straggling monoecious shrub to 1.5 m tall, the young branchlets densely clothed with malpighiaceous hairs; leaves narrowly to broadly elliptic or rarely narrowly obovate, 1–6.5 cm long, acute and mucronate at the apex, the margins subentire to shallowly serrate with gland-tipped teeth, more or less woolly with soft malpighiaceous hairs on both sides. Racemes to 5 mm long. Staminate flowers with 4 sepals, these 1.75 mm long, bearing dense malpighiaceous hairs outside and simple hairs within; petals 4, ca. 1.5 mm long, hairy like the sepals; stamens 4, exserted, hairy. Pistillate flowers with 5 sepals, these 2.5 mm long, increasing to 3.5 mm in fruit; petals 5, minute, hairy; glands 5, squarish, almost as thick as wide; styles woolly, 3-times-forked. Capsule 3 mm long and 4.5 mm wide, villous; seeds ovoid-subglobose, ca. 2 mm in diameter, lightly ridged.

GRAND CAYMAN: *Brunt* 2158; *Proctor* 15043 (type), 27987, 48292. LITTLE CAYMAN: *Proctor* 28123, 35081, 47310. CAYMAN BRAC: *Proctor* 28941, 47329, 52155.

— Endemic; occurs chiefly in rocky woodlands, or sometimes in sandy thickets. Closely related to *Argythamnia candicans*, a widespread West Indian species.

Tragia L.

Monoecious perennial herbs with woody taproots; stems twining, pubescent with stiff stinging hairs mixed with soft hairs; similar hairs on the leaves; stipules present. Leaves alternate, petiolate, the blades mostly cordate at base. Inflorescences racemose, opposite the leaves, with pistillate flowers toward base and staminate flowers toward apex, all flowers apetalous. Calyx lobes 3–6; disc lacking. Stamens 2–6, the filaments connate toward base; pistillate minute. Pistillate flowers with subglobose ovary densely covered with stinging hairs, 3-carpellate; styles 3, spreading. Capsules lobed, explosively dehiscing. Seeds smooth, brownish-black, carunculate.

— A genus of 150 species occurring in tropical and warm-temperate regions of both hemispheres.

Tragia volubilis L., *Sp. Pl.* 2: 980 (1753).

Slender twining semi-herbaceous vine sometimes up to 4 m long, rather densely pubescent with stinging hairs. Leaves with petioles 1–3 cm long; blades oblong to ovate, 2–8 cm long, acute to acuminate at apex, the margins sharply serrate. Racemes 3–7 cm long. Capsules oblate, 6–7 mm in diameter, deeply 3-lobed. Seeds 3 mm long, pale brown mottled with red.

GRAND CAYMAN: *P. Ann van B. Stafford* et al., photo, found in area of 'The Mountain' southeast of Hutland, in secondary thickets.

— Greater and Lesser Antilles, Central and S. America.

Adelia L., nom. gen. cons.

Dioecious shrubs or trees, often with spinescent twigs, glabrous or bearing simple non-glandular hairs; leaves alternate or crowded at nodes, simple, entire, and pinnately veined. Flowers small and without petals, clustered on small woolly cushions in leaf-axils. Staminate flowers short-stalked or sessile; calyx 4–5-lobed, valvate; disc ring-like, thick and fleshy; stamens 8–17, the filaments united at least near the base; rudimentary ovary minute or absent. Pistillate flowers long-stalked; calyx 5–6-lobed, the lobes reflexed; ovary 3-celled with 1 ovule per cell; styles more or less free, deeply lacerate. Capsule 3-valved, 3-lobed, dehiscing to leave a persistent columella; seeds without a caruncle; endosperm fleshy.

— A tropical American genus of about 10 species.

Adelia ricinella L., *Syst. Nat.* ed. 10, 2: 1298 (1759). **FIG. 164.** PLATE 39.

A shrub to 3 m tall or more, rarely tree-like, with whitish bark, the ultimate branches often spine-tipped; leaves oblanceolate to obovate or elliptic, 1–6.5 cm long, rounded, blunt or acutish at the apex, glabrous above, and lightly pubescent along the midvein beneath. Staminate flowers with calyx 2–3 mm long, on pedicels about the same length; stamens

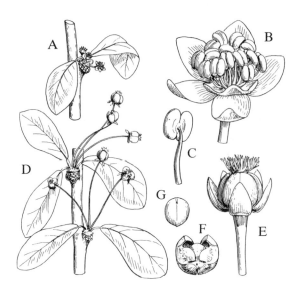

FIG. 164 **Adelia ricinella**. A, portion of stem with staminate flower. B, staminate flower, × 5. C, stamen, × 10. D, portion of stem with pistillate flowers, × ²/₃. E, pistillate flower with ovary ripening, × 4. F, coccus with seed, × 2. G, seed, × 2. (F. & R.)

8–15. Pistillate flowers on pedicels 1–5 cm long, with calyx ca. 3 mm long; ovary tomentose. Capsule deeply 3-lobed, 6–8 mm in diameter; seeds brown, globose.

GRAND CAYMAN: *Brunt* 2138; *Proctor* 15228. LITTLE CAYMAN: *Proctor* 28119. CAYMAN BRAC: *Proctor* 29115.

— West Indies south to Tobago and Curacao, mostly in dry rocky thickets and woodlands.

Bernardia Houst. ex P. Browne

Herbs or shrubs, monoecious or dioecious, bearing simple or stellate hairs; leaves alternate, with toothed margins and often with basal glands; stipules present. Flowers in unisexual, axillary, bracteate spikes, the bracts conspicuous; petals absent. Staminate flowers minute, with calyx globose and entire in bud, splitting into 3 or 4 valvate lobes at anthesis; disc absent or vestigial; stamens 3–25, with free, erect filaments and 4-celled anthers; rudimentary ovary absent or vestigial. Pistillate flowers sessile, the calyx 4–9-parted with imbricate sepals; disc annular or dissected; ovary 3-celled with a single ovule in each cell; styles 3, variously forked. Capsule 3-valved, with a persistent columella after dehiscence; seeds without a caruncle; endosperm fleshy.

— A tropical American genus of between 30 and 40 species.

Bernardia dichotoma (Willd.) Muell. Arg., *Linnaea* 34: 172 (1865).

Croton dichotomus Willd. (1805).

Bernardia carpinifolia Griseb. (1859).

A shrub up to 3 m tall, dioecious or monoecious, the branchlets tomentose with simple and stellate hairs; leaves deciduous at flowering time, more or less elliptic or rhombic-ovate, mostly 2–8 cm long, blunt to acute at the apex, the margins closely or distantly serrate or denticulate, tomentose with stellate hairs on both sides but especially beneath, and the tissue with two small round glands toward the base beneath. Spikes mostly 8–20 mm long, densely tomentose. Staminate flowers in clusters of 3–5, each cluster subtended by a bract ca. 1.5 mm long; calyx ca. 2 mm long; stamens 15–25, exserted. Pistillate flowers with calyx ca. 2.5 mm long, the sepals unequal; ovary densely tomentose. Capsule deeply 3-lobed, 7–9 mm in diameter, densely tomentose.

GRAND CAYMAN: *Hitchcock*; *Millspaugh* 1265; *Proctor* 15042. LITTLE CAYMAN: *Proctor* 35104. CAYMAN BRAC: *Proctor* 29030.

— West Indies south to Grenada, in dry thickets and woodlands.

[*Ricinus* L.

A glabrous, monoecious, wide-branching shrub with watery sap; leaves alternate, long-petiolate, the blade peltate and palmately 7–11-lobed, with serrate margins; stipules sheathing, soon falling. Inflorescences terminal or appearing axillary due to sympodial growth, paniculate, the lower flowers staminate, the distal ones pistillate. Staminate flowers with calyx globose in bud, splitting into 3–5 valvate lobes at anthesis; stamens very numerous (up to 1,000), the filaments partially connate into fascicles at the base, irregularly branched; rudimentary ovary absent. Pistillate flowers with calyx like that of the staminate ones, soon falling; ovary 3-celled, a single ovule in each cell; styles joined below, forked upwardly, with papillate branches. Capsule spiny or smooth, splitting into three 2-valved cocci; seeds with conspicuous caruncle; endosperm copious.

— A single species of African origin, now widely distributed in tropical and warm-temperate regions.

Ricinus communis L., *Sp. Pl.* 2: 1007 (1753). **FIG. 165.**

CASTOR-OIL PLANT

With characters of the genus. Plants usually 2–5 m tall; leaf-blades 10–100 cm across, the lobes acute and pinnately veined the marginal serrations more or less glandular. Capsules 12–21 mm in diameter; seeds ellipsoid, somewhat flattened, variously mottled, mostly 10–20 mm long.

GRAND CAYMAN: *Brunt* 1983; *Kings* GC 128, GC 415; *Millspaugh* 1389; *Proctor* 15236, 15237. LITTLE CAYMAN: *Kings* LC 108. CAYMAN BRAC: *Kings* CB 80, CB 81.

— Roadsides, old fields, and open waste ground. The seeds yield the castor oil of commerce.]

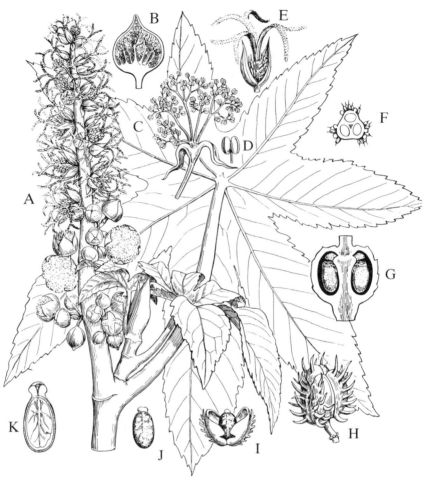

FIG. 165 **Ricinus communis**. A, upper portion of flowering branch, × $^2/_3$. B, staminate flower just before opening, × 2. C, same open, × 2. D, anther, × 8. E, pistillate flower, × 2. F, cross-section of ovary, × 4. G, ovary cut lengthwise to show ovule and obturator, × 10. H, capsule, × 1. I, coccus split open, × 1. J, seed, × 1. K, same, × 1$^1/_2$. (F. & R.)

Acalypha L.

Herbs, shrubs or rarely trees, monoecious or dioecious, bearing simple hairs or glands; leaves alternate, petiolate, simple, entire or toothed, and with pinnate or palmate venation. Flowers in terminal or axillary unisexual or bisexual spikes, the staminate flowers several at each node subtended by a minute bract, the pistillate flowers 1–3 at each node, subtended by a usually large, foliaceous, lobed bract; petals and disc absent. Staminate flowers with calyx closed in bud, valvately splitting into 4 lobes at anthesis; stamens 4–8, the filaments free or joined at the base; anther-sacs 1-celled, pendent, elongate and worm-like, opening at the apex; rudimentary ovary absent. Pistillate flowers with 3–5 imbricate calyx-lobes; ovary usually 3-celled with solitary ovules; styles free, thread-like, usually much-branched. Fruit a 3-seeded capsule; seeds with or without a caruncle; endosperm present.

— A large genus of about 400 species, the majority tropical American, with the greatest concentration of species in the Caribbean region. *Acalypha hispida* Burm. f., 'red puss-tail', from the Indonesian area, and *A. amentacea* Roxb. subsp. *wilkesiana* (Muell. Arg.) Fosberg, 'Joseph's coat', from the Pacific islands, are shrubs often cultivated in gardens.

KEY TO SPECIES

1. Inflorescences bisexual; leaves usually less than 1.5 cm long with very short petioles: **A. chamaedrifolia**

1. Inflorescences unisexual; leaves mostly 2–10 cm long, the petioles often longer than the blades: **A. alopecuroidea**

Acalypha chamaedrifolia (Lam.) Muell. Arg. in DC., *Prodr.* 15 (2): 879 (1866).

A small subwoody herb with woody taproot, the stems usually less than 15 cm long, erect or procumbent, the young branches pubescent. Leaves elliptic, oblong or ovate, 0.5–1.5 cm long or sometimes a little longer, blunt to acute at the apex, the margins crenate-dentate, pubescent or glabrate. Spikes 1–2 cm long, terminal or in upper axils, with 1–5 pistillate flowers near the base; staminate flowers red; pistillate flowers with bracts ca. 3 mm long, divided at apex into 7–9 triangular lobes, each bract enclosing 2 flowers. Capsule 3-lobed, 1.6 mm long; seeds ellipsoidal.

CAYMAN BRAC: *Proctor 28919.*

— Florida and the West Indies, on sandy banks or exposed rocky hillsides.

Acalypha alopecuroidea Jacq., *Collect.* 3: 196 (1789).

Erect annual herb to about 35 cm tall, the branches pubescent. Leaves long-petiolate, the blades ovate, 1.5–7 cm long, acuminate at the apex, the margins sharply serrate except toward base, glabrous or sparingly pubescent. Staminate spikes axillary, very short and inconspicuous, 2–9 mm long. Pistillate spikes terminal, 2–5 cm long, densely flowered and plume-like, tipped with a bristle-like appendage terminating in an abortive flower; bracts hairy, each segment with a long apical bristle. Capsules solitary within each bract, 1.3–1.4 mm long.

GRAND CAYMAN: *Hitchcock*; *Kings GC 89*; *Proctor 15292.*

— Widely distributed in tropical America, a weed of fields, roadsides and open waste places.

Gymnanthes Sw.

Glabrous shrubs or trees, usually monoecious; leaves alternate, entire or toothed, with short non-glandular petioles; stipules present. Inflorescences spike-like, axillary, protected by a conspicuous bud, bisexual, with usually only 1 basal pistillate flower per spike; calyx small or absent; petals absent; disc absent. Staminate flowers with rudimentary calyx of 1 or 2 small sepals, or absent; stamens 2 or 3 in lateral flowers of a cymule, 3–5 in central flowers; filaments free or joined at the base; rudimentary ovary absent. Pistillate flowers usually 1 at the lowest node of the inflorescence; calyx minute, of 2 or 3 reduced sepals;

ovary sessile or stalked, 3-celled with a solitary ovule in each cell; styles free or joined at the base, simple, slender and recurved. Fruit a capsule with a persistent, 3-winged columella; seeds subglobose, smooth, and with a caruncle.

— A small Caribbean genus of 12 species.

Gymnanthes lucida Sw., *Prodr.* 96 (1788). **FIG. 166**. PLATE 40.

CRAB BUSH

An evergreen shrub or small tree; leaves leathery and glossy, more or less oblanceolate, 2.5–10 cm long, obtuse at the apex, the margins entire or with a few low teeth, the veins prominulous on both sides, and the tissue often with 1 or 2 glands near the base beneath. Spikes 1–3 cm long, densely flowered. Staminate flowers 3 to each bract. Pistillate flower with short pedicel greatly elongating in fruit; sepals scale-like, minute, not all attached at the same level; ovary stalked above the sepals, the stalk short at first but elongating to 1.5–2 cm long in fruit, appearing as a continuation of the pedicel. Capsules deeply 3-lobed, nearly 1 cm in diameter.

GRAND CAYMAN: *Kings GC 243*; *Lewis 3863*; *Proctor 15031, 15032*. LITTLE CAYMAN: *Proctor 35188*. CAYMAN BRAC: *Proctor 29016*.

— Florida and the West Indies, in dry rocky woodlands. The wood is heavy, hard, close-grained, and capable of receiving a beautiful polish; it is occasionally made into canes.

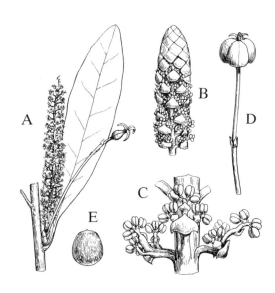

FIG. 166 **Gymnanthes lucida**. A, leaf and inflorescence, × 1. B, upper part of young inflorescence, × 2. C, part of inflorescence with three bracts and staminate flowers in their axils, × 10. D, capsule, × 1. E, seed, × 2. (F. & R.)

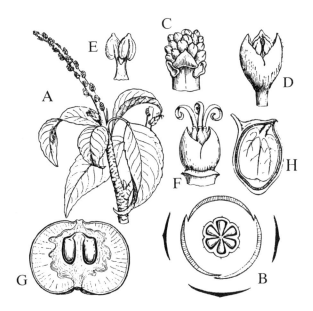

FIG. 167 **Hippomane mancinella**. A, branchlet with leaves and inflorescence, × ²/₃. B, diagram of pistillate flower. C, cluster of staminate flowers, × 3. D, staminate flower just opening, × 12. E, stamens, × 14. F, pistillate flower with three styles cut off, × 3. G, section of fruit, × 1. H, section of seed, × 4. (F. & R.)

Hippomane L.

Glabrous, monoecious shrubs or trees with poisonous milky latex; leaves alternate, long-petiolate, the petioles with a single gland at the apex, the blades cordate and pinnately veined, the margins toothed or spiny. Stipules present, soon falling. Inflorescence a terminal, bisexual spike with thickened rhachis, the bracts biglandular; petals and disc absent. Staminate flowers several to numerous in dense glomerules at each of the distal nodes; calyx 2–3-lobed, the lobes imbricate; stamens 2 with cohering filaments; rudimentary ovary absent. Pistillate flowers 1 or few, sessile, solitary at lower nodes of the spike; calyx 3-lobed; ovary 6–9-celled (rarely less) with 1 ovule in each cell; styles 2 or 3, joined at base, recurved, undivided. Fruit a small apple-like drupe, with yellowish or reddish fleshy exocarp, the bony endocarp with numerous blunt projections; seeds flattened-elongate, without a caruncle; endosperm present.

— A tropical American genus of 3 species, one of them widespread, the others confined to Hispaniola.

Hippomane mancinella L., *Sp. Pl.* 2: 1191 (1753). **FIG. 167**. PLATE 40.

MANCHINEEL

An evergreen shrub, or tree to 10 m tall or more; leaves with cordate-ovate blades usually 2–7 cm long or sometimes longer, the apex sharply acute or subacuminate, the margins distantly serrate, the tissue shiny green above, the finely reticulate venation prominulous. Spikes usually 3–9 cm long, the rhachis dark red or purple. Fruit globular, 2.5–3.5 cm in diameter; seeds 4–6 mm long.

GRAND CAYMAN: *Brunt* 1789, 1790; *Hitchcock*; *Kings* GC 114; *Proctor* 15064. LITTLE CAYMAN: *Proctor* 28100. CAYMAN BRAC: *Proctor* 29133.

— Florida, West Indies, and continental tropical America, in coastal thickets and woodlands. The wood of this tree is prised for cabinet-work and construction, but because of its poisonous sap it is hazardous to saw unless dry. The fruits are likewise poisonous, and children should be warned to shun them.

Euphorbia L.

Herbs, shrubs or trees, sometimes succulent, with copious milky latex, monoecious or rarely dioecious; leaves simple and alternate, opposite or whorled, in succulent forms soon-falling or sometimes apparently absent; stipules present or absent, sometimes glandular. Inflorescence a cyathium (aggregate flower), the 5 cup-like lobes alternating at their tips with 4 or 5 glands, these with or without appendages. Staminate flowers in 4–5-cymes, the subtending bracteoles partly fused to the involucre or reduced or absent; calyx and corolla absent; stamen 1. Pistillate flowers terminal, solitary; calyx of 3–6 united sepals or absent; ovary 3-celled with a single ovule in each cell; styles 3, free or joined at the base, usually forked. Fruit a capsule or rarely a drupe; seeds more or less ovoid, angled or terete, smooth or variously roughened; caruncle present or absent.

— A world-wide genus of about 1,200 species, extremely diverse in appearance. The name *Euphorbia* is said to have been first applied to a succulent member of this genus by King Juba of Mauritania (reigned B.C. 25–A.D. 18) in honour of his doctor, Euphorbus, the new plant and the worthy physician both being of notably fleshy build. 'Euphorbos' is a Greek adjective meaning 'well-stuffed'.

KEY TO SPECIES

1. Low herbaceous leafy plants:

  2. Perennial rosette herb with woody taproot; leaves crowded, alternate, nearly sessile, 2–8 mm long; plant of sandy sea-beaches: **E. trichotoma**

  2. Annual weak-stemmed herb with fibrous roots; leaves widely separated, alternate and opposite, long-petiolate, the blades 10–25 mm long or more; weed of gardens and nurseries: **[E. graminea]**

1. Plants shrubby or tree-like, the stems leafless or with minute scattered leaves on new growth:

  3. Straggling shrub to 1.5 m tall; branches angled; flowers yellow; seeds white: **E. cassythoides**

  3. Stout erect shrub or small tree to 7 m tall or more; branches terete; flowers greenish; seeds dark brown: **[E. tirucalli]**

Euphorbia trichotoma Kunth in H. B. K., *Nov. Gen. & Sp. Pl.* 2: 60 (1817).

Glabrous, pale green, perennial herb with a woody taproot, the branches not usually more than 20 cm long, spreading or ascending; leaves numerous and crowded, nearly sessile, mostly oblanceolate or narrowly ovate, 2–8 mm long, the margins entire to minutely erose or serrulate. Cyathia in terminal 3-branched clusters or solitary, the involucre 2 mm high, the glands obreniform, 1 mm broad, yellow. Capsule broadly 3-lobed, 3–4 mm in diameter; seeds whitish, 1.5 mm long, smooth.

GRAND CAYMAN: *Kings* GC 23a; *Proctor* 11972, 15215; *Sachet* 442. LITTLE CAYMAN: *Proctor* 35178. CAYMAN BRAC: *Millspaugh* 1185, 1232; *Proctor* 15309, 28918, 47803.

— Florida, Bahamas, Cuba, and Mexico, on sandy sea-beaches.

[*Euphorbia graminea* Jacq., *Sel. Stirp. Amer.* 151 (1763). PLATE 40.

A straggling procumbent, decumbent, or weakly erect annual or short-lived perennial herb up to 50 cm tall with slender laxly branching stems, glabrate or with scattered soft hairs; branching irregularly dichotomous; stipules minute. Lower leaves alternate, the upper ones opposite, all with long slender petioles and blades variable in shape, broadly ovate to oblong or lanceolate, mostly 1–4(–7) cm long, acute to rounded at the apex, cuneate to rounded at the base, the margins entire, the tissue very thin and glabrous to sparingly pubescent. Cymes terminal, often long-stalked with slender elongate branches; involucres very small, in forks of the cyme and terminating the lax ultimate branches. Capsules small; seeds tuberculate.

GRAND CAYMAN: *Proctor* 52126, 52141, 52145, a weed becoming common in gardens and plant nurseries. First found by Joanne Ross.

— Southern Mexico to Costa Rica and in northern S. America.]

Euphorbia cassythoides Boiss., *Cent. Euph.* 20 (1860).

Arthrothamnus cassythoides (Boiss.) Millsp. (1909).

A straggling, pale-barked shrub up to 1.5 m tall or more, the pale green branches leafless, more or less whorled, and sharply 6–7-angled, with gummy-resinous nodes; vestigial leaf-scales few, opposite. Cymes terminal, few-flowered, the yellow cyathia 1.5 mm long; styles hairy. Capsules ovoid; seeds white, 3-angled, foveolate.

GRAND CAYMAN: *Brunt* 2173; *Kings* GC 364; *Proctor* 47276, 47830, 47858.

— Bahamas, Cuba, and Hispaniola, in dry rocky or sandy woodlands, rare. Herbarium specimens of this species have been mistaken for *Rhipsalis baccifera* (Cactaceae).

[*Euphorbia tirucalli* L., *Sp. Pl.* 1: 452 (1753).

A stout shrub or small tree, the green, smooth, terete, mostly naked branches alternate or clustered in brush-like masses, the ultimate ones 5–7 mm thick and bearing scattered, alternate, narrow leaves 6–12 mm long, these soon falling. Inflorescences of small, sessile, terminal clusters, the cyathia 3 mm long. Ovary sessile at first, becoming exserted on a pedicel-like stalk 8 mm long in fruit. Capsule obtusely 3-lobed; seeds dark brown, smooth, 4 mm long.

GRAND CAYMAN: *Brunt* 2028.

— Indigenous to tropical Africa but widely cultivated in warm countries, often becoming naturalised.]

[*Euphorbia milii* Ch. des Moulins (*E. splendens* Bojer) from Madagascar, the 'crown-of-thorns' or 'crucifixion plant', is cultivated in Grand Cayman (*Kings* GC 306); as are the red-leaved shrub *Euphorbia cotinifolia* L. and probably other ornamental species.]

Poinsettia Graham

Monoecious herbs or shrubs with milky latex; leaves alternate and opposite, petiolate, those subtending the inflorescence often brightly coloured or whitish at least toward the base; stipules minute or lacking. Inflorescence a cyathium, several or many of these clustered in terminal dichasia or pleiochasia; lobes 5, usually more or less fringed, the involucral glands cup-like, usually 1 but up to 5 in central cyathia. Staminate flowers few to many, naked, with a single stamen. Pistillate flowers terminal, solitary, naked; ovary 3-celled with 1 ovule per cell; styles 3, united at the base, bifid. Fruit a 3-lobed dehiscent capsule; seeds ovoid, the caruncle vestigial or absent.

— An American genus of about 12 species. *P. pulcherrima* (Willd. ex Kl.) Graham, the ornamental poinsettia, is widely grown for its showy bracts.

KEY TO SPECIES

1. Cyathial gland circular at the apex; bracts green, often pale or purple-mottled at base but never red; seeds angled: *P. heterophylla*

1. Cyathial gland flattened and 2-lipped; bracts green or red at the base; seeds not or scarcely angled: *P. cyathophora*

Poinsettia heterophylla (L.) Kl. & Gke. in Monatsb., *Akad. Berlin* 1859: 253 (1859). FIG. 169.

Euphorbia geniculata Ort. (1797).

An annual or sometimes persisting herb, usually less than 50 cm tall; leaves alternate below, opposite above, the membranous blades pandurate, obovate or elliptic, rarely lanceolate, the margins entire or with 1–few broad teeth, glabrous or pubescent, those subtending the inflorescence-cluster often pale or purple-mottled at the base. Cyathia glabrous, the gland with a slightly flared round opening. Capsule ca. 5 mm broad; seeds up to 2.5 mm long, tuberculate.

GRAND CAYMAN: *Kings GC 145; Proctor 15079.*

— Southeastern U.S.A. and tropical America, a widely distributed and common weed of roadsides and fields.

Poinsettia cyathophora (Murray) Kl. & Gke. in Monatsb., *Akad. Berlin* 1859: 253 (1859).

Euphorbia heterophylla of authors, not L. (1753).

BLEEDING HEART, STARLIGHT, GROUND DOVE BERRY

An annual or sometimes persisting herb up to 0.8 m tall or more; leaves alternate below, opposite above, the membranous blades pandurate, ovate, lanceolate or sublinear, the margins entire or broadly toothed, glabrous or sparingly pubescent, those subtending the inflorescence-cluster often bright red at base or throughout. Cyathia glabrous or sparingly short-pubescent, the gland flattened and 2-lipped. Capsules ca. 5 mm broad; seeds up to 3 mm long, tuberculate.

GRAND CAYMAN: *Brunt 2033, 2058; Kings GC 17, GC 139; Maggs II 59; Proctor* 15020. LITTLE CAYMAN: *Kings LC 51.* CAYMAN BRAC: *Kings CB 45; Proctor 52158.*

— Eastern U.S.A. and tropical America, also in the Old World, a common weed of waste places and disturbed areas.

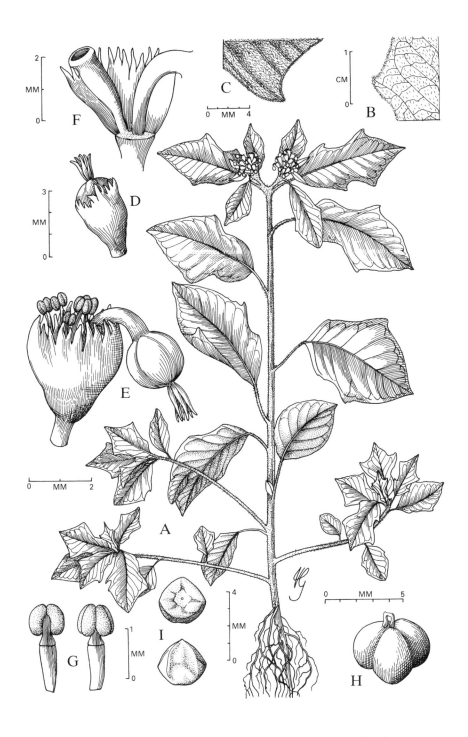

FIG. 168 **Poinsettia heterophylla.** A, habit, B, portion of leaf showing upper (dorsal) surface. C, portion of leaf showing lower (ventral) surface. D, immature cyathium. E, mature cyathium at anthesis. F, cyathium cut open to show gland. G, two views of stamen. H, capsule. I, two views of seed. (G).

Chamaesyce S. F. Gray

Monoecious herbs or subshrubs, annual or persisting, prostrate to ascending or erect, with milky latex throughout; stems with main axis aborting within 2 nodes of the cotyledons, the secondary axes few to many, rarely (in prostrate species) rooting at the nodes. Leaves opposite, simple, the blade oblique or unequal-sided at base and with entire or serrate margins; stipules present, often conspicuous. Inflorescence a cyathium, solitary, or several clustered in cymules at nodes; cyathia 5-lobed, with 4 glands (sometimes a fifth vestigial one) alternating with the lobes and each bearing a petal-like appendage (rarely the appendage obsolete). Staminate flowers few to many, naked, with 1 stamen. Pistillate flowers terminal, solitary, naked; ovary 3-celled with 1 ovule in each cell; styles 3, free or united at base, partly bifid. Fruit an exserted capsule; seeds ovoid, angled or terete, smooth or variously sculptured; caruncle absent.

— A world-wide genus of about 250 species, the majority in tropical America.

KEY TO SPECIES

1. Capsules glabrous:

  2. Plants erect or ascending:

    3. Plants herbaceous; leaf-margins toothed; cyathia in dense, stalked, glomerate clusters; capsule less than 1.4 mm long: **C. hypericifolia**

    3. Plants woody; leaf-margins entire; cyathia solitary in uppermost leaf-axils; capsule ca. 2 mm long: **C. mesembryanthemifolia**

  2. Plants prostrate or decumbent:

    4. Stems numerous, fine and dense, rarely exceeding 0.5 mm diam. and 6 cm in length, from a woody rootstock; stipules less than 0.5 mm long; leaves serrulate throughout: **C. torralbasii**

    4. Stems few, rather open, often up to 2–3 mm in diameter and usually 10–30 cm long, the rootstock not or scarcely woody; stipules often 1 mm long; leaves minutely serrulate at apex only: **C. blodgettii**

1. Capsules more or less pubescent:

  5. Cyathia solitary at leafy nodes, never in stalked glomerate clusters:

    6. Stems pubescent at least on one side; ovary and capsule pubescent on the angles: **C. prostrata**

    6. Stems glabrous; ovary and capsule sparsely and minutely strigulose on the sides: **C. bruntii**

  5. Cyathia in stalked, glomerate clusters:

    7. Stems branching at base but seldom near the tip; inflorescences both lateral and terminal; plants mostly rather robust and ascending: **C. hirta**

    7. Stems branching freely to near the tip; inflorescences always terminal on branches; plants mostly low and decumbent: **C. ophthalmica**

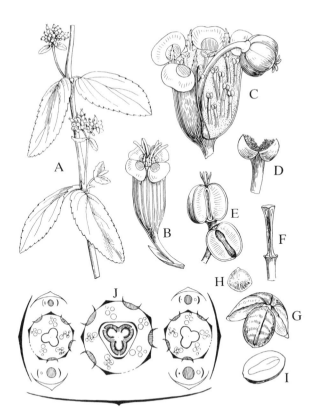

FIG. 169 **Chamaesyce hypericifolia.** A, part of flowering branch, × ²/₃. B, cyathium, × 16. C, cyathium cut open, × 24. D, stamen, much enlarged. E, capsule, × 10. F, columella of capsule, much enlarged. G, coccus with seed, much enlarged. H, seed, × 10. I, seed cut lengthwise, much enlarged. J, diagram of partial inflorescence. (F. & R.)

Chamaesyce hypericifolia (L.) Millsp., *Field Mus. Bot.* 2: 302 (1909); Burch, *Rhodora* 68: 160–163 (1966). **FIG. 170.**

Euphorbia glomerifera (Millsp.) L. C. Wheeler (1939); Adams (1972).

CHICK WEED

An annual herb, the ascending or erect stems to 40 cm tall and 3 mm in diameter at the base; leaves glabrous, ovate-elliptic to elliptic-obovate, somewhat falcate, 1.5–3.5 cm long, the apex acute, the margins serrulate. Cyathia glabrous, in short-stalked axillary and terminal glomerules. Capsules glabrous, subspherical, to 1.2 mm in diameter; seeds ovoid, 4-sided, to 0.8 mm long.

GRAND CAYMAN: *Brunt* 1721, 1931; *Hitchcock*; *Kings* GC 296; *Lewis* GC 39; *Maggs* II 60; *Millspaugh* 1291, 1304; *Proctor* 15109. LITTLE CAYMAN: *Proctor* 35185. CAYMAN BRAC: *Proctor* 28994.

— West Indies and the warmer parts of continental America, a common weed of fields, pastures, and clearings.

Chamaesyce mesembryanthemifolia (Jacq.) Dugand, *Phytologia* 13: 385 (1966).

Chamaesyce buxifolia (Lam.) Small (1903).

TITTIE MOLLY

A subwoody herb or miniature shrub, erect or sometimes decumbent, often 30 cm tall or more; leaves more or less overlapping the stem, glabrous and often glaucous, ovate to elliptic, 5–12 mm long, with entire margins; stipules whitish, united, to 1 mm long, more or less fringed. Cyathia solitary at upper nodes, glabrous outside, densely bearded within; appendages white. Capsules glabrous, subspherical, 2–2.5 mm in diameter; seeds broadly ovoid, 1.3 mm long.

GRAND CAYMAN: *Brunt* 1882, 2011, 2070; *Hitchcock*; *Kings* GC 23, GC 272; *Lewis* GC 23; *Millspaugh* 1262; *Proctor* 11973; *Sachet* 402, 441; *Sauer* 3319, 4107. LITTLE CAYMAN: *Kings* LC 93; *Proctor* 28070; *Sauer* 4166. CAYMAN BRAC: *Kings* CB 26; *Millspaugh* 1180, 1196, 1233; *Proctor* 29077.

— Florida, West Indies, and coasts of Central and northern S. America, frequent to abundant on sandy seashores or in pockets of coastal rocks. The latex of this species is reputed to be effective in removing warts or wart-like skin eruptions.

Chamaesyce torralbasii (Urban) Millsp., *Field Mus. Bot.* 2: 412 (1916).

A prostrate, glabrous, perennial herb, the wiry, densely branched stems with noticeably thickened nodes; leaves short-petiolate, obliquely ovate or rotundate, 2–3.5 mm long or more (larger in Cuban specimens), obtuse at the apex and shallowly cordate at the base, with thickened denticulate margins. Cyathia few, terminal or solitary in the uppermost leaf-axils; appendages 4, minute, subequal, entire; styles forked below the middle. Capsules glabrous; seeds 4-angled, glabrous (or reported elsewhere to be minutely hairy).

LITTLE CAYMAN: *Proctor* 47293. CAYMAN BRAC: *Proctor* 29042.

— Cuba and possibly elsewhere. Cayman Brac plants were found "in pockets of exposed limestone pavement".

Chamaesyce blodgettii (Engelm.) Small, *Fl. S.E. U.S.* 712 (1903).

An annual glabrous herb, the branches prostrate, ascending, or rarely erect, to 30 cm long; leaves oblong or oblong-oval, 4–10 mm long, minutely serrulate at apex; stipules broadly triangular, often more than 1 mm long, ciliate. Cyathia solitary in upper leaf-axils, glabrous outside, bearded within; appendages entire or 2–3-crenate. Capsules glabrous, 3-cornered-globular, 1.5–2 mm in diameter; seeds 4-angled, ca. 1 mm long.

GRAND CAYMAN: *Hitchcock*; *Kings* GC 372; *Millspaugh* 1257, 1258, 1314, 1333; *Proctor* 15123, 15240; *Sachet* 385. LITTLE CAYMAN: *Proctor* 28145, 28168. CAYMAN BRAC: *Proctor* 29076.

— Florida, West Indies and C. America, a weed of clearings, open waste ground, and second-growth woodlands.

Chamaesyce prostrata (Ait.) Small, *Fl. S.E. U.S.* 713 (1903).

Annual herb with prostrate, minutely pubescent, much-branched stems, often forming mats. Leaves oblong or broadly obovate, 4–7 mm long, puberulous or glabrate on both sides, the margins more or less serrulate; stipules broadly triangular, ciliate. Cyathia solitary in leaf-axils, glabrous or puberulous outside, bearded within; appendages crenulate. Capsules 3-angled, pilose on the angles, ca. 1 mm in diameter; seeds pink, sharply 4-angled and with numerous transverse ridges.

GRAND CAYMAN: *Correll & Correll* 51001; *Proctor* 11971.

— Widespread in warm-temperate and tropical regions, a weed of open waste ground and sandy clearings.

Chamaesyce bruntii Proctor, *Sloanea* 1: 2 (1977). PLATE 40.

A prostrate perennial herb with woody taproot, the wiry glabrous stems densely branched. Leaves purplish, short-petiolate, obliquely ovate or broadly oblong, 2–5 mm long, glabrous, broadly obtuse at the obscurely toothed apex, subcordate at base; stipules inconspicuous, consisting of 2 or 3 linear segments 0.3–0.5 mm long. Cyathia solitary in the upper leaf-axils, the 4 subequal appendages lunate-orbicular, ca. 0.3 mm wide. Capsules broadly 3-lobed, strigose-pubescent, ca. 1 mm in diameter; seeds 4-angled, smooth, 0.8–0.9 mm long.

LITTLE CAYMAN: *Proctor* 28146 (type), 35083. (Another collection, *Proctor* 35090, differs from typical *C. bruntii* in possessing sparse, lax pubescence on stems and leaves).

— Endemic; occurs in sandy clearings and natural sandy glades.

Chamaesyce hirta (L.) Millsp., *Field Mus. Bot.* 2: 303 (1909).

DOVE WEED

Annual pubescent herb, the mostly erect stems up to 35 cm tall, clothed with multicellular hairs. Leaves ovate to lanceolate or rhombic, 1–3.5 cm long, pubescent especially on the under surface, the margins serrate; stipules lacerate, to 1 mm long. Cyathia in stalked, dense, terminal and axillary clusters, strigulose throughout; appendages present or absent. Capsules strigulose, ca. 1 mm in diameter; seeds reddish, wedge-shaped with sharp angles, ca. 0.8 mm long.

GRAND CAYMAN: *Millspaugh* 1278; *Proctor* 11970. LITTLE CAYMAN: *Proctor* 35181. CAYMAN BRAC: *Kings* CB 50; *Millspaugh* 1175; *Proctor* 29103.

— A pantropical weed of open waste ground.

Chamaesyce ophthalmica (Pers.) D. G. Burch, *Ann. Missouri Bot. Gard.* 53: 98 (1966).

An annual pubescent herb, the freely branching stems mostly low and decumbent up to 20 cm long, clothed with multicellular hairs; leaves oblong-lanceolate, more or less rhombic, mostly 5–18 mm long, pubescent on both sides, the margins finely serrate; stipules bifid with narrow lobes, 0.5–0.8 mm long. Cyathia in stalked, dense clusters, these always terminal on leafy branches, strigulose throughout; appendages absent or minute. Capsules strigulose, ca. 1 mm in diameter; seeds reddish, sharply quadrangular, 0.6–0.7 mm long.

GRAND CAYMAN: *Millspaugh* 1292, 1298; *Proctor* 31052. LITTLE CAYMAN: *Proctor* 35138. CAYMAN BRAC: *Millspaugh* 1213; *Proctor* 28024, 29336.

— Florida and the West indies, a common weed.

[*Pedilanthus tithymaloides* ssp. *parasiticus* Kl. & Garcke, the monkey fiddle, is frequently cultivated. GRAND CAYMAN: *Kings* GC 118.]

RHAMNACEAE

Trees or shrubs, rarely herbs, sometimes scandent, often with spines, rarely with tendrils; leaves simple, alternate or opposite, often 3–5-nerved from the base, the margins entire or serrate; stipules small and deciduous or sometimes modified into spines. Flowers small, usually perfect, mostly in small axillary cymes, green or yellowish. Calyx with more or less obconical tube and 4–5 triangular, valvate lobes; petals 4–5 or none, inserted in throat of the calyx and often smaller than its lobes, often hooded or infolded, at the base sessile or clawed. Stamens 4–5, inserted with the petals and often concealed by them; anthers oblong or 2-lobed, opening by slits. Disc perigynous. Ovary sessile, free or immersed in the disc, superior or somewhat connate with the calyx-tube, usually 3-celled, or cells sometimes 2 or 4; style erect with capitate or 3-lobed stigma; ovules usually 1 in each cell, erect from base of the cell. Fruit usually a 3-coccous capsule or drupe with 1–3-celled stone, the seeds solitary in the cells; seeds often with aril at the base; endosperm fleshy, scant, or rarely none, the embryo large, the cotyledons flat or plano-convex.

— A cosmopolitan family of about 50 genera and 600 species.

Colubrina L. C. Rich.

Trees or shrubs, sometimes scrambling; leaves alternate, pinnate-nerved or sometimes 3-nerved from the base, the margins entire or toothed; stipules small, soon falling. Calyx 5-lobed, the lobes keeled on the inside, the tube forming a cupule confluent with the fruit. Petals 5, hooded and clawed. Stamens 5 with short, slender filaments. Disc thick, 5- or 10-lobed. Ovary immersed in the disc and confluent with it, 3-celled; style 3-lobed or 3-branched. Fruit a subglobose 3-coccous capsule; seeds flattish-ellipsoid with scant endosperm.

— A genus of about 15 species, all but one confined to the American tropics.

KEY TO SPECIES

1. Leaves entire, pinnate-nerved:
    2. Leaves densely tomentose on both sides: **C. cubensis**
    2. Leaves glabrous or glabrescent on upper side, at least when mature:
        3. Leaves beneath (and twigs) woolly ferruginous-tomentose at least when young; calyx-cupule covering nearly half the capsule; seeds 4 mm long: **C. arborescens**
        3. Leaves glabrescent beneath; calyx-cupule covering hardly one-third the capsule; seeds 5 mm long: **C. elliptica**
1. Leaves serrate, 3-nerved at base, glabrous: **C. asiatica**

Colubrina cubensis (Jacq.) Brongn., *Ann. Sci. Nat.* sér. 1, 10: 369 (1827). PLATE 40.

CAJON

A shrub to 3 m tall or more, rarely a small tree, the young branches, leaves, and inflorescence densely velvety-tomentose; leaves oblong or elliptic 2–7 cm long, rounded or acute at the apex, the pinnate venation strongly raised beneath and channelled on the upper side. Cymes stalked, longer than the petioles; flowers yellow-green; calyx densely pubescent, the lobes ca. 2 mm long; petals equal in length to the calyx. Fruit globose, ca. 7 mm in diameter.

GRAND CAYMAN: *Brunt* 1813; *Correll & Correll* 51046; *Hitchcock*; *Howard & Wagenknecht* 15027; *Kings GC* 390; *Millspaugh* 1256; *Proctor* 15016. LITTLE CAYMAN: *Proctor* 35186. CAYMAN BRAC: *Millspaugh* 1150, 1230; *Proctor* 15314.

— Florida, Bahamas, Cuba, and Hispaniola, in dry rocky woodlands. Reports of *Colubrina arborescens* (Mill.) Sarg. (under other names) were apparently based on misidentifications of *C. cubensis*.

The leaves of this species are sometimes used to make tea.

Colubrina arborescens (Mill.) Sarg., *Trees & Shrubs* 2: 167 (1911). PLATE 41.

Tree up to 10 m tall or more. Leaves ovate-lanceolate to oblong-elliptic, 3.5–12 cm long, 1.5–6 cm broad, rounded at base with about 6 pairs of rather prominent side-nerves. Common peduncle usually at least 5 mm long, stout, rusty tomentose; flowers 4–5-merous, greenish. Capsule subglobose, black, 7–8 mm long; seeds shiny black.

GRAND CAYMAN: *Proctor* 47369, 49349, 50729.

— Florida, Mexico to Honduras, Bahamas, Greater and Lesser Antilles.

Colubrina elliptica (Sw.) Brizicky & W. L. Stern, *Trop. Woods* no. 109: 95 (1958). FIG. 170. PLATE 41.

Colubrina reclinata (L'Her.) Brongn. (1827).

WILD GUAVA

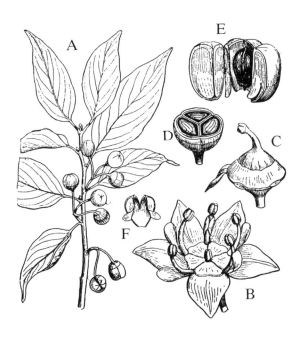

FIG. 170 **Colubrina elliptica**. A, branchlet with leaves and fruit, × ²/₃. B, flower, × 7. C, unripe fruit with one calyx-lobe still attached, × 4. D, ripe fruit cut across, × 2. E, ripe fruit splitting open, × 2. F, endocarp of one coccus after splitting, × 1¹/₃. (F. & R.)

Shrub or small tree 2–10 m tall or more, with slender, finely pubescent branches; leaves of thin texture, broadly elliptic or ovate, 2–10 cm long, blunt to subacuminate at the apex, glabrous above, minutely puberulous beneath, and often bearing a marginal gland on either or both sides near the base; petioles finely pubescent. Cymes short-stalked, few-flowered; flowers pale green; calyx puberulous, the lobes 1.3 mm long; petals slightly shorter. Fruit glabrous, 6–8 mm in diameter.

GRAND CAYMAN: *Correll & Correll 50994*; *Hitchcock*; *Proctor 15162*. LITTLE CAYMAN: *Proctor 28114*. CAYMAN BRAC: *Proctor 29137*.

— Florida, West Indies, Yucatan, Guatemala and Venezuela, in dry rocky woodlands.

[*Colubrina asiatica* (L.) Brongn., *Ann. Sci. Nat.* sér. 1, 10: 369 (1827).

A bushy shrub with long, trailing glabrous branches; leaves with petioles mostly 1–1.5 cm long, these puberulous in one side, the glabrous blades glossy green, ovate-acuminate, 3–8 cm long. Cymes shorter than the petioles, glabrous; flowers greenish-cream. Capsules 7–9 mm in diameter.

GRAND CAYMAN: *Brunt 1633*; *Correll & Correll 51032*; *Kings GC 349*; *Proctor 15191*; *Sachet 399, 433*. LITTLE CAYMAN: *Proctor 28159*; *Sauer 4172*. CAYMAN BRAC: *Proctor 29105*.

— Native to Eastern Africa, Southeast Asia, tropical Australia and the Pacific Islands. Introduced to the Cayman Islands and becoming a serious invasive threat.]

[*Ziziphus mauritiana* Lam., or cooly plum, from southern Asia and Africa, has been planted on Grand Cayman and is likely to become naturalised. It is a small tree with spiny, tomentose branches; the oblong to oval leaves are mostly 4–5 cm long, strongly 3–5-nerved from the base, light green and glabrate on upper side and whitish-tomentose beneath. The fleshy drupes are edible.]

VITACEAE

More or less woody vines with copious watery sap, the stems often swollen at the nodes, climbing by means of tendrils that are either sterile peduncles or sometimes simple branches of flowering peduncles. Leaves alternate, simple or digitately 3–5-foliolate or pedate, rarely bipinnate; stipules present. Flowers regular, perfect or unisexual, in racemes or cymose panicles usually borne opposite the leaves; calyx small, entire or with 4–6 teeth or lobes; petals 4–5, valvate, soon falling. Stamens 4–5, opposite the petals, inserted at base of the disc or between its lobes; anthers free or connate, short, 2-celled, opening inwardly. Disc various or none. Ovary usually immersed in the disc, 2–5-celled, the cells with 1 or 2 ascending anatropous ovules; style short or none, the stigma capitate or discoid. Fruit a berry, 1–6-celled, the cells 1–2-seeded; seeds with cartilaginous endosperm at the base of which is the short embryo.

— A chiefly tropical and subtropical family of about 12 genera and 700 species. *Vitis vinifera* L. is the grape, cultivated in most warm and warm-temperate countries. Over 25,000,000 metric tons of wine are made throughout the world from grapes every year. Dried grapes are called raisins.

Cissus L.

Herbaceous or woody plants generally climbing by means of tendrils; leaves simple or 3-foliolate. Flowers in more or less umbellate cymes, usually opposite a leaf; calyx short, subentire; petals 4, often adnate to each other and falling away like a cap at anthesis; disc 4-lobed, adnate to base of the ovary. Ovary 2-celled, each cell with 2 ovules. Berries 1–4-seeded, not edible; seeds ovoid or obtusely 3-cornered.

— A genus of about 200 species, widely distributed in tropical regions.

KEY TO SPECIES

1. Leaves simple:	C. verticillata
1. Leaves 3-foliolate:	
2. Flowers greenish, greenish-cream or whitish; leaflets usually 1–3 cm long, deeply and closely toothed above the middle:	C. trifoliata
2. Flowers bright red; leaflets mostly 2–6 cm long, subentire, with minute distant teeth:	C. microcarpa

Cissus verticillata (L.) Nicholson & Jarvis, *Taxon* 33: 727 (1984). PLATE 41.

Cissus sicyoides L. of Adams (1972).

Glabrous subwoody vine; leaves simple, long-petiolate, the blades narrowly to broadly ovate, mostly 4–9 cm long, subacuminate at the apex, broadly wedge-shaped, truncate or shallowly cordate at the base, the margins with distant minute teeth. Cymes shorter than the opposing leaves, 2–3-forked. Flowers small, greenish-cream. Berries obovoid-globose, black 8–10 mm in diameter.

GRAND CAYMAN: *Proctor* 15297. LITTLE CAYMAN: *Proctor* 35221. CAYMAN BRAC: *Proctor* 29083.

— Florida, West Indies, and continental tropical America, climbing on trees, rocks and fences.

Cissus trifoliata (L.) L., *Syst. Nat.* ed. 10, 2: 897 (1759). PLATE 41.

A glabrous, somewhat fleshy subwoody vine; leaves trifoliate, the leaflets obovate-wedge-shaped, deeply and closely toothed above the middle, usually 1–3 cm long. Cymes longer than the opposing leaves. Flowers greenish-cream, long-stalked, the pedicels up to 8 mm long. Berries ovoid-globose, black, 6–7 mm long, mucronate.

GRAND CAYMAN: *Brunt* 2103; *Kings* GC 151; *Proctor* 15128, 27938; *Sachet* 432. LITTLE CAYMAN: *Kings* LC 99; *Proctor* 28031. CAYMAN BRAC: *Proctor* 28932.

— Florida, West Indies, and northern S. America, in dry rocky thickets and on fences and old stone walls. This species has been used in folk medicine for treating coughs and back-ache.

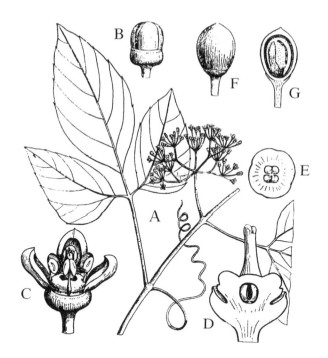

FIG. 171 **Cissus microcarpa**. A, leaf and inflorescence, × ²/₃. B, flower-bud, × 4. C, flower, × 5. D, flower with petals removed, cut lengthwise, × 11. E, cross-section of ovary, × 11. F, fruit, × 2. G, same cut lengthwise, × 2. (F. & R.)

Cissus microcarpa Vahl, *Ecolog.* 1: 16 (1796). **FIG. 171**. PLATE 41.

PUDDING WITH, DAFFODIL

A subwoody scrambler or vine with few tendrils, the branches variously 4-winged or -angled; leaves glabrous, trifoliate, the terminal leaflet subrhombic-elliptic, 3–6 cm long or more, the lateral ones obliquely ovate and smaller, all blunt to subacuminate at the apex, the margins distantly and minutely mucronate-serrulate. Cymes red, many-flowered, shorter than the opposing leaves, the branches and pedicels puberulous. Flowers red, the pedicels 3–5 mm long. Berries ovoid-globose, 7–8 mm in diameter, purple.

GRAND CAYMAN: *Kings* GC 119. LITTLE CAYMAN: *Proctor* 35117. CAYMAN BRAC: *Kings* CB 95; *Proctor* 28947.

— Cuba, Jamaica, Central and S. America, in dry or moist woodlands, climbing on rocks and trees. The Cayman plants as described above are somewhat transitional toward *Cissus caustica* Tussac, a species of Cuba and Hispaniola.

SURIANACEAE

Shrubs; leaves alternate, simple, entire and usually of thick texture; stipules minute or absent. Flowers regular, perfect or polygamous, solitary or in small panicles, these axillary or subterminal; calyx 5-lobed with persistent imbricate lobes; petals 5 or absent. Stamens 10, with free slender filaments. Ovary 1-celled or of 5 distinct 1-celled carpels, with a free style springing from the base of each; ovules 2 in each cell, ascending from the base. Fruit a nut, dry drupe, or of indehiscent achene-like carpels, these 1-seeded; seeds without aril and with little or no endosperm.

— A small family consisting of one monotypic pantropical genus and one other (*Stylobasium*) of 2 species occurring in southwestern Australia.

Suriana L.

A seashore shrub clothed with mixed capitulate and acicular hairs; leaves densely alternate; stipules absent. Flowers perfect, subterminal, solitary or in small, few-flowered panicles; petals clawed, yellow. Stamens unequal in length, those opposite the petals shorter and sometimes without anthers. Ovary of 5 distinct 1-celled carpels. Fruiting carpels achene-like, surrounded by the persistent calyx; seeds with thick, horseshoe-shaped embryo.

— A genus of 1 pantropical species.

Suriana maritima L., *Sp. Pl.* 1: 284 (1753). **FIG. 172.** PLATE 42.

JUNIPER

Characters of the genus. Bushy, 1–2 m tall, rarely more; leaves linear-spatulate, 1–3 cm long. Sepals ovate-lanceolate, 6–10 mm long; petals erose at the apex, shorter than the calyx. Ripe carpels 4–5 mm long, finely pubescent.

GRAND CAYMAN: *Brunt* 2009, 2064; *Hitchcock*; *Kings* GC 22, GC 264; *Millspaugh* 1253; *Proctor* 15122; *Sachet* 398. LITTLE CAYMAN: *Kings* LC 85; *Proctor* 28077. CAYMAN BRAC: *Kings* CB 57; *Millspaugh* 1151; *Proctor* 28923.

— Common on sandy or rocky tropical seashores.

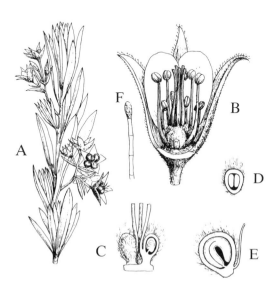

FIG. 172 **Suriana maritima.** A, branch with inflorescence, × ²/₃. B, flower with part of the calyx and corolla and one stamen removed, × 4. C, pistil cut lengthwise, × 6. D, cross-section of a carpel, × 6. E, ripe achene (nutlet) cut lengthwise, × 2. F, hail from calyx, much enlarged. (F. & R.)

SAPINDACEAE

Shrubs or trees with watery sap, or sometimes woody or herbaceous vines climbing by means of tendrils; leaves alternate, mostly pinnately or bipinnately compound, sometimes 3- or 1-foliolate; stipules absent in most genera. Flowers usually small and white, mostly polygamo-dioecious, regular or irregular, variously arranged in racemes, panicles or corymbs, axillary or sometimes terminal. Sepals 4–5, rarely none, mostly free and often unequal, imbricate; petals 4–5 or none, equal or unequal, often scaly or bearded within; disc complete in regular flowers or represented by 2 or 4 glands in irregular flowers. Stamens usually 8, usually hypogynous and inserted within the disc; anthers oblong, 2-lobed, or linear. Ovary superior, mostly 3-celled, entire or lobed, with 1–2 ovules in each cell, attached to the axis; style terminal or basal between the ovary-lobes, simple or divided, with simple stigma. Fruit a capsule or indehiscent and drupe-like, berry-like, leathery or consisting of 2–3 samaras; seeds usually with or sometimes without aril; endosperm none, the embryo thick and often folded or spirally twisted.

— A large, chiefly tropical family of about 150 genera and 2,000 species.

KEY TO GENERA

1. Plants climbing by tendrils: — **Cardiospermum**

1. Plants erect, never with tendrils:
    2. Leaves compound:
        3. Leaves 3-foliolate, having a terminal leaflet; fruits drupe-like:
            4. Plant wholly glabrous; petioles marginate; leaves with closely parallel venation: — **Hypelate**
            4. Plant puberulous at least on inflorescence; petioles not marginate; leaf-venation not closely parallel: — **Allophylus**
        3. Leaves 2–4-foliolate or more, without a terminal leaflet:
            5. Fruit a 1-seeded drupe (sometimes in *Sapindus* with more than 1 seed); sepals and petals 4 or 5:
                6. Petiole and rachis usually marginate or winged, at least in young plants; fruits more than 1 cm in diameter:
                    7. Sepals and petals 4; fruit 2 cm in diameter or more, green, with edible pulp: — **[Melicoccus]**
                    7. Sepals and petals 5 (rarely 4); fruit 1.2 cm in diameter, yellowish-olive, the very thin pulp not edible: — **Sapindus**
                6. Petiole and rachis never marginate or winged; fruits ca. 1 cm in diameter or less, black when ripe: — **Exothea**
            5. Fruit a 3-seeded capsule, the large black seeds each attached to a pale-yellow fleshy aril; sepals and petals 5: — **[Blighia]**
    2. Leaves simple; fruit a winged capsule: — **Dodonaea**

Cardiospermum L.

Climbing herbs or shrubs with paired tendrils arising from the peduncles; branches slender, grooved or ribbed; leaves biternate, the leaflets crenate or serrate, often with pellucid dots or lines in the tissue. Flowers white, irregular, polygamodioecious, with jointed pedicels, in axillary racemes or corymbs; sepals 4–5, concave, the outer ones smaller; petals 4, two with crested scale which has a bearded appendage pointing downward, the other two with a scale which has a wing-like crest on the back. Disc-glands 2, opposite the petals with the appendage. Stamens 8, the filaments unequal. Style 3-lobed. Capsule of 3 inflated membranous lobes, splitting loculicidally; seeds globose, blue-black, often arillate at the base; cotyledons large, transversely folded.

— A pantropical genus of 12 species.

KEY TO SPECIES

1. Capsule subglobose, puberulous, 3–4 cm in diameter; seed with a heart-shaped, distinctly bilobed hilum: C. halicacabum

1. Capsule top-shaped or subglobose, glabrous, less than 3 cm in greatest diameter; seed with a semicircular hilum: C. corindum

Cardiospermum halicacabum L., *Sp. Pl.* 1: 366 (1753).

A mostly herbaceous vine, the stems glabrous or puberulous, 5–6-ribbed; leaves 3–10 cm long, with ovate or lanceolate leaflets, variously toothed or lobed, puberulous. Inflorescence umbel-like, long-stalked; flowers mostly 4–5 mm long. Seeds 4–5 mm in diameter, with white hilum nearly as broad as the seed.

GRAND CAYMAN: *Hitchcock*; *Kings GC 47.* CAYMAN BRAC: *Millspaugh* 1165, 1200, 1201.

— Pantropical, thickets and woodlands.

Cardiospermum corindum L., *Sp. Pl.* ed. 2, 1: 526 (1762).

Herbaceous to subwoody vine, the stems puberulous, 5–7-ribbed; leaves 2–7 cm long, with lanceolate to ovate leaflets, incised or crenate, puberulous. Inflorescence umbel-like, long-stalked; flowers mostly 3–5 mm long. Seeds 2.5–4 mm in diameter, with white semicircular hilum.

GRAND CAYMAN: *Millspaugh* 1309; *Proctor* 15181. LITTLE CAYMAN: *Proctor* 28125, 35096, 35180. CAYMAN BRAC: *Proctor* 29017, 29049.

— Pantropical, in thickets and woodlands.

Allophylus L.

Erect shrubs or small trees; leaves often long-petiolate, trifoliolate or 1-foliolate, the leaflets often with pellucid dots or lines in the tissue. Inflorescence an axillary raceme or panicle; flowers very small, polygamo-dioecious; sepals 4, opposite in pairs, concave, broadly imbricate, the 2 outer ones smaller; petals 4, each with a small 2-lobed scale within. Disc lobed or of 4 glands. Stamens 8. Ovary commonly 2-locular and deeply 2-

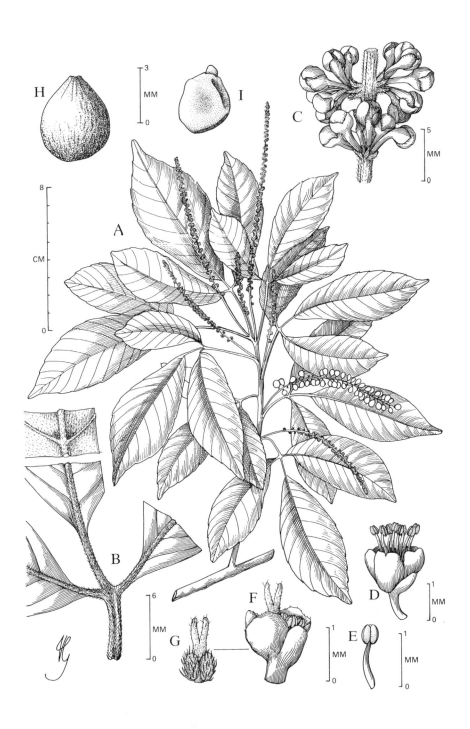

FIG. 173 **Allophylus cominia** var. **caymanensis.** A, habit. B, segments of leaf. C, section of inflorescence with flower-buds. D, staminate flower. E, stamen. F, perfect pistillate flower just opening. G, pistil, showing 2-lobed ovary and 2-branched style. H, fruit. I, seed. (G.)

lobed, united by the 2–3-lobed style; ovules solitary; only 1 ovary-lobe usually developing to form a fruit. Fruit a single indehiscent drupe-like coccus (rarely 2 cocci), dry or fleshy, usually obovoid or globose; seed erect, on a very short fleshy aril.

— A pantropical genus of perhaps 175 species.

Allophylus cominia (L.) Sw., *Nov. Gen. & Sp. Pl.* 62 (1788) var. **caymanensis** Proctor, *Sloanea* 1: 2 (1977). **FIG. 173.**

A bushy shrub to 3 m tall; young branches puberulous or glabrate, with raised lenticels. Leaves long-petiolate, trifoliate, the leaflets elliptic or oblanceolate to narrowly obovate, the central one 5–10 cm long, short-acuminate at the apex, the margins distantly crenate-serrate, glabrous or nearly so on the upper side, minutely puberulous or glabrate beneath and with small tufts of white hairs (domatia) in the vein-axils. Inflorescence often equal in length to or exceeding the leaves, simple or with 1 or 2 short branches, puberulous. Flowers greenish-cream, ca. 1 mm long, the stamens exserted. Fruit more or less globose, 4–4.5 mm in diameter.

GRAND CAYMAN: *Correll & Correll* 51021; *Hitchcock*; *Proctor* 27979, 31039, 31053 (type). LITTLE CAYMAN: *Proctor* 35219. CAYMAN BRAC: *Fawcett*; *Proctor* 28965, 29346.

— The variety is endemic, growing in rocky thickets and woodlands. Var. *cominia*, which is widely distributed in the West Indies and C. America, is usually much more densely pubescent (a variable character), has larger leaves, inflorescences much more branched, and larger fruits.

[*Melicoccus* P. Br.

Glabrous trees; leaves abruptly pinnate, the leaflets in 2 or 3 subopposite pairs, subsessile, entire. Racemes simple or paniculate, terminal on lateral branchlets, bearing numerous flowers. Flowers polygamo-dioecious; calyx deeply 4–5-lobed; petals 4 or 5, roundish or obovate. Disc 4–5-lobed. Stamens 8. Ovary 2–3-celled; stigma peltate, 2–3-lobed. Fruit a drupe; seed enclosed in a large, pulpy, edible aril.

— A tropical American genus of 2 species.

Melicoccus bijugatus Jacq., *Enum. Syst. Pl. Carib.* 19 (1760).

GENIP

Tree to 12 m tall or more, with smooth grey bark; leaves deciduous annually, the young leaves appearing with the flowers in March or April; petiole and rhachis flat, often winged; leaflets elliptic or ovate-elliptic, 5–15 cm long, subacuminate at apex; margins often undulate. Staminate inflorescence often longer than the leaves, much-branched; pistillate inflorescence shorter and less branched. Open flowers 6–8 mm in diameter, scented. Fruits green, 2–3 cm in diameter, with edible pulp.

GRAND CAYMAN: *Brunt* 2012, 2060; *Kings* GC 162; *Proctor* 15193. CAYMAN BRAC: *Proctor* 29109.

— Native to continental tropical America from Nicaragua to Surinam, widely planted elsewhere and becoming naturalised. It is a handsome shade-tree; the wood is hard and heavy, suitable for general construction purposes.]

Sapindus L.

Trees with pinnate leaves, these lacking a terminal leaflet, the rachis commonly winged; leaflets 3 or more pairs; stipules lacking. Inflorescence axillary racemes or thyrsoid panicles. Flowers regular, unisexual or bisexual (plants polygamo-dioeceous). Sepals 2 or 5, imbricate; petals 4 or 5, each with scale at base; disc annular; stamens 8 to 10, inserted on disc; ovary 3-lobed, with 1 ovule in each locule; style slender, 2–4-branched at apex. Fruit drupe-like, globose or 2- or 3-lobed, with fully developed carpels or one or two aborted. Seed globose, without aril.

— A small genus of 10 species found in warm and tropical Asia and in the neotropics.

Sapindus saponaria L., *Sp. Pl.* 1: 367 (1753).

SOAP-BERRY TREE

Tree sometimes up to 15 m tall; leaflets glabrous, 3–6 pairs, elliptic, oblong or lanceolate, 7–18 cm long, up to 3.5 cm broad or more, rounded to acute or acuminate at apex, equal- or unequal-sided. Panicles up to 30 cm long with numerous small flowers. Sepals 1–2 mm long; petals shorter, white, hairy. Ripe fruit yellow to yellowish-olive; seed globose, black.

LITTLE CAYMAN: *Platts* et al.; also reported from Cayman Brac.

— Florida, Mexico and C. America, Bahamas, Greater and Lesser Antilles. Introduced into tropical Asia and Africa.

Exothea Macfad.

Trees or shrubs with alternate, pinnately compound leaves, these with 2(–3) pairs of leaflets and no terminal leaflet; stipules lacking. Inflorescence a small loose panicle. Flowers small, slightly irregular, unisexual or bisexual (plants dioeceous or polygamous). Sepals 5, imbricate, pubescent on both sides, persistent and reflexed in fruit. Petals 5, short-clawed without scales. Stamens 7 or 8, those of pistillate flowers shorter than the petals. Ovary 2-locular, with 2 or 3 ovules in each locule; style short. Fruit a globose purple drupe; seed without aril.

— A small genus of 3 species in Florida, C. America, and the West Indies.

Exothea paniculata (A. L. Juss.) Rodlk. ex Durand, *Index Gen. Phan.* 81 (1888). PLATE 42.

Tree up to 20 m tall; leaves petiolate, with 2, 4, or 6 alternate leaflets, these oblong to elliptic-obovate, 5–13 cm long, the apex acute, obtuse, or emarginate, narrowed or nearly sessile at base, the margins entire, the upper surface somewhat glossy, the lower surface less so and pale-green. Panicles terminal and axillary, up to 14 cm long, pubescent. Sepals ovate, 3.5 mm long; petals oblong-obovate, 3 mm long, white pubescent; disc annular or lobed, pubescent; stamens borne inside the disc, the fertile stamens 3 mm long, glabrous, the sterile stamens 1.5 mm long; ovary slightly pubescent. Drupes globose, purplish-black, 10 mm in diameter or slightly more.

CAYMAN BRAC *Proctor* 47320, 47806.

— Florida, Guatemala, Belize, Bahamas, Greater and Lesser Antilles.

[*Blighia* K. D. Koenig

Trees; leaves large, even-pinnate, with 3–5 pairs of entire, glabrous leaflets. Flowers polygamous in axillary racemes or panicles; calyx 5-parted; petals 5, with a scale at the base of each about half as long as the petal. Disc ring-like, somewhat 8-lobed; stamens 8, inserted within the disc. Ovary 3-celled, with 1 ovule in each cell. Fruit a somewhat fleshy 3-valved capsule; seeds large, black, surrounded at base with a large, whitish or yellowish, fleshy aril.

— A tropical African genus of 6 species.

Blighia sapida K. D. Koenig in Koenig & Sims, *Ann. Bot.* 2: 571, t. 16, 17 (1806).

AKEE

A tree to 10 m tall, the young branchlets yellowish-tomentose; leaves 15–30 cm long, the 3–5 pairs of leaflets opposite or subopposite, glabrous above, puberulous beneath, the midrib and veins prominent. Racemes shorter than the leaves, tomentose, the flowers greenish-white, stalked, fragrant; calyx 2.5–3 mm long; petals 4 mm long, each with wide basal scale. Disc tomentose. Fruits 7–10 cm long, red, glabrate outside, densely tomentose within; aril edible when ripe.

GRAND CAYMAN: *Kings* GC 346. CAYMAN BRAC: *Proctor* 29110.

— Introduced to Jamaica from W. Africa in the 1780s, and since widely planted there and to a lesser extent in other West Indian islands and elsewhere, often becoming naturalised. The fleshy arils are poisonous unless the capsules are ripe (i.e. split open) when picked, but are wholesome and nutritious if properly prepared. They are usually served with salt codfish, or sometimes with bacon.]

Dodonaea L.

Shrubs or trees, usually viscid; leaves simple or rarely pinnate; stipules absent. Flowers unisexual or polygamo-dioecious, whitish, pale green or yellowish, in axillary or terminal racemes, corymbs or panicles. Sepals 2–5, valvate or narrowly imbricate; petals absent. Disc lacking in the staminate flower, short and stalk-like in the pistillate. Stamens 5–8, the anthers nearly sessile, linear-oblong, obtusely 4-angled. Ovary sessile, 3–6-angled, 3–6-celled; style 3–6-lobed at the apex; ovules 2 in each cell. Fruit a somewhat papery capsule, 2–6-angled, the angles acute to broadly winged, with 1–2-seeded cells; seeds without aril.

— A chiefly Australian genus of more than 50 species.

Dodonaea viscosa (L.) Jacq., *Enum. Syst. Pl. Carib.* 19 (1760). **FIG. 174.** PLATE 42.

A shrub up to 3 m tall, with very viscid foliage, the branches reddish-brown, longitudinally ridged; leaves simple, mostly oblanceolate or narrowly oblong, usually 4–11 cm long and 1.5–4 cm broad above the middle, gradually narrowed to the base, a petiole virtually absent. Flowers pale green, dioecious, in small lateral corymbs; sepals 3 mm long. Capsules usually 3-celled and 3-winged, 1.5–2.5 cm wide, deeply notched at the apex.

GRAND CAYMAN: *Kings* GC 386; *Millspaugh* 1264; *Proctor* 11964. LITTLE CAYMAN: *Kings* LC 69; *Proctor* 28078.

— Pantropical, in various habitats; many variants have been described, but the Cayman plants seem to approximate the typical variety.

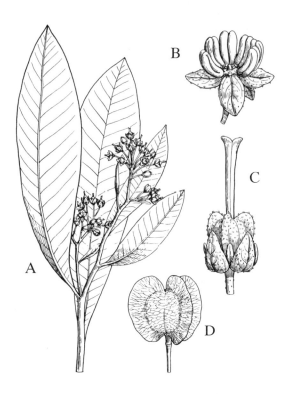

FIG. 174 **Dodonaea viscosa**. A, branchlet with leaves and flowers, × ²/₃. B, staminate flower with a sepal pressed down and a stamen removed, × 4. C, pistillate flower, × 4. D, fruit, × 1. (F. & R.)

Hypelate Sw.

A glabrous shrub or small tree; leaves alternate, trifoliate, with narrowly margined petioles. Flowers small, white, polygamo-monoecious, in axillary panicles; sepals 5, imbricate, soon falling; petals 5. Stamens 8, inserted on the disc, shorter and imperfect in the female flower. Ovary 3-celled, rudimentary in the male flowers; ovules 2 in each cell. Fruit a thin-fleshed, 1-seeded drupe; seed without endosperm, the cotyledons folded irregularly.

— A monotypic genus of the Florida keys and West Indian islands.

Hypelate trifoliata Sw., *Nov. Gen. & Sp. Pl.* 61 (1788). **FIG. 176**. PLATE 43.

POMPERO, PLUMPERRA, WILD CHERRY

Shrub or small tree to 8 m tall or more; leaves with petioles 1–3 cm long, the oblanceolate to obovate sessile leaflets mostly 1–4 cm long; venation closely parallel and prominulous. Panicles exceeding the leaves; sepals 2.5–3 mm long, petals 2 mm long. Fruits ellipsoid, 6–8 mm long, brown or black.

GRAND CAYMAN: *Brunt* 2195; *Kings* GC 377, GC 380; *Proctor* 15218. LITTLE CAYMAN: *Proctor* 28102, 35129. CAYMAN BRAC: *Matley*; *Kings* CB 92; *Proctor* 29025.

— Distribution of the genus, frequent in dry rocky woodlands. The wood is said to be very heavy, hard, close-grained, and a rich dark brown; it is very durable in contact with the soil, and is used for posts, in ship-building and for the handles of tools.

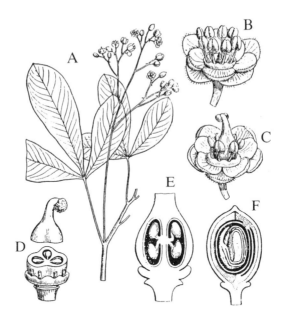

FIG. 175 **Hypelate trifoliata.** A, portion of branch with leaves and flowers, × ²/₃. B, staminate flower, × 3. C, pistillate flower, × 3. D, ovary cut across, × 6. E, ovary cut lengthwise, × 12. F, fruit cut lengthwise, × 6. (F. & R.)

BURSERACEAE

Shrubs or trees, more or less resinous or aromatic; leaves alternate, simple or usually odd-pinnate, sometimes 3-foliolate, usually not pellucid-dotted; stipules absent. Flowers small, perfect or polygamo-dioecious, in racemes or panicles. Calyx 3–5-lobed, the lobes imbricate or valvate; petals 3–5, free or rarely united toward the base, imbricate or valvate, alternating with the sepals. Disc ring- or cup-shaped, free or adnate to the calyx-tube. Stamens twice as many as the petals or rarely the same number, inserted at the base or margin of the disc; filaments free; anthers subglobose or oblong, 2-locular. Ovary superior and free, 2–5-locular, 3-angled, with short style and undivided or 2–5-lobed stigma; ovules 2 in each locule, attached to the axis above the middle, usually pendulous. Fruit capsule-like or drupe-like, dehiscent or indehiscent, containing 2–5 hard nutlets; seeds with membranous walls; endosperm lacking.

— A pantropical family of about 17 genera and 500 species.

Bursera L.

Shrubs or trees, often with thin papery bark, and with aromatic resinous sap; leaves deciduous, variously pinnate or sometimes 1-foliolate, the leaflets opposite, entire or serrate, and with rhachis naked or winged. Flowers polygamous or perfect; calyx 3–4-lobed or -parted, imbricate; petals 3–4, valvate. Disc ring-like. Stamens 6–10, subequal, inserted at base of the disc. Ovary 3–5-celled, the stigma 3–5-lobed. Fruit an ovoid, more or less 3-angled capsule, 2–3-valved, with 1–several hard nutlets.

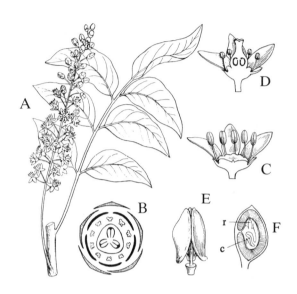

FIG. 176 **Bursera simaruba**. A, portion of branch with small leaf and inflorescence, × ²/₃. B, diagram of perfect flower. C, staminate flower, × 4. D, fertile flower, × 4. E, drupe, × 1. F, stone cut lengthwise, × 1¹/₂; c, cotyledons; r, radicle. (F. & R.)

— An American genus of about 80 species, the majority in Mexico; all are capable of yielding a fragrant resin called 'copal', used (in Mexico and C. America chiefly) as incense, in domestic medicine, as an ingredient of varnish and for numerous other purposes.

Bursera simaruba (L.) Sarg., *Gard. & For.* 3: 260 (1890). **FIG. 176.** PLATE 43.

BIRCH, RED BIRCH

A tree to 15 m tall or more, the young bark greenish, the old bark light reddish, peeling off in paper-like sheets; leaves petiolate, pinnate with usually 5–7 unequal-sided leaflets, these more or less pubescent when young, becoming glabrous with age, lance-oblong to broadly ovate, mostly 4–10 cm long, acuminate at the apex. Flowers white, in racemes appearing before or with the new leaves; staminate flowers 5-parted, with 10 stamens; perfect flowers 3-parted, with 6 stamens. Fruits 8–12 mm long, dark red.

GRAND CAYMAN: *Brunt* 1816, 2023; *Hitchcock*; *Proctor* 15027, 15139; *Sachet* 324. LITTLE CAYMAN: *Proctor* 28046. CAYMAN BRAC: *Proctor* 28957.

— Florida, West Indies, Central and northern S. America, common in dry woodlands and along roadsides. This tree is often used for living fenceposts, and to a partial extent its distribution is artificial for this reason. The wood is light and soft but firm, and is used somewhat for making crates, boxes and match-sticks.

ANACARDIACEAE

Shrubs or trees, often containing more or less poisonous resin or sap; leaves alternate or very rarely opposite, and simple, 1–3-foliolate, or odd-pinnate; stipules absent or the lowest leaflet stipule-like. Flowers perfect or polygamo-dioecious, usually regular, in panicles or rarely racemes; calyx 3–7-lobed, rarely spathe-like or irregularly rupturing; petals usually 3–7, free, sometimes persistent. Disc usually ring-like. Stamens commonly twice as many as the petals, rarely otherwise, inserted at base of the disc; filaments free; anthers opening inwardly. Ovary 1-celled or rarely 2–5-celled; styles 1–3; ovules solitary, more or less pendulous. Fruit usually drupe-like and indehiscent, superior and free or surrounded either by the enlarged base of the calyx or disc, or sometimes located at the top of a fleshy receptacle formed from the base of the calyx and top of the pedicel, 1–5-celled, the flesh frequently oily or with caustic sap; seeds with little or no endosperm, the cotyledons fleshy.

— A widely distributed family of about 65 genera and 600 species. The cashew tree, *Anacardium occidentale* L., belongs to this family; as do poison ivy and poison sumac (both *Toxicodendron*), two notorious N. American plants.

KEY TO GENERA

1. Leaves simple:

    2. Leaves obovate to elliptic, rounded to emarginate at apex; stamens 7–10; fruit a curved (reniform) drupe 2–3.5 cm long borne on a pyriform fleshy hypocarp (= receptacle + pedicel):   **[Anacardium]**

    2. Leaves oblong-lanceolate, acute at apex; stamens 5; fruit a large ovoid to subreniform drupe more than 5 cm long without an enlarged hypocarp: **[Mangifera]**

1. Leaves pinnate:

    3. Ovary 1-locular; fruits not over 1.5 cm long, not edible, the plants with poisonous sap:

        4. Leaflets 3–7, entire; a glabrous tree to 10 m tall: **Metopium**

        4. Leaflets usually 11–17, with toothed margins; often unbranched pubescent shrub to 2 m tall: **Comocladia**

    3. Ovary 2–5-locular; fruits plum-like, 2.5 cm long or more, edible, the plants not poisonous: **Spondias**

[*Anacardium* L.

Trees or shrubs with alternate, simple, petiolate leaves, the tissue glabrous, coriaceous. Inflorescence terminal or sometimes in upper leaf-axils, paniculate or sometimes corymbiform. Flowers unisexual (male) or bisexual (plants polygamous). Sepals 5, erect; petals 5, often reflexed at maturity; disc lacking. Stamens 7–10, unequal, 1 (or rarely 2) longer and stouter than the others, the latter sometimes sterile; filaments connate at base forming a short tube; anthers basifixed. Ovary sessile, ovoid, with 1 locule and 1 ovule; style lateral, with minute stigma. Fruit a kidney-shaped, laterally compressed, nut-like drupe, borne on a fleshy, pyriform hypocarp.

— A neotropical genus of 10 species.

485

Anacardium occidentale L., *Sp. Pl.* 1: 383 (1753).

CASHEW

Tree to 12 m tall (but not often over 5 m) with wide-spreading branches and glabrous twigs. Leaves with petioles up to 2 cm long, the blades obovate to broadly elliptic, 7.5–15 cm long, 4.5–12 cm broad, the apex rounded or slightly emarginate, the base cuneate to obtuse, with 8–13 pairs of lateral nerves. Panicles 10–18 cm long, longer than the uppermost leaves; bracts ovate to ovate-lanceolate, acuminate, 3–5 mm long, pubescent. Flowers fragrant, crowded at ends of inflorescence branches; pedicels 2–5 mm long; calyx lobes 3–5 mm long; petals linear-lanceolate, 7–15 mm long, at first cream-white with red-stripes, turning all red; stamens 2–12 mm long. Fruit 1.9–2.7 cm long, olive-brown; ripe hypocarp ('cashew apple') red or yellow.

GRAND CAYMAN: *Proctor* 47822. Rare in pastures near North Side.

— Native to the continental neotropics but cultivated and established world-wide in tropical countries.]

[*Mangifera* L.

Trees; leaves alternate, simple and entire. Flowers polygamous, in terminal panicles; calyx 4–5-parted, deciduous, with imbricate sepals; petals 4–5, imbricate and spreading. Disc of 4–5 fleshy lobes, alternate with the petals. Stamens 1 or 4–5, inserted within or on the disc, only 1 (rarely more) fertile and larger than the others. Ovary free, sessile, 1-celled, with lateral style and simple stigma. Drupe fleshy, the stone compressed and sometimes 2-valved.

— A genus of about 30 species chiefly in tropical Asia.

Mangifera indica L., *Sp. Pl.* 1: 200 (1753). PLATE 44.

MANGO

A tree of 10–15 m with a dense, rounded or spreading crown; leaves glabrous, oblong-lanceolate, 10–20 cm long. Flowers greenish-white, in large panicles; sepals 2.5 mm long; petals 5 mm long; fertile stamens 1 or 2. Fruits edible, varying greatly in size, shape and colour, comprising many horticultural variants.

GRAND CAYMAN: *Brunt* 1991; *Hitchcock*; *Proctor* 15185. CAYMAN BRAC: *Proctor* (sight record).

— Originally from India, but now cultivated in various forms throughout the tropics for its edible fruits, and often becoming naturalised. In Grand Cayman, trees can be found in areas remote from habitations wherever seeds have been casually tossed aside.]

Metopium P. Br.

Glabrous trees with caustic sap; leaves petiolate, odd-pinnate, the leaflets coriaceous, entire. Flowers small, greenish, in axillary panicles; sepals 5, imbricate; petals 5, imbricate. Disc ring-like. Stamens 5, with short, subulate filaments. Ovary 1-celled, the style short, the stigma 3-lobed. Fruit a small drupe.

— A chiefly West Indian genus of 3 species.

Metopium toxiferum (L.) Krug & Urban in Urban, *Bot. Jahrb.* 21: 612 (1896). PLATE 44.

POISON TREE

A tree to 10 m tall or more, with wide-spreading branches and poisonous sap; leaves 15–30 cm long, with 3–7 stalked leaflets, these with blades ovate to suborbicular or obovate, 3–10 cm long, often obtuse or notched at the apex and narrowed or unequally cordate at the base, bright shining green especially on the upper side. Panicles many-flowered, sometimes exceeding the leaves. Drupe oblong, orange-yellow, 1–1.5 cm long.

GRAND CAYMAN: *Hitchcock*. LITTLE CAYMAN: *Proctor 28128*.

— Florida, Bahamas, Cuba, Hispaniola and Puerto Rico, in dry rocky or sandy woodlands. A different species occurs in Jamaica.

Comocladia L.

Shrubs or small trees with poisonous sap, this turning blackish on exposure to air; trunk or main stem slender, often unbranched or few-branched, the leaves crowded in a palm-like rosette toward the top; leaves alternate, odd-pinnate, the leaflets rather numerous and opposite to subopposite, entire or toothed, reduced in size toward base of the leaf. Panicles axillary or appearing to be terminal, usually shorter than the leaves; flowers minute, crowded, polygamous, sessile or subsessile; calyx 3–4-cleft, persistent, with imbricate lobes; petals 3–4, red, imbricate. Disc 3-lobed. Stamens inserted in notches of the disc. Ovary 1-locular, with 3 stigmas. Fruit an oblong-ellipsoid or ovoid drupe.

— A genus of about 20 species, occurring chiefly in the West Indies and Mexico.

Comocladia dentata Jacq., *Enum. Syst. Pl. Carib.* 16 (1760). PLATE 44.

MAIDEN PLUM

A slender shrub rarely more than 2 m tall, often unbranched, the younger parts puberulous; leaves with 11–17 leaflets, these ovate to oblong, 3–13 cm long, the margins sharply toothed, the under-surface puberulous and with prominently reticulate venation. Panicles 20–25 cm long; flowers 3-parted. Drupes ovoid, 7–8 mm long.

GRAND CAYMAN: *Kings GC 115; Lewis 3612, 3822; Proctor 15017, 27977*. LITTLE CAYMAN: *Proctor 35183*.

— Cuba and Hispaniola, in dry thickets, pastures and old fields, a noxious weed wherever it occurs.

Spondias L.

Small to large trees; leaves alternate, odd-pinnate, the leaflets mostly opposite. Flowers small, polygamous, in axillary or terminal panicles; calyx small, 4–5-lobed, deciduous; petals 4–5, valvate in bud. Disc cup-shaped, crenate. Stamens 8–10, inserted below the disc. Ovary sessile, free, 3–5-celled; styles 3–5, connivent above. Fruit a fleshy drupe, the stone large and 1–5-celled.

— A small genus of about 8 species, widely distributed in tropical regions at least in cultivation.

KEY TO SPECIES

1. Inflorescences terminal (rarely in upper axils), up to 50 cm long, many-flowered; flowers white to greenish-white; calyx lobes not imbricate; leaflets 5–12 cm long: [S. mombin]

1. Inflorescences axillary, rarely more than 5 cm long, few flowered; flowers red to purple; calyx lobes imbricate; leaflets mostly 2–4.5 cm long: S. purpurea

[*Spondias mombin* L., *Sp. Pl.* 1: 371 (1753).

HOG PLUM

Tree often up to 25 m tall (Cayman plants much smaller) with smooth grey to brown bark often armed with stout prickles on the trunk. Leaves 12–40 cm long, pinnate with 9–19 leaflets; petioles 1.5–8.2 cm long; leaflets elliptic, ovate-elliptic, or oblong-elliptic, mostly 5–12 cm long, acuminate at the apex, cuneate to obliquely cuneate at base, the margins usually entire, the surfaces puberulous on veins beneath and on midrib above. Flowers white, fragrant, on pedicels 1–3 mm long; calyx lobes triangular, 0.5 mm long; petals oblong-lanceolate, 2–3.5 mm long. Stamens 9 or 10; filaments to 2.5 mm long. Fruits ellipsoid to obovoid, 3–4 cm long, orange to yellow when ripe.

GRAND CAYMAN: *Proctor* 49325. Rare in woodlands, Spotts area.

— Mexico to S. America; introduced and often becoming naturalised elsewhere; cultivated throughout the tropics.]

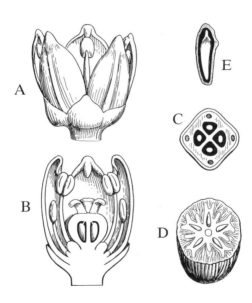

FIG. 177 **Spondias purpurea**. A, staminate flower, × 6. B, pistillate flower cut lengthwise, × 6. C, cross-section of ovary, × 9. D, endocarp of a related species *S. mangifera*, enlarged. E, cell of same with embryo, enlarged. (F. & R.)

Spondias purpurea L., *Sp. Pl.* ed. 2, 1: 613 (1762). **FIG. 177.**

PLUM

A low, spreading, deciduous tree to 5 m tall or more, with arching lateral branches, the bark smooth and greyish; leaves glabrous, with 5–13 or more opposite or alternate leaflets, these elliptic-oblong to obovate, 1.5–4 cm long, acute and mucronate at the apex, and unequal-sided at the base. Panicles small and few-flowered, produced when the branches are leafless at nodes on the older wood. Flowers red-purple, the petals 3 mm long. Fruits red or yellow, 3–3.5 cm long, edible.

GRAND CAYMAN: *Brunt* 2018. CAYMAN BRAC: *Proctor* 29111.

— Widespread in tropical America; often planted as a 'quickstick' fence or for its edible fruits, so that its true natural range would be difficult to ascertain. It is probably not indigenous in the Cayman Islands, but nevertheless can be found growing wild in rocky thickets.

[Other species of *Spondias*, such as *S. dulcis* Parkinson, the June plum, may perhaps occur as cultivated trees in the Cayman Islands, but have not been encountered during the preparation of this present book.]

SIMAROUBACEAE

Shrubs or trees, the bark usually very bitter and containing oil-sacs; leaves alternate or rarely opposite, pinnate or rarely 3- or 1-foliolate; stipules usually lacking. Flowers regular, dioecious or polygamous, sometimes perfect, in mostly axillary panicles or racemes, rarely in spikes or the flowers solitary; calyx 3–5-lobed; petals 3–5 (rarely none), imbricate or valvate. Disc ring-like, cup-like, or elongate to form a gynophore, entire or lobed. Stamens inserted at base of the disc, as many or twice as many as the petals; filaments free; anthers 2-celled, opening inwardly. Ovary 2–5-lobed, rarely entire, 1–5-celled; styles 2–5, free or sometimes fused at the apex into a cap-like stigma; ovules 1 or 2 in each cell, attached at the inner angle. Fruit a drupe, capsule or samara; seeds with or without endosperm.

— A pantropical family of about 20 genera and 120 species.

Alvaradoa Liebm.

Shrubs or small trees with bitter juice; leaves unequally pinnate, with numerous small alternate leaflets. Flowers dioecious, small, numerous in narrow spreading or drooping racemes, these axillary and terminal; calyx of 5 valvate sepals; petals 5 or lacking. Staminate flowers with large, deeply 5-lobed disc; stamens 5, alternate with the sepals, inserted between the lobes of the disc. Pistillate flowers without stamens; ovary 2–3-celled with only 1 cell fertile; styles 2–3, free and recurved; ovules 2 in each cell. Fruit a 2–3-winged samara.

— A tropical American genus of 5 or 6 species.

Alvaradoa amorphoides Liebm., *Nat. For. Kjoebenhavn Vid. Medd.* 100 (1853). **FIG. 178. PLATE 44.**

WILD SPANISH ARMADA

A deciduous shrub to 3 m tall or more, rarely a small tree; branchlets puberulous; leaves clustered near the ends of the branches, with 19–51 membranous leaflets, these oval or oblong, 1–2.5 cm long, rounded at the apex, puberulous on both sides or glabrous on the

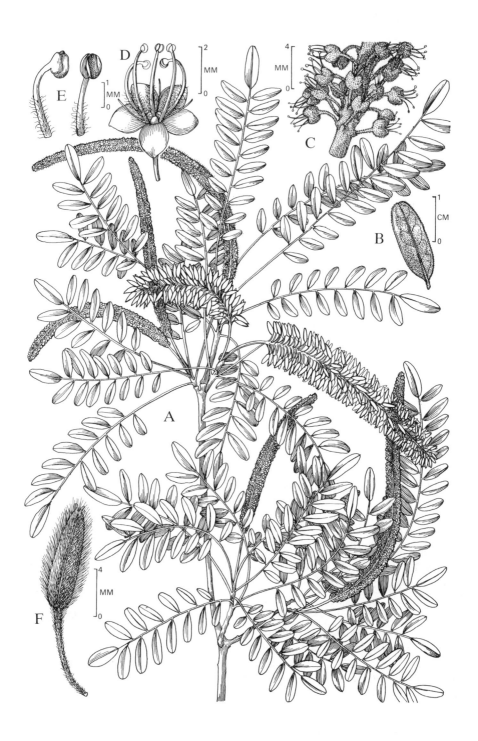

FIG. 178 **Alvaradoa amorphoides**. A, habit. B, single leaflet. C, portion of flowering raceme. D, flower. E, two stamens. F, fruit. (G.)

upper side, noticeably pale or whitish beneath. Flowers on slender pedicels, greenish, the staminate racemes up to 25 cm long, the pistillate ones much shorter, dense and plume-like. Samaras narrowly lance-oblong, dull reddish, 1–1.5 cm long, densely pilose and ciliate with long slender spreading hairs.

GRAND CAYMAN: *Brunt* 2192; *Hitchcock*; *Kings* GC 320; *Millspaugh* 1282; *Proctor* 15198, 15229. LITTLE CAYMAN: *Proctor* 35093.

— Florida, Bahamas, Cuba, Central and S. America, in dry rocky woodlands. The wood is said to be valuable as fuel because it burns slowly and for a long time.

RUTACEAE

More or less aromatic shrubs or trees, rarely herbs, occasionally scandent, sometimes armed with prickles; leaves alternate or opposite, usually compound, occasionally 1-foliolate or simple, nearly always more or less pellucid-dotted with oil-glands; stipules absent. Flowers perfect, polygamous or dioecious, in axillary or terminal cymes, panicles, racemes, spikes or clusters, rarely solitary, the parts usually in 4s or 5s, rarely 3s. Sepals imbricate or rarely none; petals imbricate or sometimes united. Stamens as many or twice as many as the petals, rarely more numerous; filaments free or united below, inserted on a hypogenous disc or sometimes adnate to the corolla-tube; anthers 2-celled, often gland-tipped, opening inwardly. Ovary usually 4–5-locular or the carpels free at the base and united in the styles or stigmas, or altogether free and 1-locular; ovules usually 2 (rarely 4) in each carpel. Fruit various, of follicles, or a samara, drupe, or berry; seeds solitary or several in each cell, with or without endosperm.

— A widely distributed family of about 120 genera and 900 species, chiefly of warm climates, especially numerous in South Africa and Australia.

KEY TO GENERA

1. Ovary 2–5-lobed; fruit of 1–5 cocci; leaves pinnately compound with 5 or more leaflets:	Zanthoxylum
1. Ovary not lobed; fruit a drupe or berry; leaves simple or if compound not more than 3-foliolate:	
2. Leaves simple; stamens numerous (20–60):	[Citrus]
2. Leaves 3-foliolate; stamens 4–10:	
3. Flowers solitary in the leaf-axils, the parts in 3s; plants armed with spines:	[Triphasia]
3. Flowers in terminal and axillary panicles, the parts in 4s or 5s; plants unarmed:	Amyris

Zanthoxylum L.

Trees or shrubs, often more or less prickly; bark aromatic; leaves alternate, even- or odd-pinnate, or rarely 1-foliolate; leaflets opposite or alternate, frequently inequilateral, crenulate or entire, usually glandular-punctate (at least along the margin), and the rhachis often winged. Plants dioecious or polygamous; flowers small, white to greenish-yellow, unisexual and/or bisexual, in axillary short spikes or clusters or in terminal or axillary panicles. Calyx 3–5-cleft, deciduous or persistent, or rarely lacking; petals 3–5(–10). Staminate flowers with 3–5 hypogynous stamens. Pistillate flowers without stamens or with scale-like staminodes; carpels 1–5, distinct or partially united, on an elevated fleshy gynophore, each 1-locular with 2 pendulous ovules in each locule. Fruits dry, of separate 2-valved follicles, each with 1 shining black seed; seeds with fleshy endosperm, at maturity often hanging from the carpels on slender funicles.

— A pantropical genus of about 215 species, a few occurring in temperate regions. It is here considered to include the genus *Fagara* L., with which it intergrades. (See Brizicky, *J. Arnold Arb.* 43: 80–83 (1962).)

KEY TO SPECIES

1. Inflorescence and rhachis of the leaves puberulous with stellate hairs; floral parts in 4s or 5s; leaves densely pellucid-punctate; prickles absent: **Z. flavum**

1. Inflorescence and other parts glabrous; floral parts in 3s; leaves not pellucid-punctate; stems often prickly: **Z. coriaceum**

Zanthoxylum flavum Vahl, *Eclog.* 3: 48 (1807). FIG. 179. PLATE 44.

Fagara flava (Vahl) Krug & Urban (1896); Adams (1972).

SATINWOOD

A small tree up to 7 m tall or more; leaves 10–30 cm long, usually with an odd leaflet; leaflets 5–11, opposite, elliptic-oblong to broadly ovate, 4–9 cm long, acute or somewhat acuminate at the apex, often inequilateral at the base, with lightly crenate margins. Sepals open in bud, stellate-puberulous, 0.5 mm long; petals greenish-white, glandular, 3–4 mm long. Fruiting cocci obovate-roundish, 4–6 mm long.

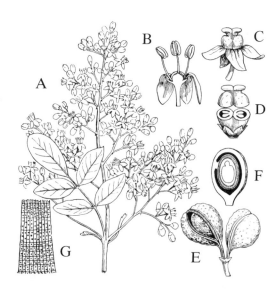

FIG. 179 **Zanthoxylum flavum.** A, inflorescence with a small leaf, × ²⁄₃. B, staminate flower cut lengthwise, × 3. C, pistillate flower, × 3. D, pistil cut across, × 4. E, fruit of 2 cocci, one open showing the seed, × 2. F, coccus cut lengthwise showing the embryo, × 2. G, cross-section of small portion of wood. (F. & R.)

GRAND CAYMAN: *Proctor* 15131. LITTLE CAYMAN: *Proctor* 35115.

— Florida, Bermuda, and the West Indies, often in sandy woodlands. The wood is prised for cabinet-work and furniture, as it has a yellow, satiny luster, a rippled grain, and takes a high polish.

Zanthoxylum coriaceum A. Rich., *Ess. Pl. Cub.* 326, t. 34 (1841). PLATE 45.

SHAKE HAND

A glabrous shrub or small tree to 5 m tall, usually armed at least on young branches with sharp spines. Leaves with or more often without an odd leaflet, 7–15 cm long; leaflets 4–9 or rarely more, opposite and oblong, elliptic or obovate, mostly 3–6 cm long, blunt, rounded or emarginate at the apex, the margins entire, dark green and very shiny on the upper surface and with prominulous venation. Inflorescence a dense corymbose terminal panicle (rarely axillary). Flowers greenish cream; sepals 0.5–0.7 mm long; petals 2.5–4 mm long; ovary of 3 carpels. Fruiting cocci roundish-ellipsoid, rough, 5–6 mm long, minutely apiculate.

GRAND CAYMAN: *Kings* GC 338; *Millspaugh* 1274; *Proctor* 11982. LITTLE CAYMAN: *Proctor* 35105, 35216. CAYMAN BRAC: *Proctor* 29013.

— Florida, Bahamas, Cuba, and Hispaniola, in dry rocky woodlands at low elevations. Adams (1972: 386) ascribed two other species to the Cayman Islands, namely *Zanthoxylum cubensis* P. Wils. and *Z. spinosum* (L.) Sw. Both of these are very similar to *Z. coriaceum*, and these records may be due to misidentifications. *Z. cubense* can be distinguished by having larger leaves, more open panicles, and 1-carpellate flowers. *Z. spinosum*, which is a Jamaican endemic, has thinner leaf-tissue and smaller white flowers with hooded petals. Identifications are often difficult in this group of species.

[*Citrus aurantiifolia* (Christm.) Swingle, the lime; C. **aurantium** L., the sour orange or Seville orange; C. **paradisi** Macf., the grapefruit; C. **sinensis** (L.) Osbeck, the sweet orange, and perhaps other citrus fruits, are quite often planted. All are indigenous to southeastern Asia. Only the first is likely to become naturalised in the Cayman Islands.]

[*Triphasia trifoliata* (Burm.f.) P. Wilson, a Chinese species, was reported by Hitchcock in 1893 as being naturalised on Grand Cayman. This record has not been substantiated by subsequent collectors.]

Amyris L.

Unarmed shrubs or trees, glabrous or pubescent; leaves opposite or alternate, 1–3-foliolate or pinnately compound; leaflets opposite, with numerous pellucid dots in the tissue. Flowers small, white, perfect, in terminal and axillary panicles; pedicels 2-bracteate; floral parts in 4s or 5s; calyx urn-shaped, toothed, persistent; petals imbricate. Stamens twice as many as the petals, inserted on the disc; filaments thread-like. Ovary 1-celled, the style short or lacking, the stigma capitate; ovules 2, pendulous. Fruit a 1-seeded drupe, usually black; seeds without endosperm; cotyledons glandular.

— A tropical American genus of about 18 species.

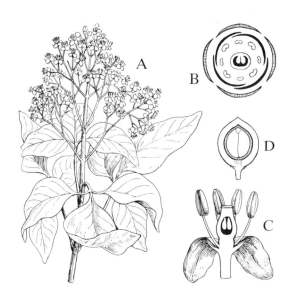

FIG. 180 **Amyris elemifera**. A, branchlet with leaves and staminate inflorescence, × ²/₃. B, diagram of perfect flower. C, perfect flower cut lengthwise, × 6. D, drupe cut lengthwise, × 2. (F. & R.)

Amyris elemifera L., *Syst. Nat.* ed. 10 (2): 1000 (1759). **FIG. 180**. PLATE 45.

CANDLEWOOD, WHITE CANDLEWOOD, TORCHWOOD

A shrub to 3 m tall or more, sometimes tree-like, glabrous throughout; leaves 3–5-foliolate, the leaflets stalked, lanceolate to broadly ovate, 2–6 cm long, acuminate, the margins entire or crenulate, the tissue densely dotted with resinous glands. Drupe globular, 5–8 mm in diameter.

GRAND CAYMAN: *Brunt* 2001; *Kings* GC 342; *Proctor* 15163. LITTLE CAYMAN: *Proctor* 28101. CAYMAN BRAC: *Kings* CB 93; *Proctor* 29014.

— Florida, West Indies and C. America, in rocky woodlands. The wood is hard, heavy, strong and close-grained; it is extremely durable and can take a brilliant polish. Most of the Cayman examples seen are too small to be of much use. Because of its high resin content, *Amyris* wood will burn when it is green.

MELIACEAE

Trees or shrubs, the wood often scented; leaves alternate, usually pinnate, sometimes 2-pinnate, 3-pinnate, or 1–3-foliolate; leaflets entire or rarely serrate; stipules absent. Inflorescence an axillary or terminal panicle; flowers regular, perfect or rarely polygamo-dioecious. Calyx 4–6-lobed, the lobes imbricate; petals 4–6, imbricate, convolute or valvate in bud, free or sometimes adnate to lower part of the stamen-tube. Stamens 8–10 (or 4–6 in *Cedrela*); filaments united to form an entire, toothed, or lobed tube, rarely free; anthers sessile or stalked, inserted within the mouth of the stamen-tube or on its margin, 2-celled. Disc ring-like or columnar, free or adnate to the stamen-tube or ovary. Ovary

2–5-celled, with usually 2 or more ovules in each cell; stigma disc-like or capitate. Fruit a dehiscent capsule, or rarely a drupe or berry; seeds solitary or numerous in each cell, sometimes winged, with or without endosperm.

— A pantropical family of about 45 genera and 800 species, some of them valuable sources of timber.

KEY TO GENERA

1. Leaves pinnate; leaflets entire; fruit a capsule:

    2. Leaflets mostly 3 pairs; seeds not winged, partly enclosed by a red aril; capsules less than 1.5 cm long, not woody: **Trichilia**

    2. Leaflets 4–8 pairs; seeds winged and without aril; capsules more than 3 cm long, woody:

        3. Capsules 8–10 cm long, opening at the base; anthers sessile on the stamen-tube: **Swietenia**

        3. Capsules 3–4 cm long, opening at the apex; anthers on filaments free above their attachment to the column: **Cedrela**

1. Leaves 2–3-pinnate; leaflets serrate; fruit a drupe: **[Melia]**

Trichilia L.

Trees or shrubs; leaves odd-pinnate or rarely 1–3-foliolate; leaflets opposite or alternate, sometimes pellucid-dotted. Panicles axillary or terminal; flowers whitish or yellowish; calyx cup-like, 4–5-lobed; petals 4–5, rarely 3, free or nearly so, imbricate or valvate. Stamens usually 8 or 10; filaments broadly winged and united more or less into a tube; anthers inserted in notches at apex of the filaments, erect and exserted. Ovary 2–3-celled, more or less immersed in a disc, with usually (rarely 1) ovules in each cell. Fruit a 2–3-celled capsule, dehiscent from the apex; seeds 1 or 2 in each cell, subtended by a fleshy red aril; endosperm lacking.

— A rather large genus of about 200 species occurring in tropical America and Africa.

KEY TO SPECIES

1. Panicles corymbose, at ends of the branches; petals 5–8 mm long; capsules tomentose: **T. glabra**

1. Panicles umbel-like, crowded, axillary; petals 3–3.5 mm long; capsules glabrous: **T. havanensis**

Trichilia glabra L., *Syst. Nat.* ed. 10, 2: 1020 (1759). PLATE 45.

A small tree to 8 m tall or more; leaves mostly with 3 pairs of leaflets and a terminal one, these elliptic or ovate, 3–7 cm long, bluntly acuminate at the apex, glabrous or nearly so, a membranous expansion occurring in the axils of the nerves beneath, this often minutely hairy. Inflorescence puberulous, usually 4–10 cm long; ovary pubescent. Capsules globular, greenish, tomentose, 1–1.5 cm long.

GRAND CAYMAN: *Proctor* 11983, 47779. LITTLE CAYMAN: *Proctor* 28028, 35095, 35101. CAYMAN BRAC: *Proctor* 15327, 29019, 47815.

— Jamaica, the Swan Islands and Cozumel, in rocky thickets and woodlands.

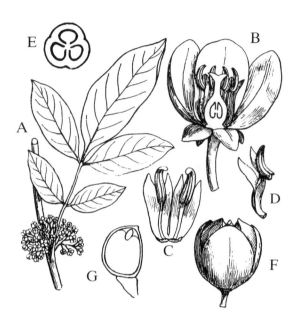

FIG. 181 **Trichilia havanensis**. A, leaf and inflorescence, × ¹/₂. B, flower cut lengthwise, × 7. C, two stamens, × 10. D, stamen, side view, × 10. E, cross-section of ovary, × 14. F, capsule, × 2. G, long.section of seed, × 2. (F. & R.)

Trichilia havanensis Jacq., *Enum. Syst. Pl. Carib.* 20 (1760). **FIG. 181**. PLATE 45.

A shrub or small tree to 5 m tall or more; leaves mostly with 3 pairs of leaflets and a terminal one, these mostly elliptic, 3–10 cm long or more, bluntly acuminate at the apex, glabrous except near the point of attachment, the nerve-axils beneath without a membranous expansion. Inflorescence less than 3 cm long; ovary glabrous. Capsule globular, dark brown, glabrous, ca. 1 cm long.

GRAND CAYMAN: *Brunt* 1935; *Hitchcock*; *Proctor* 49356.

— Cuba, Jamaica, and continental tropical America, in a variety of habitats. Cayman examples were found in moist thickets near George Town.

[*Melia* L.

Trees or shrubs; leaves pinnate or 2–3-pinnate with a terminal leaflet. Panicles axillary, ample, with numerous flowers; flowers perfect, violet or purple. Calyx 5–6-lobed, imbricate in bud; petals 5–6, convolute, spreading. Stamen-tube long and slender, toothed at the apex; anthers 10–12, sessile within the apex of the tube. Disc ring-like. Ovary 5–6-celled with 2 ovules in each cell; style long and slender, the capitate stigma 5–6-lobed. Fruit a somewhat fleshy drupe, the stone 1–6-celled with 1 seed in each cell; seeds with fleshy endosperm.

— A genus of 10 or 12 species indigenous to tropical Asia and Australia.

Melia azedarach L., *Sp. Pl.* 1: 384 (1753).

LILAC

A bushy ornamental shrub, sometimes arborescent, the young growing parts minutely stellate-puberulous; leaves 2-pinnate or nearly 3-pinnate, up to 40 cm long, the pinnae opposite in 2–5 pairs and a terminal one; ultimate leaflets lanceolate, 2–6 cm long, with serrate margins, glabrous. Sepals 1.5–2 mm long, puberulous; petals 7–9 mm long, pale violet or white; stamen-tube ca. 6 mm long, deep purple. Drupe ellipsoid, ca. 1.5 cm long, yellow, the stone deeply grooved.

GRAND CAYMAN: *Kings GC 297*; *Lewis 15*; *Maggs II 57*; *Millspaugh 1354.* CAYMAN BRAC: *Proctor 29124.*

— Native of tropical Asia, but now cultivated and naturalised throughout the tropics. The fruits are said to be poisonous, and together with the dried leaves are used in India to protect clothing etc. from insect attack. The 'seeds' are used to make beads.]

Swietenia L.

Medium-sized to very large trees with dark red wood; leaves even-pinnate, glabrous, the leaflets opposite, entire, and strongly unequal-sided. Panicles axillary and subterminal, much shorter than the leaves; flowers small, white, polygamomonoecious; calyx 5-lobed, the lobes

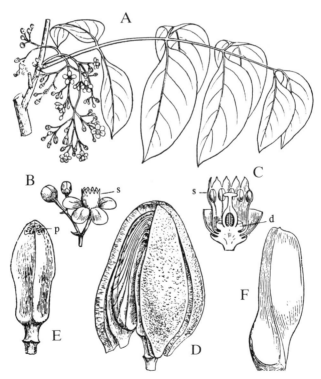

FIG. 182 **Swietenia mahagoni.** A, leaf and inflorescence, × ¹/₂. B, buds and open flower, × 1¹/₂; s, staminal tube. C, perfect flower cut lengthwise, × 5; **s**, staminal tube; **d**, disk. D, fruit, ripe and splitting open (one valve removed), × ¹/₂. E, central axis of fruit, × ¹/₂; p, attachment-points of seeds. F, seed, × ²/₃. (F. & R.)

imbricate; petals 5, convolute in bud. Stamen-tube urn-shaped and with 10 teeth, the anthers attached internally between the teeth. Disc saucer-shaped. Ovary 5-celled with 10–14 ovules in each cell, attached to the central axis; stigma disc-like. Fruit a woody capsule, 5-valved and dehiscent from the base, leaving a persistent 5-winged central axis; seeds numerous, each with a terminal oblong wing, overlapping downward in 2 rows in each cell; endosperm fleshy.

— A tropical American genus of 3 species, all notable for their extremely valuable wood.

Swietenia mahagoni Jacq., *Enum. Syst. Pl. Carib.* 20 (1760). **FIG. 182.** Plate 46.

MAHOGANY

A tree to 10 m tall or more, with a rounded crown; leaves up to 15 cm long, the leaflets mostly in 4 pairs, recurved-lanceolate or -ovate, 3–6 cm long, sharply acuminate. Calyx ca. 1 mm long or less, with rounded glabrous lobes; petals and stamen-tube ca. 3 mm long. Seeds brown, ca. 6 cm long.

GRAND CAYMAN: *Brunt* 1650; *Hitchcock*; *Kings* GC 246; *Proctor* 15135, 27989, 49365. LITTLE CAYMAN: *Kings* LC 5; *Proctor* 28179. CAYMAN BRAC: *Proctor* (sight record).

— Florida and the West Indies only; reports of this species from Central and S. America pertain to *Swietenia macrophylla* G. King. In the Cayman Islands, *S. mahagoni* occurs chiefly in rocky woodlands, where it could formerly be found with massive trunks up to 1 m in diameter or more. Nearly all of these large, very old trees have long since been cut.

Cedrela L.

Large trees with aromatic reddish wood; leaves even-pinnate, the leaflets in many pairs, opposite or subopposite, entire and more or less unequal-sided. Panicles large, terminal; calyx 4–5-lobed; petals 4–5, erect, keeled on the inner side toward the base, the keel adherent to the disc. Disc elevated and forming a column, 4–6-lobed. Stamens 4–6, adherent to the column and extending free beyond it. Ovary sessile at the apex of the column, 5-celled, with 8–12 ovules in each cell, in two series pendent from the axis; stigma disc-like. Fruit a woody capsule, 5-valved and dehiscent apically almost to the base, separating from a persistent 5-winged axis; seeds pendulous, each with a terminal wing, overlapping downward in 2 rows in each cell; endosperm scanty.

— A tropical American genus of 15 species.

Cedrela odorata L., *Syst. Nat.* ed. 10 (2): 940 (1759). **FIG. 183.** Plate 46.

Cedrela mexicana M. J. Roem. (1846).

CEDAR

A tree up to 18 m tall or more, with a long straight trunk; leaves deciduous, renewed at time of flowering, usually 30–50 cm long, the leaflets in 7 or 8 pairs, narrowly oblong-ovate, 5–11 cm long or more, more or less long-acuminate. Panicles often 30–35 cm long, open and lax; flowers unpleasantly scented, greenish-cream; calyx ca. 1.5 mm long; petals ca. 6 mm long, densely puberulous. Seeds light brown, 2 cm long.

GRAND CAYMAN: *Kings* GC 337; *Proctor* 31033. CAYMAN BRAC: *Proctor* 29368.

— West Indies and continental tropical America, in various habitats. The rather soft but durable, handsome wood is used for many purposes, and is one of the most valuable of all tropical timbers. It is virtually impervious to attacks by insects.

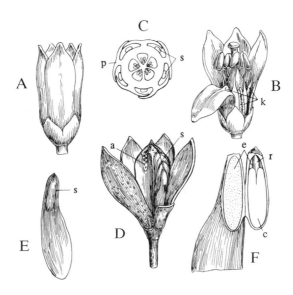

FIG. 183 **Cedrela odorata**. A, flower, × 4. B, same with one petal removed; **k**, keel of petal attached to the ovary. C, same cut across, enlarged; **s**, filaments; **p**, petal. D, capsule with one valve cut away, × ²/₃; **s**, seeds; **a**, placenta with seeds removed. E, seed, × 1; **s**, position of embryo. F, seed cut lengthwise, × 2; **c**, cotyledon; **e**, endosperm; **r**, radicle. (F. & R.)

ZYGOPHYLLACEAE

Shrubs, trees, or annual or perennial herbs, frequently more or less resinous; leaves opposite, or alternate by abortion of one of each pair, petiolate, even-pinnate or in some cases simple or digitately compound; stipules small or minute. Flowers solitary or clustered in the leaf-axils, regular or nearly so, perfect; sepals usually 5 and free, usually imbricate; petals 5, free, imbricate, valvate or convolute. Stamens 10, free, in two series, the outer ones larger than the inner; anthers opening inwardly. Ovary of 2–5 united carpels, sessile or short-stalked, the styles united into one, terminated by a compound stigma; ovules 1–several in each carpel, pendulous or ascending. Fruit a capsule, or else separating into few or several often spiny nutlets, these 1–few-seeded; seeds with or without endosperm, the cotyledons fleshy.

— A widespread tropical and subtropical family of about 20 genera and 200 species.

KEY TO GENERA

1. Creeping or decumbent herbs:

    2. Leaflets mostly in 6–9 pairs; sepals deciduous; fruit of 5 spiny nutlets: **Tribulus**

    2. Leaflets mostly in 3–4 pairs; sepals persistent; fruit of 10–12 small tuberculate nutlets surrounding the persistent beak-like style-axis: **Kallstroemia**

1. Erect evergreen tree: **Guaiacum**

Tribulus L.

Prostrate annual or perennial herbs; leaves opposite, one of each pair alternately smaller than the other or sometimes lacking; stipules obliquely lanceolate. Flowers solitary on long axillary peduncles; sepals soon falling; petals 5, yellow or orange, much larger than the sepals. Stamens 10, the 5 inner ones shorter and glandular at the base. Ovary sessile, 5-celled, surrounded at the base by an urn-shaped 10-lobed disc; ovary-cavities divided by transverse partitions into 3–5 compartments with 1 ovule in each. Fruit depressed, 5-angled, consisting of 5 bony nutlets which ultimately separate, each bearing spines, and each containing 3–5 seeds; seeds without endosperm.

— A pantropical genus of about 10 species.

Tribulus cistoides L., *Sp. Pl.* 1: 387 (1753). **FIG. 184.** PLATE 47.

BUTTERCUP, JIM CARTER WEED

Herb with prostrate stems up to 50 cm long or more, densely pubescent; leaves white-pubescent beneath; leaflets oblong, 4–17 mm long, unequal-sided; stipules 4–9 mm long. Peduncles arising from axils of the shorter leaves and longer than them; sepals 7–9 mm long; petals yellow, obovate, 1.5–2.5 cm long. Fruit 6–9 mm long, each nutlet with 2 longer spines at the top and 2 shorter ones near the base.

GRAND CAYMAN: *Brunt* 2008; *Kings GC 225*; *Proctor* 15157. LITTLE CAYMAN: *Kings* LC 57; *Proctor* 28172. CAYMAN BRAC: *Proctor* 29120.

— Southern U. S. A. south to Guatemala, West Indies and S. America, usually in open sandy ground at low elevations.

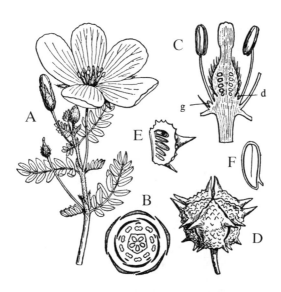

FIG. 184 **Tribulus cistoides.** A, flowering branch, × ²/₃. B, floral diagram. C, flower cut lengthwise, calyx and corolla removed, × 4, showing pistil with hypogynous disk (d), two stamens, and one staminal gland (g). D, fruit somewhat enlarged. E, coccus cut lengthwise, enlarged. F, seed cut lengthwise, enlarged. (F. & R.)

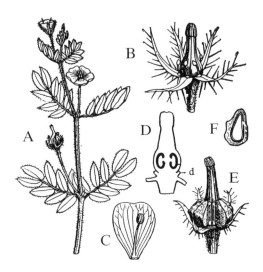

FIG. 185 **Kallstroemia maxima**. A, flowering branch, × ²/₃. B, flower with the petals and five larger stamens adhering to them removed, × 5. C, petal with its adhering stamen, × 2. D, pistil cut lengthwise, × 5; d, hypogynous disk. E, fruit with calyx, two cocci detached, × 3. F, coccus cut lengthwise, × 2. (F. & R.)

Kallstroemia Scop.

Prostrate or ascending, annual or perennial herbs; leaves opposite, one of a pair often alternately smaller than the other or sometimes lacking; stipules linear. Flowers solitary on axillary peduncles; sepals 5–6, persistent; petals 4–6, orange or yellow, soon falling; stamens 10–12, of unequal size, the smaller ones glandular at the base. Ovary sessile, 10–12-celled, with 1 pendulous ovule in each cell; styles united into a beak-like column persistent in fruit. Fruit of 10–12 hard, tuberculate, indehiscent, 1-seeded nutlets, at maturity these falling away from the persistent central axis; seeds without endosperm.

— A tropical American genus of 12 or more species.

Kallstroemia maxima (L.) Torr. & Gray, *Fl. N. Amer.* 1: 213 (1838). **FIG. 185.**

PARSLEY

A trailing herb, the branched stems up to 50 cm long or more, sparsely pubescent; leaves glabrate; leaflets elliptic or oblong-elliptic, somewhat unequal in size, the apical ones largest, 1–2 cm long, unequal-sided; stipules 4–5 mm long. Peduncles shorter or longer than the leaves; sepals 4–6 mm long; petals cream or pale yellowish, 6–8 mm long. Fruit 8–9 mm long, the nutlets glabrous, 4–5 mm long, tuberculate-ridged.

GRAND CAYMAN: *Brunt* 2113; *Kings* GC 403; *Millspaugh* 1342; *Proctor* 27939.

— Southern U. S. A., West Indies and continental tropical America, a common weed of roadsides, fields and open waste ground.

[*Guaiacum* L.

Trees with hard, heavy, resinous wood and smooth bark; stipules minute, soon falling. Leaflets 2 or 3 pairs, entire, inequaelateral at least at base. Flowers pedunculate in distal leaf-axils; sepals 4 or 5, broadly ovate, unequal; petals 4 or 5, obovate, blue or purple (rarely white), slightly clawed. Stamens 8–10 on an inconspicuous disc, with filiform filaments. Ovary short-stalked, 2–5-carpellate with 2–5 lobes and locules; ovules 8–10 in each locule. Fruit a 2–5 winged capsule, septicidal; seed solitary, enclosed in a fleshy aril.

— A neotropical genus of 6 species.

Guaiacum officinale L., *Sp. Pl.* 1: 381 (1753).

LIGNUM VITAE

Small dense-crowned tree with mottled smooth bark, up to 8 m tall. Leaflets mostly 2 pairs, sessile, obovate-elliptic, usually 1.5–3.5 cm long and about 2.5 cm broad. Peduncles 1-flowered, few or clustered in distal leaf-axils. Sepals tomentose, 4–5 mm long; petals puberulous, ca. 10 mm long; stamens 10. Capsules 1.5–2 cm long with orange arils.

GRAND CAYMAN *van B. Stafford* s.n., found growing wild in forest behind the University College of the Cayman Islands south of George Town. This handsome tree is occasionally cultivated in the Cayman Islands but has not been previously reported as a wild plant. It is a common species in dry woodlands of other West Indian Islands.

— Bahamas, Greater and Lesser Antilles, C. America, Venezuela and Colombia; introduced in the Old World tropics.]

OXALIDACEAE

Herbs with fibrous roots or sometimes aggregates of scaly bulbs, shrubs or trees. Leaves alternate, pinnately or digitately compound, or simple by suppression of leaflets; stipules present or absent. Flowers mostly bisexual and actinomorphic, solitary cymose-umbellate, racemose, or paniculate; sepals 5, free or connate, imbricate; petals 5, free or united, or basally free and connate above, contorted in bud, rarely absent in cleistogamous flowers. Stamens 10 or sometimes 5 reduced to staminodes and petaloid, basally connate in 2 unequal series, the shorter ones opposite the petals; anthers 2-loculate, opening lengthwise. Ovary 5-locular with one or more ovules in each locule; styles 5, free, the stigmas capitate. Fruit a loculicidal capsule or a berry. Seeds sometimes arillate; endosperm fleshy.

— A family of 7 genera and about 900 species, most of them in the genus *Oxalis*.

Oxalis L.

Characters of the family; herbs or subshrubs with fibrous roots, rootstocks, or scaly bulbs. Leaves 3-foliate, the leaflets obcordate. Flowers often heterostylous, with petals yellow, pink, purple or rarely white. Fruit always capsular; seeds flattened, longitudinally grooved or transversely ridged or foveolate.

Oxalis corniculata L., *Sp. Pl.* 1: 435 (1753).

Fibrous-rooted annual herb with erect flowering branches and spreading or decumbent rooted branches. Leaves digitately 3-foliate with petioles up to 4 cm long, pilose, the leaflets obcordate, 8–18 mm long and broad, puberulous, the margins ciliate, inflorescence with peduncles 2–4.5 cm long; pedicels to 1.5 cm long; sepals 3–4 mm long; petals yellow, 7–8 mm long. Capsules oblong, 9–12 mm long, beaked. Seeds 1.3 mm long, brownish-red.

GRAND CAYMAN: *Burton* s.n., found on the grounds of MRCU; *Proctor 48680*, found on Arthur Hunter estate near Bats Cave.

— A pantropical weed extending into temperate regions.

ERYTHROXYLACEAE

Glabrous shrubs or trees; leaves alternate, simple, and entire; stipules present, often overlapping on young branches. Flowers small, solitary or clustered in the leaf-axils or at defoliated nodes, regular, perfect, often heterostylous; pedicels bracteate and usually angled. Calyx with 5 imbricate lobes, persistent; petals 5, free, deciduous, usually with a 2-lobed appendage (the ligule) on the inner side. Stamens 10 in 2 series, their filaments united below into a tube. Ovary superior, 3-celled, mostly with 2 of the cells sterile, the fertile cell with 1 (rarely 2) pendulous ovule; styles 3, usually each ending in a small capitate stigma. Fruit a drupe, usually 1-seeded; seed with or without farinaceous endosperm.

— A pantropical family of 2 genera and over 200 species, all but one in the genus *Erythroxylum*.

Erythroxylum P. Browne

With characters of the family. The best-known member of this genus (and family) is *Erythroxylum coca* Lam. of the Andean region in S. America, from whose leaves is obtained the drug cocaine.

KEY TO SPECIES

1. Leaves beneath usually with a distinct often paler central area bounded by 2 longitudinal lines; pedicels (at least in fruit) up to 8 mm long:	E. areolatum
1. Leaves beneath without a distinct contrasting area; pedicels less than 4 mm long:	
2. Leaf-blades usually less than 2 cm long, noticeably pale beneath; fruits 5–6 mm long:	E. rotundifolium
2. Leaf-blades 3–5 cm long, not paler beneath; fruits 8–9 mm long:	E. confusum

Erythroxylum areolatum L., *Syst. Nat.* ed. 10, 2: 1035 (1759). **FIG. 186.** PLATE 47.

SMOKE WOOD

A shrub or small tree to 8 m tall; leaves broadly elliptic, the apex often shallowly notched (retuse); venation finely reticulate and prominulous, especially on the upper side; deciduous, the flowers appearing with the new leaves in April and May. Flowers white, numerous, fragrant, heterostylous; pedicels up to 7 mm long, enlarged toward the apex. Calyx 1.5–2 mm long; petals 2.5–3 mm long. Drupes red, subtended by the persistent staminal tube and calyx.

503

FIG. 186 **Erythroxylum areolatum.** A, habit, flowering branch. B, fruiting branch. C, portion of flowering branch. D, flower. E, stamen. F, ovary, styles and capitate stigmas. G, fruit. **Erythroxylum rotundifolium.** H, leaf. I, portion of flowering branch (buds only). J, flower. K, flower with petals removed. L, sepal. M, ovary, styles and stigmas. N, fruit. (G.)

GRAND CAYMAN: *Brunt* 1754, 2000, 2153; *Kings* GC 204a, GC 333; *Proctor* 11984, 15296, 48242, 49317, 49344. LITTLE CAYMAN: *Proctor* 35092. CAYMAN BRAC: *Proctor* 49327.

— Bahamas, Greater Antilles, Mexico and northern C. America, in rocky woodlands. The wood is reddish-brown, hard, heavy, fine-textured and very durable.

Erythroxylum rotundifolium Lunan, *Hort. Jam.* 2: 116 (1814). **FIG. 186**. PLATE 47.

A shrub to 3 m tall or more; leaves thin, obovate to roundish, the apex often shallowly notched, the venation prominulous on upper side but not as finely reticulate as in *E. areolatum*, and apparently not deciduous all at one time, the flowers appearing among mature leaves at various seasons but especially in July. Flowers white, solitary or few together; pedicels usually not over 4 mm long; calyx ca. 1 mm long; petals 1.5–2.5 mm long. Drupes red.

GRAND CAYMAN: *Brunt* 2185; *Correll & Correll* 51029; *Proctor* 15175, 15205, 49366. LITTLE CAYMAN: *Kings* LC 63; *Proctor* 28106, 28187, 47300, 48228. CAYMAN BRAC: *Proctor* 29022.

— Bahamas, Greater Antilles and Yucatan, in rocky thickets and woodlands. None of the Cayman species of *Erythroxylum* contains any cocaine.

Erythroxylum confusum Britton in Britton & Wilson, *Bahama Fl.* 190 (1920). PLATE 47.

Shrub or small tree with trunk up to 10 cm in diameter and 8 m tall, deciduous. Leaves alternate; petioles 4–9 mm long; blades obovate to oblong-elliptic, rounded at apex, narrowed or cuneate at base, dull green on upper side, somewhat lighter beneath, neither prominently reticulate-veined nor areolate. Flowers appearing with the new leaves, solitary or fascicled in leaf axils, the pedicels much shorter than the fruits and rarely as long as the petioles. Calyx ca. 2 mm long. Stamens 8–10; styles 2 or 3. Drupe oblong, scarlet, 8–9 mm long.

GRAND CAYMAN: *Proctor* 48669, 49319, 49344, 50748.

— Bahamas, Cuba, and Jamaica.

MALPIGHIACEAE

Mostly trees or shrubs, often scandent, the pubescence most often of 'malpighiaceous' hairs, i.e. hairs attached at or near the middle, thus 2-armed or flattened T-shaped; these hairs in some species stiff and sharp, easily detached, and very irritating to the skin; leaves usually opposite and simple, often glandular either on the petiole, near the midrib, or on the leaf-margins; stipules present. Flowers mostly perfect and regular, in axillary or terminal racemes or panicles, rarely solitary; calyx 5-lobed, often bearing large, sessile glands; petals 5, usually concave, narrowed below into a claw. Stamens 5 or 10, the filaments more or less united at the base. Ovary superior, of 3 free carpels or these more or less united into a 3-celled ovary with 1 ovule in each cell; styles 3, free or united into a single style with 3-lobed stigma. Fruit a drupe containing one 3-celled stone or 1–3 separate 1-celled stones, or else a capsule or samara.

— A pantropical family of perhaps 60 genera and about 850 species, best represented in the American tropics. Several S. American species of this family are known to contain alkaloids with hallucinogenic properties.

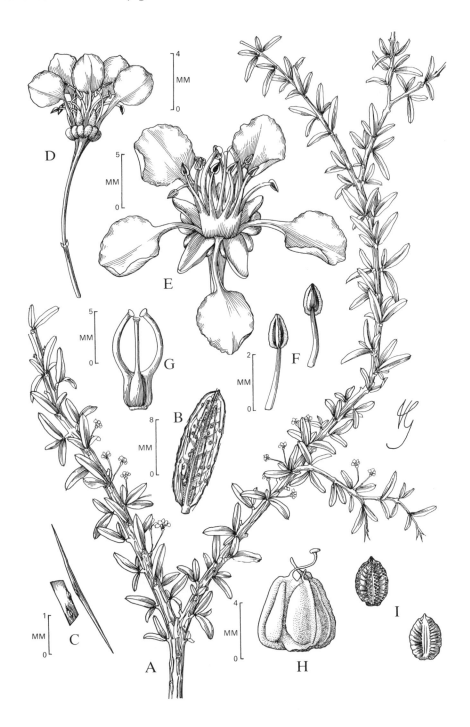

FIG. 187 **Malpighia cubensis**. A, habit. B, underside of single leaf, showing 2-armed hairs and scars where others have broken off. C, a single 2-armed hair and its point of attachment. D, habit of flower; the portion of stalk above the bracteoles is the pedicel. E, flower enlarged. F, two stamens. G, ovary, style and stigmas. H, fruit. I, two views of stone. (G.)

KEY TO SPECIES

1. Flowers pink or white; styles free; drupe containing crested stones; plant (in our
   single species) densely clothed with stinging hairs: **Malpighia**

1. Flowers yellow; styles united into 1; drupe containing smooth stones; plant without
   stinging hairs: **Bunchosia**

Malpighia L.

Erect shrubs or small trees; leaves without glands, sometimes with stinging hairs, the margins entire or toothed; stipules minute, deciduous. Flowers in axillary or terminal umbellate clusters or corymbs, rarely solitary; calyx with 6–10 glands; petals pink, red or white, unequal, often fringed or ciliate at the apex. Filaments and ovary glabrous; filaments and styles free. Ovary 3-lobed. Fruit a red or orange 3-stoned drupe, each stone with 3–5 dorsal crests.

— A tropical American genus of about 45 species.

Malpighia cubensis Kunth in H. B. K., *Nov. Gen. & Sp.* 5: 145 (1821). **FIG. 187.** PLATES 47 & 48.

Malpighia angustifolia of Hitchcock (1893) not L. (1762).

LADY HAIR

A shrub to 2 m tall; leaves narrowly oblong, 1.5–2.5 cm long or more, obtuse or acute, clothed on both sides with numerous yellowish, stiff, stinging malpighiaceous hairs. Flowers 1 or few in a glabrous umbellate cluster, the pedicels 7–8 mm long, enlarged toward the apex; calyx 2.5–3 mm long, completely covered by the glands; petals white or pale pink, 6–7 mm long, long-clawed. Drupes subglobose, 8–9 mm in diameter.

GRAND CAYMAN: *Brunt* 2073; *Hitchcock*; *Howard & Wagenknecht* 15025; *Kings* GC 113; *Proctor* 15045, 48254, 49362, 52068.

— Cuba, in rocky or sandy thickets.

[*Malpighia emarginata* Sessé & Moç. ex DC. (often called *M. punicifolia*) the West Indian 'Cherry', occurs as a cultivated plant. It is a large, glabrous shrub with pink flowers. The red, edible fruits are one of the richest known sources of vitamin C.]

Bunchosia L. C. Rich.

Shrubs or small trees, glabrate or pubescent; leaves usually with 2 glands on the lower surface a little above the base near the midrib; stipules minute, deciduous. Flowers in axillary racemes; calyx with 8–10 glands; petals yellow. Filaments glabrous, united at the base. Ovary 2–3-celled; styles united, terminating in a triangular stigma. Fruit a red, 2–3-stoned drupe, the stones smooth and uncrested.

— A tropical American genus of about 40 species.

Bunchosia media (Dryand. ex Aiton) DC., *Prodr.* 1: 581 (1824). PLATE 48.

Bunchosia swartziana Griseb. (1859).

A shrub to 4 m tall or more, rarely a small tree; leaves glabrous or nearly so, elliptic, 3.5–6 cm long or more, bluntly acuminate. Racemes usually shorter than the leaves;

pedicels puberulous; calyx ca. 2 mm long; petals 7–9 mm long, long-clawed. Drupes somewhat ellipsoid and 2–3-lobed, 8 mm long, orange.

GRAND CAYMAN: *Brunt* 2183; *Proctor* 49363.

— Cuba, Jamaica and Hispaniola, in dry rocky woodlands.

POLYGALACEAE

Herbs, shrubs, or trees, sometimes woody vines, often with glands in the leaf-tissue, flowers and fruits; leaves alternate, opposite or whorled, simple and entire, with very short petioles; stipules absent. Flowers perfect, irregular (zygomorphic), in racemes or panicles; sepals 5, free or the lower 2 united, the two lateral ones much larger than the others and petaloid, forming the 'wings'; petals 3 or rarely 5, the anterior one forming a boat-shaped 'keel'. Stamens usually 8, the filaments united for most of their length into a 'sheath', this split on the upper side and usually united at the base with the keel or the upper petals or both; anthers mostly confluently 1-celled, opening by a subterminal pore. Ovary superior, usually 1–2-celled with 1 style and a 2-lobed often fringed stigma; ovules solitary or rarely 2–6, pendulous. Fruit a capsule, drupe or samara; seeds usually pubescent, with an aril and with endosperm.

— A cosmopolitan family of 12 genera and about 800 species.

Polygala L.

Erect herbs, shrubs, or trees; leaves alternate, opposite, or whorled. Flowers usually small and white, pink, or purple, in racemes or rarely umbellate; petals 2 upper ones and a boat-shaped keel, the latter clawed, sometimes 3-lobed and usually with an apical beak or crest. Ovary 2-celled with solitary ovules; style slender, bent, more or less excavate at the apex, the stigma 2-lobed. Fruit an equally or unequally 2-celled capsule, often winged or marginate, compressed contrary to the partition, splitting at the margin.

— A very widely distributed variable genus of more than 500 species.

Polygala propinqua (Britton) Blake, *Contr. Gray Herb.* n.s. 47: 16, t. 1, fig. 5 (1916). PLATE 48.

A shrub to 2.5 m tall or more, with pale, brittle, densely puberulous branchlets; leaves broadly elliptic or oval, 2–3.5 cm long, glabrate above, minutely punctate-strigillose beneath. Inflorescences axillary, very short, few-flowered, puberulous; flowers whitish or greenish-white; sepals 1–1.3 mm long, the wings up to 1.5 mm long; keel ca. 3 mm long. Capsule minutely pubescent or subglabrous, transversely oblong, lobed about 1/3 its length, the lobes divergent, 7–7.5 mm long, 9.5–11 mm wide; seeds 5.3 mm long.

LITTLE CAYMAN: *Proctor* 28104, 35087, 47299.

— Cuba, in rocky or sandy thickets or woodlands.

Note: Guala *et al.* in *Kew Bull.* **57**: 237 (2002) identified a Cayman shrubby *Polygala* (ascribed in the first edition of this flora to *P. propinqua* (Britton) Blake) as *P. penaea* L., a species known from Hispaniola and Puerto Rico. Furthermore, they cited material from all three Cayman Islands; I have not seen most of these specimens. Nevertheless, I stand by my original identification of the Little Cayman material as *P. propinqua*, and believe (on phytogeographic grounds) that the presence of *P. penaea* in these islands is doubtful.

ARALIACEAE

Trees, shrubs, or rarely herbs, sometimes climbing. Leaves mostly alternate, simple or compound, sometimes with stellate hairs or pellucid-dotted; stipules usually present, often adnate to the dilated petiole-base. Flowers bisexual or often unisexual, the plants then polygamous or dioecious, actinomorphic, in heads, umbels or racemes. Perianth biseriate, the parts mostly in 5s. Calyx entire or toothed. Corolla of usually valvate, hook-tipped free or sometimes united petals, these soon falling. Glandular, epigynous disc present, usually confluent with the style-bases. Stamens usually as many as the petals and alternate with them, the filaments inserted on the disc; anthers 2-locular, opening lengthwise. Ovary inferior, with 1 or more locules; styles as many as the carpels, free or connate, or the stigmas sessile; ovules solitary in each locule. Fruit a berry or drupe, or the carpels separating as distinct pyrenes. Seeds with copious endosperm and small embryo.

— A mostly tropical family of about 50 genera and 1,200 species, chiefly woody.

Dendropanax Decne. & Planch.

Shrubs or trees, mostly glabrous. Leaves simple petiolate, entire or sometimes lobate when juvenile; petioles unequal in length, sometimes geniculate; stipules minute or absent. Inflorescence simple or branched umbels, sometimes many umbels in racemes. Calyx obconical or cup-shaped, the short limb denticulate; petals usually 5, free, cucullate, valvate, usually fleshy and greenish-white. Stamens usually 5, inflexed in bud; disc fleshy, short-conical, confluent with the styles; styles mostly free or more usually connate to the middle or beyond; ovary thick-walled, the locules and ovules 5–9. Fruit subglobose or ellipsoid, usually 5-grooved when dry, the exocarp fleshy, black or dark purple, the seeds compressed or trigonous.

— A genus of about 80 species in tropical America and eastern Asia.

Dendropanax arboreus (L.) Decne. & Planch., *Rev. Hort.* 4 (3): 107 (1854). PLATE 48.

Small glabrous tree to 10 m tall or more; bark smooth, buff-grey with white slash, exuding small drops of brownish-orange sap. Leaves faintly aromatic when crushed, mostly elliptic to ovate-elliptic, up to 15 cm long or more and to 10 cm broad, acuminate at apex, cuneate to rounded at base, glossy on upper surface. Inflorescence terminal; peduncle 1.5–8 cm long; pedicels 6–8 mm long; flowers green turning yellowish; petals 4–6. Fruits 5–6-angled when dry, 5–7 mm long.

GRAND CAYMAN: *Proctor* 47266, found along the Mastic Trail; and *Proctor* 47864, from forest behind the University College of the Cayman Islands S. of George Town.

— Greater Antilles and the continental neotropics; absent from the Bahamas and the Lesser Antilles.

UMBELLIFERAE

Annual or perennial herbs with hollow stems, rarely subwoody or arborescent; leaves alternate or rarely opposite, the petiole usually dilated at the base into a sheath; blades simple or usually compound, often more or less finely dissected; stipules usually lacking. Inflorescence umbellate or sometimes head-like; umbels simple or more often compound, terminal or lateral, sometimes numerous in a panicle, the individual umbellate clusters often subtended by an involucre of bracts. Flowers small, regular or sometimes irregular, perfect or often polygamomonoecious, protandrous (staminate elements maturing before the pistillate); calyx represented by 5 small teeth around the upper edge of the ovary, or these absent; petals 5, inserted at the apex of the ovary, equal or the outer ones enlarged. Stamens 5, free, the filaments inflexed in bud. Ovary inferior, 2-locular, crowned by a conspicuous disc; styles 2, distinct; ovules solitary in each locule, pendulous, anatropous. Fruit dry, crowned by the persistent disc and styles, and consisting of 2 indehiscent 1-seeded 'mericarps', these eventually separating; mericarps bearing longitudinal ribs sometimes extended into wings, with oil-tubes between or under them, rarely lacking; seeds with cartilaginous endosperm and small embryo.

A world-wide family of more than 200 genera and 2,850 species, best represented in temperate regions and on tropical mountains. The family includes numerous important cultivated plants, chiefly used for food or condiment. Among the best-known are celery (*Apium graveolens* L.), carrot (*Daucus carota* L.), parsley (*Petroselinum crispum* (Mill.) Nyman), and anise (*Pimpinella anisum* L.). Some other members of this family are poisonous, among the most notorious being *Conium maculatum*, the poison hemlock, famous in the history of Greece and other Mediterranean countries; it is supposed to be the plant by which Socrates was put to death.

KEY TO GENERA

1. Leaves and umbels simple:	Centella
1. Leaves and umbels compound:	[Anethum]

Centella L.

Perennial herbs with prostrate, rooting stems bearing clusters of long-petiolate leaves at the nodes; leaves with simple blades, palmately veined, cordate at the base. Inflorescence of simple, subcapitate, few-flowered umbels on long, axillary peduncles; subtending bracts 2; calyx-teeth absent; styles short, hair-like. Fruits laterally compressed, prominently ribbed; oil-tubes lacking.

— A widespread genus of about 20 species, the majority in Africa.

Centella asiatica (L.) Urban in Mart., *Fl. Bras.* 11, pt. 1: 287 (1879). **FIG. 188.**

Hydrocotyle asiatica L. (1753).
Centella erecta (L.f.) Fernald (1944).

A creeping, glabrous herb with somewhat fleshy stems and leaves; leaves with petioles up to 22 cm long or more, usually much longer than the blades; blades broadly ovate to orbicular, mostly 2–5 cm long, rounded at the apex, shallowly cordate at the base, the margins repand-dentate. Peduncles shorter than the leaves; umbels head-like, 2–4-flowered, the flowers sessile or subsessile. Fruits 3–4 mm long and slightly broader, ribbed and reticulate.

GRAND CAYMAN: *Brunt* 1952, 1976; *Proctor* 27978.

— Pantropical, usually in swampy meadows or areas of impeded drainage.

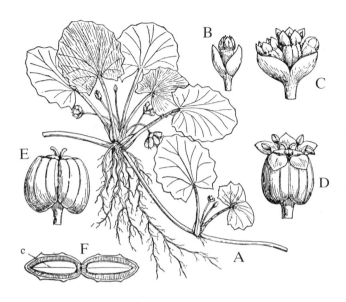

FIG. 188 **Centella asiatica.** A, portion of plant bearing flowers and fruit, × 1. B, young inflorescence, × 2¹/₂. C, inflorescence with lateral flowers developed, × 2¹/₂. D, flower, × 4. E, fruit, × 4. F, cross-section of fruit, × 7; c, cotyledons. (F. & R.)

[*Anethum graveolens* L., dill, has been recorded as cultivated and possibly escaped near George Town, Grand Cayman (*Kings* GC 203, GC 207). Both the foliage and the fruits of this species are much used for flavouring food, especially in the U. S. A. (e.g. dill pickles). It cannot be expected to thrive in the climate of the Cayman Islands.]

LOGANIACEAE

Herbs, shrubs, vines or trees; leaves opposite or rarely whorled, simple, entire or toothed; stipules present or the leaf-bases connected by a transverse line. Flowers regular and perfect, usually arranged in dichotomous cymes or in panicles; calyx 4–5-parted, usually short, with imbricate segments; corolla gamopetalous, 5-lobed. Stamens 5, inserted in the corolla-tube and alternate with the lobes; filaments short; anthers 2-celled, opening inwardly. Ovary superior, usually 2-celled; ovules usually many in a cell, attached to the axis; style simple or forked. Fruit a 2-celled, many-seeded capsule, rarely berry-like or drupe-like and indehiscent; seeds variable, sometimes winged; endosperm usually copious.

— A rather heterogeneous family of about 32 genera and nearly 800 species, widespread in tropical and warm-temperate regions. Some authors divide it into 6 smaller families. Among well-known plants of this complex are species of *Buddleja*, widely planted for ornament, and the numerous species of *Strychnos*, whose seeds yield the poisonous alkaloid strychnine.

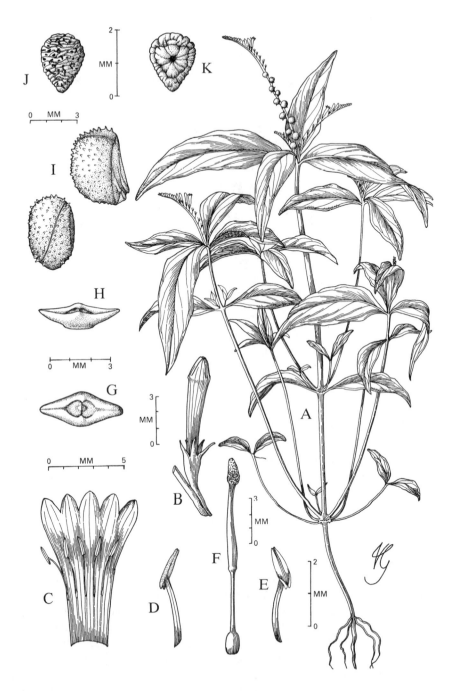

FIG. 189 **Spigelia anthelmia**. A, habit. B, flower-bud. C, corolla opened out to show stamens. D, E, two views of stamen. F, ovary, style and stigma. G, H, two views of persistent base of fruit. I, two views of fruit. J, K, two views of seed (G.)

Spigelia L.

Annual or perennial herbs, glabrous or pubescent, the stems terete or 4-angled. Leaves entire, decussate or in whorls at the top of the stems. Inflorescence usually terminal, consisting of 1–several 1-sided spikes or sometimes short spikes situated in the forks of branches, the individual flowers sessile or nearly so. Calyx 5-parted, with 2 or more linear glands at the inner base of each lobe; corolla funnel-shaped. Style filiform, simple, jointed near the middle, the upper portion quickly deciduous. Capsule 2-lobed, circumscissile above the persistent, cup-shaped base; seeds few, top-shaped, ellipsoid or ovoid, variously sculptured.

— An American genus of about 50 species.

Spigelia anthelmia L., *Sp. Pl.* 1: 149 (1753). **FIG. 189**.

A small erect annual herb, simple or few-branched, seldom more than 30 cm tall; lower leaves opposite, the upper ones larger and in a whorl of 4 (actually 2 closely decussate pairs) subtending the inflorescence, 3–9 cm long or more, minutely scabrid on the upper side and margins. Spikes 1–several in a terminal cluster, each spike usually 3–10 cm long, with 10–20 flowers, rarely less. Calyx-lobes 2–3 mm long; corolla usually pale pink, 5–10 mm long. Capsule 3–5 mm long, 4–6 mm broad, finely muricate; seeds 12–15 per capsule.

GRAND CAYMAN: *Brunt* 1940, 1949; *Kings* GC 86. CAYMAN BRAC: *Proctor* 28949.

— Florida, West Indies and continental tropical America; naturalised in tropical Africa and Indonesia, a weed of fields and open waste ground. The cut and wilted plant is extremely poisonous and may cause fatalities among livestock; the fresh plant is apparently less dangerous.

GENTIANACEAE

Annual or perennial herbs or small shrubs, glabrous throughout, the sap usually bitter. Leaves opposite and entire, often connate at the base or connected by a transverse line; true stipules lacking; rarely, the leaves reduced to scales and the whole plant lacking chlorophyll. Flowers regular and perfect, usually in a dichasial cyme; calyx 4–6-toothed or -lobed, imbricate or open in bud; corolla gamopetalous, 4–6-lobed, contorted in bud and persistent around the fruit. Stamens as many as the corolla-lobes and alternate with them, inserted in the throat or tube of the corolla. Ovary superior, usually 1-celled, with 2 parietal placentas each bearing numerous ovules; style simple. Fruit usually a 2-valved capsule, many-seeded; seeds various; endosperm usually copious.

— A world-wide family of about 80 genera and 900 species.

KEY TO GENERA

1. Green-leafed herbs with violet-blue flowers 2–3.5 cm long:	Eustoma
1. Small saprophytic herbs without chlorophyll and with minute, scale-like leaves; flowers white or creamy, less than 5 mm long:	Voyria

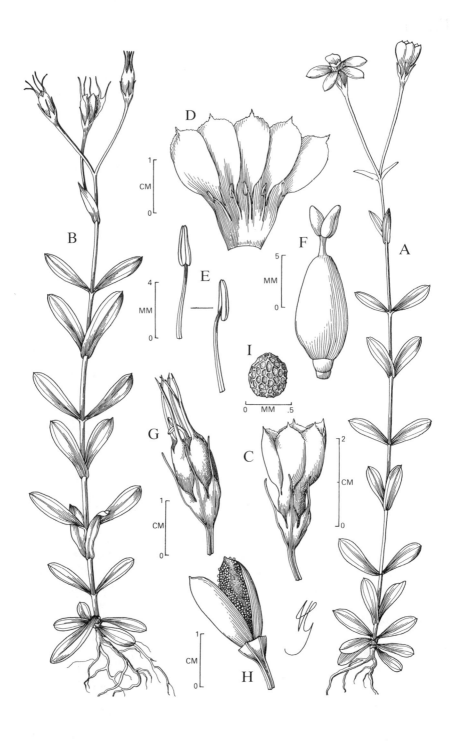

FIG. 190 **Eustoma exaltatum**. A, habit with flowers. B, habit with fruits. C, flower. D, corolla opened out. E, two stamens. F, ovary, style and 2-lamellate stigma. G, fruit. H, capsule after dehiscence. I, seed. (G.)

Eustoma Salisb.

Erect annual herbs; leaves sessile and often clasping. Flowers few, rather large, long-stalked, and blue, purplish or white; calyx deeply 5–6-parted, the segments narrow and keeled; corolla bell-shaped with short tube, the limb deeply 5–6-lobed. Stamens with hair-like filaments and oblong, versatile anthers. Style persistent in fruit, the stigma 2-lamellate. Seeds small, very numerous, foveolate.

— A small American genus of 3 species.

Eustoma exaltatum (L.) Salisb., *Parad. Lond.*, t. 34 (1806). FIG. 190. PLATE 48.

Erect herb, more or less glaucous throughout, usually 50 cm tall or less, simple or sparingly branched; stems terete; leaves oblong or oblanceolate, 2–6 cm long, rounded or obtuse and apiculate at the apex. Flowers light violet-blue, a deep violet blotch at the base within, 2–3.5 cm long; calyx-segments lance-linear, 1–1.5 cm long, united only at the base; corolla constricted below the lobes. Capsule oblong, 1.5 cm long, rounded at the apex.

GRAND CAYMAN: *Brunt* 1831, 1916; *Proctor* 27927.

— Southern U. S. A., West Indies, and continental tropical America, most often in sandy, brackish or subsaline situations at low elevations.

Voyria Aublet

Erect, saprophytic, perennial herbs without chlorophyll. Stems most often simple (sometimes few-branched), terete, solitary or a few clustered together; roots coral-like or like a minute birds nest. Leaves small, scale-like, opposite, somewhat connate at the base, or the lower ones sometimes alternate. Inflorescence a few–30-flowered dichasium, or else the flowers terminal and solitary. Flowers variously coloured, erect (or rarely nodding), (4–)5(–7)-merous with short or long pedicels; calyx tubular to campanulate, lobed, persistent; corolla actinomorphic, salverform to funnel-shaped, much exceeding the calyx, marcescent, the lobes contorted, spreading, or recurved, rarely erect. Stamens included, rarely exceeding the corolla tube, the filaments elongate or very short; anthers free or often coherent just below the stigma, introrse. Ovary 2-carpellate, unilocular; style filiform, gradually widened toward the base; stigma often more or less capitate or weakly 2-lobed; ovules numerous. Capsule fusiform to globose, more or less dehiscent or indehiscent. Seeds numerous, globose to filiform, in some species with two hair-like projections; endosperm present.

— A genus of 19 species, 18 of them widely distributed in the neotropics and one in West Tropical Africa.

Voyria parasitica (Schlect. & Cham.) Ruyters & Maas, *Acta Bot. Neerl* 30: 143 (1981).

Erect glabrous herb, light buff in colour, up to 15 cm tall or more. First flower terminal, later flowers secund on inflorescence-branches; flowers 4–5 mm long; capsules ellipsoid to globose, 2.5–4 mm in diameter; seeds filiform or fusiform. The plants grow in deeply shaded forest humus.

GRAND CAYMAN: *Proctor* 47793, found in Arlington Executive Park area; and *Proctor* 47863, found along the Mastic Trail.

— Florida, Bahamas, Greater Antilles and from southern Mexico to Honduras.

APOCYNACEAE

Herbs, shrubs or trees, or often herbaceous or woody vines, usually containing milky latex; leaves opposite or whorled (rarely alternate), simple and entire; stipules present or absent. Flowers perfect, regular or nearly so, twisted in bud, in cymes or panicles, or sometimes solitary. Calyx usually 5-lobed, often with glandular appendages within; corolla more or less funnel-shaped, 5-lobed. Stamens 5, inserted on the corolla-tube, often connivent around the stigma but the filaments usually free; anthers 2-celled. Ovary superior, 2-celled, the carpels free or united; style simple with a large stigma; ovules 1–many in each carpel. Fruit of 2 distinct follicles, or the carpels more or less united to form a capsule, drupe, or berry; seeds naked, or plumed at the apex, or with a papery wing, or sometimes with an aril; endosperm present.

— A cosmopolitan family best represented in the tropics, with about 200 genera and more than 2,000 species. Few are of any economic value, but some are grown for ornament. The white latex of several species is known to be poisonous, whereas in other species it is edible.

Species commonly cultivated in the Cayman Islands include *Allamanda cathartica* L. and *A. hendersonii* Bull., both called yellow allamanda; *Nerium oleander* L., the oleander; two species of *Plumeria* (see below); *Strophanthus gratus* (Wall. & Hook.) Baill., the blue allamanda (*Kings* GC 300); and perhaps more than one species of *Tabernaemontana*.

KEY TO GENERA

1. Plants erect, never vine-like; anthers free from the stigma, the locules without basal appendages:

  2. Leaves alternate, crowded; shrubs:

    3. Flowers white or pink; branches thick; leaves deciduous during dry seasons: **Plumeria**

    3. Flowers yellow; branches slender; leaves not deciduous: **[Thevetia]**

  2. Leaves distinctly opposite or whorled; herbs or shrubs, the branchlets slender; flowers borne in leaf-axils, solitary or in cymes:

    4. Herbs with white or pink flowers; mature follicles less than 3 mm thick: **[Catharanthus]**

    4. Shrubs or small trees:

      5. Leaves opposite in pairs; flowers greenish-yellow; fruit of separate follicles 10 mm thick or more: **Tabernaemontana**

      5. Leaves in whorls of 4; flowers white; fruit subglobose, drupe-like: **Rauvolfia**

1. Plants scandent or vine-like; anthers attached to the stigma, their locules with basal prolongations:

  6. Calyx bearing glandular appendages inside the lobes; corolla cream or yellow:

    7. Corolla cream and with a cylindrical tube; sap not or scarcely milky: **Echites**

    7. Corolla bright yellow and with a funnel-shaped tube; sap distinctly milky: **Pentalinon**

  6. Calyx without glandular appendages inside the lobes; corolla white: **Rhabdadenia**

Plumeria L.

Shrubs or trees with stout, thick branches exuding abundant milky latex when cut; leaves crowded, alternate, petiolate; flowers in terminal, stalked cymes. Calyx 5-parted nearly to the base; corolla salver-shaped, without appendages within, equally 5-lobed. Stamens inserted near base of the corolla tube, included, the anthers free. Ovary of 2 distinct carpels, these with many ovules. Follicles 2, distinct, thick and wide-divergent; seeds many, winged at the base; endosperm fleshy.

— A tropical American genus of about 40 described species.

Plumeria obtusa L., *Sp. Pl.* 1: 210 (1753). PLATE 49.

WILD JASMINE

A shrub or small tree to 4 m tall or more, glabrous throughout; leaves oblong to oblong-oblanceolate or obovate, 7–20 cm long including the petioles, and blunt, rounded or notched at the apex. Inflorescence of 4–5 branches in a dense subumbellate cluster at the apex of the long peduncle; calyx ca. 3 mm long; corolla white with a yellow eye, the spreading obovate lobes 1.5–2 cm long. Follicles 10–16 cm long and to 1.4 cm thick.

GRAND CAYMAN: *Brunt* 1764; *Hitchcock*; *Lewis GC 1*; *Proctor* 11988, 15194. LITTLE CAYMAN: *Kings LC 8*; *Proctor* 28043. CAYMAN BRAC: *Millspaugh* 1229; *Proctor* 15315, 28909, 29066.

— Bahamas, Cuba, Hispaniola, Jamaica and the Swan Islands, mostly in dry, rocky, coastal thickets.

[*Plumeria rubra* L., the jasmine or frangipani in both pink and cream forms, and *P. pudica* Jacq., a species with large dense heads of white flowers, are traditionally planted in Cayman Island graveyards. The first-named is a native of C. America now cultivated in all tropical countries; the second is said to have been introduced from the Corn Islands at a fairly recent date.]

[*Thevetia* L.

Shrubs or small trees with alternate rather crowded leaves and white latex; interpetiolar glands usually present. Inflorescence cymose, few–several-flowered; calyx 5-parted with equal lobes, these bearing many glands within; corolla tubular-campanulate to salverform, the tube orifice bearing 5 small villous scales; lobes obovate, spreading. Anthers separate, with slightly enlarged connective. Ovary apocarpous, fleshy, with annular nectary; ovules 2–4. Fruits syncarpous, flattened; seeds 2–4, large, naked.

— A tropical American genus of 9 species.

Thevetia peruviana (Pers.) Schum. in Engler & Prantl, *Nat. Pflanzen fam.* 4 (2): 159 (1895).

Glabrous shrub or small tree to 7 m tall. Leaves sessile, the blades linear, 10–15 cm long, 0.5–1 cm broad, the apex obtuse, the base cuneate-attenuate, the surfaces glossy. Cymes terminal, few-flowered, the pedicels 2–4 cm long. Calyx lobes triangular, 5 mm long, with numerous glands within at base. Corolla bright yellow, tubular-campanulate above basal tube; tube 1–1.5 cm long, the expanded throat 6–7 cm long, 3–3.5 cm in diameter, the lobes 1–2 cm long, overlapping to the right. Fruits falcate, 2 × 3 cm, truncate above.

GRAND CAYMAN: *Proctor* 47821, 49324. Naturalised in secondary thickets at Further Ground, North Side and just north of main highway at Spotts.

— S. America; widely cultivated and naturalised in the Greater and Lesser Antilles.]

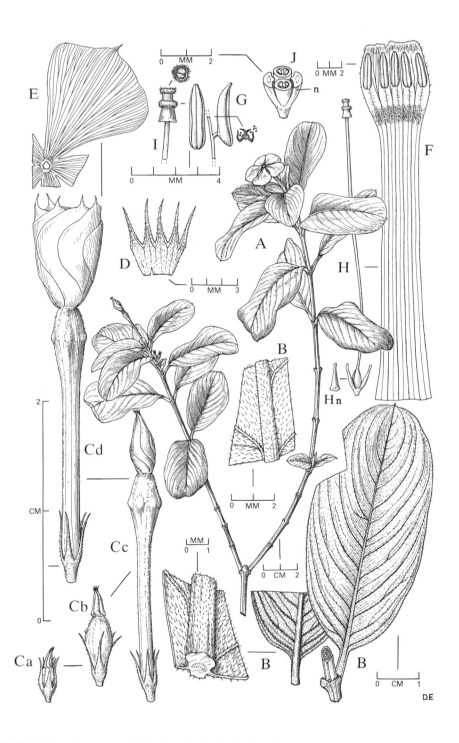

FIG. 191 **Catharanthus roseus**. A, habit. B, B, B, details of leaf. Ca, b, c, d, stages in development of flower. D, calyx spread out. E, portion of corolla-limb. F, inside of corolla-tube spread out. G, details of anther. H, pistil; Hn, nectary (disk-gland). I, details of stigma. J, cross-section through carpels; n, nectary. (D.E.)

[*Catharanthus* G. Don

Erect, subwoody perennial herbs or low shrubs. Flowers 1–4 in the leaf-axils; calyx 5-cleft; corolla with narrowly tubular throat and flat, salver-shaped limb, the throat pubescent and thickened in the apical part. Stamens inserted in the apical part of the corolla-tube, included; anthers free. Carpels 2, distinct but slightly cohering, alternating with 2 disc-glands (nectaries); ovules many in each carpel. Fruit of 2 narrowly cylindrical, many-seeded follicles; seeds without appendages.

— A small tropical genus of 8 species, all but one in Madagascar.

Catharanthus roseus (L.) G. Don, *Gen. Syst.* 4: 95 (1837). **FIG. 191.**

Vinca rosea L. (1753).
Lochnera rosea (L.) Reichb. (1828).

PERIWINKLE, BURYING-GROUND FLOWER, RAMGOAT ROSE

An erect herb usually less than 50 cm tall, more or less finely pubescent throughout. Leaves oblong or elliptic-oblong, mostly 2–7 cm long, obtuse and apiculate at the apex. Flowers in axillary pairs, subsessile; calyx-lobes 3–5 mm long; corolla white or pink, or white with a pink eye, the tube ca. 2.5 cm long, the lobes broadly obovate, 1.5 cm long. Follicles pubescent, longitudinally ribbed, 1.5–3.5 cm long, 2–3 mm thick.

GRAND CAYMAN: *Brunt* 2118; *Kings* GC 51, GC 221, GC 222; *Millspaugh* 1331; *Proctor* 15093. LITTLE CAYMAN: *Proctor* 35197. CAYMAN BRAC: *Kings* CB 60; *Proctor* 28988.

— Originally described from Madagascar, now cultivated and escaping in nearly all warm countries.]

Tabernaemontana L.

Shrubs or trees, usually glabrous or nearly so. Flowers in axillary cymes; calyx 5-parted almost to the base, bearing numerous glandular appendages within; corolla with salver-shaped limb. Anthers not connivent, free from the stigma. Ovary of 2 distinct carpels, with or without a basal ring-like nectary; ovules many. Fruit of 2 rather broad and fleshy follicles; seeds embedded among the fleshy arils.

— A tropical American genus of about 50 species

Tabernaemontana laurifolia L., *Sp. Pl.* 1: 210 (1753). PLATE 49.

A small tree to 7 m tall or more; leaves elliptic or elliptic-oblong, 7–17 cm long, indistinctly subacuminate, with petioles 1–2 cm long. Inflorescences congested, much shorter than the leaves, subumbellate; pedicels 2–3 mm long; calyx-lobes 3 mm long, imbricate; corolla greenish-yellow, the tube 15 mm long, the lobes narrow and 10 mm long. Ovary surrounded at the base by a low, ring-like nectary. Follicles 3–4.5 cm long.

GRAND CAYMAN: *Brunt* 2147, 2184; *Correll & Correll* 51006; *Howard & Wagenknecht* 15031; *Proctor* 11990.

— Jamaica; a doubtful record from Guatemala is probably not the same. This is normally a species of rocky coastal thickets and dry woodlands on limestone.

Rauvolfia L.

Shrubs or small trees; nodes with small interpetiolar deciduous stipules and persistent glands confined to leaf-axils or extending up the petioles. Leaves whorled, 3–5 at each node, sometimes unequal in size. Inflorescence terminal or lateral, 2–4-times dichotomously branched, few–many-flowered. Flowers sessile or pedicellate; calyx campanulate, 5-lobed, without glands within. Corolla white salverform, funnel-shaped, urn-shaped, or campanulate, the lobes 5, equal, overlapping to the left, glabrous outside, variously pubescent within. Stamens epipetalous; anthers free, included. Carpels 2 separated or variously syncarpous, 2-locular; ovules 1 or 2 per locule; basal disc annular; stigma cylindrical. Follicles free or variously united and drupe-like, or only one carpel developing; seeds stout or flat.

— A pantropical genus of about 100 species.

Rauvolfia nitida Jacq., *Enum. Syst. Pl. Carib.* 14 (1760); *Select. Strip. Amer. Hist.* 47 (1763).

Shrub or small tree to 15 m tall, with glands in leaf-axils only. Leaves mostly in whorls of 4 with petioles 5–10 mm long; blades ovate-elliptic to oblong-elliptic, 3–15 cm long, 3–5 cm broad, the apex acute to short-acuminate, the base attenuate, the upper surface somewhat glossy and with conspicuous secondary veins. Inflorescences few–many-flowered; flowers with pedicels 2–5 mm long, 1.5 mm in diameter, glabrous outside, villous in upper half within; lobes broadly ovate, 3–4 mm long. Fruits subglobose, 10–12 × 15–18 mm, black; seeds 2.

GRAND CAYMAN: "in primary dry forest", reported by Guala *et al.* in *Kew Bull.* **57**: 236 (2002).

— Bahamas, Greater and Lesser Antilles.

Echites P. Br.

Slender, twining, somewhat woody vines with colourless, watery sap; leaves opposite, without glands. Flowers in axillary cymes, rarely solitary; calyx 5-parted almost to the receptacle, each lobe bearing within at the base a single, often dissected glandular appendage; corolla salver-shaped, without appendages in the tube. Anthers connivent and adherent to the stigma, the cells appendaged at the base, 2-lobed. Ovary of 2 distinct carpels with many ovules, surrounded at the base by 5 more or less distinct nectar-glands; stigma spindle-shaped. Fruit of two distinct, divergent follicles; seeds numerous, plumed at the apex.

— A tropical American genus of 7 species.

Echites umbellatus Jacq., *Enum. Pl. Carib.* 13 (1760). **FIG. 192.** PLATE 50.

NIGHTSHADE

A glabrous, suffrutescent vine, scrambling over bushes or trees to a height of 5 m; leaves elliptic to very broadly ovate, 4–10 cm long, acute or subacuminate at the apex and often more or less recurved or folded. Inflorescences 2–7-flowered, stalked; calyx-lobes scarious, sharply acute, 3–4 mm long; corolla-tube 3–4 cm long, dilated below the middle and above spirally twisted and gradually constricted; lobes obliquely obovate, ca. 2 cm long. Follicles usually 10–20 cm long.

GRAND CAYMAN: *Brunt* 2022; *Kings* GC 173; *Lewis* 3845; *Millspaugh* 1243; *Proctor* 15267; *Sachet* 427. LITTLE CAYMAN: *Kings* LC 1, LC 86. CAYMAN BRAC: *Kings* CB 24; *Proctor* 29021.

— Florida, Bahamas, Greater Antilles except Puerto Rico, the Swan Islands, and the Caribbean coasts of continental America, in dry thickets and woodlands.

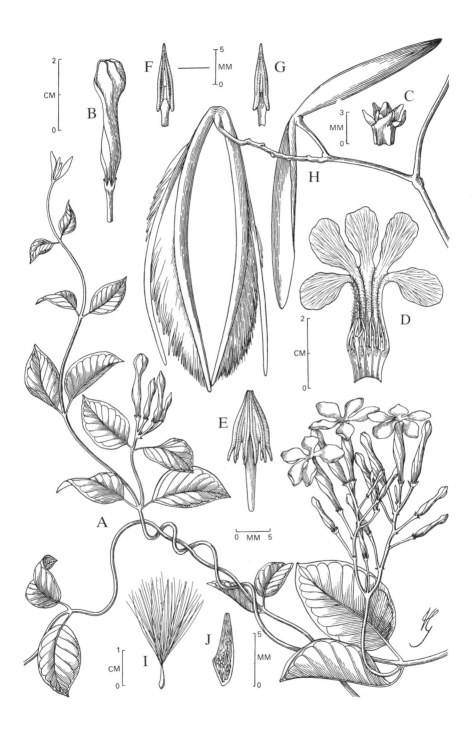

FIG. 192 **Echites umbellata**. A, branch with leaves and flowers. B, flower-bud. C, calyx showing appendages. D, corolla opened to show stamens. E, stamens showing connivent anthers. F, G, two views of anthers. H, fruiting branchlet. I, seed with plume. J, seed enlarged, with plume removed. (G.)

FIG. 193 **Pentalinon luteum**. A, branch with leaves, flowers and fruits. B, calyx with one lobe removed. C, corolla opened out to show stamens. D, enlarged portion of same. E, stamen, dorsal view. F, stamen with appendage, ventral view; a, appendage. G, ovary, style and stigma. H, seed with plume. I, seed enlarged, with plume removed. (G.)

Pentalinon Muell. Arg.

Woody or suffrutescent vines; leaves usually opposite and without glands. Flowers large, yellow, in axillary or subterminal cymes; calyx 5-parted nearly to the receptacle, the imbricate lobes bearing glandular appendages within at the base; corolla funnel-shaped, without appendages within. Anthers connivent and adherent to the stigma, the base enlarged and narrowly 2-lobed, appendaged at the apex. Ovary of 2 distinct carpels with many ovules, surrounded at the base by 5 more or less distinct nectar-glands; stigma spindle-shaped. Fruit of 2 distinct follicles; seeds numerous, dry, beaked and hairy at the apex.

— A tropical American genus of 2 species.

Pentalinon luteum (L.) Hansen & Wunderlin, *Taxon* 35: 167 (1986). **FIG. 193**.

Echites andrewsii Chapm. (1860).
Urechites lutea of Proctor (1984).

YELLOW NIGHTSHADE

A usually pubescent slender vine; leaves oblong to broadly ovate, 1.5–6 cm long or more, obtuse at the apex, the margins often revolute, dark green on the upper side and pale green beneath. Cymes few-flowered; flowers on slender pedicels 0.8–1.5 cm long; calyx-lobes lance-acuminate, mostly 7–9 mm long; corolla-tube 8–12 mm in diameter, the obovate lobes 2–3 cm long. Follicles linear, 10–15 cm long.

GRAND CAYMAN: *Kings* GC 91, GC 91B, GC 401; *Lewis* GC 4; *Maggs* II 67; *Millspaugh* 1373; *Proctor* 15089; *Sachet* 428.

— Florida and the West Indies, in thickets, pastures and scrublands. This species is known to be poisonous to livestock.

Rhabdadenia Muell. Arg.

Slender, glabrous woody vines; leaves opposite, not glandular. Flowers solitary or few in axillary or subterminal umbellate cymes; calyx 5-parted almost to the receptacle, rather broad, lacking glandular appendages within; corolla funnel-shaped, lacking appendages within. Anthers connivent and adherent to the stigma, their bases enlarged and narrowly 2-lobed. Ovary of 2 distinct carpels with many ovules, surrounded at the base by 5 more or less distinct nectar-glands; stigma spindle-shaped. Fruit of 2 distinct, terete follicles; seeds numerous, beaked and hairy at the apex.

— A small genus of 3 tropical American species.

Rhabdadenia biflora (Jacq.) Muell. Arg. in Mart., *Fl. Bras.* 6 (1): 175 (1860). **FIG. 194**. PLATE 50.

Echites paludosa Vahl (1798).

An often high-climbing woody vine with slender branches; leaves oblong or elliptic-oblong, 4–11 cm long, obtuse and apiculate at the apex, noticeably pale beneath. Cymes 1–5-flowered, on peduncles equalling or exceeding the leaves in length; pedicels 7–13 mm long; calyx-lobes ovate-oblong, usually 4–6 mm long; corolla pure white, the tube 1.5–2 cm long, the lobes obovate, 2–2.5 cm long. Filaments densely pubescent. Follicles slender, terete, 9–14 cm long.

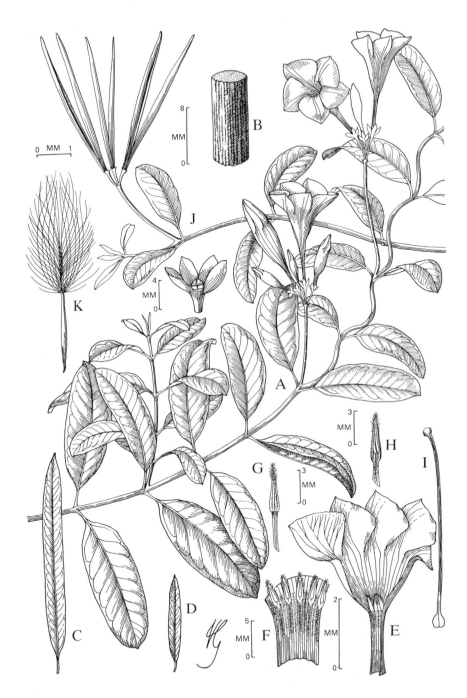

FIG. 194 **Rhabdadenia biflora.** A, branch with leaves and flowers. B, section of stem. C, D, variations of leaf-shape. E, corolla opened to show stamens. F, portion of same, enlarged. G, H, two views of anther. I, ovary and style. J, fruiting branch. K, seed with plume. (G.)

GRAND CAYMAN: *Brunt* 1773, 1801; *Hitchcock*; *Kings* GC 199, GC 348; *Lewis* 3846; *Maggs* II 65; *Proctor* 15046; *Sachet* 449. LITTLE CAYMAN: *Kings* LC 34; *Proctor* 35116.

— Florida, Bahamas, Greater Antilles and the Caribbean coasts of Central and S. America, a characteristic species of mangrove swamps.

ASCLEPIADACEAE

Mostly perennial herbs, erect, scandent or twining, or sometimes woody shrubs (rarely arborescent), usually with milky sap; leaves opposite or rarely whorled; stipules poorly developed or lacking. Inflorescence a usually umbel-like or raceme-like axillary cyme; flowers of very complicated structure, perfect, regular; calyx with very short or no tube, the 5 lobes imbricate or open in bud; corolla 5-lobed, the lobes twisted or valvate; usually a corona present, this simple or consisting of 5 or more scales or lobes, adnate to the corolla-tube or to the staminal column, highly varied in structure. Stamens 5, inserted at or near the base of the corolla, the filaments flat, short, and usually joined to form a tube, united with the stigma to form a 'gynostegium'; anthers 2-celled, opening inwardly, often extended at the base and tipped with a membranous appendage; pollen aggregated in waxy or granular masses called 'pollinia', these usually solitary in each anther-cell, and agglutinated with the stigma by 5 glandular 'corpuscles' which extrude the pollinia after dehiscence of the anthers. Ovary superior, of 2 distinct carpels, surrounded by the stamen-tube; styles 2, distinct below the stigma; stigma 1, peltately dilated and forming a more or less pentagonal disc; ovules numerous in each carpel, pendulous and overlapping on the placentas. Fruit of 2 follicles, or often only 1 developing; seeds numerous, compressed, and usually bearing a tuft of long soft white silky hairs; endosperm thin and cartilaginous; embryo large, with flat cotyledons.

— A family of perhaps 130 genera and 2,000 species, widely distributed in temperate and tropical regions. There has been much disagreement about the classification of this group, as the almost incredible complexity of the flowers has led to the proposal of numerous genera whose validity is difficult to evaluate. Except perhaps for the orchids, the flowers of Asclepiadaceae are the most highly specialised of any group of plants.

KEY TO GENERA

1. Plants erect, not vine-like:

  2. Herb with lanceolate, petiolate leaves; corona-segments with a horn-like process within: **Asclepias**

  2. Shrub, sometimes of tree-like habit, the very broad leaves almost sessile with clasping bases; corona-segments spurred near the base: **[Calotropis]**

1. Trailing or twining vines:

  3. Flowers greenish or white; calyx lobes less than 4 mm long:

    4. Corolla glabrous, less than 6 mm broad when expanded; corona simple: **Metastelma**

    4. Corolla pubescent, 10 mm broad or more when expanded; corona double: **Sarcostemma**

  3. Flowers purple or violet, large; calyx lobes more than 10 mm long: **[Cryptostegia]**

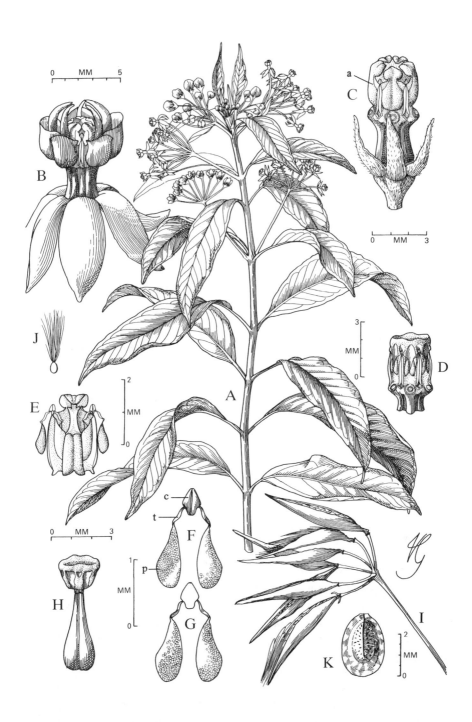

FIG. 195 **Asclepias curassavica.** A, upper portion of stem with leaves and flowers. B, flower. C, flower with corolla and corona removed; **a**, anther. D, same with anthers removed to show pollinia. E, part of same enlarged. F, G, two views of pollinia; **c**, corpuscle; **t**, translator; **p**, pollinium. H, pistil. I, fruiting inflorescence. J, seed with plume. K, seed enlarged, the plume removed. (G.)

Asclepias L.

Usually erect, perennial herbs; leaves opposite or whorled. Flowers in umbels on axillary or terminal peduncles; calyx 5–10-glandular within at the base; corolla deeply 5-lobed, the lobes reflexed; scales of the corona 5, attached to the stamen-tube, erect and concave-hooded, ligulate within. Pollinia pendulous from apex of the anther-cell. Stigma flat, 5-angled or obtusely 5-lobed. Follicles acuminate, smooth or rarely spiny-tuberculate.

— A chiefly N. American genus of about 120 species.

Asclepias curassavica L., *Sp. Pl.* 1: 215 (1753). **FIG. 195.**

RED TOP, HIPPA CASINI

Erect herb usually 30–60 cm tall with mostly simple, somewhat pubescent stems; leaves opposite or sometimes in whorls of 3, lanceolate, mostly 6–12 cm long, long-acuminate at the apex and attenuate to the short-petioled base, glabrous or nearly so. Umbels 1–several in the upper axils; pedicels 1–2 cm long, pubescent. Corolla scarlet or bright red, the reflexed segments 6–7 mm long; corona yellow, the hoods erect, 4–5 mm long, shorter than the horns. Follicles narrowly spindle-shaped, 6–9 cm long, glabrous.

GRAND CAYMAN: *Hitchcock*; *Kings* GC 152, GC 328; *Millspaugh* 1323; *Proctor* 15142.

— Florida, West Indies and continental tropical America, a common weed of roadsides and pastures. This species is alleged to be poisonous to livestock; it has been used in folk medicine to induce vomiting.

[*Calotropis* R. Br.

Shrubs or small trees; leaves broad, opposite and nearly sessile. Flowers in terminal or axillary umbellate cymes; calyx with several to many glands within at the base; corolla 5-cleft with broad lobes not reflexed; scales of the corona 5, fleshy, attached to the stamen-tube, lobed or toothed above and short-spurred at the base. Pollinia pendulous from the anther-cells. Stigma sharply 5-angled. Follicles thick, blunt, and rather rough.

— A genus of 6 species occurring in tropical Africa and Asia.

Calotropis procera (Ait.) R. Br. in Ait.f., *Hort. Kew.* ed. 2, 2: 78 (1811). **FIG. 196.** PLATE 50.

FRENCH COTTON

A shrub or small tree to 5 m tall, the younger parts deciduously white-felted, becoming glabrous with age; leaves oblong-obovate or rotund, 8–18 cm long, up to 12 cm broad, cuspidate at the apex, narrowly cordate-clasping at the base, with very short, stout petiole. Cymes stalked, few–many-flowered; pedicels 1–3 cm long; calyx woolly; corolla bell-shaped, 2–3 cm broad when expanded, white, the fleshy triangular lobes violet within; corona deep violet; stigma pale green. Follicles greenish, swollen, 8–12 cm long, minutely woolly puberulous.

GRAND CAYMAN: *Proctor* 15121. CAYMAN BRAC: *Proctor* (sight record).

— Apparently native of tropical Africa, but planted and naturalised in most warm countries, especially in dry places. The latex contains a bitter substance and is somewhat poisonous; the silky floss from the seeds can be used for stuffing pillows or for weaving cloth.]

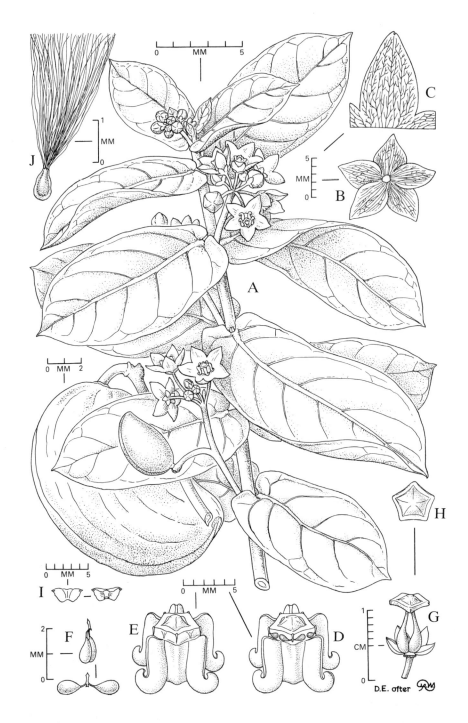

FIG. 196 **Calotropis procera**. A, branch with leaves, flowers and fruit. B, calyx with corolla removed. C, portion of corolla. D, corona (c) and gynostegium with pollinia in position after dehiscence of the anthers. E, same with pollinia removed. F, pollinia before and after dehiscence. G, carpels capped by peltate stigma. H, stigma see from above. I, long.-section through stigma. J, seed with plume.(St.)

Metastelma R. Br.

Slender perennial vines, herbaceous or somewhat woody; leaves opposite, small or sometimes reduced to scales, the stem green. Flowers very small, in sessile or stalked axillary umbel-like cymes; calyx 5-glandular within at the base, or sometimes without glands; corolla bell-shaped or subrotate, the lobes valvate, often bearded within; corona of 5 scales, these distinct or united, sometimes variously dissected or with internal projections, rarely lacking. Pollinia pendulous, often compressed. Stigma flat at the apex or apiculate. Follicles narrowly acuminate, usually smooth.

— A genus of perhaps 150 species of tropical and subtropical America, many of them poorly understood because the complex flowers are too small to be studied without a microscope.

KEY TO SPECIES

1. Leaves subsessile, linear, up to 6 cm long or more; corolla-lobes acuminate-attenuate, 4 mm long:	M. palustre
1. Leaves petiolate, oblong to ovate, 2 cm long or less; corolla-lobes obtuse, 2 mm long:	M. picardae

Metastelma palustre (Pursh) Schltr. in Urban, *Symb. Ant.* 1: 258 (1899).

Cynanchum angustifolium Pers. (1805).
Cynanchum salinarum (Wr. ex Griseb.) Alain (1955).

A trailing glabrous vine; leaves narrowly linear, 2–8 cm long, usually less than 3 mm broad, sharply acute at the apex. Peduncles 1.5–3 cm long; cymes up to 14-flowered, forming an umbel-like head; calyx-lobes narrowly lanceolate, ca. 2 mm long; corolla greenish often tinged dull rose or purplish, the lobes ovate with attenuate tips; corona-lobes 1.5–2 mm long, notched. Anther-wings 1 mm long. Follicles 4.5–7 cm long and ca. 5 mm thick.

GRAND CAYMAN: *Proctor* 31044, 48262, 48286, 52075, 52093.
— Southeastern U. S. A., Bahamas, and Cuba, in various habitats, but most often in somewhat saline situations near the sea. The Cayman plants were growing on bare sand beside the sea, southwest of Rum Point and in open salinas near Head of Barkers.

Metastelma picardae Schltr. in Urban, *Symb. Ant.* 1: 248 (1899).

Metastelma schlechtendalii of Millspaugh (1900), not Dcne. in DC. (1844).
Cynanchum picardae (Schltr.) Jiménez (1960).

A very slender, rather high-climbing vine, glabrous or nearly so, forming tangles; leaves oblong to ovate or lance-ovate, mostly 0.8–2 cm long, minutely apiculate at the apex, rounded at the base, the delicate petioles 1.5–4 mm long. Cymes subsessile, 2–5-flowered; calyx-lobes oblong, obtuse, 0.5–0.8 mm long; corolla greenish-cream, the lobes oblong, obtuse, ca. 2 mm long, very minutely puberulous within; corona-lobes spatulate; gynostegium sessile. Follicles narrowly spindle-shaped, 3–5 cm long and ca. 4 mm thick.

LITTLE CAYMAN: *Proctor* 35125, 47302. CAYMAN BRAC: *Millspaugh* 1197.
— Hispaniola, chiefly in dry thickets and scrublands near the sea. The Cayman Brac plants were collected "climbing over shrubbery at southwest point".

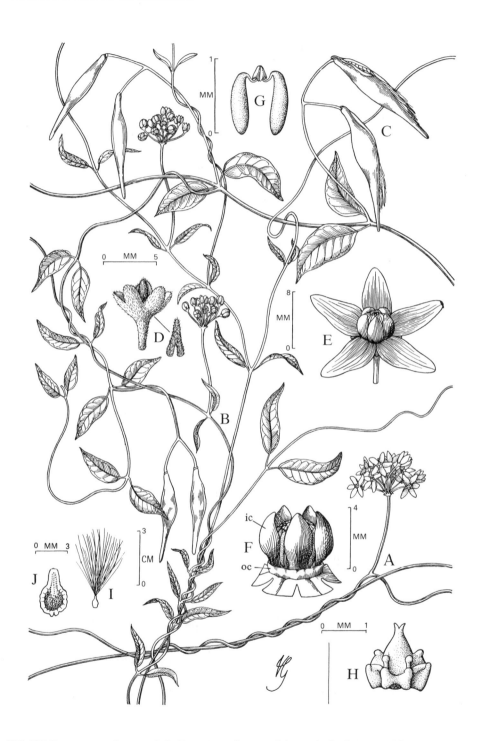

FIG. 197 **Sarcostemma clausum.** A, B, C, portions of stem with leaves, buds, flowers, and fruits. D, calyx. E, flower. F, central part of flower, enlarged; **oc**, outer corona; **ic**, scale of inner corona. G, pollinia. H, gynostegium; s, stigma. I, seed with plume. J, seed enlarged, plume removed. (G.)

Sarcostemma R.Br.

Herbaceous or suffrutescent vines, often glaucous; leaves opposite. Flowers in axillary, stalked umbels, usually white or greenish-white; calyx minutely 5-glandular within; corolla broadly bell-shaped or subrotate, shallowly or deeply 5-lobed, the lobes often contorted; outer corona ring-like, entire, adnate to the base of the corolla; inner corona of 5 scales adnate to the base of the stamen-tube, their blades free, broad and flat to concave or sac-like. Pollinia oblong or elongate, pendulous. Stigma flat, protuberant, or bearing a short forked beak. Follicles acuminate, smooth.

— A pantropical genus of about 35 species.

Sarcostemma clausum (Jacq.) Schult. in L., *Syst. Veg.* ed. nov. 6: 114 (1820). **FIG. 197.** PLATES 50 & 51.

A twining, often high-climbing vine, the stems green and glabrous; leaves glabrous, somewhat fleshy, almost linear to elliptic, mostly 2–4 cm long, acuminate or cuspidate at the apex. Umbels long-stalked, the peduncles stout, 3.5–7.5 cm long, few–many-flowered; flowers on slender pubescent pedicels; calyx densely pubescent; corolla white, the lobes 6–8 mm long, pubescent outside. Follicles 5–6.5 cm long, pubescent or glabrate.

GRAND CAYMAN: *Brunt* 1767, 1857, 1960; *Hutchings* (IJ); *Kings GC* 211; *Proctor* 15176. LITTLE CAYMAN: *Proctor* 35121. CAYMAN BRAC: *Proctor* 29370.

— West Indies and continental tropical America, common in thickets and woodlands especially along the borders of mostly brackish or subsaline swamps.

[*Cryptostegia* R. Br.

Glabrous, high-climbing vines; leaves broad, opposite. Flowers large, in terminal cymes; calyx deeply 5-parted; corolla funnel-shaped with short tube, the lobes twisted; corona-scales 5, pointed, entire or 2-lobed. Stamens with filaments short and joined only at the base, the anthers connivent around the convex stigma; pollen granular, the grains cohering in small masses. Follicles thick, woody, divergent, ribbed and 3-winged.

— A genus of 2 known species, the following and one from Madagascar. Some authors place this and some other genera with granular pollen and free filaments in a separate family, the Periplocaceae.

Cryptostegia grandiflora R.Br., *Bot. Reg.* t. 435 (1820).

A woody, trailing or climbing vine; leaves oblong-elliptic, 5–10 cm long, short-acuminate at the apex, on rather stout petioles ca. 1 cm long. Inflorescence few-flowered, puberulous; calyx-lobes lance-acuminate; corolla light violet outside, whitish within, 5–6 cm long. Follicles 9–12 cm long.

GRAND CAYMAN: *Brunt* 1853. CAYMAN BRAC: *Proctor* (sight record).

— Said to be a native of India, but widely cultivated for ornament in many tropical countries, often escaping and becoming naturalised. The latex is capable of yielding a considerable amount of rubber, and has been used for this purpose. It is reported to be poisonous to livestock.]

[*Stephanotis floribunda* Brongn. is or has been cultivated in Grand Cayman (*Kings GC* 241). It is a vine with fragrant white tubular flowers in umbels.]

SOLANACEAE

Herbs, shrubs, vines or sometimes trees; leaves alternate or sometimes in unequal pairs, simple or pinnately compound; stipules lacking. Flowers perfect, regular or nearly so, solitary or in cymes, borne outside the axils or between the paired leaves; calyx usually 5-lobed or -cleft; corolla gamopetalous, mostly 5-lobed, the lobes more or less folded in bud. Stamens usually 5, inserted inside the corolla-tube and alternate with the lobes; anthers 2-celled, variously dehiscent. Ovary superior, usually 2-celled, with numerous ovules on thick axile placentas; style slender and simple. Fruit a capsule or berry; seeds numerous, with fleshy endosperm.

A world-wide, chiefly tropical family of about 90 genera and 2,200 species. A few are economically important, e.g. 'Irish' potato (*Solanum tuberosum* L.) (native of the Andean region of S. America), tomato (*S. lycopersicum* L.), tobacco (*Nicotiana tabacum* L.) and others.

KEY TO GENERA

1. Flowers more than 15 cm long; ovary 4-celled; unarmed scrambling woody vine: **Solandra**

1. Flowers less than 5 cm long; ovary 2-celled; erect herbs or shrubs, or if scrambling or vine-like then armed with spines:
   - 2. Calyx inflated around the fruit: **Physalis**
   - 2. Calyx not inflated:
     - 3. Flowers solitary, axillary or in forks of the stem; corolla not folded in bud: **Capsicum**
     - 3. Flowers in racemes or cymose panicles:
       - 4. Corolla spreading, the tube rotate or else shorter than the lobes: **Solanum**
       - 4. Corolla tubular, the very short lobes much shorter than the tube: **Cestrum**

Physalis L.

Annual or perennial herbs, rarely somewhat woody, glabrous or with simple, branched or stellate hairs; leaves simple, alternate or clustered, petiolate, the margins entire to wavy or toothed. Flowers usually solitary at forks of the stem or arising laterally, remote from the leaf-axils; calyx 5-lobed, in fruit inflated like a papery bladder around the fruit; corolla bell-shaped, usually more or less yellow, and often with dark spots at the base. Stamens 5, inserted near the base of the corolla-tube; anthers splitting lengthwise. Style very slender with truncate or capitate stigma. Fruit a few–many-seeded berry; seeds roundish or kidney-shaped, more or less flattened.

— A cosmopolitan genus of about 100 species, most abundant in the American tropics.

Physalis angulata L., *Sp. Pl.* 1: 183 (1753).

A tender, branched herb to 50 cm tall or more, glabrous or with scattered minute hairs on the young parts; leaves long-petiolate, the blades lance-ovate to broadly ovate, mostly 2.5–9 cm long, shortly acuminate, the margins nearly entire to irregularly incised- or undulate-toothed. Corolla pale yellow, indistinctly spotted at the base within, or unspotted, 6–12 mm long. Fruiting calyces 10-angled or 10-ribbed, 2–3.5 cm long, on slender stalks 1–2.5 long; berry 10–12 mm in diameter.

FIG. 198 **Capsicum baccatum.** A, habit, × 1^1/$_2$. B, flower, × 3. C, corolla cut open, showing stamens, × 9. D, calyx and pistil, × 9. E, fruit, × 3. F, seed, × 6. (St.)

GRAND CAYMAN: *Brunt* 2032; *Hitchcock*; *Kings* GC 208.

— Southern U. S. A., West Indies and continental tropical America, a frequent weed of fields, clearings and waste places.

Capsicum L.

Annual or perennial herbs or shrubs with forking stems; leaves often in unequal pairs, simple, entire or nearly so. Flowers solitary or sometimes in small clusters, the peduncles arising from forks of the stem or from leaf-axils; calyx 5-toothed or subentire; corolla usually white or cream, the tube very short, the 5 lobes spreading, imbricate. Stamens 5, inserted in the throat of the corolla-tube; anthers opening lengthwise. Ovary 2- or rarely 3-celled; stigma club-shaped. Berries red, yellow or green, erect or nodding, usually pungent-flavoured.

— A genus of an uncertain number of species, perhaps 30, all native to tropical America. *Capsicum annuum* L. is cultivated in many forms, including the condiment red pepper and such non-pungent varieties as sweet pepper and pimiento.

Capsicum baccatum L., *Mant. Pl.* 47 (1767). **FIG. 198.**

BIRD PEPPER, WILD PEPPER

Bushy herb or shrub to 2 m tall or more, glabrous or minutely pubescent, leaves narrowly ovate, acute or acuminate, 1.5–5 cm long or more, petiolate. Pedicels solitary or sometimes paired, 1–2 cm long, somewhat thickened toward the apex. Calyx 1–2 mm long, almost truncate or very shallowly 5-lobed. Corolla cream, ca. 6 mm across when expanded. Anthers greenish-blue. Berries usually red, ovoid or globose, 5–10 mm long or sometimes longer.

GRAND CAYMAN: *Kings* GC 236.

— Throughout tropical America, and naturalised in the Old World tropics; frequent in thickets and along roadsides; sometimes cultivated. The fruits are very hot to the taste.

Solanum L.

Herbs or shrubs, sometimes scandent or climbing, unarmed or spiny, and often bearing stellate hairs; leaves alternate, simple and entire or variously lobed or pinnatifid, sometimes bearing prickles. Flowers white to violet, solitary or in simple or compound cymes, or sometimes in umbel-like clusters, the peduncles arising laterally more less remote from the leaves. Calyx bell-shaped, 4–5-lobed; corolla rotate, the tube very short, the 4 or usually 5 lobes spreading or recurved, folded in bud. Stamens 5, inserted near the base of the corolla-tube, the filaments usually short, the anthers relatively long and bright yellow, connivent around the style, opening by more or less terminal pores or longitudinal slits. Ovary usually 2-celled; stigma small, capitate or obscurely 2-lobed. Fruit a globose berry, fleshy or leathery; seeds numerous and more or less flattened.

— A very large tropical and subtropical genus of more than 1,000 species.

KEY TO SPECIES

1. Plants glabrous or bearing simple hairs; spines lacking:

   2. Herbs; corolla white, not over 6 mm across when expanded; berries black: — **S. americanum**

   2. Shrubs; corolla pale blue-violet, more than 15 mm across when expanded; berries blue: — **S. havanense**

1. Plants more or less clothed with stellate hairs; spines nearly always present on stems and leaves:

   3. Erect shrubs with straight spines:

      4. Leaves lanceolate, less than 4 cm broad; inflorescence an unbranched, elongate raceme; corolla violet, ripe berries bright red, 6–7 mm in diameter: — **S. bahamense**

      4. Leaves ovate, up to 10–15 cm broad; inflorescence short, branched; corolla white; ripe berries yellow-green, 10–15 mm in diameter: — **S. torvum**

   3. A trailing or scrambling shrub with recurved spines: — **S. lanceifolium**

Solanum americanum Mill., *Gard. Dict.* ed. 8 (1768). PLATE 51.

Solanum nodiflorum Jacq. (1789).

An erect annual herb to 1 m tall, minutely puberulous throughout with simple hairs; leaves long-petiolate, the blades membranous, narrowly to broadly ovate, 2–8 cm long, acuminate, the margins entire or with 1–few short lobes or teeth near the base; petioles up to 2 cm long. Peduncles lateral, 1–2 cm long; flowers 3–10 in an umbel-like cluster; pedicels 7–10 mm long; calyx 1 mm long; corolla-lobes 2 mm long, longer than the tube, spreading or recurved. Berries black, 5–10 mm in diameter.

GRAND CAYMAN: *Brunt 2056; Hitchcock; Millspaugh 1351; Proctor 15277.* CAYMAN BRAC: *Kings CB 65; Proctor 29079.*

— A pantropical weed chiefly of open waste ground.

Solanum havanense Jacq., *Enum. Syst. Pl. Carib.* 15 (1760). PLATE 51.

An erect or somewhat straggling shrub 1–2 m tall, of smooth texture, minutely puberulous with simple hairs on the young parts; leaves elliptic, 2–5 cm long or more, obtuse at the apex and acuminate at the base. Inflorescence lateral or subterminal, few-flowered on peduncles less than 1 cm long; pedicels 5–6 mm long; calyx deeply 5-lobed, ca. 4 mm long; corolla-limb broadly spreading, the shallow lobes much shorter than the united portion. Berries ca. 15 mm in diameter.

GRAND CAYMAN: *Brunt 2200.*

— Cuba, Jamaica and Hispaniola, in rocky woodlands most frequently near the sea.

Solanum bahamense L., *Sp. Pl.* 1: 188 (1753). FIG. 199.

A more or less spiny shrub usually 1–2 m tall, minutely stellate-puberulous throughout; spines straight and slender, 3–8 mm long, scattered on stems and leaves; leaves lanceolate to oblong, 3–11 cm long, acute or obtuse at the apex, the margins entire or somewhat wavy. Cymes simple and raceme-like, shorter than or sometimes exceeding the leaves; pedicels 6–12 mm long; calyx ca. 1.5 mm long; corolla blue-violet or rarely white, with very short tube and narrow strap-like lobes 6–8 mm long. Berries red, 6–8 mm in diameter.

GRAND CAYMAN: *Brunt 1779, 1997; Correll & Correll 51014; Kings GC 317; Proctor* 11978. LITTLE CAYMAN: *Kings LC 33.* CAYMAN BRAC: *Proctor 29035.*

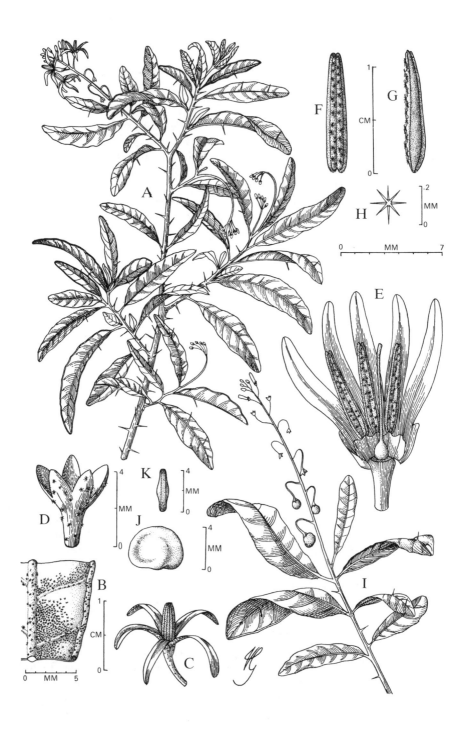

FIG. 199 **Solanum bahamense.** A, branch with leaves and flowers. B, enlarged portion of leaf, showing stellate hairs. C, flower. D, calyx. E, flower cut open to show stamens and pistil. F, G, two views of anther. H, stellate hair. I, branch with fruits. J, K, two views of seed. (G.)

— Florida, Bahamas and the Greater Antilles except Puerto Rico; doubtfully distinct from *S. racemosum* Jacq. of Puerto Rico and the Lesser Antilles. This species grows in rocky thickets and woodlands. Outside our area it tends to be very polymorphic, varying widely in the presence or absence of spines and other characters, but the Cayman population is quite uniform.

Solanum torvum Sw., *Nov. Gen. & Sp. Pl.* 47 (1788).

Solanum ficifolium Ortega (1800).

SUSUMBER

Shrub 1–3 m tall, covered with stellate pubescence; prickles rather few. Leaves petiolate with broadly ovate blades, these usually few-lobed, the lobes and apex acute to acuminate, the margins otherwise entire; petioles 1.5–5 cm long. Inflorescence extra-axillary, cymose; pedicels 5–8 mm long; calyx ca. 4 mm long, pubescent, the lobes ovate, acute; corolla ca. 1 cm long, densely stellate-puberulous outside the ribs, the lobes ovate-lanceolate. Anthers linear, 7 mm long. Berry globose, 10–15 mm in diameter.

GRAND CAYMAN: *Proctor* 48250, found in open waste ground south of George Town.
— A weed of tropical American origin, but now pantropical in distribution.

Solanum lanceifolium Jacq., *Coll.* 2: 286 (1789).

Solanum scabrum Vahl (1798), not Mill. (1768).

A very prickly scrambling or scandent shrub with stems up to 4 m long, more or less stellate-tomentose throughout; spines short, rather thick and recurved, 1–5 mm long, scattered on stems, petioles and midribs of the leaves beneath; leaves lanceolate to ovate, 4–20 cm long, petiolate. Inflorescence a contracted cymose raceme usually less than 6 cm long; pedicels 6–12 mm long, elongating in fruit; calyx 1–2 mm long, enlarging in fruit to 3–5 mm; corolla white, with very short tube and narrow strap-like lobes 7–12 mm long, stellate-pubescent on the outside. Berries red or orange, 8–10 mm in diameter.

GRAND CAYMAN: *Hitchcock*; *Kings GC 317*; *Proctor 15276*.
— Cuba, Hispaniola, Tortola and the Lesser Antilles, appearing to intergrade with a series of other species in Central and S. America, where it is a noxious weed. The Cayman plants were collected in moist swales near George Town.

Other species of *Solanum* may yet be recorded from the Cayman Islands, including *S. erianthum* D. Don. In view of their ranges, the apparent absence of these species from Cayman is surprising.

Cestrum L.

Shrubs or small trees; leaves alternate, simple and entire. Flowers in paniculate cymes, these often contracted, usually axillary; calyx 5-lobed or -toothed; corolla salver-shaped or more often funnel-shaped with long slender tube and short, spreading lobes. Stamens usually 5, inserted in the corolla-tube, included; filaments often pilose below and sometimes with a tooth-like appendage; anthers small, splitting lengthwise. Ovary 2-celled, usually short-stalked; ovules few. Fruit a small berry; seeds oblong, smooth.

— A tropical American genus of about 150 species.

Cestrum diurnum L., *Sp. Pl.* 1: 191 (1753). PLATE 51.

Shrub 2–5 m tall, or sometimes a small tree to 10 m tall, mostly glabrous; leaves oblong to ovate or elliptic, cuneate at base, blunt to acuminate at the apex, up to 15 cm long and 6.5 cm broad, the petiole up to 2.5 cm long. Flowers fragrant at night; calyx varying from less than 3 mm long to 8 mm long; corolla greenish-white to cream. Berries purplish to deep blue or black.

— A number of varieties have been described from various parts of the American tropics. Two of these occur in the Cayman Islands:

1. Peduncles 0.5–1 cm long, shorter than or not much exceeding the
   petioles; flowers pedicellate in loose simple (rarely branched) racemes;
   calyx distinctly lobed; berries black: **a. C. diurnum** var. **venenatum**
1. Peduncles 2.5–5 cm long, much longer than the petioles; flowers
   sessile in tight heads or short spikes; calyx truncate or nearly so;
   berries purple: **b. C. diurnum** var. **marcianum**

a. *Cestrum diurnum* var. *venenatum* (Mill.) O. E. Schulz in Urban, *Symb. Ant.* 6: 263 (1909).

A shrub or small tree to 7 m tall or more, glabrous or minutely puberulous on the young parts; leaves thin, oblong to elliptic, 5–11 cm long, acute or subacuminate at the apex, often somewhat folded. Calyx 3.5–4.5 mm long, the rounded lobes ciliolate; corolla cream or white with cream lobes, the slender tube ca. 10 mm long, the obtuse lobes minutely puberulous around the margin. Berries ellipsoid, 6–7 mm long.

GRAND CAYMAN: *Proctor* 15013. CAYMAN BRAC: *Kings* CB 17; *Millspaugh* 1192; *Proctor* 15320, 29033.

— Jamaica and the Swan Islands, in rocky woodlands.

b. *Cestrum diurnum* var. *marcianum* Proctor, *Sloanea* 1: 3 (1977).

A shrub 1.5 m tall, minutely woolly puberulous on the youngest parts; leaves pale green, narrowly ovate-oblong, 5–8 cm long, blunt to acutish at the apex; petioles 3–5 mm long. Calyx 2.5–3 mm long, truncate or nearly so, the limb ciliolate; corolla white, the tube 10–11 mm long, the obtuse lobes minutely puberulous. Berries subglobose, 5–6 mm in diameter.

GRAND CAYMAN: *Proctor* 15294 (type).

— Cuba. Cayman plants were found in thickets bordering a stony pasture. This differs from all other variants of *Cestrum diurnum* in its long-stalked, spicate or capitate inflorescence and in some other details, but resembles *C. diurnum* var. *portoricense* in the small, nearly truncate calyx.

Cestrum laurifolium L'Her. was recorded from Grand Cayman by Fawcett, but his specimen (if he took one) has not been located; it seems very likely to have been a misidentification of *C. diurnum* var. *venenatum*. True *C. laurifolium* has sessile inflorescences, and therefore presumably could not be confused with *C. diurnum* var. *marcianum*.

Solandra Sw.

Woody, high-climbing or scrambling vines; leaves alternate, simple and entire. Flowers very large, solitary, terminal; calyx tubular and somewhat inflated, membranous, 2–5-cleft; corolla funnel-shaped, the tube cylindrical below and more or less abruptly much-expanded above, at the apex divided into 5 broad, imbricate, spreading lobes. Stamens 5, inserted at the base of the corolla-tube, included, the declined anthers oblong. Ovary imperfectly 4-celled; ovules many; stigma subcapitate. Fruit a large pulpy berry.

— A tropical American genus of about 10 species.

Solandra longiflora Tussac, *Fl. Antill.* 2: 49, t. 12 (1818). PLATE 51.

A glabrous, scrambling woody vine; leaves petiolate, elliptic or obovate, 5–12 cm long, acute at the apex. Calyx 6–7 cm long; corolla opening white, turning creamy yellow, 17–20 cm long or more, the narrow tubular part much longer than the calyx, the lobes finely undulate-toothed. Berry globose, ca. 4 cm in diameter.

CAYMAN BRAC: *Proctor* 29343.

— Cuba and Hispaniola, in various habitats; the Cayman Brac plants were found in dense woodland along the brink of north-facing limestone cliffs.

CONVOLVULACEAE

Annual or perennial herbs, twining vines, shrubs, or rarely trees, sometimes parasites with elongate slender stems and leaves reduced to scales; sap often milky; leaves alternate, simple and entire, lobed or variously dissected; stipules absent. Flowers axillary, solitary or in cymose heads or panicles, regular or nearly so, usually soon withering. Calyx of usually 5 free sepals, these imbricate, equal or unequal, usually persistent and often enlarging in fruit. Corolla gamopetalous and tubular, funnel-, bell- or salver-shaped, the more or less expanded limb 5-lobed or -toothed, or almost entire, twisted and often plaited or folded in bud, the folds indicated in the expanded corolla by longitudinal lines or stripes. Stamens 5, inserted in the base of the corolla-tube and alternating with its lobes; anthers 2-celled, splitting lengthwise. Disc ring-shaped or absent. Ovary superior, sessile, mostly 2–3-celled with solitary ovules or 1-celled with 4 ovules; style simple or forked, or styles 2; stigma capitate or 2-lobed. Fruit a usually 1–4-celled capsule, dehiscent or indehiscent, often more or less enclosed by the persistent calyx; seeds erect, glabrous or hairy, with scanty but hard endosperm.

— A nearly cosmopolitan family with about 50 genera and 1,800 species, particularly well represented in tropical America and Asia.

KEY TO GENERA

1. Ovary and fruit deeply 2-lobed or the carpels distinct; small creeping plant with minute flowers:	Dichondra
1. Ovary and fruit entire, not lobed:	
2. Styles 2, distinct, with elongate stigmas:	Evolvulus
2. Style 1, entire:	
3. Stigmas elliptic or oblong, flattened; flowers rather small and white:	Jacquemontia
3. Stigmas globose or capitate; flowers if white much larger:	
4. Filaments glandular at the base; anthers often spirally twisted at or soon after anthesis; pollen smooth; corolla white or yellow:	Merremia
4. Filaments glabrous or pilose at the base but not glandular; anthers never spirally twisted; pollen spiny; corolla not yellow:	Ipomoea

Dichondra J. R. & G. Forst.

Slender perennial herbs, usually creeping and rooted at the nodes, more or less clothed with both simple and 'malpighiaceous' hairs; leaves small, roundish-cordate or reniform, entire and long-petiolate. Flowers minute, stalked, solitary in the leaf-axils; sepals subequal, distinct; corolla broadly bell-shaped, deeply 5-lobed. Stamens shorter than the corolla, inserted just below the sinuses of the lobes; anthers small, nearly globose. Ovary 2-lobed, each lobe a single carpel with 2 ovules; styles 2, arising between the ovary-lobes; stigmas capitate. Fruit of 1 or 2 capsules, each usually 1-seeded; seeds subglobose, smooth.

— A nearly pantropical genus of about 12 species or fewer.

Dichondra repens J. R. & G. Forst., *Char. Gen. Pl.* 39, t. 20 (1776). **FIG. 200.**

A prostrate, creeping herb, forming mats, the delicate individual stems mostly 15–30 cm long, minutely appressed-hairy; leaves long-petiolate, the blades more or less reniform, mostly 3–12 mm broad, sparsely and minutely hairy on both surfaces. Peduncles shorter than the petioles; sepals oblong-spatulate, ca. 2 mm long; corolla yellowish, ca. 2 mm long; capsules hairy.

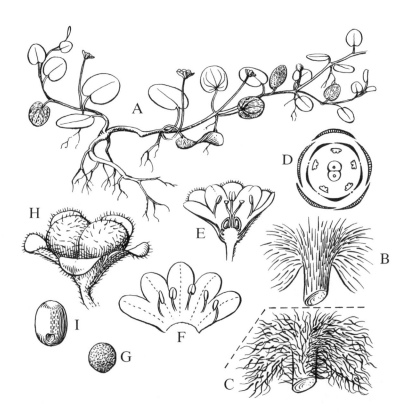

FIG. 200 **Dichondra repens.** A, habit × ¹/₂. B, leaf-base enlarged to show arrangement of hairs, × 8. C, same showing hairs on leaf from New Zealand, × 8. D, floral diagram. E, long section of flower, × 4. F, corolla spread out to show stamens, × 4. G, pollen grain, × 800. H, capsule, × 4. I, seed, × 4. (St.)

LITTLE CAYMAN: *Proctor* 28193.

— Widespread in tropical and warm-temperate America, occurring in a number of varieties. The single Cayman specimen is not adequate for identification with any particular variety. In Little Cayman, this species grows in sandy clearings and is rare.

Evolvulus L.

Annual or perennial herbs or small shrubs, the stems slender, creeping or erect, not twining; leaves small, simple and entire. Flowers axillary, solitary or in stalked, 1–several-flowered clusters, rarely in terminal spikes; sepals equal or subequal; corolla small, blue or white, usually rotate, funnel- or salver-shaped, the limb obscurely 5-angled but unlobed, folded in bud. Stamens 5, inserted at the mouth of the corolla-tube. Ovary 2-celled, each cell with 2 ovules, or rarely 1-celled with 4 ovules; styles 2, free or slightly joined at the base, each 2-forked at the apex. Capsule 2–4-valved, 1–4-seeded; seeds glabrous, smooth or minutely warty.

— A genus of more than 100 species, widely distributed in tropical and warm-temperate regions.

KEY TO SPECIES

1. Small erect shrub; leaves few, minute, more or less scale-like:	E. squamosus
1. Trailing or creeping herbs; leaves numerous, conspicuous:	
2. Leaves orbicular or obovate, notched at the apex; peduncles very short or obsolete, the pedicel scarcely exceeding the petioles:	E. nummularius
2. Leaves narrowly elliptic or oblong, acute and mucronate at the apex; peduncles hair-like, often equalling or exceeding the leaves in length:	E. convolvuloides

Evolvulus squamosus Britton, *Bull. New York Bot. Gard.* 3: 499 (1905). PLATE 51.

CRAB BUSH

A knee-high, brushy shrub with numerous stiff, slender, grey-green branchlets, these glabrate or sparsely pubescent; leaves minute, lance-linear, 1–2 mm long, sericeous. Flowers solitary or paired in the upper axils, the peduncles nearly obsolete, the sericeous pedicels 2–4 mm long. Sepals acuminate, ca. 2 mm long; corolla white, 7–9 mm broad when expanded, puberulous in stripes to each lobe. Capsule glabrous, 2 mm in diameter.

LITTLE CAYMAN: *Kings* LC 35, LC 76a; *Proctor* 28087, 28130.

— Bahamas. Our plants grow in dry sandy clearings or open glades of dry woodlands. This species is apparently the sole host of the diminutive land-snail *Cerion nanus*, endemic to Little Cayman.

Evolvulus nummularius (L.) L., *Sp. Pl.* ed. 2, 1: 391 (1762).

Perennial creeping herb, rooting at most of the nodes; leaves distichous, 5–17 mm long. Flowers mostly solitary; peduncles nearly obsolete, the bracteoles located at base of the pedicel, the latter 2–6 mm long; sepals ca. 2.5 mm long; corolla white or rarely pale blue, 5–7 mm broad when expanded. Capsule glabrous, 2–3 mm in diameter.

GRAND CAYMAN: *Kings* GC 247 (not seen).

— Pantropical, a frequent weed of grazed pastures, lawns, and open waste ground.

Evolvulus convolvuloides (Willd.) Stearn, *Taxon* 21: 649 (1972).

Evolvulus glaber Spreng. in L. (1824).

Trailing or ascending herb, often suffrutescent at the base or growing from a woody taproot, the slender branches usually rooting at some of the nodes, finely sericeous when young; leaves 5–28 mm long, glabrate. Flowers 1 or 2 at the apex of filiform peduncles up to 20 mm long; pedicels 1.5–4 mm long; sepals ca. 4 mm long; corolla pale blue, 7–10 mm broad. Capsule glabrous, ca. 2.5 mm in diameter.

GRAND CAYMAN: *Brunt* 2154; *Proctor* 27967.

— Florida, West Indies and northern S. America; the Cayman plants were found in open, brackish, seasonally flooded ground with *Salicornia*.

Jacquemontia Choisy

Trailing or twining vines, herbaceous or subwoody; leaves usually entire, sometimes toothed or lobed. Flowers axillary in small, stalked, umbel-like cymes or heads; sepals 5, equal or unequal; corolla white, blue or violet, bell-shaped or somewhat rotate, the limb 5-angled. Stamens 5, mostly shorter than the corolla, the filaments sometimes dilated at the base. Ovary 2-celled, each cell with 2 ovules; style filiform, with 2 stigmas. Capsule globose, 2-celled, 4-seeded; seeds glabrous, tuberculate or hairy.

— A pantropical genus of about 60 species.

Jacquemontia havanensis (Jacq.) Urb., *Symb. Ant.* 3: 342 (1902). PLATE 52.

Jacquemontia jamaicensis (Jacq.) Hall. f. (1899).

A slender, creeping or climbing, tangled, subwoody vine, the young stems and leaves puberulous; leaves sublinear to narrowly or broadly oblong, mostly 5–25 mm long, the apex acute or notched with a mucro. Inflorescence 1–few-flowered; sepals ca. 2 mm long; corolla white, ca. 1 cm long. Capsules 3–4 mm long; seeds ovoid, narrowly marginate, rough.

GRAND CAYMAN: *Brunt* 1819; *Kings* GC 149, GC 354; *Proctor* 11969, 15171. LITTLE CAYMAN: *Kings* LC 96; *Proctor* 28088. CAYMAN BRAC: *Kings* CB 41; *Millspaugh* 1198, 1205; *Proctor* 29024.

— Florida, West Indies, Yucatan, and the Turneffe Islands of Belize, frequent in dry rocky thickets and scrublands.

Merremia Dennst. ex. Endl., nom. cons.

Small or large vines; leaves simple and entire to variously lobed or palmately compound. Flowers axillary, solitary or in more or less condensed, stalked cymes; sepals subequal, often enlarging in fruit; corolla bell- or funnel-shaped, usually white or yellow. Stamens and style included. Ovary 2–4-celled with 2 ovules in each cell; style filiform with a globose or biglobose stigma. Capsule 2–4-celled, dehiscing lengthwise by valves, the pericarp thin and fragile; seeds glabrous or pubescent.

— A pantropical genus of about 60 species.

KEY TO SPECIES

1. Leaves simple and entire, deeply cordate; inflorescence umbel-like, few–many-flowered and dense; corolla yellow:	**M. umbellata**
1. Leaves palmately divided:	
2. Leaves with 5 serrulate, sessile leaflets; corolla ca. 2 cm long, cream:	**M. quinquefolia**
2. Leaves with 5–7 coarsely sinuate-dentate lobes; corolla ca. 4 cm long, white with purple eye:	**M. dissecta**

Merremia umbellata (L.) Hall. f., *Bot. Jahrb.* 16: 552 (1893).

Ipomoea polyanthes Roem. & Schult. (1819).
I. mollicoma Miq. (1850).

A trailing or twining herbaceous vine with glabrate stems and milky sap; leaves long-petiolate, the blades narrowly triangular to broadly ovate, mostly 3–9 cm long, long-acuminate, glabrate or puberulous. Inflorescence 3–18-flowered, on stout peduncles 3–5 cm long; pedicels ca. 1 cm long, thickened at the apex. Sepals 6–8 mm long; corolla bright yellow, ca. 2.5 cm long. Anthers not twisted. Capsule 8–10 mm in diameter, partly enclosed by the persistent calyx; seeds velvety-pubescent.

GRAND CAYMAN: *Correll & Correll* 51003; *Lewis* 3833; *Millspaugh* 1322; *Proctor* 15148.
— Pantropical, frequent in thickets and old fields.

Merremia quinquefolia (L.) Hall. f. in Engler, *Bot Jahrb.* 16: 522 (1893). PLATE 52.

Trailing or twining herbaceous vine, the stems up to 4 m long, slender and sparingly pilose. Leaflets elliptic to lanceolate, pointed at both ends, mostly 2–6 cm long, of thin texture. Inflorescence mostly 3–6-flowered; corolla 1.5–2 cm long. Capsules 8–10 mm in diameter, depressed-globose, dehiscing into 4 valves; seeds 4, carinate, brown, covered with scale-like hairs.

GRAND CAYMAN: *Proctor* 52137, from corner of South Sound Road and dyke road in low weedy vegetation, found by Joanne Ross.
— Widespread in the neotropics.

Merremia dissecta (Jacq.) Hall. f., *Bot. Jahrb.* 16: 552 (1893). **FIG. 201.**

A trailing or twining herbaceous vine with hirsute stems; leaves palmately divided almost to the base, the 7–9 segments coarsely sinuate-dentate, glabrous except for the long, hairy petioles. Flowers solitary or 2–3 together on long, stout, often hairy peduncles; pedicels 1–2 cm long, much thickened at the apex. Anthers spirally twisted. Capsules ca. 15 mm in diameter, surrounded by the enlarged, fleshy calyx; seeds black, glabrous.

GRAND CAYMAN: *Brunt* 1722, 2016; *Proctor* 15081. CAYMAN BRAC: *Kings* CB 84; *Proctor* 29097.
— Southern U. S. A., West Indies and continental tropical America, along the borders of thickets, overgrown fields, and open waste places.

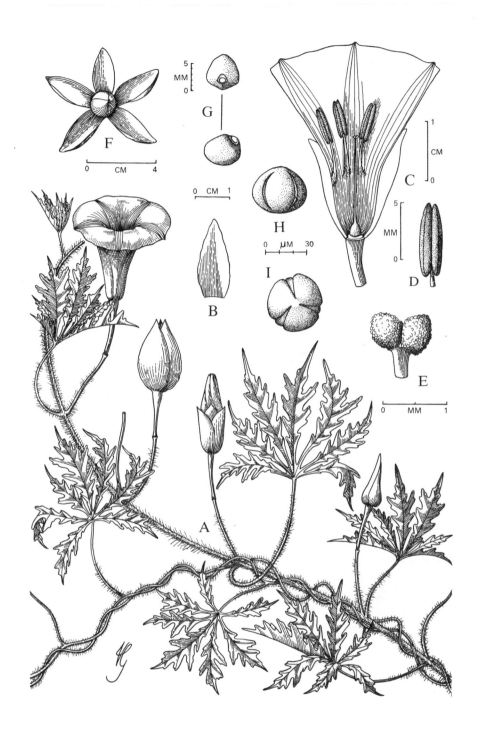

FIG. 201 **Merremia dissecta**. A, branches with leaves, flowers, and immature fruit. B, sepal. C, flower cut open to show stamens and pistil. D, anther. E, stigma. F, capsule with persistent calyx. G, two views of seed. H, I, two views of pollen grain. (G.)

PLATE 1

1. SCHIZAEACEAE *Anemia adiantifolia* (AS)

2. POLYPODIACEAE *Pteridium aquilinum* var. *caudatum* (KDG)

3. POLYPODIACEAE *Pteris longifolia* var. *bahamensis* (AS)

4. POLYPODIACEAE *Pteris longifolia* var. *bahamensis* (AS)

5. POLYPODIACEAE *Acrostichum aureum* (MC)

6. POLYPODIACEAE *Acrostichum aureum* (MC)

PLATE 2

1. POLYPODIACEAE *Nephrolepis biserrata* (AS)

2. POLYPODIACEAE *Nephrolepis multiflora* (AS)

3. POLYPODIACEAE *Thelypteris kunthii* (KDG)

4. POLYPODIACEAE *Thelypteris augescens* (AS)

5. POLYPODIACEAE *Thelypteris augescens* (AS)

6. POLYPODIACEAE *Polypodium polypodioides* (AS)

PLATE 3

1. POLYPODIACEAE *Polypodium polypodioides* (FJB)

2. POLYPODIACEAE *Polypodium heterophyllum* (FR)

3. POLYPODIACEAE *Polypodium phylitidis* (AS)

4. ZAMIACEAE *Zamia integrifolia* (male) (AS)

5. ZAMIACEAE *Zamia integrifolia* (female) (AS)

6. CYPERACEAE *Rhynshospora colorata* (MC)

PLATE 4

1. GRAMINEAE (POACEAE) *Dactyloctenium aegyptium* (KDG)

2. GRAMINEAE (POACEAE) *Pharus glaber* (AS)

3. GRAMINEAE (POACEAE) *Pharus glaber* (AS)

4. BROMELIACEAE *Tilandsia recurvata* (FJB)

5. BROMELIACEAE *Tilandsia utriculata* (KDG)

6. BROMELIACEAE *Tilandsia flexuosa* (MC)

7. BROMELIACEAE *Tilandsia setacea* (KDG)

PLATE 5

1. BROMELIACEAE *Tilandsia balbisiana* (FJB)

2. BROMELIACEAE *Tilandsia balbisiana* (FJB)

3. BROMELIACEAE *Tilandsia bulbosa* (MC)

4. BROMELIACEAE *Hohenbergia caymanensis* (MC)

5. BROMELIACEAE *Hohenbergia caymanensis* (AS)

6. BROMELIACEAE *Hohenbergia caymanensis* (flower spike) (MC)

PLATE 6

1. PALMAE (ARECACEAE) *Roystonia regia* (AS)

2. PALMAE (ARECACEAE) *Roystonia regia*

3. PALMAE (ARECACEAE) *Thrinax radiata* (MC)

4. PALMAE (ARECACEAE) *Thrinax radiata* (in fruit) (MC)

5. PALMAE (ARECACEAE) *Coccothrinax proctorii* (KDG)

6. PALMAE (ARECACEAE) *Coccothrinax proctorii* (KDG)

PLATE 7

1. AMARYLLIDACEAE *Hymenocallis latifolia* (MC)

2. AMARYLLIDACEAE *Hymenocallis latifolia* (MC)

3. AGAVACEAE *Agave caymanensis* (KDG)

4. AGAVACEAE *Agave caymanensis* (KDG)

5. AGAVACEAE *Agave caymanensis* (inflorescence)

PLATE 8

1. ORCHIDACEAE *Vanilla claviculata* (FJB)

2. ORCHIDACEAE *Pleurothallis caymanensis*

3. ORCHIDACEAE *Pleurothallis caymanensis* (CR)

4. ORCHIDACEAE *Encyclia kingsii*

5. ORCHIDACEAE *Encyclia kingsii* (inflorescence) (MC)

6. ORCHIDACEAE *Encyclia phoenicia* (MC)

7. ORCHIDACEAE *Encyclia phoenicia* (flower) (MC)

PLATE 9

1. ORCHIDACEAE *Prosthechea boothiana* (AS)

2. ORCHIDACEAE *Prosthechea cochleata* (FJB)

3. ORCHIDACEAE *Myrmecophila thomsoniana*

4. ORCHIDACEAE *Myrmecophila thomsoniana* (AS)

5. ORCHIDACEAE *Myrmecophila thomsoniana* (KDG)

PLATE 10

1. ORCHIDACEAE *Myrmecophila thomsoniana* var. *minor*
2. ORCHIDACEAE *Dendrophylax fawcettii* (AS)
3. ORCHIDACEAE *Dendrophylax fawcettii* (KDG)
4. ORCHIDACEAE *Oeceoclades maculata* (AS)
5. CANELLACEAE *Canella winterana* (MC)
6. CANELLACEAE *Canella winterana* (KDG)

PLATE 11

1. Lauraceae *Licaria triandra* (AS)

2. Lauraceae *Licaria triandra* (AS)

3. Lauraceae *Ocotea coriacea* (AS)

4. Lauraceae *Ocotea coriacea* (AS)

5. Piperaceae *Peperomia pseudopereskiifolia* (KDG)

6. Piperaceae *Peperomia simplex* (AS)

7. Piperaceae *Piper amalago* (AS)

PLATE 12

1. NYMPHACEAE *Cissampelos pareira*

2. NYMPHACEAE *Cissampelos pareira*

3. PAPAVERACEAE *Argemone mexicana* (FJB)

4. ULMACEAE *Celtis trinervia* (AS)

5. ULMACEAE *Celtis trinervia* (AS)

6. ULMACEAE *Celtis iguanaea* (AS)

PLATE 13

1. MORACEAE *Ficus citrifolia* (AS)

2. MORACEAE *Ficus citrifolia* (AS)

3. MORACEAE *Ficus aurea* (KDG)

4. MORACEAE *Ficus aurea* (AS)

5. PHYTOLACCACEAE *Rivina humilis* (MC)

6. PHYTOLACCACEAE *Trichostigma octandrum* (AS)

7. NYCTAGINACEAE *Pisonia aculeata* (JR)

PLATE 14

PLATE 15

1. CACTACEAE *Nopalea cochenillifera* (KDG)

2. CACTACEAE *Nopalea cochenillifera* (MC)

3. CACTACEAE *Consolea macrantha*

4. CACTACEAE *Consolea macrantha*

5. CACTACEAE *Pilosocereus royenii* (AS)

6. CACTACEAE *Pilosocereus royenii* (MC)

PLATE 16

1. CACTACEAE *Harrisia gracilis* (FR)

2. CACTACEAE *Harrisia gracilis* (FR)

3. CACTACEAE *Selenicereus grandiflorus* (DB)

4. CACTACEAE *Selenicereus grandiflorus* (FJB)

5. AIZOACEAE *Sesuvium portulacastrum* (KDG)

6. AMARANTHACEAE *Blutaparon vermiculare* (KDG)

7. AMARANTHACEAE *Blutaparon vermiculare* (KDG)

PLATE 17

1. BATACEAE *Batis maritima* (MC)

2. PLUMBAGINACEAE *Limonium companyonis*

3. CLUSIACEAE *Clusia rosea* (FJB)

4. CLUSIACEAE *Clusia rosea* (KDG)

5. TILIACEAE *Corchorus siliquosus* (KDG)

6. STERCULIACEAE *Helicteres jamaicensis* (MC)

PLATE 18

1. STERCULIACEAE *Waltheria indica* (FJB)

2. STERCULIACEAE *Neoregnellia cubensis* (FJB)

3. STERCULIACEAE *Neoregnellia cubensis* (FJB)

4. STERCULIACEAE *Melochia tomentosa* (AS)

5. STERCULIACEAE *Sterculia apetala*

6. MALVACEAE *Malvaviscus arboreus* var. *cubensis* (AS)

7. MALVACEAE *Hibiscus pernambucensis* (AS)

PLATE 19

1. MALVACEAE *Hibiscus pernambucensis* (MC)

2. MALVACEAE *Kosteletzkya pentasperma* (AS)

3. MALVACEAE *Thespesia populnea* (KDG)

4. SALICACEAE *Banara caymanensis*

5. SALICACEAE *Banara caymanensis* (MC)

6. SALICACEAE *Casearia staffordiae* (AS)

PLATE 20

1. SALICACEAE *Casearia staffordiae* (AS)

2. SALICACEAE *Casearia guianensis* (AS)

3. SALICACEAE *Casearia hirsuta* (AS)

4. SALICACEAE *Casearia aculeata* (AS)

5. SALICACEAE *Casearia odorata*

PLATE 21

1. SALICACEAE *Casearia odorata* (MC)

2. SALICACEAE *Zuelania guidonia* (AS)

3. SALICACEAE *Xylosma bahamense*

4. SALICACEAE *Xylosma bahamense* (FJB)

5. TURNERACEAE *Turnera triglandulosa* (MC)

6. TURNERACEAE *Turnera triglandulosa* (MC)

PLATE 22

1. PASSIFLORACEAE *Passiflora suberosa* (AS)

2. PASSIFLORACEAE *Passiflora cupraea* (KDG)

3. PASSIFLORACEAE *Passiflora cupraea* (AS)

4. CUCURBITACEAE *Cionosicyos pomiformis* (JFB)

5. CUCURBITACEAE *Cionosicyos pomiformis* (AS)

6. CUCURBITACEAE *Momordica charantia* (FJB)

PLATE 23

1. CUCURBITACEAE *Momordica charantia* (MC)

2. CAPPARACEAE *Capparis ferruginea* (AS)

3. CAPPARACEAE *Capparis ferruginea* (KDG)

4. CAPPARACEAE *Capparis cynophallophora* (AS)

5. CAPPARACEAE *Capparis cynophallophora* (AS)

6. CAPPARACEAE *Capparis flexuosa* (FJB)

PLATE 24

1. CAPPARACEAE *Capparis flexuosa* (FJB)

2. CRUCIFERAE *Cakile lanceolata* (FJB)

3. SAPOTACEAE *Sideroxylon foetidissimum* (AS)

4. SAPOTACEAE *Sideroxylon foetidissimum* (AS)

5. SAPOTACEAE *Sideroxylon salicifolium* (FJB)

PLATE 25

1. Sapotaceae *Sideroxylon horridum*

2. Sapotaceae *Sideroxylon horridum* (MC)

3. Theophrastaceae *Jacquinia proctorii* (KDG)

4. Theophrastaceae *Jacquinia proctorii* (AS)

5. Theophrastaceae *Jacquinia keyensis* (AS)

6. Theophrastaceae *Jacquinia keyensis* (KDG)

PLATE 26

1. MYRSINACEAE *Myrsine acrantha* (AS)

2. MYRSINACEAE *Myrsine acrantha* (AS)

3. CHRYSOBALANACEAE *Chrysobalanus icaco* (MC)

4. CHRYSOBALANACEAE *Chrysobalanus icaco* (MC)

5. LEGUMINOSAE (FABOIDEAE) *Crotalaria verrucosa*

6. LEGUMINOSAE (FABOIDEAE) *Crotalaria verrucosa* (KDG)

PLATE 27

1. LEGUMINOSAE (FABOIDEAE) *Tephrosia cinerea* (KDG)

2. LEGUMINOSAE (FABOIDEAE) *Abrus precatorius* (FJB)

3. LEGUMINOSAE (FABOIDEAE) *Centrosema virginianum* (MC)

4. LEGUMINOSAE (FABOIDEAE) *Clitoria ternatea* (MC)

5. LEGUMINOSAE (FABOIDEAE) *Clitoria ternatea* (KDG)

6. LEGUMINOSAE (FABOIDEAE) *Erythrina velutina* (AS)

PLATE 28

1. Leguminosae (Faboideae) *Mucuna pruriens*

2. Leguminosae (Faboideae) *Mucuna pruriens* (FJB)

3. Leguminosae (Faboideae) *Canavalia rosea* (FJB)

4. Leguminosae (Faboideae) *Dalbergia brownei* (AS)

5. Leguminosae (Faboideae) *Dalbergia ecastaphyllum* (AS)

6. Leguminosae (Faboideae) *Dalbergia ecastaphyllum* (AS)

PLATE 29

1. LEGUMINOSAE (FABOIDEAE) *Piscidia piscipula* (AS)

2. LEGUMINOSAE (FABOIDEAE) *Piscidia piscipula* (AS)

3. LEGUMINOSAE (CAESALPINIOIDEAE) *Caesalpinia bonduc* (MC)

4. LEGUMINOSAE (CAESALPINIOIDEAE) *Haematoxylum campechianum* (MC)

5. LEGUMINOSAE (CAESALPINIOIDEAE) *Haematoxylum campechianum* (KDG)

6. LEGUMINOSAE (CAESALPINIOIDEAE) *Senna ligustrina* (AS)

PLATE 30

1. LEGUMINOSAE (CAESALPINIOIDEAE) *Chamaecrista lineata* (AS)

2. LEGUMINOSAE (CAESALPINIOIDEAE) *Chamaecrista lineata* (AS)

3. LEGUMINOSAE (CAESALPINIOIDEAE) *Bauhinia divaricata* (FJB)

4. LEGUMINOSAE (CAESALPINIOIDEAE) *Bauhinia divaricata* (MC)

5. LEGUMINOSAE (MIMOSOIDEAE) *Adenanthera pavonina* (KDG)

6. LEGUMINOSAE (MIMOSOIDEAE) *Mimosa pudica* (AS)

PLATE 31

1. LEGUMINOSAE (MIMOSOIDEAE) *Leucaena leucocephala* (MC)

2. LEGUMINOSAE (MIMOSOIDEAE) *Leucaena leucocephala* (MC)

3. LEGUMINOSAE (MIMOSOIDEAE) *Calliandra cubensis* (AS)

4. THYMELEACEAE *Daphnopsis occidentalis* (FJB)

5. THYMELEACEAE *Daphnopsis americana* (AS)

6. THYMELEACEAE *Daphnopsis americana* (FJB)

PLATE 32

1. MYRTACEAE *Calyptranthes pallens* (AS)

2. MYRTACEAE *Calyptranthes pallens* (MC)

3. MYRTACEAE *Psidium guajava* (KDG)

4. MYRTACEAE *Myrcianthes fragrans* (MC)

5. MYRTACEAE *Myrcianthes fragrans* (AS)

6. MYRTACEAE *Eugenia axillaris* (KDG)

7. MYRTACEAE *Eugenia axillaris* (FR)

PLATE 33

1. COMBRETACEAE *Terminalia eriostachya* (AS)
2. COMBRETACEAE *Terminalia eriostachya* (AS)
3. COMBRETACEAE *Terminalia catappa* (MC)
4. COMBRETACEAE *Conocarpus erectus* (KDG)
5. COMBRETACEAE *Conocarpus erectus* (MC)

PLATE 34

1. Combretaceae *Laguncularia racemosa* (KDG)

2. Rhizophoraceae *Rhizophora mangle* (KDG)

3. Rhizophoraceae *Rhizophora mangle* (KDG)

4. Olacaceae *Schoepfia chrysophylloides*

5. Olacaceae *Schoepfia chrysophylloides* (KDG)

6. Loranthaceae *Dendropemon caymanensis*

PLATE 35

1. Viscaceae *Phoradendron* spp. (FJB)

2. Apodanthaceae *Pilostyles globosa*

3. Celastraceae *Elaeodendron xylocarpum* var. *attenuatum* (AS)

4. Celastraceae *Elaeodendron xylocarpum* var. *attenuatum* (AS)

5. Celastraceae *Crossopetalum rhacoma* (AS)

6. Celastraceae *Crossopetalum caymanense*

7. Celastraceae *Crossopetalum caymanense* (AS)

PLATE 36

1. CELASTRACEAE *Crossopetalum caymanense* (AS)

2. CELASTRACEAE *Gyminda latifolia* (AS)

3. CELASTRACEAE *Gyminda latifolia* (AS)

4. BUXACEAE *Schaefferia frutescens* (AS)

5. BUXACEAE *Buxus bahamensis* (FJB)

6. EUPHORBIACEAE: Phyllanthoideae: *Savia erythroxyloides* (AS)

7. EUPHORBIACEAE: Phyllanthoideae: *Savia erythroxyloides* (FJB)

8. EUPHORBIACEAE: Phyllanthoideae: *Astrocasia tremula* (AS)

PLATE 37

1. EUPHORBIACEAE: Phyllanthoideae *Securinega acidoton*

2. EUPHORBIACEAE: Phyllanthoideae *Chascotheca neopeltandra*

3. EUPHORBIACEAE: Phyllanthoideae *Chascotheca neopeltandra* (AS)

4. EUPHORBIACEAE: Phyllanthoideae *Phyllanthus caymanensis*

5. EUPHORBIACEAE: Phyllanthoideae *Phyllanthus nutans* (AS)

6. EUPHORBIACEAE: Phyllanthoideae *Phyllanthus angustifolius* (MC)

PLATE 38

1. EUPHORBIACEAE: Phyllanthoideae *Phyllanthus angustifolius* (FJB)

2. EUPHORBIACEAE: Oldfieldioideae *Picrodendron baccatum* (MC)

3. EUPHORBIACEAE: Euphorbioideae *Jatropha divaricata*

4. EUPHORBIACEAE: Euphorbioideae *Jatropha divaricata*

5. EUPHORBIACEAE: Euphorbioideae *Croton linearis* (AS)

6. EUPHORBIACEAE: Euphorbioideae *Croton lucidus* (FJB)

PLATE 39

1. EUPHORBIACEAE: Euphorbioideae *Croton lucidus* (AS)

2. EUPHORBIACEAE: Euphorbioideae *Croton nitens*

3. EUPHORBIACEAE: Euphorbioideae *Argythamnia proctorii* (FJB)

4. EUPHORBIACEAE: Euphorbioideae *Argythamnia proctorii* (JR)

5. EUPHORBIACEAE: Euphorbioideae *Adelia ricinella*

PLATE 40

1. EUPHORBIACEAE: Euphorbioideae *Gymnanthes lucida* (FJB)

2. EUPHORBIACEAE: Euphorbioideae *Hippomane mancinella* (KDG)

3. EUPHORBIACEAE: Euphorbioideae *Hippomane mancinella* (KDG)

4. EUPHORBIACEAE: Euphorbioideae *Euphorbia graminea* (JR)

5. EUPHORBIACEAE: Euphorbioideae *Chamaesyce bruntii*

6. EUPHORBIACEAE: Euphorbioideae *Chamaesyce bruntii*

7. RHAMNACEAE *Colubrina cubensis* (FJB)

PLATE 41

1. RHAMNACEAE *Colubrina arborescens*
2. RHAMNACEAE *Colubrina elliptica* (AS)
3. VITACEAE *Cissus trifoliata*
4. VITACEAE *Cissus verticillata*
5. VITACEAE *Cissus verticillata*
6. VITACEAE *Cissus microcarpa* (FJB)

PLATE 42

1. SURINACEAE *Suriana maritima* (MC)

2. SAPINDACEAE *Allophylus cominia* var. *caymanensis* (AS)

3. SAPINDACEAE *Allophylus cominia* var. *caymanensis* (AS)

4. SAPINDACEAE *Exothea paniculata* (KDG)

5. SAPINDACEAE *Exothea paniculata* (KDG)

6. SAPINDACEAE *Dodonaea viscosa* (AS)

PLATE 43

1. SAPINDACEAE *Hypelate trifoliata* (MC)

2. SAPINDACEAE *Hypelate trifoliata* (MC)

3. BURSERACEAE *Bursera simaruba*

4. BURSERACEAE *Bursera simaruba* (KDG)

5. BURSERACEAE *Bursera simaruba* (FJB)

6. BURSERACEAE *Bursera simaruba* (FJB)

PLATE 44

1. ANACARDIACEAE *Mangifera indica* (KDG)

2. ANACARDIACEAE *Mangifera indica* (KDG)

3. ANACARDIACEAE *Metopium toxiferum* (FJB)

4. ANACARDIACEAE *Comocladia dentata* (KDG)

5. ANACARDIACEAE *Comocladia dentata* (FJB)

6. SIMAROUBACEAE *Alvaradoa amorphoides* (FJB)

7. RUTACEAE *Zanthoxylum flavum*

PLATE 45

1. Rutaceae *Zanthoxylum coriaceum*

2. Rutaceae *Zanthoxylum coriaceum* (FJB)

3. Rutaceae *Amyris elemifera* (AS)

4. Rutaceae *Trichila glabra* (AS)

5. Rutaceae *Trichila glabra* (AS)

6. Rutaceae *Trichila havanensis*

PLATE 46

1. RUTACEAE *Swietenia mahagoni* (MC)

2. RUTACEAE *Swietenia mahagoni* (KDG)

3. RUTACEAE *Swietenia mahagoni* (MC)

4. RUTACEAE *Cedrela odorata* (MC)

5. RUTACEAE *Cedrela odorata* (KDG)

6. RUTACEAE *Cedrela odorata*

PLATE 47

1. RUTACEAE *Tribulus cistoides*

2. RUTACEAE *Erythroxylum areolatum* (FR)

3. RUTACEAE *Erythroxylum areolatum* (FJB)

4. RUTACEAE *Erythroxylum rotundifolium* (MC)

5. RUTACEAE *Erythroxylum confusum* (AS)

6. MALPIGHIACEAE *Malpighia cubensis* (AS)

PLATE 48

1. MALPIGHIACEAE *Malpighia cubensis*
2. MALPIGHIACEAE *Bunchosia media* (AS)
3. POLYGALACEAE *Polygala propinqua*
4. POLYGALACEAE *Polygala propinqua*
5. ARALIACEAE *Dendropanax arboreus* (AS)
6. ARALIACEAE *Dendropanax arboreus* (AS)
7. GENTIANACEAE *Eustoma exaltatum* (AS)

PLATE 49

1. Apocynaceae *Plumeria obtusa* (KDG)

2. Apocynaceae *Plumeria obtusa* (AS)

3. Apocynaceae *Tabernaemontana laurifolia*

4. Apocynaceae *Tabernaemontana laurifolia* (MC)

5. Apocynaceae *Tabernaemontana laurifolia* (MC)

PLATE 50

1. APOCYNACEAE *Echites umbellatus* (FJB)

2. APOCYNACEAE *Rhabdadenia biflora* (MC)

3. APOCYNACEAE *Rhabdadenia biflora* (KDG)

4. ASCLEPIADACEAE *Calotropis procera* (MC)

5. ASCLEPIADACEAE *Calotropis procera* (KDG)

6. ASCLEPIADACEAE *Sarcostemma clausum* (FJB)

PLATE 51

1. ASCLEPIADACEAE *Sarcostemma clausum* (KDG)

2. SOLANACEAE *Solanum americanum*

3. SOLANACEAE *Solanum havanense* (AS)

4. SOLANACEAE *Solanum havanense* (AS)

5. SOLANACEAE *Cestrum diurnum* (AS)

6. SOLANACEAE *Cestrum diurnum* (AS)

7. SOLANACEAE *Solandra longiflora* (AS)

8. CONVOLCULACEAE *Evolvulus squamosus* (FJB)

PLATE 52

1. CONVOLCULACEAE *Jacquemontia havanensis* (KDG)

2. CONVOLCULACEAE *Merremia quinquefolia* (JR)

3. CONVOLCULACEAE *Ipomoea quamoclit* (AS)

4. CONVOLCULACEAE *Ipomoea pes-caprae* (FJB)

5. BORAGINACEAE *Argusia gnaphalodes* (MC)

6. BORAGINACEAE *Argusia gnaphalodes* (MC)

PLATE 53

1. BORAGINACEAE *Tournefortia astrotricha* var. *astrotricha* (KDG)

2. BORAGINACEAE *Tournefortia astrotricha* var. *subglabra* (AS)

3. BORAGINACEAE *Heliotropium humifusum* (MC)

4. BORAGINACEAE *Ehretia tinifolia* (MC)

5. BORAGINACEAE *Bourreria venosa* (MC)

6. BORAGINACEAE *Cordia sebestena* (AS)

PLATE 54

1. BORAGINACEAE *Cordia sebestena* (AS)

2. BORAGINACEAE *Cordia gerascanthus* (FJB)

3. BORAGINACEAE *Cordia gerascanthus* (AS)

4. BORAGINACEAE *Cordia laevigata* (KDG)

5. BORAGINACEAE *Cordia laevigata* (KDG)

6. BORAGINACEAE *Cordia brownei* (AS)

PLATE 55

1. BORAGINACEAE *Cordia globosa* (AS)

2. BORAGINACEAE *Cordia globosa*

3. BORAGINACEAE *Rochefortia acanthophora* (AS)

4. BORAGINACEAE *Rochefortia acanthophora* (AS)

5. BORAGINACEAE *Rochefortia acanthophora* (AS)

6. VERBENACEAE *Stachytarpheta jamaicensis* (FJB)

7. VERBENACEAE *Lantana bahamensis* (KDG)

PLATE 56

1. VERBENACEAE *Lantana bahamensis* (KDG)

2. VERBENACEAE *Lantana involucrata* (FJB)

3. VERBENACEAE *Citharexylum spinosum* (MC)

4. VERBENACEAE *Citharexylum spinosum* (AS)

5. VERBENACEAE *Aegiphila caymanensis* (FJB)

6. VERBENACEAE *Petitia domingensis* (AS)

7. VERBENACEAE *Petitia domingensis* (MC)

PLATE 57

1. VERBENACEAE *Avicennia germinans* (KDG)

2. VERBENACEAE *Avicennia germinans* (KDG)

3. LABIATAE *Salvia occidentalis* (AS)

4. LABIATAE *Salvia caymanensis* (MC)

5. LABIATAE *Salvia caymanensis* (MC)

6. OLEACEAE *Forestiera segregata* (AS)

7. OLEACEAE *Forestiera segregata* (AS)

PLATE 58

1. OLEACEAE *Chionanthus caymanensis* (AS)

2. OLEACEAE *Chionanthus caymanensis*

3. OLEACEAE *Chionanthus caymanensis* (AS)

4. SCROPHULARIACEAE *Agalinis kingsii*

5. MYOPORACEAE *Bontia daphnoides* (AS)

6. BIGNONIACEAE *Crescentia cujete* (AS)

7. BIGNONIACEAE *Crescentia cujete* (AS)

8. BIGNONIACEAE *Crescentia cujete* (KDG)

PLATE 59

1. BIGNONIACEAE *Catalpa longissima*

2. BIGNONIACEAE *Tabebuia heterophylla* (MC)

3. BIGNONIACEAE *Tabebuia heterophylla* (FJB)

4. BIGNONIACEAE *Tecoma stans* (KDG)

5. ACANTHACEAE *Blechum pyramidatum* (KDG)

6. ACANTHACEAE *Asystasia gangetica* (KDG)

7. ACANTHACEAE *Asystasia gangetica* (KDG)

PLATE 60

1. ACANTHACEAE *Ruellia nudiflora* (FJB)

2. GOODENIACEAE *Scaevola plumieri* (MC)

3. GOODENIACEAE *Scaevola taccada* (KDG)

4. GOODENIACEAE *Scaevola taccada* (KDG)

5. RUBIACEAE *Exostema caribaeum*

6. RUBIACEAE *Randia aculeata* (MC)

7. RUBIACEAE *Randia aculeata* (MC)

PLATE 61

1. RUBIACEAE *Scolosanthus roulstonii* (AS)

2. RUBIACEAE *Scolosanthus roulstonii* (AS)

3. RUBIACEAE *Strumpfia maritima* (FJB)

4. RUBIACEAE *Erithalis fruticosa*

5. RUBIACEAE *Chiococca alba* (AS)

6. RUBIACEAE *Chiococca alba* (FJB)

7. RUBIACEAE *Antirhea lucida* (AS)

8. RUBIACEAE *Antirhea lucida* (AS)

PLATE 62

1. RUBIACEAE *Guettarda elliptica*

2. RUBIACEAE *Guettarda elliptica* (AS)

3. RUBIACEAE *Hamelia cuprea* (KDG)

4. RUBIACEAE *Morinda royoc* (KDG)

5. RUBIACEAE *Morinda royoc* (KDG)

6. RUBIACEAE *Psychotria nervosa* (MC)

7. RUBIACEAE *Psychotria nervosa* (KDG)

8. RUBIACEAE *Faramea occidentalis* (MC)

PLATE 63

1. COMPOSITAE *Vernonia divaricata* (KDG)

2. COMPOSITAE *Baccharis dioica* (AS)

3. COMPOSITAE *Pluchea odorata* (AS)

4. COMPOSITAE *Iva imbricata*

5. COMPOSITAE *Iva imbricata* (AS)

6. COMPOSITAE *Iva cheiranthifolia*

7. COMPOSITAE *Iva cheiranthifolia*

PLATE 64

1. COMPOSITAE *Borrichia arborescens* (FJB)

2. COMPOSITAE *Verbesina caymanensis* (WP)

3. COMPOSITAE *Verbesina caymanensis*

4. COMPOSITAE *Salmea petrobioides*

5. COMPOSITAE *Pectis caymanensis* (MC)

6. COMPOSITAE *Pectis caymanensis*

7. COMPOSITAE *Emilia fosbergii* (JDM)

Ipomoea L.

Creeping or twining, sometimes erect herbs or shrubs, rarely trees; leaves entire, angled, lobed, or palmately or pinnately divided. Flowers often large, soon-withering, solitary or in stalked axillary clusters, rarely aggregated in terminal panicles. Sepals 5, equal or unequal, persistent and often enlarged in fruit; corolla bell- or funnel-shaped, the limb entire or 5-angled, usually plaited. Stamens included or sometimes exserted; pollen spiny. Ovary 2–4-celled; style simple; stigma capitate. Capsule 2–4-celled, 2–4-seeded; seeds glabrous or often pubescent.

— A genus of between 400 and 500 species, widely distributed in tropical and warm-temperate regions. *Ipomoea batatas* (L.) Lam., the sweet potato, is one of the most important food-crops of tropical countries. Other species are often grown for ornament.

KEY TO SPECIES

1. Corolla bright red or scarlet, with exserted stamens and style:

   2. Leaves pinnate with narrow linear segments: — I. quamoclit

   2. Leaves entire or usually somewhat angled or lobed: — I. hederifolia

1. Corolla various colours or white, but never bright red; stamens included (except in *I. macrantha*, with large white flowers):

   3. Leaf-blades deeply 3-lobed or else digitately divided into 3–7 leaflets, cut nearly to junction of the petiole:

      4. Corolla 1.5–2 cm long: — I. triloba

      4. Corolla 4–7 cm long:

         5. Sepals and ovary glabrous; wild plants with twining or climbing stems:

            6. Sepals ca. 6 mm long: — [I. cairica]

            6. Sepals 10–20 mm long: — I. indica var. acuminata

         5. Sepals either ciliate or else pilose at the base:

            7. Trailing cultivated vine with large edible tuber; corolla pinkish-mauve: — [I. batatas: see p. 552]

            7. Twining wild vine; corolla pink: — I. passifloroides

   3. Leaf-blades entire or somewhat lobed but not divided nearly to junction of the petiole:

      8. Sepals 15–20 mm long:

         9. Corolla pink, mauve, or bluish-purple, with funnel-shaped tube not over 5 cm long; sepals attenuate at the apex: — I. indica var. acuminata

         9. Corolla white, with cylindrical tube 6–7 cm long; sepals broad at the apex; flowers opening at night: — I. violacea

      8. Sepals less than 15 mm long:

         10. Sepals and ovary glabrous; wild plants:

            11. Stems trailing on sand, rooting at nodes; leaves rather thick or leathery, often notched at apex:

               12. Leaves broadly oblong to suborbicular, rarely twice as long as broad; corolla pink or purple: — I. pes-caprae

               12. Leaves oblong or narrowly triangular, 2–3 times as long as broad; corolla white: — I. imperati

            11. Stems climbing or twining over bushes or fences, not rooting at nodes; leaves thin, acuminate at the apex: — I. tiliacea

         10. Sepals ciliate and ovary hairy; plants cultivated for edible tubers: — [I. batatas, p. 552]

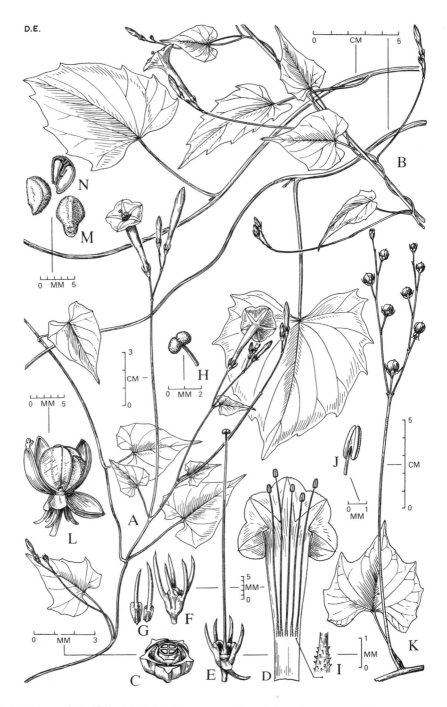

FIG. 202 **Ipomoea hederifolia.** A, B, habit. C, cross-section through base of flower. D, corolla spread out to show stamens. E, calyx with pistil. F, calyx alone. G, separated sepals. H, stigma. I, base of filament. J, anther. K, fruiting inflorescence. L, capsule after dehiscence. M, two views of seed. N, section through seed. (D.E.)

Ipomoea quamoclit L., *Sp. Pl.* 1: 159 (1753). PLATE 52.

A slender, glabrous, annual twiner; leaves oblong to ovate in outline, 4–9 cm long, pinnately divided into linear segments mostly 1 mm wide or less, a small finely divided accessory leaf resembling a stipule arising at the base of each regular leaf. Peduncles 1–6-flowered; sepals 4–6 mm long, mucronulate; corolla 2.5–4 cm long with flat, lobate limb. Ovary 4-celled. Capsules ovoid, ca. 10 mm long.

CAYMAN BRAC: *Proctor 29367.*

— Southeastern U. S. A., West Indies and continental tropical America, introduced in the Old World tropics; the Cayman Brac specimens were found in roadside thickets.

Ipomoea hederifolia L., *Syst. Nat.* ed. 10, 2: 925 (1759). **FIG. 202.**

Ipomoea coccinea of Griseb. (1862), not L. (1753).

A slender, annual twiner, glabrous or nearly so; leaves long-petiolate, the simple, cordate blades very broadly ovate, up to about 7 cm long and broad, acuminate at the apex and with margins entire or somewhat angular and with a few broad, lobe-like teeth. Flowers few in elongate dichasial cymes; sepals ca. 3 mm long with a subulate tooth of about the same length or longer arising from the back of each; corolla 2–4 cm long, the limb obscurely 5-lobed. Ovary 4-celled. Capsules globose, 6–7 mm in diameter.

GRAND CAYMAN: *Brunt 1688.*

— Eastern and southern U. S. A., West Indies and continental tropical America, in open waste ground and moist thickets.

Ipomoea triloba L., *Sp. Pl.* 1: 161 (1753). **FIG. 203.**

A slender annual twiner, the usually glabrous stems mostly trailing; leaves long-petiolate and with blades cordate and usually deeply 3–5-lobed or rarely entire, more or less ovate or sagittate in outline, mostly 2–5 cm long, the apex acuminate. Flowers solitary or few, clustered on a rather long peduncles; sepals ca. 6 mm long, acuminate, hairy; corolla pink or rose. Ovary 2-celled. Capsules subglobose, ca. 4 mm in diameter.

GRAND CAYMAN: *Brunt 1941, 2109; Hitchcock; Kings GC 85, GC 330; Millspaugh 1385; Proctor 27936, 27955; Sachet 373, 419.* LITTLE CAYMAN: *Proctor 35141, 35184.* CAYMAN BRAC: *Proctor 28977.*

— Southeastern U. S. A., West Indies and continental tropical America; naturalised in the Old World tropics; a weed of sandy ground, moist pastures, clearings and thickets.

[*Ipomoea cairica* (L.) Sweet, *Hort. Brit.* 287 (1827).

A slender, nearly glabrous trailing or climbing vine, the stems becoming subwoody. Leaves palmately divided into 5 lanceolate to ovate segments, the middle one longest and 1.5–4 cm long. Flowers 1–10 in umbel-like cymes on a short peduncle; sepals ovate, acute, ca. 6 mm long; corolla lavender or light mauve, 4–5 cm long. Capsules ca. 10 mm in diameter; seeds hairy.

GRAND CAYMAN: *Proctor 11967.*

— Native of tropical Asia and Africa, now introduced and naturalised in most warm countries.]

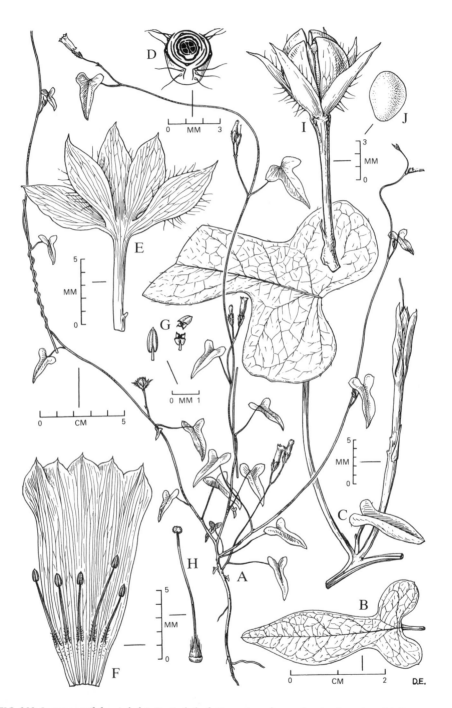

FIG. 203 **Ipomoea triloba.** A, habit. B, single leaf. C, portion of stem showing insertion of inflorescence. D, cross-section through base of flower. E, calyx spread out. F, corolla spread out, showing stamens. G, anther, whole and cut crosswise. H, pistil. I, capsule. J, seed. (D.E.)

Ipomoea indica (Burm.) Merrill, *Interpr. Rumph. Herb. Amboin.* 445 (1917). var. *acuminata* (Vahl) Fosberg, *Bot. Notiser* 129: 38 (1976).

> *Ipomoea acuminata* (Vahl) Roem. & Schult. in L. (1819).
>
> *I. cathartica* Poir. in Lam. (1816).
>
> *I. jamaicensis* of Hitchcock (1893), not G. Don (1837).

Slender perennial vine climbing to 5 m or more, the young stems more or less puberulous. Leaves long-petiolate, the blades cordate, ovate, mostly 4–8 cm long, the margins varying from entire to deeply 3-lobed, acuminate at the apex. Flowers solitary or few in a small, loose cluster; peduncles rather stout, 2–4 cm long, appressed-puberulous; sepals lance-acuminate, unequal, mostly 15–20 mm long; corolla mostly 5.5–7 cm long. Capsules subglobose, 10–12 mm in diameter; seeds puberulous.

GRAND CAYMAN: *Brunt* 2140; *Hitchcock*; *Lewis* 3830, 3844a; *Millspaugh* 1244, 1246, 1372, 1381, 1403; *Proctor* 15107, 15108, 27980; *Sachet* 420. LITTLE CAYMAN: *Kings* LC 19; *Proctor* 35112. CAYMAN BRAC: *Kings* CB 74; *Millspaugh* 1227, 1235.

— Pantropical; common in roadside thickets and along the borders of moist woodlands.

Ipomoea passifloroides House, *Ann. New York Acad. Sci.* 18: 230 (1908).

Slender perennial twining vine, densely short-pilose. Leaves narrowly ovate, 4–7 cm long, deeply 3-lobed, 7-nerved from the base. Peduncles short, 5 cm long or less, 2–5-flowered; bracts ovate, obtuse, 6–8 mm long; pedicels densely hirsute; sepals subequal, 7–9 mm long, pilose outside toward the base, glabrous within; corolla bright pink, 4–5 cm long. Capsules glabrous, 8–10 mm in diameter; seeds pubescent.

GRAND CAYMAN: *Proctor* 47346, found in rocky scrub woodland ca. 3.5 km due NW of East End Village, and *Proctor* 47356, found along the Canaan Road SSE of Hutland.

— Cuba, known only from the Sierra Maestra region.

Ipomoea violacea L. *Sp. Pl.* 1: 161 (1753). **FIG. 204.**

> *Ipomoea macrantha* Roem. & Schult. in L. (1819).
>
> *Ipomoea tuba* (Schlecht.) G. Don (1837).

COWSLIP

Perennial, glabrous, creeping or high-climbing vine (to 12 m or more), with white latex, the older stems becoming subwoody. Leaves rather fleshy, long-petiolate, the blades deeply cordate, very broadly ovate, 4–15 cm long, abruptly acuminate at the apex, the margins entire or with a short, acuminate lobe on either side. Flowers nocturnal, solitary or sometimes 2 together on a stout peduncle; sepals obtuse, 20 mm long; corolla white, the limb 7–8 cm across. Capsules subglobose, 2–2.5 cm in diameter, enclosed by the enlarged fleshy sepals; seeds densely hairy on the angles.

GRAND CAYMAN: *Brunt* 1654, 1746, 1877; *Fawcett*; *Hitchcock*; *Proctor* 11968. LITTLE CAYMAN: *Kings* LC 45; *Proctor* 28045. CAYMAN BRAC: *Kings* CB 103; *Millspaugh* 1234; *Proctor* 28930.

— Pantropical, chiefly in sandy, swampy, or rocky coastal thickets, often in exposed situations. The cliffs toward the eastern end of Cayman Brac are in many places festooned with this vine.

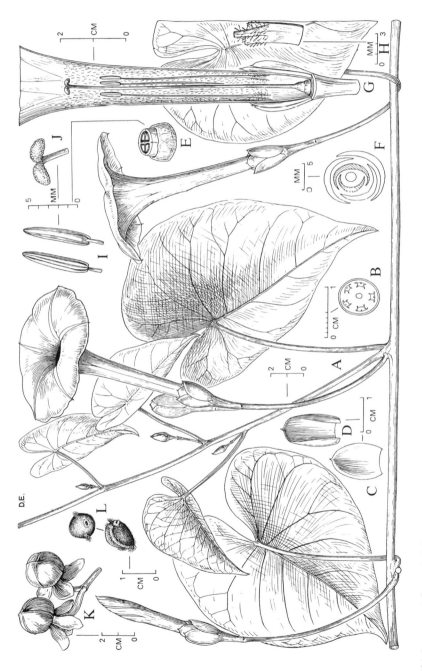

FIG. 204 **Ipomoea violacea**. A, habit. B, cross-section of stem. C, D, two sepals. E, cross-section through ovary. F, diagram of sepals. G, base of corolla opened to show stamens and pistil. H, base of filament. I, two views of anther. J, stigma. K, capsules. L, seeds. (D.E.)

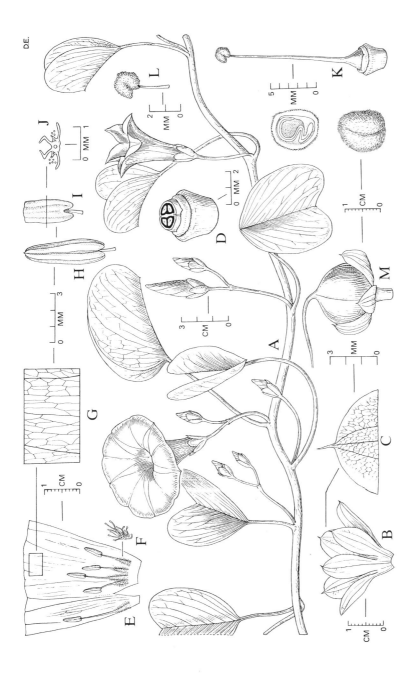

FIG. 205 **Ipomoea pes-caprae** ssp. **brasiliensis**. A, habit. B, calyx spread out. C, apex of sepal enlarged. D, cross-section through ovary. E, base of corolla spread out to show stamens. F, hairs from base of filament. G, venation of corolla-tube. H, anther, ventral view. I, portion of anther, dorsal view. J, cross-section of anther. K, pistil. L, stigma. M, capsule, N, seed. O, section through seed. (D.E.)

Ipomoea pes-caprae (L.) R. Br. in Tuckey, *Narrat. Exped. Zaire* 477 (1818) ssp. *brasiliensis* (L.) Ooststr., *Blumea* 3: 533 (1940). **FIG. 205.** PLATE 52.

BAY VINE

A perennial, subsucculent trailing vine, the glabrous stems reaching 6 m or more in length; leaves on petioles up to 15 cm long, the blades entire, mostly 5–10 cm long, broadly notched at the apex. Flowers 1–few on stout peduncles; pedicels 1–3 cm long or more; sepals ovate-oblong, apiculate, the outer 6–10 mm long, the inner 8–15 mm long; corolla 4–5 cm long. Ovary 2-celled. Capsules subglobose, ca. 1.5 cm in diameter; seeds brownish-pubescent.

GRAND CAYMAN: *Kings GC 67; Lewis GC. 24; Maggs II 58; Proctor 15062.* LITTLE CAYMAN: *Kings LC 18; Proctor 28165.* CAYMAN BRAC: *Kings CB 19; Millspaugh 1228; Proctor 28997.*

— Pantropical on sandy seashores. The subspecies *pes-caprae* is restricted to certain Old World localities.

Ipomoea imperati (Vahl) Griseb., *Cat. Pl. Cub.* 203 (1866); LaValva & Sabato, *Taxon* 32: 110–132 (1983).

Ipomoea carnosa R. Br. (1810).

Trailing herb, the glabrous stems extensively creeping on or under sand and rooting at most nodes; leaves rather long-petiolate, the leathery blades narrowly oblong or triangular, mostly 2.5–6 cm long, often slightly notched at the apex, the margins entire or sometimes irregularly lobed. Flowers solitary on short or long peduncles; sepals oblong, acute, 10–15 mm long; corolla 3.5–5 cm long, white, the tube often yellowish within and with a purplish spot at the base. Capsules globose, 10–15 mm in diameter; seeds smooth.

GRAND CAYMAN: *Millspaugh 1310.* LITTLE CAYMAN: *Proctor 28053.* CAYMAN BRAC: *Millspaugh 1222.*

— Pantropical on sandy beaches near the sea, apparently never very common.

Ipomoea tiliacea (Willd.) Choisy in DC., *Prodr.* 9: 375 (1845).

Ipomoea fastigiata (Roxb.) Sweet (1827).

HOG SLIP, WILD SLIP

Herbaceous twining vine, glabrous or nearly so, the roots sometimes with tubers, the stems climbing to 5 m high or more; leaves long-petiolate, the blades shallowly cordate, broadly ovate, 4–7 cm long or more, acuminate and with entire margins. Flowers few to many in cymes on long peduncles; sepals lanceolate, 7–10 mm long, smooth and scarious, sharply pointed; corolla light mauve to pink or purplish with paler or whitish tube and darker spot at the base within, 3.5–5 cm long. Ovary 2-celled. Capsules depressed-globose, ca. 6 mm long and 8 mm broad; seeds nearly black, glabrous.

GRAND CAYMAN: *Brunt 1716, 1936; Hitchcock; Kings GC 75, GC 299; Lewis 3844b.*

— Pantropical, frequent in roadside thickets, along borders of woodlands, and in overgrown pastures.

[*Ipomoea batatas* (L.) Lam., the sweet potato, is or has been commonly cultivated in Grand Cayman (*Millspaugh 1290*) and Cayman Brac.]

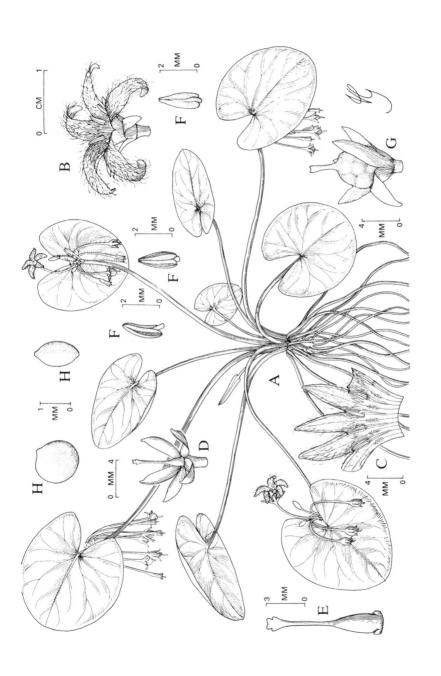

FIG. 206 Nymphoides indica. A, habit. B, flower. C, corolla opened out. D, calyx and pistil. E, ovary style and stigma. F, F, F, stamens. G, fruit. H, H, two views of seed. (G.)

MENYANTHACEAE

Perennial herbs, mostly aquatic; leaves alternate, simple or trifoliate, and entire or toothed; stipules absent. Flowers regular, perfect, and solitary or in umbel-like cymes; calyx 5–6-cleft, persistent; corolla gamopetalous, rotate or funnel-shaped, the 5 or 6 lobes induplicate-valvate in bud. Stamens equal in number to the corolla-lobes and alternate with them, inserted in the corolla-tube; filaments free; anthers 2-celled, opening lengthwise. Ovary superior or partly inferior, 1-celled, with numerous ovules on usually 2 parietal placentas; style simple, with 2–3-lobed stigma. Fruit a capsule, indehiscent, bursting irregularly, or 2-valved; seeds with copious endosperm.

— A small family of world-wide distribution, consisting of 5 genera and about 40 species.

Nymphoides Séguier

Aquatic perennial herbs with fleshy rootstocks and elongate leafy flowering stems; leaves floating, simple and deeply cordate with entire margins. Flowers in clusters closely subtended by a leaf-blade, thus appearing to arise from the top of an elongate petiole, yellow or white, the parts in 5s; corolla deeply lobed, the lobes with fringed margins and often hairy within. Stamens inserted near the base of the corolla, with appendages on the corolla between them; anthers sagittate. Capsule rupturing irregularly.

— A cosmopolitan genus of about 20 species.

Nymphoides indica (L.) Ktze., *Révis. Gen. Pl.* 2: 429 (1891). **FIG. 206.**

Nymphoides humboldtiana (Kunth) Ktze. (1891).

WATER SNOWFLAKE

Leaves glabrous, broadly ovate to nearly orbicular, 3–15 cm wide, often purplish beneath. Pedicels 2–5 cm long, erect when flowering and deflexed in fruit; calyx-lobes ca. 6 mm long; corolla 2–2.5 cm across when expanded, white, yellow at the base. Flowers heterostylous, the yellow filaments 1 or 4 mm long, the ovary and style 10 or 6 mm long. Capsules ovoid, ca. 6 mm long, many-seeded; seeds ca. 1.5 mm in diameter, smooth.

GRAND CAYMAN: *Brunt* 1793, 2165; *Kings* GC 210; *Proctor* 27956; *Sachet* 450.

— Pantropical, in ponds or small pools, sometimes on mud where water has receded. This species is presumably tolerant of somewhat brackish conditions.

HYDROPHYLLACEAE

Annual or perennial herbs, rarely shrubs, often hairy or scabrid, sometimes armed with spines; leaves alternate or rarely opposite, sometimes in basal rosettes, entire to pinnately or palmately divided; stipules absent. Flowers perfect and regular, borne in dichasial, umbel-like, or helicoid cymes, or sometimes solitary in the leaf-axils. Calyx 5-lobed, imbricate, the lobes often with appendages between; corolla gamopetalous, 5-lobed, the lobes imbricate or contorted. Stamens usually 5, inserted on the corolla-tube near its base and alternate with the lobes; filaments often dilated at the base, or subtended by appendages; anthers 2-celled, opening lengthwise. Ovary superior or partly inferior, 2-

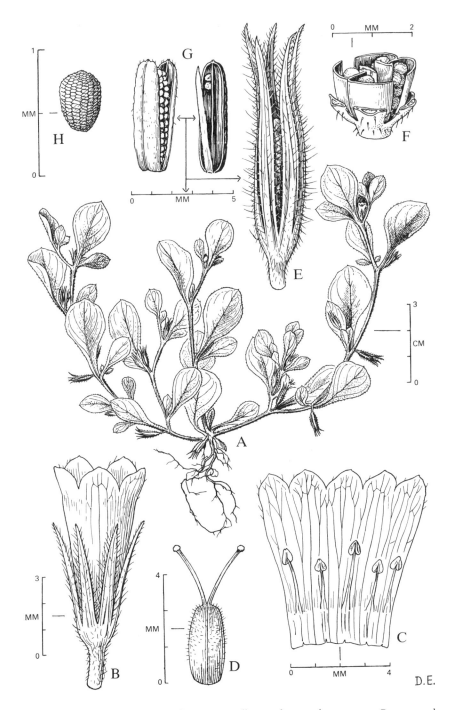

FIG. 207 **Nama jamaicensis.** A, habit, B, flower. C, corolla spread out to show stamens. D, ovary and styles. E, fruiting calyx. F, cross-section through capsule. G, two views of capsule after dehiscence. H, seed. (D.E. ex St.)

carpellate but usually 1-celled with 2 parietal placentas meeting in the centre, the ovules 4–numerous; style 1 or 2 (rarely more). Fruit a capsule, dehiscent or rarely indehiscent; seeds variously sculptured; endosperm copious or thin.

— A widespread family of 20 genera and about 270 species, best represented in N. America, absent from Australia.

Nama L., nom. cons.

Prostrate to erect, annual or perennial herbs, sometimes partly woody; leaves alternate and mostly entire, rarely toothed. Flowers solitary or paired in the axils of the upper leaves, or several in reduced lateral or terminal cymes; calyx divided nearly to the base, with narrow, subequal lobes elongating in fruit; corolla white to purple, tubular or funnel-shaped. Stamens usually included, unequally inserted or unequal in length; filaments usually glabrous, dilated or appendaged at the base. Styles 2, usually free or sometimes partly united; stigmas small, capitate. Capsules dehiscent; seeds brown and variously pitted, reticulate or smooth.

— A chiefly American genus of between 40 and 50 species.

Nama jamaicensis L., *Syst. Nat.* ed. 10, 2: 950 (1759). **FIG. 207.**

Annual herb with spreading prostrate branches, these pubescent and more or less winged by the decurrent petiole-margins; leaves obovate to spatulate, 1–6 cm long or more, obtuse at the apex. Flowers solitary or few, subsessile or stalked; sepals linear, ca. 7 mm long in flower, longer in fruit; corolla white, 5–6 mm long, narrowly funnel-shaped with lobes 1–1.5 mm long. Capsules ca. 5 mm long; seeds light brown, minutely pitted.

GRAND CAYMAN: *Hitchcock.*

— Florida, Texas, West Indies, and continental tropical America, a frequent weed of shady waste places.

BORAGINACEAE

Annual or perennial herbs, or else shrubs, trees or woody vines, usually with scabrid, setose, or hispid indument, rarely tomentose or glabrous; leaves alternate (rarely opposite or whorled), simple, entire or toothed; stipules none. Flowers usually regular, perfect or rarely polygamous by abortion, borne in usually dichotomous cymose spikes, racemes, heads or panicles. Calyx usually 5-toothed or -lobed, more or less bell-shaped, imbricate or often open in bud, usually persistent and often enlarging in fruit. Corolla gamopetalous, usually 5-lobed (rarely 6–8-lobed), tubular, funnel-, bell- or salver-shaped or subrotate, the lobes imbricate or twisted in bud, sometimes with pleats or appendages in the tube or partly closing the throat. Stamens as many as the corolla-lobes and alternating with them, inserted in the corolla-throat or -tube; filaments free; anthers 2-locular, opening lengthwise, attached above the base and more or less 2-lobed below the point of attachment. Disc ring-like, entire or 5-lobed, or apparently absent. Ovary superior, 2-locular and 2-carpellate, but often becoming falsely 4-locular at maturity, simple or deeply 4-lobed; ovules usually 4, 2 in each carpel; styles 1 or 2, terminal on simple ovaries, or sometimes arising from between the lobes on lobate ones, simple or 1–2-forked at the apex. Fruit of 4 1-seeded nutlets or a 1–4-seeded nut or drupe; seeds with or without endosperm.

— A cosmopolitan family of about 100 genera and 2,000 species.

KEY TO GENERA

1. Style 1, with solitary simple or 2-lobed stigma, or stigma sessile, then usually large and peltate or conical; flowers in simple or forked one-sided spikes, uncoiling in development; herbs or shrubs:

  2. Shrubs, erect and up to 2 m tall or more, or vine-like; fruits drupe-like, fleshy at least at first, later drying into 2 or 4 nutlets:

    3. Leaves succulent, narrowly oblanceolate, covered with silky tomentum; fruits hollowed at the base: **Argusia**

    3. Leaves not as above; fruits not hollowed at the base: **Tournefortia**

  2. Annual herbs or low, much-branched or dense, small-leaved shrubs; fruits dry, breaking up into 2 or 4 nutlets: **Heliotropium**

1. Styles 2:

  4. Styles free or united below, each simple or forked; flowers in cymose panicles or heads, these not one-sided or uncoiling in development; shrubs or trees:

    5. Stigmas 2 (styles simple):

      6. Expanded corolla 4–5 mm across, the lobes longer than the tube; ripe fruits yellow or ultimately black: **Ehretia**

      6. Expanded corolla more than 8 mm across, the tube longer than the lobes; ripe fruits red: **Bourreria**

    5. Stigmas 4 (styles forked):   **Cordia**

  4. Styles basally coherent or spreading, slender, persistent; ripe fruits red: **Rochefortia**

Argusia Amman ex Boehmer

Herbs, shrubs, or trees clothed with whitish-silky hairs. Leaves fleshy, alternate, more or less crowded, linear to oblong or obovate. Flowers in scorpioid or corymbose cymes; calyx sessile or pedicellate, the lobes cuneate to orbicular; corolla small, white, salver-shaped, the 5 valvate lobes shorter than the nearly cylindrical tube; stamens included, with short filaments and elongate anthers; style simple. Fruit drupe-like, dry and bony, hollowed at the base; carpels more or less embedded in hard, corky exocarp, at maturity the 2 nutlets ultimately separating.

— A small genus of 3 species, all confined to seacoasts or saline shores, one (herbaceous) occurring in temperate Asia, another (a tree) widespread in the Old World tropics, and the third (a shrub) occurring in the West Indies.

Argusia gnaphalodes (L.) Heine, *Fl. Nouv. Caléd.* 7: 108 (1976). **FIG. 209.** Plate 52.

Heliotropium gnaphalodes L. (1759).
Tournefortia gnaphalodes (L.) R. Br. (1819).
Mallotonia gnaphalodes (L.) Britton (1915).
Messerschmidia gnaphalodes (L.) I. M. Johnston (1935).

LAVENDER, SEA-LAVENDER

Dense, mound-like shrub to 2 m tall; leaves linear-oblanceolate, 3–6 cm long or more, 3–6 mm broad below the obtuse apex. Inflorescence stalked, scarcely exceeding the leaves; flowers in dense scorpioid cymes becoming head-like in fruit; calyx-lobes 2–3 mm long; corolla slightly exceeding the calyx, the tube pubescent outside. Fruits dark brown, glabrous, 5–6 mm long.

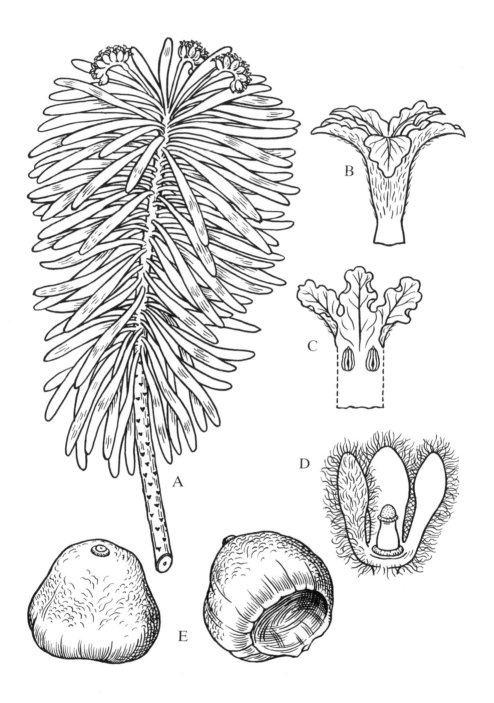

FIG. 209 **Argusia gnaphalodes.** A, branch with leaves and inflorescence, × ³/₈. B, corolla, × 6. C, corolla opened to show stamens, × 6. D, calyx with two sepals removed to show pistil, × 6. E, two views of fruit, × 6. (St.)

GRAND CAYMAN: *Fawcett*; *Hitchcock*; *Kings GC 263*; *Proctor 15047*; *Sachet 400*. LITTLE CAYMAN: *Kings LC 56*; *Proctor 28171*. CAYMAN BRAC: *Kings CB 102*; *Millspaugh 1177*; *Proctor 28931*.

— Bermuda, Florida, West Indies, and the coasts of Yucatan, Cozumel, Belize, and Venezuela, characteristically in sand at the top of sea-beaches.

Tournefortia L.

Shrubs, woody vines, or tree-like; leaves alternate, entire. Flowers in scorpioid cymes usually forked 1–several times, sometimes elongate and flexuous. Calyx 5-lobed, one lobe often longer than the others. Corolla cylindrical with 5 short spreading lobes. Stamens included; ovary 4-celled, sometimes 4-lobed; style terminal, usually forked at apex. Fruit a small entire or 4-lobed drupe, fleshy or dry, finally separating into 2–4 nutlets, these 1–2-seeded.

— A diverse pantropical genus of perhaps 100 species or more.

KEY TO SPECIES

1. Leaves less than 7 cm long; slender trailing or scrambling shrubs; corolla tube 2 mm long or less:
  2. Leaves mostly less than 3 cm long; calyx lobes ca. 1 mm long: **T. minuta**
  2. Leaves up to 9 cm long; calyx lobes 1.5–2 mm long: **T. volubilis**
1. Leaves mostly 10–20 cm long; an erect shrub to 3 m tall or more; corolla-tube 5 mm long or more: **T. astrotricha**

Tournefortia minuta Bert. ex Spreng. In L., Syst. Veg. ed. 16, **1**: 644. 1824.

Very slender-stemmed twining or scrambling shrub with stems up to 2 m long or more. Leaves narrowly lanceolate to oblong, rarely more than 3 cm long and 1 cm broad. Inflorescences often shorter than the leaves, or if branched the branches 2–4 cm long; calyx lobes subulate; corolla greenish, soon falling.

LITTLE CAYMAN: *Proctor 52173*, found in arid thickets near the east end.
— Hispaniola and Jamaica; doubtfully reported from Cuba.

Tournefortia volubilis L., *Sp. Pl.* 1: 140 (1753).

AUNT ELIZA BUSH

Usually trailing or twining, rarely erect, the younger stems puberulous; leaves lanceolate or narrowly oblong to narrowly ovate, 2–7 cm long, acute or subacuminate at the apex, glabrous or sparsely puberulous. Inflorescence-branches very slender, curved or flexuous, up to 8 cm long but usually much shorter; calyx ca. 1 mm long; corolla pale greenish, the lobes subulate, ca. 1 mm long. Ripe fruits white, usually with black spots, glabrous, 2–3 mm in diameter.

GRAND CAYMAN; *Brunt 1996, 2132*; *Kings GC 140*; *Proctor 15155, 15170*. LITTLE CAYMAN: *Kings LC 12*; *Proctor 28112, 35102*. CAYMAN BRAC: *Proctor 29018*.
— Florida, Texas, West Indies and continental tropical America, chiefly in rocky thickets and woodlands; a variable species over much of its range, but the Cayman population is quite uniform.

Tournefortia astrotricha DC., *Prodr.* 9: 520 (1845). PLATE 53.

An erect or sometimes arborescent shrub 1.5–5 m tall, with stout branches, sometimes of open or straggling habit; leaves elliptic but often more or less asymmetric or folded, usually of rather harsh texture, acute or subacuminate at the apex. Flowers odourless or somewhat fragrant, in helicoid one-sided cymes, these often becoming elongate and pendent; calyx ca. 2.5 mm long; corolla cream or whitish turning pink. Ripe fruits pure white, unspotted, rather fleshy.

— The Cayman plants of this species have been assigned to two varieties, as follows:

1. Leaves densely short-pubescent beneath; corolla ca. 5 mm long; fruits to
   10 mm in diameter: var. **astrotricha**
1. Leaves glabrous beneath or with a few scattered hairs; corolla ca. 6 mm
   long; fruits 6–9 mm in diameter: var. **subglabra**

Tournefortia astrotricha var. *astrotricha*. PLATE 53.

GRAND CAYMAN: *Kings GC 155; Proctor 11999, 15041.* CAYMAN BRAC: *Proctor* 28946.
— Jamaica and the Swan Islands, in dry rocky thickets and woodlands. The Swan Islands plants are intermediate in pubescence between the two varieties.

Tournefortia astrotricha var. *subglabra* Stearn, *Jour. Arnold Arb.* 52: 633 (1971). PLATE 53.

GRAND CAYMAN: *Brunt* 2080; ?*Fawcett*; ?*Kings* GC 343. Fawcett's record of *T. cymosa* L. (= *T. glabra* L.) probably pertains to this variety, but his specimen has not been found for examination.
— Jamaica, in habitats like those of the preceding variety.

Heliotropium L.

Annual or perennial herbs or sometimes low shrubs; leaves alternate or subopposite, rarely whorled, always simple and entire. Flowers usually in one-sided scorpioid spikes, these simple or 1–2-forked, or sometimes the flowers solitary and internodal. Calyx of 5 teeth or lobes, often unequal; corolla small, with cylindrical tube and spreading 5-lobed limb. Stamens included; anthers obtuse, mucronate or short-appendaged. Ovary 4-celled, the terminal stigma sessile or borne on a distinct style, the apex usually bearing a conical or elongate sterile appendage, this often bifid. Fruit dry, entire or lobed, at maturity separating into 2–4 bony nutlets, the nutlets 1–2-seeded; endosperm thin.

— A widely distributed, diverse genus of about 150 species, sometimes not easily distinguishable from *Tournefortia*.

KEY TO SPECIES

1. Plant glabrous, succulent, and usually somewhat glaucous: H. curassavicum

1. Plant pubescent, the leaves not succulent:

   2. Flowers borne on more or less elongate, distinctly scorpioid spikes; annual
   herbs, sometimes persisting and slightly woody at the base:

      3. Corolla white, 2 mm long; fruits 2 mm long, the nutlets rounded and not
      ribbed: H. angiospermum

3. Corolla pale blue, 4–5 mm long; fruits 3–4 mm long, the nutlets pointed
   and strongly ribbed: **H. indicum**

2. Flowers solitary in the axils or on short, terminal, non-scorpioid spikes;
   depressed or erect shrubs:
   4. Erect shrub to 1 m tall or more; leaves up to 10 mm long or more; flowers
      in short terminal spikes: **H. ternatum**
   4. Prostrate, dense, mat-like shrub not more than a few cm high and ca. 15 cm
      across; leaves 2–3 mm long; flowers solitary in the upper axils: **H. humifusum**

Heliotropium curassavicum L., *Sp. Pl.* 1: 130 (1753).

Annual succulent herb of pale blue-green colour, the branches decumbent-ascending; leaves sessile, linear-oblanceolate or narrowly spatulate, mostly 1–4 cm long and 2–6 mm broad. Flowers sessile or nearly so in simple or twin scorpioid cymes 1–4 cm long or more; calyx 1–2 mm long; corolla white, the tube ca. 2 mm long. Anthers sagittate, sessile. Fruit 2 mm in diameter, separating into 4 nutlets.

GRAND CAYMAN: *Brunt* 1830; *Kings GC 322*; *Proctor* 15137.

— Widespread in tropical America in brackish or saline soils, mostly at or near sea-level.

Heliotropium angiospermum Murray, *Prodr. Stirp. Goett.* 217 (1770). **FIG. 209**, H–I.

Heliotropium parviflorum L. (1771).

SCORPION TAIL, BASTARD CHELAMELLA

Erect, loosely branched herb, at length becoming somewhat woody toward the base, and up to 1 m tall but often lower, the stems sparsely clothed with appressed hairs. Leaves alternate or subopposite, petiolate, the blades lanceolate or narrowly ovate to elliptic, mostly 2–9 cm long, acute at the apex, the upper surface with scattered hairs, the lower pubescent on midrib and veins. Cymes simple or once-forked, up to 12 cm long in fruit. Stigma sessile. Fruit glabrous but covered with minute scales.

GRAND CAYMAN: *Brunt* 2029; *Hitchcock*; *Lewis* 3856; *Millspaugh* 1287; *Proctor* 15159. LITTLE CAYMAN: *Kings* LC 41; *Proctor* 28058. CAYMAN BRAC: *Kings* CB 67; *Proctor* 29121.

— Widespread in tropical America, frequent along roads and paths, in pastures and in open waste ground.

Heliotropium indicum L., *Sp. Pl.* 1: 130 (1753). **FIG. 209**, J–K.

SCORPION TAIL

Erect, annual herb to 1 m tall or more, the stems and leaves more or less hispid; leaves petiolate, the blades broadly ovate, 4–6 cm long or often longer, acute at the apex and abruptly contracted at the base, the margins somewhat wavy. Flowers in simple scorpioid spikes up to 15 cm long in fruit, but much shorter while flowering. Stigma borne on an evident style. Fruit glabrous.

GRAND CAYMAN: *Kings GC 275*.

— A pantropical weed, probably of American origin.

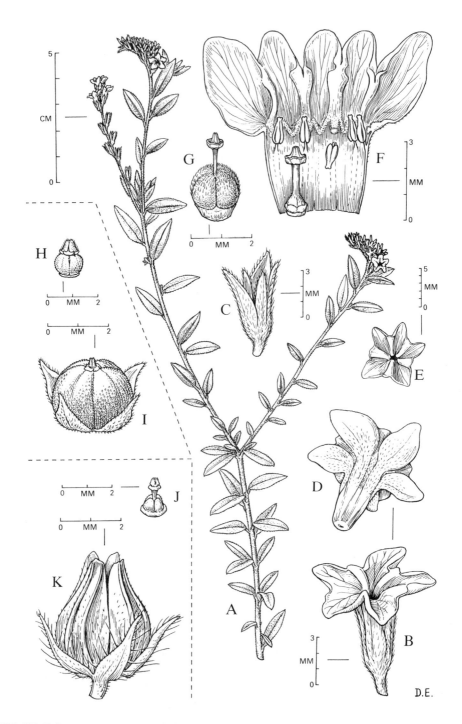

FIG. 209 **Heliotropium ternatum**. A, habit. B, flower. C, calyx. D, corolla, ventral view. E, corolla, view of limb. F, corolla spread out, showing stamens and pistil. G, fruit. H, **H. angiospermum**, pistil. I, fruit. J, **H. indicum**, pistil, K, fruit. (D.E. ex St.)

Heliotropium ternatum Vahl, *Symb. Bot.* 3: 21 (1794). **FIG. 209**, A–G.

A mostly erect, bushy shrub up to 1.5 m tall, densely white-pubescent throughout; leaves rigid, alternate, opposite or in whorls of 3, sharply acute at the apex, the hairs on the upper side with pustulate bases, the margin usually revolute. Flowers few in short terminal spikes; calyx 3–4 mm long, white-strigose; corolla white with yellow eye, the tube 4 mm long, the expanded limb mostly 3–4 mm across. Nutlets subglobose, black.

GRAND CAYMAN: *Brunt* 2007, 2199; *Proctor* 15195.

— West Indies and continental tropical America, in dry sandy or rocky thickets.

Heliotropium humifusum Kunth in H. B. K., *Nov. Gen. & Sp.* 3: 85, t. 205 (1818). PLATE 53.

A dense, mat-like, suffrutescent shrub, diffusely branched, clothed throughout with stiff white strigose hairs mostly with pustulate bases; leaves densely overlapping, rigid, lance-oblong, acute at the apex and 1-nerved, attached to the stem by a broad base. Calyx strigose, 2–3 mm long; corolla white with yellow eye, the tube scarcely exceeding the calyx, the expanded limb 2–3 mm across or more. Stigma capitate on a short style. Fruit subglobose, enclosed by the calyx, the nutlets hispidulous.

GRAND CAYMAN: *Crosby, Hespenheide & Anderson* 37 (GH, MICH): *Kings* GC 59, GC 123; *Proctor* 15159, 52241; *Sauer* 4214 (WIS). LITTLE CAYMAN: *Kings* LC 48, LC 78; *Proctor* 28129, 35137. CAYMAN BRAC: *Proctor* 35154.

— Cuba and Hispaniola, usually in gravel-filled pockets of dry, exposed limestone, or in sandy clearings. Cuban plants of this affinity growing on serpentine appear to represent a different species, though usually identified as being the same.

Ehretia P. Br.

Trees or shrubs; leaves alternate and petiolate, the blades entire, serrate or toothed. Flowers small, in terminal panicles or corymbose cymes; calyx of 5 segments imbricate or open in bud; corolla with short or cylindrical tube, the 5 lobes spreading or recurved. Stamens usually exserted. Ovary imperfectly or completely 4-celled, with ovules attached laterally; style terminal, bifid, the 2 stigmas club-shaped or capitate. Fruit a globose or subglobose drupe, at maturity the stone separating into two 2-seeded or four 1-seeded nutlets; seeds with scant endosperm and ovate cotyledons.

— A pantropical genus of about 50 species, the majority in the Old World. Some of them are valued for their timber.

Ehretia tinifolia L., *Syst. Nat.* ed. 10, 2: 936 (1759). **FIG. 210**. PLATE 53.

A glabrous arborescent shrub or small tree to 10 m tall or more; leaves broadly elliptic or oblong-elliptic, mostly 7–15 cm long, acutish at the apex, the midrib and primary veins prominent on the underside. Panicles many-flowered; calyx ca. 1.5 mm long, the rounded lobes ciliate; corolla white or creamy-white, the tube 1 mm long, the reflexed or revolute lobes ca. 2.5 mm long. Fruits ca. 5 mm in diameter, with 2 nutlets.

GRAND CAYMAN: *Correll & Correll* 51037, 51047; *Proctor* 15044, 31038, 49254, 52136.

— Greater Antilles except Puerto Rico, the Swan Islands, Mexico and Honduras, in thickets and woodlands.

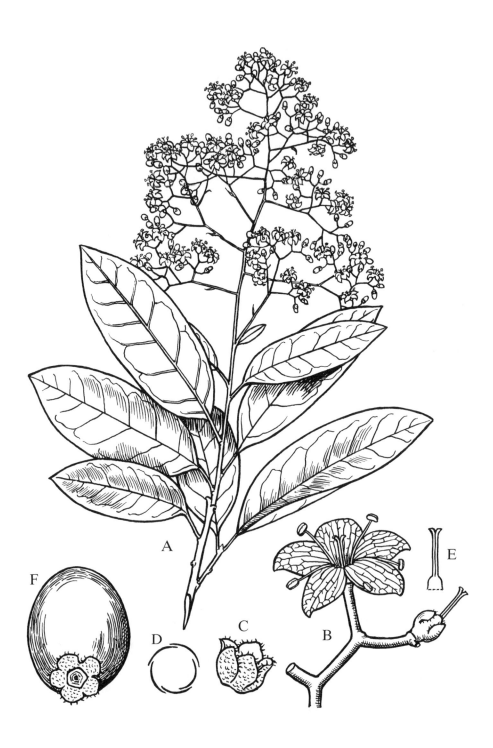

FIG. 210 **Ehretia tinifolia**. A, branch with leaves and flowers, × $^3/_8$. B, portion of inflorescence with flower, × 6. C, calyx, × 9. D, diagram of calyx. E, pistil, × 6. F, fruit, × 6. (St.)

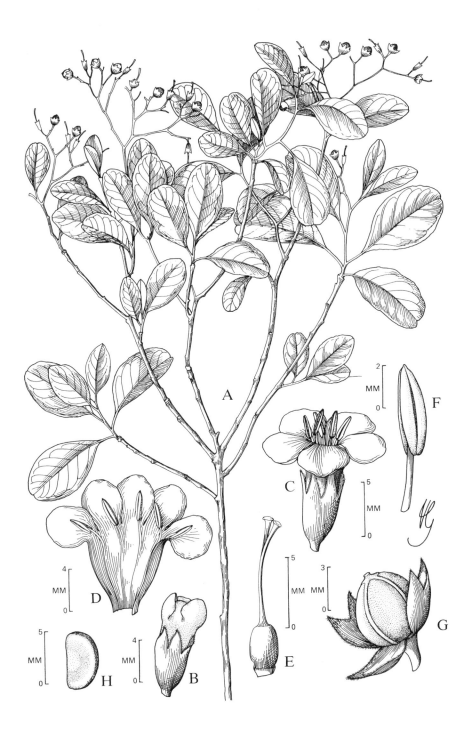

FIG. 211 **Bourreria venosa**. A, habit. B, flower-bud. C, flower. D, dissected corolla. E, pistil, showing ovary, style and stigmas. F, anther. G, fruit. H, seed. (G.)

Bourreria P. Br., nom. cons.

Shrubs or small trees; leaves alternate, petiolate and entire. Flowers usually rather many in terminal corymbose cymes; calyx closed in bud, at anthesis splitting into 2–5 valvate teeth or lobes; corolla white, the 5 (or rarely 6) lobes imbricate in bud, the tube usually short, the throat widely dilated, and the lobes broad and spreading. Stamens borne in the corolla-tube, included or exserted, the filaments glabrous or villous near the base. Ovary 4-celled, the style terminal and more or less bifid at the apex, with truncate or subcapitate stigmas; ovules laterally attached. Fruit a thin-fleshed drupe enclosing 4 bony nutlets, these usually ridged on the back; endosperm fleshy.

— A tropical American genus of between 15 and 20 species.

Bourreria venosa (Miers) Stearn, *Jour. Arnold Arb.* 52: 625 (1971). **FIG. 211**. PLATE 53.

A shrub or small tree to 6 in tall or more, glabrous or nearly so; leaves broadly elliptic to obovate or sometimes nearly rotund, 2–10 cm long. Inflorescence few–many-branched, glabrous or puberulous; calyx 6–7 mm long, the lobes often splitting unequally; corolla ca. 10 mm long, the lobes ca. 4 mm long and broad. Style more or less distinctly 2-branched, the branches up to 2.5 mm long but often less or obsolete. Ripe fruits red, 5–8 mm in diameter.

GRAND CAYMAN: *Brunt* 1715, 1804(?), 1814, 2069; *Correll & Correll* 51044; *Fawcett*; *Hitchcock*; *Kings* GC 310, GC 426; *Proctor* 11995; *Sachet* 438(?). LITTLE CAYMAN: *Proctor* 28149, 35144. CAYMAN BRAC: *Proctor* 29032.

— Jamaica and the Swan Islands, in sandy or rocky thickets and woodlands. Perhaps not really distinct from *B. succulenta* Jacq., with undivided capitate stigma, which has a wide range from Florida and Mexico through the West Indies to Venezuela.

Cordia L.

Shrubs or trees, often roughly pubescent, the hairs simple, branched or stellate; leaves alternate or mostly so, petiolate, the margins entire or toothed. Flowers in panicles, cymes, spikes or heads, sessile or stalked; calyx tubular to bell-shaped, and grooved, striate or smooth, usually 5-toothed or 3–10-lobed, usually persistent, often enlarging in fruit; corolla usually white or creamy, funnel-shaped to subrotate, with 5 or more lobes. Stamens equally or unequally inserted in the corolla-tube, included or shortly exserted. Ovary 4-celled, the style terminal and 2-lobed or 2-branched, the branches each 2-lobed or 2-branched; ovules 1–4, erect, laterally attached. Fruit a drupe with a hard 1–4-celled stone; seeds without endosperm.

— A pantropical genus of about 250 species.

KEY TO SPECIES

1. Inflorescence branched and open, corymbose or paniculate:

    2. Calyx 7–15 mm long in flower, enlarging in fruit; corolla 15–55 mm long; fruits not red:

        3. Leaves rough; corolla orange or scarlet: **C. sebestena**

        3. Leaves smooth; corolla white, persistent and turning brown, enclosing the fruit: **C. gerascanthus**

2. Calyx 3–4 mm long; corolla 10–15 mm long or across; fruits with viscid juice:	
4. Leaves entire; ripe fruits red:	**C. laevigata**
4. Leaf-margins dentate; fruits white:	**[C. dentata]**
1. Inflorescence dense and unbranched:	
5. Inflorescence a cylindrical spike; leaves densely pubescent beneath, on upper side scabrous but without hairs:	**C. brownei**
5. Inflorescence a globose head; leaves sparsely pubescent beneath and minutely glandular, on upper side with numerous pustulate hairs:	**C. globosa**

Cordia sebestena L., *Sp. Pl.* 1: 190 (1753) var. *caymanensis* (Urb.) Proctor, *Sloanea* 1: 3 (1977). PLATES 53 & 54.

Cordia caymanensis Urb., *Symb. Ant.* 7: 344 (1912).

BROADLEAF

Shrub or small tree to 10 m tall, very rough-hispid throughout; leaves very broadly elliptic or ovate to rotund, up to 30 cm long (but often much less), shortly subacuminate at the apex, the margins subentire or often the apical third sharply serrate, the base truncate or broadly cuneate; petioles up to 4 cm long. Inflorescence corymbose; calyx tubular, 10–12 mm long, with short, broad, unequal lobes; corolla salver-shaped, the tube 1.5–2 cm long, the limb 2–3 cm across. Stamens included. Flowers heterostylous, the styles once-forked if included or twice-forked if exserted. Ripe fruits white, acuminate, ca. 3 cm long.

GRAND CAYMAN: *Brunt* 1651; *Hitchcock*; *Kings* GC 63; *Lewis* 3823; *Maggs* II 55; *Millspaugh* 1261, 1358 (type of var.); *Proctor* 15169; *Sachet* 401, 406, 439. LITTLE CAYMAN: *Kings* LC4; *Proctor* 28047. CAYMAN BRAC: *Kings* CB 9; *Millspaugh* 1223; *Proctor* 29037.

— The endemic variety *sebestena* is widely distributed in coastal thickets from Florida through the West Indies and along the eastern coasts of continental tropical America. Var. *caymanensis* differs from var. *sebestena* in a suite of overlapping characters, including the (on average) larger, usually serrate leaves that are never cordate at the base, and in the smaller calyx and corolla. The rough, sandpaper-like leaves have traditionally been used to polish tortoise-shell. Var. *sebestena* is widely cultivated for ornament in tropical countries of both hemispheres.

Cordia gerascanthus L., *Syst. Nat.* ed. 10, 2: 936 (1759). PLATE 54.

SPANISH ELM

An evergreen or sometimes deciduous tree to 10 m tall or more, the young branchlets glabrous; leaves narrowly to broadly elliptic or ovate, 5–15 cm long or more, acuminate at the apex; petioles of young leaves with scattered spreading hairs. Inflorescence rather densely corymbose, the branches puberulous; flowers fragrant; calyx 7–10 mm long, striate-grooved, hispidulous; corolla salver-shaped, the expanded limb 2–2.5 cm across, persistent in fruit and acting as a parachute for fruit-dispersal. Fruits oblong, enclosed by the calyx.

GRAND CAYMAN: *Brunt* 1805; *Howard & Wagenknecht* 15029; *Millspaugh* 1273, 1300; *Proctor* 15299. LITTLE CAYMAN: *Proctor* 28126. CAYMAN BRAC: *Kings* CB 1; *Proctor* 15321.

— Greater Antilles, Mexico, C. America and Colombia, in rather dry, rocky woodlands. The wood is useful for general construction.

Cordia laevigata Lam., *Tabl. Encycl. Méth Bot.* 1: 422 (1792). PLATE 54.

Cordia nitida Vahl. (1793).

C. collococca of Hitchcock (1893), not L. (1759).

A small tree to 12 m tall, the young branchlets and petioles finely rusty-pubescent; leaves elliptic to obovate, mostly 5–18 cm long, blunt or short-acuminate at the apex, the upper surface shining. Inflorescence corymbose, puberulous; calyx globose in bud, broadly bell-shaped after anthesis, 3–5-lobed; corolla creamy white. Fruits mostly 8–12 mm in diameter, red when ripe.

GRAND CAYMAN: *Correll & Correll 51016; Hitchcock.* CAYMAN BRAC: *Proctor 29082.*

Greater Antilles and the Virgin Islands, the Swan Islands and perhaps C. America, in rocky woodlands usually on limestone. The C. American plants are alleged to have creamy-white instead of red fruits and might better be considered a different species. Lesser Antillean plants identified as *Cordia laevigata* differ in the tubular calyx and other details, and likewise are probably not conspecific.

[*Cordia dentata* Poiret in Lam., *Encycl.* 7: 48 (1806).

Cordia alba of Adams (1972), not (Jacq.) R. & S. (1819).

DUPPY CHERRY

Tree to 10 m tall or more with pubescent branches. Leaves with petioles 1–2 cm long, the blades broadly elliptic to orbicular, 4–12 cm long, 4–7 cm broad, the apex rounded to abruptly short-acute, the base abruptly cuneate, the margins irregularly toothed, the upper surface slightly scabrous with stiff short hairs, the lower surface glabrate. Inflorescence terminal, cymose-scorpioid; peduncle 4–5 cm long, pubescent. Calyx turbinate, 3 mm long, 10-ribbed, 3-lobed, strigose-pubescent. Corolla tube 3–4 mm long, the limb spreading, shallowly 5-lobed, the lobes 4–6 mm long, white. Drupe ellipsoid, 1 cm in diameter, white, with gelatinous-viscid flesh; endocarp bony.

GRAND CAYMAN: *Proctor 47791*, found in a vacant lot in the town of George Town; probably not indigenous.

— Greater and Lesser Antilles, Trinidad, and northern S. America.]

Cordia brownei (Friesen) I. M. Johnst., *Jour. Arnold Arb.* 31: 177 (1950). **FIG. 212.** PLATE 54.

A shrub usually 1–3 m tall with densely puberulous branchlets; leaves lanceolate to oblong or oblong-elliptic, rarely ovate, 2–8 cm long, blunt to acutish at the apex, the margins irregular. Spikes 1.5–3.5 cm long, on puberulous peduncles 2–4 cm long; flowers at apex of spike opening first, then in succession downwardly; calyx 2.5–3 mm long, very hairy (Cayman Brac) or hairs virtually lacking (Grand Cayman); corolla white, with small intermediate lobes between the main lobes. Ripe fruits subglobose, red, 3–4 mm in diameter.

GRAND CAYMAN: *Brunt 2136; Correll & Correll 50997; Kings GC 126; Proctor 11998,* 15268, 27985; *Sachet 382.* LITTLE CAYMAN: *Proctor 28142, 35126.* CAYMAN BRAC: *Proctor 29063, 35158.*

— Jamaica, in rocky thickets and woodlands.

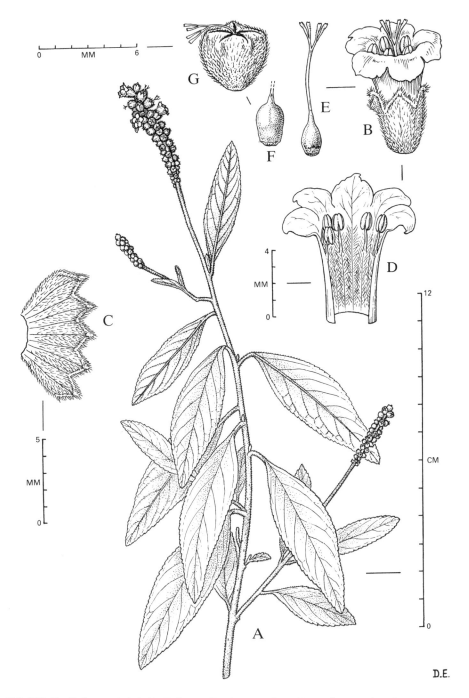

FIG. 212 **Cordia brownei**. A, habit. B, flower. C, calyx spread out. D, corolla opened to show stamens. E, pistil. F, ovary. G, fruit enclosed by calyx, with style protruding. (D.E. ex St.)

569

Cordia globosa (Jacq.) Kunth, H. B. K., *Nov. Gen. & Sp.* 3: 76 (1818) var. *humilis* (Jacq.) I. M. Johnst., *Jour. Arnold Arb.* 30: 98, 117 (1949). PLATE 55.

BLACK SAGE

A somewhat straggling shrub mostly 1–1.5 m tall, the young branchlets pustulate-puberulous and minutely glandular; leaves lance-elliptic or ovate, mostly 1–4 cm long, acute at the apex, the margins sharply toothed, the upper surface with numerous white pustulate hairs, the lower side with non-pustulate hairs on the nerves and veins and the whole surface densely clothed with minute glistening yellow glands. Inflorescence 0.6–1.5 cm in diameter on peduncles 0.5–2 cm long; calyx hispid, the lobes with hair-like tips 1–2 mm long; corolla white, 5–7 mm long. Ripe fruits ovoid, red, partly enclosed in the calyx.

GRAND CAYMAN: *Brunt* 2013, 2101; *Correll & Correll* 51030; *Fawcett*; *Hitchcock*; *Kings GC* 126a; *Proctor* 12000, 15187; *Sachet* 414.

Var. *humilis* occurs in Florida, the Bahamas, Greater Antilles, and from Mexico to Panama, in sandy or rocky thickets and woodlands. Var. *globosa* occurs in the Lesser Antilles and northern S. America. Var. *globosa* differs from var. *humilis* in having non-glandular leaves and larger inflorescences, the individual flowers with longer, more hispid calyx-processes.

Rochefortia Sw. (1788); Lefor (1968)

A genus with a single species described below.

Rochefortia acanthophora (DC.) Griseb., *Fl. Br. W.I.*: 482 (1862). PLATE 55.

GREENHEART EBONY

Ehretia acanthophora DC. (1845).

Shrub or tree 2–6 m tall with profuse multiple branching, the distal branches drooping; leaves rigid, blade 2–15 mm long, shortly petioled to subsessile; fruit red with ripe, subglobose, about 3.5 mm in diameter, 4-seeded.

Flowers August, fruits September.

GRAND CAYMAN: *P. Ann van B. Stafford s.n*; *Proctor.*
— Greater Antilles and Antigua, in rocky woodlands.

VERBENACEAE

Herbs, shrubs or trees, or sometimes woody vines; branches often 4-angled, sometimes armed with thorns; leaves opposite or whorled, simple or palmately compound, the margins entire or variously toothed, lobed or incised; stipules absent. Flowers usually perfect, in axillary or terminal spikes, racemes, cymes or panicles, or sometimes in heads. Calyx 4–5-toothed or -lobed, persistent and often enlarging in fruit; corolla gamopetalous, regular or irregular, funnel-or salver-shaped, 4–5-lobed, often more or less 2-lipped. Stamens usually 4 (2 long and 2 short) or rarely only 2, the filaments inserted in the corolla-tube; anthers 2-celled, opening lengthwise; staminodes sometimes present. Ovary superior, usually 2-carpellate but often 4-lobed and usually 2–5-celled (rarely 1-celled), with 1 or 2 ovules in each cavity, basally or laterally attached to the placentas; style terminal and solitary. Fruit a drupe with 1 or more stones, or a dry schizocarp separating into 2 or 4 nutlets.

— A rather large, widely distributed family of about 100 genera and 2,600 or more species. Several are important sources of timber, e.g. *Tectona grandis* L. f., or teak. Among the species cultivated for ornament are *Holmskioldia sanguinea* Retz., or Chinese hat, several species of *Clerodendrum* and *Petrea volubilis* L.

KEY TO GENERA

1. Inflorescence a spike, raceme, or head:
  2. Plants herbaceous:
    3. Fertile stamens 2; calyx closely appressed to the rhachis and more or less sunk in depressions: **Stachytarpheta**
    3. Fertile stamens 4; calyx not appressed to the rhachis:
      4. Inflorescence a loose spike with scattered flowers; fruiting calyx globose, bearing minute hooked hairs: **Priva**
      4. Inflorescence a dense head; fruiting calyx not globose: **Lippia**
  2. Plants woody:
    5. Plants erect or arching but not scrambling or vine-like; wild species:
      6. Inflorescence a more or less dense head:
        7. Fruit dry; calyx distinctly 2–4-toothed or -cleft: **Lippia**
        7. Fruit a small fleshy drupe; calyx with margin truncate or very obscurely toothed: **Lantana**
      6. Inflorescence a spike or raceme:
        8. Calyx in flower smooth, in fruit dry and cup-shaped, subtending the fruit: **Citharexylum**
        8. Calyx in flower ribbed, in fruit fleshy, contracted at the apex and enclosing the fruit: **Duranta**
    5. Plant a scrambling woody vine, cultivated for its showy racemes of purple flowers with bluish calyces: **[Petrea p. 581]**
1. Inflorescence a cyme or panicle:
  9. Corolla regular; plants not spiny:
    10. Leaves of rather smooth texture on both sides, minutely gland-dotted beneath, glabrous or with unicellular hairs chiefly on the midrib; style-branches filamentous, 4 mm long or more; drupe with up to 4 1-seeded nutlets: **Aegiphila**
    10. Leaves with raised densely reticulate venation beneath, densely pubescent with pluricellular hairs; style-branches ca. 1 mm long; drupe with 1–4-celled stone: **Petitia**
  9. Corolla irregular; plants spiny: **Clerodendrum**

Stachytarpheta Vahl, nom. cons.

Annual or perennial herbs or low shrubs; leaves opposite and toothed or serrate. Flowers in terminal spikes, sessile or partly immersed in the rhachis, each flower solitary in the axil of a bract. Calyx tubular, 5-toothed, the teeth equal or unequal; corolla with cylindrical tube and 5 spreading lobes. Stamens 2, inserted above the middle of the corolla-tube and included; staminodes 2, small and inconspicuous; anther-cells divergent and opening in

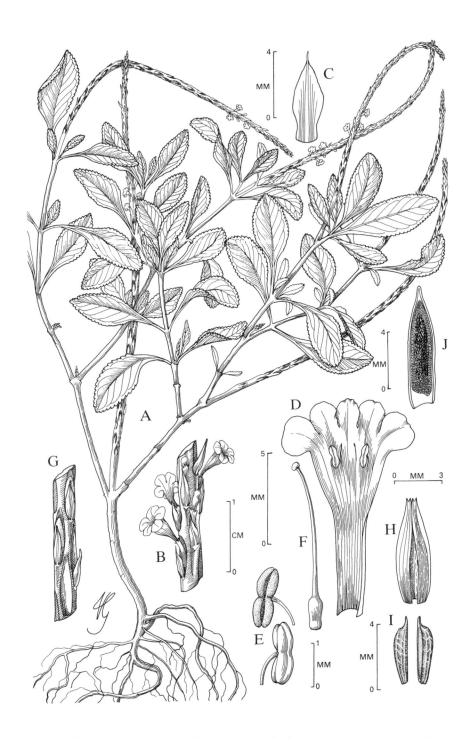

FIG. 213 **Stachytarpheta jamaicensis.** A, habit, B, portion of inflorescence showing bracts and flowers. C, bract. D, corolla cut open to show stamens. E, two views of anther. F, pistil. G, portion of fruiting spike. H, fruiting calyx. I, seeds. J, seed, enlarged. (G.)

one continuous line. Ovary 2-celled, with 1 ovule attached laterally near the base of each cell; stigma subcapitate. Fruit dry, oblong, enclosed in the persistent calyx, eventually separating into 2 1-seeded nutlets; seeds without endosperm.

— A tropical American genus of between 30 and 40 species.

Stachytarpheta jamaicensis (L.) Vahl, *Enum. Pl.* 1: 206 (1804). **FIG. 213**. PLATE 55.

VERVINE

Annual herb, the stems decumbent or ascending to 1 m tall, glabrous throughout or with a few scattered hairs; leaves oblong-elliptic to ovate, mostly 2–9 cm long, obtuse at the apex, abruptly long-cuneate at the base, the margins coarsely crenate-serrate. Spikes glabrous, stiff, 10–50 cm long, up to 5 mm thick or more; bracts lanceolate or narrowly ovate, 4–6 mm long; calyx 5–7 mm long; corolla blue-violet or violet, 8–10 mm long.

GRAND CAYMAN: *Brunt* 1901; *Hitchcock*; *Kings* GC 121, GC 290, GC 291; *Lewis* 3831, GC 36; *Maggs* II 53; *Millspaugh* 1340; *Proctor* 11981; *Sachet* 403; LITTLE CAYMAN: *Proctor* 28048. CAYMAN BRAC: *Millspaugh* 1173; *Proctor* 28990.

— Florida, the West Indies and continental tropical America, introduced elsewhere, a weed of open waste places and dry sandy clearings and thickets.

Priva Adans.

Annual or perennial herbs, often with rough pubescence; leaves opposite, thin, the margins serrate or toothed. Flowers small, in terminal racemes, the stems often branching dichasially below the inflorescence so that the latter appears to arise from a fork. Calyx tubular at anthesis, 5-ribbed, the ribs ending in small teeth, persistent and enlarging to enclose the fruit completely except for a minute orifice. Corolla obliquely 5-lobed. Stamens 4, included, inserted in 2 pairs at different levels; anther-cells parallel or slightly divergent. Ovary 2-celled; stigma 2-lobed. Fruit dry and surrounded by the enlarged, bladdery calyx, consisting at maturity of 2 nutlets, these usually variously spined, ridged or roughened on the dorsal surface.

— A pantropical genus of about 20 species.

Priva lappulacea (L.) Pers., *Synops. Pl.* 2: 139 (1806). **FIG. 214**.

OLD LADY COAT TAIL

Ascending to erect annual herb often 50–75 cm tall, with quadrangular pubescent stems; leaves with petioles up to 2 cm long or more, the blades more or less ovate and mostly 3–8 cm long, acute to acuminate at the apex, the margins crenate-serrate, rough-hispidulous on the upper side. Racemes lax, up to 20 cm long; flowers short-stalked; calyx at anthesis 2–3 mm long, in fruit 5–6 mm long; corolla pale bluish or whitish, 4–5 mm long. Nutlets 3–3.5 mm long, curved-spiny on the dorsal side.

GRAND CAYMAN: *Brunt* 1933; *Hitchcock*; *Kings* GC 87, GC 153; *Millspaugh* 1286. CAYMAN BRAC: *Proctor* 29366.

— Florida, the West Indies and continental tropical America; introduced in the Old World tropics; a weed of open waste ground, fields and roadsides.

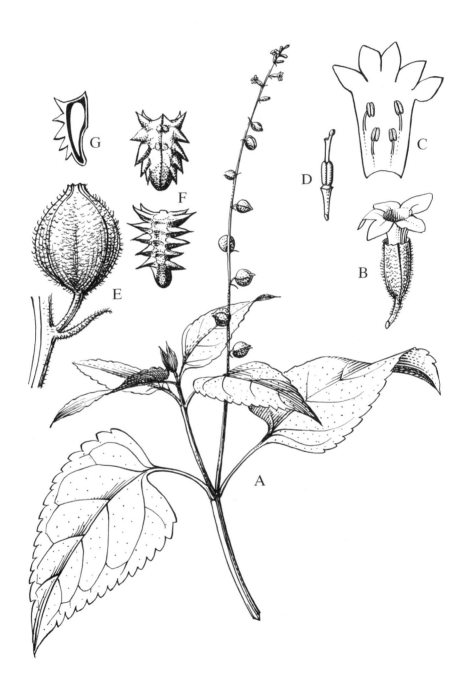

FIG. 214 **Priva lappulacea**. A, habit, × ¹/₂. B, flower, × 3. C, corolla spread out to show stamens, × 3. D, pistil, × 3. E, fruit, × 3. F, two views of seed, × 3. G, section through seed, × 3. (St.)

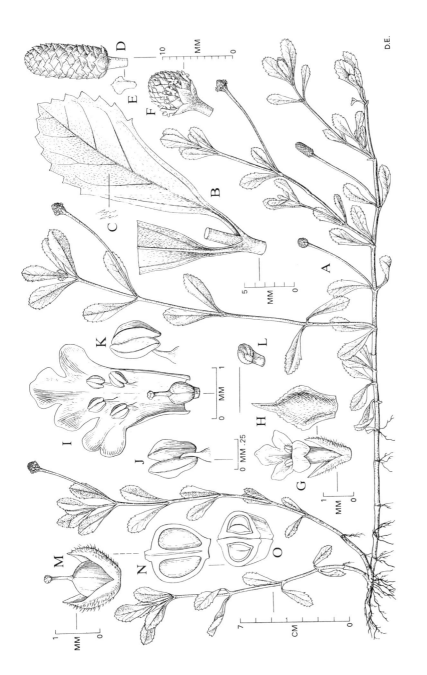

FIG. 215 **Lippia nodiflora**. A, habit. B, node with leaf. C, hairs from surface of leaf. D, mature spike. E, bract from spike. F, young spike with flowers. G, flower. H, bract. I, corolla spread open to show stamens and pistils. J, K, two views of anther. L, stigma. M, fruit and enclosing calyx-lobes. N, O, long. and cross-section of fruit. (D.E.)

Lippia L.

Herbs, shrubs, or small trees, sometimes aromatic; leaves opposite or in whorls of 3, entire or toothed. Flowers small, sessile, in axillary stalked heads or dense cylindrical spikes, often with conspicuous bracts; calyx small, thin, 2–4-toothed; corolla with cylindrical tube and 4 spreading lobes, one of them larger than the others. Stamens 4, in 2 pairs at different levels, usually included. Ovary 2-celled, each cavity with 1 basal ovule; stigma obscurely 2-lobed, oblique or recurved. Fruit small, dry, enclosed by the persistent calyx, at maturity usually separating into 2 hard nutlets; seeds without endosperm.

— A genus of nearly 200 species, the majority of them in tropical America.

KEY TO SPECIES

1. Flower-heads nearly sessile or on peduncles usually 1 cm long or less; shrubby plant with strongly aromatic foliage:	L. alba
1. Flower-heads on elongate peduncles much longer than the leaves and up to 9 cm long; trailing herb, sometimes with ascending or erect flowering branches, not aromatic:	L. nodiflora

Lippia alba (Mill.) N. E. Br., Britton & Wilson, *Sci. Surv. Porto Rico & Virg. Is.* 6: 141 (1925).

SAGE

Straggling or arching aromatic shrub up to 1 m tall or more, puberulous or tomentellous throughout; leaves opposite or in whorls of 3, oblong elliptic or ovate, mostly 1–5 cm long, blunt or broadly acute at the apex, the margins serrulate, the venation prominently raised on the lower surface and channeled on the upper. Flower-heads subglobose to shortly cylindrical, 0.5–1.2 cm long; bracts ovate, 3–5 mm long; corolla pale rosy-lavender with tube 3–3.5 mm long.

GRAND CAYMAN: *Proctor* 15189. CAYMAN BRAC: *Kings* CB 51.

— West Indies, Texas and Mexico south to Argentina, usually in sandy or gravelly thickets near the sea. This species is occasionally cultivated for use in making tea.

Lippia nodiflora (L.) Michx., *Fl. Bor. Amer.* 2: 15 (1803). **FIG. 215.**

Phyla nodiflora (L.) Greene (1899).

MATCH HEAD

Creeping perennial herb, very minutely strigillose with malpighiaceous hairs throughout (appearing glabrous except under a strong lens); leaves oblanceolate to obovate or spatulate, mostly 2–4.5 cm long, blunt or acutish at the apex, long-cuneate at the base, the apical half with serrulate margins. Flower-heads globose at first, becoming cylindrical and up to 1.5 cm long; bracts purplish and closely overlapping, 2–3 mm long; corolla pale pink or lavender-whitish with tube 2–3 mm long.

GRAND CAYMAN: *Brunt* 1770; *Hitchcock*; *Kings* GC 61; *Lewis* GC 16; *Millspaugh* 1365; *Proctor* 15014; *Sachet* 423. LITTLE CAYMAN: *Kings* LC 72. CAYMAN BRAC: *Proctor* 29088.

— Pantropical in moist pastures, ditches and sandy clearings.

Lantana L.

Erect, scrambling or climbing shrubs, always more or less aromatic, the stems 4-angled, ribbed or subcylindrical, sometimes armed with prickles; leaves opposite or whorled, more or less toothed. Flower in dense, stalked, axillary heads or contracted spikes, with or without a subtending involucre of bracts; inner bracts (subtending individual flowers) also present. Calyx membranous, truncate, sinuate or minutely 2–4-toothed, enlarging in fruit; corolla with slender, often curved tube and 4–5 lobes, one larger than the others. Stamens 4, in 2 pairs inserted at different levels, included; anthers ovate, parallel-celled. Ovary 2-celled with 1 ovule in each cavity. Fruit a small, globose, juicy drupe, the stone 2-celled or separating into 2 1-seeded nutlets; seeds without endosperm.

— A mostly tropical American genus of perhaps 50 species, many of them not clearly differentiated, thus often difficult to identify. The following treatment of the Cayman species is not very satisfactory and cannot "work" outside the local geographic context. More field observations are needed on these plants.

KEY TO SPECIES

1. Flowers yellow or orange, changing to orange, red or mauve; inflorescence usually not subtended by a conspicuous involucre of bracts, the outer bracts seldom over 2 mm broad:
  2. Stems armed with numerous prickles; flowers changing to mauve, the corolla-tube ca. 10 mm long, the limb 6–7 mm across: **L. aculeata**
  2. Stems unarmed or nearly so; corolla never mauve, the tube 4–8 mm long, the limb 3–6 mm across:
    3. Corolla tube 5–8 mm long:
      4. Lower surface of the leaves thinly pubescent or almost glabrous; calyx as broad as long: **L. camara**
      4. Lower surface of the leaves with numerous short hairs; calyx longer than broad: **L. urticifolia**
    3. Corolla tube ca. 4 mm long: **L. bahamensis**
1. Flowers white, pale pink or lavender with yellow eye; inflorescence subtended by broad outer bracts 2–5 mm wide, forming an involucre:
  5. Leaves rounded or blunt at the apex; tertiary venation irregular on the lower leaf-surface: **L. involucrata**
  5. Leaves acute at the apex; tertiary venation demarcating regular oblong areoles on the lower surface: **L. reticulata**

Lantana aculeata L., *Sp. Pl.* 2: 627 (1753).

A straggling shrub to 1 m tall but frequently less, with strong pungent odour; young branchlets glandular, sparingly pubescent and beset with scattered sharp recurved prickles. Leaves ovate or broadly ovate, 3–7 cm long, acuminate at the apex, the margins coarsely crenate-serrate. Peduncles up to 6 cm long, thickened toward the apex. Flower-heads ca. 2.5 cm in diameter. Drupes blue, ca. 4 mm in diameter.

LITTLE CAYMAN: *Proctor* 35179. CAYMAN BRAC: *Proctor* 28913.

— With a wide but uncertain range in the tropics; in other regions said to intergrade with *Lantana camara* (see next), and by many recent authors not considered a distinct species or even a variety. However, the Cayman population of *L. aculeata* is so strikingly different from 'normal' *L. camara* (as represented in the Cayman Islands) that it makes no sense in the local context to lump them together. No intergrades have been observed locally. Common on the north side of Cayman Brac in sandy thickets; rare on Little Cayman.

Lantana camara L., *Sp. Pl.* 2: 627 (1753).

WHITE SAGE

An erect, mostly unarmed shrub to 2 m tall or more, with aromatic odour different from that of *Lantana aculeata*; young branchlets glabrate and glandular. Leaves narrowly to broadly ovate or oblong-ovate, mostly 2.5–6 cm long or more, blunt to acute at the apex, the margins crenate-serrate. Peduncles up to 5 cm long, slightly thickened at the top. Flower-heads 1.5–2 cm in diamter, corolla yellow changing to orange or red, the tube usually 6–8 mm long. Drupes blue-black, ca. 3 mm in diameter.

GRAND CAYMAN: *Brunt 2057*; *Hitchcock*; *Howard & Wagenknecht 15028*; *Kings GC 314*; *Lewis GC 41*; *Millspaugh 1330, 1332*; *Sachet 368*. LITTLE CAYMAN: *Proctor 28035*. CAYMAN BRAC: *Millspaugh 1202, 1215*.

— Florida, the West Indies and continental tropical America; Cayman plants occur chiefly in rocky thickets and woodlands.

Lantana urticifolia Mill., *Gard. Dict.* ed. 8 (1768).

Lantana arida Britton (1910).

SWEET SAGE

Aromatic bushy shrub to 2 m tall; young branchlets scabrid and more or less pubescent. Leaves broadly ovate, mostly 2–6 cm long (rarely more), acute to acuminate at the apex, densely and softly pubescent on the under side, scabrid with short pustular hairs on the upper (adaxial) side. Peduncles mostly 2–4 cm long, thickened at the top. Flower-heads 1–1.5 cm in diameter (or more); corolla deep yellow, the tube usually 5–7 mm long. Drupes blue-black, 3–4 mm in diameter.

GRAND CAYMAN: *Kings GC 112*; *Proctor 15009*.

— Widely distributed in the West Indies, usually in rocky thickets or dry woodlands on limestone.

Lantana bahamensis Britton, *Bull. N.Y. Bot. Gard.* 3: 450 (1905). PLATES 55 & 56.

An erect unarmed shrub to 1 m tall or more, the young branchlets glandular and rather densely puberulous, the foliage very aromatic; leaves oblong-lanceolate or elliptic, blunt to acute at the apex, the margins crenulate-serrulate. Peduncles 0.5–2 cm long, very slender (almost filiform). Flower-heads 0.6–1.5 cm in diameter; corolla-tube ca. 4 mm long. Drupes shining black, ca. 3 mm in diameter.

GRAND CAYMAN: *Proctor 15230*.

— Bahamas and Cuba, in dry rocky thickets. This might be considered a much-reduced variant of *Lantana camara*, and more collections and field observations are necessary to establish its true relationship with that variable species.

Lantana involucrata L., *Cent. Pl.* 2: 22 (1756). PLATE 56.

ROUNDLEAF SAGE, BITTER SAGE

An erect, unarmed, aromatic shrub up to 1.5 m tall, glandular and puberulous throughout, the young branches stiff and nearly terete; leaves broadly elliptic or ovate to obovate or nearly orbicular, mostly 1–4.5 cm long, abruptly cuneate at the base, the margins finely crenulate, the petioles up to 10 mm long. Peduncles up to 6 cm long, rarely very short (3 mm or less). Flower-heads 0.8–1.7 cm in diameter (including the bracts); corolla-tube 2–3 mm long. Drupes light to deep purple or mauve, 3–4 mm in diameter.

GRAND CAYMAN: *Brunt* 1638, 1752, 1811; *Correll & Correll* 51027; *Kings* GC 339; *Lewis* 3854; *Millspaugh* 1252; *Proctor* 15090; *Sachet* 413, 426. LITTLE CAYMAN: *Kings* LC 38; *Proctor* 28059. CAYMAN BRAC: *Millspaugh* 1218; *Proctor* 28914.

— Florida, West Indies, Yucatan and Belize, and northern S. America, chiefly in sandy coastal thickets.

Lantana reticulata Pers., *Synops. Pl.* 2: 141 (1806).

Lantana stricta of Hitchcock (1893), not Sw. (1788).

Similar to the preceding species and often scarcely distinguishable from it. The single Cayman record (GRAND CAYMAN: *Hitchcock*) needs further confirmation. Occurs elsewhere chiefly in the Greater and Lesser Antilles, in rocky thickets.

Citharexylum L.

Shrubs or trees with quadrangular or striate stems, usually unarmed or rarely spiny; leaves opposite or whorled, entire or toothed, usually with 1 or 2 glands at the base of the blade. Flowers in axillary and terminal, more or less elongate, spikes or racemes; calyx at anthesis tubular or bell-shaped, the margin truncate or 5–7-toothed or -lobed, becoming enlarged and cup-shaped in fruit; corolla funnel-shaped or salver-shaped, white or yellowish, the 5 lobes subequal, usually pubescent in the throat. Stamens 4, in 2 pairs at different levels, included, a fifth stamen represented by a rudimentary staminode. Ovary incompletely 4-celled, each cavity with a single lateral ovule; stigma 2-lobed. Fruit a juicy, berry-like drupe containing two 2-seeded nutlets.

— A genus of more than 100 species in tropical and subtropical America.

Citharexylum spinosum Sp. *Pl.* 2: 265 (1753). PLATE 56.

A glabrous shrub or small tree to 5 m tall, the branchlets closely striate-grooved; leaves rather long-petiolate, the blades elliptic or oblong-elliptic to ovate, mostly 5–9 cm long or sometimes longer, rounded and often notched at the apex or sometimes abruptly short-acuminate, the upper surface shining and with prominulous venation. Spikes often curved or lax, mostly 4–10 cm long; flowers sessile, fragrant; calyx 3–4 mm long, with broadly triangular lobes; corolla white, with funnel-shaped tube 4–6 mm long. Drupes red at first, turning black, subglobose, 8–12 mm in diameter.

GRAND CAYMAN: *Correll & Correll* 51000. LITTLE CAYMAN: *Proctor* 28135. CAYMAN BRAC: *Proctor* 29001.

— Florida, the West Indies and northern S. America, in rocky thickets and woodlands.

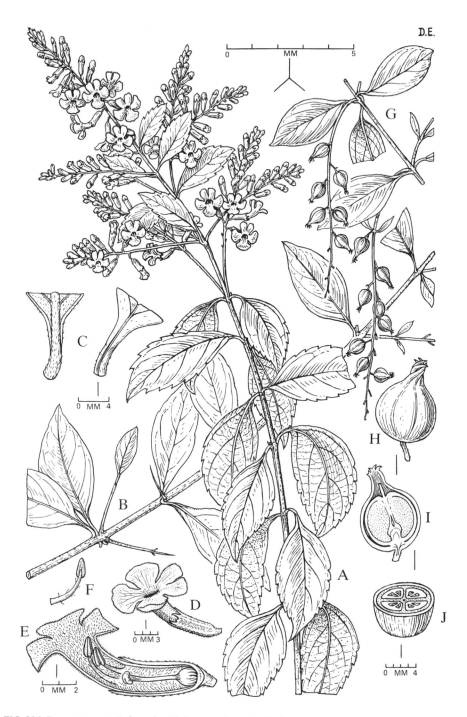

FIG. 216 **Duranta erecta**. A, branch with leaves and terminal inflorescence. B, portion of older branch with spines. C, two views of leaf-base. D, flower. E, section through flower to show stamens and pistil. F, stamen. G, portion of fruiting branch. H, fruit. I, long-section of fruit. (St.)

Duranta L.

Shrubs, sometimes arborescent, and in some species armed with spines; branches sometimes arching or pendent; leaves opposite or whorled, entire or toothed. Flowers in terminal and axillary racemes; calyx in flower tubular or bell-shaped, truncate and 5-ribbed, each rib ending in a small tooth, enlarging to enclose the fruit; corolla salver-shaped, the 5 lobes equal and spreading or oblique, usually pubescent in the throat. Stamens 4, in 2 pairs at different levels, included; anthers with parallel cells. Ovary imperfectly 8-celled, composed of four 2-celled carpels, each cavity with 1 ovule; stigma obliquely subcapitate. Fruit a fleshy drupe enclosed by the enlarged calyx and containing 4 2-seeded nutlets; seeds without endosperm.

— A tropical American genus of about 35 species.

Duranta erecta L., *Sp. Pl.* 2: 637 (1753). **FIG. 216.**

Duranta repens of Adams (1972).

A shrub to 3.5 m tall or more, the glabrate or puberulous branches often arching or trailing, unarmed or sometimes with short axillary spines; leaves glabrous or thinly puberulous, opposite or a few of them alternate, elliptic or obovate, 2–6 cm long, rounded, blunt or acute at the apex, the margins entire or obscurely crenate-serrate toward the apex. Racemes axillary and unbranched or terminal and branched, 2–6 cm long or sometimes much longer. Flowering calyx 3–4 mm long. Corolla blue-violet or rarely white, the tube ca. 6 mm long. Fruits orange, pear-shaped, 6–9 mm long.

GRAND CAYMAN: *Hitchcock*; *Kings* GC 369; *Proctor* 15006.

— Florida, West Indies and continental tropical America; introduced in the Old World tropics; sometimes cultivated as an ornamental. The Cayman plants occur wild in rocky woodlands.

[*Petrea volubilis* L., a woody ornamental vine, is cultivated in Grand Cayman.]

Aegiphila Jacq.

Shrubs or small trees, sometimes scandent, the stems terete or more or less 4-angled; leaves opposite or rarely whorled, simple, the petioles jointed at or near the base. Flowers usually polygamo-dioecious, in axillary or terminal cymes, these paniculate, corymbose or head-like. Calyx bell-shaped or tubular, the apex truncate and entire or with 4–5 teeth or lobes, becoming enlarged and thickened in fruit. Corolla funnel- or salver-shaped with cylindrical tube, the 4–5 lobes somewhat unequal or oblique. Stamens 4–5, equal or nearly so, exserted in the male flowers. Ovary incompletely 4-celled, with 1 laterally attached ovule in each cavity; stigma of 2 linear branches, exserted in the female flowers. Fruit a 1–4-seeded drupe; seeds without endosperm.

— A tropical American genus of perhaps 150 species.

KEY TO SPECIES

1. Stems, underside of leaves, and inflorescence densely clothed with velvety brownish hairs; calyx truncate with 4 minute teeth; corolla-tube 13 mm long or more: **A. caymanensis**

1. Stems glabrous or nearly so; calyx with 4 or 8 wavy lobes, rarely merely 4-toothed; corolla-tube 5–8 mm long: **A. elata**

Aegiphila caymanensis Moldenke, *Fedde, Repert. Sp. Nov.* 33: 118 (1933). PLATE 56.

A velvety-tomentose shrub; leaves oblong-lanceolate, 5–9 cm long, 2–3.8 cm broad, acute to acuminate at the apex, the under-surface (in addition to hairs) densely clothed with minute glistening yellow glands; fine venation obscure. Cymes terminal and in the upper axils, laxly few-flowered; pedicels 6–9 mm long, glandular; calyx ca. 4 mm long, glandular; corolla white, fading cream. Fruit oblong, ca. 7 mm long.

GRAND CAYMAN: *Hitchcock* (type collected Jan. 17, 1891); *Correll & Correll* 51005, coll. Nov. 11, 1979, near Farm Road NW of East End Village; *Mrs. P. Ann van B. Stafford* s.n., coll. March 2, 2005 near Jasmin Lane, Spotts area. Long thought to be extinct, this rare species has been rediscovered by Mrs. Stafford and will now be propagated for survival.

— Endemic.

Aegiphila elata Sw., *Nov. Gen. & Sp. Pl.* 31 (1788).

An erect or straggling glabrate shrub to 2 m tall or more; leaves oblong-ovate, 6–15 cm long, 3–6 cm broad, short-acuminate at the apex, more or less densely gland-dotted beneath but the dots not glistening yellow; fine venation evident or prominulous. Inflorescence a terminal cymose panicle, often densely flowered; pedicels 2–6 mm long, densely puberulous; calyx 3–4 mm long, glandular and sparingly appressed-puberulous; corolla pale yellow. Fruit subglobose, yellow, 9 mm long or more.

GRAND CAYMAN: *Brunt* 1751; *Kings GC* 60; *Millspaugh* 1281; *Rothrock* 158, 235.

The West Indies and continental tropical America, along margins of sandy or rocky thickets and woodlands.

Aegiphila martinicensis Jacq. was attributed to the Cayman Islands by Moldenke in "The known geographic distribution of the members of the Verbenaceae. . .", p. 46, 1949, based upon a literature reference that was probably erroneous. *A. martinicensis* is similar to *A. elata* and the two have sometimes been confused.

Petitia Jacq.

Unarmed shrubs or trees; leaves opposite, simple, entire and long-petiolate. Flowers small, rather numerous in axillary cymose panicles; calyx bell-shaped, subtruncate or 4-toothed, not enlarging in fruit; corolla short, salver-shaped, with 4 spreading equal imbricate lobes. Stamens 4, inserted near the top of the corolla-tube, the ovate anthers almost sessile. Ovary 2-celled, with 2 ovules in each cavity; style shortly forked at the apex. Fruit a small drupe containing a single 2–4-seeded stone.

— A small genus of 2 species occurring in the Bahamas and the Greater Antilles.

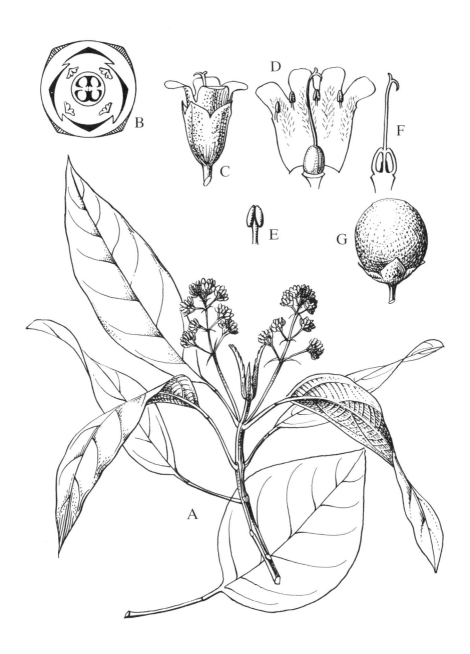

FIG. 217 **Petitia domingensis**. A, branch with leaves and flowers, × ²/₃. B, floral diagram. C, flower, × 4. D, corolla spread open to show stamens and pistil, × 4. E, anther, × 8. F, long. section of pistil, × 4. G, fruit, × 4. (St.)

Petitia domingensis Jacq., *Enum. Syst. Pl. Carib.* 12 (1760). **FIG. 217.** PLATE 56.

FIDDLEWOOD

A small tree to 10 m tall or more, sometimes smaller and shrubby, the younger parts densely and minutely brownish-puberulous; leaves with petioles 1.5–4 cm long, the blades lance-oblong or elliptic-oblong and mostly 7–14 cm long, acuminate at the apex, glabrate on the upper surface, beneath softly tomentellous and also densely coated with minute glistening yellow glands. Calyx puberulous, 2 mm long; corolla cream or greenish-white, the tube slightly longer than the calyx, the lobes 1 mm long. Fruits subglobose, blackish or red, 3.5–4 mm in diameter.

GRAND CAYMAN: *Brunt* 1756, 1924-a; *Correll & Correll* 51012; *Hitchcock*; *Kings GC* 107, GC 409; *Proctor* 15087. CAYMAN BRAC: *Millspaugh* 1164; *Proctor* 15324.

— Distribution of the genus, frequent or common in rocky woodlands. The wood is hard, heavy ad strong; it is used for furniture and general construction.

Clerodendrum L.

Mostly shrubs or trees, sometimes herbs, the stems erect, arching or scandent; leaves opposite, subopposite or whorled, and simple, entire or toothed; petioles jointed at or near the base, when breaking off often leaving a raised or spine-like persistent base. Flowers more or less irregular, in cymose panicles, terminal and often in the upper axils; calyx usually bell-shaped, 5-toothed or -lobed, in fruit enlarging and subtending or enclosing the fruit. Corolla often with long straight or curved tube, the 5 subequal or unequal lobes spreading. Stamens 4, in 2 pairs inserted at different levels, long-exserted, with ovoid or oblong anthers. Ovary imperfectly 4-celled with one lateral ovule in each cavity; style shortly 2-branched. Fruit a drupe with thin, fleshy exocarp, the bony endocarp ultimately separating into 2 or 4 nutlets; seeds without endosperm.

— A large pantropical genus of about 350 species, most abundant in Asia and Africa.

Clerodendrum aculeatum (L.) Schldl., *Linnaea* 6: 750 (1831).

CAT CLAW

A shrub up to 2 m tall or more, forming dense thickets, with puberulous and spiny arching or straggling branches; leaves opposite or occasionally a few alternate, or often in 3s or clustered, puberulous especially beneath. Cymes on peduncles 0.8–1.5 cm long, few-flowered, axillary or in dense terminal clusters; calyx 5-lobed, the lobes reflexed at anthesis; corolla white. Filaments purple, unequal. Fruit splitting into 2 hard nutlets at maturity.

Represented in the Cayman Islands by two varieties:

1. Leaves elliptic or lanceolate, up to 6 cm long or more and 2 cm broad;
    corolla-tube more than 15 mm long; fruits 5–7 mm in diameter: var. **aculeatum**
1. Leaves narrowly lanceolate or lance-oblong, mostly less than 3 cm long
    and up to 0.8 cm broad; corolla-tube 10 mm long or less; fruits ca. 4 mm
    in diameter: var. **gracile**

Clerodendrum aculeatum var. *aculeatum*

GRAND CAYMAN: *Correll & Correll 50993; Hitchcock; Kings* GC 133, GC 148; *Millspaugh* 1380; *Proctor* 15153. LITTLE CAYMAN: *Proctor* 28111. CAYMAN BRAC: *Proctor* 29135.

— Bermuda, the West Indies and Mexico to Venezuela and the Guianas, in dry rocky or gravelly thickets.

Clerodendrum aculeatum var. *gracile* Griseb. ex Moldenke in *Carib. Forester* 2: 13 (1940).

GRAND CAYMAN: *Proctor* 27941.

— Cuba, in dry thickets. The peduncles of this variety are often 1-flowered, and all the parts are of a more slender appearance than in var. *aculeatum.*

AVICENNIACEAE

Shrubs or small trees of mangrove swamps, the roots sending up erect pencil-like aerial pneumatophores; leaves opposite and simple; stipules absent. Flowers perfect, sessile in the leaf-axils or in terminal spikes or heads, each flower subtended by 3 imbricate bracteoles. Calyx of 5 imbricate sepals, almost free, persistent. Corolla gamopetalous, with short tube and 4 almost equal lobes. Stamens 4, inserted near the base of the corolla-tube, shortly exserted; anthers 2-celled, opening lengthwise inwardly. Ovary superior, incompletely 4-celled, with a free-based placenta having 4 pendulous ovules attached to the apex; style terminal, with 2 stigmatic lobes. Fruit a compressed, leathery, asymmetric, 1-seeded capsule, eventually splitting by 2 valves; seed with folded cotyledons, germinating in the fruit.

— A single genus with about 11 species, widely distributed on tropical and subtropical seashores. This group is included in the Verbenaceae by many authors.

Avicennia L.

With the characters of the family.

Avicennia germinans (L.) L., *Sp. Pl.* ed. 3, 2: 891 (1764). **FIG. 218.** PLATE 57.

Avicennia nitida Jacq. (1760).

BLACK MANGROVE

A shrub or tree to 15 m tall or more, the young branchlets and peduncles very minutely puberulous; pneumatophores up to 30 cm long; leaves oblong-elliptic or lance-oblong, 3–13 cm long, blunt or acute at the apex, the lower surface pale greyish-green. Spikes simple or 3-branched, densely flowered toward the apex. Calyx puberulous and ciliate. Corolla white with yellow eye, the lobes densely puberulous on both sides. Style 3–4 mm long. Capsules 2.5–4 cm long.

GRAND CAYMAN: *Brunt* 2043; *Kings* GC 255, GC 270; *Proctor* 15212; *Sachet* 459. LITTLE CAYMAN: *Kings* LC 28; *Proctor* 28073.

— Seacoasts of subtropical and tropical America, also in W. Africa.

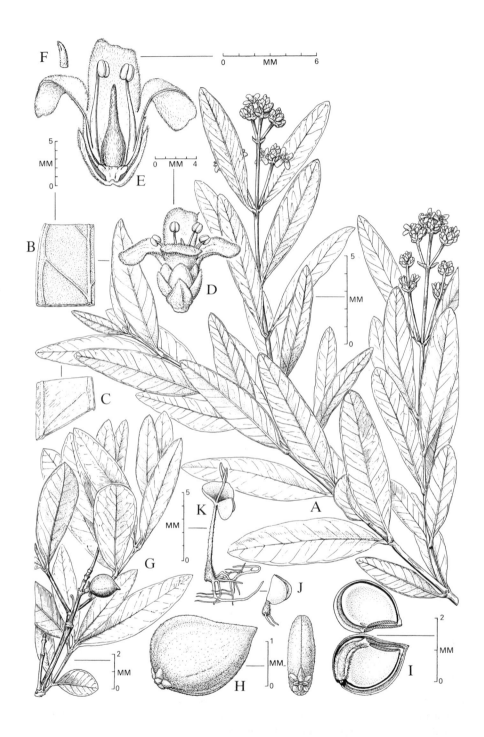

FIG. 218 **Avicennia germinans**. A, branch with leaves and flowers. B, C, details of leaf. D, flower. E, long.-section of flower, showing stamens and pistil. F, apex of style. G, fruiting branch. H, two views of fruit. I, fruit split open. J, K, stages in development of seedling. (D.E.)

LABIATAE

Herbs or shrubs, rarely trees, often aromatic and usually with 4-angled stems; leaves opposite or whorled, simple, and entire, crenate, serrate or lobed, often glandular; stipules absent. Flowers mostly perfect, rarely unisexual, solitary in the leaf-axils or more often in compact cymes, these axillary or terminal, often condensed into whorls spaced on the axes of racemes or panicles, or crowded into a spike or head; bracts small or large; bracteoles often present. Calyx tubular or bell-shaped, regular or 2-lipped, basically 5-lobed but the upper 3 teeth or lobes often united. Corolla gamopetalous, tubular at the base, the limb basically 5-lobed but usually 2-lipped, the 2 upper lobes usually to form an entire or notched hood, or rarely absent; the 3 lower lobes partly or wholly united to form the lower lip. Stamens usually 4 in 2 pairs inserted at different levels of the corolla-tube, or sometimes only 2; anthers 2-locular or sometimes 1-locular by abortion, opening lengthwise inwardly. A ring-shaped or unilateral disc or gland present. Ovary superior, of 2 bilobed carpels, becoming more or less 4-lobed and 4-locular, each cavity with 1 basal, erect ovule; style slender, central and usually gynobasic (i.e. arising basally between the lobes of the ovary); stigma 2-lobed or entire. Fruit of 4 free or paired, dry, 1-seeded nutlets, usually enclosed by the persistent calyx; seeds with little or no endosperm, the embryo straight and with flat cotyledons.

— A large, cosmopolitan family of about 200 genera and 3,200 species.

KEY TO GENERA

1. Calyx regular or nearly so, truncate and with 5 equal teeth:	**Hyptis**
1. Calyx distinctly lobed and 2-lipped or irregular:	
2. Stamens 2; corolla usually blue in Cayman species:	**Salvia**
2. Stamens 4; corolla whitish	**Ocimum**

Hyptis Jacq., nom. cons.

Erect herbs, sometimes shrubby; leaves opposite, usually toothed. Flowers in short-stalked axillary clusters, terminal racemes or panicles of contracted more or less whorled cymules. Calyx 10-nerved, truncate and bearing 5 subequal bristle-like teeth, or rarely equally 5-lobed; corolla 2-lipped, the upper lip 2-lobed, the lower one 3-lobed with the lateral lobes deflexed and the central lobe concave or sac-like. Stamens 4, slightly exserted; filaments glabrous; anthers 2-celled. Nutlets ovoid or ellipsoid, smooth or rough.

— A large tropical American genus of about 400 species.

KEY TO SPECIES

1. Calyx-tube in fruit 5–7 mm long, with glandular and non-glandular hairs; foliage strongly aromatic:	**H. suaveolens**
1. Calyx-tube in fruit 2.5–3 mm long, puberulous (the hairs not glandular) and with sessile glands; foliage not strongly aromatic:	**H. pectinata**

Hyptis suaveolens (L.) Poit., *Ann. Mus. Hist. Nat. Paris* 7: 472, t.29, f.2 (1806).

Erect annual herb up to 1 m tall, clothed throughout with gland-tipped hairs, the stems becoming subwoody at the base; leaves long-petiolate, the blades broadly ovate to orbicular, 2–6 cm long or more, the margins crenate-serrate or doubly serrate; pubescent on both sides. Cymes short-stalked in the upper axils or forming a terminal raceme; corolla pale blue or whitish, the tube 4–6 mm long. Nutlets brown, ribbed, 3–4 mm long.

GRAND CAYMAN: *Proctor* 15263. CAYMAN BRAC: *Millspaugh* 1154.

— Widespread in the American tropics, mostly a weed of open waste ground; introduced into the Old World tropics.

Hyptis pectinata (L.) Poit., *Ann. Mus. Hist. Nat. Paris* 7: 474, t.30 (1806).

Erect shrub-like herb up to 2 m tall or more, puberulous throughout with minute non-glandular hairs interspersed with minute sessile glands; leaves long-petiolate, the blades lanceolate to broadly ovate and 2–6 cm long (longer toward base of the main stem), the margins irregularly crenate-dentate. Cymes subsessile, often forked, axillary among the upper leaves and also forming simple or branched terminal racemes; corolla whitish, the tube 1.5–2 mm long. Nutlets black, smooth, 0.5 mm long.

GRAND CAYMAN: *Hitchcock*; *Millspaugh* 1341.

— Pantropical, a weed of thickets and waste places.

Salvia L.

Herbs or sometimes shrubs, erect or decumbent; leaves opposite, serrate or subentire. Flowers sessile or shortly stalked in few-flowered whorls at the nodes of terminal spikes, racemes or panicles. Calyx 2-lipped, the lips subequal and not spiny, in our species clothed with spreading, gland-tipped hairs; corolla strongly 2-lipped, the upper lip entire to 2-lobed, the lower lip spreading and more or less 3-lobed. Stamens 2; anthers 1-celled, the connective very much elongated, on one branch ascending and bearing the anther-cell at the apex, the other descending and more or less flattened. Nutlets smooth, usually becoming mucilaginous when wet.

— A very widely distributed genus of more than 700 species. Culinary sage is *Salvia officinalis* L., a native of the Mediterranean region. *S. splendens* Ker-Gawl. or red salvia is a scarlet-flowered species often cultivated for ornament.

The peculiar structure of the anthers in this genus causes them to function as a sort of lever; a bee, pushing into the flower in search of nectar, comes into contact with the sterile arm of the anther, thus raising it and causing the other end to descend onto the bee's back and dust it with pollen. The flowers are protandrous, i.e. the staminate part matures before the pistillate, thus ensuring cross-pollination. When the pistil matures, the style bends down and places the stigma in position to be touched first by a visiting insect.

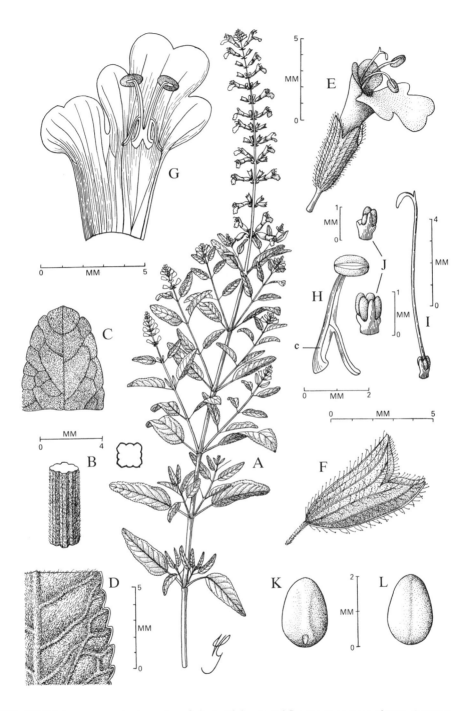

FIG. 219 **Salvia caymanensis.** A, portion of plant with leaves and flowers. B, sections of stem. C, upper side of leaf. D, lower side of leaf. E, flower. F, calyx. G, upper part of corolla cut open to show stamens. H, anther; c, connective. I, pistil. J, two ovaries. K, L, two views of nutlet. (G.)

589

KEY TO SPECIES

1. Mature calyces 2.8–3.5 mm long, not gaping in fruit: — **S. occidentalis**

1. Mature calyces more than 5 mm long, obviously open and gaping:

   2. Annual decumbent-ascending to erect herb with mostly pale blue or sometimes white flowers; leaf-blades broadly ovate, puberulous but not white-woolly beneath: — **S. serotina**

   2. Erect slender shrub to 1 m tall with deep blue flowers; leaf-blades narrowly lance-oblong, densely white-puberulous beneath: — **S. caymanensis**

Salvia occidentalis Sw, Nov. Gen. & Sp. Pl. 14 (1788). PLATE 57.

A straggling, diffusely branched herb with ascending flowering-branches, the young parts sparsely pubescent; leaf-blades rhombic-ovate, mostly 1–5 cm long, acute at the apex, cuneate at the base, the margins coarsely serrate. Racemes slender and elongate, up to 12 cm long or more. Calyx at anthesis ca. 2 mm long; corolla-tube ca. 2.5 mm long. Nutlets ca. 2 mm long.

GRAND CAYMAN: *Hitchcock*; *Proctor* 15039. LITTLE CAYMAN: *Kings* LC 61. CAYMAN BRAC: *Kings* CB 3; *Millspaugh* 1186.

— Throughout tropical America, common along roadsides, in open waste places, and in various other disturbed habitats.

Salvia serotina L., Mant. Pl. 25 (1767).

A decumbent to erect herb usually less than 30 cm tall, more or less puberulous throughout; leaf-blades mostly 1.5–3 cm long and up to 2.5 cm broad, obtuse or subacute at the apex, truncate-subcordate at the base, the margins crenate. Racemes 5–12 cm long. Calyx at anthesis 3–4 mm long; corolla tube ca. 3 mm long. Nutlets ca. 2 mm long.

GRAND CAYMAN: *Hitchcock*.
— Florida, the West Indies and C. America, a weed of pastures and cultivated fields.

Salvia caymanensis Millsp. & Uline ex Millsp., *Field Mus. Bot.* 2: 94 (1900). **FIG. 219.** PLATE 57.

Salvia serotina var. *sagittaefolia* Millsp. (1900).

A small, stiffly erect shrub with canescent-puberulous branchlets; leaf-blades mostly 1–3.5 cm long and usually not over 1 cm broad, acutish at the apex, broadly short-cuneate at the base, the margins crenulate. Racemes 1.5–10 cm long. Calyx at anthesis 4–5 mm long; corolla-tube 4–6 mm long. Nutlets 1.9–2 mm long.

GRAND CAYMAN: *Brunt* 2081; *Kings* GC 422; *Millspaugh* 1295 (type), 1391; *Proctor* 27969.

— Endemic, in sandy thickets and clearings. Sometimes considered merely a local variant of *Salvia serotina*, but falls outside the normal variability of that species; in habit it is especially distinctive.

Ocimum L.

Aromatic herbs or low shrubs; leaves opposite, usually serrate, the tissue finely dotted with pellucid glands. Flowers small, in 4–10-flowered whorls arranged in terminal racemes or panicles, the pedicels usually recurved. Calyx 2-lipped, the upper lip broad, the edges wing-like and decurrent along the tube, the lower lip 4-lobed, the lobes ending in bristly teeth. Corolla 2-lipped, the upper lip 4-lobed, the lower entire. Stamens 4, in 2 pairs, exserted. Nutlets smooth or wrinkled, often mucilaginous when moistened.

A genus of about 150 species, widely distributed in tropical and warm-temperate regions. *Ocimum basilicum* L., or basil, is a well-known culinary herb.

KEY TO SPECIES

1. Plants annual; fruiting calyx 7–8 mm long:	O. micranthum
1. Plants perennial; fruiting calyx 4–5 mm long:	[O. sanctum]

Ocimum micranthum Willd., *Enum. Hort. Berol.* 630 (1809). **FIG. 220.**

PIMENTO BASIL

A somewhat bushy herb usually less than 50 cm tall, thinly puberulous on the younger parts; leaf-blades broadly ovate-elliptic, mostly 2–6 cm long, short-acuminate at the apex,

FIG. 220 **Ocimum micranthum.** A, habit, × ¹/₂. B, flower, × 4. C, underside of calyx, × 4. D, corolla spread open to show stamens, × 4. E, style, × 4. F, fruiting calyx, × 4. G, nutlet, × 2¹/₂. (St.)

cuneate at the base, the margins subentire or obscurely serrate. Racemes compact, mostly under 6 cm long, the flowers in whorls of 6. Corolla whitish with pale violet mottling, ca. 4 mm long. Nutlets black, 1–2 mm long.

GRAND CAYMAN: *Correll & Correll* 51042; *Kings* GC 213; *Millspaugh* 1266. LITTLE CAYMAN: *Kings* LC 6.

— Florida, the West Indies and continental tropical America, common in fields, thickets and open waste ground. The pleasant, aromatic odour of this species is due to the essential oil methyl cinnamate, which can be used like citronella as a mosquito-repellent.

[*Ocimum sanctum* L., *Mant. Pl.* 1: 85 (1767).

A bushy herb or subshrub usually more than 70 cm tall, puberulous and thinly pilose on the younger parts; leaves long-petiolate, the blades mostly 2–5 cm long, acute at the apex, the margins coarsely serrate. Racemes simple or paniculate, up to 10 cm long, the flowers in whorls of 4–8. Corolla ca. 3 mm long. Nutlets brown, 1–1.5 mm long.

GRAND CAYMAN: *Kings* GC 135.

— Native of the Old World tropics, now sparingly naturalised in the West Indies and continental tropical America.]

OLEACEAE

Trees or shrubs, often scandent or vine-like; leaves usually opposite, rarely alternate or whorled, simple or pinnately compound; stipules absent. Flowers regular, perfect or unisexual, borne in terminal or axillary racemes, cymes or panicles. Calyx 4-many-lobed, valvate or rarely absent; corolla gamopetalous or rarely with free petals, usually 4-lobed or rarely with more. Stamens 2 or rarely 4, hypogynous or inserted on the corolla; anthers 2-celled, opening lengthwise; disc absent. Ovary superior, 2-celled, usually with 2 ovules in each cavity, pendulous or ascending on axile placentas; style simple or absent; stigma capitate or bifid. Fruit a berry, drupe, capsule or samara; seeds with fleshy endosperm (rarely none) and straight embryo.

— A widespread family of perhaps 30 genera and about 600 species, occurring in both temperate and tropical regions. The most important species in this family is *Olea europea* L., the olive, the fruit being eaten when pickled, and also the source of olive-oil.

KEY TO GENERA

1. Plants creeping or vine-like, with pinnate leaves; corolla-tube elongate; fruit of 2 berry-like lobes:	[Jasminum]
1. Plants erect, with simple leaves; corolla-tube very short, or petals free or lacking; fruit a simple drupe:	
2. Flowers unisexual, dioecious, in small axillary clusters; calyx and corolla absent, or minute and soon falling:	Forestiera
2. Flowers perfect, in terminal and axillary panicles; calyx and corolla both present and well-developed:	Chionanthus

[*Jasminum* L.

Erect shrubs or often vines; leaves usually opposite and simple, 3-foliolate or odd-pinnate. Flowers solitary or usually in cymes, terminal or in the upper axils, often fragrant and showy; calyx 4–9-toothed or -lobed; corolla salver-shaped with cylindrical tube and 4–5 or more imbricate lobes. Stamens 2, included. Ovary 2-celled with 2 (rarely 3–4) ovules in each cavity, these laterally attached near the base. Fruit a double berry, or one of the carpels sometimes abortive; seeds usually solitary, without endosperm.

— A genus of about 200 species, in the tropical and warm-temperate regions of the Old World. Several are widely cultivated for their showy or fragrant flowers.

Jasminum fluminense Vell., *Fl. Flumin.* 10 (1825).

STAR OF BETHLEHEM

A scrambling shrub or vine, the younger parts pubescent; leaves 3-foliolate; leaflets broadly ovate, 1.5–5 cm long, acute or acuminate at the apex, the margins entire, and with tufts of woolly hairs in the nerve-axils beneath. Cymes on pubescent peduncles; flowers fragrant at night; corolla white, 5–9-lobed, the tube ca. 1.5 cm long. Berries black, 5–8 mm in diameter.

GRAND CAYMAN: *Brunt* 1888; *Proctor* 15076.

— Native of tropical Africa, widely naturalised in the American tropics. In Grand Cayman, this species is grown in gardens, has escaped and has become naturalised in roadside thickets.]

Forestiera Poir.

Dioecious trees or shrubs; leaves mostly opposite, simple, entire or toothed, usually deciduous during dry seasons. Flowers small, 1–few in small fascicles in the leaf-axils or at nodes on old wood; calyx deeply 4–6-lobed or absent; corolla none or rarely 1 or 2 small free petals present. Stamens 2 or 4. Ovary 2-celled, each cavity with 2 pendulous ovules. Fruit a 1-seeded drupe.

— A genus of about 15 species occurring in southern U. S. A., Mexico, northern C. America and the West Indies.

Forestiera segregata (Jacq.) Krug & Urb. in Engl., *Bot. Jahrb.* 15: 339 (1892). PLATE 57.

A diffusely branched shrub to 2 m tall or more, rarely tree-like; branchlets stiff and minutely puberulous; leaves glabrous, narrowly elliptic, oblanceolate or obovate, mostly 2–6.5 cm long, obtuse at the apex, minutely gland-dotted beneath. Inflorescence subtended by sessile and clawed, ciliate bracteoles 2–2.5 mm long. Flowers yellowish-green, fragrant. Staminate flowers pedicellate, lacking calyx and corolla, consisting of 2–4 naked stamens with filaments 4–6 mm long. Pistillate flowers undescribed. Drupe obliquely spindle-shaped, acute, bluish-purple, ca. 7 mm long.

GRAND CAYMAN: *Brunt* 2152; *Proctor* 15063, 15066, 15138. LITTLE CAYMAN: *Proctor* 35134.

— Florida, Bermuda, the Bahamas, the Greater Antilles, Virgin Islands and Antigua, in dry rocky thickets and woodlands.

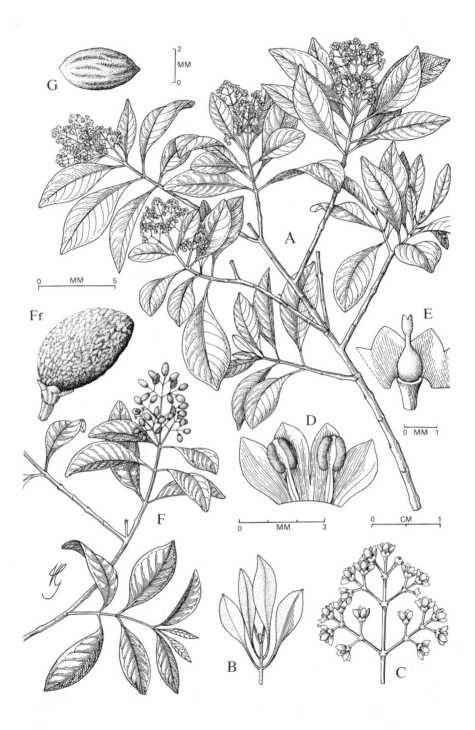

FIG. 221 **Chionanthus caymanensis.** A, habit, branch with leaves and flowers. B, young leafy branchlet. C, portion of flowering inflorescence. D, corolla opened out to show stamens. E, calyx (opened out) and pistil. F, fruiting branch; Ff, fruit. G, seed. (G.)

Chionanthus L.

Trees or sometimes shrubs; leaves opposite, entire and more or less coriaceous. Flowers in terminal and axillary panicles, the branches racemose, thyrsoid, or cymose, or the inflorescence contracted to an umbel or head. Calyx small, 4-parted or -toothed; corolla of 4 free or nearly free valvate petals, these linear or oblong. Stamens 2 or rarely 4, borne on the base of the petals. Ovary 2-celled with 2 ovules in each cavity, attached laterally near the apex and pendulous. Fruit an ovoid, oblong or subglobose drupe with thin flesh and hard endocarp, usually containing a solitary seed.

— A pantropical and warm-temperate taxon of more than 150 species.

Chionanthus caymanensis Stearn, *Bot. Notiser* 132: 58 (1979). **FIG. 221. PLATE 58.**

IRONWOOD

A shrub or small tree to 10 m tall, with ashy-grey bark, glabrous throughout, the young shoots clothed with numerous minute white waxy scales. Leaves with marginate petioles 0.5–1 cm long; blades obovate, 2–8 cm long, 1–2.5(–3) cm broad above the middle, abruptly short-acuminate at the apex, long-cuneate at the base and decurrent on the petiole, very minutely gland-dotted on both surfaces; domatia absent. Panicles 3–6 cm long, many-flowered; flowers white, fragrant; calyx rugose, ca. 1 mm long with triangular lobes; petals oblong-obovate, 2 mm long, joined at the base. Anthers ellipsoid, 1–1.4 mm long. Style 1 mm long; stigma bifid. Drupes obliquely ellipsoid, 6–8 mm long.

GRAND CAYMAN: *Proctor* 27958, 48303, 49228. LITTLE CAYMAN: *Proctor* 28116, 28183 (type), 47312. CAYMAN BRAC: *Proctor* 48209.

— Endemic.

SCROPHULARIACEAE

Herbs, shrubs or vines, rarely trees; leaves alternate, opposite or whorled, always simple, margins entire, toothed or lobed; stipules absent. Flowers perfect, more or less zygomorphic, solitary, clustered, or in racemes or panicles. Calyx nearly truncate or 4–5-toothed or -divided, persistent; corolla gamopetalous, tubular to broadly bell-shaped, often 2-lipped, with 4, 5 or more imbricate lobes. Stamens usually 4 (rarely 2 or 5), in 2 pairs inserted at different levels in the corolla-tube; anthers usually 2-celled, opening lengthwise. Disc usually present. Ovary superior, typically 2-celled; ovules on large axile placentas; style solitary, often persistent, the stigma usually more or less 2-lobed. Fruit usually a capsule variously dehiscent, or rarely a berry; seeds usually numerous, small and smooth, angled or winged; endosperm fleshy, the embryo small.

— A large, cosmopolitan family of about 200 genera and 3,000 species.

KEY TO GENERA

1. Leaves alternate:	Capraria
1. Leaves opposite or whorled:	
2. Flowers in lax, peduncled cymes; corolla red:	[Russelia]

2. Flowers solitary or clustered in the leaf-axils; corolla not red:

   3. Leaves (at least the larger ones) serrate or toothed; corolla less than 6 mm long:

      4. Flowers subsessile; corolla tubular, the lobes shorter than the tube: **Stemodia**

      4. Flowers on slender pedicels longer than the calyx; corolla rotate, the lobes longer than the tube: **Scoparia**

   3. Leaves entire, linear; corolla more than 8 mm long:

      5. Stems erect; leaves linear-subulate; corolla pink: **Agalinis**

      5. Stems prostrate, often forming mats; leaves spatulate or obovate; corolla blue or white: **Bacopa**

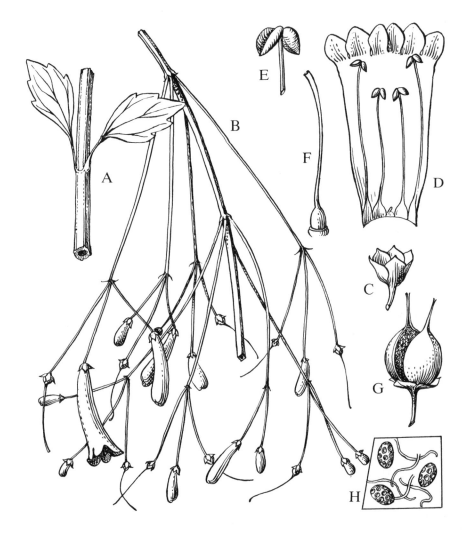

FIG. 222 **Russelia equisetiformis**. A, portion of stem with pair of leaves, × ⁴/₅. B, portion of inflorescence, × ⁴/₅. C, calyx, × 1¹/₂. D, corolla spread open to show stamens. E, anther, × 4. F, pistil, × 1¹/₂. G, capsule, × 1¹/₂. H, seeds, × 6. (St.)

[*Russelia* Jacq.

Shrubs with striate or angled stems; leaves opposite or whorled, sometimes reduced to more scales. Flowers in simple or compound cymes; calyx deeply 5-lobed, the lobes ovate; corolla tubular or narrowly funnel-shaped, the 5 lobes much shorter than the tube, somewhat unequal. Stamens 4, the anther-cells divergent. Capsule glabrous, ovoid or globose, septicidally dehiscent; seeds small, ellipsoid, brown or black, and variously reticulate, pitted or ridged.

— A genus of more than 50 species occurring naturally in Mexico and C. America; several are widely cultivated in warm countries.

Russelia equisetiformis Cham. & Schlecht., *Linnaea* 6: 377 (1831). **FIG. 222**.

Russelia juncea Zucc. (1832).

A lax, arching, much-branched shrub, the slender green branches striate and glabrous; leaves up to 7 in a whorl or obsolete, when present lanceolate to ovate, up to 2 cm long, with toothed margins, and gland-dotted beneath. Pedicels ca. 1 cm long; sepals 2 mm long. Corolla crimson with tube up to 2.5 cm long. Capsules broadly ovoid, ca. 5 mm long.

GRAND CAYMAN: *Brunt* 1714; *Hitchcock*; *Kings* GC 80; *Proctor* 15188. LITTLE CAYMAN: *Proctor* 35201 (cult.). CAYMAN BRAC: *Proctor* (sight record).

— Mexico and C. America, cultivated and naturalised elsewhere. In Grand Cayman, this species now grows wild along sandy roadsides.]

Stemodia L., nom. cons.

Herbs or low shrubs, mostly glandular-pubescent and aromatic; leaves opposite or whorled, serrate or toothed. Flowers solitary in the axils or in terminal, often leafy-bracted spikes or racemes; calyx 5-parted, the segments imbricate, equal and nearly free. Corolla with nearly cylindrical tube, the limb 2-lipped, the upper lip notched or entire, the lower 3-lobed. Stamens 4; anthers-cells distinct, stalked. Stigma usually 2-lobed. Capsule 2-valved, the valves 2-cleft; seeds striate or reticulate.

— A pantropical genus of about 30 species.

Stemodia maritima L., *Syst. Nat.* ed. 10, 2: 1118 (1759).

Perennial sprawling herb with erect branches, the older stems becoming somewhat woody; leaves lanceolate, mostly 0.6–2 cm long, acute at the apex, sessile and clasping-subcordate at the base, the margins sharply serrate. Flowers solitary; calyx 2–3 mm long; corolla 5.5 mm long, pale blue or white. Capsules elongate-ovoid, 2.5 mm long; seeds minutely punctate, apiculate.

GRAND CAYMAN: *Brunt* 1915, 2076, 2077; *Hitchcock*; *Kings* GC 65; *Proctor* 15036, 47824, 48284, 50731.

— Bahamas, Greater Antilles and S. America, in low moist ground, sandy clearings or seasonally flooded grasslands near the sea.

Capraria L.

Perennial shrubby herbs; leaves alternate, toothed. Flowers solitary or paired (rarely 4 together) in the leaf-axils, usually long-stalked; calyx of 5 narrow, almost equal sepals; corolla white, bell-shaped, 5-lobed, the lobes subequal. Stamens 4 (rarely 5); anther-cells divergent, confluent. Stigma dilated. Capsule longitudinally grooved and 4-valved; seeds reticulate.

— A tropical American genus of 4 species.

Capraria biflora L., *Sp. Pl.* 2: 628 (1753).

Stems erect, to 1 m tall or more, usually hairy; leaves oblanceolate or narrowly elliptic, mostly 1.5–4 cm long, acute at the apex, narrowed to the point of attachment, sharply serrate on the distal half. Pedicels slender and flexuous, up to 15 mm long; sepals 4–6 mm long; corolla ca. 1 cm long. Capsule oblong-ovate, 4–6 mm long; seeds light brown, 0.5 mm long.

GRAND CAYMAN: *Brunt* 2075, 2130; *Hitchcock*; *Kings* GC 65a; *Millspaugh* 1364; *Proctor* 27940.

— Throughout tropical and subtropical America, common along roadsides and ditches, in open waste ground and in damp pastures.

Scoparia L.

Erect, branched herbs or low shrubs; leaves opposite or whorled, gland-dotted and with entire or serrate margins. Flowers solitary or paired (rarely 4) in the leaf-axils, usually long-stalked; calyx 4–5-parted, the segments imbricate, nearly free; corolla white, nearly rotate, 4-lobed with subequal lobes, densely bearded in the throat. Stamens 4; anther-cells distinct, parallel or divergent. Style pubescent, club-shaped with truncate or notched stigma. Capsule 2-valved, membranous; seeds numerous, angular.

— A tropical American genus of 20 species, one of them (the Cayman species) also in the Old World tropics and subtropics.

Scoparia dulcis L., *Sp. Pl.* 1: 116 (1753). **FIG. 223**.

A bushy annual herb with a taproot, up to 50 cm tall or more, glabrous throughout; leaves opposite or whorled, linear-oblanceolate or narrowly elliptic, mostly 0.5–4 cm long, at least the larger ones serrate in the distal half. Pedicels filiform, 5–8 mm long; calyx-lobes oblong, 1.5–2 mm long; corolla 3–4 mm across when expanded, with reflexed lobes. Capsules ovoid-globose, ca. 3 mm long; seeds brown, 0.3–0.4 mm long.

GRAND CAYMAN: *Proctor* 27970.

— Widespread in the tropics and subtropics; Cayman plants were found in sandy thickets and clearings.

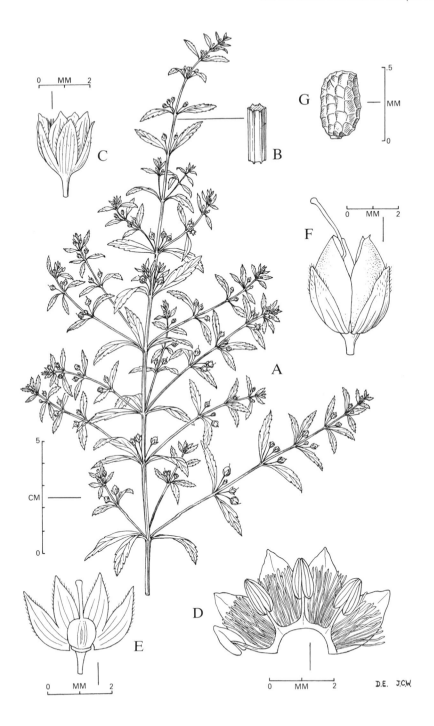

FIG. 223 **Scoparia dulcis**. A, habit. B, section of stem. C, flower. D, corolla spread open to show stamens and hairs. E, calyx with pistil. F, capsule. G, seed. (D.E. & J.C.W. ex St.)

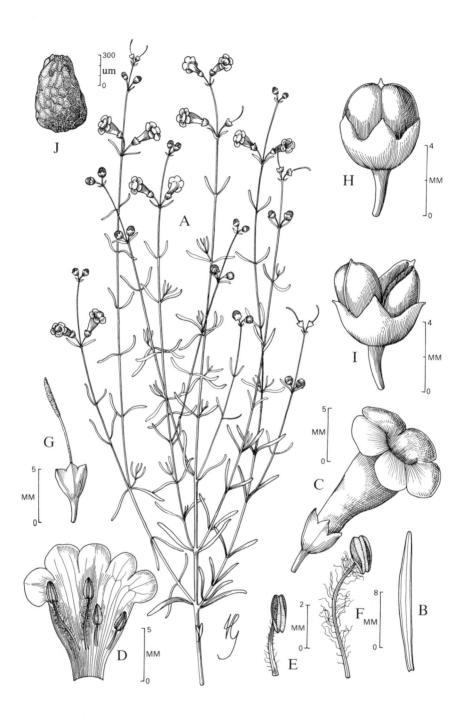

FIG. 224 **Agalinis kingsii**. A, habit. B, leaf. C, flower. D, corolla opened out. E, F, stamens. G, flower with corolla and stamens removed, showing calyx, style and stigma. H, fruit. I, fruit after dehiscence. J, seed. (G.)

Agalinis Raf., nom. cons.

Erect hemiparasitic terrestrial herbs; leaves opposite, narrow, and sessile. Flowers solitary in the bract-axils of loose racemes; calyx campanulate, 5-toothed or -lobed; corolla campanulate or funnel-shaped, the tube broad, the limb 5-lobed and somewhat 2-lipped, the posterior (upper) lobes inflexed in bud. Stamens 4, included, the filaments pubescent; anthers 2-celled. Capsule ovoid or globose, 2-valved; seeds numerous, angled.

— An American genus of about 60 species.

Agalinis kingsii Proctor, *Sloanea* 1: 3 (1977). **FIG. 224**. PLATE 58.

Slender annual subglabrous herb to 50 cm tall or more; leaves linear-subulate, up to 4 cm long and 2 mm broad, often incurved; margins rough-edged. Racemes very lax and few-flowered; pedicels ascending, 6–10 mm long; calyx-tube ca. 2.5 mm long, the lobes triangular; corolla pink, campanulate, 10–15 mm long, the lobes ciliate. Capsules globose, ca. 4 mm long; seeds wedge-shaped, 0.7–0.9 mm long.

GRAND CAYMAN: *Kings GC 257* (type MO, isotype BM); *Proctor 47271, 48285.*

— Endemic, collected "in mangrove swamps on the drier land" at Forest Glen, near North Side. The species has since been found E. of Duck Pond Bight and S. of the Salina Reserve. *Agalinis kingsii* differs from *A. albida* Britton & Pennell of Cuba and Jamaica in its very much larger leaves, longer pedicels and pink flowers. It differs from the related *A. purpurea* (L.) Pennell of the U. S. A. and Cuba in its glabrous leaves, longer pedicels, smaller corollas and smaller capsules.

Bacopa Aublet

Herbs, mostly with prostrate stems (a few with erect stems). Leaves opposite or whorled, sessile or stalked, often fleshy, usually punctate. Flowers axillary, pedicellate, often with 2 bracts at apex of pedicel; calyx 5-parted, the lobes unequal, the outer larger; corolla zygomorphic with cylindrical tube or campanulate and slightly bilabiate, the lobes spreading with the upper one emarginate, the lower 3-toothed or -lobed, blue, purple or white. Stamens 4. Capsule globose, dehiscent; seeds small, numerous.

— A genus of about 40 species, the majority in the neotropics.

Bacopa monnieri (L.) Pennell, *Proc. Acad. Nat. Sci. Phil.* 98: 94 (1946).

Stems creeping, rooting at the nodes, glabrous, fleshy, up to 50 cm long or more, much branched and forming mats. Leaves subsessile, spatulate or cuneate-obovate, 6–12 mm long or more, 2–5 mm broad, rounded at apex, the margins entire or nearly so, the veins obscurely pinnate, the tissue punctate. Pedicels mainly in alternate axils, 10–18 mm long with 2 bracts at the apex; largest calyx segments ovate, 6–7 mm long, acute; corolla campanulate, 10 mm broad, pale blue, mauve or white. Stamens 4. Capsule ovoid, 4 mm long, pointed; seeds pointed at one end.

GRAND CAYMAN: *Clifford 214*; *Guala 1940*; *Proctor 50720, 50736.* Occurs in wet cattle pastures, damp lawns and beside fresh or brackish pools.

— Florida, Mexico, and widespread in both the New and Old World tropics.

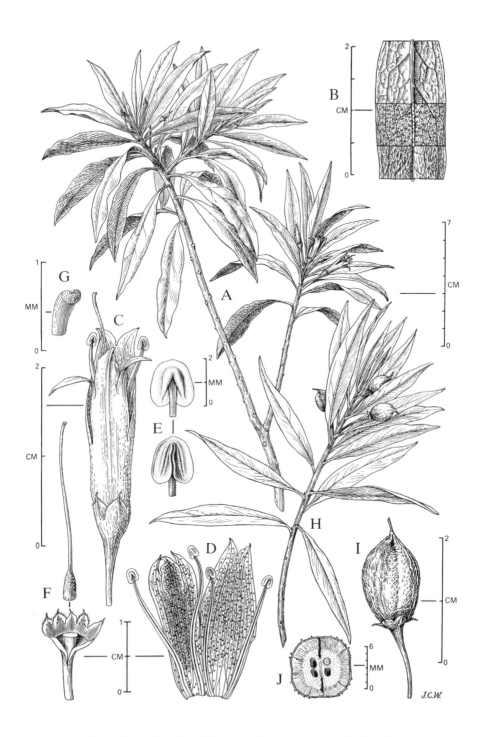

FIG. 225 **Bontia daphnoides.** A, branch with leaves and flowers. B, details of leaf. C, flower. D, upper part of corolla spread open to show stamens. E, two views of anther, F, calyx and pistil. G, stigma. H, fruiting branch. I, fruit. J, cross-section of fruit. (J.C.W.)

MYOPORACEAE

Shrubs or trees; leaves alternate or opposite, simple and entire, often glandular; stipules absent. Flowers perfect, zygomorphic, solitary or in axillary cymose clusters; calyx 5-lobed, persistent; corolla gamopetalous, 5-lobed, the limb oblique or 2-lipped. Stamens 4 (rarely 5), in 2 pairs inserted at different levels of the corolla-tube, sometimes accompanied by a staminode; anthers 2-celled, opening lengthwise. Ovary superior, usually 2-celled (rarely 3–10-celled); ovules 2–8 in each cavity, paired and pendulous on axile placentas; style terminal with simple stigma. Fruit a drupe; seeds small, with scant endosperm.

— A chiefly Australasian family of 5 genera and about 180 species, a single representative occurring in the West Indies.

Bontia L.

A glabrous, bushy, often arborescent shrub; leaves alternate, somewhat fleshy. Flowers solitary in the leaf-axils; calyx-segments imbricate; corolla with cylindrical tube and 2-lipped limb, the upper lip 2-lobed, the lower 3-lobed and recurved, the middle lobe densely bearded. Ovary 2-celled with 4 ovules in each cavity. Drupe ovoid-acuminate.

— A monotypic genus of the Caribbean area.

Bontia daphnoides L., *Sp. Pl.* 2: 638 (1853). **FIG. 225.** PLATE 58.

Plants up to 3 m tall or more; leaves narrowly lanceolate, mostly 4–12 cm long, acuminate at the apex, finely gland-dotted, the venation (except the midvein) obscure. Peduncles 1–2.5 cm long; calyx-lobes ovate-acuminate, 3–5 mm long, with hair-like tip; corolla dull yellow blotched with purple, ca. 2 cm long, gland-dotted. Drupes 1–1.5 cm long, yellowish when ripe, crowned by the persistent elongate style.

GRAND CAYMAN: *Brunt* 2055, 2177; *Hitchcock*. LITTLE CAYMAN: *Proctor* 28175. CAYMAN BRAC: *Proctor* 29136.

— West Indies and north coast of S. America, in subsaline coastal thickets and dry woodlands on limestone.

BIGNONIACEAE

Trees, shrubs, or vines, rarely herbs; leaves opposite or rarely alternate, simple, trifoliolate, pinnately compound or sometimes digitate; if trifoliolate, the terminal leaflet often modified to a tendril; stipules absent. Flowers perfect, zygomorphic, often large, usually in cymes or racemes, sometimes solitary or clustered. Calyx bell-shaped, usually 5-toothed or -lobed, sometimes truncate or spathe-like; corolla gamopetalous, 5-lobed, sometimes 2-lipped. Stamens 4 (rarely 2), inserted in the corolla-tube; anthers 2-celled, opening lengthwise, the cells often widely divergent; 1–3 staminodes sometimes present. A hypogynous, ring-like or cup-shaped disc present. Ovary superior, 2-celled, with numerous ovules borne on 2 axile placentas in each cavity, or the ovary 1-celled with 2 bifid parietal placentas; style terminal, simple with 2-lobed stigma. Fruit indehiscent or else a 2-valved capsule; seeds without endosperm, those from capsular fruits often winged.

— A pantropical family of about 120 genera and 800 species, the great majority in tropical America.

KEY TO GENERA

1. **Leaves simple:**

   2. Fruit a large round indehiscent capsule; seeds not winged: **Crescentia**

   2. Fruit an elongate dehiscent capsule; seeds winged: **Catalpa**

1. **Leaves compound:**

   3. Leaves digitately 3–5-foliolate, the leaflets clothed with very minute peltate scales: **Tabebuia**

   3. Leaves pinnate:

      4. Leaflets entire; calyx spathe-like; corolla scarlet, very large: **[Spathodea, p. 606]**

      4. Leaflets serrate; calyx regular; corolla yellow: **Tecoma**

Crescentia L.

Small trees, the branches with prominent nodes; leaves alternate or mostly fascicled on dwarf shoots in the axils of older, fallen leaves. Flowers solitary or several in a cluster from nodes on the old wood (or occasionally axillary); calyx leathery, closed in bud, splitting into 2 or 5 lobes at anthesis; corolla broadly bell-shaped with swollen tube and an oblique, 5-lobed limb. Stamens 4, in 2 pairs at different levels, included or slightly exserted. Ovary 1-celled. Fruit globose or ovoid, often very large, with a hard shell; seeds numerous, flattened, borne in pulp on spongy placentas.

— A tropical American genus of 5 species.

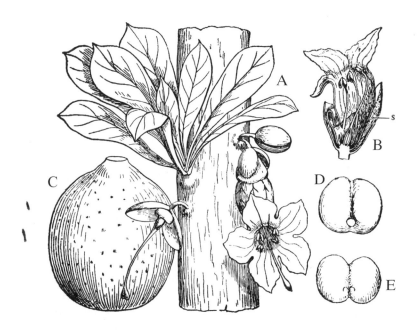

FIG. 226 **Crescentia cujete.** A, portion of woody stem bearing flowers and a cluster of leaves, × $^1/_2$. B, flower cut lengthwise, × $^1/_2$; s, staminode. C, fruit, very much reduced. D, seed, × $^3/_4$. E, embryo, × $^3/_4$; the radicle is concealed by the auricle of the cotyledon. (St.)

Crescentia cujete L., *Sp. Pl.* 2: 626 (1753). **FIG. 226.** Plate 58.

A tree up to 10 m tall (but often less), with stout branchlets; leaves oblanceolate or narrowly obovate, mostly 2–15 cm long, variable in length in the same fascicle or on the same branch. Calyx up to 2 cm long, deeply 2-lobed; corolla 4–6 cm long, greenish-cream, often purple-veined. Fruits up to 25 cm in diameter (often less).

GRAND CAYMAN: *Brunt* 2100; *Kings* GC 204; *Proctor* 15295.

— Florida, West Indies and continental tropical America, in thickets, savannas, pastures and along roadsides. The fruits, called 'calabashes' (from the Spanish 'calabazo', meaning a gourd), have traditionally been used in rural areas throughout its range as containers for water and other liquids; the shell is so tough that water can even be boiled in it. Less well known is the fact that the wood of this tree is exceedingly tough and durable, suitable for making furniture and the handles of tools.

Catalpa Scop.

Trees with simple, petiolate, opposite or whorled leaves. Inflorescence terminal, paniculate; calyx deeply bilabiate; corolla broadly campanulate, bilabiate, 5-lobed, the lobes undulate, the three lower ones longer. Fertile stamens 2, the others sterile or rudimentary. Ovary oblong, the ovules multiseriate in each locule; disc annular-pulvinate. Capsule elongate, slender, nearly cylindrical, 2-locular; seeds many, the two acute wings on each seed formed from fused hairs; body of the seed sometimes long silky-pubescent.

— A genus of 10 species, 4 in eastern Asia, 2 in eastern N. America, and 4 in the Antilles.

Catalpa longissima (Jacq.) Dum. Cours., *Bot. Cult.* 2: 190 (1802).

Tree to 25 m tall or more. Leaves with slender petioles mostly 1–3.5 cm long, the blades usually ovate-lanceolate, mostly 5–12 cm long, 2–4.5 cm broad, the apex long-acute to acuminate, the base rounded. Calyx campanulate, 2-lobed to the base, 4–7 mm long; corolla mostly white but pinkish on the lobes and yellow in the throat with fine purple lines, 25–30 mm long; stamens included. Fruit very slender, mostly 30–75 cm long, 2–4 mm in diameter; seeds 7–10 mm wide including the tapering wings, silky pubescent.

GRAND CAYMAN: *Proctor* 47845, found at NE border of forest behind the Community College, S of George Town.

— Hispaniola, Jamaica; introduced in the Lesser Antilles.

Tabebuia DC.

Trees or shrubs; leaves opposite or nearly so, usually 3–5-foliolate, sometimes 1-foliolate or simple. Flowers in terminal or axillary cymes or cymose panicles; calyx tubular or bell-shaped, toothed or shallowly lobed, sometimes more or less 2-lipped; corolla more or less funnel-shaped, the 5 spreading, slightly unequal lobes rounded and variously toothed, undulate or ruffled. Stamens 4, in 2 pairs inserted at different levels, a short staminode also usually present. Ovary 2-celled. Fruit an elongate, linear, lengthwise-dehiscent capsule; seeds with 2 membranous whitish wings.

— A tropical American (chiefly West Indian) genus of more than 100 species. Several are frequently planted for ornament, e.g. *Tabebuia rosea* (Bertol.) DC., the pink poui and *T. rufescens* J. R. Johnst., the yellow poui (Jamaica).

Tabebuia heterophylla (DC.) Britton, *Ann. Missouri Bot. Gard.* 2: 48 (1915). PLATE 59.

Tabebuia pentaphylla of Hitchcock (1893), not Hemsl. (1882).
Tabebuia riparia (Raf.) Sandwith (1944).

WHITEWOOD

A shrub or small tree to 5 m tall or more; leaflets usually 5- (sometimes 3-)stalked, more or less coriaceous and variable in shape, elliptic or oblong-elliptic to oblanceolate or obovate, mostly 2–8 cm long, the apex rounded or blunt to acutish, the venation finely reticulate. Flowers petiolate, solitary or in small cymose clusters on peduncles 1–3 cm long; calyx 9–12 mm long, slightly 2-lipped; corolla 4–7 cm long, light pink with yellow throat; stigma spatulate. Capsules 7–12 cm long, rarely longer, minutely lepidote like the leaves, beaked at the apex.

GRAND CAYMAN: *Brunt* 1955, 2133; *Correll & Correll* 50999; *Hitchcock*; *Howard & Wagenknecht* 15023; *Kings GC* 132; *Lewis GC* 11, *GC* 57; *Proctor* 15055, 15227; *Sachet* 430. LITTLE CAYMAN: *Kings LC* 14; *Proctor* 28060. CAYMAN BRAC: *Kings CB* 73; *Millspaugh* 1214; *Proctor* 29054; *Sauer* 4139.

— Greater Antilles, the Virgin Islands and northern Lesser Antilles, in dry rocky thickets and woodland on limestone. The wood is used for building cat-boats and schooners.

[*Spathodea campanulata* Beauv., the African tulip tree, is planted as an ornamental in Grand Cayman (*Brunt* 2019; *Kings GC* 95).]

Tecoma Juss.

Erect shrubs or trees; leaves opposite, odd-pinnate or rarely simple; flowers in terminal racemes or panicles; calyx more or less bell-shaped, 5-toothed or -lobed; corolla funnel- or bell-shaped, the limb 5-lobed and slightly 2-lipped, the lobes nearly equal. Stamens 4, in 2 pairs inserted at different levels, included or exserted. Ovary 2-celled. Capsule linear-elongate, lengthwise-dehiscent, many-seeded, the seeds winged.

— A tropical American genus of about 16 species.

Tecoma stans (L.) Kunth in H. B. K., *Nov. Gen. & Sp.* 3: 144 (1819). **FIG. 227**. PLATE 59.

SHAMROCK, COW-STICK or HEMLOCK (Cayman Brac).

A shrub to 3 m tall or more; leaves with mostly 5 or 7 leaflets (rarely simple or 3-foliolate), these lanceolate or narrowly ovate, mostly 3–10 cm long, the terminal leaflet the largest, all acuminate and sharply serrate, puberulous beneath chiefly along the midrib and principal veins. Flowers in a short raceme; corolla 4–5 cm long. Capsules 11–20 cm long, glabrous.

GRAND CAYMAN: *Brunt* 2072; *Kings GC* 154; *Lewis* 3851; *Proctor* 15085. CAYMAN BRAC: *Proctor* 28987.

— Florida, West Indies, and continental tropical America, introduced and naturalised in the Old World tropics. The Cayman plants are common in sandy or rocky thickets and along roadsides.

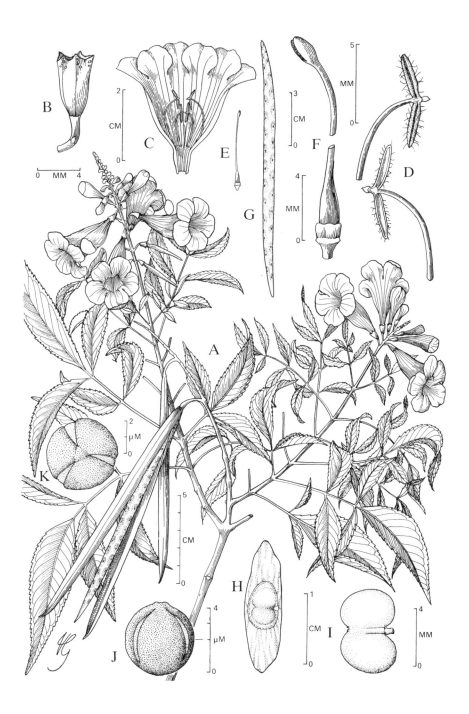

FIG. 227 **Tecoma stans**. A, branch with leaves, flowers and fruits. B, calyx. C, corolla cut open to show stamens. D, two views of anther. E, pistil. F, same, enlarged. G, one valve of fruit, inside surface. H, seed. I, seed with wings removed. J, K, two views of pollen grain. (G.)

ACANTHACEAE

Herbs, shrubs or trees, sometimes vines, the shoots often angled and swollen above the nodes; leaves opposite or rarely in whorls of 3, simple, entire or toothed, the epidermis often with cystoliths. Flowers perfect, zygomorphic or nearly regular; solitary, clustered or cymose in the leaf-axils, or in terminal spikes, racemes, cymes or panicles, often with large bracts. Calyx 4–5-lobed, imbricate or valvate, or reduced to a ring; corolla gamopetalous, 5-lobed and nearly regular or 2-lipped, sometimes 1-lipped. Stamens mostly 4, in 2 pairs inserted at different levels of the corolla-tube, sometimes reduced to 2; 1 or more staminodes sometimes present; anthers 1–2-celled, opening lengthwise. Disc present. Ovary superior, 2-celled, with 2–many ovules in each cavity on axile placentas; style simple, with capitate or lobed stigma. Fruit usually a capsule, often elastically dehiscent, the valves rupturing and recurving explosively, flinging the seeds out, the action aided by the hook-like and hardened ovule-stalks; seeds usually flat, with little or no endosperm.

— A pantropical and subtropical family of about 250 genera and 2,500 species. Many ornamental shrubs and vines belong to this family, which otherwise has little economic importance.

KEY TO GENERA

1. Flowers in spikes or racemes:

    2. Bracts large and conspicuous, completely concealing the calyx: **Blechum**

    2. Bracts minute, shorter than the pedicels: **[Asystasia]**

1. Flowers in panicles: **Ruellia**

Blechum P. Br.

Herbs with 4-angled, often straggling stems; leaves entire, wavy or crenate, usually pubescent. Inflorescence a dense terminal spike, the flowers partly concealed by large, overlapping bracts; flowers solitary or paired in the bract-axils, each subtended by 2 narrow bracteoles longer than the calyx. Calyx 5-parted, with linear segments; corolla narrowly tubular, with 5 spreading, nearly equal lobes. Stamens 4, inserted in the upper part of the corolla-tube; anthers oblong, with parallel cells. Capsule ovoid, pointed 6–16-seeded.

— A tropical American genus of 10 species.

Blechum pyramidatum (Lam.) Urb., *Repert. Spec. Nov. Regni Veg.* 15: 323 (1918). **FIG. 228**. PLATE 59.

Blechum brownei Juss. of this flora ed. 1.

An annual decumbent herb with ascending branches up to 50 cm tall or more, often rooting at the lower nodes; leaves narrowly ovate or elliptic, mostly 1.5–5 cm long, sparingly pubescent and with numerous linear cystoliths. Spikes 3–10 cm long, or rarely some flowers solitary in the leaf-axils; corolla pale violet or whitish, ca. 15 mm long. Capsules 6–7 mm long, 12–16-seeded.

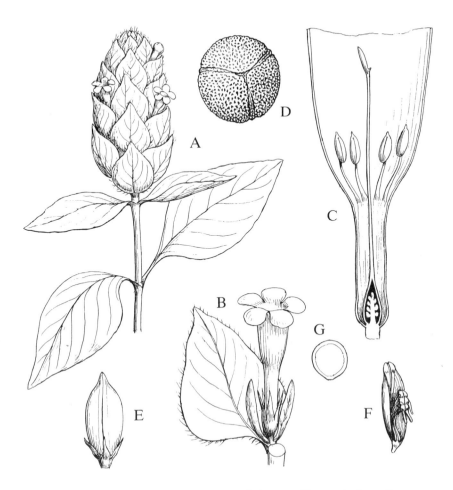

FIG. 228 **Blechum pyramidatum.** A, habit, × 1. B, single flower and bract, × 3. C, corolla cut open to show stamens and pistil, × 6. D, pollen grain greatly enlarged. E, capsule, × 3. F, capsule after dehiscence, × 3. G, seed, × 6. (St.)

GRAND CAYMAN: *Brunt* 1900; *Hitchcock*; *Kings* GC 88; *Lewis* 3860; *Millspaugh* 1328; *Proctor* 15075. CAYMAN BRAC: *Millspaugh* 1174; *Proctor* 29112.

— West Indies and continental tropical America.

[*Asystasia* Blume

Asystasia gangetica (L.) T. Anders. has been found becoming naturalised along roadsides and in clearings on Cayman Brac (*Proctor* 29339). This is a trailing or straggling herb, sometimes scrambling over bushes to a height of 2 or 3 m, with ovate leaves and elongate, one-sided racemes of pale yellow or dull purplish flowers, the corolla 3.5–4 cm long. A native of tropical Asia introduced as a garden plant, this species is obviously a recent escape from cultivation. PLATE 59.]

FIG. 229 **Ruellia tuberosa**. A, habit. B, cross-section of stem. C, part of upper surface of leaf. D, enlarged portion of same, showing cystoliths. E, corolla cut open to show stamens. F, two anthers. G, pistil, cut through style. H, fruiting branch. I, fruit, closed and open. J, seed. K, pollen grain. (G.)

Ruellia L.

Perennial herbs or shrubs with 4-angled stems; leaves entire or sometimes crenate. Flowers rather large, solitary or clustered in the leaf-axils, or in axillary and terminal cymose panicles; bracts narrow, and the bracteoles minute. Calyx deeply 5-lobed, the segments long and narrow; corolla funnel-shaped with the tube narrow at the base, the 5 spreading lobes nearly equal. Stamens 4, in 2 pairs, included. Style with recurved apex, the stigma simple or of 2 unequal lobes. Capsule cylindrical, pointed at the ends, usually many-seeded; seeds flattened, ovate or orbicular, attached by their edges.

— A large pantropical and subtropical genus of about 250 species.

KEY TO SPECIES

1. Leaves linear or narrowly lanceolate, less than 1 cm broad:	[R. brittoniana]
1. Leaves elliptic or ovate, 1.5–4 cm broad:	
2. Roots tuberous-thickened; corolla 4.5–6 cm long; capsule subcylindrical; seeds orbicular, more than 20 per locule:	R. tuberosa
2. Roots fibrous; corolla 3–4 cm long, capsule ellipsoid; seeds flattened-lenticular, 8–18 per locule:	R. nudiflora

[*Ruellia brittoniana* Leonard, *Jour. Wash. Acad. Sci.* 31: 96, f.1 (1941).

An erect herb with purplish stems; leaves 5–12 cm long or more, narrowly acuminate, the cystoliths aggregated beneath to form an irregular zig-zag pattern. Inflorescences few-flowered or the flowers sometimes solitary; corolla violet-blue, 4–5 cm long, the tube puberulous. Capsules 2–2.5 cm long, glabrous except for a patch of hairs on each side of the beak, about 20-seeded; seeds nearly orbicular, 2–2.5 mm in diameter.

GRAND CAYMAN: *Brunt* 1944.

— Native of Mexico, cultivated and naturalised in many warm countries. The Cayman plants were found along sandy roadsides at West Bay.]

Ruellia tuberosa L., *Sp. Pl.* 2: 635 (1753). **FIG. 229.**

Ruellia clandestina L. (1753).

HEART BUSH, DUPPY GUN

Erect herb with numerous thickened roots; stems more or less pilose; leaves mostly 3–10 cm long, obtuse or acute, the minute linear cystoliths all scattered in a random pattern. Inflorescences few-flowered; corolla violet-blue, 4.5–6 cm long, the tube puberulous. Capsules ca. 2 cm long, glabrous throughout, 10–15-seeded, dehiscing explosively when wet; seeds nearly orbicular, ca. 2 mm in diameter.

GRAND CAYMAN: *Brunt* 1970; *Hitchcock*; *Kings GC* 103; *Maggs* II 56; *Millspaugh* 1388; *Proctor* 15094; *Sachet* 408. CAYMAN BRAC: *Proctor* 28976, 35223.

— West Indies and northern S. America, usually in pastures and open waste ground or along roadsides. The local name, heart bush, refers to the use of its roots as a medicine for heart disease.

Ruellia nudiflora (Engelm. & Gray) Urban, *Symb. Ant.* 7: 382 (1912). PLATE 60.

Erect or decumbent perennial herb with fibrous roots, the stems simple or sparsely branched, usually puberulous or short-pilose. Leaves petiolate, the blades oblong-elliptic, ovate or lance-oblong, mostly 4–10 cm long, rounded to subacute at the apex, short-cuneate at the base, the margins often undulate, densely puberulous, short-pilose or sometime nearly glabrous. Inflorescences terminal and axillary, pedunculate, composed of short dichotomous cymes; bracts small; pedicels usually 3–10 mm long; calyx of chasmogamous (normal reproductive) flowers 7–17 mm long, of cleistogamous flowers 4–11 mm long, glandular-puberulous, the segments linear-acuminate; corolla light violet, the lobes rounded or often emarginate. Capsules 1.5–2 cm long, densely puberulous; seeds brown.

GRAND CAYMAN: first found by *Mrs. Joanne Ross* along Walkers Road south of George Town; *Proctor 52096* (from the same locality).

— Texas to Arizona, Mexico, Belize to Costa Rica, Cuba, and Hispaniola, a variable often weedy species.

GOODENIACEAE

Perennial herbs or shrubs with watery sap; leaves alternate, rarely opposite or basal only, simple and entire or toothed, rarely pinnatifid; stipules absent. Flowers perfect, usually zygomorphic, solitary or in cymes, racemes or heads arising from the leaf-axils; Calyx shortly 5-lobed, truncate or sometimes obsolete; corolla 5-lobed and 1–2-lipped, split down one side. Stamens 5, free or shortly adnate to the base of the corolla; anthers free or coherent around the style, 2-celled, opening lengthwise inwardly. Ovary inferior to superior, mostly 1–2-celled, with 1 or more erect or ascending ovules in each cavity, basal or on axile placentas; stigma simple or 2–3-branched, surrounded by a cup. Fruit a drupe, nut or capsule; seeds with fleshy endosperm.

— A chiefly Australasian family of about 12 genera and 300 species.

Scaevola L., nom. cons.

Rather fleshy herbs or shrubs; leaves alternate or subopposite, usually entire. Flowers in axillary cymes or dichasia; calyx 5-lobed or the lobes nearly obsolete; corolla white or blue, bearded within, and with winged lobes. Stamens free. Ovary inferior, 2-celled, with 1 ovule in each cavity. Fruit a drupe with 2-seeded stone.

— A genus of more than 80 species chiefly found in Polynesia and the Australian region. Two species occur naturally in the West Indies, one widespread, the other Cuban.

KEY TO SPECIES

1. Leaves less than 9 cm long; calyx entire or barely 5-lobed, 1 mm long; fruits black:	S. plumieri
1. Leaves often more than 15 cm long; calyx distinctly 5-lobed, the segments 2–5 mm long; fruits white:	[S. sericea]

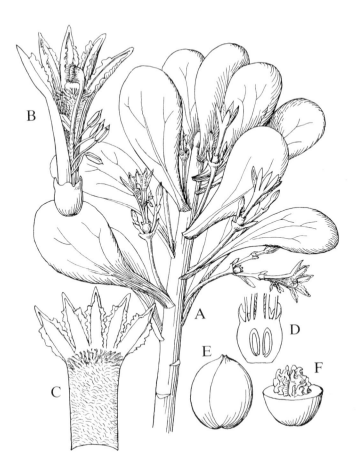

FIG. 230 **Scaevola plumieri**. A, end of branch with leaves and flowers, × ²/₃. B, flower, × 2. C, corolla cut open, × 2. D, ovary with calyx, cut lengthwise, × 3. E, drupe, × 1. F, drupe with upper half cut away exposing the stone, × 1. (F. & R.)

Scaevola plumieri (L.) Vahl., *Symb. Bot.* 2: 36 (1791). **FIG. 230**. PLATE 60.

BAY BALSAM

A nearly glabrous, much-branched shrub to 1.5 m tall; leaves obovate, 4–8 cm long, rounded at the apex and of fleshy texture. Flowers in stalked dichasia; corolla ca. 2.5 cm long, greenish and glabrous on the outside, the lobes white within. Drupes ellipsoid or subglobose, black, mostly 10–15 mm long, with rugose endocarp.

GRAND CAYMAN: *Brunt* 2063; *Proctor* 15214. LITTLE CAYMAN: *Kings* LC 66; *Proctor* 28025.

— Florida, the West Indies, and the Caribbean coast of C. America, also along the coasts of tropical Africa, always at the top of sandy sea-beaches.

[*Scaevola taccada* (Gaertn.) Roxb., *Fl. Ind.*, ed (1820) 2: 146 (1824). PLATE 60.

Scaevola sericea Vahl. *Symb. Bot.* 2: 37 (1791).

Erect bushy shrub sometimes to 3 m tall; branchlets white with white pith; leaf-scars conspicuous, semilinear. Leaves spirally arranged, often densely crowded; blades oblong-obovate, broadly rounded at apex, tapering downward to a short broad petiole; axils with tufts of white hairs. Peduncles and cyme-branches varying from densely pubescent to glabrous; calyx segments linear to linear-lanceolate, acutish. Corolla 5-lobed, at first white then becoming yellowish, the tube 10–15 mm long, pubescent inside and outside; base of segments long-ciliate. Stamens soon withering. Style ca. 2 cm long. Drupe white, ca. 1 cm in diameter; seeds 0–2.

GRAND CAYMAN: *Proctor* 47842, 48264, 48682. LITTLE CAYMAN: *Proctor* 48234 (cultivated).

— Widely distributed on Pacific islands from Hawaii to Malaya. Introduced as a salt-tolerant ornamental on many West Indian islands and rapidly becoming naturalised; in many localities it has become a serious invasive weed.]

RUBIACEAE

Herbs, shrubs or trees; leaves opposite or whorled, simple and entire or rarely toothed; stipules often sheathing the stem, sometimes divided into linear segments or reduced to glandular hairs, sometimes expanded and leaf-like. Flowers regular, usually perfect, solitary or in cymes or panicles, sometimes condensed to heads or glomerules, or rarely in spikes, with or without bracts. Calyx of 4–6 free or united sepals, or rarely the limb truncate. Corolla gamopetalous, mostly 4–6-lobed (rarely more), the lobes valvate, imbricate or twisted. Stamens as many as the corolla-lobes and alternate with them, inserted in the corolla-tube; anthers 2-celled, opening lengthwise. Ovary mostly inferior, crowned by a disc, usually 2-celled, the placentation various, rarely 1- or several-celled; ovules 1–many in each cavity; style usually slender, often forked toward the apex. Heterostyly is present in many genera. Fruit a capsule, berry, drupe or schizocarp; seeds usually with plentiful endosperm.

— A very large cosmopolitan family of more than 450 genera and about 6,000 species, most numerous in the tropics. Among the many economically important species of this family, the most valuable are those producing coffee (*Coffea* spp.) and quinine (*Cinchona* spp.). Ornamental species planted in Cayman gardens include *Ixora coccinea* L. and species of *Pentas*.

KEY TO GENERA

1. Plants woody; shrubs or (rarely) trees:
   2. Fruit a capsule; flowers solitary:
      3. Leaves thin, 3–5 cm long; corolla 5-lobed, white or cream with tube 25–30 mm long: **Exostema**
      3. Leaves thick and fleshy, less than 1 cm long; corolla 4-lobed, yellow, with tube 5–6 mm long: **Rhachicallis**
   2. Fruit a drupe or berry:
      4. Flowers sessile, solitary or several in the leaf-axils:
         5. Plants more or less spiny:

6. Corolla 5-lobed; fruits black with hard outer rind, 8–12 mm in diameter: **Randia**

6. Corolla 4-lobed; fruits white, soft, ca. 4 mm in diameter: **Catesbaea**

5. Plants without spines:

7. Small erect shrub; leaves more or less obovate, 6–9 mm long; flowers pale yellow: **Scolosanthus**

7. Trailing or arching shrub; leaves linear to lanceolate, 10–30 mm long; flowers pink: **Ernodea**

4. Flowers several or many in a cyme, panicle, raceme, spike or head:

8. Leaves (or many of them) in whorls of 3:

9. Low, dense shrub; leaves linear and sessile, 1–2 cm long and not over 2 mm broad; corolla pink, the tube ca. 1.5 mm long: **Strumpfia**

9. Tall shrub; leaves elliptic or ovate and long-petiolate, 3–10 cm long and up to 4 cm broad; corolla yellow, the tube ca. 25 mm long: **Hamelia**

8. Leaves opposite, never in 3s:

10. Inflorescence a panicle:

11. Panicles axillary; fruits 5–10-seeded: **Erithalis**

11. Panicles terminal; fruits 2-seeded: **Psychotria**

10. Inflorescence not a panicle:

12. Plants climbing or scrambling; individual flowers stalked: **Chiococca**

12. Plants erect and shrubby or tree-like:

13. Flowers pedicellate, in umbellate clusters of 3 on branches of compound corymb; fruits black, 1-seeded: **Faramea**

13. Flowers sessile on inflorescence-branches; fruits if black with more than 1 seed:

14. Flowers on the branches of a forked cyme; fruit a drupe:

15. Leaves and inflorescence glabrous; calyx-teeth 5, persistent; fruit 2-seeded: **Antirhea**

15. Leaves and inflorescence pubescent; calyx-teeth 2, deciduous; fruit 3–6-seeded: **Guettarda**

14. Flowers in dense globose heads; fruit a fleshy compound berry: **Morinda**

1. Plants herbaceous:

16. Stems delicate, creeping, rooting at the nodes, the stipules minute or obsolete; fruit a many-seeded capsule; corolla 4–5-lobed: **Hedyotis**

16. Stems usually erect, not rooting at the nodes, the stipules evident; fruit 2-seeded; corolla 4-lobed:

Exostema (Pers.) L. C. Rich.

Shrubs or trees; leaves opposite, thin-textured; stipules small and soon falling. Flowers solitary in the leaf-axils, or in terminal corymbs or panicles; calyx cylindrical or top-shaped, 5-lobed; corolla more or less salver-shaped with long slender tube and 5 linear reflexed lobes. Stamens 5, borne near the base of the corolla-tube; anthers linear, attached by the base, exserted. Ovary 2-celled, with numerous ovules in each cavity; stigma club-shaped or capitate. Fruit a 2-valved capsule, somewhat leathery or woody, splitting lengthwise; seeds numerous, broadly winged.

— A tropical American genus of about 50 species.

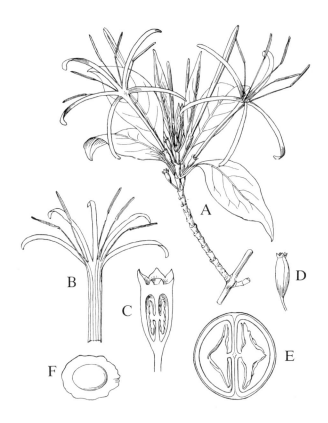

FIG. 231 **Exostema caribaeum.** A, branchlet with leaves and flowers, × ²/₃. B, upper part of corolla cut open, × ²/₃. C, ovary with calyx cut lengthwise, × 4. D, capsule, × ²/₃. E, cross-section of capsule, × 4. F, seed × 4. (F. & R.)

Exostema caribaeum (Jacq.) Schult. in L., *Syst. Veg.* ed. nov. 5: 18 (1819). **FIG. 231.** PLATE 60.

BASTARD IRONWOOD

A shrub to 3 m tall or slender tree to 6 m, the branchlets glabrous and with enlarged nodes; leaves elliptic or broadly elliptic, mostly 3–7 cm long, up to 3 cm broad, acute, glabrous or puberulous in the vein-axils beneath. Pedicels 4–8 mm long; calyx-lobes triangular, 0.5–1 mm long; corolla greenish-white, turning cream or pale yellow, the lobes narrowly linear and ca. 3.5 cm long. Anthers 1.5–2 cm long. Capsules woody, 1.2–1.7 cm long; seeds suborbicular, 2.5 mm in diameter, winged all around.

GRAND CAYMAN: *Correll & Correll* 51045; *Proctor* 15249. LITTLE CAYMAN: *Proctor* 28140.

— Florida, West Indies, Mexico and C. America, in dry rocky woodlands.

Rhachicallis DC.

A small, intricately branched shrub; leaves opposite, densely overlapping; stipules sheathing, persistent. Flowers sessile, the base enclosed by the stipular sheath; calyx 4-toothed with smaller accessory teeth between the main ones; corolla salver-shaped with slender cylindrical tube. Stamens 4, short, included. Ovary half-superior, 2-celled, each cavity with numerous ovules on a peltate placenta; style slightly 2-lobed. Capsule subglobose, 2-valved, septicidally dehiscent; seeds angular, pitted.

— A monotypic West Indian genus.

Rhachicallis americana (Jacq.) Ktze., *Revis. Gen. Pl.* 1: 281 (1891). **FIG. 232**.

JUNIPER

Stems flexible, canescent, mostly less than 1 m tall; leaves linear to oblong or ellipsoid, mostly 2–8 mm long; stipules mucronate and ciliate. Corolla deep yellow with a red eye, the tube 5–6 mm long, puberulous, the lobes densely pubescent on the outside. Capsules 3 mm long.

GRAND CAYMAN: *Correll & Correll* 51023; *Hitchcock*; *Kings* GC 396; *Lewis* GC 37; *Proctor* 15199; *Sachet* 447. LITTLE CAYMAN: *Proctor* 28055. CAYMAN BRAC: *Kings* CB 58; *Millspaugh* 1178; *Proctor* 28926.

— Bahamas, Greater Antilles (except Puerto Rico) and Yucatan, on exposed limestone rocks beside the sea.

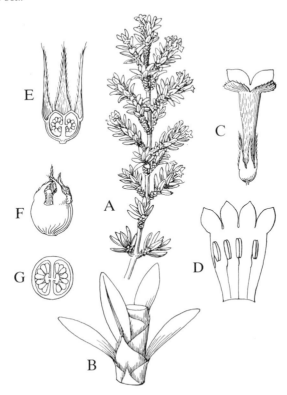

FIG. 232 **Rhachicallis americana.** A, branch with leaves and flowers, × ²/₃. B, small portion of branch showing leaves and stipules, × 4. C, flower, × 4. D, corolla cut open, × 4. E, ovary with calyx cut lengthwise, × 8. F, fruit, × 4. G, cross-section of fruit, × 4. (F. & R.)

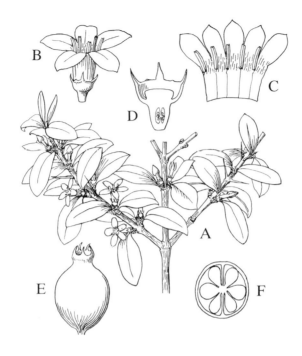

FIG. 233 **Randia aculeata**. A, portion of plant, × ²/₃. B, flower, × 2. C, corolla cut open, × 2. D, ovary with calyx, cut lengthwise, × 4. E, berry, × 2. F, cross-section of berry, × 2. (F. & R.)

Randia L.

Shrubs or trees, erect or scandent, sometimes with axillary or extra-axillary spines; leaves opposite; stipules interpetiolar, often sheathing. Flowers axillary or lateral, solitary or in clusters or corymbs; calyx-tube ovoid, obovoid or top-shaped, the limb tubular, cup- or bell-shaped, and truncate or 4-toothed or -lobed, the lobes often foliaceous. Corolla bell-, funnel-, or salver-shaped, 5-lobed (rarely 4–6-lobed), twisted in bud. Stamens usually 5, inserted in the throat or mouth of the corolla, included or exserted. Ovary usually 2-locular, with few to many ovules; style simple or bifid. Fruit a 2-locular berry with many compressed seeds immersed in pulp.

— A pantropical genus of perhaps 300 species, mostly in the Old World.

Randia aculeata L., *Sp. Pl.* 2: 1192 (1753). **FIG. 233. PLATE 60.**

LANCEWOOD

A shrub to 2.5 in tall or more, glabrous or nearly so, with stiff horizontal or ascending branches occasionally armed with a few stout spines; leaves deciduous, narrowly to broadly obovate or nearly orbicular, mostly 1–5 cm long, acutish to rounded at the apex. Flowers fragrant, sessile or nearly so; corolla-tube green, 4–7 mm long, hairy within; lobes 4–6, white, 3.5–4 mm long. Berries ellipsoid or globose, 8–12 mm long.

GRAND CAYMAN: *Brunt 2071*; *Correll & Correll 51015*; *Kings GC 105*; *Millspaugh 1319*; *Proctor 11986, 15156*; *Sachet 429*; *Sauer 4104*. LITTLE CAYMAN: *Proctor 28041, 35120, 35172*. CAYMAN BRAC: *Millspaugh 1203*; *Proctor 28952*.

— Florida, the West Indies and Mexico to Venezuela, in dry rocky thickets and woodlands.

Catesbaea L.

Shrubs with more or less spiny stems; leaves opposite or fascicled on short lateral spurs; stipules small, soon falling. Flowers small or large, solitary in the axils, short-stalked; calyx with 4 awl-shaped lobes; corolla white, funnel- or bell-shaped, glabrous within, the 4 lobes broad, valvate in bud. Stamens 4, inserted at the base of the corolla-tube, included or shortly exserted. Ovary 2-locular with few or many ovules; stigma bifid. Fruit a white (rarely black) berry, crowned by the persistent calyx-lobes; seeds compressed, angled or round; endosperm fleshy.

— A West Indian genus of about 10 species.

Catesbaea parviflora Sw., *Nov. Gen. & Sp. Pl.* 30 (1788). FIG. 234.

A shrub usually 1 m tall or less (rarely up to 2.5 m), densely branched, the young branches arching, puberulous, the spines rather few or sometimes apparently lacking; leaves densely clustered, glabrous, rigidly leathery, obovate to orbicular, 4–10 mm long. Flower nearly sessile; corolla 5–6 mm long. Berries white, ovoid, ca. 2 mm long.

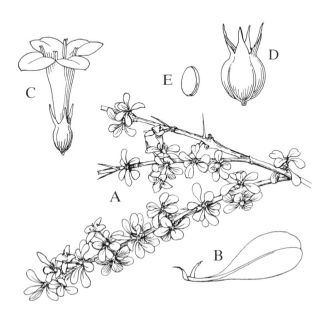

FIG. 234 **Catesbaea parviflora**. A, portion of plant with flowers, × ²/₃. B, leaf with stipules, × 4. C, flower, × 2. D, berry, × 4. E, seed, × 4. (F. & R.)

GRAND CAYMAN: *Brunt* 2005; *Kings GC* 254; *Proctor* 15168, 15217. LITTLE CAYMAN: *Proctor* 35107. CAYMAN BRAC: *Proctor* 29065.

— Florida, Bahamas, Cuba, Jamaica and Puerto Rico in dry rocky scrublands and thickets. The Cayman plants differ from those of Jamaica in their smaller fruits, more resembling those of Florida and the Bahamas in this respect. However, this size differential is overlapping and not wholly consistent. Black-fruited plants from Puerto Rico and the northern Lesser Antilles (sometimes included in a broad concept of *C. parviflora*) are here considered to represent a different species, *C. melanocarpa* Krug & Urb.

Scolosanthus Vahl

Spiny or rarely unarmed shrubs; leaves opposite or fascicled, coriaceous, short-petiolate or sessile; stipules minute, interpetiolar, connate, forming a short sheath around the stem, persistent. Flowers bisexual, 4-merous, actinomorphic, in axillary fascicles or solitary; hypanthium obovoid or obconical, crowned by a small 4-lobed calyx; corolla narrowly funnel-shaped (rarely bell-shaped), the tube longer than the lobes, the lobes imbricate in bud, erect or spreading at anthesis. Stamens 4, included, the filaments connate at the base; ovary inferior, 2-carpellate, each carpel with one pendulous ovule; style filiform, bilobed. Fruit a more or less compressed-globose fleshy drupe containing 1 or 2 pyrenes, each pyrene with a single seed.

— A genus of about 23 described species confined to the Bahamas and Greater Antilles. Of these, 1 is credited to the Bahamas, 9 to Cuba, 9 to Hispaniola, 2 to Jamaica and 3 to Puerto Rico. All are endemic to their respective islands except for 2 shared by Hispaniola and Puerto Rico (*Scolosanthus densiflorus* and *S. versicolor*, the latter also occurring in the Virgin Islands). The only two entirely non-spiny species are *S. howardii* of Jamaica and *S. roulstonii* of Grand Cayman.

Scolosanthus roulstonii Proctor, sp. nov. GRAND CAYMAN: bluff at Little Salt Creek, 19 Nov. 2005, P. Ann van B. Stafford s.n. (IJ, holotype). PLATE 61.

Unarmed nearly glabrous shrub to 1 m, tall or more with light brown bark, the youngest twigs very minutely puberulous. Leaves fasciculate often on short spurs along branches, mostly 6–9 mm long, leathery, glabrous, obovate or very broadly obovate or rarely elliptic, dark glossy green above, paler and longitudinally striate beneath, acute at apex, tapering to the nearly sessile base. Flowers pale yellow, few in sessile clusters or solitary on leaf-spurs; calyx lobes minute; corolla tube 3–3.5mm long. Stamens and style included. Fruit cream-white, broadly ellipsoid, 3–3.5 mm long, 2–2.5 mm thick, crowned by the 4 minute calyx lobes.

This species is very different from *Scolosanthus bahamensis*, having larger glabrous (rather than papillose- puberulous) leaves that are acute (rather than obtuse) at the apex; in addition, the corolla lobes are sharply acute (rather than suborbicular). This new Cayman species differs from *S. multiflorus* of Jamaica in its shorter stature and smaller leaves (0.6–0.9 cm vs. 2–4.5 cm long), shorter corolla tube (3–3.5 mm vs. 7 mm) and smaller fruits (3–3.5 mm vs. 5 mm long). Similar types of differences separate it from the other members of the genus, but in particular this new Cayman species differs in its entire absence of spines; a characteristic shared only with *S. howardii* of Jamaica.

Scolosanthus roulstonii Proctor, sp. nov. a *S. bahamensi* foliis majoribus glabris (non papilloso-puberulis) ad apicem acutis (non obtusis), lobis corollae argute acutis (non suborbicularibus) differt. A *S. multifloro* statura breviore, foliis minoribus 0.6–0.9 cm (non 2–4.5 cm) longis, tubo corollae breviore 3–3.5 mm (non 7 mm) longo et fructibus minoribus 3–3.5 mm (non 5 mm) longis differt.

Collected (*Proctor 47275*) ca. 3.5 km due NW of East End Village in rocky scrub woodland; and along the High Rock Quarry road ca. 1 km S of the 'T' intersection at its north end (*Proctor 48255*). Other specimens have also been seen.

Strumpfia Jacq.

A small, densely branched shrub, the branches with very short nodes and rough with the rigid, persistent, spreading stipules; leaves in whorls of 3, rigidly leathery; stipules sheathing. Flowers in short axillary racemes; calyx 5-lobed, the lobes persistent; corolla deeply 5-parted, the tube very short. Stamens 5, borne near the base of the corolla-tube; anthers subsessile, erect, joined by their connectives to form a column surrounding the style. Ovary 2-celled with 1 ovule in each cavity; style glabrous but surrounded by a ring of hairs at the base. Fruit a small fleshy drupe containing a 1–2-celled stone; seeds oblong, with fleshy endosperm.

— A monotypic West Indian genus.

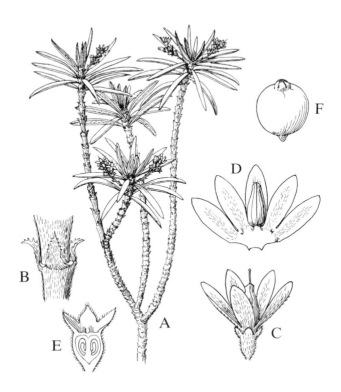

FIG. 235 **Strumpfia maritima.** A, portion of plant, × ²/₃. B, stipules, × 4. C, flower, × 4. D, corolla cut open, showing connate anthers. E, ovary with calyx cut lengthwise, × 8. F, drupe, × 4. (F. & R.)

Strumpfia maritima Jacq., *Enum. Syst. Pl. Carib.* 28 (1760). **FIG. 235.** PLATE **61.**

Low, flat-topped or mound-like shrub mostly less than 1 m tall, the young branches finely and densely white-tomentose; leaves sessile, linear, 1–2.5 cm long, 1–3 mm broad with revolute margins, puberulous on the upper side and densely white-tomentellous beneath. Racemes shorter than the leaves, few-flowered; corolla pink, the tube ca. 1.5 mm long. Drupes globose, white, ca. 3 mm in diameter.

GRAND CAYMAN: *Brunt* 2010; *Correll & Correll* 51022; *Fawcett*; *Hitchcock*; *Kings* GC 124; *Lewis* GC 2; *Proctor* 15200; *Sachet* 393; *Sauer* 3312. LITTLE CAYMAN: *Proctor* 28095; *Sauer* 4190. CAYMAN BRAC: *Kings* CB 98; *Millspaugh* 1169; *Proctor* 29038.

— Florida, the West Indies and Yucatan, in dry rocky scrublands usually near the sea.

Erithalis P. Br.

Shrubs or rarely small trees, mostly glabrous; leaves opposite; stipules short, sheathing, persistent. Flowers in axillary or rarely terminal corymbose panicles; calyx-tube globose or ovoid, the limb more or less cup-shaped and truncate or 5–10-toothed, persistent on the fruit.

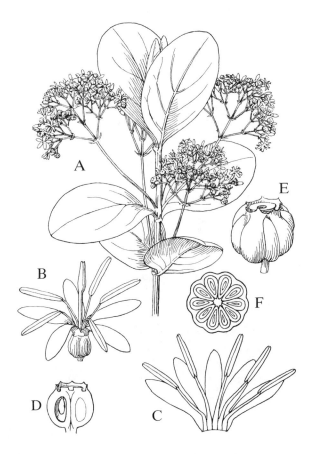

FIG. 236 **Erithalis fruticosa.** A, end of branch with leaves and flowers, × ²/₃. B, flower, × 6. C, corolla cut open, × 6. D, ovary with calyx cut lengthwise, × 12. E, drupe, × 6. F, cross-section of drupe, × 6. (F. & R.)

Corolla with short tube and 5–10 lobes, these narrow and recurved. Stamens 5–10, inserted at the base of the corolla; filaments united at the base. Ovary 5–22-celled, with 1 pendulous ovule in each cavity. Fruit a globose drupe of 5–22 bony nutlets; seeds oblong, compressed.

— A West Indian genus of about 10 species.

Erithalis fruticosa L., *Syst. Nat.* ed. 10, 2: 930 (1759). **FIG. 236**. PLATE 61.

BLACK CANDLEWOOD

A shrub usually 1–2 m tall but sometimes up to 5 m, nearly glabrous throughout; leaves leathery, elliptic or narrowly obovate to rotund, mostly 2.5–6.5 cm long, obtuse at the apex, the margins often minutely revolute, the venation somewhat obscure. Panicles exceeding the leaves. Flowers fragrant, corolla white or cream, with usually 4–5 lobes, these ca. 4 mm long. Drupes purple-black, 2–2.5 mm in diameter.

GRAND CAYMAN: *Brunt* 1763, 1818, 1917, 2059, 2068; *Correll & Correll* 51034; *Fawcett*; *Hitchcock*; *Kings* GC 335; *Millspaugh* 1251; *Proctor* 15092; *Sachet* 407. LITTLE CAYMAN: *Kings* LC 100; *Proctor* 28097. CAYMAN BRAC: *Proctor* 29064.

— Florida, West Indies, and the east coast of C. America, in sandy or rocky thickets and scrublands, or sometimes on limestone cliffs.

Chiococca P. Br.

Glabrous shrubs, often trailing or climbing; leaves opposite, somewhat leathery and shining; stipules broad, with a sharp point, persistent. Flowers in axillary simple or compound racemes; calyx 5-toothed or -lobed, persistent; corolla funnel-shaped with glabrous throat, 5-lobed, the lobes valvate in bud. Stamens 5, inserted near the base of the corolla-tube, the filaments connate at the base, often pubescent; anthers linear, attached at the base, usually included. Ovary 2- (rarely 3-)locular, with solitary pendulous ovules; stigma entire or shortly 2-lobed. Fruit a flattened, leathery white drupe containing 2 nutlets; seeds compressed; endosperm fleshy.

— A tropical American genus of perhaps 20 species.

KEY TO SPECIES

1. Leaves mostly 3–6 cm long or more; racemes often longer than the leaves:	C. alba
1. Leaves mostly 1–3 cm long; racemes shorter than the leaves:	C. parvifolia

Chiococca alba (L.) Hitchc., *Rep. Missouri Bot. Gard.* 4: 94 (1893). **FIG. 237**. PLATE 61.

Chiococca parvifolia of Hitchcock (1893), as to Cayman specimen, not Wullshl. ex Griseb. (1861).

Trailing, scrambling or climbing shrub with elongate stems to 6 m long or more; leaves ovate, lanceolate or narrowly elliptic, more or less acuminate at the apex. Racemes up to 7 cm long, often branched; corolla yellow, the tube 3–5 mm long, the lobes reflexed. Drupes 4–5 mm in diameter.

GRAND CAYMAN: *Brunt* 1701, 1780, 2197; *Hitchcock*; *Sachet* 378. LITTLE CAYMAN: *Proctor* 28118. CAYMAN BRAC: *Proctor* 28953.

— Florida, West Indies, Mexico and C. America, in rocky thickets and woodlands.

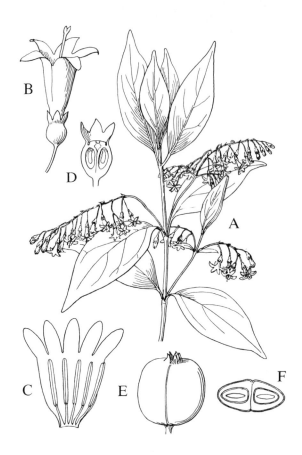

FIG. 237 **Chiococca alba.** A, branch with leaves and flowers, × ²/₃. B, flower, × 4. C, corolla cut open, × 4. D, ovary with calyx cut lengthwise. E, drupe, × 4. F, cross-section of drupe, × 4. (F. & R.)

Chiococca parvifolia Wullschl. ex Griseb., *Fl. Brit. W. I.* 337 (1861).

A climbing shrub like the last species; leaves narrowly to broadly elliptic or ovate, acute or blunt at the apex. Racemes up to 2 cm long, seldom branched but sometimes paired in the axils; corolla pale yellow, the tube ca. 3 mm long, the lobes mostly erect. Drupe 4–5 mm in diameter.

CAYMAN BRAC: *Proctor* 29000.

— Florida and the West Indies, especially Trinidad and Tobago, in rocky thickets and woodlands.

Antirhea Commers.

Shrubs or trees, glabrous or pubescent; leaves opposite; stipules interpetiolar, soon falling. Flowers perfect or polygamous, usually sessile along the upper side of the branches of a 1–2-forked cyme, or rarely solitary. Calyx truncate or irregularly 4–5-toothed or -lobed, persistent. Corolla with cylindrical or funnel-shaped tube and 4–5 short lobes, these imbricate in bud. Stamens 4–5, inserted in the throat of the corolla-tube, the filaments short, the anthers included or shortly exserted. Ovary 2–10-locular with solitary pendulous ovules; stigma capitate or 2–3-lobed. Fruit a small, fleshy drupe with 2–10-celled stone; seeds elongate, without endosperm.

— A rather widespread tropical genus of about 40 species, occurring in the West Indies, Madagascar, tropical Asia and Australia. Some species yield valuable timber.

Antirhea lucida (Sw.) Benth & Hook. f. in Benth. & Hook.f., *Gen. Pl.* 2: 100 (1873). **FIG. 238**. PLATE 61.

A glabrous tree 5–10 m tall with light grey bark; leaves oblong-elliptic or broadly elliptic, 4–8 cm long or more, obtuse or acutish at the apex, shining light green on the upper side; stipules acuminate, 6–8 mm long. Inflorescence 1-forked with curved branches; flowers fragrant; calyx ca. 3 mm long with 5 oblong, minutely ciliate lobes; corolla cream-white, bell-shaped, with tube 3.5–5 mm long. Drupes ca. 1 cm long.

GRAND CAYMAN: *Brunt* 1807. LITTLE CAYMAN: *Proctor* 35080, 35211.

— Bahamas, Greater Antilles, St. Croix and the Swan Islands, in rocky thickets and woodlands.

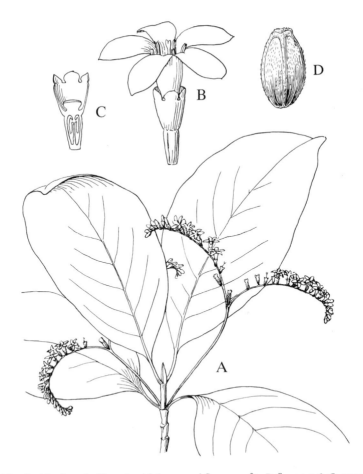

FIG. 238 **Antirhea lucida**. A, end of branch with leaves and flowers, × ²/₃. B, flower, × 6. C, ovary with calyx cut lengthwise, × 6. D, drupe, × 2. (F. & R.)

Guettarda L.

Shrubs or small trees; leaves opposite or sometimes in whorls of 3; stipules interpetiolar, soon falling. Flowers perfect or polygamo-dioecious, sessile or subsessile on the branches of axillary forked cymes. Calyx with tubular limb truncate at the apex or irregularly toothed, deciduous. Corolla salver-shaped with long, cylindrical tube sometimes curved, and 4–9 obtuse lobes, these imbricate in bud. Anthers sessile or subsessile in the upper part of the corolla-tube, included. Ovary 4–9-locular, with solitary pendulous ovules; stigma capitate, entire or short-lobed. Fruit a usually globose drupe with thin flesh and hard, 4–9-celled stone; seeds without endosperm.

— A genus of more than 80 species, of which 60 occur in tropical America, 20 in New Caledonia, and 1 is widespread on tropical coasts.

Guettarda elliptica Sw., *Nov. Gen. & Sp. Pl.* 59 (1788). PLATE 62.

A shrub to 3 m tall, or rarely a small tree; leaves ovate, more or less elliptic or narrowly obovate, mostly 1–7 cm long, obtuse or acute at the apex, puberulous on both sides. Peduncles slender, 0.5–2.5 cm long; inflorescences subcapitate, the cyme-branches short and few-flowered; corolla whitish, the tube 5 mm long or more, puberulous. Drupes dark red turning blackish, 4–8 mm in diameter, 3–6-seeded.

GRAND CAYMAN: *Brunt* 2004; *Proctor* 15173, 15210. LITTLE CAYMAN: *Proctor* 28136, 35106, 35114, 35119. CAYMAN BRAC: *Proctor* 28954, 29027.

— Florida, the West Indies (except the Lesser Antilles) and Mexico to Venezuela, in dry, rocky thickets and woodlands. The Grand Cayman plants have much smaller leaves and shorter peduncles than those of Little Cayman and Cayman Brac.

Hamelia Jacq.

Shrubs or trees; leaves thin, opposite or whorled and rather long-petiolate; stipules interpetiolar, soon falling. Flowers sessile or short-stalked on the branches of 2–3-forked or compound cymes. Calyx with 5 persistent lobes. Corolla yellow or red, tubular or narrowly bell-shaped, 5-angled at least in bud. Stamens 5, inserted near the base of the corolla-tube; filaments short; anthers linear, attached at the base, included. Disc prominent, persistent, forming a protuberance on the apex of the fruit. Ovary 5-locular, with numerous ovules on axile placentas; stigma narrow, entire. Fruit a 5-lobed, 5-locular berry; seeds numerous, small, variously angled or tuberculate.

— A tropical American genus of about 40 species.

Hamelia cuprea Griseb., *Fl. Brit. W. I.* 320 (1861). PLATE 62.

A shrub to 3 m tall or more, glabrous throughout or nearly so; leaves in whorls of 3, ovate or elliptic, 4–11 cm long (including petioles to 2.5 cm), acuminate at the apex, often recurved-plicate; stipules triangular-attenuate, to 2 mm long. Inflorescence laxly few-flowered; pedicels 2–7 mm long; corolla bright yellow streaked with orange, ca. 2.5 cm long, constricted above the base. Berries ovoid, 5–7 mm long, ripening reddish black.

GRAND CAYMAN: *Kings GC* 345; *Proctor* 15140. LITTLE CAYMAN: *Proctor* 35123. CAYMAN BRAC: *Proctor* 29011, 29072.

— Cuba, Jamaica and Hispaniola, in rocky thickets and woodlands.

Morinda L.

Usually glabrous shrubs or trees, or sometimes woody vines; leaves opposite or rarely in whorls of 3; stipules more or less sheathing. Flowers in dense globose heads, these stalked or subsessile, axillary or terminal, solitary or several in an umbel. Calyx-limb short and truncate or minutely toothed, persistent. Corolla usually white, salver-shaped, with usually 5 lobes (rarely 3–7), valvate in bud. Stamens as many as the corolla-lobes, inserted near the top of the tube, the anthers included or exserted. Ovary 2–4-locular with solitary ascending ovules; style with 2 linear arms. Fruit a fleshy compound berry (syncarp) formed by union of the enlarged calyx-tubes, and containing numerous 1-seeded nutlets; seeds with fleshy endosperm.

— A pantropical genus of about 80 species, the majority Indian and Malayan.

KEY TO SPECIES

1. Leaves less than 10 cm long and 3 cm broad; peduncles less than 5 mm long; syncarps not over 2 cm in diameter: **M. royoc**

1. Leaves up to 30 cm long or more and 6–15 cm broad; peduncles 10–25 mm long; syncarps to 10 cm long: **[M. citrifolia]**

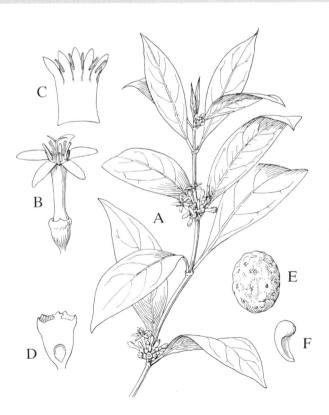

FIG. 239 **Morinda royoc**. A, branch with leaves and flowers, × ²/₃. B, flower, × 3. C, corolla cut open, × 2. D, ovary with calyx cut lengthwise, × 6. E, fruit, × ²/₃. F, nutlet, × 2. (F. & R.)

Morinda royoc L., *Sp. Pl.* 1: 176 (1753). **FIG. 239.** Plate 62.

YELLOW ROOT, RHUBARB ROOT

An erect or straggling shrub to 2 m tall; leaves lanceolate or narrowly elliptic, mostly 3–10 cm long, subacuminate at the apex; stipules cuspidate, inconspicuous. Corolla-tube ca. 5 mm long, the 5 lobes shorter. Syncarps nearly spherical, yellowish when ripe.

GRAND CAYMAN: *Brunt* 1649, 1812, 1926; *Hitchcock*; *Kings* GC 64; *Millspaugh* 1280, 1359; *Proctor* 15050.

— Florida, Bahamas, Cuba, Jamaica, Hispaniola, Mexico to Venezuela, Curacao and Aruba, in sandy or rocky thickets or along the borders of pastures.

[*Morinda citrifolia* L., *Sp. Pl.* 1: 176 (1753).

MULBERRY, HOG APPLE, NONI

An erect shrub or small tree to 10 m tall; leaves broadly elliptic, acute or cuspidate at the apex, shining bright green; stipules membranous, up to 1.5 cm long. Corolla-tube 9–10 mm long, the lobes ca. 4 mm long. Syncarps ovoid-ellipsoid, somewhat asymmetric or irregular, creamy-translucent when ripe and of foetid odour.

GRAND CAYMAN: *Brunt* 1982; *Proctor* 15117, 31050. LITTLE CAYMAN: *Kings* LC 112. CAYMAN BRAC: *Proctor* 29094.

— Native of tropical Asia and Australia, now widely naturalised in the American tropics. It frequently occurs in coastal thickets, as the fruits are apparently dispersed by floating in the sea. In India, this species has been extensively cultivated for the production of a dye called 'al', used for dyeing cloth various shades of red. This substance is obtained chiefly from the bark of the roots. The wood is said to be hard and durable.]

Psychotria L., nom. cons.

Shrubs or sometimes small trees, rarely herbs or climbers; leaves usually opposite, rarely in whorls of 3 or 4; stipules free or more or less united, persistent or deciduous. Flowers often dimorphic (heterostylous), usually in terminal corymbs, cymes or panicles; calyx 5-toothed; corolla cylindrical or funnel-shaped, straight, glabrous or hairy within, usually 5-lobed (rarely 4–6-lobed), the lobes valvate in bud. Stamens the same number as the corolla-lobes, inserted in the upper part of the tube; anthers included or exserted. Ovary 2-celled with solitary ovules erect from the base; style 2-armed. Fruit a berry, or a drupe with 2 nutlets, these often dehiscing longitudinally on the ventral side; seeds convex and smooth or ribbed on one side, and flat or concave and smooth on the other; endosperm fleshy or cartilaginous.

— A very large, pantropical genus of more than 1500 species.

KEY TO SPECIES

1. Erect shrub with cylindrical branchlets; leaves long-acuminate at apex, the nerves prominent beneath; corolla tube 2–3 mm long; fruits ellipsoid:	P. nervosa
1. Semiscandent shrub with 4-angled branchlets; leaves obtuse or acute at apex, the nerves not prominent; corolla tube 5 mm long; fruit compressed-obovoid:	P. microdon

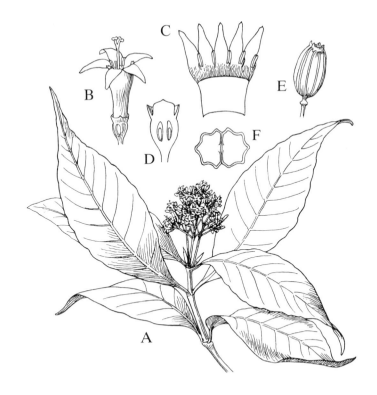

FIG. 240 **Psychotria nervosa**. A, end of branch with leaves and flowers, × ²/₃. B, flower, × 4. C, corolla cut open, × 4. D, ovary with calyx cut lengthwise, × 8. E, fruit, × 2. F, cross-section of fruit, × 8. (F. & R.)

Psychotria nervosa Sw, *Nov. Gen. & Sp. Pl.* 43 (1788). **FIG. 240.** PLATE 62.

Psychotria undata Jacq. (1798).

STRONG BACK, KIDNEY BUSH

A shrub up to 2.5 m tall, rarely arborescent and up to 6 m, the young branchlets glabrate or reddish-pubescent; leaves narrowly elliptic or elliptic, mostly 5–16 cm long, acuminate at the apex, glabrate or puberulous beneath, the 9–16 pairs of side-nerves prominent; stipules fused and sheathing in bud, splitting on one side and soon falling. Panicles sessile, pubescent or glabrous; corolla white, the tube 2–3 mm long. Drupes ellipsoid, red, 6–7 mm long.

GRAND CAYMAN: *Brunt* 1978; *Correll & Correll* 51043 (pubescent form); *Hitchcock*; *Kings* GC 316; *Lewis* GC 33a; *Proctor* 11987, 15001, 15007; *Sachet* 376.

— Florida, the West Indies and continental tropical America, variable; Cayman plants grow in rocky woodlands. Two rather different forms occur: (1) shrubby plants with pubescent stems, leaves and inflorescences, the latter rather compact; and (2) taller, more arborescent plants (e.g. *Brunt* 1978) that are nearly glabrous and have a more open type of inflorescence. Form 1 has been called *Psychotria undata*, whereas form 2 resembles typical *P. nervosa* as it occurs in Jamaica. However, the distinctions are unstable and break down when the whole range of plants in this complex is considered.

629

Psychotria microdon (DC.) Urban, *Symb. Ant.* 9: 599 (1928).

Semiscandent or scrambling shrub up to 2 m tall or more; lateral branches decussate, short or elongate, the twigs greyish and somewhat 4-angled. Leaves with blades obovate to oblanceolate, 5–10 cm long, 2–4 cm broad, acute to obtuse at both apex and base, the margins entire, puberulous along the veins beneath; petioles 0.5–2 cm long; stipules ovate, 1.5 mm long, soon falling. Flowers with parts in 5s, nearly sessile, in terminal corymbs borne on the ends of lateral branches; hypanthium cup-shaped, 1.5–2 mm long, glabrous, green and finely toothed. Corolla white, bell-shaped, with tube ca. 5 mm long, the blades spreading, 3–3.5 mm long. Stamens included, with anthers lanceolate, ca. 2 mm long; style filiform. Ripe fruits red, 5–6 mm long.

GRAND CAYMAN: *Proctor* 48648-a, found in the Forest Glen area W of The Mastic. The specimen is sterile; more material is needed.

— Cuba, Hispaniola, Puerto Rico, Virgin Islands, Lesser Antilles, northern S. America and along the Pacific Coast to Peru.

Faramea Aublet

Shrubs or small trees with opposite leaves and interpetiolar awned stipules. Inflorescences terminal, pedunculate, corymbose with umbellate flower-clusters; calyx cupular, truncate or 4-toothed; corolla white, the tube cylindrical or funnel-shaped, the 4 lobes narrow, erect or recurved, glabrous. Stamens 4, inserted in upper part of corolla-tube. Style glabrous, with 2 linear branches. Ovary 2-locular, each locule with 1 ovule. Fruit a leathery 1-seeded berry; seed globose or reniform, grooved on ventral side.

— A neotropical genus of 120 species.

Faramea occidentalis (L.) A. Rich., *Mem. Rub.* 95 (1830). PLATE 62.

Glabrous tree up to 10 m tall; stipules 2–4 mm long bearing filiform appendages 3–5 mm long. Petioles ca. 5 mm long; leaf-blades oblong, oblong-lanceolate to elliptic, 7–20 cm long, 4–10 cm broad, the apex abruptly acuminate-caudate, the base obtuse or roundish, glabrous. Flowers very fragrant, aggregated in 3s in loose pedunculate compound corymbs, heterostylous; pedicels 5–20 mm long; calyx cylindrical; corolla long-pointed in bud, the cylindrical tube 8–10 mm long, the linear lobes 8–10 mm long. Stamens slightly exserted in short-styled forms, included in long-styled forms; anthers 4 mm long, apiculate. Short styles 3 mm long; long styles 12–14 mm long. Fruit depressed-globose, 8–10 mm in diameter, black, 1-seeded.

GRAND CAYMAN: *Proctor* 47862, 48660; found in The Mastic and Forest Glen areas.
— Greater and Lesser Antilles, Mexico, C. America, Trinidad, Tobago and northern S. America.

Hedyotis L.

Small herbs or subshrubs, the stems erect, creeping or diffuse; leaves opposite; stipules entire or variously toothed or cut, rarely minute or apparently absent. Flowers solitary or cymose in the leaf-axils, often dimorphic (heterostylous); calyx 4-toothed or -lobed;

corolla funnel- or salver-shaped, mostly 4-lobed, the lobes valvate in bud. Stamens usually 4, inserted above the middle of the corolla-tube; anthers included or exserted. Ovary 2-celled with many ovules on axile placentas; style 2-armed. Fruit a several–many-seeded globose or 2-lobed capsule, dehiscing at the top or all the way down; seeds flattened or angled.

— A pantropical subtropical or warm-temperate genus of about 300 species. This concept includes the species often separated as the genera *Houstonia* and *Oldenlandia*.

Hedyotis callitrichoides (Griseb.) W. H. Lewis in *Rhodora* 63: 222 (1961).

Oldenlandia callitrichoides Griseb. (1862).

A delicate, creeping herb forming small mats, the glabrous stems 0.1–0.2 mm in diameter; leaves thin, petiolate, very broadly ovate or orbicular, 1–3mm long, obtuse at the apex, sparsely puberulous on the upper side; stipules minute or obsolete. Flowers solitary on delicate peduncles 4–5 mm long; calyx 4–5-lobed, the acuminate lobes hairy. Corolla white, 1.5–2 mm long, 4–5-lobed. Capsules oblong, 1.5–2 mm long; seeds flattened.

GRAND CAYMAN: *Hitchcock*.

— Bahamas, Cuba, Jamaica, Hispaniola, Guadeloupe, Yucatan, Trinidad and northern S. America; reported from tropical Africa. The Cayman plants were found "growing on stone in a shallow well".

Ernodea Sw.

Suberect or trailing slender shrubs; leaves opposite, crowded and narrow; stipules sheathing, persistent. Flowers solitary and sessile in the leaf-axils; calyx with short tube, the limb 4–6-parted with narrow persistent segments. Corolla with long cylindrical tube and 4–6 narrow reflexed lobes, these valvate in bud. Stamens as many as the corolla-lobes, inserted at the top of the tube and long-exserted. Disc fleshy. Ovary 2-celled with solitary axile ovules; style long-exserted and with capitate stigma. Fruit a dry-fleshed drupe with two 1-seeded, plano-convex nutlets.

— A West Indian genus of 7 known species, 6 of them endemic to the Bahamas.

Ernodea littoralis Sw., *Nov. Gen. & Sp. Pl.* 29 (1788). **FIG. 241.**

GUANA BERRY

Stems 4-angled, trailing or ascending to 1 m long or more; leaves rigid, glabrous, linear to lanceolate, mostly 1–3 cm long, spine-tipped. Corolla pale pink, the tube 8–10 mm long. Drupes yellow, ellipsoid, 4–5 mm long, crowned by the persistent calyx-lobes.

GRAND CAYMAN: *Brunt* 1635, 1815; *Hitchcock*; *Kings* GC 373; *Millspaugh* 1254; *Proctor* 15051. LITTLE CAYMAN: *Kings* LC 17; *Proctor* 28096. CAYMAN BRAC: *Millspaugh* 1194; *Proctor* 28999.

— Florida, the West Indies, Yucatan and Honduras, mostly in dry, rocky scrublands or thickets near the sea.

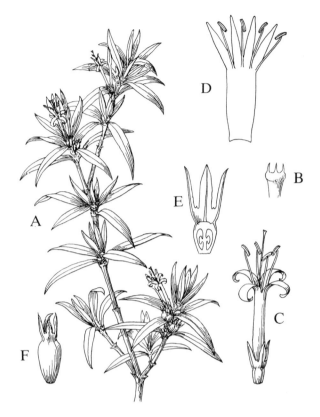

FIG. 241 **Ernodea littoralis**. A, branch with leaves, flowers and fruits, × 2/3. B, stipule, × 2. C, flower, × 2. D, corolla cut open, × 2. E, ovary with calyx cut lengthwise, × 4. F, fruit, × 2. (F. & R.)

Spermacoce L.

Annual or perennial herbs with 4-angled stems; leaves opposite; stipules united with the petioles to form a bristly sheath. Flowers several to many, sessile and glomerate in the leaf-axils. Calyx 4-lobed; corolla funnel-shaped, 4-lobed, the lobes valvate in bud. Stamens 4, inserted near the base of the corolla-tube; anthers included. Ovary 2-celled with solitary ovules; stigma 2-lobed. Fruit a hard-shelled capsule, one or both valves opening from the top; seeds oblong, with horny endosperm.

— A pantropical genus of more than 150 species, now construed to include the plants long separated in the genus *Borreria*.

KEY TO SPECIES

1. One valve of the capsule always remaining closed; flowers in axillary clusters:

    2. Stems and leaves minutely scabrid; capsules setulose: **S. confusa**

    2. Stems and leaves hirsute with white hairs; capsules densely long-hairy: **S. tetraquetra**

1. Both valves of the capsule opening; flowers in terminal heads:

3. Leaves elliptic, ovate-elliptic, or elliptic-oblanceolate with about 5 pairs of distinct lateral veins, up to 2 cm broad or more, and lacking axillary leaf-clusters: **S. assurgens**

3. Leaves linear or narrowly lanceolate, the lateral veins obscure, and mostly less than 1 cm broad, with axillary clusters of smaller leaves giving a verticillate appearance: **S. verticillata**

Spermacoce confusa Rendle, *J. Bot.* 74: 12 (1936); Nicolson, *J. Arnold Arb.* 58: 446–447 (1977). **FIG. 242.**

Annual herb with erect stems to 45 cm tall or more (often less), scabrid on the angles; leaves narrowly lanceolate, lanceolate, or rarely narrowly ovate, mostly 2–5 cm long. Corolla white, ca. 2 mm long. Fruits subglobose or ellipsoid, 2.5 mm long.

GRAND CAYMAN: *Hitchcock*; *Millspaugh* 1302, 1338; *Kings GC 303*; *Proctor* 15260.

— Southeastern U. S. A., the West Indies and continental tropical America, common in open waste places, also frequently a weed of cultivated fields.

FIG. 242 **Spermacoce confusa.** A, portion of plant with flowers and fruits, × ²/₃. B, flower subtended by bract, × 8. C, corolla cut open, × 8. D, fruit, × 8. E, seed, × 8. (F. & R.)

Spermacoce tetraquetra A. Rich. in Sagra, *Hist. Cub.* 11: 29 (1850).

Annual erect or sprawling herb like the last species but not scabrid, and clothed nearly all over with long whitish hairs; leaves lanceolate or oblong-lanceolate, mostly 1.5–3 cm long. Corolla white, ca. 2 mm long. Fruits globose, 2 mm long.

CAYMAN BRAC: *Proctor* 29335.

— Bahamas, Cuba and Jamaica, in sandy clearings and open waste ground, appearing after rains.

Spermacoce assurgens Ruiz & Pavon, *Fl. Peruv.* 1: 60, t.92, fig. c (1798).

Borreria laevis of many modern authors, not *Spermacoce laevis* Lam. (1791).

Decumbent or erect annual or persisting herb, the stems glabrous or nearly so and up to 50 cm tall or more; leaves lanceolate or narrowly elliptic to elliptic, mostly 1–4 cm long, glabrate and rather strongly nerved. Corolla white, hairy, ca. 4 mm long. Capsules puberulous, 2.5–3 mm long.

GRAND CAYMAN: *Brunt* 1892, 1932, 2096; *Hitchcock*; *Proctor* 15105; *Sachet* 418. LITTLE CAYMAN: *Proctor* 35199. CAYMAN BRAC: *Proctor* 28989.

— Florida, West Indies, and continental tropical America, introduced in Hawaii, a weed of pastures, roadsides and open waste ground.

Spermacoce verticillata L., *Sp. Pl.* 1: 102 (1753).

Erect glabrous herb, woody at base, the branches angled; stipules sheathed, with setae 3–5 mm long. Leaves appearing verticillate with clusters of axillary leaves nearly as large as the primary leaves, the latter 2–6 cm long and mostly 0.3–1(–1.2) cm broad, the apex acute to acuminate, the base narrowed, rough on the margins and on the midrib beneath. Flower-heads to 12 mm in diameter, terminal (or a few lateral), dense; calyx lobes 2, narrow, to 1 mm long; corolla 1.5–2 mm long, white, the lobes ovate, acute. Capsules oblong to subglobose, 1 mm in diameter, glabrous, each coccus bearing a calyx lobe; seeds ellipsoid, 1 mm long, dark brown, foveolate.

GRAND CAYMAN: *Proctor* 52103, found by Mrs. Joanne Ross, near S end of Bobby Thompson Way (formerly Halfway Pond Road).

— Widespread in the neotropics; also in Africa.

[CAPRIFOLIACEAE

Lonicera japonica Thunb., the honeysuckle (*Kings GC* 336), and *Sambucus simpsonii* Rehder, the elder (*Hitchcock*), have been recorded from Grand Cayman, presumably on the basis of cultivated plants. The writer has seen no evidence that they still occur, or that they ever escaped or persisted outside cultivation.]

COMPOSITAE/ASTERACEAE

Herbs, shrubs, vines, or small trees; leaves alternate, opposite or whorled, simple or variously toothed, lobed or divided; stipules absent. Flowers (florets) small and crowded into heads (capitula) surrounded by an involucre of one or more series of free or connate bracts (phyllaries); rarely the heads compound, consisting of several 1–few-flowered capitula; heads solitary or arranged in spikes, racemes, cymes, corymbs or panicles (just as if they were individual flowers). The receptacle from which the florets arise are usually convex, sometimes conical or cylindrical, or occasionally concave, and may be pitted or smooth and naked, scaly or hairy. Florets of 1 or 2 kinds in each head: perfect, unisexual or neuter, or rarely the heads dioecious; outer florets often ligulate (ray-florets), the inner ones tubular and without ligules (disc-florets), or all the florets may be ligulate or all tubular. Calyx superior, represented by a pappus of persistent or sometimes soon-falling hairs, bristles, awns or scales crowning the ovary, or sometimes reduced to a ring or absent. Corolla gamopetalous, with a long or short tube, 4–5-lobed and regular (disc-florets) or ligulate (rarely 2-lipped) and irregular (ray-florets). Stamens 5 (rarely 4), inserted in the corolla-tube; filaments free; anthers connate into a tube surrounding the style, rarely free, 2-celled and opening lengthwise, often appendaged at each end. Ovary inferior, 1-celled with a solitary basal ovule; style of the perfect or pistillate flowers usually 2-armed, the arms smooth, papillose or hairy and of various shapes. Fruit a 1-celled, 1-seeded achene, indehiscent and usually dry, sometimes beaked, often crowned by the persistent pappus; seeds without endosperm.

— A cosmopolitan family with an estimated 950 genera and more than 20,000 species, one of the largest plant families. All of the Cayman genera have the disc-florets (or all the florets) regular and 4–5-lobed; none have 2-lipped florets. For convenience in so large a group, the genera of Compositae are arranged in series called Tribes, of which 8 occur in the Cayman Islands. These can be characterised as follows:

TRIBE I. VERNONIEAE. Leaves usually alternate. Heads discoid. Corolla white or purple. Anthers sagittate at the base. Style-arms more or less cylindrical, hairy on the back. Pappus usually of bristles or scales. Genera: **Cyanthillium** and **Vernonia** (Lepidoploa)

TRIBE II. EUPATORIEAE. Leaves usually opposite, the upper sometimes alternate. Heads discoid. Corolla purple, white or whitish. Anthers blunt at the base. Style-arms long, subcylindrical, hairy. Pappus usually of bristles. Genera: **Ageratum**, **Chromolaena** and **Koanophyllon**

TRIBE III. ASTEREAE. Leaves usually alternate. Heads radiate or disciform (heterogamous), or sometimes by suppression of the ray-florets homogamous. Receptacle usually naked. Corolla of the disc usually yellow, of the ray the same colour or different. Anthers blunt at the base. Style-arms flattened, appendaged. Genera: **Aster**, **Conyza** and **Baccharis**

TRIBE IV. INULEAE. As in Tribe III, but anthers tailed at the base and the style-arms linear, without appendages. Genus: **Pluchea**

TRIBE V. HELIANTHEAE. Leaves opposite or alternate. Heads usually radiate, sometimes disciform, or by suppression of the rays homogamous and discoid. Receptacle scaly. Corolla of the disc usually yellow, of the rays often of the same colour, sometimes different. Anthers entire at the base or with two very short points. Style-arms truncate or appendaged. Pappus usually of awns or scales, often lacking, rarely bristly. Genera: **Ambrosia**, **Iva**, **Isocarpha**, **Borrichia**, **Parthenium**, **Thymophylla**, **Melanthera**, **Eclipta**, **Sphagneticola**, **Eleutheranthera**, **Verbesina**, **Spilanthes**, **Salmea**, **Synedrella**, **Bidens**, **Flaveria** and **Tridax**

TRIBE VI. HELENIEAE. Leaves opposite or alternate. Heads usually radiate, sometimes disciform, or by suppression of the rays discoid; phyllaries in one series. Receptacle naked. Corolla of the disc usually yellow. Anthers obtuse at the base. Style-arms truncate or appendaged. Pappus usually of scales. Genera: **Porophyllum**, **Helenium** and **Pectis**

TRIBE VII. SENECIONEAE. Leaves usually alternate. Heads usually radiate, less often disciform or discoid. Receptacle usually naked, sometimes scaly. Disc corollas red to yellow. Anthers with an apical appendage, often sagittate at the base. Style-arms usually truncate, with or without an appendage. Pappus of bristles. Genera: **Erechtites** and **Emilia**

TRIBE VIII. LACTUCEAE. Leaves spiral; heads entirely ligulate; latex ducts present; mostly herbs (a few pachycaul trees); receptacle naked, fimbrilliferous or scaly; anthers sagittate at base; style-arms long and thin; pappus bristly or scaly, rarely absent. Genera: **Launaea** and **Youngia**

Other Tribes occur in various parts of the world, but are not represented in the Cayman Islands.

KEY TO GENERA

1. **Heads discoid, the florets all tubular and regular:**
  2. Herbs:
    3. Leaves alternate (in *Porophyllum* both alternate and opposite):
      4. Phyllaries in 2–several series, distinctly imbricate:
        5. Heads in loose, open cymose panicles: Cyanthillium
        5. Heads in dense corymbs: Pluchea
      4. Phyllaries in 1 main series, essentially valvate:
        6. Phyllaries glabrous: Porophyllum
        6. Phyllaries pubescent:
          7. Florets cream or greenish-yellow; disc-florets perfect; marginal florets pistillate: Erechtites
          7. Florets crimson; all florets perfect: [Emilia]
    3. Leaves opposite:
      8. Leaves simple:
        9. Heads several or many in long-stalked corymbs: Ageratum
        9. Heads solitary (rarely 2 or 3 on a common peduncle):
          10. Peduncles and phyllaries glabrous; some of the leaves alternate: Porophyllum
          10. Peduncles pubescent; phyllaries pubescent or at least ciliate:
            11. Leaves more or less toothed; receptacle flat or slightly convex:
              12. Phyllaries shorter than the florets; pappus of 2–4 awn-like bristles: Melanthera
              12. Phyllaries foliaceous and much longer than the florets; pappus a ciliate cupule: Eleutheranthera
            11. Leaves entire; receptacle conical: Spilanthes
      8. Leaves compound (or at least some of them):
        13. Heads unisexual, less than 3 mm in diameter, the pistillate ones 1-flowered, the staminate ones in more or less elongate racemes: Ambrosia

13. Heads bisexual, up to 10 mm or more in diameter always multiple-flowered, never in racemes: **Bidens**

2. Shrubs, or plants woody at least near the base:
  14. Leaves alternate:
    15. Heads on one-sided, simple or forked, spike-like cymes: **Vernonia** (Lepidoploa)
    15. Heads in corymbs or glomerate clusters:
      16. Leaves less than 3 cm long, viscid-resinous and gland-dotted: **Baccharis**
      16. Leaves mostly 4–10 cm long or more, not viscid-glandular:
        17. Leaves entire, densely tomentose beneath; pappus of numerous white bristles: **Pluchea**
        17. Leaves toothed, glabrate or sparingly puberulous beneath; pappus of 2 stiff awns: **Verbesina**
  14. Leaves opposite or mostly so (a few alternate in *Isocarpha*):
    18. Heads in leafy racemes: **Iva**
    18. Heads in corymbs or glomerules:
      19. Pappus of numerous bristles; receptacle without scales:
        20. Leaves crenate-toothed; achenes 3–4 mm long: **Chromolaena**
        20. Leaves mostly entire; achenes ca. 1.5 mm long: **Koanophyllon**
      19. Pappus absent or of 2 or 3 awns; receptacle scaly:
        21. Achenes oblong and 5-angled; pappus absent; plants woody only at the base: **Isocarpha**
        21. Achenes flat; pappus of 2 awns; plants woody throughout: **Salmea**

1. **Heads with at least some outer florets ligulate (the rays sometimes very inconspicuous):**
  22. Leaves alternate:
    23. All florets ligulate:
      24. Achenes grey, glabrous, 3.5–4 mm long: **Launaea**
      24. Achenes red-brown, pubescent toward apex, 1.5 mm long: **Youngia**
    23. Both ligulate and discoid florets present:
      25. Leaves simple; pappus of numerous bristles:
        26. Leaves rather few, narrowly oblanceolate, up to 10 mm broad or more; heads relatively few in open cymose panicles: **Aster**
        26. Leaves very numerous, linear, those on the stem less than 3 mm broad except near the base; heads very numerous in rather dense thyrsoid panicles: **Conyza**
      25. Leaves pinnatifid or pinnatisect:
        27. Pappus of 2 recurved awns: **Parthenium**
        27. Pappus of numerous 3-awned scales: **Thymophylla**
  22. Leaves opposite:
    28. Erect shrub; leaves thick and leathery, often silvery-tomentose: **Borrichia**
    28. Erect, decumbent or mat-like herbs; leaves neither leathery nor silvery-tomentose:
      29. Achenes without a pappus:
        30. Heads solitary at tips of axillary peduncles; flowers white, the corollas 1.5–2 mm long: **Eclipta**
        30. Heads tightly aggregated in sessile glomerules at stem forks and in leaf axils, rarely short pedunculate; flowers greenish-yellow, the corollas 3 mm long: **Flaveria**
      29. Achenes with a pappus:

31. Pappus cup-like: **Sphagneticola**
31. Pappus of bristles, awns or scales:
 32. Heads sessile in the leaf-axils and stem-forks; achenes of 2 kinds, the outer flat and with comb-like wings, the inner angular and crowned with a pappus of 2 or 3 bristles: **Synedrella**
 32. Heads stalked; achenes all alike:
  33. Leaves or leaflets broad with toothed margins; heads more than 5 mm across; plants not aromatic:
   34. Pappus of 2 awns: **Bidens**
   34. Pappus of many bristles: **Tridax**
  33. Leaves linear or very small, entire or bristle-margined; heads less than 5 mm across, longer than broad; plants often aromatic:
   35. Receptacle conical; ligules bright orange, up to 10 mm long; plants ill-smelling: **Helenium**
   35. Receptacle flat, small, ligules yellow, mostly less than 5 mm long; plants often pleasantly aromatic: **Pectis**

TRIBE I. VERNONIEAE

Cyanthillium Blume

Annual herbs with puberulous erect stems. Leaves simple, alternate, and pinnately veined. Inflorescences terminal, corymbiform or corymbiform-paniculate. Heads discoid with bell-shaped ca. 3-seriate involucres, the involucral bracts narrowly lanceolate, somewhat overlapping, sharply attenuate at tips, pilose, glandular, and reflexed in fruits; receptacle naked. Flowers bisexual, the corollas tubular or bell-shaped, shortly 5-lobed. Anthers included, basally spurred; styles hispidulous in upper half, the branches lavender. Achenes nearly cylindrical, not ribbed, appressed-pilose, pustulate; pappus biseriate, white, the outer series of scales, the inner of numerous fine bristles longer than the achene.

— A genus of 5 species with pantropical distribution.

Cyanthillium cinereum (L.) H. Rob., *Proc. Boil. Soc. Wash.* 103: 252 (1990).

Erect herb up to 50 cm tall or more, the stems puberulous; leaves rhombic-ovate or obovate, mostly 1.5–4 cm long, obtuse at the apex, long-cuneate at the base, the margins sinuate-toothed or subentire. Heads ca. 15-flowered; involucral bracts pubescent, linear-oblanceolate with sharply acuminate tips. Achenes short-bristly, 1.5 mm long; pappus white, the inner bristles ca. 3 mm long.

GRAND CAYMAN: *Brunt* 1930; *Kings GC* 370; *Proctor* 15280. CAYMAN BRAC: *Proctor* 29091.

— A pantropical weed of pastures, roadsides and open waste places.

Vernonia Schreb.

Herbs or shrubs; leaves alternate, rarely opposite or whorled, entire or toothed, pinnate-veined. Heads discoid, solitary or in cymes, corymbs or panicles; phyllaries imbricate in several series; receptacle naked and smooth or minutely pitted. Florets 5-lobed. Anthers sagittate at the base. Stlye-arms linear or awl-shaped and usually terete, papillose or short-hairy. Achenes cylindrical or top-shaped, angled or ribbed; pappus usually 2-seriate, the inner series of bristles, the outer usually of shorter bristles or scales.

— A large pantropical to warm-temperate genus of more than 1,000 species.

Vernonia divaricata Sw., *Fl. Ind. Occ.* 3: 1319 (1806). **FIG. 243.** PLATE 63.

CHRISTMAS BLOSSOM

An erect shrub to 2 m tall or more, the young branches finely white-tomentose; leaves narrowly to broadly ovate, mostly 3–11 cm long, acuminate at the apex and rounded at the base, the margins entire, slightly scabrid-roughened on the upper side and puberulous beneath. Inflorescence usually of 2–several curved, spike-like branches; involucres 5–6 mm long, the phyllaries in 5 series. Florets light purple, ca. 6 mm long. Achenes appressed-silky, ca. 1 mm long; pappus white, the inner bristles ca. 3 mm long.

GRAND CAYMAN: *Brunt* 1924; *Correll & Correll* 51020; *Hitchcock*; *Proctor* 15018. LITTLE CAYMAN: *Proctor* 28124, 35146. CAYMAN BRAC: *Millspaugh* 1161; *Proctor* 29023, 29352.

— Jamaica; frequent in thickets, along the edges of clearings and in dry rocky woodlands.

Editor's note: *Lepiodoploa divaricata* is an invalid name and the original name for this species *Vernonia divaricata* as used in the First Edition of the flora should be retained.

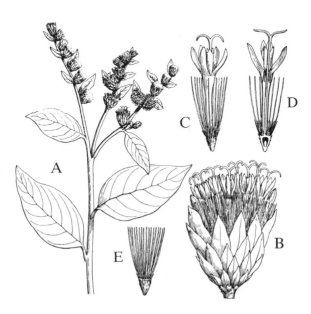

FIG. 243 **Vernonia divaricata**. A, branch with inflorescence, × ²⁄₃. B, flowering head, × 5. C, floret, × 6. D, floret cut lengthwise, × 6. E, achene and pappus, × 5. (F. & R.)

TRIBE II. EUPATORIEAE

Ageratum L.

Annual or perennial herbs, or sometimes shrubs; leaves opposite or sometimes a few upper ones alternate. Heads discoid, many-flowered, in corymbs or panicles; phyllaries in 2–3 series, narrow and equal or nearly so; receptacle naked or scaly. Florets 5-lobed. Anthers obtuse at the base and with a terminal appendage. Style-arms elongate, exserted, terete. Achenes 5-angled and ribbed; pappus of 5 or more scales, these awned or not, free or united at the base, rarely reduced to a cup.

— A mostly tropical American genus of about 35 species.

KEY TO SPECIES

1. Perennial salt-tolerant herb of seaside locations; inflorescence long-pedunculate; flower-heads bright blue; achenes crowned by a cup-like pappus:	**A. littorale**
1. Annual weedy herb of gardens and waste ground; inflorescence short-pedunculate; flower-heads lavender or white; achenes crowned by a pappus of 5 or 6 awned scales:	**A. conyzoides**

Ageratum littorale A. Gray, *Proc. Amer. Acad.* 16: 78 (1881).

A decumbent-ascending herb with glabrous stems; leaves rhombic-ovate or elliptic, mostly 1.5–4 cm long including the slender petioles, acute at the apex and long-cuneate at the base, the margins crenate-serrate. Heads in small, dense corymbs at the apex of long, erect peduncles; involucres 3–4 mm long, the phyllaries minutely ciliate. Florets bright lavender-blue, ca. 2.5 mm long. Achenes glabrous, ca. 2 mm long; pappus a very short, crown-like cup.

GRAND CAYMAN: *Kings GC 176*; *Proctor 15037, 15145, 31049*.

— Florida Keys and Cape Sable, also on the Turneffe Islands, Belize; grows on beaches and coastal sands.

Ageratum conyzoides (L.) L., *Sp. Pl.* 2: 839 (1753).

Erect annual herb usually to 50 cm (rarely 100 cm) tall. Leaves with petioles 1–3.3 cm long; blades of lower leaves ovate, of upper leaves obovate to elliptic-oblong, 3–10 cm long, 1.9–7 cm broad, the apex acute, the base obtuse or cuneate, with crenate, ciliate margins, 3-nerved from base, and pilose with yellow glands on both surfaces to nearly glabrous. Inflorescence of 4–18 heads in terminal cymose clusters; involucres bell-shaped, the phyllaries biseriate, oblong, 3–4.7 mm long; receptacle convex, naked. Flowers 50–80 per head. Achenes 5-angled, 1.2–2 mm long, black.

GRAND CAYMAN: *Proctor 52144*.

— Native of the neotropics, but now a pantropical weed.

Chromolaena DC.

Erect or sprawling perennial herbs or shrubs, the stems often pubescent. Leaves simple, usually opposite, the blades often 3-nerved from the base. Inflorescence usually corymbose, the heads discoid 10–40-flowered; involucres constricted-cylindrical, the bracts ca. 4–7-seriate, overlapping, 3–5-veined, broadly acute to rounded at apex, deciduous; receptacle elongate, flat or conical on top, glabrous. Flowers bisexual, the corollas actinomorphic, tubular, and shortly 5-lobed, the lobes often glandular. Anthers usually included, the style-branches elongate, linear and exserted, the upper half of each style-branch with a papillose sterile appendage. Achenes narrowly obconical, usually 5-ribbed, sometimes glandular; pappus of numerous bristles about equal in length to the corolla and achene.

— A genus of 166 species widely distributed in the Western Hemisphere from the U. S. A. to Argentina.

Chromolaena odorata (L.) R. M. King & H. Rob., *Phytologia* 20: 204 (1970).

Eupatorium odoratum L. of this Flora, ed 1.

A mostly erect, short-lived shrub to 2 m tall, the younger stems puberulous or scabrid; leaves petiolate, ovate or broadly ovate, mostly 2.5–8 cm long, acuminate at the apex and abruptly cuneate at the base, the margins broadly crenate-toothed, puberulous or glabrate on the upper side, densely puberulous beneath. Corymbs 2–3-branched; phyllaries in about 5 series, up to 7 mm long, green-tipped. Florets lavender-whitish, ca. 5 mm long. Achenes scabrous on the angles, 3–4 mm long.

GRAND CAYMAN: *Hitchcock*; *Proctor* 15293; *Rothrock* 166. CAYMAN BRAC: *Millspaugh* 1187; *Proctor* 29108.

— Southern U. S. A., the West Indies and continental tropical America, common along roadsides, in pastures, clearings and in rocky thickets. This species has become an invasive weed in parts of Africa and Malaysia.

Koanophyllon Arruda

Shrubs or small trees (rarely vines) with terete, more or less striate stems. Leaves opposite (rarely alternate), petiolate, the blades ovate to broadly lanceolate or elliptic, the apex blunt to acute, the base acute, truncate or cordate, the margins mostly entire to serrate, the venation pinnate or 3-nerved, the surfaces with few to many hairs (but rarely densely pubescent) and with sparse to numerous glandular punctations. Inflorescences pyramidally paniculate to corymbose; involucral bracts 7–16, slightly imbricate in 2–4 more or less unequal series, mostly spreading at maturity, some of them deciduous; receptacle flat to slightly convex, glabrous. Flowers 5–ca. 20 in a head, the corollas usually whitish to greenish-yellow, funnel-shaped from a broadly cylindrical basal tube, the lobes broadly triangular and bearing numerous clustered short-capitate glands. Anther appendages often wider than long; style-bases not enlarged, glabrous, the branches broadened and smooth apically, without glands. Achenes prismatic, 5-ribbed; pappus of ca. 30–35 long, scabrid bristles.

— A Western Hemisphere genus of 114 species ranging from Florida, the West Indies, southwestern U. S. A. and Mexico through C. America and the Andes Region of S. America to Brazil and Argentina.

Koanophyllon villosum (Sw.) R. M. King & H. Rob., *Phytologia* 39: 265 (1975).

Eupatorium villosum of this Flora, ed. 1.

A bushy shrub to 2 m tall, more or less velvety-pubescent throughout; leaves petiolate, deltate-ovate, mostly 2–7 cm long, blunt to acute at the apex, truncate at the base, the margins entire or obscurely toothed. Heads numerous in dense corymbs; phyllaries 2–2.5 mm long. Florets whitish, 2.5–3 mm long, with glandular tube. Achenes sparsely puberulous, ca. 1.5 mm long.

GRAND CAYMAN: *Brunt* 1809, 1889; *Correll & Correll* 51010; *Hitchcock*; *Kings* GC 350; *Millspaugh* 1401; *Proctor* 11992; *Rothrock* 169; *Sachet* 366. LITTLE CAYMAN: *Proctor* 35143. CAYMAN BRAC: *Kings* CB 71; *Proctor* 28963.

— Florida and sporadically throughout the West Indies, including the Swan Islands; common in clearings and in thickets on limestone.

TRIBE III. ASTEREAE

Aster L.

Mostly perennial herbs (rarely annual); leaves alternate, and entire, toothed or divided. Heads radiate, usually in corymbs or panicles, rarely solitary or in racemes; involucre bell-

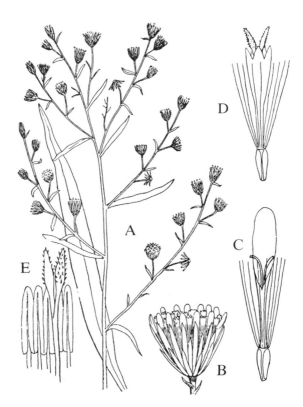

FIG. 244 **Aster subulatus.** A, portion of plant, × 2/3. B, head, × 4. C, ray-floret with achene, × 12. D, disk-floret with achene, × 8. E, anthers and apex of style, × 32. (F. & R.)

shaped, the phyllaries imbricate in 3–4 series; receptacle flat or convex. Ray-florets in 1 or 2 series, usually pistillate and fertile; disc-florets 5-lobed, perfect, and usually fertile. Anthers obtuse at the base. Style-arms of the disc-florets flattened, crowned with a short or long papillose appendage. Achenes usually compressed, 1–4-ribbed or ribless; pappus of bristles in 2 or 3 series.

— A cosmopolitan genus of more than 500 species.

Aster subulatus Michx., *Fl. Bos. Amer.* 2: 111 (1803). **FIG. 244.**

Aster exilis of this Flora, ed. 1.

Erect annual herb up to 50 cm tall or more, glabrous or minutely puberulous; leaves elongate, narrowly oblanceolate, mostly 5–15 cm long, acuminate at the apex and long-tapered to the semi-clasping base, the margins entire or minutely serrulate. Heads 5–8 mm broad in flower; phyllaries 2–5 mm long, glabrous or nearly so. Ray-florets white or very pale mauve; disc-florets light yellow, ca. 3.5 mm long. Achenes thinly pubescent, 1–1.5 mm long; pappus of white bristles ca. 3 mm long.

GRAND CAYMAN: *Kings* GC 419.

— Southeastern U. S. A., Bahamas, Cuba, Jamaica and Mexico to Belize, chiefly along roadsides and on moist disturbed soil.

Conyza Less., nom. cons.

Erect annual or perennial herbs; leaves usually in a basal rosette and alternate and numerous on the stems, entire, toothed, or deeply divided. Heads radiate, solitary or in corymbs or panicles; involucre bell-shaped, with phyllaries in 1–3 series; receptacle flat and naked, pitted. Ray-florets pistillate, in 2 or more series, with filiform corollas and very short ligules; disc-florets 4–5-toothed. Achenes narrow, with pappus of 2–3 rows of bristles.

— A chiefly temperate and subtropical genus of about 60 species.

Conyza canadensis (L.) Cronquist, *Bull. Torr. Bot. Club* 70: 632 (1943).

Erigeron canadensis L. (1753).
Conyza canadensis var. *pusilla* (Nutt.) Cronquist (1943).

Erect weedy herb to 1 m tall or more; leaves on the stem linear or very narrowly oblanceolate and up to 5 cm long, the margins usually sparsely ciliate, the surfaces sometimes puberulous. Branches of the panicle raceme-like; heads mostly 3–5 mm across when in flower; phyllaries mostly 2–3 mm long, glabrous except for the minutely ciliolate tips. Ray-florets ca. 2.5 mm long, the minute ligule white; disc-florets yellow, 4-toothed, ca. 2.5 long. Achenes 1 mm long; pappus-bristles 3 mm long.

GRAND CAYMAN: *Brunt* 1720; *Kings* GC 261, GC 351; *Millspaugh* 1259. LITTLE CAYMAN: *Proctor* 28162. CAYMAN BRAC: *Proctor* 28921, 52166.

— Temperate N. America and the West Indies, a common weed of roadsides, pastures, and open waste ground. Also widely naturalised in the Old World.

Baccharis L.

Dioecious shrubs; leaves alternate, occasionally reduced to scales, entire, toothed or lobed. Heads discoid with oblong to hemispheric involucre; phyllaries in several series; receptacle flat or convex, naked, sometimes minutely pitted. Staminate florets tubular with 5-lobed limb and rudimentary styles, their arms tipped with an ovoid pubescent appendage; anthers obtuse and entire at the base. Pistillate florets slender with 5-toothed or entire limb; style-arms smooth, exserted, truncate or club-shaped. Achenes somewhat compressed, 5–10-ribbed, developed only in pistillate heads; pappus of the staminate florets consisting of a few short bristles, of the pistillate flowers numerous in 1 or more series.

— A New World genus of about 400 species, the majority tropical.

Baccharis dioica Vahl, *Symb. Bot.* 3: 98, t.74 (1794). Plate 63.

A densely branched, round-topped shrub 1 m tall or more, the young branchlets angled and resinous; leaves crowded, obovate, mostly 1.5–2 cm long or more, stiff, glabrous and viscid, the apex minutely cuspidate. Heads sessile in dense clusters, the staminate ca. 5 mm long and 20-flowered, the pistillate 7 mm long and 50-flowered; florets white. Achenes glabrous, 1–2 mm long; pappus dull whitish.

LITTLE CAYMAN: *Proctor* 35190. CAYMAN BRAC: *Proctor* 29046.

— Florida and the West Indies to Montserrat, in rocky thickets and dry rocky scrublands.

TRIBE IV. INULEAE

Pluchea Cass.

Erect herbs or shrubs; leaves alternate, entire or toothed, rarely pinnatifid. Heads discoid, heterogamous, usually in terminal corymbose cymes; involucre ovoid or bell-shaped, the phyllaries in few or several series; receptacle flat, naked. Outer florets in several rows, pistillate and fertile; central florets usually few, perfect but sterile (functionally staminate). Corolla of the perfect florets tubular, 5-lobed, their anthers sagittate at the base with tailed auricles, their styles entire or forked, papillose. Corolla of the pistillate flowers filiform, 3-toothed or -lobed; style with 2 linear arms. Achenes 4–5-angled; pappus a single series of several or many scabrid bristles.

— A pantropical genus of about 50 species.

KEY TO SPECIES

1. Annual herb with lanceolate leaves mostly less than 6 cm long (rarely longer); florets bright red-purple; achenes 1 mm long:	*P. odorata*
1. Coarse shrub with elliptic leaves usually 6–15 cm long; florets dull light pink; achenes 0.5 mm long:	*P. carolinensis*

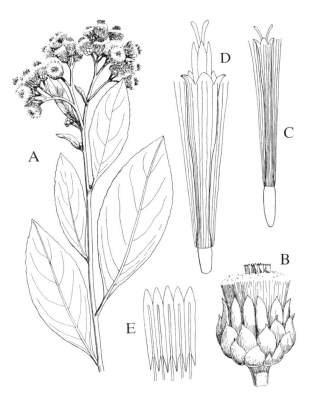

FIG. 245 **Pluchea carolinensis.** A, end of branch with inflorescence, × ²/₃. B, head, × 4. C, pistillate floret, × 12. D, perfect floret, × 12. E, anthers × 12. (F. & R.)

Pluchea odorata (L.) Cass., *Dict. Sci. Nat.* 42: 3 (1826). PLATE 63.

> *Pluchea purpurascens* (Sw.) DC. (1836).

Aromatic herb to 1 m tall (rarely more), the stems puberulous; leaves mostly 2–6 cm long and 0.5–2 cm broad, at least the upper ones subsessile, acute or obtuse at the apex, the margins obscurely toothed with glandular serrations, puberulous and glandular on both sides. Corymbs mostly 2–4 cm across; heads 4–5 mm long. Achenes glandular; pappus-bristles ca. 3 mm long.

> GRAND CAYMAN: *Brunt* 1662; *Kings* GC 357; *Maggs* II 51; *Proctor* 15261; *Sachet* 460.
> — Southeastern U. S. A., West Indies, Mexico and C. America, chiefly in marshes and moist sandy ground.

Pluchea carolinensis (Jacq.) G. Don in Sweet, *Hort. Brit.* ed. 3: 330 (1839). **FIG. 245.**

> *Pluchea odorata* of many authors, not (L.) Cass. (1826).
> P. *symphytifolia* of this Flora, ed. 1.

A bushy shrub up to 3 m tall, the stems whitish-tomentose; leaves all petiolate, the blades elliptic, acute or blunt at the apex, the margins entire or with a few obscure glandular teeth, minutely puberulous and glandular on the upper (adaxial) surface, densely woolly tomentose and glandular beneath. Corymbs mostly 8–15 cm across; heads 5–6 mm long. Achenes strigillose; pappus-bristles ca. 3 mm long.

GRAND CAYMAN: *Hitchcock*; *Kings* GC 357.

— Florida, the West Indies, Mexico to Venezuela, and islands of the Pacific; occurs in brackish thickets and open waste ground.

TRIBE V. HELIANTHEAE

Ambrosia L.

Monoecious (rarely dioecious) annual or perennial herbs, sometimes woody at the base; leaves opposite or alternate (rarely in whorls of 3), entire to pinnately cut or dissected. Staminate heads in spikes or racemes, the pistillate solitary or in glomerules at the base of the staminate raceme. Staminate involucre hemispheric or saucer-shaped, 5–12-lobed; receptacle flat, naked or scaly; florets funnel-shaped, 5-toothed; anthers scarcely coherent, mucronate-tipped; style undivided, capitate-penicillate. Pistillate involucre closed, usually with small tubercles or prickles near the top and contracted above into a short beak surrounding the style; corolla absent; style deeply divided into 2 long, narrow arms. Achenes broad and thick, without a pappus, closely surrounded by the persistent, hardened involucre.

— A mostly American genus of between 35 and 40 species. The wind-borne pollen of some species is a principal cause of hay-fever.

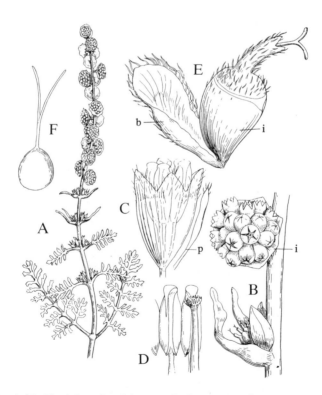

FIG. 246 **Ambrosia hispida.** A, branch with leaves and inflorescence, × ²/₃. B, a glomerule of pistillate heads (below) and a staminate head, × 4; **i**, involucre. C, staminate floret detached with scale (**p**), × 12. D, anthers and style from staminate floret, × 12. E, pistillate head with bract (**b**), × 12; **i**, involucre. F, ovary and style, × 8. (F. & R.)

Ambrosia hispida Pursh, *Fl. Amer. Sept.* 2: 743 (1814). **FIG. 246**.

GERANIUM, RUNNING WORMWOOD

A perennial, bitter-aromatic, trailing herb with ascending branches to about 35 cm tall or more, canescent-tomentose throughout; leaves ovate or broadly triangular in outline, mostly 2.5–5 cm long, bipinnatifid or the lower ones tripinnatifid. Racemes up to 8 cm long; staminate heads 6–20-flowered, 2–4 mm across, the involucres clothed with short hairs having pustulate bases; florets cream, densely glandular. Pistillate heads 3–10 in a cluster; achenes black, 2.5–3 mm long.

GRAND CAYMAN: *Brunt* 1963; *Hitchcock*; *Kings* GC 260, GC 319; *Proctor* 15154; *Sachet* 443. LITTLE CAYMAN: *Kings* LC 25, LC 26; *Proctor* 28068. CAYMAN BRAC: *Proctor* (sight record).

— Florida, the West Indies and C. America, mostly in sandy clearings near the sea; sometimes cultivated.

Iva L.

Annual or perennial herbs, or shrubs; leaves opposite or the upper ones alternate, simple and often of thick texture. Heads nodding, in spikes, racemes or panicles, rarely solitary in the leaf-axils. Involucre hemispheric or cup-shaped, composed of few, rounded bracts; receptacle scaly, the chaff-like linear or spatulate scales enveloping the florets. Marginal florets 1–6, pistillate, fertile, their corollas short and tubular, or none; styles 2-armed. Disc-florets perfect in structure but functionally staminate only, their corollas funnel-shaped and 5-lobed; anthers entire at the base, scarcely coherent, tipped with mucronate appendages; styles undivided, dilated at the apex. Achenes compressed, obovoid, glabrous; pappus lacking.

— An American genus of about 15 species.

KEY TO SPECIES

1. Plant glabrous; leaves mostly alternate; involucres 5–6 mm broad:	I. imbricata
1. Plant pubescent; leaves mostly opposite; involucres 3–4 mm broad:	I. cheiranthifolia

Iva imbricata Walt., *Fl. Carol.* 232 (1788). PLATE 63.

Glabrous fleshy perennial herb or subshrub with woody roots and simple or sparingly branched stems often sprawling or decumbent, rarely more than 50 cm tall. Leaves sessile or nearly so, oblong or linear-oblong to narrowly spatulate, the larger seldom more than 3 cm long and 0.6–0.8 cm broad, more or less obtuse and mucronulate, usually entire and obscurely 3-nerved. Heads greenish-white, nearly sessile or very short-pedunculate in the leaf axils, the upper ones longer than their subtending leaves; involucres broadly bell-shaped, the 6–9 suborbicular bracts subimbricate in 2 series, ca. 4 mm long and 2 mm broad. Fructiferous flowers 2–4 per head, their corollas tubular; staminate flowers much more numerous. Achenes obovoid, ca. 4 mm long, granular-resinous.

GRAND CAYMAN: *Proctor* 48273, 52074. Found only at Head of Barkers.

— Coastal southeastern U. S. A., Bahamas and Cuba.

Iva cheiranthifolia Kunth in H. B. K., *Nov. Gen. & Sp.* 4: 276 (1820). PLATE 63.

An aromatic shrub to 2 m tall, with arching canescent branches; leaves opposite and alternate, very narrowly elliptic or narrowly oblong-elliptic, mostly 1–6 cm long or more, 0.3–1.5 cm broad, sharp-pointed at the apex, 3-nerved from the base, the margins entire, finely puberulous and gland-dotted on both sides. Heads numerous, short-stalked, 3–5 mm across; phyllaries 3–5 nearly orbicular, puberulous. Achenes ca. 2.5 mm long.

GRAND CAYMAN: *Brunt* 1769, 2082; *Hitchcock*; *Lewis* 3829; *Proctor* 27928, 27953, 52132.

— Bahamas and Cuba; common in brackish thickets and along the margins of swamps.

Isocarpha R. Br.

Perennial herbs, sometimes woody or shrub-like at the base; leaves alternate or opposite, simple and entire or toothed. Heads discoid, solitary or clustered at the ends of long peduncles; involucre of rather few scales in 2–3 series; receptacle conical, with chaffy scales enveloping the florets. Florets all perfect and fertile, 5-lobed; anthers obtuse at the base; style-arms terete, slender, hairy. Achenes 4–5-angled, truncate at the apex and without pappus.

— A genus of 10 tropical and warm-temperate American species.

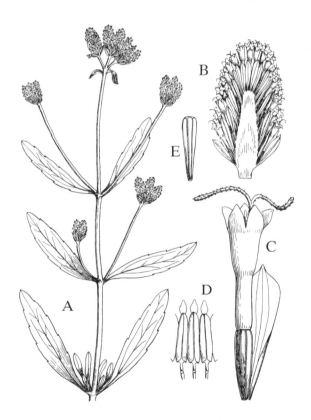

FIG. 247 **Isocarpha oppositifolia.** A, portion of plant, × $^2/_3$. B, head cut lengthwise, × 4. C, floret, × 12. D, anthers, × 14. E, ripe achene, × 8. (F. & R.)

Isocarpha oppositifolia (L.) Cass., *Dict. Sci. Nat.* 24: 19 (1822). **FIG. 247**.

A suffrutescent herb, the older stems woody and often decumbent, the young stems ascending and pubescent, up to 30 cm tall or more; leaves mostly opposite, often with smaller ones clustered in the axils, narrowly oblanceolate, mostly 1.5–4.5 cm long, obtuse at the apex, the margins mostly entire, pubescent and glandular on both sides, triplinerved. Heads 6–10 mm long, few in a small, dense cluster; phyllaries 2-nerved and mucronate, 3–4 mm long. Florets white, 2.5 mm long. Achenes 5-angled, dark, 2 mm long.

CAYMAN BRAC: *Kings* CB 54; *Proctor* 29044.

— Bahamas, Cuba, Jamaica and Texas to Venezuela, Trinidad and Tobago; occurs in dry, rocky thickets and scrublands.

Borrichia Adans.

Shrubs, often of fleshy texture; leaves opposite, simple and entire or toothed, 1–3-nerved. Heads solitary, terminal or in forks of the branches, radiate. Involucre hemispherical, the phyllaries leathery and in 2–3 series; receptacle slightly raised, with rigid concave scales surrounding the florets. Ray-florets pistillate, the ligules subentire or 2–3-toothed. Disc-florets perfect and fertile, 5-toothed; anthers entire at the base or nearly so; style-arms swollen and hairy above. Achenes oblong, 3–5-angled; pappus a short, toothed cup.

— A small tropical American genus of 7 species.

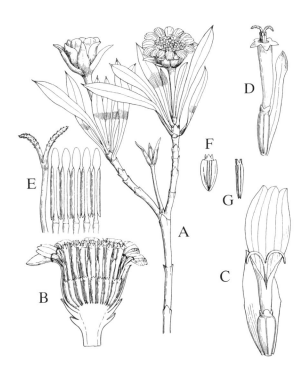

FIG. 248 **Borrichia arborescens**. A, portion of plant, × ²/₃. B, head cut lengthwise, × 2. C, ray-floret, × 4. D, disk-floret, × 4. E, anthers and style, × 8. F, achene of ray-floret, × 3. G, achene of disk-floret, × 3 (F. & R.)

Borrichia arborescens (L.) DC., *Prodr.* 5: 489 (1836). **FIG. 248.** PLATE 64.

BAY CANDLEWOOD

A rather succulent shrub to 1.5 m tall, occurring in two forms, one whitish-canescent, the other nearly glabrous throughout; leaves linear-oblanceolate to narrowly obovate, 2–7.5 cm long, sharply mucronate at the apex, the margins entire. Peduncles mostly 1–2.5 cm long; heads of heavy, leathery texture, 1–2 cm in diameter or more; ligules yellow, 6–9 mm long; disc-florets 6 mm long. Achenes 3.5–4 mm long; pappus-cup ca. 1 mm long.

GRAND CAYMAN: *Brunt* 1687, 1850, 1884, 1973; *Hitchcock*; *Kings* GC 35, GC 38, GC 268, GC 269, GC 395; *Millspaugh* 1239, 1242, 1247; *Proctor* 11989; *Sachet* 392. LITTLE CAYMAN: *Kings* LC 40; *Proctor* 28026. CAYMAN BRAC: *Kings* CB 86; *Proctor* 28927.

— Florida, West Indies and Yucatan; common on rocky seacoasts, sometimes in sandy or brackish thickets.

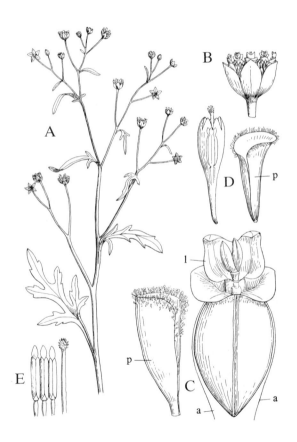

FIG. 249 **Parthenium hysterophorus.** A, portion of plant, × ²⁄₃. B, head, × 6. C, ray-floret with achene and receptacular scale, × 18; **p**, scale; **1**, ligule; **a, a**, awns. D, disk-floret with scale (**p**), × 18. E, anthers and style of disk-floret, × 24. (F. & R.)

COMPOSITAE (HELIANTHEAE): Thymophylla

Parthenium L.

Herbs or shrubs; leaves alternate, entire to pinnately cut. Heads radiate, small, in terminal panicles; involucre bell-shaped, of few broad phyllaries in 2 or 3 series; receptacle more or less raised, bearing scales surrounding the disc-florets. Ray-florets few, pistillate, their ligules short, broad, 2-toothed and persistent. Disc-florets perfect but only staminate in function, funnel-shaped and 5-toothed; anthers entire at the base; styles unbranched. Achenes compressed, keeled on the inner face each surrounded by a phyllary and 2 concave receptacular scales, crowned by the persistent ligule and a pappus of 2–3 recurved awns.

— A chiefly Mexican genus of 15 species.

Parthenium hysterophorus L., *Sp. Pl.* 2: 988 (1753). **FIG. 249**.

An annual, diffusely branched herb up to 50 cm tall or more, the stems more or less pubescent; leaves elliptic or obovate in outline, mostly 2–12 cm long, bipinnatifid, pubescent on both sides and densely gland-dotted beneath. Heads stalked, 3–5 mm in diameter, in loose open panicles; phyllaries in 2 series, ovate, clothed with minute, appressed, yellowish hairs. Scales of the receptacle with glandular margins. Florets and ligules white. Achenes black, 2 mm long, hairy at the top; pappus of 2 slender recurved awns.

GRAND CAYMAN: *Brunt* 2020; *Kings* GC 45, GC 302; *Lewis* 3838, 3850; *Proctor* 15290.

— Throughout the American tropics and subtropics, and introduced into the Old World; a weed of roadsides and open waste ground.

[*Thymophylla* Lagasca

Annual or perennial herbs; leaves opposite or alternate, the blades linear-filiform or pinnatisect with linear-filiform lobes and with scattered glands. Inflorescences terminal, the heads solitary, pedunculate, radiate, heterogamous; involucral bracts 8–10, strongly connate for 2/3 of their length, glandular; receptacle naked. Ray florets pistillate, fertile, yellow; disc florets bisexual, yellow. Achenes obconical to cylindrical; pappus of 10 small scales, variously aristate or dissected.

— A genus of about 25 species of southwestern U. S. A., Mexico and C. America.

Thymophylla tenuiloba (DC.) Small, *Fl. S.E. U.S.* 1295 (1903); Howard (1989).

Dyssodia tenuiloba (DC.) B. L. Robinson (1913).

Small diffusely branched perennial to 20 cm tall with puberulous stems. Leaves alternate, elliptic in outline, 1–2 × 1 cm, pinnatisect into 9–12 filiform spinulose-tipped segments bearing oval glands. Peduncles 3–4 cm long; involucres top-shaped, with 12–15 bracts, connate to near the ciliate or erose tips, each with 1–4 glands. Ray-flowers 11–15, the ligules 3–4 mm long, bright yellow; disc-flowers 50–100. Achenes obpyramidal, black, 2.5–3 mm long, striate, sparsely pubescent; pappus of up to 120 small scales, these 3-awned.

GRAND CAYMAN: *Burton* 267, recorded as "adventitious in a plant nursery".

— Native of Texas and Mexico; introduced and more or less naturalised in Florida, the Bahamas, the Lesser Antilles, and also in Africa and Asia.]

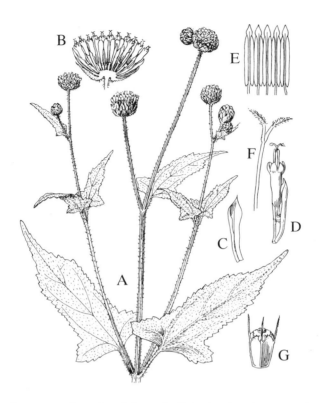

FIG. 250 **Melanthera aspera**. A, portion of plant, × ²/₃. B, head cut lengthwise, × 2. C, receptacular scale, × 4. D, floret with scale, × 4. E, anthers, × 8. F, style-arms, × 8. G, achene, × 4. (F. & R.)

Melanthera Rohr

Perennial herbs or undershrubs; leaves opposite, simple but toothed or sometimes hastate-lobed. Heads solitary or rarely paired on long terminal or axillary peduncles, discoid (in our species) with all the florets perfect and fertile, or radiate, the ray-florets pistillate or neuter, and the disc-florets perfect. Involucre hemispheric, the phyllaries in 2–3 series; receptacle convex, scaly, the scales concave. Disc-florets tubular, 5-lobed; anthers black, obtuse or minutely sagittulate at the base; style-arms with an acute, hairy appendage. Achenes more or less 3-angled; pappus of 2–4 stiff, ciliate, awn-like bristles, these soon falling, absent from the achenes of ray-florets.

— A genus of about 50 species in tropical America and Africa.

Melanthera aspera (Jacq.) L. C. Rich. ex Spreng., *Neue Entdeck.* 3: 40 (1822). **FIG. 250.**

SOFT LEAF (so-called in jest, as the leaves are harsh, not soft)

A stiff bushy or scrambling herb to 1 m tall or more, the stems and leaves of rough, harsh texture; leaves long-petiolate, the blades broadly ovate or triangular-hastate, 3–12 cm long, acuminate at the apex, the margins coarsely crenate- or serrate-toothed, often with a sharp, short lobe on either side at the base. Peduncles up to 8 cm long; heads up to 1 cm in diameter. Achenes 3 mm long; pappus-bristles 1–3 mm long.

LITTLE CAYMAN: *Kings* LC 22; *Proctor* 28050, 28054. CAYMAN BRAC: *Proctor* 29123.

— Florida, Bahamas, Cuba, Jamaica and the Yucatan Peninsula; in sandy thickets and clearings at low elevations, mostly near the sea.

Eclipta L., nom. cons.

Herbs; leaves opposite, simple, entire or toothed. Heads radiate, mostly solitary on axillary peduncles; peduncles 1–several in an axil; involucre campanulate, of few herbaceous phyllaries in 2 series; receptacle flat or slightly raised, scaly. Ray-florets pistillate or sterile, the ligule small and inconspicuous. Disc-florets perfect and fertile, the corolla tubular and 4–5-toothed; anthers obtuse at the base; style-arms with a short, obtuse appendage. Achenes cylindrical or slightly compressed, rugose; pappus absent or of 2 short awns.

— A small but widespread tropical genus of 3 or 4 species.

Eclipta prostrata (L.) L., *Mant. Pl.* 2: 286 (1771).

Eclipta alba of this Flora, ed. 1.

An erect or decumbent herb up to 50 cm tall or more, the stems strigose with basally swollen hairs. Leaves sessile, very narrowly elliptic or narrowly lanceolate, mostly 2–7 cm long, bluntly acuminate at the apex and long-cuneate at the base, the margins entire or minutely toothed, the surfaces bearing strigose hairs. Peduncles 1–6 cm long; heads 6–9 mm across; phyllaries ovate, acute, 3.5–4 mm long; ray-florets with linear ligules ca. 2 mm long. Achenes glabrous, rugose, black, 2 mm long; pappus of 2 very short awns joined by a fimbriate flange.

GRAND CAYMAN: *Brunt* 1843; *Kings* GC 276; *Proctor* 15275. CAYMAN BRAC: *Proctor* 28975.

— A pantropical weed of roadside ditches and low moist ground.

Sphagneticola Hoffm.

Procumbent perennial herbs, puberulous to pubescent, sometimes succulent, the stems often rooting at the nodes. Leaves simple, opposite, sessile or short petioled. Inflorescence terminal (but often appearing axillary through lateral displacement), 1–few-headed. Heads radiate, many-flowered, long pedunculate; involucre weakly 2–(3-)seriate, the bracts foliaceous or the inner ones foliaceous only toward the apex; receptacle scaly. Ray-flowers pistillate, yellow or orange; disc-flowers bisexual, yellow or orange. Anthers black. Style-branches erect or slightly reflexed. Achenes black and tuberculate at maturity; pappus fimbriate, obscured by a corky crown-shaped collar.

— A small genus of 4 species, 3 of them occurring in the lowland neotropics and subtropics, the fourth native to southeastern Asia and adjacent Pacific islands.

Sphagneticola trilobata (L.) Pruski in Acevedo-Rodriguez, *Fl. St. John* 114 (1996). FIG. 251.

Wedelia trilobata of this Flora, ed. 1.

MARIGOLD

Herb with trailing stems rooting at the nodes, the branches and tips ascending, sparsely hairy or nearly glabrous throughout; leaves sessile or nearly so, obovate or oblong-

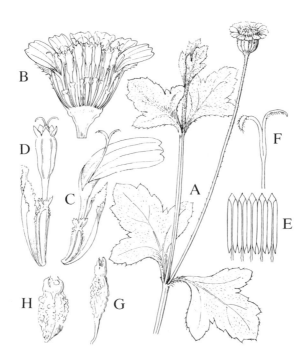

FIG. 251 **Sphagneticola trilobata**. A, branch with leaves and flower-head, × ²/₃. B, head cut lengthwise, × 2. C, ray-floret with scale, × 4. D, disk-floret with scale, × 4. E, stamens × 8. F, style-arms, × 8. G, achene of ray-floret, × 4. H, achene of disk-floret, × 4. (F. & R.)

obovate, 3–12 cm long, often strongly 3-lobed but sometimes merely toothed or subentire. Peduncles up to 5 cm long or more; heads mostly 1.5–2 cm broad; outer phyllaries ca. 12 mm long. Ligules yellow, 7–12 mm long, 3-toothed at the apex. Disc-florets 5 mm long. Achenes rugose, 4–5 mm long; pappus of irregular, short, united scales.

GRAND CAYMAN: *Brunt* 1908; *Hitchcock*; *Kings* GC 315; *Lewis* GC 46; *Maggs* II 50; *Millspaugh* 1245; *Proctor* 15151; *Sachet* 440. LITTLE CAYMAN: *Kings* LC 50.

— Florida, the West Indies, continental tropical America, Africa and Hawaii; common along roadsides, in damp pastures and in sandy thickets and clearings near the sea.

Eleutheranthera Poit. ex. Bosc.

Annual herbs; leaves opposite, simple, subentire or toothed. Heads discoid or rarely with a few minute, sterile ray-florets, solitary on short axillary or terminal peduncles, 1 or 2 peduncles from an axil. Involucre bell-shaped, of few foliaceous scales; receptacle convex, scaly, the scales surrounding the florets. Florets few, perfect, 5-toothed; anthers free, sagittulate at the base; style-arms linear, hairy on the back. Achenes thick, smooth or rough; pappus a small, ciliate cup and sometimes 2 or 3 short awns.

— A genus consisting of 1 pantropical species and 1 in Madagascar.

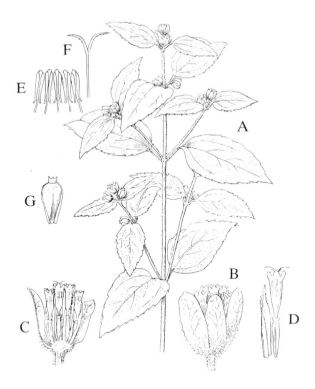

FIG. 252 **Eleutheranthera ruderalis**. A, portion of plant, × ²/₃. B, flowering head, × 4. C, same cut lengthwise, × 4. D, floret with scale, × 6. E, anthers, nearly free, × 16. F, style-arms, × 16. G, ripe achene, × 4. (F. & R.)

Eleutheranthera ruderalis (Sw.) Sch. Bip., *Bot. Zeit.* 24: 165 (1866). **FIG. 252.**

An erect bushy herb to about 50 cm tall, the stems pubescent; leaves narrowly ovate or ovate, mostly 1.5–4 cm long, acute at the apex, the margins crenate-serrate, pubescent on both sides and gland-dotted beneath. Peduncles to 1 cm long or more; phyllaries to 7 mm long; receptacle scales ca. 4 mm long, folded around the florets. Florets few, yellow, 5-toothed. Achenes obovoid, puberulous, 3 mm long.

GRAND CAYMAN: *Kings GC 304*.

— A pantropical weed of moist fields and pastures.

Verbesina L.

Herbs or shrubs, sometimes arborescent; leaves opposite or alternate, usually toothed or lobed. Heads discoid (in our species) with all the florets perfect; or radiate, the ray-florets pistillate or sterile and the disc-florets perfect; solitary or few on long peduncles, or numerous in corymbs or corymbose panicles. Involucre hemispheric or bell-shaped, the scales in 2–6 series or sometimes apparently 1-seriate; receptacle convex or conical, bearing concave or folded scales. Disc-florets tubular and 5-lobed; anthers obtuse or

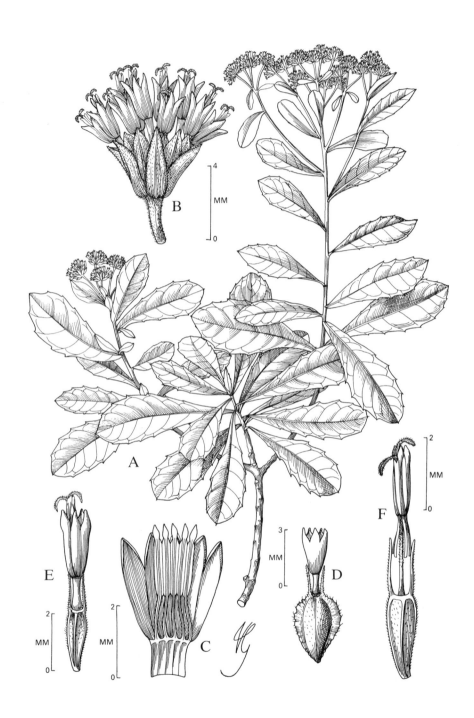

FIG. 253 **Verbesina caymanensis**. A, branch, habit. B, flowering head. C, single floret cut open to show the stamens. D, floret with winged achene. E, floret with unwinged achene. F, same, corolla partly removed. (G.)

minutely sagittulate at the base; style-arms with an acute, hairy appendage. Achenes compressed, with marginate or winged edges; pappus of 2 awns.

— A tropical American genus of about 150 species.

Verbesina caymanensis Proctor, *Sloanea* 1: 4 (1977). **FIG. 253.** PLATE 64.

Shrub to 1 m tall or more, the branchlets striate and sparsely puberulous; leaves alternate, sessile, oblanceolate to narrowly obovate, 3–12 cm long, up to 4 cm broad, obtuse or acute at the apex, the margins broadly and shallowly serrate in the upper part, the serrations sharply gland-tipped; the surface thinly scabridulous on the upper side and minutely hispid or glabrescent beneath. Heads in small terminal corymbose panicles; phyllaries pubescent, acute, 4–5 mm long; receptacle-scales mucronate; florets 4.5–5 mm long. Achenes narrowly flattened-obovoid, 3.5–4.5(–5) mm long, blackish and hispidulous, with a broad, whitish, notched and rugose wing along each side (or in some achenes the wings lacking); awns 1 or 2 (rarely 3), ca. 3 mm long, often unequal, ciliate.

CAYMAN BRAC: *Proctor* 29073, 29361 (type).

— Endemic; confined to limestone cliffs and rocky thickets near Spot Bay. *Verbesina caymanensis* differs from *V. propinqua*, its closest Jamaican counterpart, in various details of the leaves, florets and achenes. These species belong to a subgroup of *Verbesina* that by some authors has been treated as a separate genus *Chaenocephalus*. This subgroup has a remarkable distribution, with 7 species occurring in Jamaica and 6 others found in Ecuador, Peru and Argentina; the phytogeographic significance of such a distribution is not clear.

Spilanthes Jacq.

Annual or sometimes perennial herbs; leaves opposite, entire or toothed. Heads long-stalked, solitary or in loose, open cymes, discoid (in our species) with all the florets perfect and fertile, or radiate, the ray-flowers pistillate and the disc-florets perfect. Involucre short, bell-shaped; phyllaries in 1 or 2 series; receptacle convex, conical or subcolumnar, bearing folded scales. Ray-florets, when present, with short inconspicuous ligules. Disc-florets tubular, with enlarged 4–5-lobed limb; anthers obtuse at the base; style-arms truncate. Achenes compressed, wingless pappus of 1–3 bristle-like awns, or absent.

— A pantropical genus of about 60 species.

Spilanthes urens Jacq., *Enum. Syst. Pl. Carib.* 28 (1760).

Herb with trailing, nearly glabrous stems, often rooting at the nodes, the flowering-branches ascending to 20 or 30 cm (often less); leaves sessile, very narrowly to narrowly elliptic, mostly 2–9 cm long, acutish at the apex, triplinerved, glabrate or puberulous beneath. Heads discoid, solitary on long terminal peduncles, 0.8–1.3 cm in diameter; phyllaries 4–5 mm long; florets numerous, white, 2 mm long. Achenes obovate-oblong, 2.5 mm long, ribbed and thinly pubescent; pappus of 2 incurved, broad-based awns.

GRAND CAYMAN: *Brunt* 1883, 1923; *Correll & Correll* 51051; *Howard & Wagenknecht* 15024; *Kings GC* 321; *Lewis* 3836; *Proctor* 15152; *Sachet* 452. LITTLE CAYMAN: *Proctor* 28069.

— The West Indies (except the Bahamas) and continental tropical America; frequent along moist roadsides, in seasonally flooded pastures and in damp sandy clearings near the sea.

Salmea DC., nom. cons.

Erect or climbing shrubs; leaves opposite and entire, toothed or lobed. Heads discoid, in terminal or axillary cymose panicles or corymbs; involucre bell- or top-shaped; phyllaries in 2–5 series, membranous or dry and papery; receptacle conical or columnar, bearing folded scales. Florets tubular, enlarged above, 5-toothed; anthers shortly sagittulate at the base; style-arms papillose at the top. Achenes flattened, dark, and with or without a pale border, usually ciliate; pappus of 2 subequal awns.

— A genus of 12 species in the West Indies, Mexico and C. America.

Salmea petrobioides Griseb., *Fl. Brit. W. I.* 375 (1861). PLATE 64.

A densely branched glabrous shrub to 1.5 m tall or more; leaves somewhat fleshy, narrowly to broadly obovate, mostly 2–5 cm long, rounded, acutish or shortly apiculate at the apex, the margins entire, the lateral venation obscure. Heads in dense terminal corymbs; involucre glutinous, narrowly bell-shaped, ca. 4 mm long, the phyllaries dry and long-persistent. Florets white, 2.5 mm long. Achenes black, ca. 2 mm long, glabrous.

GRAND CAYMAN: *Brunt* 1995; *Correll & Correll* 51035; *Hitchcock*; *Lewis* 3826; *Millspaugh* 1404; *Proctor* 11991. LITTLE CAYMAN: *Proctor* 35222. CAYMAN BRAC: *Millspaugh* 1231.

— Bahamas and Cuba, in sandy or rocky thickets mostly near the sea.

Synedrella Gaertn., nom. cons.

Annual herbs; leaves usually opposite, serrate or subentire. Heads sessile, radiate; involucre oblong, of few bracts, the outer leaf-like; receptacle small, bearing flat chaffy scales among the florets. Ray-florets pistillate, fertile, each with a short, broad ligule. Disc-florets perfect and fertile, tubular and 4-toothed; anthers minutely sagittulate; style-arms with hairy appendages. Achenes compressed, dimorphic, those of the ray-florets crested-winged, those of the disc-florets with narrow entire wings and 2 awns.

— A tropical American genus of 2 species, one of wide distribution, the other a native of Ecuador.

Synedrella nodiflora (L.) Gaertn., *Fruct. & Sem. Pl.* 2: 456, t.171, f.7 (1791). FIG. 254.

Ucacou nodiflorum (L.) Hitchc. (1893).

An erect herb to 50 cm tall or more, sometimes flowering when very small; stems appressed-pubescent; leaves ovate, 1.5–6 cm long or more, acute at the apex, the margins obscurely serrate, appressed-pubescent on both sides and triplinerved. Heads 1–several crowded together in the leaf-axils, each with 2 green phyllaries 7.5–10 mm long and several scarious inner ones. Ray-florets few, with yellow ligules ca. 2 mm long; disc-florets 6–10, 2 mm long. Achenes 3.5–4 mm long.

GRAND CAYMAN: *Brunt* 1934; *Hitchcock*; *Kings GC* 54, GC 305; *Millspaugh* 1276.

— A common weed throughout tropical America; naturalised in the Old World tropics; frequent along roadsides, in pastures and in open waste places.

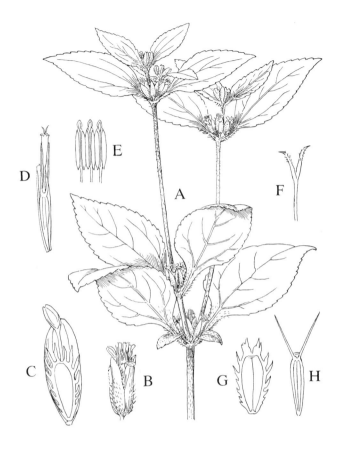

FIG. 254 **Synedrella nodiflora**. A, upper part of plant, × ²/₃. B, head in flower, × 2. C, ray-floret with scale, × 4. D, disk-floret with scale, × 4. E, stamens, × 14. F, style-arms, × 14. G, achene of ray-floret, × 4. H, achene of disk-floret, × 4. (F. & R.)

Bidens L.

Herbs, shrubs, or sometimes vines; leaves mostly alternate, simple and toothed, or variously divided. Heads usually radiate (rarely discoid only), solitary or few at ends of branches, sometimes in panicles; involucre more or less bell-shaped, of rather few phyllaries in 1 or 2 series; receptacle flat or slightly raised, bearing flat scales subtending the florets. Ray-florets in 1 series, usually neuter and sterile. Disc-florets perfect and fertile, tubular and 5-toothed; anthers entire at the base or minutely sagittulate; style-arms hairy and appendaged at the top. Achenes compressed or 4-angled, usually oblong or linear-oblong; pappus of 2–4 (rarely 6) awns usually upwardly or downwardly barbed, rarely smooth or lacking.

— A cosmopolitan genus of about 230 species.

KEY TO SPECIES

1. Leaves bipinnate; outer phyllaries linear-subulate, soon reflexed, not ciliate; achenes mostly 4-awned; rays small, yellow: **B. cynapiifolia**

1. Leaves pinnate or simple; outer phyllaries spatulate, ciliate; achenes mostly 2–3-awned; rays conspicuous, white (sometimes absent): **B. alba var. radiata**

Bidens cynapiifolia Kunth in H. B. K., *Nov. Gen. & Sp.* 4: 235 (1820).

Bidens bipinnata of Hitchcock (1893), not L. (1753).

SPANISH NEEDLE

An erect or somewhat diffuse herb to 1 m tall or more, the spreading branches glabrate and striate-angled; leaves pinnate to bipinnate, the leaflets thin, ovate or lanceolate, the terminal one the largest and up to 5 cm long, caudate-acuminate at the apex, the margins serrate. Peduncles mostly 3–6 cm long; ray-florets 4 or 5. Achenes narrowed at the apex; awns 2–2.5 mm long, retrorsely barbed.

GRAND CAYMAN: *Hitchcock*; *Kings* GC 362; *Proctor* 15074. CAYMAN BRAC: *Proctor* 29348.

— The West Indies and continental tropical America; a common weed of roadsides, fields and open waste ground.

Bidens alba (L.) DC., *Prodr.* 5: 605 (1836) var. *radiata* (Sch. Bip.) Ballard & Melchert in *Phytologia* 32: 291–298 (1975).

Bidens pilosa L. var. *radiata* of most recent authors; Adams (1972).

SPANISH NEEDLE

Erect herb to 1 m tall (usually less), the young stems angled and puberulous. Leaves simple, or pinnate with 1–3 pairs of lateral leaflets; leaflets ovate or lanceolate, the terminal one up to 7 cm long or more (but often much smaller), acuminate at the apex, the margins coarsely serrate. Peduncles mostly 1–4.5 cm long; ray-flowers usually 5 or 6, the ligules 8-nerved; disc-florets light yellow. Achenes black, thinly hispidulous; awns up to 3 mm long, retrorsely barbed.

GRAND CAYMAN: *Hitchcock*; *Kings* GC 73; *Lewis* GC 50; *Millspaugh* 1276; *Proctor* 15073; *Sachet* 492. LITTLE CAYMAN: *Kings* LC 60.

— A pantropical weed of roadsides, fields, and waste places.

[*Cosmos caudatus* Kunth, a herb with dissected leaves and showy pink-rayed heads, is grown in gardens and may occasionally escape. (GRAND CAYMAN: *Kings* GC 418; *Millspaugh* 1353). *Cosmos* differs from *Bidens* in its beaked achenes.]

Flaveria Juss.

Annual herbs with divaricate branching. Leaves opposite, decussate, simple, sessile, with entire or dentate margins, strongly 3-nerved. Inflorescences terminal or axillary, the heads in dense cymes or glomerules, radiate or discoid, heterogamous; involucral bracts 2–5 in

2 or 3 series; receptacle convex, naked. Ray flowers 1 or absent, pistillate, the ligule (if present) yellow, inconspicuous; disc flowers 1–15, bisexual, the corollas tubular below, bell-shaped toward apex, 5-lobed, yellow. Achenes obovoid, 8–10-ribbed, glabrous, black; pappus lacking.

— A Western Hemisphere genus of 21 species, several of them occurring as weeds in open waste ground.

Flaveria trinervia (Spreng.) Mohr, *Contr. U.S. Nat. Herb.* 6: 810 (1901).

Erect (rarely prostrate) annual seldom over 30 cm tall, the stems glabrous. Leaves sessile and usually connate, but sometimes with petioles up to 1.5 cm long; blades lanceolate, oblanceolate, or elliptic, 3–10(–15) cm long and 1–4 cm broad, the apex acute to subobtuse, the base narrowed, the margins serrate or denticulate. Heads greenish-yellow or whitish, 1-flowered, aggregated in glomerules; involucral bracts 3–4.5 mm long, strongly nerved; corollas 2–3 mm long. Achenes subclavate, 2–2.6 mm long.

GRAND CAYMAN: *Proctor 47350*, in open waste ground, George Town; *Joanne Ross s.n.*, in vicinity of Cayman Kai.

— Southern U. S. A., Bahamas, Greater Antilles, Barbados and elsewhere in the neotropics; also widespread in the Old World tropics.

Tridax L.

Perennial herbs with decumbent stems; leaves opposite and toothed or incised. Heads radiate, solitary on long terminal or axillary peduncles; involucre ovoid or hemispheric, its bracts subequal in 2–3 series; receptacle flat or convex, bearing chaffy scales subtending the disc-florets. Ray-florets pistillate, with 3-lobed or -toothed ligules. Disc-florets perfect and fertile, tubular and 5-lobed; anthers auricled or sagittate at the base; style-arms subulate-appendaged. Achenes compressed, silky-hairy; pappus of many plumose bristles or scales.

— A tropical American genus of more than 25 species.

Tridax procumbens L., *Sp. P.* 2: 900 (1753).

Stems decumbent-ascending, hairy; leaves lanceolate to ovate, mostly 2.5–6.5 cm long, acute to acuminate, coarsely serrate or toothed, rough-hairy. Peduncles up to 20 cm long; heads 7–12 mm in diameter, the phyllaries pubescent. Ray-florets 3–6, the ligules cream or pale yellow and 2.5–5 mm long; disc-florets light yellow. Achenes of the ray-florets ca. 2.5 mm long with pappus 3 mm long; achenes of disc-florets 2 mm long with pappus 6 mm long.

GRAND CAYMAN: *Brunt 1896a, 2015*; *Proctor 27942*. CAYMAN BRAC: *Kings CB 64*; *Proctor 15313*.

— Florida, the West Indies and continental tropical America; a weed of roadsides, pastures and open waste places. Introduced in Africa and elsewhere in the Old World tropics.

TRIBE VI. HELENIEAE

Porophyllum Guett.

Glabrous, glandular herbs; leaves opposite and alternate, entire or crenate, often glaucous. Heads discoid, solitary or in open corymbs on terminal peduncles; involucres subcylindrical or narrowly bell-shaped, of few bracts in 1 series; receptacle naked. Florets with slender, elongate tube and a narrowly bell-shaped, 5-lobed limb; anthers obtuse at the base; style-arms linear, hairy at the top. Achenes linear, striate; pappus of rough or barbed bristles in 1 or 2 series.

— A tropical to warm-temperate American genus of about 50 species.

Porophyllum ruderale (Jacq.) Cass., *Dict. Sci. Nat.* 43: 56 (1826).

STINKING BUSH

A bitterly aromatic, erect, glaucous herb up to 70 cm tall or more; leaves long-petiolate, the blades thin and broadly elliptic to rotund, 1–3 cm long or more. Peduncles 2–4 cm long, swollen at the top; phyllaries 5, 17–20 mm long and 2–3.5 mm broad; florets with slender tube 12.5 mm long, green, the somewhat deflexed limb purplish-red. Achenes 8–8.5 mm long, hispidulous; pappus of numerous upwardly barbed bristles ca. 10 mm long.

GRAND CAYMAN: *Brunt* 1627; *Proctor* 15305.

— West Indies and continental tropical America; an occasional weed of sandy roadsides and open waste ground.

Helenium L.

Annual or perennial caulescent herbs, usually with taproots. Leaves alternate, essentially sessile, the lowest ones often pinnately lobed, the upper ones usually not, all beset numerous minute droplets of a resin-like secretion. Peduncles terminal, 1-headed; receptacle usually globose or prolate, naked or with a few short bristles. Involucral bracts ca. 16 in 2 series, lanceolate to sublanceolate, usually pubescent and resinous, usually reflexed at maturity; ray flowers absent or present, if present about 8, pistillate or not, fertile or infertile, the rays yellow, apically 3-lobed, often reflexed; disc usually globose or prolate; disc flowers numerous, bisexual, fertile the corollas usually yellow with 5 moniliform-pubescent triangular lobes. Achenes obpyramidal with 4 or 5 angles; pappus of 5 translucent scales each of which is prolonged into an awn-like tip, or else the entire scale is narrow and awn-like.

— A genus of about 40 species, most of them in western N. America.

Helenium amarum (Raf.) H. Rock, *Rhodora* 39: 151 (1957).

BITTER SNEEZEWEED

Taprooted, bushy-branched annual mostly 10–30 cm tall. Leaves mostly linear but the lowermost pinnately lobed, the latter mostly withered at anthesis. Heads numerous; rays about 8; disc flowers entirely yellow. The entire plant is odoriferous and bitter and imparts an unpleasant taste to the milk of cows that eat it.

LITTLE CAYMAN: *Proctor* 47305, found along gravelly roadside near Grape Tree Point.

— Southeastern U. S. A., west to Texas.]

Pectis L.

Annual or perennial glandular herbs, usually aromatic; leaves opposite, usually linear and often with stiff marginal hairs (cilia) below the middle. Heads radiate, small, solitary or in corymbs, sessile or on axillary or terminal peduncles; involucre cylindrical or bell-shaped, the bracts glandular and few in 1 series; receptacle naked. Ray-florets pistillate and fertile, with entire or 3-toothed ligules. Disc-florets perfect and fertile, tubular with enlarged, regular or irregular, 5-lobed limb; anthers obtuse at the base; style-arms short and obtuse. Achenes linear, striate; pappus of few to many bristles or scales.

— A tropical to temperate American genus of about 70 species.

KEY TO SPECIES

1. Plants erect, not or scarcely aromatic; leaves up to 7 cm long with scattered small glands beneath; pappus of 2–4 spreading, stiff, glabrous spines:	P. linifolia
1. Plants prostrate, forming mats, strongly aromatic; leaves less than 1.5 cm long, the relatively large glands mostly in 2 rows beneath; pappus of erect setulose bristles:	P. caymanensis

Pectis linifolia L., *Syst. Nat.* ed. 10, 2: 1221 (1759).

Erect, diffusely branched, annual, glabrous herb up to 70 cm tall or more; leaves lance-linear, 2–7 cm long, attenuate at both ends, and with 1–2 pairs of basal bristles. Heads 6–9-flowered on filiform, bracteolate peduncles 1–2.5 cm long; phyllaries purplish, 4.5–5.5 mm long. Ligules yellow with dark veins, ca. 1 mm long. Achenes 4.5 mm long, hispidulous toward the top.

CAYMAN BRAC: *Proctor* 29036, 29354.

— West Indies, southwestern U. S. A. to northern S. America, and the Galapagos Islands. The Cayman Brac plants were found scattered in rocky woodlands, not common.

Pectis caymanensis (Urb.) Rydb., *N. Amer. Fl.* 34: 204 (1916). PLATE 64.

Pectis cubensis of Hitchcock (1893), not Griseb. (1866).

P. cubensis var. *caymanensis* Urb. (1907).

TEA BANKER, MINT

Mat-like perennial herb, subwoody at the base and with a woody taproot, the stems often pinkish; leaves oblong-linear or very narrowly lanceolate, 4–12 mm long, minutely scabrid toward the apex and sharply mucronate, and with 4–6 pairs of long cilia near the base. Peduncles mostly 5–10 mm long; ligules yellow, more or less longitudinally nerved. Achenes dark brown, minutely striate.

Occurs in two varieties, which can be distinguished as follows:

1. Stems glabrous, seldom more than 12 cm long; phyllaries ciliolate, ca. 3 mm long; ligules ca. 3 mm long; achenes 2–2.5 mm long, strigose with reddish hairs: var. **caymanensis**
1. Stems sparingly hispidulous in lines, up to 25 cm long or more; phyllaries glabrous, ca. 6 mm long; ligules ca. 5 mm long; achenes 3–3.2 mm long, glabrous or minutely white-strigillose toward the base: var. **robusta**

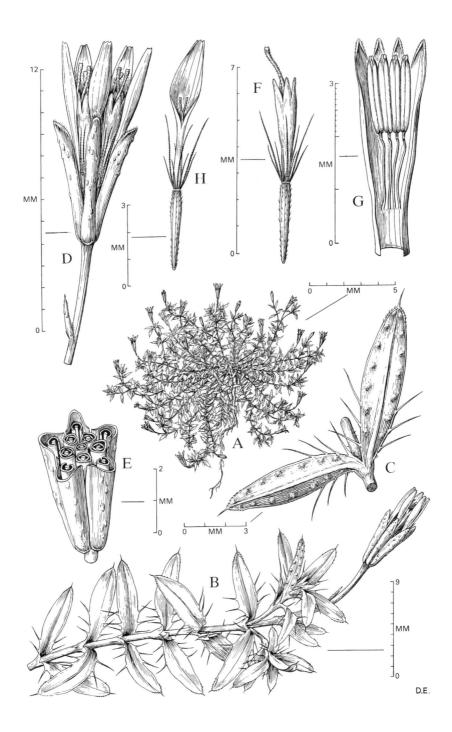

FIG. 255 **Pectis caymanensis var. caymanensis.** A, habit. B, flowering branch. C, pair of leaves. D, head. E, section through head. F, disk-floret with achene. G, corolla of disk-floret cut open to show stamens. H, ray-floret with achene. (D.E.)

Pectis caymanensis var. *caymanensis* FIG. 255.

GRAND CAYMAN: *Brunt* 1761; *Correll & Correll* 51054; *Hitchcock*; *Howard & Wagenknecht* 15026; *Kings* GC 58; *Millspaugh* 1279 (type); *Proctor* 11993. LITTLE CAYMAN: *Proctor* 28186, 35193. CAYMAN BRAC: *Proctor* 29080, 35155.

— Cuba; occurs in sandy clearings or soil-filled pockets of exposed limestone. Frequently used to make a pleasantly aromatic tea.

Pectis caymanensis var. *robusta* Proctor, *Sloanea* 1: 4 (1977).

GRAND CAYMAN: *Proctor* 31023 (type); *Sachet* 386.

— Endemic; found growing in gravelly sand near the sea. This variety is generally larger and coarser in appearance than var. *caymanensis*.

[*Tagetes erecta* L., the African Marigold, and **T. patula** L., the French marigold, are grown in gardens. In spite of their common names, both are natives of Mexico.]

TRIBE VII. SENECIONEAE

Erechtites Raf.

Annual or perennial herbs; leaves alternate, simple and entire to pinnatifid. Heads discoid but heterogamous, the outer florets (of 2 or more series) pistillate, the inner ones perfect, all fertile. Involucre cylindrical, of numerous narrow bracts in 1 series, but often with a few very small accessory ones (bracteoles) outside; receptacle flat, naked. Outer florets filiform. Inner florets tubular and 5-toothed; anthers obtuse at base; style-arms of the perfect florets truncate. Achenes oblong; pappus of numerous long, silky bristles.

— A genus of about 15 species, chiefly occurring in N. America, Australia and New Zealand.

Erechtites hieracifolia (L.) Raf. ex DC., *Prodr.* 6: 294 (1838). **FIG. 256.**

Erect, glabrate or somewhat hairy herb up to 1 m tall; leaves sessile, linear-lanceolate to narrowly oblong, 5–17 cm long, often auricled at the base, and with toothed or coarsely serrate-lobed margins. Heads clustered or in open corymbs; phyllaries 9–12 mm long. Achenes ribbed, slightly hairy, ca. 3 mm long; pappus of white, soft, smooth hairs 10–12 mm long.

GRAND CAYMAN: *Kings* GC 72.

— N. America, Greater Antilles, northern S. America, Hawaii and central Europe; a weed of roadside banks and pastures. The distribution given is that of var. *hieracifolia*; other varieties occur in various regions but have not been found in the Cayman Islands.

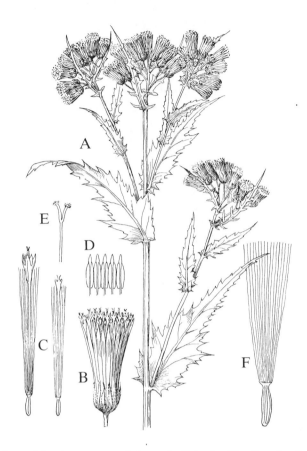

FIG. 256 **Erechtites hieracifolia**. A, upper part of plant, × ²/₃. B, head, × 2¹/₄. C, perfect (left) and pistillate florets, × 4. D, anthers, × 10. E, style-arms, × 10. F, achene with pappus, × 4. (F. & R.)

[*Emilia* Cass.

Annual or perennial herbs; leaves alternate and entire; toothed or pinnatifid. Heads discoid and with all florets perfect, on long peduncles, solitary or few in a lax corymb; involucre cylindrical or narrowly campanulate, the narrow phyllaries in 1 series with no bracteoles outside; receptacle flat, naked, tuberculate. Florets tubular with an elongate, cylindrical, 5-lobed limb; anthers obtuse at the base; style-arms with a short appendage. Achenes oblong, subterete or ribbed; pappus of numerous silky bristles.

— A genus of about 30 species in the tropics of Asia and Africa, two being naturalised in the New World.

Emilia fosbergii Nicolson, *Phytologia* 32: 34 (1975).

Emilia coccinea of many authors, not (Sims) Sweet (1839).
Emilia javanica of Adams (1972), not (Burm. f.) Robins. (1908).

A short-lived annual herb up to 60 cm tall or more (but often much less), nearly glabrous or thinly pubescent. Leaves (except the lowest) sessile and clasping, oblong, oblanceolate or obovate, 4–8 cm long, acute, and with irregularly toothed or serrate margins. Heads few in loose corymbs; involucre 10–14 mm long; florets crimson or scarlet, 11–14.5 mm long, mostly numbering 40–65 in a head. Achenes 3.5–4 mm long, hispidulous on the angles; pappus of soft white hairs ca. 7 mm long.

GRAND CAYMAN: *Brunt* 2117; *Kings* GC 205.

— Now pantropical, a common weed of fields and waste places.]

TRIBE VIII. LACTUCEAE

Launaea Cass.

Erect annual herbs, the stems glabrous or glaucous. Leaves in a basal rosette, and/or cauline, sharply toothed or lobed, the venation pinnate. Inflorescence paniculate, with numerous heads, the peduncles often branched. Heads few–many-flowered, cylindrical, the outer bracts 4 or 5, ovate, shorter than the oblong inner ones; receptacle flat, naked. Corolla yellow, glabrous or pubescent. Achenes prismatic or slightly flattened with truncate apex, ribbed, shiny or rugose, glabrous or puberulous; pappus of numerous white bristles, these monomorphic or dimorphic.

— A widespread tropical genus of about 80 species.

Launeae intybacea (Jacq.) Beauverd, *Bull. Soc. Bot. Genère* ser. 2, 2: 114 (1900).

Erect glabrous herb up to 1.2 m tall, sparingly branched, containing white latex. Basal and lower stem leaves obovate to oblanceolate in outline, 8–20 cm long, 4–7 cm broad, runcinate-dentate or irregularly coarsely dentate, ciliate. Upper leaves few and small. Mostly narrowly lanceolate. Heads numerous, pedunculate; outer involucral bracts ovate, 2–4 mm long with scarious margins; inner ones lanceolate to oblong, 8–12 mm long with short white hairs at apex. Flowers yellow. Achenes 3.5–4 mm long, ribbed and muricate; pappus of soft white hairs 4–5 mm long.

GRAND CAYMAN: *Proctor* 47370, a roadside weed near Queen Elizabeth II Botanic Park.

— A nearly cosmopolitan weed.

[*Youngia* Cass.

Rosette herbs, usually annual, with a spreading tuft of basal leaves and erect stems with scattered alternate leaves. Basal leaves lyrate, pinnatifid and membranous in texture. Inflorescence paniculate or corymbose, the heads small. Involucre cylindrical, its bracts biseriate; receptacle flat, naked. Flowers 8–20, the ligules 4- or 5-dentate. Anthers with linear acute basal auricles. Achenes small, oblong, subterete or angular, attenuate at both ends, ribbed with disc at apex; pappus uniseriate, of numerous fine hairs.

— A genus of about 30 species of temperate and tropical Asia.

Youngia japonica (L.) DC., *Prodr.* 7: 194 (1838).

Annual rosette herb, the basal leaves lanceolate-oblong or narrowly obovate, 3–6 cm long, 2–5 cm broad, runcinate-pinnatifid with a large terminal segment, the margins dentate. Erect stems up to 25 cm tall or more, with rather few reduced cauline leaves with rounded or acute apex and tapering base, glabrous, the texture membranous. Inflorescence scapose, diffusely cymose-paniculate. Involucres 4 mm long with 6 or 7 bracts, those of the outer series ovate, 1.2 mm long, those of the inner series linear-oblong, to 5 mm long, glabrous; heads 20-flowered, yellow. Achenes oblong-ellipsoid, 1.5 mm long, tapered at each end, many ribbed, brown; pappus-bristles white, 4 mm long.

GRAND CAYMAN: *Walls* s.n. (BM), coll. April 1992, det. by C. D. Adams.

— Native of Japan and Eastern Asia, now established as a weed in many tropical countries.]

New record added after completion of Dr Proctor's original manuscript.

ARISTOLOCHIACEAE

Herbaceous or woody vines, rarely herbs or shrubs, often somewhat aromatic. Leaves alternate, petiolate, with or without pseudostipules developed from axillary buds; blades entire or lobed, oblong, ovate, or pandurate, more or less cordate, the venation palmate or pinnate. Flowers axillary, solitary or in clustered racemes, zygomorphic, perfect epigynous; calyx conspicuous, gamosepaloue, variously inflated, often constricted towards base into a neck, expanding at apex into 1- or 2-lobed limb. Corolla absent. Stamens 5 or 6, the anthers sessile and adnate to the styles. Ovary inferior, 5- or 6-locular with ovules axile and numerous; styles 6, fleshy, marginally connate, with more or less capitate stigma. Fruit a dehiscent capsule; seeds numerous; vertically compressed in 5 or 6 rows.

— A mostly tropical or warm-temperate family of seven genera and 410 species. The majority of species belong in the genus *Aristolochia*.

Aristolochia L.

Essentially with characters of the family. A genus of more than 350 species, a single one only recently discovered in the Cayman Islands.

Aristolochia odoratissima L. *Sp. Pl.* (ed. 2) 2: 1362 (1763).

Glabrous annual vine from perennial rhizome; pseudostipules absent; petioles 2–3 cm long; leaf blades hastate or subpandurate, (rarely ovate), 6.5–11 × 3.5–7 cm, deeply cordiform or hastate at base, acuminate at apex, the venation palmate, the undersurface minutely whitish-puberulous. Flowers solitary, the perianth geniculate, yellowish with purple network, the tube to 7 cm long, the limb abruptly spreading, 10–13 × 4–6 cm. Capsules curved-cylindric, 7–10 cm long, c. 1 cm in diameter, 5-ribbed, pointed at both ends, dehiscence basal; seeds flattened, 2 × 3 mm, 1 mm thick.

GRAND CAYMAN: Lower Valley area, *Stafford s.n.*, May 2007 (IJ), leaves and capsules.

— Apparently indigenous, this species was discovered because it is the obligate larval host of a particular small butterfly. The capture of specimens of this butterfly led to the search for the plant host, which was successful, and for which Mrs Ann Stafford is to be congratulated.

ADDENDUM

Encyclia fucata (Lindl.) Britton & Millsp., *Bahama Flora* 91 (1920).

> *Epidendrum fucatum* Lindl. (1838).

Epiphytes or lithophytes with short rhizomes and ovoid to ovoid-fusiform pseudobulbs 2–5.5 cm long, 1–2.3 cm thick. Leaves 1–2 from apex of pseudobulb, conduplicate at the base, rigid, coriaceous, linear-ligulate, obtuse, 15–37 cm long. Inflorescence from apex of pseudobulbs, up to 100 cm long including the scape, diffuse, suberect panicle with branches 8–62 cm long, many flowered. Flowers with sepals and petals greenish-brown, yellow-brown or yellow, lip white with a reddish spot on the mid lobe, resupinate; sepals elliptic-oblanceolate, oblanceolate, acute to obtuse, 10–12.5 mm long; petals oblanceolate-spatulate, obtuse to acute, 10–11.5 mm long; lip trilobed 8.5–9.5 mm long, lateral lobes falcate, oblong, rounded, 6–8 mm long, midlobe elliptic to obovate, obtuse-rounded to emarginated, uncommonly apiculate, 4–4.5 mm long, 2–3 mm wide, narrowing to 1–1.3 mm wide at isthmus below the lateral lobes. Column wingless. Ovary smooth, with pedicel 8–9 mm long; capsules smooth, ellipsoid, 16–28 mm long; pedicels slender, 10–20 mm long.

GRAND CAYMAN: *Sue Gibbs*, photographs. CAYMAN BRAC: *Sue Gibbs*, photographs.
— Cuba and Bahamas, a variable and widespread species. The description above is based on photographs sent by Sue Gibbs and material from Cuba. The discovery of these plants brings into question the reports of *E. kingsii* on these two islands. These are two very closely related species, and with more material of *E. kingsii*, we may be able to determine whether the two are actually distinct.

APPENDICES

APPENDIX I: ADDITIONS TO THE CAYMAN FLORA

*Families new to the flora

116 species in 54 families are new to the *Flora of the Cayman Islands*; 7 families are new to the Flora: Schizaeaceae, Agavaceae, Plumbaginaceae, Apodanthaceae, Oxalidaceae, Araliaceae and Aristolochiaceae.

This integrated list compiled from the following sources:

- Proctor in *Kew Bulletin* 51: 483–505 (1996).
- Guala, Burton, Proctor & Clifford in *Kew Bull.* 57: 235–237 (2002).
- Collections with Joanne Ross, P. Ann van B. Stafford et al., Nov. 2002–Jul. 2003.

*SCHIZAEACEAE
Anemia adiantifolia

POLYPODIACEAE
Macrothelypteris torresiana
Nephrolepis exaltata
Pityrogramma calomelanos
Polypodium aureum
[*Pteris vittata*]
Thelypteris hispidula var. *versicolor*

HYDROCHARITACEAE
Halophila baillonis
Halophila engelmannii

CYMODOCEACEAE
Halodule beaudettei

CYPERACEAE
Cyperus surinamensis
Eleocharis rostellata

GRAMINEAE
Axonopus compressus
Cynodon nlemfuensis
Digitaria ciliaris
Digitaria setigera
Oplismenus hirtellus
Panicum repens
Paspalum arundinaceum
Paspalum notatum
Paspalum setaceum
Pharus glaber
[*Rottboellia cochinchinensis*]
Urochloa decumbens
Urochloa subquadripara
[*Zoysia tenuifolia*]

BROMELIACEAE
Tillandsia festucoides

ARACEAE
Syngonium podophyllum

*AGAVACEAE
Furcraea hexapetala
Agave caymanensis

ORCHIDACEAE
Beloglottis costaricensis
Brassavola nodosa
Cyclopogon cranichoides
Cyclopogon elatus
Cyrtopodium punctatum
Dendrophylax porrectus
Epidendrum nocturnum.
Oeceoclades maculata
Polystachya concreta
Tropidia polystachya

LAURACEAE
Licaria triandra

PIPERACEAE
Peperomia obtusifolia
Peperomia pseudopereskiifolia
Peperomia simplex

NYCTAGINACEAE
Pisonia margaretiae sp. nov.

CACTACEAE
Epiphyllum phyllanthus var. *plattsii*
var. nov.

AMARANTHACEAE
Alternanthera sessilis

POLYGONACEAE
Polygonum hydropiperoides

*PLUMBAGINACEAE
Limonium companyonis

STERCULIACEAE
Neoregnellia cubensis
[*Sterculia apetala*[

MALVACEAE
Hibiscus lavateroides

SALICACEAE
Banara caymanensis
Casearia staffordiae sp. nov.

TURNERACEAE
Turnera triglandulosa

CUCURBITACEAE
Cionosicyos pomiformis
Fevillea cordifolia

CRUCIFERAE
Lepidium virginicum

THEOPHRASTACEAE
Jacquinia proctorii
LEGUMINOSAE — Faboideae
Aeschynomene americana
Aeschynomene sensitiva
Christia vespertilionis
[*Desmodium gangeticum*]
Erythrina velutina
[*Indigofera spicata*]
[*Sesbania sericea*]

LEGUMINOSAE — Caesalpinioideae
Hymenaea courbaril

LYTHRACEAE
Ammannia coccinea

THYMELAEACEAE
Daphnopsis americana

MYRTACEAE
Eugenia biflora
Eugenia foetida
Pimenta dioica
{*Syzygium jambos*}

ONAGRACEAE
Ludwigia affinis

COMBRETACEAE
Terminalia eriostachya var. *margaretiae* var. nov.

VISCACEAE
Phoradendron trinervium

*APODANTHACEAE
Pilostyles globosa var. *caymanensis* var. nov.

CELASTRACEAE
Maytenus jamaicensis

EUPHORBIACEAE
[*Euphorbia graminea*]
Jatropha divaricata
Margaritaria nobilis
Phyllanthus tenellus
Tragia volubilis
Securinega acidoton
RHAMNACEAE
Colubrina arborescens

SAPINDACEAE
Exothea paniculata
Sapindus saponaria

ANACARDIACEAE
Anacardium occidentale
[*Spondias mombin*]

*OXALIDACEAE
Oxalis corniculata

ZYGOPHYLLACCAE
[*Guaiacum officinale*]

ERYTHROXYLACEAE
Erythroxylum confusum

*ARALIACEAE
Dendropanax arboreus

GENTIANACEAE
Voyria parasitica

APOCYNACEAE
Rauvolfia nitida
Thevetia peruviana

SOLANACEAE
Solanum torvum

CONVOLVULACEAE
Ipomoea passifloroides
Merremia quinquefolia

BORAGINACEAE
Cordia dentata
Tournefortia minuta

SCROPHULARIACEAE
Bacopa monnieri

BIGNONIACEAE
Catalpa longissima

ACANTHACEAE
Ruellia nudiflora

GOODENIACEAE
[*Scaevola sericea*]

RUBIACEAE
Faramea occidentalis
Psychotria microdon
Spermacoce verticillata

COMPOSITAE
Ageratum conyzoides
Flaveria trinervia
Helenium amarum
Iva imbricata
Launaea intybacea
[*Thymophylla tenuiloba*]
Youngia japonica

ARISTOLOCHIACEAE
Aristolochia odoratissima

APPENDIX II: NEW TAXA IN THIS VOLUME

Agave caymanensis Proctor	p. 183, 184.
Pisonia margaretiae Proctor	p. 254.
Epiphyllum phyllanthus var. *plattsii* Proctor	p. 256, P14.
Casearia staffordiae Proctor	p. 316, P19, P20.
Terminalia eriostachya var. *margaretiae* Proctor	p. 414, P33.
Pilostyles globosa var. *caymanensis* Proctor	p. 428.
Scolosanthes roulstonii Proctor	p. 620, 621.

GLOSSARY

Abaxial On the side (of an organ) away from the axis, as the lower surface of a leaf.

Abortive Imperfectly developed, not fully developed at maturity, as abortive stamens with filaments only.

Abruptly pinnate Said of a compound leaf without a terminal leaflet.

Acaulescent Stemless; having the stem very short and often underground, thus appearing stemless; opposite of **caulescent**.

Achene A small, dry, indehiscent, 1-seeded fruit with a thin tight pericarp, as the individual fruit of a composite flower or of the Amaranthaceae.

Acicular Needle-shaped.

Acid Characterised by an excess of free hydrogen ions; opposite of **alkaline**.

Acidulous Slightly acid.

Acroscopic Facing or directed toward apex, e.g. of a fern-frond.

Acuminate Tapering to an apex, the sides concave along the taper.

Acute Tapering to a sharp apex, the sides straight along the taper.

Adaxial On the side (of an organ) toward the axis, as the upper surface of a leaf.

Adnate United with another (unlike) part, as the stamens to the corolla.

Adpressed Same as **oppressed**.

Adventitious Said of plants recently introduced to an area where they are not indigenous; organs arising from abnormal positions, as buds near a wound or roots from a stem or leaf.

Aerial Living above the surface of the ground or water.

Aggregate fruit A cluster of ripened ovaries traceable to separate pistils of the same flower and inserted on a common receptacle.

Albumen An obsolete name for endosperm.

Alkaline Having the ability to neutralise acids; having a pH of more than 7; opposite of **acid**.

Alternate Any arrangement of leaves or other parts not opposite or whorled; placed singly at different heights along an axis or stem.

Alternation of generations 1. The growth of reproductive bodies into structures which differ from those from which they are reproduced. 2. The occurrence in one life-cycle of two or more modes of reproduction which are differently produced and which differ morphologically. Usually the sexual forms (gametophytes) and the asexual forms (sporophytes) alternate, the sporophytes containing double the number of chromosomes found in the gametophytic generation.

Anatropous Applied to ovules or seeds which grow in an inverted position; having an ovule bent over in growth so that the micropyle is near the base of the funicle, with the body of the ovule united with the funicle.

Androgynophore Hermaphrodite stipe.

Andromonoecious Refers to species that have bisexual and male flowers on the same plant.

Angiosperm A plant with seeds enclosed in an ovary or pericarp.

Angulate More or less angular (not curved or rounded), having angles or corners usually of a determinate number.

Annual A plant of one year's (or one growing-season's) duration, completing its life-cycle in that period.

Annular In the form of a ring; arranged in a circle.

Annulus (plural: **annuli**) A rim of thickened cells around a fern sporangium, functioning as an elastic mechanism in shedding spores.

Anther The pollen-producing part of a stamen.

Antherozoid One of the motile male cells produced in an antheridium; the sperm cell of a fern.

Anthesis The process or time of opening of a flower-bud and the expansion of the flower-parts.

Anthocarp A fruit formed by the fusion of part or all of the floral parts with the fruit itself.

Apex (plural: **apices**) The tip of an organ, the extreme end or point farthest from the point of attachment; the growing point of a stem or root.

Apical Pertaining to the apex.

Apiculate Terminating in a short, sharp, somewhat flexible point, but not a spine.

Apogamous Developed without fertilisation; development of an embryo without the fusion of gametes.

Appendage An attached subsidiary or secondary part, as hairs, prickles or leaves of a stem.

Appressed Flattened against underlying or adjacent tissues; pressed down or against.

Approximate Close together but not united, as leaves along a stem; opposite of distant.

Aquatic Living in water; growing naturally in water or under water.

Arboreous Tree-like or pertaining to trees.

Arborescent Attaining the size or character of a tree.

Archegone The egg cell of a fern, produced in an archegonium.

Arcuate Curved; bent in an arc.

Areolate Marked out into small spaces enclosed by anastomosing veins; reticulate.

Areole A space surrounded by anastomosing veins; a pit, spot or small raised area bearing hairs, glochids or spines (or all three), as in many cacti.

Arid Dry, having little rainfall.

Aril An extra covering of part or all of a seed, which is an outgrowth from the hilum, and which may be more or less soft and fleshy, or else dry and bony.

Arillate Bearing an aril.

Aristate Bearing a terminal bristle or awn.

Articulate Jointed.

Ascending Directed obliquely upward, but not truly erect.

Asymmetric Not equally bilateral.

Attenuate Tapering gradually and narrowly.

Auricle An ear-shaped appendage or part, as ear-like lobes at the base of leaves or small lobes at the summit of the sheath in many grasses.

Auriculate Having an ear-like lobe or appendage.

Awl-shaped Narrow and tapering to a point; sharp-pointed from a broader base; **subulate**.

Awn A bristle-like appendage, especially on the glumes of grasses.

Axil The angle between a leaf-petiole and the stem on the upper (acroscopic) side; any angle between an organ and its axis.

Axile placentation A condition in which the ovules are borne on a central axis in the ovary.

Axillary Pertaining to an axil, or located in or attached in an axil.

Axis (plural: **axes**) The main or central line of development of a plant or part of a plant; the main stem.

Barbed Beset with rigid points or short bristles, these usually reflexed; pertaining to a hair or other straight process which is armed with one or more teeth that point backward.

Basiscopic Facing or directed toward the base, e.g. of a fern-frond.

Bearded Bearing or furnished with long or stiff hairs.

Berry Any simple fruit having a pulpy or fleshy pericarp, usually with several or many seeds embedded in the pulp.

Bi- In combination, 2, twice, or doubled, as biauriculate (having two ear-like appendages), bicrenate (having two rounded teeth), or bilobed (having 2 lobes).

Biennial A plant which completes its life-cycle in two growing seasons, usually fruiting during the second season.

Bifid Forked at least halfway to the base.

Bifurcate Forked at the tip.

Bilamellate Having two flat lobes or divisions.

Bilateral Having two sides which are equal.

Binomial nomenclature The system of applying two Latin names, a generic name and a specific name, to designate kinds of plants or animals.

Bipinnate Twice pinnately compound; doubly pinnate.

Bipinnatifid Twice pinnatifid, with the divisions of a pinnatifid leaf themselves pinnatifid.

Bipinnatiform Pertaining to a leaf-like structure having bipinnatifid architecture.

Biseriate In two rows or series.

Bisexual Having both stamens and pistils in the same flower or inflorescence.

Bitemate Pertaining to a leaf with three main divisions each having three leaflets.

Bract A reduced, scale-like leaf subtending a flower or branch of an inflorescence; a leaf or scale in whose axil an inflorescence, flower, or floral organ is produced. Sometimes applied to any leaf subtending a flower.

Bracteole A small bract or bractlet; a secondary bract, i.e. borne on a petiole, pedicel, or other secondary axis.

Bristle A rigid hair.

Bud An unopened flower; an undeveloped shoot or stem; an undeveloped axis covered with the rudiments of leaves.

Bulb A large bud, usually subterranean, made up of a short, thick stem with roots at the base and having a number of membranous or fleshy overlapping scale-like leaves.

Bulbil A deciduous bud usually formed on an aerial part of a plant, as on old inflorescences of Agave species. The term may also be applied to small subterranean offsets of bulbs.

Bulblet A little bulb, esp. the small deciduous buds formed in the axils of leaves, on the fronds of certain ferns, and in the inflorescence of a few grasses.

Bulbous Bulb-shaped.

Bullate Blistered or puckered; pertaining to a leaf with inflated convexities on the upper surface and corresponding cavities beneath; swollen or bubble-like.

Bur A fruit, seed, or head bearing hooked or barbed appendages.

Buttress A knee-like or plank-like outgrowth developing from the trunk-base of some kinds of trees.

Caespitose Growing or aggregated in tufts or small clumps.

Callose Hard and thick in texture.

Callosity A thickened or raised area; an area of tissue firmer than the surrounding tissue.

Callus A hard protuberance; in grasses, the swelling at the point of insertion of the lemma or palea; a roll of new tissue developed around a wound.

Calyculus A group of small bracts that resembles a calyx.

Calyptra A lid or hood

Calyx (plural: **calyces**) The outer whorl of the perianth or floral envelope, usually green in colour and composed of sepals; the sepals considered collectively.

Calyx lobe One of the free, projecting parts of a united calyx.

Calyx tube The tube or cup of a calyx in which the sepals are united.

Cambium The thin layer of formative tissue beneath the bark of gymnospennous and dicotyledonous trees and shrubs, from which new wood and bark originate; a sheath of generative tissue usually located between the xylem and the phloem; the tissue from which secondary growth arises in stems and roots.

Campanulate Bell-shaped.

Canescent Covered with greyish downy pubescence.

Canaliculate Having one or more longitudinal grooves or channels

Capitate Borne in heads; head-shaped.

Capitulate Borne in very small heads.

Capitulum A dense inflorescence consisting of sessile flowers.

Capsule A dry, dehiscent fruit resulting from the maturation of a compound (multi-carpellate) ovary.

Carpel A simple pistil or one unit of a compound pistil. A carpel usually consists of three parts: the ovary (a swollen basal portion containing the ovules, which after fertilisation develop into seeds), the style (a usually narrow prolongation of the apex of the ovary), and the stigma (the specialised tip of the style on which pollen lodges and germinates).

Carpellate Pertaining to, or bearing carpels.

Cartilaginous Tough and firm but not bony, like gristle.

Caruncle A protuberance or appendage at or near the hilum of a seed.

Caryopsis The achene of a grass, differing from the achene of Compositae in being derived from a superior ovary.

Cataphyll Any rudimentary scale-like leaf which precedes the foliage-leaves, or (in Phyllanthus) subtends a deciduous branchlet; a bud-scale or rhizome-scale.

Catkin A pendulous, scaly spike with many simple, usually unisexual flowers; it usually falls as a unit after the pollen or seeds have been shed.

Caudate Bearing a tail-like appendage.

Caudex (plural: **caudices**) The stem or trunk of a tree-fern; the perennial base of an otherwise herbaceous, short-lived plant; the main axis of any woody plant.

Caulescent Having an obvious main stem above the ground.

Cauliflorous Bearing flowers on old stems; flowering on the trunk of a woody plant, or on specialised spurs from it, or on the larger branches.

Cauline Pertaining to a stem or axis (as contrasted with **basal**).

Cell A microscopic unit of protoplasm, surrounded by a membrane, and in plants usually by a more or less rigid cell wall, the fundamental unit of structure and function of all living organisms. The term is also applied to the pollen-bearing cavity of an anther and to the ovule-bearing cavity of an ovary.

Cellular Composed of cells; containing small, enclosed spaces of similar shape and size.

Chartaceous Papery or firmly tissue-like in texture; having the texture of writing-paper.

Chasmogamous Of or relating to a flower that opens to allow for pollination (the opposite of cleistogamous).

Chlorophyll The green colouring matter in the cells of plants, which enables them to synthesize carbohydrates from carbon dioxide and water. It is a mixture of two green and two yellow pigments (chlorophyll a, chlorophyll b, carotin and xanthophyll).

Chromosome One of the minute, rod-like bodies, usually definite in number in the cells of a given species, into which the chromatin (deeply staining material) of a cell nucleus becomes condensed during the process of cell division. The chromosomes carry the genes (units of heredity) in linear sequence, and by splitting lengthwise provide the mechanical basis for equal allocation of hereditary material to the daughter cells.

Cilia Minute protoplasmic threads acting as organs of motion which propel gametes and many unicellular organisms from place to place. The term is also applied to marginal hairs on leaves and other parts of plants.

Ciliate Fringed with hairs on the margins.

Ciliolate Fringed with minute hairs; minutely ciliate.

Cinereous Ash-coloured; light grey.

Circumscissile Opening or dehiscing by a transverse line around a fruit or an anther, the apical part usually being shed like a lid.

Clathrate Like a lattice; net-like.

Clavate Club-shaped, slender at the base and gradually thickening upward.

Claw The slender, tapered base or stalk-like part of a petal or sepal.

Cleft Divided into lobes separated by narrow or acute sinuses which extend more than halfway to the midrib.

Cleistogamous Small flowers self-fertilising without opening.

Clinandrium (plural: **clinandria**) In orchids, a cavity in the column between the anther sacs which often contains the stigmatic surface.

Clonal Pertaining to a clone.

Clone The vegetatively produced progeny of a single individual.

Coalescent United by growth, the fusing or organic cohesion of similar parts or organs.

Coccus (plural: **cocci**) A 1-seeded carpel into which compound fruits split when ripe. The term is also used for a bacterium having spherical or nearly spherical form, and for the spore mother cell of certain hepatics.

Coherent Having two or more similar parts touching but not fused.

Colony A collection of organisms of the same kind growing together in close association.

Columella A small column or central axis; the persistent sterile central axis in some fruits, around which the carpels are arranged.

Column The structure formed by the fusion of stamen, style and stigma in the flowers of orchids; the lower undivided part of the awns of certain species of Aristida; the axis or central pillar of a capsule; the structure formed by the fusion of stamen filaments in the flowers of Malvaceae.

Columnar Shaped like a pillar.

Compound Consisting of 2 or more similar parts, as a leaf divided into leaflets, or a fruit or pistil made up to several carpels.

Concolorous Uniform in colour, having 1 colour only.

Conduplicate Folded together lengthwise, as certain leaves or certain petals in a bud.

Cone An inflorescence or fruit covered with overlapping scales.

Confluent Blended together, as in ferns, those sori which spread so that they join those adjacent to them, or in flowers, those with anther-lobes which are united at the summit of the filament and diverge from that point.

Congeneric Belonging to the same genus.

Congeners Two or more species belonging to the same genus.

Conifers A general term used for cone-bearing trees; a tree or shrub belonging to the Coniferae.

Conjugate Coupled, as a pinnate leaf with two leaflets; joined or arranged in pairs.

Connate United with like structures during the process of formation; joined together, as confluent filaments or opposite leaves which are united at the base.

Connective The part of a filament which serves to connect the anthers; the stalk connecting the separated lobes of an anther.

Connivent Coming together or converging, but not organically connected.

Conspecific Belonging to the same species.

Contorted Twisted together; convoluted.

Convoluted Same as contorted.

Cordate Heart-shaped, with the notch at the base.

Coriaceous Of leathery texture.

Corm A solid, fleshy, underground base of a stem, usually bulb-like in shape, covered with thin membranes.

Corolla A collective term for the petals of a flower, whether separate or fused. The corolla usually more or less surrounds the stamens and pistil, and is subtended or encircled by the calyx.

Corona A more or less petaloid structure which develops between the corolla and the stamens, as in Asclepiadaceae.

Corpuscle A small body connecting the pollen masses in Asclepiadaceae, by means of which the pollen becomes attached to insects to accomplish cross-pollination. The term is also used for any minute body within a cell that has definite form and function.

Corticate Having a cortex or a similar specialised outer layer.

Cortex Rind or bark.

Corymb A more or less flat-topped indeterminate flower-cluster in which the outer flowers have the longest pedicels and open first.

Corymbose With flowers arranged in corymbs.

Costa A ridge or midrib of a frond or leaf.

Costules The midribs of the subdivisions (segments or pinnules) of a fern frond, or of other compound leaves.

Cotyledon A seed leaf; a leaf-like organ, developed within the seed and in which nourishment for the young plant is usually stored. The number of cotyledons produced may be 1 (monocotyledonous plants), 2 (dicotyledonous plants), or several (many gymnosperms).

Crateriform Cup-shaped; crater-like.

Crenate Pertaining to leaf-margins with broad rounded teeth, these separated by narrow sinuses.

Crenulate Minutely crenate.

Cristate Bearing an elevated appendage resembling a crest; bearing elevated and toothed ridges. The term is often applied to leaves whose apices are abnormally forked, expanded, and twisted.

Crownshaft The smooth green sheath, formed by the connate expanded petioles of certain species of palm (as in *Roystonea*), which encloses the terminal bud. Palms with a crownshaft produce their inflorescences below the leaves.

Crustaceans Crust-like, or forming a brittle crust.

Cryptogams Plants which reproduce by methods other than by flowers and seeds.

Culm The stem of a grass or sedge, in grasses usually hollow except at the nodes. The term is not applied to the rhizome or other underground stems.

Cuneate Wedge-shaped, with the narrow point at the base of the leaf-blade, petal, or other structure.

Cupule A cup-shaped involucre of bracts adherent at least by their bases.

Cuspidate Tapering abruptly to a sharp point.

Cyathium (plural: **cyathia**) An inflorescence reduced to look like a single flower; in the Euphorbiaceae, a cup-shaped involucre containing individually free male (staminate) flowers and usually a single female flower (gynoecium).

Cyme A flat-topped, usually few-flowered inflorescence in which the central terminal flowers open in advance of the outer ones. A **helicoid cyme** has the lateral branches always on the same side. A **scorpioid cyme** has the lateral branches occurring alternately on opposite sides.

Cymose Bearing or flowering in cymes.

Cymule A reduced or very small cyme; a cymose cluster.

Cystolith A small concretion of calcium carbonate (lime), usually acicular or elliptic in shape, occurring in the epidermal cells of certain plants, and usually visible as a minute raised whitish line.

Cytological Pertaining to **cytology**, the study of the internal structure, function, and life-history of cells.

Deciduous Falling off at certain seasons or stages of growth, such as leaves, petals, sepals, flowers, etc. This term is commonly applied to trees which are not evergreen.

Declinate Bent downward or forward, as the stamens in many flowers; directed downward from the base.

Decompound Pertaining to a leaf that is twice or more compound; repeatedly divided.

Decumbent Reclining at the base, but curved or bent upward toward the apex, said of stems lying on the ground but rising at the tip.

Decurrent Said of a leaf whose margins extend downward as ridges or narrow wings along the stem; any organ extending along the side of another.

Decussate A term applied to leaves or branches arranged in pairs, with each successive pair at right angles to the next pair, or at least growing out at a different angle.

Deflexed Bent sharply downward or outward.

Defoliated Pertaining to a node, stem, or branch from which the leaves have fallen.

Dehiscent Pertaining to an anther or fruit which opens naturally by valves, slits or pores.

Deltate Triangular; term used in preference to 'deltoid' for flat structures such as leaves. (See also **ovate** versus **ovoid**, etc.).

Dentate Having a toothed margin, the teeth usually rather coarse and pointing outward.

Denticulate Finely toothed, or minutely dentate.

Depauperate Reduced, undeveloped, impoverished or dwarfed.

Dibrachiate Branched twice with widely spreading arms.

Dichasium (plural: **dichasia**) A cyme with 2 lateral axes; a determinate inflorescence in which the first flower to open lies between 2 lateral flowers, the pedicels of the lateral flowers often elongating to form a false dichotomy, the central flower remaining sessile.

Dichasial branching Pertains to a compound dichasium in which the lateral flowers are replaced by branchlets that either terminate in secondary dichasia, or else in turn (rarely) give rise to tertiary dichasia.

Dichotomous Successively branched into more or less equal pairs; branching by repeatedly forking in pairs.

Dicot An abbreviation for 'dicotyledonous plant'.

Dicotyledon A plant with two seed-leaves or cotyledons.

Diffuse Loosely branching or spreading; widely or loosely spreading.

Digitate Having parts which diverge from the same point like the fingers of a hand; palmately divided.

Dimerous Having each whorl of two parts.

Dimidiate Unequally divided or lop-sided, as a leaf or leaflet with the principal vein ('mid-vein') much nearer to one margin than the other.

Dimorphic Occurring in two forms.

Dioecious Having staminate and pistillate flowers on different plants; the opposite of **monoecious**.

Diploid Having the number of chromosomes normally occurring in the somatic or vegetative cells of a given species, which is twice the number found in the gametes or the haploid generation of the same species. In many groups of plants the basic diploid chromosome number may be multiplied through hybridisation or other factors, producing **polyploid** plants of various degrees (**triploids, tetraploids,** etc.).

Disc (disk) 1. Any flat, round growth. 2. A more or less fleshy torus developing from the receptacle within the calyx, or within the points of attachment of the corolla and stamens, and surrounding the base of the pistil; this structure is sometimes interpreted as representing coalesced nectaries or staminodes. 3. The flattened or conical receptacle in the heads of Compositae. 4. The flattened, clinging tip of some tendrils. 5. The base of a pollinium. 6. The expanded base of the style in Umbelliferae. 7. In a bulb, the solid base of the stem around which the scales are arranged.

Disc-floret One of the tubular flowers making up the central part of the heads of Compositae, as distinguished from the **ray-florets** which occur around the margin of the heads in many species.

Disciform (discoid) Shaped like a disc; round and flat.

Disk Alternative spelling for **disc**.

Dispersal The act of dispersing or scattering.

Dissected Deeply divided, or cut into many segments.

Distal Remote from the point of attachment; farthest from the axis; toward the apex.

Distant Widely scattered, said of leaves not closely attached or aggregated; remote; the opposite of **approximate**.

Distichous In two vertical ranks along an axis, producing leaves or flowers in two opposite rows.

Diurnal Functioning during the day; said of flowers that open by day and close at night; the opposite of **nocturnal**.

Divergent Said of two organs or structures that incline away from each other from the same point of attachment.

Domatium (plural: **domatia**) A small cavity or other form of structure on a plant, in which live minute insects or mites, apparently in symbiosis with the plant.

Dormant Said of organisms or parts of organisms that are not actively functioning, as for example buds which do not develop unless stimulated to growth by special conditions.

Dorsal Pertaining to the outer or back surface of an organ; the surface away from the axis. See also **abaxial**.

Dorsiventral With a distinct upper and lower surface.

Drupe A fleshy, single-seeded, usually indehiscent fruit; a stone-fruit.

Drupelet A little drupe.

Echinate Bearing or covered with spines or bristles; prickly.

Ecological Pertaining to the relation of organisms to their environment.

Ecologist A student of ecology.

Ecology The study of organisms in relation to their environment.

Egg-cell The female gamete.

Ellipsoid A solid or 3-dimensional body which is elliptic in section.

Elliptic A figure about twice as long as wide, widest in the middle, and rounded at both ends.

Emarginate With a shallow notch at the apex.

Embryo The rudimentary plant within a seed.

Endemic Occurring in one limited locality or region only; confined to a particular area and found nowhere else.

Endocarp The inner layer of a pericarp or fruit-wall.

Endoparasite A parasite that lives within another organism.

Endosperm A multicellular, usually starchy or oily, nutritive tissue formed inside the seeds of many flowering plants, separate from the embryo; in gymnosperms, the term is applied to the prothallus of the female gametophyte.

Entire Pertaining to leaf-margins without teeth or marginal divisions; having a continuous, even margin.

Epicalyx A secondary calyx, as a group of bracts forming a calyx-like structure beneath the true calyx; an involucre formed of bracts.

Epicarp The outer layer of a pericarp or fruit-wall; sometimes called exocarp.

Epidermis The outermost layer of cells of a plant leaf or stem.

Epigynous Having the calyx, corolla, and stamens growing from the top of an inferior ovary.

Epipetric Growing on rocks, **epipetrous**.

Epipetrous Alternative spelling of **epipetric**.

Epiphytic Growing on other plants, but not parasitically.

Equitant Folded over as if astride; said of leaves that are folded together lengthwise in two ranks.

Erose With an irregular, uneven margin; gnawed or jagged.

Evolution The natural process through which organisms have acquired their characteristic structure and functions; the orderly development of species from pre-existing ones through differential natural selection of random variations.

Excentric One-sided.

Excurrent Directed outward. The term is also used to denote a leaf-vein that projects beyond the margin, or an axis that remains central and undivided, the other parts being regularly disposed around it.

Exine The integral wall of a spore, which may or may not be surrounded by a **perispore**.

Exocarp The outer layer of the pericarp or fruit-coat.

Exserted Projecting beyond, as stamens from a perianth; opposite of **included**.

Extrorse Facing or opening outward, away from the axis.

Falcate Sickle-shaped; curved like a sickle.

Farinaceous 1. Having a surface with a mealy coating. 2. Containing starch.

Farinose Covered with a mealy powder.

Fascicle A close cluster or bundle of flowers, leaves, stems, or roots.

Fasciculate In close clusters or bundles; arranged in fascicles.

Fermented Referring to organic substances transformed through the action of enzymes (catalysts produced by living cells), as the production of alcohol from sugar by yeasts.

Ferruginous Rust-coloured, reddish-brown.

Fertile Said of spore-bearing fern-fronds, pollen-bearing stamens, and seed-bearing fruits; opposite of **sterile**.

Fertilisation The union of male and female reproductive cells.

Fiber (fibre) A thread, or thread-like structure; elongate, thick-walled cells many times longer than wide.

Fibrillose Covered with minute fibers; having a lined appearance as if composed of fine fibers.

Fibrous Composed or covered with tough, string-like fibers.

Filament The stalk of a stamen.

Filament-tube A tubular structure formed from coalesced filaments.

Filamentous Formed of fine fibers.

Filiform Thread-like; hair-like.

Fimbriate Having a fringed margin, the fringe composed of processes hair like at the apex but more than one cell wide at the base.

Fimbrilliferous Bordered by very small hair-like processes.

Flabellate Fan-shaped; narrow at the base and much broader at the apex.

Flaccid Limp; lacking rigidity.

Flexuous Bent alternately in different directions; zigzag.

Floccose Bearing or covered with tufts of woolly hairs.

Flora 1. An aggregate term referring to all the plants occurring in a country or particular area. 2. A catalogue or descriptive account of the plants growing in a country or particular area.

Floral Pertaining to flowers or the parts of flowers.

Floret An individual flower of diminutive size, as in the grasses and Compositae.

Floristic Pertaining to a flora.

Flower A structure bearing stamens or pistils, or both, and usually having a calyx and corolla; the structure concerned with the production of seeds in the angiosperms.

Fluted Marked by alternating ridges and groove-like depressions.

Foetid Having an offensive smell.

Foliaceous Leaf-like; bearing leaves.

Foliage The leaves of a plant referred to collectively.

Foliolate Having leaflets; usually used as a suffix, with such prefixes as bi-, tri-, etc., or 2-, 3-, etc.

Follicle A dry, unilocular, capsular fruit that dehisces longitudinally by a suture (slit) on one side only, as in fruits of Asclepiadaceae.

Foveolate Marked with a regular uniform pattern like a honeycomb, or having numerous shallow pits in such a pattern; also said of a membrane having small perforations in such a pattern.

Free central placenta A placenta running through the centre of a unilocular ovary and attached at the ends only.

Frond The leaf of a fern, which differs from a typical leaf in bearing reproductive organs (sporangia) on its surface or margins.

Fruit The structure which develops from the ovary of an angiosperm after fertilisation, with or without additional structures formed from other parts of the flower.

Fugacious Withering or dropping off early.

Fungus A thallophyte lacking chlorophyll.

Funicle The stalk on which an ovule is borne.

Fusiform Spindle-shaped; swollen in the middle and narrowing gradually toward each end.

Galea A helmet-shaped petal, corolla-lip, calyx or sepal.

Gamete A reproductive cell of either sex prior to fertilisation.

Gametophyte The phase in the life-cycle of plants which bears the sex organs and gives rise to the gametes.

Gamopetalous Having the corolla-parts (petals) fused by their edges, at least at the base, to form a tube.

Gamosepalous Having a calyx composed of connate sepals.

Gelatinous Jelly like.

Genera Plural of genus.

Generation The individuals of a species equally remote from a common ancestor.

Generic Pertaining to genera.

Genetic Pertaining to the inheritance of characters by means of genes.

Geniculate Bent abruptly like a knee.

Genus (plural: genera) In classification, the principal subdivision of a family, a more or less closely related and definable group of plants comprising one or more species. The generic name is the first word of a binomial used to designate a particular kind of plant or animal. Large genera are frequently divided for convenience into subgenera, but in such cases the same generic name is still used for all the species.

Germination The beginning of growth from a spore or seed.

Gibbous Swollen on one side, often at the base.

Glabrate Nearly glabrous, bearing only a few scattered hairs.

Glabrescent Becoming glabrous with age.

Glabrous Lacking hairs, bristles, or scales.

Gland A small structure, prominence, pit, or appendage which usually secretes such substances as mucilage, oil, or resin. Similar non-secretory structures are often also called glands.

Glandular Pertaining to or having glands.

Glaucous Covered with an extremely fine whitish or bluish substance that is easily rubbed off.

Globose Spherical; globular.

Glochid A small barbed hair or bristle.

Glochidiate Beset with glochids; bearing glochids.

Glomerate Densely clustered; in a dense, compact cluster or head.

Glomerule A head-like cyme; a cyme of almost sessile (usually small) flowers condensed to form a head or capitate cluster.

Glume A chaff-like bract; specifically, one of the two empty bracts at the base of a grass spikelet.

Glutinous Covered with a sticky exudation.

Gonophore A stalk elevating the stamens and pistils; see also **gynophore**.

Granulate Finely roughened as if with grains of sand.

Granulose Same as granulate.

Gregarious Pertaining to a species whose individuals tend to occur together in groups.

Gymnosperm A plant whose seeds are not enclosed in an ovary or pericarp.

Gynobasic style A style which arises from the base of the ovary.

Gynoecium (plural: **gynoecia**) The pistils collectively of a flower.

Gynophore A stalk or prolonged axis bearing the ovary at its apex.

Gynostegium The staminal crown in an *Asclepias* flower.

Habitat The kind of locality in which a plant grows.

Halophyte A plant that grows in a saline habitat or that tolerates salt.

Halophytic Growing in saline soil or in salt water.

Hastate Shaped like an arrowhead, but with the basal lobes diverging nearly at right angles to the midline of the axis; halberd-shaped.

Hastula The terminal (apical) part of the petiole of a palm leaf; also called a **ligule**.

Haustorium The food-absorbing sucker of a parasitic plant.

Head A dense cluster of sessile or nearly sessile flowers terminating an axis (**peduncle**); a shortened spike reduced to a globular form; a **capitulum**.

Helicoid cyme A scorpioid inflorescence produced by the suppression of successive axes on the same side causing the sympodium to be spirally twisted.

Herb Any seed plant without a woody stem.

Herbaceous Not woody; dying at the end of a growing season.

Herbarium A collection of pressed, dried, identified plant specimens systematically arranged.

Heterogamous 1. Bearing two or more kinds of flowers (staminate, pistillate, hermaphroditic, and/or neutral) in a single inflorescence, as in the heads of Compositae. 2. Producing two kinds of gametes.

Heterosporous Producing two kinds of spores, **microspores** and **megaspores**.

Heterostylous Having styles of two or more distinct forms, or of different lengths.

Hilum The scar on a seed where the funicle was attached.

Hippocrepiform Horseshoe-shaped.

Hirsute Bearing long, rather coarse hairs.

Hispid Having bristly or stiff hairs.

Hispidulous Minutely hispid.

Holotype The particular single specimen cited by an author as the basis of a new species or other taxon. The identity of a holotype precisely fixes the application of a particular name.

Homogamous 1. Having all the flowers of an inflorescence alike, whether all staminate, pistillate, hermaphroditic or neuter. 2. Having the anthers and stigmas mature at the same time.

Homosporous Producing spores that are all alike.

Humus Decaying organic matter in the soil; the mold or soil formed by the decomposition of vegetable matter.

Hyaline Translucent or transparent.

Hybrid The offspring of two different varieties, species, or genera; a cross between two different kinds of related plants.

Hygroscopic Readily absorbing moisture from the atmosphere; altering form, size or position through changes in humidity.

Hypanthium The tube-like extension of a receptacle on which the calyx, corolla, and stamens are borne; a cup-like or shortly tubular structure formed by the fusion of the calyx-tube with the ovary wall, bearing the calyx-lobes, corolla and stamens at the apex.

Hypocarp A fleshy enlargement of the receptacle, or for the stem, below the proper fruit, as in the cashew.

Hypogynium The swollen or perianth like structure subtending the ovary in *Scleria* and some other Cyperaceae.

Hypogynous Having the calyx, corolla, and stamens at the base or below the free superior ovary.

Imbricate Overlapping like shingles on a roof.

Incumbent 1. Said of an anther attached to the inner face of its filament. 2. Said of a cotyledon with its back lying against the radicle.

Indehiscent Said of a fruit that remains closed after it is ripe; not opening by any regular process; opposite of **dehiscent**.

Indeterminate 1. The growth of a stem, branch or shoot not limited or stopped by the development of a terminal bud. 2. the indefinite branching of a floral axis because of the absence of a terminal bud. 3. An inflorescence with axillary buds which continue to develop indefinitely.

Indigenous Native to a country or particular area; not introduced.

Indument Any hairy covering or pubescence.

Indumentum Same as **indument**.

Induplicate With edges folded in, turned inward, or folded cross-wise.

Indurate Hardened.

Indusium (plural: **indusia**) A scale-like shield or covering overarching the sori of many ferns, originating as an outgrowth of the epidermis. In some ferns, as *Adiantum*, the sori are protected by indusioid organs that structurally are not true indusia; these may be designated by such terms as **indusial flap**, **indusial flange** or **false indusium**.

Inferior Said of one organ when it is below another, such as an **inferior ovary**, which has the perianth located on top.

Inflexed Turned in at the margins.

Inflorescence An arrangement of flowers on a stem or axis; a cluster of flowers or a single flower.

Infra-areolar Within an areole.

Inframarginal Near but not at the margin; submarginal.

Inframedial Located nearer to the midvein than to the margin.

Interfoliar Situated between two opposite leaves, as the stipules of many Rubiaceae.

Internode The part of a stem between two nodes.

Interpetiolar Between the petioles.

Intrapetiolar Inside or beneath the petiole (= **subpetiolar**), or between the petiole and the stem (= **intrafoliaceous**).

Introrse Turned inward, towards the axis.

Involucre A cluster of modified leaves or bracts at the base of a flower cluster or capitulum.

Involute Having the upper surface of a leaf rolled inward.

Irregular Said of flowers that are asymmetrical, having parts of the same whorl (sepals or petals, or both) different in size and/or shape.

Keel A central dorsal ridge; the two united petals of a papilionoid flower, which together form a boat-shaped structure.

Lacerate Irregularly cleft or cut; a margin having a torn appearance.

Laciniate Cut into deep narrow segments; cut into pointed lobes separated by deep, narrow, irregular incisions.

Lamellate Composed of thin plates or scales; thinly stratified.

Lamina The expanded part of a leaf or frond; a leaf-blade.

Lanceolate Lance-shaped; usually applied to rather narrow leaves broadest below the middle and about 3 times longer than wide. The term is used, often in abbreviated form, to denote shapes intermediate between lanceolate and some other shape, as **lance-oblong**, **lance-linear**, etc.

Lanate Clothed with soft woolly hairs.

Lanose Covered with long and loosely tangled hairs.

Lateral On or at the side.

Latex Milky sap, as in Asclepiadaceae or some Euphorbiaceae.

Lax Loose; not rigid.

Leaf A lateral appendage arising from the node of a stem and subtending a bud; it is usually of expanded shape and green, being the chief organ of photosynthesis in most flowering plants.

Leaflet One of the subdivisions of a compound leaf.

Legume 1. A plant of the family Leguminosae. 2. A fruit formed from a single carpel opening by two sutures.

Lemma The lower of the two bracts immediately subtending or enclosing the floret of a grass.

Lenticel A corky spot on young bark that serves as a path of gas exchange between the atmosphere and internal tissues of the stem.

Lenticular Lens-shaped; shaped like a biconvex lens.

Lepidote Covered with minute scurfy scales.

Ligulate Strap-shaped; also said of organs that possess a ligule.

Ligule 1. The strap-shaped corolla-lobe of the ray-flowers in Compositae. 2. A membranous appendage projecting from the top of the leaf-sheath in grasses. The term also has other meanings not relevant to this book.

Limb The expanded part of a gamopetalous corolla, as distinct from the tube or throat.

Linear Line-like; long and narrow with parallel or nearly parallel margins.

Lingulate Tongue-shaped.

Lithophyte A plant that grows on rock and derives its nourishment chiefly from the atmosphere.

Lobate Divided into, or bearing lobes.

Lobe A rounded division of a plant organ; one of the segments of a leaf, leaflet, calyx, corolla, etc., that is divided to about the middle.

Locule One of the compartments or 'cells' of an ovary or anther.

Loculicidal Said of a fruit that dehisces along a suture about midway between the partitions separating the carpels.

Lodicule A small scale outside the stamens in the flowers of grasses.

Lunate In the shape of a half-moon; crescent-shaped.

Malpighiaceous Pertaining to hairs attached at or near the middle, with two horizontal points oriented in opposite directions, especially characteristic of the Malpighiaceae.

Mangrove A salt-tolerant tree or shrub, especially of such genera as *Rhizophora*, *Avicennia* and *Laguncularia*.

Marginate Having a margin of distinctive structure or colour, forming a well-defined border.

Marcescent Withering but not falling off, as a blossom that persists on a twig after flowering.

Medial Located at or near the middle.

Megasporangium (plural: **megasporangia**) A sporangium containing only megaspores.

Megasporophyll The carpel of an angiosperm.

Mentum An extension of the foot of the column in some orchids, in the shape of a projection in front of the flowers.

Mericarp One of the seed-like ripened carpels in the fruit of Umbelliferae.

Micropyle 1. A minute opening at the apex of an ovule through which the pollen tube often grows to reach the egg cell. 2. The corresponding opening in the integument of a seed, through which water may enter when the embryo plant starts to develop.

Microsporangium (plural: **microsporangia**) A sporangium containing microspores; a pollen sac.

Midrib, midvein The main rib or central vein of a leaf or leaf-like structure.

Miocene The geological period between the Oligocene and the Pliocene.

Moniliform Resembling a string of beads.

Monocot An abbreviation for 'monocotyledonous plant'.

Monoecious Having separate staminate and pistillate flowers on the same plant.

Monograph An exhaustive systematic account of a particular genus, family or group of organisms.

Monolete Said of a fern spore having a single surface line or minute ridge separating two more or less flat areas, the line marking the narrow zone where 4 spore faces end, the adjoining flat areas being where two neighbouring spores were in contact; occurring in cases where 4 spores are radially arranged about the zone-line in a disc-like uniaxial tetrad.

Monotypic Said of a taxon containing or comprised of but a single element, as a genus with but one species or a family with but one genus.

Montane Pertaining to mountains.

Mordant A chemical solution used to fix (render permanent) a dye, as in tissues prepared for microscopic study.

Morphological Pertaining to morphology.

Morphology The study of form and structure; also used to designate the structure of an organism as contrasted with its physiology or classification.

Motile Capable of self-movement.

Mottled Marked with spots or irregular patches of different colours or shades.

Mucilage A gummy or gelatinous plant secretion, usually belonging to the amylose group of hydrocarbons.

Mucilaginous Composed of, or covered with mucilage; slimy.

Mucro A short, sharp terminal point or tip.

Mucronate Having a relatively blunt apex ending abruptly in a mucro.

Mucronulate Slightly or minutely mucronate.

Multifid Cleft into several or many lobes or segments.

Multiple fruit A fruit derived from a cluster of flowers, i.e. the ripened ovaries are traceable to the pistils of separate flowers, as in the pineapple.

Multiseriate Arranged in many series, whorls, or rows.

Muricate Rough with many short, sharp points on the surface.

Muriculate Minutely or finely muricate.

Mycorrhiza (plural: **mycorrhizae**) A specialised rootlet or root-like structure in which the cells are permeated with or coated by a symbiotic fungus.

Naked 1. Lacking a covering, such as a perianth, scales, or pubescence. 2. Said of an ovary that is not enclosed in a pericarp, or of a seed that develops when such a naked ovary ripens.

Natural history The study of plant and animal life.

Naturalist A student of plant and animal life.

Naturalised Said of a plant species introduced from another region that becomes established, maintains itself, and reproduces successfully in competition with the indigenous vegetation.

Nectar A sweet secretion produced by flowers.

Nectary A nectar-secreting gland.

Neotropical Pertaining to the tropical parts of the New World, i.e. the West Indies, C. America and the tropical parts of S. America.

Neuter Pertaining to flowers that lack both stamens and pistil.

Nocturnal Occurring or functioning during the night; the opposite of **diurnal**.

Node The point on a stem where a leaf or leaves are normally borne; a joint.

Nodose Having prominent or swollen nodes.

Nodule A small hard knot or rounded body.

Non-motile Not capable of self-movement.

Nucleus (plural: **nuclei**) A dense, complex spheroidal body, surrounded by a membrane, occurring in the cytoplasm of a living cell; it contains substances that determine the transmission of hereditary characters and that become organised in minute bodies called chromosomes just prior to cell division.

Nut A hard or bony, dry, indehiscent fruit derived from two or more carpels enclosed in a dense pericarp and usually containing one seed. The term is loosely used for any hard, dry, one-seeded fruit.

Nutlet 1. A small nut. 2. A one-seeded portion of a fruit which fragments as it matures, as in some Boraginaceae.

Ob- In combination, inversely or oppositely, as obovate, reversed ovate, i.e. having the broader portion toward the apex, or **obcordate**, inversely heart-shaped, i.e. having the notch at the apex.

Oblate Having the form of a compressed sphere.

Oblong Longer than broad, with the margins nearly parallel.

Obpyriform Pear-shaped with the broad end uppermost.

Obreniform Inversely reniform

Obsolete Not evident; rudimentary.

Obturator A plug or special structure closing an opening, as the small body accompanying the pollen masses of orchids and asclepiads which close the opening of the anthers, or the cushion-like structure (called a **caruncle**) closing the micropyle in Euphorbiaceae and often persisting as a protuberance on the seed.

Obtuse Blunt-pointed.

Ocrea (plural: **ocreae**) A nodal sheath formed by the fusion of two stipules, as in many Polygonaceae.

Ocreola (plural: **ocreolae**) A small ocrea.

Odd-pinnate Said of a pinnate leaf with an odd number of leaflets, i.e. pinnate with a single terminal leaflet; imparipinnate.

Oedema Accumulation of excessive fluid in tissues.

Oligocene The geological period between the Eocene and the Miocene.

Operculate Having a cap or lid.

Orbicular Circular in outline; round and flat; orbiculate.

Oval Broadly elliptic with the width greater than half the length.

Ovary The ovule-bearing part of the pistil.

Ovate Egg-shaped and flat, with the broader end at the base.

Ovule The structure that becomes a seed after fertilisation.

Pachycaul Having or characterised by a thick primary stem and few or no branches.

Palea The upper of the two bracts immediately subtending or enclosing the floret of a grass.

Palmate With veins, lobes, or divisions radiating from a common point.

Pandurate Fiddle-shaped; constricted at or near the middle.

Panicle A branched or compound raceme; in more casual meaning, an irregular compound inflorescence with pedicillate flowers.

Paniculate With flowers arranged in panicles.

Pantropical Distributed generally throughout tropical regions.

Papilla (plural: **papillae**). A minute nipple-shaped projection.

Papillate, **papillose** Having papillae on the surface.

Pappus The modified calyx in the florets of Compositae, consisting of scales, bristles, or plumes of various shape attached to the apex of the achene.

Paraphyses Sterile hairs occurring among the sporangia in ferns.

Parasitic Deriving nourishment at the expense of another organism.

Parietal Attached to or lying near and more or less parallel with a wall, as the placenta when it arises from the peripheral wall of a carpel.

Partite A suffix meaning divided or cleft, as 2-partite, 3-partite, etc.

Pedicel The stalk of a flower.

Pedicellate Having a pedicel; borne on a pedicel.

Peduncle A primary flower-stalk supporting two or more flowers, or a solitary flower if it is the remnant of a cluster.

Pedunculate Pertaining to or borne upon a peduncle.

Pellucid Wholly or partly transparent; translucent. Applied especially to various dots or lines in leaves which contain internal oil-glands that allow the passage of light.

Peltate Pertaining to a leaf-blade which is attached to its petiole somewhere on the lower surface instead of by the margin; derived from the latin 'pelta' meaning a small shield, hence shield-shaped.

Pendulous Hanging downward; drooping.

Penicillate Resembling a little brush; having a terminal tuft of hairs.

Penninerved Pinnately veined or nerved.

Perennial Continuing to live from year to year, as contrasted with **annual**.

Perfect Said of flowers having both stamens and pistil in functioning condition.

Perianth A collective term designating both the calyx and corolla considered together, especially if they are of similar colour and texture.

Pericarp The wall of a mature ovary or fruit.

Perigynous Having the perianth parts and stamens borne on or arising from the periphery of the ovary (not beneath it), with the calyx lobes, corolla, and stamens arising on or near the rim of the hypanthium. To be contrasted with **epigynous** and **hypogynous**.

Perisperm That part of the endosperm (stored nutrient material) of a seed that lies outside the embryo, or which surrounds the embryo.

Perispore The husk-like outer covering which surrounds the exine of a spore; however, some fern spores wholly lack a perispore.

Persistent Retaining its place, shape, or structure; remaining attached after the growing period; not deciduous.

Petal A unit of the inner floral envelope or corolla of a polypetalous flower, usually white or variously coloured, seldom green.

Petaloid Resembling a petal.

Petiole The stalk of a leaf.

Petiolule The stalk of a leaflet in a compound leaf.

Phyllanthoid A type of branching, found in the genus *Phyllanthus* (Euphorbiaceae), in which the penultimate axes have spiral phyllotaxy with leaves modified as cataphylls which subtend deciduous, floriferous, distichous-leaved ultimate axes (or in some cases leafless and flattened phylloclades). The stems, in other words, are differentiated into persistent, flowerless, 'leafless' long-shoots, and deciduous, floriferous, 'leafy' short-shoots.

Phyllary An involucral bract, usually several to many in number, subtending the flower-head in Compositae.

Phylloclade A green flattened or rounded stem functioning as a leaf.

Phyllotaxy The system of leaf arrangement on an axis or stem.

Phytogeography The science or study of plant distribution.

Pilose Clothed with long, soft hairs.

Pinna (plural: **pinnae**) One of the primary divisions of a compound leaf, or (especially) of a compound fern frond.

Pinnate Pertaining to a compound leaf or frond with a single primary axis, as contrasted with **palmate**. The term is sometimes used as a suffix preceded by a number indicating the degrees of division, as **2-pinnate**, **3-pinnate**, etc.; the same meaning may also be conveyed by a prefix of latin derivation, as **bipinnate** or **tripinnate**, etc. If the term is used alone, without any prefix, it may mean 1-pinnate. A form such as '1–4-pinnate' indicates the extent of variation in the degrees of cutting of a leaf or frond.

Pinnatifid Pinnately cut more than halfway to the axis into **segments**.

Pinnatisect Cut to the midrib or rhachis in a pinnate pattern.

Pinnule A secondary (or tertiary, etc.) leaflet or pinna of a decompound leaf or frond. Usually distinguished from a segment by being attached to the next higher axis only by the base of its own axis, the tissue margin being entirely free.

Pistil The female organ of a flower, consisting when complete of an ovary, style and stigma. A simple pistil consists of a single carpel; a compound pistil consists of 2 or more carpels, these usually fused together.

Pistillate Pertaining to a flower having a pistil but no stamens.

Pistillode A rudimentary pistil in a staminate flower.

Pith The spongy tissue often occurring in the centre of a dicotyledonous stem; also found in some gymnosperms.

Pitted Marked with small or minute depressions.

Placenta The part of the ovary or carpel where the ovules are attached.

Pleiochasium (plural: **pleiochasia**). A cymose inflorescence in which each branch bears more than two lateral branches.

Plicate Folded or plaited.

Plumose Resembling a plume or feather.

Plumule The primary leaf-bud of an embryo.

Pluricellular Composed of two or more cells.

Pneumatophore An aerial structure which grows vertically upward from roots embedded in mud, composed of spongy tissue presumed to function as an aerating organ.

Pollen The dusty or sticky grains produced in the stamens of flowers. Each pollen grain is in fact a male gametophyte which, on contact with a suitable stigma, germinates to extend a 'pollen tube' ultimately to an ovule. Fertilisation is effected when one of the male nuclei enters the ovule and unites with the female nucleus.

Pollen-sac The cavity of an anther containing the pollen.

Pollen-tube See above under **pollen**.

Pollinia The pollen masses of the orchids and milkweeds.

Pollination The process by which pollen travels from an anther to a stigma.

Pollinium A coherent mass of pollen, as in orchids and asclepiads.

Polygamous Bearing perfect and unisexual flowers on the same plant.

Polymorphic With several or various forms; variable in structural characters.

Polyploid Having more than the standard number of chromosomes in its nuclei, usually expressed in terms of a definite multiple of the base number, as **tetraploid** meaning with four times the number of chromosomes found in the gametic cells of a normal 'diploid' population.

Polystichous Having leaves borne in many rows or series and spreading in many directions.

Pore A small or minute aperture.

Prickle Outgrowth of the epidermis, can easily be snapped off.

Prismatic 1. Shaped like a prism. 2. With many colours, like a rainbow.

Procumbent Prostrate but not taking root at the nodes; trailing.

Proliferous Bearing offshoots; often used to denote leaves or fronds which produce small root-bearing buds or bulbils that afford a means of vegetative reproduction.

Prominulous Minutely raised, as the veins of a leaf.

Prop root A root produced above the ground which serves as a prop or support to the plant, as in mangroves.

Prostrate Lying flat on the ground.

Protandrous With anthers maturing before the pistil in the same flower.

Prothallus (plural: **prothalli**). The gametophyte of a fern.

Pruinose Bearing a waxy-powdery secretion on the surface.

Pseud-, pseudo- A prefix meaning false or spurious.

Pseudobulb The thickened or bulb-like stems of certain orchids, usually borne above the ground or other substrate, and usually bearing one or more leaves at the apex.

Pseudocarp A false fruit; a fruit derived from parts other than the ovary, as in some Rosaceae, in which the 'fruit' is chiefly composed of the greatly enlarged receptacle.

Pteridophyte One of the four main divisions of the plant kingdom according to some systems of classification, including the ferns and several other more or less remotely related groups which have a somewhat similar lifecycle.

Puberulous Minutely pubescent.

Pubescent Covered with soft, straight, short hairs.

Pulvinate Having a pulvinus.

Pulvinus A swelling at the base of a petiole or petiolule which frequently acts as a centre of sensitivity, irritability, or movement in a leaf.

Punctate Marked with dots, translucent or otherwise.

Punctulate Minutely punctate.

Pustular, pustulate Blister-like; covered with small blister-like prominences.

Pyrene A small, hard, stone-like seed in a drupe or similar fruit.

Pyriform Pear-shaped.

Quandrangular Having four sides.

Quadrinomial A 4-unit epithet applied to an organism or population, designating genus, species, variety or subspecies, and form, as *Pteridium aquilinum* var. *caudatum* forma *glabratum*, or *Pteridium aquilinum caudatum glabratum*.

Raceme A simple, elongate, indeterminate inflorescence bearing a number of pedicellate flowers, the pedicels of approximately equal length.

Racemose Having flowers in racemes.

Rachilla (plural: **rachillae**) Same as Rhachilla. In grasses and sedges, the axis that bears the florets.

Rachis see **Rhachis**

Radiate Spreading from or arranged around a common centre.

Radicle The embryonic root of a germinating seed.

Raphe The continuation of a funicle adnate to the side of an ovule, usually evident as a raised line or ridge.

Rank A row, especially a vertical row, often used as a suffix, as 2-ranked, 3-ranked, etc.

Ray-floret A ligulate flower in the flower-head of Compositae.

Receptacle In flowering plants, the more or less enlarged or elongated apex of the pedicel; in Compositae, the enlarged apex of the peduncle.

Reflexed Abruptly curved or bent downward or backward.

Regular Having flower parts of the same kind all alike in size and shape; radially symmetrical.

Remote Widely spaced; scattered; not close together.

Reniform Kidney-shaped.

Repand Undulate or wavy; having a slightly undulating or sinuous margin.

Replum A thin wall dividing a fruit into two chambers, formed by an ingrowth from the placenta and thus not a true part of the carpel walls.

Resipunate Upside down, inverted by the twisting of the pedicel, as the flowers of orchids.

Reticulate Netted; forming a network.

Retrorse Bent or turned backward or downward.

Retuse Having a shallow notch at an obtuse apex.

Revolute With the margins rolled under toward the lower (abaxial) side, as the margin of a leaf.

Rhachilla (plural: **rhachillae**) Same as **rachilla**.

Rhachis 1. Any axis bearing lateral appendages or organs. 2. In compound leaves or fronds, the portion of the main axis to which the primary divisions or pinnae are attached. Sometimes spelled 'rachis'.

Rhizome An underground stem that gives rise to roots and aerial stems, distinguished from a true root by the presence of nodes, buds, or leaves, the latter sometimes reduced to scales.

Rhizophore A naked branch which grows down into the soil and develops roots from the apex.

Rind A tough outer layer, as on some fleshy fruits; sometimes used to designate any outer skin.

Root The usually underground part of a plant which supplies it with water and dissolved mineral nutrients, in structure always lacking nodes and leaves; the absorbtive, anchoring, and storage organ of vascular plants.

Rootlet A little root; a small branch of a root.

Rootstock An underground stem or rhizome; sometimes applied especially to an erect rhizome, as in some ferns.

Rosette A cluster of spreading or radiating basal leaves.

Rostellum A small beak; especially, a narrow extension of the upper edge of the stigma in some orchid flowers.

Rosulate In the form of a rosette.

Rotate Wheel-shaped with flat and spreading parts.

Rotund Rounded in outline; nearly orbicular but slightly inclined toward the oblong.

Rucinate Sharply pinnatifid or incised, the lobes pointing downward.

Rugose Wrinkled in appearance, as a leaf-surface with sunken veins.

Rugulose Minutely rugose.

Ruminate Having a crumpled appearance, or appearing as if chewed.

Russet Reddish or dull red.

Saccate Shaped like a pouch or little bag; sac-shaped.

Sagittate Shaped like an arrowhead, with prominent basal lobes pointing or curving downward.

Sagittulate Minutely sagittate.

Salverform, salver-shaped Said of a corolla composed of a slender tube abruptly expanding into a flat, rotate limb.

Samara A dry, indehiscent, one-seeded, winged fruit.

Saprophyte A plant which derives all its nourishment from dead animal or vegetable matter.

Scabrid Referring to a surface with scattered or intermittent roughness.

Scabridulous Minutely roughened.

Scabrous Referring to a surface that is rough with a covering of very short, stiff, bristly hairs, scales or points.

Scale 1. A usually small, dry, thin or membranous epidermal appendage, differing from a hair in being two or more cells in width. 2. A small, thin, semitransparent bract or leaf-like structure, usually appressed, found on apical buds, bulbs, at the base of shoots, and on rhizomes and other organs; see also **cataphyll**.

Scandent Climbing or scrambling over rocks or other plants without the aid of tendrils, but often with adherent roots or rootlets.

Scape A leafless peduncle arising from the ground, either entirely naked or else at the most bearing scales or bracts.

Scapiform, scapose Resembling or bearing a scape.

Scarious Thin, dry, and membranous, and usually not green.

Schizocarp A dry compound fruit which splits apart at maturity into two or more single-seeded segments known as mericarps.

Scorpioid Coiled like a scorpion's tail, said of a coiled (circinnate) inflorescence in which the flowers are 2-ranked, being borne alternately on opposite sides of the axis.

Scrambling Said of a plant with weak, elongate stems that grow over other plants or any kind of support, but do not twine or have the aid of tendrils or aerial roots.

Scrub A type of vegetation composed of low and often densely packed bushes.

Scurfy Covered with minute bran-like scales.

Secondary axis A branch of a main axis.

Secondary thicket, secondary vegetation Vegetation which comes up naturally after cutting, fire, or other disturbing factors.

Secund Arranged on one side only; unilateral

Seed A mature, fertilised ovule, containing an embryo and some form of stored food material, and protected by a seed-coat or testa.

Segment A division or part of a leaf or other organ that is cleft or divided but not truly compound.

Semi- A prefix meaning partly, to some extent, incompletely, etc.

Sepal One of the parts of the calyx.

Sepaloid Sepal-like, resembling a sepal.

Septate Divided by partitions; having **septa**.

Septicidal Said of a capsule that dehisces along the partitions.

Septum (plural: **septa**) A partition; a cross-wall; a dividing membrane.

Seriate Arranged in one or more rows or series.

Sericeous Clothed with a silky pubescence; covered with closely appressed fine soft hairs.

Serrate Having a saw-toothed margin, the teeth inclined toward the apex.

Serration A saw-like notch; a tooth of a serrate margin.

Serrulate Minutely serrate.

Sessile Lacking a stalk, as a leaf without a petiole.

Seta (plural: **setae**) A bristle or bristle-like structure; a needle-shaped process.

Setaceous Bearing setae; covered with bristles.

Setiform Bristle-shaped.

Setose Covered with bristles.

Setulose Minutely setose.

Sheath Any more or less tubular structure surrounding a part, as the basal part of a grass leaf surrounding the culm.

Shrub A relatively low, usually several-stemmed, woody plant; a bush.

Silicula A dry fruit that opens along two lines and has a central persistent partition; it is as broad, or broader, than it is long, as in the Cruciferae.

Silique A 2-valved capsular fruit in which the valves dehisce from a frame (the **replum**) on which the seeds grow, and across which a false partition is formed.

Sinuate Having a deeply wavy margin.

Sinus (plural: **sinuses**). The notch between two lobes or segments.

Solitary Single; only one in the same place.

Sorus (plural: **sori**). In ferns, a cluster of sporangia.

Soriferous Bearing sori.

Spadix The thick or fleshy flower-spike of certain plants, as in the Araceae, surrounded or subtended by a **spathe**.

Spathe The bract or pair of bracts surrounding or subtending a flower cluster or spadix.

Spathulate, spatulate Shaped like a spatula, oblong with an attenuated base.

Species (plural: **species**) A term used in classification to denote a group or population of closely similar, mutually fertile individuals which show constant differences from allied groups that are more or less reproductively isolated.

Specific name The name of a species; the second part of a binomial name.

Specimen A plant, or portion of a plant, prepared and preserved for study; a preserved sample intended to show the characteristics of a species or other taxon.

Spermatozoid A motile or free-swimming male gamete.

Spher- A prefix meaning round.

Spicate Spike-like; arranged in or having spikes.

Spike An unbranched, simple, elongate inflorescence bearing sessile or subsessile flowers.

Spikelet 1. A secondary spike, i.e. the part of a compound inflorescence which is itself a spike. 2. The ultimate unit in the inflorescence of grasses, consisting of two **glumes** and one or more florets each of which is subtended by a **lemma** and **palea**.

Spine Any sharp, rigid process or outgrowth, usually a modified branch, but also sometimes a modified stipule, petiole, or other part; a thorn.

Spinose Bearing a spine or spines.

Spinulose Minutely spinose.

Sporangium (plural: **sporangia**). A minute capsule or sac containing spores.

Spore A minute, unicellular, reproductive body, analogous in function to a seed but lacking an embryo. In ferns, the spores of many or most species are produced through the process of **meiosis**, by which the number of chromosomes in the nucleus is reduced by half; germination of a fern spore gives rise to the gametophyte generation in the life-cycle of these plants.

Sporophyll A spore-bearing leaf or frond.

Sporophyte The asexual spore-producing plant or generation in plants having an alternation of generations.

Spur 1. A tubular elongation of the base of a petal or of a gamopetalous corolla. 2. An extension of the base of a leaf beyond its point of attachment. 3. A short branch in many trees on which flowers and fruits are borne. 4. A short branch borne from the axil of a scale-leaf or cataphyll and bearing the true foliage leaves of the plant.

Stamen The pollen-bearing organ of a flower, typically consisting of a filament and an anther, or the anther sometimes sessile.

Staminal cup, column or tube Structures formed by various degrees of fusion of the filaments of stamens, as the staminal column of Malvaceae.

Staminate Having, producing, or consisting of stamens; having stamens and no pistil, as the staminate flowers of dioecious plants.

Staminode A sterile stamen or stamen-like structure; a structure occupying the position of, and often shaped like, a stamen, but lacking a fertile anther.

Standard The upper, broad, often erect to recurved petal in many leguminous flowers.

Stellate Star-shaped; having hairs or scales with branches or points radiating from a centre.

Stem The main axis of a plant, bearing leaves and flowers, as contrasted with a root, which bears neither of these structures.

Sterile Barren or non-reproductive, as a fern-frond without sporangia or a flower lacking a pistil.

Stigma (plural: **stigmas** or **stigmata**) The part of a pistil (usually the apex) which is receptive to pollen grains and on which they germinate.

Stigmatic Pertaining to the stigma.

Stipe The stalk or petiole of a fern-frond; the stalk beneath an ovary.

Stipel The stipule of a leaflet in a compound leaf.

Stipitate Having a stipe.

Stipular Having stipules, or relating to them.

Stipular spine A spine representing a modified stipule, or having the position of a stipule.

Stipule A more or less leafy appendage at the base of the petiole in many plants.

Stolon An elongate creeping stem (usually representing a modified basal branch) which roots at the nodes and often gives rise to new plants at some nodes, or at its tip, or both.

Stoloniferous Producing stolons.

Strand 1. A thread or fiber. 2. The zone of exposed beach above the high tide line.

Striate Marked with fine, linear, parallel lines.

Striation A fine, linear marking; a minute elongate ridge or furrow; one of a pattern of fine lines.

Strigillose Minutely strigose.

Strigose Bearing appressed, sharp, straight, stiff hairs.

Strobilus (plural: **strobili**) 1. A cone-like cluster of sporophylls. 2. An inflorescence or ovule-bearing structure made up largely of more or less imbricated scales, and usually of a somewhat conical shape. The term is also written as **strobile**.

Strophiole An appendage at the hilum of some seeds; a caruncle.

Strychnine A poisonous alkaloid used medicinally in small doses as a nerve stimulant.

Stylar column The column in orchid flowers.

Style The elongate part of a pistil, bearing the stigma at its apex; lacking in flowers having a sessile stigma.

Sub- A prefix meaning less than, almost, approaching, etc., as **subcordate**, meaning slightly cordate, **subcylindric**, meaning not quite circular in cross-section, or **submarginal**, meaning almost at the margin.

Subfamily In classification, a category beneath that of family and above that of genus.

Subgenus In classification, a category beneath that of genus and above that of species; a subgeneric name is often used for one or more species that form a natural grouping within a genus, but does not become part of a binomial epithet.

Substrate The material in or on which a plant is rooted.

Subtend To be attached beneath and close to, as a bract below a flower.

Subterranean Underground; occurring beneath the surface of the soil.

Subtropical Inhabiting or characterising regions bordering the tropics.

Subulate Awl-shaped.

Succulent Juicy or fleshy; having tissues thickened to conserve moisture.

Suffrutescent A stem which is woody and perennial at the base, but has an upper herbaceous portion that dies back at the end of a growing season. The term is also sometimes used to mean 'slightly shrubby'.

Sulcate Grooved lengthwise.

Superior Said of one organ when it is above another, such as a **superior ovary**, which has the perianth attached beneath it.

Supra-axillary Borne above the axil.

Supramedial Said of a sorus which is slightly closer to the margin than to the midvein.

Suture 1. A line of fusion or union. 2. A line along which dehiscence may occur.

Symbiosis The living together of dissimilar organisms, with benefit to one or both.

Sympodial Having a stem made up of a series of superposed branches arranged so as to appear like a simple axis; this is caused by the self-pruning and withering of the terminal bud at the end of each period of growth, so that successive growth-episodes must originate from lateral buds.

Syncarp A multiple or fleshy aggregate fruit.

Syncarpous 1. Composed of two or more united carpels. 2. Bearing an aggregate fruit.

Synonym One of two or more scientific names applied to the same taxon, one of which is correct and the others incorrect under the International Rules of Nomenclature. The term is commonly used to designate only the incorrect names.

Syntype One of two or more specimens (not duplicates of the same number) of equal rank on which a species, subspecies, variety, or form is based, in the absence of a designated holotype.

Taproot A central or leading root which penetrates deeply into the ground without dividing; a prolonged and relatively thick primary root.

Tawny Dull brownish-yellow.

Taxon (plural: **taxa**) In classification, a term applied to any coherent element, population, or grouping, regardless of level, as **order**, **family**, **genus**, **species**, etc.

Taxonomic Pertaining to the systematic classification of organisms.

Taxonomy The science of classification.

Tegument A natural outer covering; an integument.

Tendril A thread-like process or extension by which a plant grasps an object and clings to it for support; this structure may represent a modified stem, leaf, leaflet, or stipule.

Terete Circular in cross-section.

Terminal Pertaining to the end or apex.

Terrestrial Growing on the ground.

Tertiary Of the third order or rank, as **tertiary veins**, which are branches of secondary veins, which in turn arise from the primary vein or midvein.

Testa The outer coat or integument of a seed, usually hard and brittle.

Tetrahedral Having or made up of four faces.

Tetraploid Said of an organism whose cell nuclei contain twice the normal (diploid) number of chromosomes expected in members of its particular taxon; it is called tetraploid because it has four times the gametic number.

Thallus (plural: **thalli** or **thalluses**) A plant body undifferentiated into stem, root or leaf.

Thallophyte A plant lacking differentiation into roots, stems, leaves, and without internal vascular tissues.

Thorn A modified branch in the form of a sharp, woody spine.

Thyrse A compact or contracted panicle with the main axis indeterminate but the secondary and ultimate axes cymose and determinate.

Thyrsoid Resembling a thyrse.

Tissue An aggregate of cells similar in structure or function.

Tomentellous Minutely tomentose.

Tomentose Covered with matted, soft, woolly hairs.

Tomentum A covering of matted, soft, woolly hairs; wool-like pubescence.

Tortuous Marked by repeated twists, bends, or turns.

Torose Knobby; having a cylindrical body abruptly swollen at intervals.

Torulose The diminutive of torose.

Torus The receptacle of a flower; the apical portion of a floral axis on which the parts of a flower are inserted.

Trapezoidal Having four unequal sides.

Tri- A prefix meaning three, as **trichotomous** meaning three-forked or three-branched, **trifid** meaning a three-cleft apex, **trifoliate** meaning with three leaflets, **trigonous** meaning three-angled, etc.

Tribe A taxon consisting of one or more genera forming a natural group within a family, but not equivalent to a subfamily.

Trichome Any hair-like outgrowth of the epidermis.

Trigonous, trigonal Three-angled, the faces between planes.

Trimerous Having flower parts, such as petals, sepals, and stamens, in sets of three.

Triplinerved Said of a leaf with three main nerves or ribs, the lateral two arising from the median one above the base, contrasting with a trinerved leaf in which the three nerves all arise from the same point at the base of the lamina.

Triploid Said of an organism whose cell nuclei contain three times the normal gametic number of chromosomes, or one and one half the normal diploid number; such organisms usually arise by hybridisation between a diploid and a tetraploid.

Triquetrous Three-edged, the faces between concave.

Tropical Strictly, the life-zone on either side of the equator where the day and night are of about equal length and the temperature is rather uniform throughout the year with a mean daily temperature of about 27°C (= 80°F); in this sense, such climatic variation as occurs in the zone is a result of fluctuations in rainfall. In a broader sense, the tropical zone is the region of the Earth's surface immediately north and south of the equator, and bounded approximately by the Tropic of Cancer (23.5 °N) and the Tropic of Capricorn (23.5 °S); within this very large area many climatic differences occur and a large number of life-zones are represented. Therefore the term 'tropical' can mean different things according to the context in which it used.

Trullate With widest axis below middle and with straight margins; ovate but margins straight and angled below middle, trowel-shaped.

Truncate Ending abruptly; having a base or apex that is nearly straight across as if cut off.

Tuber A relatively short, thickened rhizome with numerous buds capable of vegetative propagation, as in the 'Irish' potato.

Tubercle A wart-like or knob-like projection.

Tuberculate Bearing tubercles.

Tufted Clustered or bunched, the units or divisions originating from approximately the same point; caespitose.

Turbinate Top-shaped.

Turgid Swollen, as by pressure of liquid from within.

Twining Twisting together or about another part or object more or less spirally.

Ultimate Last, final, or utmost; situated at the end, as the ultimate branch of a stem; farthest from the base.

Umbel An indeterminate, often flat-topped inflorescence consisting of several or many pedicellate flowers arising from a common point of attachment.

Umbellate Arranged in umbels, or pertaining to umbels.

Umbelliform In the shape of an umbel.

Undershrub A low-growing woody plant.

Undulate Wavy.

Undulate-toothed A wavy margin with teeth.

Uni- A prefix meaning, one, as **uniseriate** meaning in one row, or **unisexual** meaning of one sex.

Urceolate Pitcher-shaped; hollow and contracted at the mouth.

Urn-shaped In the shape of a vase or pitcher; **urceolate**.

Utricle 1. A bladdery, 1-seeded, usually indehiscent fruit, as in the Urticaceae. 2. Any small membranous sac or bladder-like body.

Vaginate Sheathed; surrounded by a sheath.

Valvate Having or pertaining to valves; opening by valves, as a capsule. The term is also applied to flower-buds whose segments meet exactly without overlapping.

Valve One of the segments into which the walls of a fruit separate by dehiscence; one of the parts of a dehiscent pericarp. The term is also applied to a partially detached flap of an anther wall in Lauraceae.

Variegated Marked with different colours or tints in spots, streaks, or patches.

Variety A morphological variant or variant group within a species, differing from other variants of the same species by one or more minor characteristics.

Vascular Furnished with vessels or ducts through which fluid passes.

Vascular bundle A strand of conductive tissue composed of xylem and phloem cells, these sometimes separated by a cambium, and sometimes also accompanied by sclerenchymatous (fibrous) supporting tissue.

Vascular plants Plants containing vascular tissues, such as ferns and seed plants.

Vegetation The total aggregation of plant-life in a particular area.

Vegetative reproduction Non-sexual reproduction; reproduction by means of non-sexual structures such as bulbils, stolons or tubers.

Vein A strand of vascular tissue in a leaf-blade.

Veinlet A minute vein.

Velamen The spongy, multiple epidermis that covers the aerial roots of epiphytic orchids and certain other plants and is capable of absorbing atmospheric moisture.

Velamentous With parchment-like sheath or layer of air-cells on roots of some orchids.

Venation Pertaining to the arrangement or pattern of veins.

Ventricose Swollen on one side.

Vermifuge A medicine for eliminating worms.

Versatile anther An anther attached near the middle and turning freely on its support.

Verrucose Covered with small wart-like projections.

Verticillate Whorled.

Vestigial Rudimentary; partly formed, often minute structure located where normal, fully formed parts ought to be, or once existed.

Villous Clothed with long, soft, weak but straight hairs.

Vine 1. The plant which bears grapes, *Vitis vinifera*. 2. Any trailing or climbing plant.

Virgate Shaped like a wand or rod, straight, long and slender.

Viscid Having a sticky coating or secretion.

Viscidium sticky substance in flowers used to collect pollen or pollinia

Whorl The cyclic arrangement of appendages at a node.

Wing 1. Any thin, often dry and membranous expansion attached to an organ, as on petioles or fruits. 2. A lateral petal of a leguminous flower.

Xylem A complex tissue in the vascular system of higher plants, functioning chiefly for the conduction of fluids upward, but also for support and storage, and typically making up the woody part of a plant stem.

Zoology The study of animal life.

Zygomorphic Bilaterally symmetrical; divisible into similar halves in only one plane.

Zygote The cell resulting from the fusion of two gametes; a fertilised egg cell before its nucleus has divided to initiate growth.

BIBLIOGRAPHY

GENERAL

The literature pertaining directly to Cayman Islands vegetation is scanty, and consists chiefly of lists or short reports. Other literature, dealing with aspects of the geology, zoology and history of these islands is more extensive, but not relevant to this book. The following botanical references have been consulted in addition to general taxonomic works and special papers or monographs on particular groups of plants:

Fawcett, W. (1888). Cayman Islands. *Kew Bull.* 1888, 160–163.

Fawcett, W. (1889). Report by the Director of Public Gardens and Plantations on the Cayman Islands. *Bull. Bot. Jamaica* 11, 6–7.

Guppy, H. B. (1917). *Plants, Seeds and Currents in the West Indies and Azores.* Williams and Norgate, London.

Hitchcock, A. S. (1893). List of plants collected in the Bahamas, Jamaica and Grand Cayman. *Ann. Rep. Missouri Bot. Gard.* 4, 47–179.

Hitchcock, A. S. (1898). List of cryptogams. *Ann. Rep. Missouri Bot. Gard.* 9, 111–120.

Millspaugh, C. F. (1900). Plantae Utowanae. *Field Mus. Bot.* Ser. 2, 3–133.

Proctor, G. R. (1980). Checklist of the plants of Little Cayman. *Atoll Res. Bull.* 241, 71–80.

Savage English, T. M. (1913). Some notes on the natural history of Grand Cayman. *Handbook of Jamaica for 1912*, 598–600.

Savage English, T. M. (1913). Some notes from a West Indian coral island. *Kew Bull.* 1913, 367–372.

Swaby, C. and Lewis, C. B. (1946). Forestry in the Cayman Islands. *Development and Welfare in the West Indies*, Bull. no. 23. Barbados.

FLORAS

The following standard West Indian floras were frequently consulted:

Acevedo-Rodriguez, P. (1996). Flora of St. John, U.S. Virgin Islands. *Mem. N.Y. Bot. Gard.* 78: 1–581.

Adams, C. D. (1972). *Flowering Plants of Jamaica.* University of the West Indies, Mona.

Alain, H. (1962). *Flora de Cuba, Vol. 5.* Puerto Rico University, Rio Piedras.

Britton, N. L. & Millspaugh, C. F. (1920). *Bahama Flora.* New York, 1–695.

Britton, N. L. & Wilson, P. (1923–1930). *Botany of Porto Rico and the Virgin Islands. Sci. Surv. of Porto Rico and the Virgin Islands, Vols. 5–6.* Acad. Sci., New York.

Corell, D. S. & Corell, H. B. (1982). *Flora of the Bahama Archipelago.* J. Cramer, Vaduz.

Fawcett, W. & Rendle, A. B. (1910–1936). *Flora of Jamaica. Vols. 1. 3, 4, 5, & 7.* British Museum (Natural History), London.

Gooding, E. G. B., Loveless, A. R., & Proctor, G. R. (1965). *Flora of Barbados.* H.M.S.O., London.

Grisebach, A. H. R. (1859–64). *Flora of the British West Indian Islands.* Lovell Reeve, London.

Howard, R. A. (and collaborators). *Flora of the Lesser Antilles.* Vol. 1, 1974; Vol. 2, 1977; Vol. 3, 1979; Vol. 4, 1988; Vol. 5, 1989; Vol. 6, 1989. Harvard University, Boston.

León, H. (1946–1957). *Flora de Cuba.* Vol. 1–4. P. Fernandez, Havana, Cuba.

Urban, I. (1898–1928). *Symbolae Antillanae.* Vols. 1–9. Williams and Norgate, London.

PHYTOGEOGRAPHY

Burton, F. J. (2008a). *Threatened Plants of the Cayman Islands: the Red List*. Royal Botanic Gardens, Kew.

Burton, F. J. (2008b). Vegetation classification for the Cayman Islands. In: F. J. Burton, *Threatened Plants of the Cayman Islands: the Red List*. Royal Botanic Gardens, Kew.

Burton, F. J. (1994). Climate and tides of the Cayman Islands. In: M. A. Brunt & J. E. Davies (eds), *The Cayman Islands: Natural History and Biogeography*, pp. 51–60. Kluwer Academic Publishers, Dordrecht.

Jones, B., Hunter, I. G. & Kyser, K. (1994). Stratigraphy of the Bluff Formation (Miocene–Pliocene) and the newly defined brac formation (Oligocene), Cayman Brac, British West Indies. *Caribbean Journal of Science* 30(1–2): 30–51.

Jones, B. & Hunter, I. G. (1990). Pleistocene paleogeography and sea levels on the Cayman Islands, British West Indies. *Coral Reefs* 9: 81–91.

Muhs, D. R., Bush, C. A., Stewart, K. C., Rowland, T. R. & Crittenden, R. C. (1990). Geochemical evidence of Saharan dust parent material for soils developed on quaternary limestones of Caribbean and western Atlantic islands. *Quaternary Research* (*Orlando*) 33(2): 157–177.

REFERENCES FOR TAXONOMIC UPDATES IN THIS SECOND EDITION

Brummitt, R. K. & Powell, C. E. (eds.) (1992). *Authors of Plant Names*. The Royal Botanic Gardens, Kew.

Hunt, D., Taylor, N. & Charles, G. (2006). *The New Cactus Lexicon*. DH books, Milborne Port.

Lewis, G., Schrire, B., Mackinder, B. & Lock, M. (2005). *Legumes of the World*. The Royal Botanic Gardens, Kew.

INDEX OF COMMON NAMES

INDEX OF BOTANICAL NAMES

Bold page numbers indicate illustrations and colour page numbers with prefix 'P' indicate plate numbers

GENERA WHOSE CLASSIFICATION IN MODERN MOLECULAR-BASED PLANT TAXONOMY DIFFERS FROM THAT USED IN THIS FLORA
(See note on p. 724)

Taxonomy used in this flora ORDER: FAMILY	Modern molecular-based taxonomy ORDER: FAMILY
	Genera

MONOCOTYLEDONES

HYDROCHARITALES: HYDROCHARITACEAE	ALISMATALES: HYDROCHARITACEAE
	Halophila, Thalassia

NAJADALES: RUPPIACEAE	ALISMATALES: RUPPIACEAE
	Ruppia

COMMELINALES: CYMODOCEACEAE	ALISMATALES: CYMODOCEACEAE
	Halodule, Syringodium

CYPERALES: CYPERACEAE	POALES: CYPERACEAE
	Cyperus, Eleocharis, Abildgaardia, Fimbristylis, Schoenoplectus, Rhynchospora, Cladium, Remirea, Scleria

CYPERALES: GRAMINEAE	POALES: POACEAE
	Andropogon, Anthephora, Arundo, Axonopus, Bambusa, Bothriochloa, Cenchrus, Chloris, Cynodon, Dactyloctenium, Digitaria, Distichlis, Echinochloa, Eleusine, Eragrostis, Imperata, Lasiacis, Leptochloa, Melinis, Oplismenus, Panicum, Paspalum, Pennisetum, Pharus, Rottboellia, Saccharum, Setaria, Sorghum, Spartina, Sporobolus, Stenotaphrum, Urochloa, Zea, Zoysia

TYPHALES: TYPHACEAE	POALES: TYPHACEAE
	Typha

BROMELIALES: BROMELIACEAE	POALES: BROMELIACEAE
	Tillandsia, Bromelia, Hohenbergia, Ananas

ARECALES: PALMAE	ARECALES: ARECACEAE
	Roystonea, Phoenix, Cocos, Thrinax, Coccothrinax

ARALES: ARACEAE	ALISMATALES: ARACEAE
	Philodendron, Syngonium

ARALES: LEMNACEAE	ALISMATALES: ARACEAE
	Lemna

LILIALES: PONTEDERIACEAE	COMMELINALES: PONTEDERIACEAE
	Eichhornia

LILIALES: DRACAENACEAE	ASPARAGALES: ASPARAGACEAE
	Sansevieria

LILIALES: ASPHODELACEAE	ASPARAGALES: XANTHORRHOEACEAE
	Aloë

LILIALES: AMARYLLIDACEAE	ASPARAGALES: AMARYLLIDACEAE
	Zephyranthes, Hippeastrum, Crinum, Hymenocallis

LILIALES: AGAVACEAE	ASPARAGALES: ASPARAGACEAE
	Yucca, Agave, Furcraea

ORCHIDALES: ORCHIDACEAE	ASPARAGALES: ORCHIDACEAE
	Vanilla, Prescottia, Beloglottis, Cyclopogon, Eltroplectris, Sacoila, Triphora, Tropidia, Pleurothallis, Bletia, Polystachya, Epidendrum, Encyclia, Prosthechea, Myrmecophila, Brassavola, Dendrophylax, Oeceoclades, Cyrtopodium, Ionopsis, Tolumnia

DICOTYLEDONES

MAGNOLIALES: CANELLACEAE	CANELLALES: CANELLACEAE

Canella

MAGNOLIALES: LAURACEAE	LAURALES: LAURACEAE

Licaria, Ocotea, Persea, Cassytha

PAPAVERALES: PAPAVERACEAE	RANUNCULALES: PAPAVERACEAE

Argemone

URTICALES: ULMACEAE	ROSALES: CANNABACEAE

Celtis, Trema

URTICALES: MORACEAE	ROSALES: MORACEAE

Ficus, Maclura, Artocarpus

URTICALES: URTICACEAE	ROSALES: URTICACEAE

Pilea

MYRICALES: MYRICACEAE	FAGALES: MYRICACEAE

Myrica

CASUARINALES: CASUARINACEAE	FAGALES: CASUARINACEAE

Casuarina

BATALES: AIZOACEAE	CARYOPHYLLALES: AIZOACEAE

Sesuvium

BATALES: PORTULACACEAE	CARYOPHYLLALES: PORTULACACEAE

Portulaca

BATALES: PORTULACACEAE	CARYOPHYLLALES: TALINACEAE

Talinum

BATALES: BASELLACEAE	CARYOPHYLLALES: BASELLACEAE

Anredera

BATALES: CHENOPODIACEAE	CARYOPHYLLALES: AMARANTHACEAE

Atriplex, Salicorna

BATALES: AMARANTHACEAE	CARYOPHYLLALES: AMARANTHACEAE

Amaranthus, Achyranthes, Iresine, Gomphrena, Lithophila (now included in Iresine), Blutaparon, Alternanthera

BATALES: BATACEAE	BRASSICALES: BATACEAE

Batis

POLYGONALES: POLYGONACEAE	CARYOPHYLLALES: POLYGONACEAE

Coccoloba, Antigonon, Polygonum

PLUMBAGINALES: PLUMBAGINACEAE	CARYOPHYLLALES: PLUMBAGINACEAE

Limonium

THEALES: CLUSIACEAE	MALPIGHIALES: CLUSIACEAE

Clusia

THEALES: CLUSIACEAE	MALPIGHIALES: CALOPHYLLACEAE

Mammea

MALVALES: TILIACEAE	MALVALES: MALVACEAE

Corchorus, Triumfetta

MALVALES: STERCULIACEAE	MALVALES: MALVACEAE

Helicteres, Waltheria, Neoregnellia (unresolved name), Melochia, Guazuma, Sterculia

MALVALES: LECYTHIDACEAE	ERICALES: LECYTHIDACEAE
Barringtonia	

VIOLALES: SALICACEAE	MALPIGHIALES: SALICACEAE
Banara, Casearia, Zuelania, Xylosma	

VIOLALES: TURNERACEAE	MALPIGHIALES: PASSIFLORACEAE
Turnera	

VIOLALES: PASSIFLORACEAE	MALPIGHIALES: PASSIFLORACEAE
Passiflora	

VIOLALES: CARICACEAE	BRASSICALES: CARICACEAE
Carica	

VIOLALES: CUCURBITACEAE	CUCURBITALES: CUCURBITACEAE
Fevillea, Melothria, Cionosicyos, Momordica, Citrullus, Sechium, Cucurbita, Cucumis	

CAPPARALES: CAPPARACEAE	BRASSICALES: CAPPARACEAE
Capparis	

CAPPARALES: CAPPARACEAE	BRASSICALES: CLEOMACEAE
Cleome	

CAPPARALES: CRUCIFERAE	BRASSICALES: BRASSICACEAE
Cakile, Lepidium	

CAPPARALES: MORINGACEAE	BRASSICALES: MORINGACEAE
Moringa	

EBENALES: SAPOTACEAE	ERICALES: SAPOTACEAE
Sideroxylon, Pouteria, Chrysophyllum, Manilkara	

PRIMULALES: THEOPHRASTACEAE	ERICALES: PRIMULACEAE
Jacquinia	

PRIMULALES: MYRSINACEAE	ERICALES: PRIMULACEAE
Myrsine	

ROSALES: CRASSULACEAE	SAXIFRAGALES: CRASSULACEAE
Bryophyllum	

ROSALES: CHRYSOBALANACEAE	MALPIGHIALES: CHRYSOBALANACEAE
Chrysobalanus	

ROSALES: LEGUMINOSAE	FABALES: FABACEAE
Crotalaria, Indigofera, Tephrosia, Gliricidia, Sesbania, Stylosanthes, Desmodium, Alysicarpus, Christia, Abrus, Centrosema, Clitoria, Teramnus, Erythrima, Mucuna, Galactia, Canavalia, Phaseolus, Vigna, Lablab, Cajanus, Rhynchosia, Moghania, Dalbergia, Piscidia, Sophora, Aeschynomene, Caesalpinia, Haematoxylum, Delonix, Cassia, Senna, Chamaecrista, Bauhinia, Hymenaea, Tamarindus, Adenanthera, Desmanthus, Mimosa, Leucaena, Acacia, Calliandra	

MYRTALES: THYMELAEACEAE	MALVALES: THYMELAEACEAE
Daphnopsis	

MYRTALES: PUNICACEAE	MYRTALES: LYTHRACEAE
Punica	

CORNALES: RHIZOPHORACEAE	MALPIGHIALES: RHIZOPHORACEAE
Rhizophora	

SANTALALES: OLACACEAE	SANTALALES: SCHOEPFIACEAE
Schoepfia	

SANTALALES: VISCACEAE	SANTALALES: SANTALACEAE

Phoradendron

RAFFLESIALES: APODANTHACEAE	CUCURBITALES: APODANTHACEAE

Pilostyles

EUPHORBIALES: BUXACEAE	BUXALES: BUXACEAE

Buxus

EUPHORBIALES: EUPHORBIACEAE	MALPIGHIALES: PHYLLANTHACEAE

Savia, Astrocasia, Securinega, Margaritaria, Chascotheca, Phyllanthus

EUPHORBIALES: EUPHORBIACEAE	MALPIGHIALES: PICRODENDRACEAE

Picrodendron

EUPHORBIALES: EUPHORBIACEAE	MALPIGHIALES: EUPHORBIACEAE

Jatropha, Aleurites, Manihot, Croton, Argythamnia, Tragia, Adelia, Bernardia, Ricinus, Acalypha, Gymnanthes, Hippomane, Euphorbia, Poinsettia, Chamaesyce, Pedilanthus tithymaloides subsp. parasiticus (now Euphorbia tithymaloides subsp. parasitica)

RHAMNALES: RHAMNACEAE	ROSALES: RHAMNACEAE

Colubrina, Ziziphus

RHAMNALES: VITACEAE	VITALES: VITACEAE

Cissus

SAPINDALES: SURIANACEAE	FABALES: SURIANACEAE

Suriana

SAPINDALES: SIMAROUBACEAE	SAPINDALES: PICRAMNIACEAE

Alvaradoa

SAPINDALES: ZYGOPHYLLACEAE	ZYGOPHYLLALES: ZYGOPHYLLACEAE

Tribulus, Kallstroemia, Guaiacum

GERANIALES: OXALIDACEAE	OXALIDALES: OXALIDACEAE

Oxalis

LINALES: ERYTHROXYLACEAE	MALPIGHIALES: ERYTHROXYLACEAE

Erythroxylum

POLYGALALES: MALPIGHIACEAE	MALPIGHIALES: MALPIGHIACEAE

Malpighia, Bunchosia

POLYGALALES: POLYGALACEAE	FABALES: POLYGALACEAE

Polygala

UMBELLALES: ARALIACEAE	APIALES: ARALIACEAE

Dendropanax

UMBELLALES: UMBELLIFERAE	APIALES: APIACEAE

Centella, Anethum

GENTIANALES: ASCLEPIADACEAE	GENTIANALES: APOCYNACEAE

Asclepias, Calotropis, Metastelma, Sarcostemma, Cryptostegia, Stephanotis

POLEMONIALES: SOLANACEAE	SOLANALES: SOLANACEAE

Physalis, Capsicum, Solanum, Cestrum, Solandra

POLEMONIALES: CONVOLVULACEAE	SOLANALES: CONVOLVULACEAE

Dichondra, Evolvulus, Jacquemontia, Merremia, Ipomoea

POLEMONIALES: MENYANTHACEAE	ASTERALES: MENYANTHACEAE

Nymphoides

POLEMONIALES: HYDROPHYLLACEAE	Unplaced asterid I: BORAGINACEAE
Nama	
LAMIALES: BORAGINACEAE	Unplaced asterid I: BORAGINACEAE
Argusia (unresolved status), Tournefortia, Heliotropium, Ehretia, Bourreria, Cordia, Rochefortia	
LAMIALES: VERBENACEAE	LAMIALES: LAMIACEAE
Aegiphila, Petitia, Clerodendrum	
LAMIALES: AVICENNIACEAE	LAMIALES: ACANTHACEAE
Avicennia	
LAMIALES: LABIATAE	LAMIALES: LAMIACEAE
Hyptis, Salvia, Ocimum	
SCROPHULARIALES: OLEACEAE	LAMIALES: OLEACEAE
Jasminum, Forestiera, Chionanthus	
SCROPHULARIALES: SCROPHULARIACEAE	LAMIALES: PLANTAGINACEAE
Russelia, Stemodia, Scoparia, Bacopa	
SCROPHULARIALES: SCROPHULARIACEAE	LAMIALES: SCROPHULARIACEAE
Capraria	
SCROPHULARIALES: SCROPHULARIACEAE	LAMIALES: OROBANCHACEAE
Agalinis	
SCROPHULARIALES: MYOPORACEAE	LAMIALES: SCROPHULARIACEAE
Bontia	
SCROPHULARIALES: BIGNONIACEAE	LAMIALES: BIGNONIACEAE
Crescentia, Catalpa, Tabebuia, Spathodea, Tecoma	
SCROPHULARIALES: ACANTHACEAE	LAMIALES: ACANTHACEAE
Blechum, Asystasia, Ruellia	
CAMPANULALES: GOODENIACEAE	ASTERALES: GOODENIACEAE
Scaevola	
RUBIALES: RUBIACEAE	GENTIANALES: RUBIACEAE
Exostema, Rhachicallis, Randia, Catesbaea, Scolosanthus, Strumpfia, Erithalis, Chiococca, Antirhea, Guettarda, Hamelia, Morinda, Psychotria, Faramea, Hedyotis, Ernodea, Spermacoce	
RUBIALES: CAPRIFOLIACEAE	DIPSACALES: CAPRIFOLIACEAE
Lonicera	
ASTERALES: COMPOSITAE	ASTERALES: ASTERACEAE
Cyanthillium, Vernonia, Ageratum, Chromolaena, Koanophyllon, Aster, Conyza, Baccharis, Pluchea, Ambrosia	

NOTE:

The taxonomy used in this flora follows Cronquist (1968): *The Evolution and Classification of Flowering Plants*.

The modern molecular-based taxonomy in this table follows Angiosperm Phylogeny Group (2009): *Botanical Journal of the Linnean Society* 161 (2): 105–121.